Research on The Management of Innovation

RESEARCH ON THE MANAGEMENT OF INNOVATION

The Minnesota Studies

Edited by

Andrew H. Van de Ven

Harold L. Angle

Marshall Scott Poole

OXFORD
UNIVERSITY PRESS
2000

OXFORD

UNIVERSITY PRESS

Oxford   New York
Athens Auckland Bangkok Bogotá Buenos Aires Calcutta
Cape Town Chennai Dar es Salaam Delhi Florence Hong Kong Istanbul
Karachi Kuala Lumpur Madrid Melbourne Mexico City Mumbai
Nairobi Paris São Paulo Singapore Taipei Tokyo Toronto Warsaw

and associated companies in
Berlin   Ibadan

Library of Congress Cataloging-in-Publication Data
Research on the management of innovation :
the Minnesota studies / edited by Andrew H. Van de Ven,
Harold L. Angle, Marshall Scott Poole.
p. cm.
Includes bibliographical references and index.
ISBN 0–19–513976–3
1. Technological innovations—Management.   2. Organizational change—Management.
I. Van de Ven, Andrew H.   II. Angle, Harold L.   III. Poole, Marshall Scott, 1951–
HD45 .R39 2000
658.5'14—dc2l   00–037470

1 3 5 7 9 8 6 4 2

Printed in the United States of America
on acid-free paper

Contents

List of Figures

List of Tables

Preface to the Paperback Edition

We are delighted that Oxford University Press is reprinting *Research on the Management of Innovation: The Minnesota Studies*. It was initially published in 1989 and subsequently taken out of print, prematurely in our opinion. It received an honorable mention in 1990 for the equivalent of an "Emmy" award in professional and scholarly publishing in the area of business and management by the Association of American Publishers in New York. It was a finalist for the 1991 Terry Book Award as making an outstanding contribution to the advancement of management knowledge by the Academy of Management. It is the original source of pioneering research by the Minnesota Innovation Research Program (MIRP) that contributed to a growing interest in understanding processes of innovation and organizational change.

Perhaps the 1989 publication was ahead of its time, for the book is increasingly cited and requested as a progenitor of current directions in innovation research and practice. Indeed, the book has spawned numerous subsequent books and journal articles from MIRP and other innovation scholars. We are grateful that Oxford University Press has agreed to reprint the book and thereby make it accessible to the growing number of readers who seek access to this book.

The Minnesota Innovation Research Program (MIRP) began in 1983 with the objective of understanding how and why innovations develop from concept to implementation. Fourteen research teams, involving more than thirty faculty and doctoral students at the University of Minnesota, conducted longitudinal studies that tracked a variety of new technologies, products, services, and programs as they developed from concept to implementation in their natural field settings. This book presents the emergent ideas, methods, and findings from the research program about the process of innovation.

Although many papers and books have been written about the management of innovation, this one is different. Most have focused on the antecedents (initiators, facilitators, inhibitors) or consequences (outcomes) of innovation. Very few have directly examined how and why innovations actually emerge, develop, grow, or terminate over time. An appreciation of the temporal sequence of events in developing and implementing new ideas is fundamental to managing innovation. Innovation managers and entrepreneurs need to know more than the starting conditions and investments that are required to achieve desired innovation outcomes. They are centrally responsible for directing the innovation process that goes on within the proverbial "black box" between inputs and outcomes. To do this, the innovation manager needs a "road map" that indicates how and why the innovation journey unfolds and what paths are likely to lead to success or failure.

As we shall see, this journey is often highly unpredictable and uncontrollable. As a result, a process theory may never reach the precision to tell managers exactly what to do and how an innovation will turn out. Nevertheless, it may produce some fundamental "laws of innovating" that are useful for explaining a broad class of processes, sequences, and paths that are central to managing the innovation journey. Empirical evidence for such a process theory can make a major contribution to improving the capabilities of managers, entrepreneurs, and policymakers to innovate.

This book is the original source for many lasting contributions to understanding processes of innovation and organizational change. Following the introductory chapters of the MIRP framework and methods, Chapter 3 presents and evaluates the Minnesota Innovation Survey. This survey instrument has been adopted by numerous researchers and consultants to measure the attitudes of participants involved in a wide variety of innovations and organizations. Where used properly, the organizations have found this survey to provide useful data for innovation participants to make periodic assessments of their progress, to reflect on their collective experiences, and to modify their action courses based on the results.

Chapter 4 develops a compelling alternative to the predominant stage model of evolution of an innovative idea. Schroeder and colleagues introduce a "fireworks" model that emphasizes that innovations do not progress smoothly through well-defined stages, but move through a fluid process where an idea seems to start off with a shock, then proliferates, is subject to setbacks and surprises, and then links with the old organization along the way until the innovation finally becomes part of the accepted order or establishes a new order. In their 1991 review of this book in *Administrative Science Quarterly*, Oscar Hauptman and Nitin Nohria noted this chapter to be among the best in the book, which provides a compelling review of seven MIRP innovations that develop and support this model.

In Chapter 5 Angle provides a systematic review of the social-psychological literature on the factors that motivate and enable individuals to innovate. It has become standard required reading for graduate courses on innovation and creativity. In Chapters 6 and 10, Ring and his colleagues introduce their initial ideas, grounded in observations of the MIRP innovations, about the informal processes of building relationships and engaging in transactions. The processes of sense making and committing may be just as important as the formal processes of negotiation, agreement, and administration. These insights subsequently led to Ring and Van de Ven's widely cited articles on the structures and processes for inter-organizational relationships published in the *Strategic Management Journal* in 1992 and *Academy of Management Review* (AMR) in 1994.

In Chapter 7, Dornblaser and associates present the novel empirical finding that assessments of innovation outcomes vary over time between resource controllers and innovation managers. They start with divergent assessments, lead to convergence in the middle period, and then diverge again at the end of the innovation development period. Dornblaser and colleagues marshal both quantitative survey data and qualitative event data to show evidence of both rational and superstitious learning as innovations develop over time. This chapter is a unique contribution to understanding shifting performance evaluations of innovations as they develop; no subsequent study has yet examined these dynamics.

Another example of the generative contributions spawned by this book is the meta-theory of innovation processes by Poole and Van de Ven in Chapter 20. Their typology of process theories was initially created to capture the different process patterns that were observed in the MIRP innovations. Expanding upon the grounded theory developed here, Van de Ven and Poole undertook a major literature review of process theories of change and development published in *Academy of Management Review* as "Explaining Development and Change in Organizations," which was selected as *AMR's* Best Paper of the Year in 1995.

Sections 3–6 of the book contain eleven chapters describing the temporal processes that unfolded in the development of the fourteen diverse innovations included in the MIRP. Numerous readers have noted their appreciation for the longitudinal case studies presented in these chapters. These chapters provide an exceptionally rich empirical database revealing dynamic processes that have previously not been observed, published, and, hence, accessible. Karl Weick observed in 1999 that "through patient, deep inquiry, these authors have captured dynamics previously undetected in organizations. In doing so they have created a major advance in organization studies that is equally compelling to practitioners and researchers." We believe that the rich longitudinal case studies in Chapters 8–18 represent the major lasting contribution of this book, for they provide a rich storehouse of revelatory cases for subsequent analysis and grounded theory building.

Providing an empirically grounded mapping of the innovation process has eluded innovation scholarship because, to date, few studies have directly examined the innovation process in real time. As a result, at the time the MIRP studies were initiated, few empirically substantiated statements about how the innovation process unfolds were available. One of MIRP's major objectives was to map empirically how innovations develop from concept to reality. As Chapters 8–18 show, these maps are based on what MIRP researchers observed, not on what they thought should have happened. Such descriptive maps represent a useful first step in developing a scientific understanding of the innovation process. They identify the temporal sequence of events, junctures, and hurdles that innovation teams and managers experience along the innovation journey. Knowing how the innovation process unfolds over time and in different situations provides useful empirical data for analyzing, inferring, and verifying prescriptions for managing the innovation process. Having taken the first step, we invite you and others to participate in the next steps. Comparative analysis of these cases provide a rich empirical base for addressing a range of questions and applying a variety of conceptual perspectives for drawing inferences about processes of innovation and change

Subsequent to the publication of this book, we have continued to pursue these further steps in data analysis, inference, and verification of the MIRP data. As field observations subsided when the innovations came to their natural conclusions, we returned to the library (as they say) to more carefully study process theories and methods relevant to understanding how and why innovation develops. At the time colleagues often asked, "What is your next new research venture?" They were often surprised by our response: "The same old one." We figured that if it took ten years to collect the data, we deserved at least ten years to analyze and make sense of the data. This included conducting literature reviews, attending conferences, and refining methods for analyzing data on theories and methods for understanding processes of organizational innovation and change.

We explored alternative methods for analyzing our rich qualitative data on the innovation process. We found that while our qualitative data generated many important insights on the innovation process, they were often limited to anecdotes and left us without capabilities to make empirical generalizations and inferences. By search, trial, and error we experimented with various methods that retained some of the sensitivity of the narrative method yet enabled us to systematically deal with larger and longer event sequence data in order to analyze complex processes and derive testable generalizations. We found it necessary both to extend traditional quantitative methods and to introduce some new approaches for diagnosing nonlinear dynamic patterns in our event sequence data on innovations.

While developing these methods, we conducted research workshops in 1993, 1995, and 1997 at the University of Minnesota, each attended by fifty to seventy researchers interested in process research methods from universities throughout the United States and Europe. These workshops made us aware of a much larger community of researchers interested in studying processes of organizational change but searching for methods to do so. Workshop participants encouraged us to document and distribute the process research methods we were developing. This resulted in the second book on MIRP, authored by Marshall Scott Poole, Andrew Van de Ven, Kevin Dooley, and Michael Holmes, *Organizational Change and Innovation Process: Theory and Methods for Research,* also being published by Oxford University Press.

While preparing this second MIRP book, we applied the new methods in a series of journal-length articles on specific innovation topics. While painful at times, the experience of preparing journal articles turned out to be very useful for learning from the scrutiny and feedback of anonymous journal reviewers. The discipline of peer reviews sharpened our thinking and methods about core components of the innovation process. It laid the basic building blocks for writing the third—and capstone—book on MIRP, *The Innovation Journey,* authored by Andrew Van de Ven, Douglas Polley, Raghu Garud, and S. Venkataraman, published by Oxford University Press in 1999. This book provides a synthesis of the collective findings from the research program and speculates about their implications for understanding and managing the process of innovation.

The MIRP studies could not have been undertaken without our research colleagues who conducted the longitudinal studies of the 14 innovations included in the program during the 1980s. Like many of the ideas spawned by the research program, many MIRP researchers have scattered throughout the globe. These scholars include:

Harold Angle, *University of Cincinnati*

David Bastien, *St. Thomas University*

John Bryson, *University of Minnesota*

J. Stuart Bunderson, *Washington University*

Yi-Ting Cheng, *National Taiwan University*

Yun-han Chu, *National Taiwan University*

Shobha Das, *Nanyang Technological University*

Kurt Dirks, *Simon Fraser University*

Kevin J. Dooley, *Arizona State University*

Bright Dornblaser, *University of Minnesota*

Raghu Garud, *New York University*

Robert Goodman, *University of Wisconsin*

David Grazeman, *University of Southern California*

Todd Hostager, *University of Wisconsin-Eau Claire*

Roger Hudson, *University of Wisconsin-Parkside*

Paula King, *Gabberts, Inc.*

Mary Knudson, *Purdue University*

Marian Lawson, *Northern Kentucky University*

Ian Maitland, *University of Minnesota*

John Kralewski, *University of Minnesota*

Tse-min Lin, *University of Texas*

Karin Lindquist, *University of Utah*

Charles Manz, *University of Massachussets, Amherst*

Alfred Marcus, *University of Minnesota*

John Mauriel, *University of Minnesota*

Linda Neumann, *University of Uruguay*

Douglas Polley, *St. Cloud State University*

Gordon Rands, *University of Western Illinois*

Michael Rappa, *North Carolina State Univesity*

Peter Ring, *Loyola Marymount University*

Nancy Roberts, *Naval Postgraduate School, Monterey*

William Roering, *Michigan State University*

Vernon Ruttan, *University of Minnesota*

Roger Schroeder, *University of Minnesota*

Gary Scudder, *Vanderbilt University*

Eric Trist (*deceased*)

Andrew Van de Ven, *University of Minnesota*

Linn VanDyne, *Michigan State University*

S. Venkataraman, *University of Virginia*

Robert Wiseman, *Michigan State University*

In addition to conducting their own innovation studies, MIRP investigators had meetings about once a month and worked together from 1983 to 1989 to develop and apply a common research framework and to share their research findings across the wide variety of innovations being studied. We fondly recall the intellectual stimulation and excitement produced by the MIRP team meetings. The MIRP framework and approach to longitudinal data collection are clearly the collective learning experiences of this large interdisciplinary group of scholars.

We feel most fortunate that Herbert Addison and Oxford University Press have agreed to reprint this book. Herb provided us tremendous encouragement, guidance, and support of this work. In his distinguished career as Oxford's executive editor of business books, Herb has made numerous contributions in selecting and publishing major works that have advanced the profession of management.

Minneapolis, Minnesota   A. H. R

Cincinnati, Ohio   H. L. A.

College Station, Texas   M. S. P.

February 2000

Preface to the Original Edition

Research on the Management of Innovation is written for people who study, advise, or manage the process of innovation. It includes a wide variety of innovations (technological, product, process, and administrative), which have been studied from different perspectives (individual, group, organizational, industry, and national) and in different settings (private, public, and not-for-profit). It address the questions, *How and why do innovations develop over time from concept to implemented reality? What processes lead to successful and unsuccessful outcomes? To what extent does knowledge about innovation processes generalize from one situation to another?* Answers to these questions are provided, based on the findings from the Minnesota Innovation Research Program (MIRP).

Since 1983, MIRP has been tracking a diverse set of innovations over time as they develop in their natural field settings. Beyond simply examining the inputs and outputs of innovations (as was typically done in past research), MIRP has been focusing on the real-time *processes* of managing innovations. These two volumes present the major conceptual models, methods, and empirical findings that are emerging from MIRP to describe and explain the innovation journey and the paths along this journey that lead to success or failure for different kinds of innovations. The results of the MIRP studies are integrated to produce a new innovation process theory as well as a "road map" for managing the innovation journey.

Similar to the historical development of MIRP (described in Chapter 1), this book represents a collective achievement. It is not simply an edited collection of chapters written by authors working independently, as is commonly true of edited works. Instead, all thirty-four authors of the twenty-one chapters are MIRP coinvestigators. They have been working together since 1983 to develop and apply a common research framework, organizing themselves into fourteen interdisciplinary research teams to study how a wide variety of innovations develop over time. Thus, this book represents the collective learning experiences of a large interdisciplinary group of investigators who have interacted and reacted to one another's accumulating longitudinal data and evolving perspectives on the management of innovation.

The managers of the innovations being studied by MIRP have also contributed significantly. Many provided MIRP researchers unusually intimate access to their innovation activities, not after their completion but as the innovations were being developed—a degree and type of access that was essential to observing how the innovation process unfolds over time. Periodically, innovation managers receive feedback on their innovations from their MIRP study teams and participate in research workshops on the overall findings emerging across all MIRP studies. These workshops offer valuable opportunities to discuss common themes, processes, and issues in the management of innovation. As a result, managers and researchers have found that they share many common experiences and problems in the management of innovation they had thought were unique to their situations. The contributions of innovation managers in these feedback sessions and workshops have been particularly gratifying, as they have clearly communicated genuine interest and involvement in the research itself. In effect, these innovation managers have become research partners rather than passive subjects. Their comments have lent numerous insights that pointed to new theoretical directions and have served as important "reality tests" for the research findings.

In addition to the contributions of these practitioners, MIRP's development has been greatly assisted over the years by many other innovation scholars. In particular, we must recognize the valuable contributions of Professor Eric Trist, who was visiting professor at the University of Minnesota in 1983 and participated in the initial critical meetings to establish the research program. Subsequently, we received useful feedback from many other scholars, who have either visited the university and attended one of the regular meetings of MIRP investigators or participated in one of MIRP's periodic research workshops with innovation managers. These colleagues have provided invaluable feedback on MIRP's methods and findings and have kept us abreast of other research and perspectives relevant to the management of innovation. In alphabetical order, they include:

Chris Argyris, Harvard University

Robert Burgelman, Stanford University

Bala Chakravarthy, University of Minnesota

Arnold Cooper, Purdue University

Kenneth Craik, University of California, Berkeley

Larry Cummings, University of Minnesota

Yves Doz, INSEAD, France

Bruce Erickson, University of Minnesota

Susan Foote, University of California, Berkeley

Edward Freeman, University of Virginia

Joseph Galaskiewicz, University of Minnesota

Jay Galbraith, University of Southern California

Luther Gerlach, University of Minnesota

George Green, University of Minnesota

J. Richard Hackman, Harvard University

Donald Hambrick, Columbia University

Gudmund Hernes, University of Bergen, Norway

Paul Hirsch, University of Chicago

George Huber, University of Texas

Paul Johnson, University of Minnesota

Rosabeth Kanter, Harvard University

John Kimberly, University of Pennsylvania

Edward Lawler, University of Southern California
Paul Lawrence, Harvard University
Huseyin Leblebichi, University of Illinois
Dorothy Leonard-Barton, Harvard University
Arie Lewin, Duke University
Ian MacMillan, University of Pennsylvania
James March, Stanford University
Bill McKelvey, University of California, Los Angeles
Robert McPhee, University of Wisconsin, Milwaukee
Ian Maitland, University of Minnesota
Robert Miles, Emory University
Henry Mintzberg, McGill University
Lawrence Mohr, University of Michigan
Mary Nichols, University of Minnesota
Paul Nutt, Ohio State University
Charles O'Reilly, University of California, Berkeley
William Ouchi, University of California, Los Angeles
Donald Pelz, University of Michigan
Thomas Peters, Skunkworks, Inc.

Jeffrey Pfeffer, Stanford University
Charles Perrow, Yale University
Andrew Pettigrew, University of Warwick
James Brian Quinn, Dartmouth College
Everett Rogers, University of Southern California
Richard Saavedra, University of Minnesota
Gerald Salancik, University of Illinois
John Slocum, Southern Methodist University
Bary Staw, University of California, Berkeley
David Teece, University of California, Berkeley
Howard Thomas, University of Illinois
Michael Tushman, Columbia University
Noel Tichy, University of Michigan
John Van Maanen, Massachusetts Institute of Technology
Karl Weick, University of Michigan
David Whetten, University of Illinois
Raymond Willis, University of Minnesota
Oliver Williamson, University of California, Berkeley

These scholars provided new perspectives and feedback that have helped us avoid tunnel vision and a "group-think" syndrome, which easily set in when such an intensive longitudinal research program is undertaken.

Funding to support the research program over the years has been provided by a major grant from the Program on Organization Effectiveness of the Office of Naval Research, under contract No. N00014-84-K-0016. In addition, for specific studies in the program, we gratefully recognize support from 3M, IBM, Honeywell, Control Data, CENEX, Dayton-Hudson, First Bank System, Bemis, Dyco Petroleum, ADC Telecommunications, Farm Credit Services, Hospital Corporation of America, the Federal Reserve Bank of Minneapolis, the Bush Foundation, McKnight Foundation, the Minnesota Agricultural Experiment Station, and the University of Minnesota. Without the broad base of support from these organizations, this longitudinal research would not have been possible. Finally, we wish to acknowledge the generous data-processing support we have received from Academic Systems Support Services at the University of Minnesota.

While the research has been underway since 1983, actual preparation of this volume began in January 1986, when MIRP investigators met several times to develop a publication strategy. Consistent with the objectives of MIRP itself, the publication strategy chosen was to strike a balance between reporting findings that are unique to individual innovation studies and those that are common to all innovation studies. Flexibility has been built into MIRP to encourage each research team to adopt diverse perspectives for investigating questions unique to the innovation they are studying, while not losing the capability to compare and generalize findings across innovations through the use of a core research framework. We believe that creativity and learning are maximized when investigators with diverse perspectives interact, applying a common framework of research concepts and methods in a variety of study sites.

The first step in implementing this publication strategy occurred in November 1986, when all MIRP investigators conducted a day-long retreat to review and discuss one another's initial chapter outlines. Having thus reached an initial appreciation of the focus, contents, methodology, and empirical data to be included in each chapter, the MIRP researchers began more detailed analysis of data from their individual innovation studies and prepared more complete drafts of their respective chapters. Concomitantly, these researchers began to share data more widely across the individual studies. The MIRP contributing authors presented and discussed completed drafts of their chapters at a two-day research conference in May 1987 with thirty-five highly recognized innovation scholars and managers. The feedback obtained at this conference, as well as that obtained from Andrew Van de Ven's doctoral seminar on innovation research that spring (where twelve students critiqued each chapter in detail), provided useful information for contributing authors' second revisions, which were completed in November 1987. As editors, we then undertook another review of the chapters, giving contributing authors detailed suggestions for undertaking a third (and in most cases, the final) revision of chapters. As contributing authors prepared final drafts in 1988, the editors drafted the last two chapters (20 and 21), which integrate and present the major findings to date from the research program by advancing a process theory of innovation and by offering many practical suggestions for managing the innovation journey. These capstone chapters, in turn, were distributed to all contributing authors, innovation managers, and other innovation scholars, and were then revised on the basis of the useful feedback obtained.

The above process of repeatedly drafting, discussing, editing, and revising chapters far exceeded our initial expectations of time and effort. The process has been very rewarding, however, because it has permitted the development of (1) more robust explanations for processes observed in our field studies of innovations; (2) a better synthesis and integration of research findings and perspectives across the diverse innovations studied by MIRP; and (3) a broadening of our initial perspectives on the management of innovation. Not only are these developments reflected in this volume but they have also provided the basis for redirecting our ongoing research program.

The preparation, editing, and revisions of original chapters by so many contributing authors represented a major coordination task. Fortunately, we were ably assisted by Linda Neumann and Susan McGuire. Linda Neumann drafted the biographical descriptions, "About the Contributing Authors", and read several drafts of each chapter—detecting the inevitable errors and suggesting many ways to improve readability. In addition, both Linda and Susan provided much administrative assistance in organizing and conducting the yearly conferences and workshops with innovation managers and scholars, as well as monthly meetings among MIRP investigators over the years. These conferences, workshops, and meetings contributed immensely to fostering an intellectually stimulating interdisciplinary forum for both researchers and managers to learn from one other about the management of innovation.

Last, but certainly not least, we are indebted to our families and loved ones. They have shared most in this undertaking and made it an exciting, growing, and enjoyable experience—both personally and professionally.

ANDREW H. VAN DE VEN
HAROLD L. ANGLE
MARSHALL SCOTT POOLE

Minneapolis
January 1989

Minnesota Innovation Research Program Members at the May 14–15, 1987, Innovation Workshop (from top left): John Mauriel, Raghu Garud, Michael Rappa, Bill Roering, Douglas Polley, Gary Scudder, Yunhan Chu. Second row: Janet Porter, David Bastien, Charles Manz, Roger Schroeder, Paula King, Nancy Roberts, Gary Seiler. Bottom row: Linda Neumann, John Bryson, Scott Poole, Andrew Van de Ven, Harold Angle, Peter Ring, Alfred Marcus, S. Venkataraman. Not pictured: Bright Dornblaser, Robert Goodman, Roger Hudson, Karin Lindquist, Todd Hostager, Mary Knudson, John Kralewski, Tse-min Lin, Ian Maitland, Gordon Rands, Vernon Ruttan, Mark Weber, Bob Wiseman.

SECTION
I

Overview of Research
Program and Methods

AN INTRODUCTION TO THE MINNESOTA INNOVATION RESEARCH PROGRAM

Andrew H. Van de Ven

Harold L. Angle

The challenge of international competition has brought with it a growing awareness that the United States is losing its innovativeness. Witness the common call for stimulating innovation that is found in bestselling books by Ouchi (1981), Magaziner and Reich (1982), Peters and Waterman (1982), Kanter (1983), and Lawrence and Dyer (1983). Underlying this concern is the belief that a significant gap has arisen between the ability of the United States to create and to implement new ideas. As a free society with diverse people and large research institutions, the United States is generally regarded as the most inventive nation in the world. However, other nations are surpassing the United States in developing and implementing its ideas. Innovation requires more than the creative capacities to invent new ideas; it requires managerial skills and talents to transform good ideas into practice. But these capabilities to manage the process of innovation remain underdeveloped. As Teece (1987: 3) argues,

> at a time when so much attention is given to innovation and entrepreneurship, it is rather pathetic that a deep understanding of the process is lacking. It is no wonder that firms and governments have difficulty trying to stimulate [and manage] innovation when its fundamental processes are so poorly understood.

We concur with Lewin and Minton (1986) that a managerial perspective on innovation is

We greatly appreciate useful comments from Scott Poole on an earlier draft of this chapter, as well as MIRP study directors for preparing the summaries of their innovation studies presented in the appendix of this chapter.

needed—one that focuses on the key problems and challenges confronting managers while they initiate and direct the development of an innovation over time. But the search for such a perspective has proven to be elusive.

In their extensive review of innovation studies commissioned by the National Science Foundation, Tornatzky et al. (1983) point out that while many studies have examined the antecedents (facilitators/inhibitors) to or consequences (outcomes) of innovation, very few have directly examined how and why innovations actually emerge, develop, grow, or terminate over time. Yet an appreciation of the temporal sequence of activities in developing and implementing new ideas is fundamental to the management of innovation. Innovation managers need to know more than the factor inputs required to achieve desired innovation outcomes. These managers are centrally responsible for directing the innovating process within the proverbial "black box" between inputs and outcomes. To do this, the innovation manager needs a "road map" that indicates how and why the innovating journey unfolds and what paths are likely to lead to success or failure.

In other words, the innovation manager needs a *process theory* that explains how and why innovations develop on the basis of the probabilistic arrangement of discrete events over time (Mohr 1982). Although such a process theory may never reach the precision to tell managers exactly what to do and how an innovation will turn out, it may produce some fundamental "laws of innovating" useful for describing and explaining a broad class of processes, sequences, and performance conditions central to the management of innovation. As March (1981) indicated,

> Typically, it is not possible to lead an organization in any arbitrary direction that might be desired, but it is possible to influence the course of events by managing the process of change . . . , particularly timing small interventions so

that the force of natural organizational processes amplifies the interventions.

Empirical verification of a process theory could make a major contribution to improving the capabilities of managers, entrepreneurs, and policymakers to innovate.

Since 1983 researchers at the University of Minnesota have been engaged in a longitudinal research program with the objective of developing such a process theory. No overarching process theory of innovation has yet emerged from the research program, nor are prospects bright in the near future. However, we believe it is timely to publish this interim report on the findings of this research program because the program has undertaken some constructive initial steps toward development of such a process theory. These steps include (1) carefully observing a wide variety of innovations in real time as they develop in their natural field settings, (2) analyzing and comparing their developmental progressions across innovations, and (3) developing and applying alternative process theories that may explain these observed developmental patterns. Since many of the innovations have not yet reached their natural conclusions (that is, been implemented, institutionalized, or terminated), many of the longitudinal field studies in the program are still ongoing. Therefore, it is important to emphasize at the outset that this is not a final report. On the contrary, we anticipate additional findings and insights as we continue to track these innovations as they develop. Even so, the research program has accumulated five years of detailed historical and real-time data on the innovations—data that provide a rich and unique opportunity to address in this report many key questions and issues that have not been systematically addressed before about the process of innovation. In particular, this interim report contains original chapters written by members of the research program that address three process questions about the management of innovation.

How and Why Do Innovations Actually Develop Over Time From Concept to Implemented Reality?

As stated before, very little is known about how innovations actually emerge, develop, grow, or terminate over time. Yet an appreciation of the temporal processes is fundamental to the management of innovation. Most innovation scholars and managers view the innovation process as a simple sequence of developmental stages (such as idea invention, design, testing, implementation, and diffusion). However, these simple phase or stage models often lack empirical validity (Poole 1983). Before valid principles about the management of innovation can emerge, there is a great need to describe, empirically, how innovations actually develop over time, and then to develop and test process theories that accurately explain these observations and the conditions when they apply. Achieving this is the first and most important issue to be addressed.

What Innovation Processes Lead to Successful and Unsuccessful Outcomes?

Unfortunately, while many innovation process models in scholarly and trade journals are prescriptive, they lack empirical validity. Too many innovation scholars and consultants have jumped to prescriptions with little or no substantiated evidence on how innovations actually develop over time and what processes are associated with success or failure. The little empirical evidence that exists is limited largely to retrospective case histories conducted after innovation outcomes were known. Even though attempts can be made to minimize bias, a priori judgments about the success or failure of innovations studied invariably filter and color one's analysis, often leading to self-fulfilling prophesies. Moreover, historical case studies tend to be too far removed in time and space to observe the real-time interactions between processes and outcomes as innovations develop over time. As will be evident, we believe the likelihood of making significant and substantiated prescriptions on the management of innovation increases when researchers (1) place themselves into the manager's frame of reference with real-time studies of innovations without knowing the outcomes of actions taken, (2) carefully observe and track innovation processes and outcomes as they occur over time, and (3) show the evidence to substantiate relationships between innovation processes and outcomes.

To What Extent Can Knowledge About Managing Innovation and Change Processes be Generalized From One Situation to Another?

To date, most research and scholarship on innovation have been narrowly defined on the one hand and fractionated into many distinct segments on the other. Most has focused on only one kind of organizational arrangement for innovation, such as internal organizational innovation (Normann 1971), or new business startups (Cooper 1979); or one stage of the innovation process, such as the diffusion stage (Rogers 1982); or one type of innovation, such as technological innovation (Utterback 1974). Although such research has provided many insights into specific aspects of innovation, such segmentation severely limits opportunities to understand the overall process of innovation from beginning to end and to generalize findings from one type of innovation to another.

To address this segmentation problem, this report presents and compares findings being obtained from a highly diverse set of innovations that were selected for longitudinal study in the research program. The set includes a variety of technological, product, process, and administrative innovations in private, public, and not-for-profit sectors. We identify and show the limits to which findings about the process of innovation can be generalized from one innovation setting to another. In so doing, we

5

hope to show the significant benefits and efficiencies that can be obtained by applying principles for innovation management from one setting to another (which heretofore have not been recognized). For example, we show that many of the processes of new business creation in new company startups apply equally well to internal corporate venturing and interorganizational joint ventures, and vice versa.

In conclusion this interim report on the research program is timely for its preliminary findings amply demonstrate the need to reexamine a number of conventional beliefs about the nature, consequences, and generality of innovation processes. We believe that answers to the three questions above can significantly contribute to our knowledge about the management of innovation—and thereby ultimately contribute to increasing U.S. innovativeness and competitiveness.

This chapter provides a background introduction to this program, the research design and the work accomplished to date, and introduces the conceptual framework for the chapters which report the findings obtained to date from the research program on the management of innovation. Finally, the chapter sets forth the plan of work designed to achieve the program's objective of developing and testing a grounded theory of the process of innovation.

BRIEF HISTORY OF THE MINNESOTA INNOVATION RESEARCH PROGRAM

In the summer of 1983 the Office of Naval Research funded our proposal to launch a longitudinal research program on the management of innovation at the Strategic Management Research Center of the University of Minnesota. As Table 1–1 indicates, this research grant sig-

TABLE 1–1. MIRP Historical Development.

Initiation (1982–83):
  Community needs assessment (1982)
  Innovation study groups (spring 1983)
  Pilot field studies (summer 1983)
  Grants from ONR and other sources

Startup (1983–85):
  Organize research program (fall–winter 1983)
  Develop framework and methods (1983–84)
  Launch innovation studies (1983–84)
  Write innovation case histories (1984)
  Refine research instruments (spring 1985)

Takeoff (1984–present):
  Real-time tracking of innovations (1984–end)
  Data analysis, reporting, and sharing:
    Biweekly meetings of researchers
    Feedback sessions to innovation teams
    Workshops with innovation managers
    Papers for journals, conferences, books

nified the culmination of two years of initiatives that led to the startup and takeoff of an interdisciplinary research program which became known as the Minnesota Innovation Research Program (MIRP).

These initiatives began with a needs assessment of the managerial community in 1982, in which meetings were conducted with over thirty chief executive officers of public and private firms to identify what they judged were the most significant problems and issues confronting their organizations in the next ten years (see Van de Ven 1982). Of all the issues surfaced in these meetings, the management of innovation was reported as the CEOs' most central concern in managing their enterprises. This concern was reflected in a variety of questions that they often raised:

> How can an organization develop and maintain a culture of innovation and entrepreneurship?
> What are the critical factors in successfully launching new organizations, joint ven-

tures with other firms, or innovative projects within large organizations over time?

How can we decrease the lead time and costly failures normally experienced in launching new businesses, products, or programs?

Van de Ven presented the results of this needs assessment to Minnesota faculty and students at one of the regularly scheduled workshops of the Strategic Management Research Center (Van de Ven 1982). This workshop stimulated a small number of faculty and students to meet further and explore what is known about these questions. In one brainstorming session, the group speculated on the statement: "The year is 2000. What innovations were developed during the 1980s that have made a significant impact on our society today?" This session led to a listing of the types of innovations we would need to cover in a comprehensive research program. Specifically, we decided to search for innovations in the fields of agriculture, electronics, health care, consumer products, education, nuclear power, government, and public- and private-sector partnerships.

News of these meetings, in turn, stimulated others to participate, and several study groups were formed (dealing with small business entrepreneurship, internal corporate innovation, and industry-level technological innovation). From these study groups, which met periodically during spring 1983, it became apparent that an appreciation of the questions and issues raised by CEOs required interdisciplinary study. They reflected a general management scope of innovation which surpassed the segmented and functionally oriented issues traditionally addressed by innovation scholars (mentioned before). Fortunately, through a "snowball" process of participation in these informal study groups, we found there were many scholars from different academic disciplines at

the University of Minnesota who were not only independently interested in studying innovation management, but also welcomed the opportunity to engage in interdisciplinary study of innovation. Van de Ven and associates (1983) therefore prepared a proposal to launch an interdisciplinary research program on the management of innovation, which the Office of Naval Research subsequently approved in fall 1983.

In the meantime, two of the study groups undertook pilot studies of new business startups and the development of advanced integrated circuits during summer 1983. These pilot studies subsequently became two of the longitudinal studies in the overall program. They were particularly useful at the time in providing an initial introduction to the prospects and problems of studying innovations in their natural field settings. They provided information that was useful in determining how to organize the research, develop a workable field research design, and gain access to field sites undertaking innovations during late 1983 and 1984. Fortunately we found a receptive community of managers that responded positively to our requests to study a variety of innovations in different sectors and industries—not after the fact of their completion, but over time, as the innovations in their organizations were in the process of development.

By the end of 1984 access and agreements were established to launch fourteen related studies of a wide variety of innovations as they developed over time in their natural field settings (see Figure 1–1). As the figure illustrates, these fourteen related studies include technological, product, process, and administrative innovations in private, public, and not-for-profit sectors. The fourteen studies were undertaken by different interdisciplinary research teams (in total consisting of fifteen faculty and nineteen doctoral students from eight different academic departments and five schools at Minnesota).

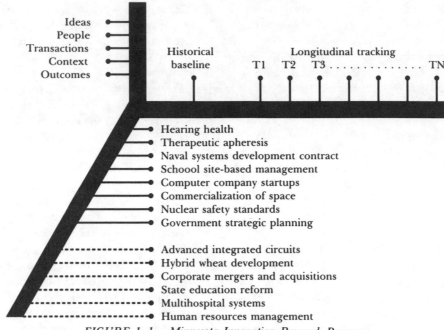

FIGURE 1–1. *Minnesota Innovation Research Program.*

In addition to the extensive field work involved in launching the innovation studies, the interdisciplinary MIRP researchers began meeting regularly (about monthly) in fall 1983 to develop a core conceptual framework to study innovations over time. The purpose of these meetings was to allow MIRP participants to explore the potential for collaborative synergy among the diverse innovation field sites and the research interests and capabilities represented by the participants. Although there were few initial explicit commitments regarding how the research would be conducted, individuals agreed to three basic principles as part of membership in MIRP: (1) the research would study the innovation process, at multiple levels, across a diverse group of organizational settings, (2) the research would be multidisciplinary, and (3) the research would be longitudinal.

We foresaw that one of the most salient opportunities for contributing to present knowledge about innovation processes would be by leveraging the efforts of each innovation study team by collecting a common core of data with a consistent conceptual framework. This would enable rigorous comparisons to be made across settings and types of innovations and would therefore provide the means to work toward development of a general process theory of innovation. Without such a common guiding framework, findings from individual innovation studies would not be comparable.

One of the most difficult tasks during 1983 and 1984 of the program was orchestrating a multistakeholders negotiation process aimed at reaching consensus on such a common framework. To bring some order to the process of developing this framework, we settled on three criteria to guide us in the framework's development: *parsimony, significance,* and *generality.* The first criterion, parsimony, was premised on the idea that each study team required sufficient slack to pursue questions unique to its individual study, in addition to those of MIRP's overall cross-study interest.

Only a simple common framework would provide the requisite flexibility and "headroom" to permit full treatment of unique aspects without overburdening either researchers or host organizations.

Simplicity alone, however, could not be sought at the expense of the other two criteria, significance and generalization. As Van de Ven and associates (1984: 4) stated, MIRP participants firmly held to the convictions that

The common framework should represent a conceptual advance to an understanding of the management of innovation;

The common framework should generalize as much as possible across the diverse innovations being studied so that we leverage opportunities for learning and insight from each study.

After many (sometimes heated) discussions, the program participants concluded that these criteria could be achieved in a common framework that centers on five basic concepts: *ideas, people, transactions, context,* and *outcomes.* (These concepts are defined and elaborated on later in this chapter.) These concepts were selected because they constitute the central factors of concern to managers in directing innovations (Van de Ven 1986). From a managerial perspective, the process of innovation consists of motivating and coordinating *people* to develop and implement new *ideas* by engaging in *transactions* (or relationships) with others and making the adaptations needed to achieve desired *outcomes* within changing institutional and organizational *contexts.* As described in Chapter 2, a significant change in these concepts represent an *event.* A systematic mapping of events over time was established as the central task for all the MIRP studies.

It was believed that this basic framework of concepts accommodated the individual requirements of each study, yet nevertheless provided an opportunity to integrate the findings across MIRP innovation studies. Summaries of the unique questions, approaches, and progress of each innovation study are presented in Appendix 1–A to this chapter. These summaries show that each innovation study, by itself, represents a significant longitudinal research effort. Moreover, each study distinguishes itself by addressing novel and important questions on the management of innovation, which will be featured in subsequent chapters.

Three overlapping research stages were undertaken to conduct the studies. The actual time periods varied somewhat for specific studies. First, from approximately January to May 1984, exploratory studies were conducted to gain access to participating organizations and to become familiar with each innovation.

Second, during summer and fall 1984, case histories and baseline data were obtained on each innovation. The case histories provided a mapping of events leading up to the initiation of longitudinal studies of the innovations. The baseline information provided a broad measure of the institutional settings in which the innovations were being developed. This information was based on a combination of published reports and other documents, interviews, and questionnaires.

Third, longitudinal tracking of the innovations began as soon as it became clear what specific aspects of each innovation should be studied over time, and access to organizational sites was obtained. Specific data collection instruments were developed during winter 1984 and revised in spring 1985. They consisted of on-site observation guides, interviews, questionnaires, and compilations of relevant documents, to enable tracking of the innovations as they developed over time. Depending on the specific innovation under study, data-collection intervals ranged from six to nine months. A more detailed description of the research methodology used to conduct the longitudinal field studies is presented in Chapters 2 and 3.

The organizations participating in the longitudinal research receive periodic research feedback both on the specific innovations in which they are engaged as well as on findings emerging across all the MIRP studies. The latter is accomplished with yearly research workshops in which MIRP researchers present and discuss their research findings with innovation managers and innovation scholars from other universities. Systematic feedback from both categories of participants has been instrumental, we believe, in our avoiding the development of tunnel vision. Bringing in external participants, as suggested by Janis (1983) helps to counter any latent tendencies toward the development of a "groupthink" syndrome, and we believe that this has been successful in the case of MIRP. The nature of the contributions on the part of the innovation managers has been particularly gratifying, as these persons have clearly sustained genuine interest and involvement in the research itself. In effect, these managers tend to be research partners, rather than simply passive subjects in someone else's research. Their comments have lent numerous insights that have pointed to new theoretical directions, as well as increasing the real-world relevance of the research. In addition to these periodic feedback sessions and research workshops, MIRP researchers have been prolific in writing papers reporting their innovation perspectives and findings for publication in journals and books, or presentation at scientific and professional conferences.

Intangible products from MIRP during the last six years have, in many ways, been as significant as these tangible products. They include the creation of an intellectually stimulating interdisciplinary forum for researchers to broaden their appreciation of innovation and to help one another learn how to formulate and study research problems. MIRP also served as a catalyst to mentor and develop many junior faculty and graduate students in longitudinal field studies. Finally, MIRP has provided the launch pad for ten doctoral dissertations either recently completed or currently being conducted on topics related to innovation management.

Before we move to the next section, which covers core concepts in the MIRP research framework, a word may be in order as to *why* we think this collaborative effort has worked so well. Angle and Hudson (1986) performed an empirical study of MIRP participants to answer the general question: "Why is it that, in an individualist culture, more than 30 faculty and students have spent years studying innovation collectively, under a single research framework?" Results of a survey conducted among both program participants and knowledgeable nonparticipants (that is, those who had an opportunity to participate but did not), called into question the conventional wisdom embedded in the question. Among both participants and nonparticipants, scores at the collectivism end of a measure of individualism-collectivism were prevalent. Moreover, advantages such as opportunity for cross-fertilization of ideas in an interdisciplinary environment and the material and intellectual support available in a programmatic undertaking predominated as reasons for continued participation in MIRP, despite the acknowledgment in the survey that such collective activity can be demanding on personal time and energy and can represent large opportunity costs. On balance, participants saw the opportunities for attaining synergy and for leveraging individual activities as clearly outweighing such costs.

CORE CONCEPTS IN MIRP RESEARCH FRAMEWORK

Because of the limited research and theory on innovation processes in the literature we decided that it is more productive to undertake a grounded-theory strategy (Glaser and Strauss

1967) to discover an innovation process theory from data systematically obtained from longitudinal research, than to test existing theories that were logically deduced from a priori assumptions that often do not fit or are not based on concrete particulars of the phenomena to be explained.

In particular, Table 1–2 compares assumptions implicit in the preexisting literature about the five core MIRP concepts, as opposed to how they were observed to behave in the MIRP field studies of innovations. Conventional wisdom tends to treat an innovation idea as a unitary phenomenon that maintains a stable identity over the time in which it is developed. Moreover, it is commonly assumed that all parties to the innovation share an essentially similar view of what the idea is. Then, there is a commonly held view that the role of the "innovator" in an organization is clearly differentiated from other roles, so that an innovator, or more likely an entire innovation team, is dedi-cated to the innovation project as its primary (if not only) responsibility. The network of other stakeholders with whom innovators must interact is also seen as fairly stable. Insofar as the context is concerned, it tends to be viewed as unitary and "real"—a source of both resources and constraints. This entire process is usually viewed as one in which there are definable stages (such as inception, development, testing, adoption, and diffusion), with this simple, cumulative series of phases resulting in a clearly interpretable outcome ("success" or "failure").

Field observations of these concepts disclosed a different reality from these rather orderly conceptions of the innovation process. Innovation ideas were found to proliferate into many ideas. There is not only invention, but there is reinvention as well, with some ideas being discarded as others are reborn. There are many persons involved in innovation, but most of these are only partially included in the inno-

TABLE 1–2. *A Comparison of the Conventional Wisdom and MIRP Observations.*

	Literature Implicitly Assumes:	But We See This:
Ideas	One invention, operationalized	Reinvention, proliferation, reimplementation, discarding, and termination
People	An entrepreneur with fixed set of full-time people over time	Many entrepreneurs, distracted, fluidly engaging and disengaging over time in a variety of organizational roles
Transactions	Fixed network of people/firms working out details of an idea	Expanding and contracting network of partisan stakeholders diverging and converging on ideas
Context	Environment provides opportunities and constraints on innovation process	Innovation process constrained by and creates multiple enacted environments
Outcomes	Final result orientation; a stable new order comes into being	Final result may be indeterminate; multiple in-process assessments and spinoffs; integration of new orders with old
Process	Simple, cumulative sequence of stages or phases	From simple to multiple progressions of divergent, parallel, and convergent paths, some of which are related and cumulative, others not

vation effort, as they are distracted by very busy schedules as they perform many other roles unrelated to the innovation. The network of stakeholders involved in transactions is constantly being revised. This "fuzzy set" epitomizes the general environment for the innovation, as multiple environments are "enacted" (Welck 1979) by various parties to the innovation. Rather than a simple, unitary, and progressive path, we see multiple tracks, spinoffs, and the like, some of which are related and coordinated, and others of which are not. Rather than a single after-the-fact assessment of outcome, we see multiple, in-process assessments. The discrete identity of the innovation may become blurred as the new and the old are integrated.

Thus, early into the research program, MIRP researchers found that it was necessary to reexamine a number of basic assumptions that managers and scholars tend to hold regarding these core concepts and their place in the management of innovation. Based on early research findings across the various innovations being studied, MIRP researchers adopted the following definitions and approaches to the core concepts for understanding the process of innovation.

Innovation Ideas

Invention is the creation of a new idea, but innovation is more encompassing and includes the process of developing and implementing a new idea. The idea may be a recombination of old ideas, a scheme that challenges the present order, or a formula or a unique approach that is perceived as new by the individuals involved (Zaltman, Duncan, and Holbek 1973; Rogers 1982). As long as the idea is perceived as new to the people involved, it is an "innovative idea," even though it may appear to others to be an "imitation" of something that exists elsewhere.

Included in this definition are both tech-

nical innovations (new technologies, products, and services) and administrative innovations (new procedures, policies, and organizational forms). Daft and Becker (1978) and others have suggested keeping technical and administrative innovations distinct. We believe that making such a distinction often results in a fragmented classification of the innovation process. Most innovations involve new technical and administrative components (Leavitt 1965). For example, Ruttan and Hayami (1984) have shown that many technological innovations in agriculture and elsewhere could not have occurred without concomitant innovations in institutional and organizational arrangements. Moreover, Damanpour and Evan (1984) identified important temporal lags between the rates of adoption of technical and administrative innovations. They found that the adoption of administrative innovations tends to stimulate the adoption of technical innovations more readily than the reverse, and that the length of the temporal lag is inversely related to organizational performance. Learning to understand the close connection between technical and administrative dimensions of innovation is a key part of understanding the management of innovation.

In Chapter 4, Schroeder et al. describe the temporal development of innovation ideas that were observed across the innovations included in MIRP. They propose a process model (see Figure 4–1) that is based on six empirically grounded observations:

1. Innovation is stimulated by "shocks," either internal or external to the organization, because when people reach a threshold of dissatisfaction with existing conditions, they will initiate action to resolve their dissatisfaction.
2. The initial idea tends to proliferate into several ideas during the innovation process. More specifically, after the onset of a simple unitary progression of activity to de-

velop an innovative idea, the innovation process proliferates into a multiple divergent progression of developmental activities.

3. During the process, unpredictable setbacks and surprises frequently arise; little learning occurs because of escalating commitments and a prevalence of mixed messages of positive and negative information that are randomly ordered over time; and mistakes "snowball" into vicious cycles that are overcome only through external interventions by investors or top management.

4. As an innovation develops, the "old" and the "new" exist concurrently, and over time they are linked together. This process of parallel and then convergent progression appears to take three forms: (a) the operating organization fundamentally changes direction as a result of the innovation, or (b) the innovation is altered to blend into the operating organization, or (c) the old and new can coexist in parallel progression with loosely coupled links between them.

5. Restructuring of the innovation unit occurs frequently during the innovation process, occasionally as a result of external interventions to alter the innovation's course of action, but more frequently to make the structural adjustments necessary to transition the innovation from one sequence of activities to another. This restructuring takes many transactional forms of changes in personnel, action teams, joint ventures, mergers or acquisitions, or altered control systems.

6. Hands-on involvement of top management or external investors occurs frequently throughout the development period. External interventions control proliferations, deal with setbacks and vicious cycles, help to link the old and the new, and restructure the organization as needed, in addition to providing general goals and resources to support the innovations.

As Schroeder et al. suggest in Chapter 4, continued real-time tracking of the MIRP innovations will enable us to extend these process observations of how innovation ideas are developed and transformed over time. However, these interim findings clearly suggest that if a process model is to be descriptively accurate and useful to the management of innovation, it needs to be more complex and fluid than the simple unitary-stage theories typically found in the literature.

Innovation Outcomes

Innovation outcomes are conventionally considered to occur at a point in time after development and implementation of the idea. Kimberly (1981) rightly points out that a positive bias pervades the study of innovation. Innovation is often viewed as a good thing because the new idea must be useful—profitable, constructive, or solve a problem. New ideas that are not perceived as useful are not normally called innovations; they are usually called "mistakes." Objectively, of course, the usefulness of an idea can be determined only after the innovation process is completed and implemented. In this sense, it is not possible to determine if work on new ideas will turn out to be "innovations" or "mistakes" until the very end of the longitudinal research. In the interim, MIRP has adopted a perceived measurement of innovation effectiveness that provides periodic assessments of the degree to which people judge the innovation process to be achieving their expectations (see Chapter 3).

By comparing these in-process outcome assessments across the MIRP innovations in Chapter 7, Dornblaser, Lin, and Van de Ven found that the criteria used by innovation managers and resource controllers shifted over time; they were different in the beginning, converged during the developmental process, and diverged in opposite and conflicting directions as innovation implementation problems arose. Not only did initially nebulous targets crystallize later into more operational criteria, but the targets themselves were often reconstructed to redirect the innovations. These changes coincided with unanticipated developmental setbacks and problems, shifting organizational priorities, as well as independent environmental events that had "spillover" effects on the innovations.

Understanding how outcome criteria shift over time demonstrate some of the central concerns of innovation participants and the directions they are likely to take in developing the next steps of an innovation. On the assumption that people act on the basis of their assessments, they will do more of what they think leads to success and less of what is perceived to lead to failure.

Dornblaser, Lin, and Van de Ven go on in Chapter 7 to examine some of these relationships between innovation actions and outcomes. Based on both quantitative and qualitative longitudinal data across the innovation studies, they show that both rational and superstitious learning occurs as innovations develop over time. In-process innovation outcomes are partially a consequence of action, partially a predictor of future actions, but often incomplete explanations of actions that occur in the development of innovations. These predictors and consequences of outcome assessments only become clear with time as developmental progressions stabilize. Even then, perceived effectiveness judgments may provide a rational basis for choosing subsequent actions to develop an innovation, but unspecified and conflicting targets or changing frames of reference often produce superstitious learning and struggles between innovation managers and resource controllers on the developmental paths of their innovations. A model of success and failure action loops is proposed in Chapter 7 to

explain these contradictory dynamics as innovations develop over time. The model not only explains how innovation outcomes are both a cause and consequence of action but also acknowledges that outcome attributions can be produced by spurious unknown factors.

One important implication of these findings is that innovation success or failure may be more usefully viewed as "by-products along the journey" than as "bottom-line" final end results (as they have typically been in the past). Although perceived effectiveness judgments during the innovation process can provide useful rationales for choices of subsequent actions, shifting and conflicting outcome criteria produce compelling opportunities for superstitious learning. Conventional wisdom suggests that one way to avoid these practices is to achieve agreement on goals and criteria among different groups. However, the prevalence of contradictory success criteria found across MIRP studies contradicts this wisdom. It suggests that contradiction and nonlinearity may be inherent in most innovative undertakings. As a consequence, a central problem in the management of innovation may be the management of paradox (Van de Ven and Poole 1988). As Cameron (1980) found, highly effective organizations were paradoxical in that they performed in contradictory ways to satisfy contradictory expectations. This is reminiscent of the way in which ambiguity in official goals is functional for complex organizations (Perrow 1961:855), in which these goals are kept "purposely vague and general" in order to accommodate the diverse operative goals of various interest groups.

The fact that in-process outcome criteria shift over time is now leading us to examine *why* and *when* they shift. In particular, we are exploring

> Escalating as well as declining commitments to the innovation idea, tasks, and transactions encountered over time as an

innovation develops, with some judged to be performed well and others not (Staw and Ross 1987);

Trial-and-error learning in a highly uncertain undertaking—naive at first, mistakes encountered, more knowing later (Lawrence and Dyer 1983; also see Chapter 11);

Perceptions of success or failure influence how much autonomy or slack investors or top management permit an innovation unit, and thus how much latitude it is has to experiment and be innovative (see Chapter 16);

Environmental awareness of what competing or substitute innovation groups are doing (imitative learning). These competing groups are also improving their technologies or services, which represent moving targets which the innovation must exceed if it is to be implemented or gain a comparative advantage (Nelson and Winter 1977);

Contextual events such as political decisions, resource allocations, or shifting administrative priorities that occur independent of the innovation itself, but significantly help or hinder the likelihood of innovation success (Pressman and Wildavsky 1974; Schelling 1978).

As these explanations suggest, innovation outcomes are dynamic, they both affect events and are affected by them.

People

Certainly, people are a central aspect in any innovation effort. Most innovations are too complex for one person to accomplish individually. A group of people needs to be recruited, organized, and directed. As these people associate with the innovation unit they apply their different skills, energy levels, and frames of reference (interpretive schemas) to the innovation ideas as a result of their backgrounds, experiences, and activities that occupy their attention. Mobilizing and directing this innovation team or unit significantly complicates the innovation management task over that of a single-person venture. Contrary to the view sometimes implicit in the literature that innovation consists of an entrepreneur who works with a fixed set of full-time people to develop a new idea, Cohen, March, and Olsen's (1972) garbage can model appears more descriptive of the process, where many stakeholders fluidly engage and disengage in the innovation process over time as their needs for inclusion, identity, and facticity dictate (Turner 1987).

The focus on people as creators and facilitators of innovation must be balanced by equivalent attention to people as inhibitors of innovation. Indeed, much of the folklore and applied literature on the management of innovation has ignored the research by cognitive psychologists and social psychologists about the limited capacity of human beings to handle complexity and maintain attention. As a consequence, one often gets the impression that inventors or innovators have superhuman creative heuristics or the ability to "walk on water." A more realistic view of innovation should begin with an appreciation of the physiological limitations of human beings (Van de Ven 1986: 594). Individuals suffer from physiological limitations that limit their ability to handle complexity. They tend to adapt unconsciously to gradually changing conditions, to overconform to group and organizational norms, and to limit their focus to repetitive activities (Van de Ven 1980). As a result, the basic people problem in managing innovation becomes, "How do individuals become attached to and invest effort in the development of innovative ideas, when people and their organizations seem mostly designed to focus on, harvest, and protect existing practices and ideas rather than to pave new directions?"

This problem is addressed by Angle in

Chapter 5. Based on an extensive and current review of research dealing with the psychology of innovation, as well as empirical data across the MIRP studies, Angle makes the fundamentally important point that it is the *organization's context* that enables and motivates innovative individual behavior. In other words, if one begins with the realistic assumption that normal people have the capability and potential to be creative and innovative (although limitedly so), then the actualization of this potential turns on whether the management structures a context that not only makes it possible to innovate but also encourages innovation.

Key motivating factors include providing a balance of intrinsic and extrinsic rewards for innovative behaviors. Pay, in itself, is a relatively weak motivator; it more often serves as a proxy for recognition. Individualized rewards tend to increase idea generation and radical innovations; whereas group rewards tend to increase innovation implementation and incremental innovations.

However, Angle emphasizes that the presence of motivating factors, by themselves, will not stimulate innovative behavior. The organization must also structure a context that enables innovation to happen. These enabling conditions include

Frequent communications among people with dissimilar viewpoints across departmental lines;

Moderate environmental uncertainty and mechanisms for focusing attention on changing conditions;

Cohesive work groups with open conflict resolution mechanisms that integrate creative personalities into the mainstream;

Structures that provide access to innovation role models and mentors;

Moderate personnel turnover; and

Psychological contracts that legitimate spontaneous innovative behavior.

Although extensive research has been conducted to substantiate the above motivating and enabling conditions for innovative behavior, very little research has examined what happens to people as they engage in an innovation over time. In particular, in Chapter 5 Angle points out there is a need to better understand the following human dynamics that were often observed in three temporal periods of the development of innovation groups studied by MIRP:

1. During the *startup period,* the dominant dynamics observed are individual recruitment and engagement in an innovation team, and the problems of the "hung jury," "acquiescent team player," and "tolerance for closure and trust," as described by Van de Ven (1985). In addition, the startup period is characterized by emotional euphoria, great expectations, and confidence among participants in the success of the innovative undertaking.

2. In the *middle period* the euphoria wanes, problems surface and the reality of the difficulty, complexity, and high risk of success have set in, and no conclusions or solutions are in sight to provide closure to the innovation effort. It is here where "hen-pecking," lack of trust or confidence in fellow workers, and leadership become manifest. Some people desert the effort and this creates problems of continuity. New people come on board without an "organizational memory." As a consequence, while the setting should be the most opportune time for learning by trial-and-error, remarkably infrequent instances of learning take place.

3. In the *ending period,* the innovation or a part of it terminates. An "end to the tunnel" comes in sight and group members find closure to their experiences. If the effort was not successful, members concoct reasons for why their ordeal was not in vain. Attributions of failure usually point to ex-

ternal "uncontrollable factors." If the effort is a "success," team celebrations emerge. Success attributions are directed to the commitment, talent, and heroic efforts of the team; the leader is embellished with a superhuman personification; and a variety of efforts are made to prevent or forestall termination of the group.

With the exception of Kanter (1983), adequate recognition or treatment of these social-psychological dynamics of group development have not been found in the innovation literature, although they have been alluded to in the literature on policy implementation (Pressman and Wildavsky 1974) and organizational development (Schein 1969). Perhaps this is because they only become apparent when undertaking real-time longitudinal field research. However, they were observed across many MIRP studies to represent some of the most "gut-wrenching" experiences for innovation participants and managers. These findings suggest that a process theory of innovation management needs to address the implementation problems described by Kanter (1983) of obtaining commitment, acceptance, and compliance through persuasion, bargaining, incentives, and power relations among people both within and outside the innovation unit.

Transactions

As we have just seen, innovation is not the enterprise of a single entrepreneur. Instead, it is a network-building effort that centers on the development of sets of ideas by people who, through transactions, become sufficiently committed to their ideas to carry them to acceptance and legitimacy. *Transactions* provide a useful unit of analysis for examining the development of a wide variety of relationships inherent to the management of innovation: (1) both collegial relationships among peers and hierarchical relationships among superiors and subordinates who engage in the development and management of an innovation, (2) proposals and commitments to obtain and allocate resources to the innovation and its subcomponents, and (3) quid pro quo arrangements with other individuals, units, and organizations to subcontract, coventure, or otherwise undertake various activities needed to develop an innovation over time.

According to Commons (1950), transactions are dynamic and go through three temporal stages: negotiations, agreements, and administration. In observing transactions in the MIRP studies, Ring and Van de Ven (Chapter 6) and Ring and Rands (Chapter 10) observe that most transactions do not follow simple linear progressions through these stages. The more novel and complex the innovative idea, the more often trial-and-error cycles of renegotiation, recommitment, and readministrations of transactions occurred. Moreover, they find that as the risk of a deal increases and trust among parties decreases, the lower the likelihood that parties will enter into or remain in a transaction; and if they do, the more costly and complex are the governance structures and contingent safeguards of the transactions. These more costly and complex procedures, in turn, increase the likelihood of transaction failure. Ring and Rands go on in Chapter 10 to empirically describe different patterns of informal processes of sensemaking, understanding, and committing that underlie each of the formal stages of negotiations, agreements, and administration.

Context

Context is the setting, or institutional environment, within which innovative ideas are developed and transacted among people. It is well known that an innovation does not exist in a vacuum and that an organization's internal structure and practices are in great measure a reflection of the amount of support or direction it can draw from its larger community. Thus, the management of innovation must not only be concerned with micro or proprietary developments of a particular innovative device or service, but also with the creation of an industry, or a macro infrastructure needed to implement or commercialize an innovation. As Rappa (Chapter 13), Knudson and Ruttan (Chapter 14), and Garud and Van de Ven (Chapter 15) report in their MIRP studies, collective action among firms in public and private sectors was found to be critical to create the social, economic, and political infrastructure a technological community needs to sustain its competing members.

For most technological innovations, this infrastructure includes institutional norms, basic scientific knowledge, financing, and a pool of competent human resources (see Chapter 15). These infrastructure resources are often initially developed as "public goods" in the public sector and appropriated by proprietary firms that transform them into "private goods" through innovation. Separate organizations often exist to provide these necessary resources for a given industry. However, these financial, educational, and research organizations are seldom easily accessible to a new industry that is emerging to commercialize an innovation. In addition, the infrastructure also requires establishing industry governance structures and procedures to regulate the behavior of competing firms, and legitimating the industry's domain in relation to other industrial, social, and political systems.

Thus, our study of the macro context of an innovation is examining (1) the role of the public sector in stimulating or inhibiting innovations in the private sector, (2) how and when this infrastructure is organized, (3) what firms cooperate to create this infrastructure, (4) how they bundle their market transactions to establish resource distribution channels (such as vendor-supplier-distributor relationships and joint ventures), and (5) what firms

emerge as industry competitors as well as cooperators.

Inherent in studying these processes is the paradox of cooperation and competition. Each firm competes to establish its distinctive position in the industry; at the same time, firms must cooperate to establish the infrastructure required for all industry participants to survive collectively. For example, it clearly benefits all firms to cooperate to set up industry standards. However, in doing so, each firm will try to ensure that standards that suit it best get institutionalized (see Chapter 15). Another key paradox is that not only do government policies act to simultaneously stimulate and retard industry development, but they also often change radically and unpredictably, thus creating an investment climate that can inhibit risk taking (Marcus 1980; Marcus, Nadel, and Merrikin 1984). An understanding of the unfolding of these and other paradoxes can offer valuable insights on how firms and government learn to cooperate to sustain themselves collectively, while at the same time compete to either perform their unique roles or carve out their distinctive position in an emerging industry.

OVERVIEW

The purpose of this introductory chapter has been to provide an introduction to the Minnesota Innovation Research Program, so that the reader understands the premises on which the remaining chapters are based. This has been accomplished by describing MIRP's objectives, history, conceptual framework, and some overall preliminary findings pertaining to the evolution of the key concepts in the framework. We have also summarized, in Appendix 1–A, the way the MIRP research is being carried out and have provided abstracts of each of the individual research projects that collectively comprise MIRP.

With this MIRP introduction as a backdrop, we now attempt to provide the reader a "roadmap" for an excursion through the remaining chapters. This is accomplished by previewing what is to be found in each of the sections. At the outset we admit that no "one best way" was found to present MIRP's findings across all studies, groups of studies, and those that pertain to individual studies. We experimented with several alternative layouts for the clusters of chapters that appear in this book. As in most complex decisions, each succeeding choice improved the situation in some respects while leaving us unsatisfied in other areas. What we have arrived at represents, we believe, the best balance in presenting the findings and perspectives pertaining to the overall program and individual innovation studies. In particular, Sections I, II, and VII focus on the overall program findings across MIRP studies, and Sections III to VI provide in-depth descriptions of individual innovation studies in MIRP.

We begin with the present section, which introduces the Minnesota Innovation Research Program. Chapter 2 describes the methodological considerations that have guided our investigation into the dynamics of organizational innovation. The chapter goes into some detail on a tracking methodology that we have found to be useful for maintaining empirical rigor in the analysis of many largely qualitative observations over time in a complex setting. Chapter 3 presents an evaluation of the Minnesota Innovation Survey (questionnaire) instrument on which many of our empirical findings have been based. The chapter applies standard psychometric procedures for evaluating tests and measurement instruments.

Whereas Section I provides an introduction to the MIRP framework and methods, Section II contains chapters that amplify the findings across the innovation studies as they pertain to and extend the core concepts in the overall MIRP framework. Some of the high-lights developed in the four chapters were introduced in the previous section. Chapter 4, focuses on the innovation *idea,* as it describes the complexity of the progression of an organizational innovation through its course of development. Chapter 5 highlights the *people* component of the MIRP framework, as it explores the psychological dynamics of creativity and innovation, both at the individual and social-system level. The *transactions* component of the framework is highlighted in Chapter 6, where formal and informal aspects of transactions are seen as complementary, rather than antithetical. The final chapter of Part II deals with innovation *outcomes* and documents the shifting in-process outcome criteria for different stakeholders over time. The chapter also proposes a process model of success and failure action loops to explain relationships between innovation actions and outcomes.

The perspective in Sections III to VI shifts to the individual studies within MIRP. The chapters in Sections III to VI are grouped according to conventional classifications of innovations. Whereas Section III examines innovation processes undertaken to create new businesses in the private sector, Section IV focuses on administrative innovations in both public and private sector organizations, and Section V deals with technological innovations in more complex organizational and industrial settings. All the studies in Sections III, IV, and V focus primarily on processes by which innovative ideas were created, developed, and implemented within the organizations undertaking the innovation. In contrast, Section VI examines the processes by which organizations adopt innovations that were developed elsewhere.

Section III consists of three chapters in one. It examines the process of new business creation in three different proprietary organizational settings: the creation of a new computer software business in a new company startup; the development of a new hearing health products business within an established, diversified company; and a joint venture among three companies to develop a new technology and related business in the health services industry. Contrary to much conventional wisdom, the research identifies many similarities in the dynamics of new startup innovations, despite the contextual and structural contrasts.

Section IV contains three chapters that examine processes of developing and implementing administrative innovations over time. Chapter 9 describes the processes by which networks of interest groups were mobilized to enact laws introducing major educational reform in Minnesota. Roberts and King emphasize the critical roles of "policy entrepreneurs" in pushing and riding educational reform ideas into good currency. Chapter 10 examines another major administrative innovation—the negotiation of a joint endeavor agreement between NASA (a large federal government bureaucracy) and 3M (an innovative private enterprise) to use the space shuttle as the transportation device to conduct experiments in low-gravity environments that may lead to efforts to commercialize space. Applying and expanding the transactions framework developed in Chapter 6, Ring and Rands develop a grounded-theory of emergent interpersonal transaction processes to explain dynamics observed in the longitudinal case study. Finally, Chapter 11 provides a longitudinal description of integration and accommodation processes in three corporate mergers or acquisitions. Although mergers and acquisitions are not new to the corporate landscape, they were found to be a common component of many technological innovations. Moreover, they tend to be novel and threatening experiences requiring significant learning on the part of individuals involved.

Section V, on technological Innovation, comprises four chapters that present longitudinal case studies of such technological innovations as: the design and production of a

highly complex weapons system by a defense contractor for the U.S. Navy; the development of gallium arsenide-based integrated circuits technology in the United States, Japan, and Western Europe; the introduction of a new hybrid wheat strain—a biological innovation that has involved numerous public and private sector organizations over its thirty years of development; and the emergence of a new industry to sustain the development of a cochlear implant device that has the potential to restore hearing in persons once considered irremediably deaf. Descriptions of the longitudinal development of these complex innovations commonly show how new technologies transgress across organizational, industry, sector, and indeed, national boundaries. No single institution commands sufficient competence or resources to control technological development. As a consequence, the management of technological innovation demands an appreciation of the simultaneous roles of cooperation and competition among firms in building an infrastructure to support technological innovation as well as proprietary entrepreneurship.

The perspective shifts sharply in the ensuing Section VI, which deals with studies of innovation adoption. Three chapters consider innovations that do not originate within the agencies or organizations that must implement them, but instead are imposed from some external source. Chapter 16 describes the organizational-effectiveness implications of two different reactions taken by U.S. nuclear power companies in response to the externally imposed nuclear safety procedures mandated by the U.S. Nuclear Power Commission. Marcus and Weber provide a very important and generalizable finding that the companies that adopted the "letter" of the mandated innovation performed less successfully than those that adopted the "spirit" of the innovation and modified it to meet their local situations. Then, Chapter 17 compares two different strategies adopted to implement site-based management, a decentralized management system in two public school districts. Contrary to expectations, the school district that implemented the innovation in "breadth" (across all schools in the district) was more successful than the school district that adopted the strategy of introducing it in "depth" within a school selected as the demonstration site. Finally, Chapter 18 chronicles the reactions of eight government organizations to an externally initiated application of private-sector style strategic planning processes.

In Section VII we return to the overall MIRP program purview, and draw some conclusions and inferences from these individual studies pertaining to innovation leadership, an emerging general process theory of innovation, and prescriptions for the management of innovation. In Chapter 19, the different forms of leadership appropriate at different innovation stages are illustrated with examples from several of the studies reported in preceding chapters, leading to the development of a double-loop model of leadership effectiveness for organizational innovation. Chapter 20 sets forth desiderata for developing a general theory of innovation process and integrates many of the separate theoretical contributions of earlier chapters in order to progress toward a metatheory of organizational innovation. Finally, Chapter 21 is written for the practitioner and distills a number of empirically supported recommendations for improving the management of innovation.

Again, it must be remembered that this is an interim statement of preliminary findings from an ongoing research program. Thus, the findings and perspectives presented should not be viewed as conclusions. Instead, they represent a significant in-process assessment of progress toward furthering our understanding of organizational innovation processes of many different types, in a wide variety of settings. For every answer we have attained, we have uncovered whole families of important new questions for further and continuing study.

APPENDIX 1–A SUMMARIES OF INDIVIDUAL MIRP INNOVATION STUDIES

Flexibility has been built into the Minnesota Innovation Research Program to enable each research team to investigate questions and issues unique to each innovation as well as to explore the emergence of patterns across various studies. Summaries of the questions, approaches, and progress of each innovation study are briefly summarized below. Although these summaries cannot do justice to the richness of each project, they show that each innovation study, by itself, represents a significant longitudinal research effort. Moreover, each study distinguishes itself by addressing novel and important questions on the management of innovation.

Medical Business Innovations
Andrew Van de Ven, Raghu Garud, and Douglas Polley, Department of Strategic Management and Organization, University of Minnesota
This study examines the processes of creating two new medical businesses in different organizational settings. Business creation is more encompassing than product innovation and more proprietary than technological innovation. Creating a new business usually entails the process of developing all the functional competencies necessary to exploit a technology into a family of related proprietary products, which, if successful, constitute a self-sustaining ongoing economic enterprise. The research project objectives are to empirically describe how two new medical businesses are being created over

time, in order to develop a process theory of business creation in different organizational settings.

One new business studied has involved the development of a line of otological products for the hearing health industry, including cochlear implants for profoundly deaf patients, hearing aids for the hearing impaired, and a related set of otological diagnostic instruments and services. This innovation has been housed within a large diversified corporation. A second new business, therapeutic apheresis, has been a joint venture between three corporations. This innovation has focused on developing a new line of medical products and diagnostic instruments for treating a variety of autoimmune diseases through the separation of pathogenic substances from blood and returning the beneficial blood components to the patient. Both business creation efforts have attempted to develop and commercialize new-to-the-world technologies and products, and both have attempted to create totally new businesses that could generate significant revenues in ten or fifteen years for the corporation involved.

Case histories were completed on the two innovations in 1984, and real-time field data collections of business development efforts have occurred since then. Seven rounds of interviews and questionnaire surveys were conducted with all key people involved in each medical business innovation in January and June 1985, January and August 1986, February and October 1987, and May 1988. In addition, the researchers regularly attend and observe monthly or bimonthly meetings of the business management teams involved in each business. After each data collection round, the innovation management teams received the survey results.

Based on publicly available records, interviews, and attendance at professional association conferences, the study also examines the emergence of new industries related to the two medical businesses. These industry analyses provide a more macro understanding of business creation and examine patterns of cooperation and competition among all major organizations worldwide that are connected with developments of the two medical business areas.

The Development of Complex Defense Contracting Innovation

Gary Scudder and Roger Schroeder Department of Operations and Management Science, University of Minnesota; Gary Seiler, Andrew Van de Ven, and Robert Wiseman, Department of Strategic Management and Organization, University of Minnesota

This study examines the management of complex innovation in the defense contracting industry. The study focuses on the management processes in developing and integrating a new naval weapons system into an organization that is currently producing the prior generation of the same weapon type. This represents a major process innovation for managing the development of complex systems across functional, organizational, and resource boundaries over time. The study examines the design-to-production transition process, a process innovation specifically designed to link together all of the functional areas required to accomplish a $500 million naval systems development contract. This process is being used to ensure that the final weapon design will be producible when introduced into the factory environment, within constraints of high quality and cost goals.

A case study of this innovation was developed in 1984 and is periodically updated to reflect changes that have been observed in the organization. Data collections occurred during November 1984, May and November 1985 that included both interviews with key program personnel and administrations of a survey questionnaire. Although funding for the study was terminated in 1986, additional interviews and observations on a less intensive basis are being used to track the innovation over time to its natural conclusion.

Site-Based Management in Public Schools

John Mauriel and Karin Lindquist, Department of Strategic Management and Organization, University of Minnesota

This study examines the implementation of an innovative participatory process known as school-based management in several public school districts. School-based management is grounded on the fundamental premise that the individual school is the basic organizational unit in a school district. This assumption challenges the widespread educational administration practice of centralizing school decisions in a central hierarchical structure. School-based management involves the relocation of strategic decisionmaking responsibility from the district superintendent to the school principal and then to school site councils, consisting of various stakeholders. As such, the intent is to institutionalize stakeholder participation in the decisions that matter most to each school's strategic operation.

Intensive longitudinal case studies are being conducted in four school districts that are implementing this innovation. Historical case analyses have been completed in two of the school districts in the fall of 1985, and the third is now underway. Baseline surveys of questionnaires and interviews have also been completed in two other school districts. In addition, a second round of data has been collected from the primary multischool research site.

New Company Startup, Adaptation, and Growth

Andrew Van de Ven and S. Venkataraman, Department of Strategic Management and Organization, University of Minnesota, Roger Hudson, College of Business Administration, University of Tennessee

This study aims to develop a process description and a theory of small company startup, adaptation, and survival or failure. Using transactions as the unit of analysis, the study integrates natural selection and organizational adaptation perspectives. It proposes that new business growth/decline is the result of the cumulative probabilities of transaction failures at the organizational population level. The study identifies the kinds of transactions in which small businesses engage, and the adaptive behavior of entrepreneurs. Strategic adaptation involves the selection of industry population niches, types of transactions, and entrepreneurial composition of the small business, which combine to define the risk of failure for a small business.

The study also examines several alternative ways to startup new businesses: (1) independent entrepreneurs starting their own firms, (2) corporate sponsorship of new businesses with equity investments and training, (3) spinoffs of internal corporate departments and ventures into stand-alone businesses, and (4) a variety of joint ventures, licensing agreements, and contractual relationships among firms to transact business. These alternative arrangements for new business startups vary from a market to hierarchical arrangements. The basic research question being examined is, "What are the developmental patterns and problems over time of different kinds of organizational arrangements for new business startups?"

During 1983, site visits were conducted to fourteen new business startups in Massachusetts, Pennsylvania, Illinois, and Minnesota. The site visits were conducted to develop case histories and collect baseline data on the initial planning and startup of seven new independent business startups, and another seven that were formed by equity investments of a corporate sponsor. A second round of site visits to the new businesses was conducted in fall of 1984 and followed up with a questionnaire survey and telephone interviews in the spring of 1985. It was found that significant performance reversals and shifts occurred between 1983 and 1985 for the fourteen firms, and three had gone out of business.

During 1986 and 1987 more of the new companies went out of business, and the research team has been tracking the companies to measure key factors and processes related to company survival, growth and demise. In order to observe these processes in depth and in real time, the research team has been tracking a particular company intensively by conducting monthly observations of key events and meetings of company principals since fall 1983.

The Commercialization of Space
Peter Ring and Gordon Rands,
Department of Strategic Management
and Organization, University of
Minnesota
This project explores the evolving set of intra- and interorganizational relationships associated with the efforts of a private-sector firm and a federal agency to coventure in the use of low-orbit space for commercial purposes.

Commencing its fourth year, the research has produced a series of interviews with all the major participants in the private-sector organization, from scientist to CEO. Survey data with the Minnesota Innovation Survey were also collected. Archival research on U.S. government policy on the commercialization of space is now nearing completion. Initial discussion with NASA officials regarding an interview schedule with major players in that agency has been interrupted by the loss of the *Challenger*. A case history of the initial two years of the coventure, from the perspective of the private-sector coventurer, has been completed and been reviewed by key players in the firm. A fourth round of survey data collection was conducted during summer 1988. Interviews with NASA officials have also been conducted.

The research team has developed a theory on the structure and process of transactions among firms. The theory explains how different governance structures and endogenous safeguards are used to structure transactions depending on the risk of the deal and trust among parties. In addition, the theory examines formal and informal processes in the development, execution, and growth or termination of transactions.

Nuclear Safety Standards Innovation Study
Alfred Marcus and Mark Weber,
Department of Strategic Management
and Organization, University of
Minnesota
This study investigates the adoption process by nuclear power companies of a set of new safety standards that are implemented and regulated by the Nuclear Power Commission. A framework for studying the innovation has been developed. Innovations may be internally generated in response to opportunities. They also may be externally imposed after accidents, scandals, or incidents shock an industry. The safety standards under consideration were externally imposed. Management has a variety of responses it can make to externally imposed changes. A typology of these responses has been developed, and the nuclear power study is examining the relationships between this typology, the dispositions of implementors, and organizational performance.

During the summer of 1985, twenty-five interviews with industry safety review managers were conducted by telephone. These in-depth telephone interviews provided an informative update of the original information concerning utility responses to NRC standards. During 1986 a survey was sent to safety review managers and twenty-two responses were received. This information is being used to do additional research in the area. A proposal to the National Science Foundation was submitted in June 1988 to continue the longitudinal research of safety incidents in the nuclear power companies.

The Development of Government Strategic Planning Systems

John Bryson, Hubert H. Humphrey Institute of Public Affairs, University of Minnesota; William Roering, College of Business Administration, University of Florida

Although strategic planning has been in vogue in private-sector organizations for over twenty years, the concept is just now gaining good currency in general purpose units of government. Because government organizations exist in a much more politicized environment, it is not at all clear how private-sector models of strategic planning can be applied to the public sector. However, the need for strategic planning has increased as public pressure mounts to cut government spending and taxes while demand continues to grow for government services. Therefore, government officials have a growing need to find ways to rethink what services they should provide and how to provide them.

A study of the development and implementation of strategic planning processes in general purpose units of government was undertaken in eight organizations. The study lasted two and one-half years and concluded in November 1987.

The Development of Gallium Arsenide Integrated Circuits

Michael Rappa, Sloan School, Massachusetts Institute of Technology, Andrew Van de Ven, University of Minnesota

This study examines the development of III–V compound semiconductor technology, which is expected to play an important role in future high-speed data processing, communications, and defense systems. The focus of the study is to understand the development of a revolutionary technological innovation with both field and bibliometric methods for tracking the processes of technological change over time.

From 1984 to 1987 a database was constructed that includes activities of more than 7,200 researchers working in about 400 R&D laboratories in eight countries on gallium arsenide integrated circuits technology since 1975. The data are being analyzed to examine several hypotheses regarding the growth of a research community, an educational infrastructure, and mobility of personnel among organizations engaged in gallium arsenide integrated circuits. The research is also examining collaboration and specialization in R&D activities among firms, and the diffusion of ideas and techniques in this emerging industry. A comparative analysis of institutional and industry activities in the United States, Japan, and Western Europe is also being conducted. In 1984 we conducted a pilot survey among forty firms in the United States and Japan that examined gallium arsenide R&D programs. The purpose of this survey was to obtain a basic understanding of the structure of these programs and perceptions of the achievements made by groups and the obstacles that still remain.

From June to September 1985, Michael Rappa conducted research in Tokyo, Japan. During this period, a comprehensive analysis was completed of the major Japanese electrical equipment manufacturers' commitment to development of this technology. The work in-

25

cluded estimates of manpower and expenditures and investigated the role of government, recent technical achievements, and plans for implementation.

The Development of Hybrid Wheat
Mary Knudson and Vernon Ruttan, Department of Agricultural and Applied Economics, University of Minnesota

Hybrid wheat development covers an extended historical period from 1950 to present and involves a number of major private and public organizations. At the onset of hybrid wheat breeding, scientists were searching for the introduction of a new technology that would enhance wheat yields. In the early 1960s private-sector breeders took (and have maintained) the lead in hybrid wheat development as a result of the publications of Johnson and Schmidt at the University of Nebraska and Wilson and Ross at Kansas State. These seminal publications made hybrid wheat development a feasible dream. Other private-sector breeders also have played key roles in the development of hybrid wheat. Among them are Cargill, DeKalb, Northrup King, Pioneer, and more recently, Monsanto, Rohm and Haas, and Shell. Public-sector breeders remained in hybrid wheat research only through the 1960s. Since then few resources have been devoted by the state experiment stations or the USDA for hybrid wheat development. In fact, North Dakota State is the only public-sector institution that still has an ongoing hybrid wheat project.

Historical and baseline studies of the development of hybrid wheat have been completed. The data include a written account of the R&D of hybrid wheat and examine case studies of several firms and universities that were once involved with hybrid wheat research. The history uses Usher's cumulative synthesis sequence theory to describe the development of hybrid wheat and Ruttan's endogenously in-

duced model of innovation to identify the endogenous factors within the wheat industry that characterize the R&D progress of hybrid wheat. This includes the identification of exogenous factors that induced the wheat industry to undertake and continue in hybrid wheat development.

In the spring of 1985 an industrywide survey was conducted of thirty-eight wheat breeders to gain information on the historical, baseline, and longitudinal phases of the hybrid wheat development study. The results of these surveys are currently being used to develop a theory that identifies the factors within an institution that contribute to the successful progress of a biological innovation.

Progress has also been made in drawing comparisons with the R&D of semidwarf wheat, a competing biological innovation. Actors have been identified who have participated in the semidwarf stay based on other studies that have already documented the story of semidwarf development. Events that were significant to semidwarf's developmental progress were identified using the Usher and Ruttan frameworks. Using the results of both the hybrid wheat and semidwarf wheat study, factors that affect R&D in the private and public sector were defined and compared.

Managerial and Organization Dynamics of Mergers and Acquisitions
David Bastien, Department of Communication, University of Wisconsin, Milwaukee; Andrew Van de Ven, University of Minnesota

Although mergers and acquisitions are not new or innovative to the corporate landscape, they tend to be precedent-setting, traumatic, and novel experiences for the individuals involved, particularly when they are members of the acquired firm. In addition, although an extensive literature and body of knowledge have developed on financial, legal, and strategic aspects of mergers and acquisitions, little systematic scientific knowledge is available on the human dynamics that unfold and the kinds of managerial behaviors that facilitate and inhibit execution of a merger. Mergers and acquisitions also provide natural field experimental settings for observing significant changes in corporate cultures, structures, systems, and styles.

Historical case studies have been conducted of five corporate mergers, and longitudinal studies are now being undertaken in another five mergers and acquisitions. The present studies include multiple sequence analysis, the analysis of cultural change (using methods derived from linguistic analysis), and detailed case histories. Based on these case studies, a typology of different kinds of mergers and acquisitions has been developed that is robust in discriminating between the differing dynamics and problems encountered in the mergers and acquisitions observed to date. Common patterns have also been observed, and the processes of individual and organization skill learning have been identified as the central postmerger or -acquisition change processes. Leadership in learning has also been examined.

The Process of Policy Innovation
Nancy Roberts, Naval Post Graduate School, Monterey, California; Paula King, Management Department, St. John's University

This study examines the role, function, and impact of policy entrepreneurs in the policy formulation and enactment process. In particular, it centers on the contribution of policy entrepreneurs to policy innovation. Policy entrepreneurs are actors in the policy community who risk personal, professional, and/or political credibility by espousing innovative policy alternatives and who actively seek to have their ideas enacted into law.

We began this longitudinal assessment of policy entrepreneurs at the Minnesota De-

partment of Education in 1983. The focus at that time was on the commissioner of education, who was espousing innovative educational policies for the state. Our study was broadened in 1984 to include other Minnesota policy entrepreneurs who were actively committed to and engaged in educational innovation. Research continuing to date has centered on ten policy entrepreneurs whose efforts have been central to enactment of innovative legislation. Our plans are to continue the study of policy entrepreneurs at least through the current legislative session, which ends in June 1988.

Hospital Organizations Innovativeness Study
John Kralewski and Bright Dornblaser, Department of Hospital and Health Care Administration, University of Minnesota

This study focuses on factors associated with the relative innovativeness of hospitals. Changes in styles of medical practice and shifts to prospective and capitation payments schemes have placed hospitals in a very vulnerable position. Demand for inpatient care is declining and payment for those receiving inpatient services is being reduced. Consequently, many hospitals are financially distressed.

Theoretically, the hospitals that succeed in making a transition to new roles and more efficient operations will be those that are best able to nurture creative ideas and translate those ideas into innovative programs. We hypothesize that hospitals that are profit oriented will be more innovative and will pursue innovations that have higher risks and potentially greater payoff. We also hypothesize that there are important structural differences within both investor-owned and not-for-profit hospital systems that contribute to their ability to innovate.

To test these hypotheses, forty-two investor-owned (for profit) and forty-one not-for-profit hospitals are being studied. Data were collected on the type of innovations being developed and implemented in these hospitals, the organizational support for innovation (the innovation climate), and the risks of any payoffs from the innovations. Data are being collected on the degree of centralization of decisionmaking and the financial performance of the hospitals. The analysis of the data is proceeding along three main lines of inquiry:

1. The innovation climate and type of innovation in investor-owned versus not-for-profit hospitals is being compared and analyzed.
2. The type of innovation and the risk of any payoff associated with these innovations in the investor-owned hospitals are being analyzed as dependent variables related to the degree of centralization of decision-making and the corporate culture of these organizations.
3. The degree of innovations of nursing units in twelve matched investor-owned and not-for-profit hospitals is being analyzed in terms of the effects of innovations on the productivity of the hospital to its final performance.

*Human Resource Management
Innovation Study*
**Harold Angle and Marian Lawson,
Department of Strategic Management
and Organization, University of
Minnesota**
This study consists of two projects, both within the same large manufacturing organization. The first study examines the effects of human resource management policies and practices, with emphasis on human and organizational impacts of a significant organizational change—the establishment of a new major headquarters site in another state. The second study is a longitudinal case study of an administrative innovation; the reorganization and re-

orientation of an organization's human resource management system in order to bring human resource management closer to the strategic management process.

Study 1: The Relocation Impact Study. The primary focus of the study is an analysis of how relocating some of a manufacturing firm's operating divisions from one region of the country to another affects the careers, work adjustment, and personal adjustment of persons involved. This study includes both persons who relocate and those who were encouraged to relocate but did not. It also includes family members. In addition to this "micro" perspective, the impacts of this innovation on organizational effectiveness are investigated.

At the beginning of the project (1985), surveys were administered to about 700 employees and 400 spouses who were involved in the relocations. Two follow-up surveys were subsequently administered:

1. A survey of persons who did not relocate and were reassigned to nonrelocating divisions that remained at the original site. The organization found new jobs within the company for virtually every employee who elected not to move with the relocating divisions (approximately 200 persons). This survey collected data from these individuals, and their new bosses, to ascertain the impacts of the organization's reassignment policies on (a) the people themselves, (b) the divisions that received them, and (c) the organization as a whole.
2. A second survey of the persons who did relocate. This was administered early in mid-1987 to ascertain changes in the effects of the relocation after persons affected have had several months to settle-in at the new site. Again, family members were surveyed in addition to the employees, themselves.

For all surveys, the identity of all respondents was known, enabling the research team to compare individual time-1/time-2 responses, and employee/spouse responses, as well as to compare performance data on organizational records with data collected in the surveys. In addition to the three surveys, supplementary data were collected by means of interviews and nominal group meetings, as well as by compilation of appropriate documentary evidence.

Although the three-year period contracted for the study has elapsed, it now appears that the project will continue as additional divisions begin to relocate. The first of these relocations will take place early in 1989.

Study 2: The Human Resource Management Innovation Study. This study documents and analyzes the planning and implementation of an innovation initiated by the organization's vice president of human resources in which the basic relationship between line management and human resource management is altered so that line managers take a more direct role in human resource management and human resource managers become true generalists whose principal focus in on business, rather than staff, concerns. Considerable data were collected at the onset of this three-year project; however (with the exception of certain documentary evidence), it was not possible to collect longitudinal process data as this innovation unfolded. Accordingly, we are only now beginning to accumulate the information needed to conclude the analysis of this case.

REFERENCES

Angle, H.L., and R.L. Hudson. 1986. *The Minnesota Innovation Research Program (MIRP): Collective Action in an Individualist Culture?* Discussion Paper 57, Strategic Management Research Center, University of Minnesota (August).

Cameron, K. 1980. "Critical Questions in Assessing Organizational Effectiveness." *Organizational Dynamics.* 9: 66–80.

Cohen, M.D., J.G. March, and J.P. Olsen. 1972. "A Garbage Can Model of Organizational Choice." *Administrative Science Quarterly* 17: 1–25.

Cooper, A. 1979. "Strategic Management: New Ventures and Small Business." In D. Schendel and C. Hofer, eds., *Strategic Management,* pp. 316–26. Boston: Little, Brown.

Commons, J. 1950. *The Economics of Collective Action.* New York: Macmillan.

Daft, R., and S. Becker. 1978. *Innovation in Organization.* New York: Elsevier.

Damanpour, F., and W.M. Evan. 1984. "Organizational Innovation and Performance: The Problem of Organizational Lag." *Administrative Science Quarterly* 29: 392–402.

Glaser, B.G., and A.L. Strauss. 1967. *The Discovery of Grounded Theory: Strategies for Qualitative Research.* Chicago: Aldine.

Janis, I.L. 1983. *Groupthink: Psychological Studies of Policy Decisions and Fiascoes,* 2d ed. Boston: Houghton Mifflin.

Kanter, R. 1983. *The Change Masters.* New York: Simon and Schuster.

Kimberly, J.R. 1981. "Managing Innovation." In P. Nystrom and W. Starbuck, eds., *Handbook of Organizational Design,* Vol. 1, pp. 84–104. Oxford: Oxford University.

Lawrence, P.R., and P. Dyer. 1983. *Renewing American Industry.* New York: Free Press.

Leavitt, H.J. 1965. "Applied Organizational Change in Industry: Structural, Technological, and Humanistic Approaches," In J. March, ed., *Handbook of Organizations,* pp. 1144–1170. Chicago: Rand McNally.

Lewin, A.Y., and J.W. Minton. 1986. "Organizational Effectiveness: Another Look, and an Agenda for Research." *Management Science* 32: 514–38.

Magaziner, I.C., and R.B. Reich. 1982. *Minding America's Business: The Decline and Rise of the American Economy.* New York: Harcourt, Brace, Jovanovich.

March, J.G. 1981. "Footnotes to Organizational Change." *Administrative Science Quarterly* 26: 563–77.

Marcus, A. 1980. *Promise and Performance: Choosing and*

Implementing an Environmental Policy. Westport, Conn.: Greenwood-Praeger.

Marcus, A., M. Nadel, and K. Merrikin. 1984. "The Applicability of Regulatory Negotiation to Disputes Involving the NRC." *Administrative Law Review* 36, 3 (Summer): 213–38.

Mohr, L.B. 1982. *Explaining Organizational Behavior: The Limits and Possibilities of Theory and Research.* San Francisco: Jossey-Bass.

Nelson, R.N., and S.G. Winter. 1977. "In Search of a Useful Theory of Innovation." *Research Policy* 6: 36–76.

Normann, R. 1971. "Organizational Innovativeness: Product Variation and Reorientation." *Administrative Science Quarterly* 16: 203–15.

Ouchi, W.G. 1981. *Theory Z.* Reading, Mass.: Addison-Wesley.

Perrow, C. 1961. "The Analysis of Goals in Complex Organizations." *American Sociological Review* 26: 854–66.

Peters, T.J., and R.H. Waterman, Jr. 1982. *In Search of Excellence: Lessons from America's Best Run Companies.* New York: Harper & Row.

Poole, M.S. 1983. "Decision Development in Small Groups, III: A Multiple Sequence Model of Group Decision Development." *Communication Monographs* 50:321–41.

Pressman, J., and A. Wildavsky. 1974. *Implementation.* Berkeley: University of California Press.

Rogers, E. 1982. *Diffusion of Innovations,* 3d ed. New York: Free Press.

Ruttan, V., and K. Hayami. 1984. "Toward a Theory of Induced Institutional Innovation." *Journal of Development Studies* 20: 203–23.

Schein, E.H. 1969. *Process Consultation: Its Role in Organization Development.* Reading, Mass.: Addison-Wesley.

Schelling, T. 1978. *Micromotives and Macrobehaviors.* New York: Norton.

Staw, B.M., and J. Ross. 1987. "Behavior in Escalation Situations: Antecedents, Prototypes, and Solutions." In L.L. Cummings and B. Staw, eds., *Research in Organizational Behavior,* Vol. 9, pp. 39–78. Greenwich, Conn.: JAI.

Teece, D.J., ed. 1987. *The Competitive Challenge: Strategies for Industrial Innovation and Renewal.* Cambridge, Mass.: Ballinger.

Tornatzky, L.G., et al. 1983. *The Process of Technological Innovation: Reviewing the Literature.* Washington, D.C.: National Science Foundation.

Turner, J.H. 1987. "Toward a Sociological Theory of Motivation." *American Sociological Review.* 52: 15–27.

Utterback, J.M. 1974. "Innovation in Industry and the Diffusion of Technology." *Science* Vol. 183, No. 4125 (February): 620–26.

Van de Ven, A.H. 1980. "Problem Solving, Planning and Innovation, Part II: Speculations for Theory and Practice." *Human Relations* 33: 757–79.

———. 1982. *Strategic Management Concerns Among CEOs: A Preliminary Research Agenda.* Presented at Strategic Management Colloquium, University of Minnesota, Minneapolis, Minnesota, October.

———. 1983. *Organizing Innovations: Research Proposal to the Program on Organizational Effectiveness of the Office of Naval Research,* Minneapolis: University of Minnesota, Strategic Management Research Center.

———. 1985. "Spinning on Symbolism: The Problem of Ambivalence." *Journal of Management* 11:101–02.

———. 1986. "Central Problems in the Management of Innovation." *Management Science* 32: 590–607.

Van de Ven, A.H., and associates. 1984. *The Minnesota Innovation Research Program.* Minneapolis: University of Minnesota, Strategic Management Research Center, Discussion Paper 10 (April).

Van de Ven, A.H., and M.S. Poole. 1988. "Paradoxical Requirements for a Theory of Organizational Change. In R. Quinn and K. Cameron, eds., *Paradox and Transformation: Toward a Theory of Change in Organization and Management,* ch. 2. Cambridge, Mass.: Ballinger.

Weick, K. 1979. *The Social Psychology of Organizing,* 2d ed. Reading, Mass.: Addison-Wesley.

Zaltman, G., R. Duncan, and J. Holbek. 1973. *Innovations and Organizations.* New York: Wiley.

2

METHODS FOR STUDYING INNOVATION PROCESSES

Andrew H. Van de Ven

Marshall Scott Poole

Little is known about how innovations emerge, develop, grow, or terminate over time. Not only has little systematic research been conducted to examine how innovations develop over time, but few process theories adequately explain the sequence of events in the innovating process. Yet an appreciation of temporal processes is fundamental to managing innovations. Given the limited research and theory on innovation processes, MIRP researchers decided to adopt a grounded-theory strategy to discover an innovation process theory (Glaser and Strauss, 1967) which avoids existing overbroad theories in the literature.

This chapter suggests methods for

generating a process theory of innovation from data systematically obtained from the MIRP longitudinal field studies of innovation. The strategies for developing grounded theory suggested by Glaser and Strauss (1967) were taken as a point of departure for research design. Developing a grounded theory of how and why innovations develop over time also requires some special methods for process analysis of longitudinal event data. In particular, we came to recognize that four requirements are necessary to study the process of innovation:

1. A clear set of concepts for selecting and describing the objects to be studied;
2. Systematic methods for observing change in the objects over time;
3. Methods for representing raw data to identify process patterns;
4. A motor or theory to make sense of the process pattern and a means of determin-

Support for this research program has been provided in part by a grant to the Strategic Management Research Center at the University of Minnesota from the Program on Organization Effectiveness, Office of Naval Research (code 4420E), under contract No. N00014-84-K-0016, as well as by other sources.

The order of authorship is arbitrary; the authors shared equally in all phases of development of this chapter.

ing whether the theory fits the observed patterns.

As Pettigrew (1985), Nisbet (1970), and Van de Ven and Poole (1988) indicate, these four requirements represent basic methodological steps in process analysis and for building "process theories" as distinct from "variance theories" that examine interrelationships among variables measured more or less at one point in time (Mohr 1982). Social scientists have developed powerful methods to examine variations (correlations and regressions) among variables measured either at one point in time or at successive times by a panel technique. Far less well developed are methods to examine how phenomena develop over time. The major methodological challenge for MIRP has been to develop concepts and methods for doing so.

After clarifying definitions of key terms implicit in the four requirements for processual analysis, this chapter discusses the various approaches that have been adopted by MIRP investigators to meet these requirements. These methods are being developed through an interactive trial-and-error process. When the MIRP field studies began, relatively simple and standard methods were adopted to collect, tabulate, and analyze the data. As the longitudinal data accumulated and became more complex, more sophisticated methods were necessary for data reduction, display, and drawing conclusions about observed innovation processes. Some of these new methods, in turn, entailed substantial "tooling up" costs of recoding the data and restructuring the data files in order to determine their utility. Thus, we forewarn the reader that some of the methods to be presented in this chapter remain to be tested because they are still in the "tooling up" stage. Presenting them here, however, will provide the reader with a useful description of MIRP's evolving methodology for developing a grounded theory of innovation processes. It

will also provide the methodological background necessary to appreciate the core themes and findings presented in subsequent chapters.

DEFINITIONS

The four requirements for process analysis derive from definitions of innovation, innovation processes, change, and processes of change. A theory of innovation is fundamentally a theory of change in a social system. While *innovation* is defined as the introduction of a new idea, the *process of innovation* refers to the temporal sequence of events that occur as people interact with others to develop and implement their innovation ideas within an institutional context. *Events* are instances when changes occur in the innovation ideas, people, transactions, contexts, or outcomes while an innovation develops over time. *Change* is an empirical observation of differences in time on one or more dimensions of an entity. All four elements in this definition are necessary. As Nisbet (1970) describes, a mere array of differences is not change but only differences. Time is also a critical element, for any differences necessarily involve earlier and later points of reference. Mobility, motion, and activity in themselves do not constitute change, although each is in some degree involved in change. Certain dimensions or categories of an entity are the objects being transformed. Change without reference to an object is meaningless.

The *process of change* adds an additional and more abstract element to this definition of change. Whereas *change* is an empirical or manifest observation, the process of change is an *inference* of a pattern of differences noted in time. Thus, processes of change, like an innovation process, are not directly observed. Instead, they are conceptual inferences about the temporal order or pattern of observed changes.

With these definitions, the relationships among the four requirements for process analysis become apparent. While the first requirement specifies the objects being investigated, the second requirement deals with empirical observations of changes in these objects. The third requirement addresses methods for inferring processes of change, and the fourth requirement is concerned with theories or conceptual motors that can explain these processes of change.

As applied to the study of innovation, MIRP addresses the first requirement with the concept of events, which are changes in the ideas, people, transactions, context, or outcomes of the innovations being studied. The second requirement entails longitudinal data collection methods and field observation of innovation events. The third requirement involves methods for identifying order and developmental paths in sequences of innovation events. The last requirement involves developing and applying theories about innovation process to explain the event patterns. The remainder of this chapter elaborates on these four requirements.

CONCEPTS FOR SELECTING AND DESCRIBING INNOVATIONS

"Generating a theory from data means that most hypotheses and concepts not only come from the data, but are systematically worked out in relation to the data during the course of the research" (Glaser and Strauss 1967:6). Of course, researchers do not approach reality as a *tabula rasa.* They must have a perspective that enables them to choose (1) a set of objects to study and (2) a set of concepts to determine what aspects of the objects being observed represent relevant data. We discuss each of these starting principles for discovering grounded theory in the MIRP studies.

Selection of Innovation Sample

In the social sciences, the comparative method is perhaps the most general and basic strategy for generating and evaluating valid knowledge. This strategy involves the study of several comparison groups chosen to differ in the scope of the population and conceptual categories of central interest to the research. Glaser and Strauss (1967:49–60) point out that the discovery of new theory from data is facilitated by maximizing differences among comparative groups (thereby maximizing differences in data) because this brings out the widest possible coverage on the ranges, types, variations and causes in conditions, structural mechanisms, processes, and consequences that are all necessary for elaboration of a theory.

Since the inception of MIRP its researchers have been interested in developing an orientation to the management of innovation that is broadly strategic as opposed to the functionally defined and technically oriented perspective that has pervaded prior studies of innovation. According to Van de Ven and associates (1984:18–19),

> Our knowledge of innovation and entrepreneurship is exceedingly narrow—usually focusing on one kind of organizational arrangement for innovation (e.g., internal innovations or new business startups), or one stage of the innovation process (e.g., the diffusion stage), or one kind of innovation (e.g., high technology). What is needed is a broader-guaged perspective that strategically examines a variety of innovations in alternative organizational settings across levels of analysis and over time.

Corresponding to this strategic management orientation is a definition of the population of innovations of interest to the MIRP researchers. Innovation was broadly defined to include both technical innovations (new technologies, products, and services) and administrative innovations (new procedures, policies, and organizational forms) in public, private, and not-for-profit organizational settings. Contrary to the belief that some of these types of innovations may not be comparable, MIRP researchers argued that technical and administrative innovations are closely interrelated and coproduced.

Daft and Becker (1978) and others have emphasized keeping technical and administrative innovations distinct. We disagree; making such distinctions often results in a fragmented classification of the innovation process. A major objective of the research is to understand the close connection between technical and administrative dimensions of innovations (Van de Ven and associates 1984: 5).

This strategic management orientation and definition of a population of innovations lead MIRP researchers to select a sample of fourteen highly diverse types of innovations, each of which would be studied using a common set of concepts described in the next section. They argued that while the maximum diversity in types of innovations increases the potential for generating new theories on the management of innovation, a core framework of common concepts permits identifying process similarities and differences across the innovations and thereby permits testing the generality of innovation process theories (Van de Ven and associates 1984: 19).

Three steps were undertaken in selecting the innovation sample. First, a brainstorming meeting was conducted in summer 1983 among the initial MIRP coinvestigators at which they speculated, "The year is 2000. What innovations were developed during the 1980s that made a significant impact on our society today?" The search for innovations was targeted for the fields of agriculture, industry, electronics, health care, education, nuclear power, government, and public- and private-sector partnerships. Although this list was not comprehensive, innovations in these areas could have a profound effect on society. More-

over, selecting innovations in these areas would ensure a sufficiently diverse sample across technical and administrative dimensions, technology/industry sectors, public and private organizations, and domestic and international arenas. Such a sample was sufficiently diverse to permit systematic examination of the degrees to which processes of innovation are similar or different in various settings.

The next step involved identifying specific innovations in these areas and co-investigators interested in or already engaged in studying them. From fall 1983 to spring 1984 meetings were conducted with managers in public and private organizations to identify and explore study access to innovations that might be planned or just under way in these areas. As potential innovation study sites arose, meetings were initiated with faculty researchers in appropriate academic departments to determine their interest in conducting a specific study and joining the research program. In a few other circumstances, university researchers who already had access to study specific innovations in MIRP's areas of interest (e.g., multi-institutional hospital organizations and nuclear power safety standards) joined the program in order to benefit from the interdisciplinary intellectual community created by MIRP.

The third step entailed introducing the researchers interested in specific studies to managers and organizational sites of the innovations and negotiating a mutually agreeable research relationship. Not all these negotiations proved successful (potential studies of deregulation of financial services industry, industrial innovations in ceramics, optical disk recording media, fashion merchandising did not materialize). Other negotiations resulted in unanticipated shifts in the innovations to study (meeting with managers to study the development of supercomputers was transformed to a study of gallium arsenide integrated circuits to the mutual interest of researchers and managers). Finally, several potential innovation studies could not be launched due to inabilities in finding qualified or interested researchers.

The results of these three steps led to launching longitudinal studies of fourteen types of innovations in the following areas: the development of hybrid wheat, advanced integrated circuits, cochlear implants and therapeutic apheresis biomedical devices, public- and private-sector ventures to conduct experiments on the space shuttle, multi-institutional hospital systems, startup of a computer hardware and software company, defense contracting of a naval weapons system, the introduction of nuclear power safety standards, strategic planning systems in local municipal governments, educational reforms at the state and local school levels, human resources management innovations, and organizational mergers and acquisitions.

Core Set of MIRP Concepts

Implicitly or explicitly, study of any change or innovation process entails examination of categories or variables that describe the innovation, the actions taken to bring it about, and the forces that influence the development of the innovation. As should be expected, different categories will produce very different substantive inquiries. As described in Chapter 1, the MIRP framework examines the development of an innovation in terms of five key concepts: ideas, people, transactions, context, and outcomes. Consistent with a grounded-theory research strategy, the meanings of these key concepts evolved over time as the MIRP field studies progressed. We began with the common assumptions and perspectives of these concepts implicit in the literature illustrated in Table 1-2. However, by observing innovations in their field studies, MIRP researchers began to appreciate richer and more dynamic perspectives of these concepts. These evolving perspectives, in turn, not only entailed new definitions of the concepts, but also revised

ways to operationalize and measure them in various innovation studies. As a consequence, these concepts came to be operationalized differently depending on the type of innovation being studied, the data collection methods employed, and the substantive focus of various MIRP research teams. These differences will become evident in subsequent chapters that report the research of particular innovations.

Whatever the concepts and organizational settings examined in a study, research on innovation processes requires a clear understanding of how change can be observed. Measurement of change necessarily implies not only a longitudinal study, but also rigorous methods for observing differences over time in the conceptual categories of the innovation being investigated.

METHODS FOR OBSERVING CHANGE OVER TIME

Most studies of innovation or change to date have been retrospective case histories conducted after the outcomes of change were known. However, it is widely recognized that prior knowledge of the success or failure of an innovation invariably biases a study's findings. Historical analysis is necessary for examining many questions and concerted efforts can be undertaken to minimize bias, but it is generally better, if possible, to initiate historical study before the outcomes of a strategic change process become known.

Moreover, time itself sets a frame of reference that directly affects our perceptions of change. As Pettigrew (1985) notes,

> the more we look at present-day events, the easier it is to identify change; the longer we stay with an emergent process and the further back we go to disentangle its origins, the more likely we are to identify continuities.

Appreciating this dilemma requires that investigators carefully design their studies in order to observe changes that are relevant to the purposes and users of their research.

Because the purpose of MIRP is to understand the management of innovation, the MIRP researchers decided to place themselves into the manager's temporal and contextual frames of reference. This entailed two overlapping stages in launching the field studies.

Once access was obtained to specific innovation study sites from fall 1983 to mid-1984, case histories and baseline data were obtained on each innovation. The case histories provided a mapping of events that led to the present longitudinal studies of innovations. The baseline information described the institutional settings in which the innovations were taking place with information being obtained from published reports, interviews and questionnaires. Case histories on each innovation were written following a standardized outline of headings and topics.

Longitudinal tracking of the innovations began as soon as it was determined what specific aspects of each innovation should be studied over time. Core data collection instruments were initially devised during the winter of 1984 and revised in spring of 1985 to track the innovation process in as many of the studies as possible. These instruments consisted of schedules for on-site observations, interviews, questionnaires, and archival records used to study the innovations as they developed in the ensuing years. Tables 2-1 to 2-3 summarize the factors measured in these instruments. These MIRP instruments, along with procedures for data collection, coding, dataset construction, and analysis are available in a *Methods Manual for Minnesota Innovation Research Program* (Van de Ven et al. 1987). A psychometric assessment of the measurement properties of the survey instrument was conducted by Van de Ven and Chu and is presented in Chapter 3.

TABLE 2–1. *Factors Measured In MIRP Survey.*

Innovation Idea
· uncertainty

People
· decision influence
· leadership
· individual competence
· time allocation

Internal Transactions
· standardization of procedures
· workload pressure
· communication frequency
· conflict frequency & resolution methods

External Group Transactions
· dependence
· formalization
· mutual influence
· perceived effectiveness
· duration of relationship
· complementarity
· consensus/conflict
· communication frequency

Organizational Context
· organizational risk taking
· freedom to express doubts
· chance of rewards & sanctions
· resource scarcity

Outcomes
· perceived effectiveness
· perceived innovation problems

In addition to obtaining regularly scheduled data with standardized interviews and questionnaires, real-time field observations were conducted in several MIRP studies in order to obtain first-hand observations of how changes in the innovations occurred over time. The repetitive surveys and interviews provided comparative-static observations of the innovation concepts being tracked over time. *Differences* between time periods on these concepts indicated what *changes* occurred in the organizational unit or program. But to understand how these changes came about, it was necessary to supplement the regularly scheduled data collection with intermittent real-time data. For ex-

TABLE 2–2. *Factors Measured From Archival Records.*

Characteristics Of Innovation Unit
- organization chart
- personnel roster
- organizational form
- type of innovation
- novelty of innovation
- developmental stage
- age

Characteristics of Innovation's Industry
- competitors of innovation unit
- structure of industry
- industry emergence process
- suppliers to innovation
- potential customers of innovation

Resources Used On Innovation
- dollars budgeted and spent
- days of outside consulting help
- proportion of resources "bootleged"

TABLE 2–3. *Factors Measured In MIRP Interviews.*

Innovation Idea
- describe idea as viewed today
- changes in idea since last time
- how and when these changes occurred

People
- list of people involved
- Changes in personnel and roles
- how and when these changes occurred

Transactions
- describe key agreements with others
- changes in agreements
- how and when these changes occurred

Internal Organizational Context
- changes in internal organization
- macro-organization factors effecting innovation
- how and when these changes occurred

Industry/Community Context
- industry factors affecting the innovation
- changes in industry/community context
- how and when these changes occurred

Outcomes
- criteria used to judge success
- grade given on these criteria (A, B, C, D)
- changes in outcome criteria or grade
- how and when these changes occurred

Final Question
- major challenges in past six months
- steps being taken to deal with problems
- significant changes expected in future

ample, in studies of the development of cochlear implant, therapeutic apheresis, Qnetics, school site-based management, and state educational reform innovations, this involved observing key committee meetings, decision or crisis events, and conducting informal discussions with key participants engaged in an innovation. Both regularly scheduled and real-time observations were necessary because, although differences scores between regularly scheduled observations identified *what* changes occurred, real-time observations at key intermittent periods were needed to understand *how* these changes occurred.

As Argyris (1968, 1985) has forcefully argued over the years, significant new methods and skills of action science are called for to conduct this kind of longitudinal real-time research. In addition, it implies significant researcher commitment and organizational access, which few researchers have achieved to date. As a consequence, very few process studies of innovation have been conducted. Our process research experience suggests that one reason that gaining organizational access has

been problematic is that researchers seldom place themselves in the manager's frame of reference to conduct their studies. (See Chapter 18 by Bryson and Roering for a useful discussion of researcher positioning.) Without observing the innovating process from a manager's perspective, it becomes difficult (if not impossible) for an investigator to understand the dynamics confronting managers who are involved in an innovation effort and thereby generate findings that are relevant to the theory and practice of innovation management. If organizational participants perceive little potential use for a study's findings, there is little to motivate their providing access and information to an investigator.

Of course, not all innovations included in MIRP could be studied in real time, nor could they adopt all the core MIRP data collection instruments on a regular basis over time. These studies tracked developments in their innovations by adopting somewhat different methods to observe changes in the core framework concepts (ideas, people, transactions, context, and outcomes).

For example, design of the study of hybrid wheat development required an appreciation of the unique temporal laws that govern the production process of a biological innovation—just as different technological and administrative laws govern the rate of development of product and service innovations. As Knudson and Ruttan describe in Chapter 14, the biological rate of hybrid wheat development entails several decades (thirty years) in order to move from basic research through technology development and to market introduction. Hence, Knudson and Ruttan adopted historical methods and "snowball" sampling and field interview techniques that were consistent with hybrid wheat's temporal biological laws of development.

In addition to examining unique characteristics of the innovation under study, a substantive research question also must adopt different data collection methods. For example, to study the emergence of a new technological paradigm, in Chapter 13 Rappa adopts a novel bibliometric methodology to observe the rates of growth in communities of researchers engaged in gallium arsenide integrated circuits in the United States, Japan, and Western Europe between 1976 and 1986. In these studies, practical considerations eliminated the possibility of conducting real-time process field observations or of administering survey instruments designed to obtain participants' recall of activities occurring during the previous six months. Instead, they used historical, bibliometric, and other archival methods to identify events in the development of the innovation under study.

METHODS TO IDENTIFY PROCESSUAL PATTERNS

Obtaining systematic observations of the innovation process over time using multiple methods quickly produces an overwhelming amount of rich raw data about an organizational innovation effort. Drawing inferential links between these data and theory requires methods for organizing and evaluating the raw data in a manner that facilitates identifying process patterns.

Although the task may seem formidable on *quantitative* data, an extensive methodology has developed to codify procedures for handling longitudinal panels of quantitative data, including procedures for constructing computer data files and for analyzing longitudinal data (Tuma and Hannan 1984). Procedures for coding and analyzing quantitative data collected in the MIRP studies are described in Chapter 3 by Van de Ven and Chu.

Far less has been written about methods for analyzing longitudinal *qualitative* data. The four basic steps outlined below have been found useful in tabulating qualitative data in a manner that helps identify process change patterns. These methods were used in a number of MIRP studies described in this book. Also described are more exact and detailed methods that were developed in the course of the project but not yet been applied to the data. These methods will be used in the next stage of MIRP analysis, which builds on the present reports (see Van de Ven et al. 1988).

Chronological Listing of Qualitative Events

The first step in tabulating qualitative data is to develop a chronological listing of events that occur in the development of an organizational innovation being investigated, as illustrated below:

Month/Year	Event	Data Source
.	.	.
:	:	:

Events require careful definition and vary with the innovation being investigated. For example, although the real-time studies of cochlear implant, therapeutic apheresis, Qnetics, and the naval defense contractor adopted fine-grained definitions of events. More coarse-grained definitions were adopted in the hybrid wheat, school site-based management, and government strategic planning studies. However, whether fine- or coarse-grained, all the MIRP studies defined *events* as instances when changes were observed to occur in either the ideas, people involved, transactions or relationships engaged in, context, or outcomes of the innovation being examined over time. Chronological listings of events were obtained by combining data collected through multiple methods and sources over time, including surveys, interviews, participant observations, archival sources, and published information.

Coding Chronological Events into Conceptual Tracks

The next step in organizing the longitudinal data into a format that facilitates identifying change processes is coding the chronological listing of events into multiple tracks that correspond to the conceptual research categories. Poole's (1983a, 1983b) multiple sequence model provides a useful descriptive system that specifies tracks used for recording process activities. Poole's method avoids the problem of preordaining the existence of stages or phases to the process, yet provides a way to identify cycles or transitions among activity tracks. In this way, it facilitates the development and testing of models or theories about innovation and change processes. Instead of assuming that change occurs through a predetermined set of phases or stages, Poole (1983a, 1983b) suggests portraying events as a set of parallel strands or tracks of activities. Phases and their sequence can then be distilled from analysis of these activities.

Each track represents a different concept or category in one's research framework. For example, following from the MIRP definition of innovation process, a minimal description of innovation requires at least five tracks:

People track	
Ideas track	
Transactions	
Context track	
Outcomes track	
Time	

For each track, a coding scheme is developed that enumerates the kinds of activities or issues occurring at each point in time, as follows:

People track: a coding of the people/groups involved in an activity, the roles and activities they perform at a given point in time, and how they formulate problems and makes decisions;

Ideas track: a coding of the substantive ideas or strategies that innovation group members use to describe the content of their innovation at a given point in time;

Transactions track: the informal and formal relationships among innovation group members, other firms, and groups involved in the innovation effort;

Context/environmental track: a coding of the

39

exogenous events outside of the innovation unit in the larger organization and industry/community that are perceived by innovation group members to affect the innovation;

Outcomes track: a coding of success criteria and ratings by innovation participants of how well the innovation is progressing and accomplishing their expectations of effectiveness at a given point in time.

For each concept track a number of different coding schemes are possible, depending on the particular questions being addressed by the researchers. For example, Ring and Rands (Chapter 10) categorized events on the transactions track with a two-tiered coding system that classified them first as belonging to a negotiation, agreement, or administration phase and, second, within each phase, as serving a sensemaking, understanding, or contracting function. This enabled Ring and Rands to study the evolution of transactions governance structures and processes. Bastien (Chapter 11) dealt with transactions in a different way. He "dissolved" this track into "decision process" and "communication process" in order to study integration processes in corporate mergers and acquisitions.

Tracking coding systems must be carefully designed to capture meaningful and important constructs necessary for testing developmental theories and hypotheses. In the case of grounded theory development, coding systems should embody the appropriate sensitizing concepts. In all events, coding systems represent refinements of the basic concepts that select out and emphasize key features, while omitting other aspects. In coding these events frequent references were made back to basic concepts with an eye to broadening or enriching the coding scheme or to supplementing it with other coding systems, as suggested by Miles and Huberman (1984). Poole, Folger, and Hewes (1987) and Folger, Hewes, and

Poole (1984) discuss the development and validation of category systems and the types of data that can be gleaned from them.

The multiple tracks yield a rich and complex depiction of development. This must be further analyzed to distill a description that can be used to test (or derive) developmental theories.

Analyzing Process Patterns or Cycles in Activity Tracks

After the chronological data are coded in these conceptual tracks, we will use two major approaches to identify and examine process patterns: phase analysis and sequence analysis. These two approaches are presently in development, and MIRP data files are being restructured to apply them. Thus, we can report here only some initial qualitative impressions in applying these approaches.

Phase Analysis. Phase analysis consists of identifying discrete phases of innovation activity and then analyzing their sequences and properties. A *phase* is a period of unified and coherent activity that serves some innovation function. For a set of n developmental tracks, a phase is defined by a meaningful set of co-occurring activities across the n tracks. So one phase for the five MIRP tracks might be "concept refinement," indicated by a change in some *idea,* occurring at a meeting of three experts (people) engaged in discussion and conflict (transactions) during a period of low resources (context) resulting in a weighting of tension and morale (outcomes). The phase would be indicated by the co-occurrence of these events as an ordered quintuple: (change in idea, experts, discussion and conflict, low resources, and high tension and high morale).

In the definition of phases n can be any number up to the total number of tracks. We might be interested only in defining phases on a single track (as Ring and Rands did with the transaction track in Chapter 10), or we might want to define phases using two, three, four, or all five tracks (obviously, definition of a phase gets progressively harder as we add more conceptual tracks). It is often the case that a given phase has multiple indicators—that is, that a set of n-tuples all correspond to the same phase. For example, "concept refinement" might be indicated by both of the following quintuples: (1) change in idea, experts, discussion and conflict, low resources, high tension and higher morale and (2) change in idea, experts, discussion and conflict, influx of resources, high tension and morale. Either of these n-tuples would be recoded into the phase designation "concept refinement." Typically, a very large set of n-tuples can be reduced to a much smaller set of phases.

This logic suggests a procedure for identifying phases. First, we code events on each single track into categories that reflect meaningful events. For example, on the ideas track, we might code events to indicate whether ideas are combined, dropped, extended, or tabled. A variety of coding systems could be applied to any track, depending on the question of interest. Once each track is coded, we could identify n-tuples of codes across tracks that indicate phases and recode these n-tuples into phases. Figure 2–1 illustrates this procedure: Part A shows sample codes for three tracks: 1a, 1b, and 1c are the codes for track 1; 2a–2c are codes for track 2; etc. Part B shows phase indicators, ordered triples of codes (one for each track) that define what phase is indicated by each set of codes across tracks; for example, a code of 1a on track 1, 2a on track 2, and 3a on track 3 would indicate phase A. Part C shows some sample track data, and indicates the phase codes derived from these data and the resulting phases. (See Poole and Roth 1989 for another illustration of phase indicator coding.) In this illustration the track data yielded a sequence A-B-C, which then dissolved into a short period of nonorganized activity (indicated by Ø) and

reemerged into an A-B sequence. This procedure is discussed in more detail below.

Phase indicators can be defined either theoretically or empirically. On the one hand, we can conceptually define meaningful combinations of codings that indicate phases (either within a single track or across tracks). Of course, there might also be combinations of codings that are not meaningful and suggest there is no coherent phase across tracks. These may correspond to periods of disorganization or anomie; they may also correspond to periods when action in one or two tracks is coherent but no overall pattern appears across tracks. There are also empirical methods for clustering track codes to determine which combinations predominate in the sample (see Poole, 1981, for example). However, these require a large number of observations and are often hard to interpret.

The process of phase identification allows us to relable the combinations of track codes into phases. This reduces the complexity of five levels to one. As mentioned above, this recoding process can be done for any n tracks or with only a single track. It is also possible to use hierarchical recoding procedures in which the tracks are recoded into two or more tracks and then recoded into a single track.

When one or more tracks of phases are identified, the next step is to identify meaningful patterns. We have focused on three sorts of patterns: developmental path, structural properties of developmental sequences, and causal relationships between paths.

One of the most useful ways to analyze developmental data is to identify the *types and structures* of sequences (Mintzberg, Raisinghani, and Theoret 1976; Nutt 1984; Poole 1983b). A subsequent section explicates a vocabulary for describing process progressions; here we mention some rudimentary methods for developing topologies. In the most common method—the "stare-at-the-charts" approach—charts are made of the developmental path and sorted into

A. *Track codes*

Track 1 codes: 1a, 1b, 1c, 1d

Track 2 codes: 2a, 2b, 2c

Track 3 codes: 3a, 3b

B. *Phase indicators (ordered triples of track codes)*

Phase A indicators: (1a, 2a, 3a), (1c, 2a, 3a), (1d, 2a, 3a), (1a, 2c, 3a)
(1c, 2c, 3a), (1d, 2c, 3a), (1a, 2a, 3b)

Phase B indicators: (1b, 2a, 3a), (1b, 2b, 3a), (1b, 2c, 3a), (1b, 2a, 3b)
(1b, 2b, 3b), (1b, 2c, 3b)

Phase C indicators: (1a, 2b, 3a), (1c, 2b, 3a), (1d, 2b, 3a), (1a, 2b, 3b)
(1c, 2b, 3b), (1d, 2b, 3b), (1a, 2c, 3b)

Nonmeaningful indicators ($\emptyset$): (1c, 2a, 3b), (1d, 2c, 3b), (1c, 2c, 3b), (1c, 2a, 3b)

C. *Track data recoded into phases*

Time	1	2	3	4	5	6	7	8	9	10	11	12	13	14	15	16	17
Track 1	1a	1a	1a	1b	1b	1a	1c	1c	1c	1c	1d	1d	1d	1d	1b	1b	1b
Track 2	2c	2c	2c	2c	2b	2b	2b	2b	2b	2b	2c	2a	2c	2c	2b	2b	2a
Track 3	3a	3a	3a	3a	3a	3a	3a	3b	3b	3b	3b	3b	3a	3a	3b	3b	3b
Phase indicators	A	A	A	B	B	C	C	C	C	C	$\emptyset$	$\emptyset$	A	A	B	B	B
Phases			A		B			C				$\emptyset$		A		B	

FIGURE 2–1. *A Sample Data Set Illustrating Phasic Recoding*

sets. This approach works well when there are only a few types of sequences and takes advantage of the researcher's ability to distinguish fine nuances in a well-understood data set. However, with complex data there is a danger that important distinctions will be missed or that types that differ due only to "error" will be distinguished. Recent methodological advances suggest more systematic ways of developing topologies. Pelz (1985) shows how the nonparametric statistic gamma can be used to establish the sequencing of phases and their degree of overlap (see Poole and Roth 1989 for an application to typology development). Abbott (1984) outlines a number of methods for statistical derivation of developmental topologies.

A number of *structural properties* of se-

quences are of interest, including cycles, breakpoints, degree of disorganization, and relative levels of key functional behaviors. Cycles and breakpoints deserve special discussion. A recurrent pattern of behavior is called a *cycle*. Cycles are identified when repetitions occur within tracks over time. *Breakpoints* are of key importance to understanding change processes because they represent transitions between cycles of activities. They indicate the pacing of activities within tracks and possible linkages between tracks of activities. Poole (1983a, 1983b) and Mintzberg, Raisinghani, and Theoret (1976) suggest four types of breakpoints: normal breakpoints or topical shifts, delays, internal disruptions, and external interrupts.

Breakpoints interrupting cycles within a

track suggest that the track is loosely coupled or operating somewhat independently of the other tracks. On the other hand, when breakpoints occur simultaneously in many tracks, those tracks are likely to be highly interdependent, and the rupture may presage major events or shifts in developmental activity. Thus, when cyclical breaks in multiple tracks occur in some coherent fashion, phases or stages similar to those in classical models may be found. However, at other points, there may be no relationship in the cyclical breaks between tracks and therefore no recognizable phases. In this case, each track is analyzed in its own right, but the entire ensemble of tracks does not yield a coherent analysis.

It is also useful to consider *causal relationships* among tracks. If there are sufficient observations, methods for establishing lagged variation in time series are very promising (as described in a section below). In most cases, however, there is less data than optimal for these applications. More informal methods of analysis are needed, and unfortunately these are currently in a rather primitive state of development (see Bastien, Chapter 11, for an example of informal causal analysis). Miles and Huberman (1984) offer several methods to display data that may facilitate identifying causal relations. Lazarsfeld's (1978) procedures for lagged causal analysis are also useful here.

Sequence Analysis. It is also useful to examine sequence order in series of related events, as done qualitatively by Schroeder et al. (Chapter 4) and Bastien (Chapter 11). These sequences illustrate the developmental trajectories of particular paths of an innovation, such as work on health accounting software by Qnetics (Chapter 8) or reorganization of offices during a merger (Bastien's Chapter 11). Separate event trajectories can be laid out for each aspect of an innovation and the linkages between them studied.

Sequence analysis is particularly useful in the study of proliferation and selection processes—that is, the study of how innovative activities grow more complex or are integrated and how some developments "die off" while others grow. These are central problems in the study of innovation. For example, Bastien (Chapter 11) examined the connections between problem trajectories and their proliferation in mergers and acquisitions. Mergers that dealt with problems sequentially had fewer difficulties than those in which several problem trajectories developed simultaneously. Schroeder et al. (Chapter 4) and Scudder et al. (Chapter 12) found that event trajectories produced a "fireworks" model of innovation, Van de Ven, Garud, Venkataraman, and Polley (Chapter 8) found many "dead ends"—idea trajectories that petered out as their firms refocused their attention on a few central projects. Early proliferation may be necessary as an experimental prologue to later focusing.

Vocabulary for Describing Processual Progressions
New concepts about developmental processes are needed to proceed further in this analysis of event sequences in the above activity tracks. Several theorists working on the problem of child development have developed conceptual schemes that are useful for the study of innovation. Flavell (1972) has developed a vocabulary for describing the types of relationship existing between two developmental events or acquisitions. Van den Daele (1969, 1974) has proposed a typology for the classification of sequences of events that complements Flavell's analysis.

Types of Relations Between Developmental Events. After an extensive review of developmental literature, Flavell (1972) posited five different types of developmental linkages: addition, substitution, modification, inclusion, and mediation. These can serve as a vocabulary for the description of innovation processes.

In *addition,* a later-emerging event or element supplements but does not replace or extinguish the earlier one. The two elements x_1 and x_2 coexist and are both equally available. For example, getting a government contract provides resources that add to privately obtained financing. Both types of resources are available to the innovators.

With *substitution* a later event or element largely replaces an earlier one. Element x_2 replaces x_1. For example, in Chapter 8 Garud and Van de Ven describe how failures experienced in the development of the cochlear implant technology led the innovation unit to shift (or substitute) its focus on the development of hearing aids and to deemphasize further developments of cochlear implants. The innovation unit's capabilities have changed fundamentally.

In *modification* a later event represents "a differentiation, generalization, or more stable (solid, reliable, efficient) version of the earlier one" Flavell (1972:345). In this case x_1 is revised to become x_2—that is, x_2 is an elaboration of x_1. For example, in Chapter 8 Polley and Van de Ven discuss how an initial idea for a therapeutic apheresis device was elaborated into two distinct Phase I and Phase II products. The innovation team's competencies and organization used to develop the first product were *generalized* and *modified* to develop the second-phase product.

Inclusion occurs when earlier events or elements "become incorporated into the later one as an integral part of a computer program" (Flavell 1972:345). In this case x_{11} and x_{12} are integrated to yield x_2. For example, in Chapter 12, Scudder et al. show how development of a complex weapon system required the creation and integration of new factory automation, materials management, software programs, and many other micro innovations.

In *mediation* an earlier event or element "represents some sort of developmental bridge or stepping stone (mediator) to the later one" (Flavell 1972:345). So x_1 develops, and once it reaches a certain point, it becomes a developmental bridge to the attainment of x_2. For example, in the enactment of state educational reform legislation, Roberts and King describe in Chapter 9 a series of key activities by policy entrepreneurs that set the stage and permitted the reform to transition from proposal to legislation to law.

The five relationships have very different implications for innovation development. Addition or inclusion relationships imply quite different developmental processes than do substitution or modification. Mediation suggests a more complex relation than that usually emphasized in writing on innovation. The five types of relationships are useful building blocks to examine larger developmental sequences and progressions of innovations over time.

Based on mathematical set theory, Van den Daele (1969, 1974) proposes a useful analytic scheme to depict qualitative progressions among temporal events. Van den Daele's models offer a way of expanding on the sequences of the relationships that Flavell outlines. Van den Daele proposes four types of progressions[1] that may describe temporal patterns in sequences of events.

Simple, Unitary Progression. This is a sequence of the form $U \rightarrow V \rightarrow W$. This model assumes that a temporal sequence of events may consist of any number of subsets, or stages but that these subsets must occur in an ordered progression. If a developmental progression has no more than one subset of events over

---

[1]Riegel (1969) considered four formal models that parallel van den Daele's. He outlined mathematical formulations for (1) branching processes, in which elements successively differentiate; (2) root models, based on progressive combination of positions; (3) jigsaw models, which show how patterns emerge from a given set of pieces; and (4) fallout models, which illustrate progressive acquisitions of parts from a predetermined store. Such formal treatments may offer useful rigor for future work.

time, it is called a simple unitary progression. This progression is consistent with descriptions of developmental sequences advanced by many theories of innovation stages.

Simple, Multiple Progressions. This model assumes that developmental processes can follow more than a single possible path. It attempts to specify alternative developmental progressions. In general terms multiple progressions can be of three forms:

Parallel	Divergent	Convergent

This second model assumes a temporal sequence of events may reflect more than one pathway at a given time in the ordered progression. For example, more than one feasible technological path to develop an innovation might be pursued in a given stage. These paths diverge from each other at a point in time, and subsets of events may occur to develop and complete the progression of activities in each pathway. Any developmental progression that has more than one subset of parallel paths at a time is called a multiple progression. Two specialized types of multiple progressions can be identified. Paths may *diverge* when more subsets of paths emerge over time in a temporally ordered collection of events. Conversely, a *convergent* multiple progression exists when there is a decrease in the number of parallel paths over time. Combinations of these two types of multiple progression can occur, and a description of how multiple progressions of events diverge, proceed in parallel, or converge over time provides a useful vocabulary for making process statements about the overall developmental pattern of an innovation over time.

Cumulative Progression (**unitary or multiple**). This model assumes that more than one stage may belong to a unit at a time. In set theory terms: $U \supset a$, $V \supset ab$, $W \supset bc$ (unitary model). For example, a multiple parallel partially cumulative model could look like this:

$$U \supset a \rightarrow V \supset a \quad b \rightarrow W \supset a \quad b \quad c$$
$$U \supset a \rightarrow V \supset b \quad\quad \rightarrow W \supset b \quad c$$
$$U \supset a \rightarrow V \supset a \quad b \rightarrow W \supset c$$

If events are cumulative, then elements found in earlier events or stages are added to and built on in subsequent events or stages. In Flavell's terms, accumulation may take the forms of addition or modification. For example, ideas developed in stage U may be used to provide the insight for a breakthrough solution in stage V. Complete cumulation means that every event from each stage is carried from its onset until the end of the developmental progression. This, of course, seldom happens because losses of memory, mistakes and detours, and terminated pathways all imply partially cumulative or substitution progressions (as illustrated in the bottom two tracks above).

Conjunctive Progression (**unitary, multiple, or cumulative**). This model posits that the elements of subsets may be related, such that aRb. Conjunctive events are causally related events, meaning that events in one pathway may trigger or influence events in other pathways of a multiple progression. Flavell's inclusion and mediation relationships represent two forms of conjunctive progressions. Of course, what is related at one time may be viewed as unrelated at another. Therefore, strict causality among events is difficult to establish. However, in the next section we describe exploratory methods being developed to identify temporal relationships among different event tracks.

Before we turn to this, it should be pointed out that these four forms of progression do not occur independently. Every developmental model makes a commitment (implicitly or explicitly) to some form of invariant sequential order, between unit variation (unitary or multiple sequence), within-unit variance (simple or cumulative structure), and in the relationship of developmental elements (conjunctive or disjunctive). For example, in Chapter 4, Schroeder et al. review the literature and show that a simple, unitary, conjunctive progression characterizes most of the stage models of innovation and organization change. However, by comparing how seven innovations actually developed over time, they present a much more complex and fluid process model. They observed innovations to unfold from a simple unitary process into multiple, divergent, parallel, and convergent progressions of events over time. Some of these paths in the multiple progression process are conjunctive (that is, related), and many appeared disjunctive (unrelated in any noticeable form of functional interdependence). Moreover, many of the paths that innovation participants perceived as being related or conjunctive at one time were often reframed as being independent or disjunctive at another time.

Causal Relationships among Event Tracks

The vocabulary presented here provided Schroeder et al. (Chapter 4), Van de Ven et al. (Chapter 8), and others robust concepts to make a qualitatively rich description of the developmental sequences they observed across innovations. However, to substantiate and further understand these developmental sequences, new analytical methods are needed to identify causal relationships among developmental paths and event tracks.

One method we are currently exploring is to code an innovation's chronological listing of qualitative events with dichotomous indicators of our conceptual tracks and then to statistically examine the time-dependent pattern of causal relations among the tracks. Although this methodology is in its formative stage, it is introduced here to indicate a potentially fruitful direction for the analysis of event sequences in future research.

Theoretically, it is possible to exhaust the information contained in a text with binary oppositions. A dichotomous indicator uses 1 to represent the presence and 0 the absence of a certain informative feature of the qualitative event. If needed, fuzzy categories can also be coded (for example, with an asterisk) and fuzzy set theory may be applicable (Zimmermann 1985). The choice of a particular set of indicators depends on the substantive problem of interest. With the indicators chosen, the listing of events is mapped into a matrix of 1s and 0s, which we call a *bit-map*. For a chronological listing of qualitative events, its bit-maps are time-dependent, meaning that the sequential order of the rows is crucial and should be taken into account when information is to be extracted, although the columns are interchangeable.

Such a bit-map can be analyzed with a variety of statistical techniques to identify substantively interpretable time-dependent patterns (or lack thereof) of the 1s and 0s. We have found that notions of pattern commonly used in quantitative time series analysis—such as "equilibrium," "explosiveness," "oscillation," "saturation," and "damping"—can be transplanted to qualitative event sequence analysis as the notions of "convergence," "divergence," "conjunction," and "vicious cycles." Efforts have just begun to operationalize these qualitative notions as informative time-dependent patterns of a bit-map. We are also investigating the possibility of applying artificial intelligence techniques, specifically the Holland classifier as implemented by Schrodt (1986) (see also Holland 1986; Frey 1986), in developing time series methods appropriate for qualitative event sequence analysis.

As an experimental implementation of these efforts to identify causal relations among conceptual tracks, we are currently analyzing bit-maps constructed from the chronological listing of qualitative events of several innovation studies. A set of five indicators was chosen to represent the presence or absence of changes in the core MIRP concepts of ideas, people, transactions, context, and outcomes. The purpose is to investigate possible causal relationships, including vicious cycles, among changes in the five fundamental tracks as events develop. A preliminary cross-correlation analysis, conceptually based on the notion of Granger causality and vector autoregression (Freeman 1983; Freeman, Williams, and Lin 1987), shows that changes in ideas and, to a lesser degree, in people are significantly associated with later changes in transactions, whereas the reverse is not true. Although statements like these are in themselves substantively interesting, other models, such as the lagged loglinear model, are also being explored. Time-dependent patterns identified in an innovation will be compared with those in other innovations in order to establish validity across innovations.

Assessment of Tracking Methodology

These techniques facilitate development of a grounded theory of innovation processes. The methodology can also be used to empirically examine existing models or theories of innovation processes. If spelled out clearly, these theories of innovation should imply definite longitudinal patterns among tracks. Thus, the track data should permit testing and comparison of a variety of innovation process theories.

There are four major *advantages or strengths* of this tracking methodology:

1. It does not presume a unitary sequence model. It lets one examine sequences in a detailed and systematic fashion.

2. It permits assessment of alternative models of innovation processes by empirical comparisons and contrasts of sequences of events.

3. It does not force one to assume that the activities are always organized. Much innovation behavior is disorganized and nondirected. The tracking method allows us to detect this as noncoherency within and across tracks.

4. It is a milieu for combining qualitative and quantitative techniques. It helps one recognize patterns in complex data. Case studies are useful but often represent a "flight" from the data.

Of course, any methodology has *limitations,* and three should be recognized with the tracking methodology proposed here.

1. Once chosen, tracks may limit what we can see and discover. This is true of any method one might employ. As Poggi stated, "A way of seeing is a way of not seeing."

2. The tracking methods may move the observer to too micro a focus. However, this danger is minimized by including conceptual tracks dealing with multiple levels of analysis, such as people, transactions, and context in the MIRP research framework. Furthermore, longitudinal study of innovations from beginning to end minimizes "local" causality, which neglects recognition of longer-term causal patterns.

3. Dividing the observed process into multiple tracks may "overdecompose" the process and lead one to miss complexities and interconnections. Yet any analytical method requires some decomposition of complex phenomena into workable conceptual categories (here, ideas, people, transactions, contexts, outcomes) that de-

47

fine the domain of the subject matter under investigation.

The objective of the tracking methodology, of course, is to enable description to be linked to theory. Otherwise it is not possible to pick out patterns. Two levels of theory can be distinguished: (1) a theory of significant event and activity combinations so we can identify phases and cross-track relationships and (2) beyond this, a theory that can explain transitions and movement through sequences—that is, a theory of the developmental process itself. It is possible that different explanatory models hold for different parts of a developmental sequence, so "switching rules" should also be considered.

Inherent in this form of data analysis is the need to tack back and forth between theory and data. In making conceptual transitions from the most concrete level of an event in the case to more general and abstract levels of overall development patterns, it is important to stay close to the data and to follow Kaplan's (1964) rules of correspondence:

Case event

⇕

Event chronology

⇕

Track maps

⇕

Phases

⇕

Overall development pattern

The advantage of making gradual transitions across these levels of abstraction is that the validity of each step can be assessed. A variety of construct validity steps can be designed to maximize validity of conceptual inferences (see Poole 1983a for an illustration).

Some issues require going beyond tracks and the data in them. For example, to verify the presence of a political/incremental decision process, it might be necessary to determine whether the persons involved viewed the process as "political" as well as assess fit to predicted patterns over time. Combination of different forms of data is crucial here. Any longitudinal analysis of innovation processes cannot go far unless it is driven by an explicit motor or theory of innovation and change processes, which is now addressed.

MOTORS OR THEORIES OF CHANGE PROCESSES

Two components of developmental theories should be distinguished—global and local models. First, there is the *global model*, which describes and explains the overall long-run developmental process. This model indicates the general trends of development over extended time periods and offers an explanation of why the long-term developmental path unfolds as it does. Examples of global theories include Usher's (Ruttan 1959) theory of technological change, Etzioni's (1963) theory of epigenesis, Darwin's theory of evolution, and Marx's dialectical theory of change.

Second, there is the *local theory of immediate action*, which describes and explains the operative processes that create developmental patterns over the short term. This model details interactions among persons, ideas, and contexts that give rise to the innovation. Examples of local models include many theories of organizational decisionmaking, Ring and Van de Ven's (Chapter 6) transaction governance model, and the theory of exogenously induced innovation developed by Marcus in Chapter 16.

The two components can be contrasted in several respects. The global model takes as its units of analysis the developmental trajectory or path, phases or stages of development, configurations of industries or groups. The local model focuses on the individual decision, idea, transaction, or episode and on the innovating group and organization and its transactions. The global model incorporates a long time frame of weeks, months, or years; the local model works with hours or days.

The global model attempts to identify the formal and final causes of innovation (Aristotle 1941): It pictures the final pattern or goal of the innovation process and elaborates a scheme of forces and processes that enform and produce this final state. For example, Usher's cumulative synthesis model depicts an invention as a product of four steps—problem perception, stage setting, act of insight, critical revision—each of which may take years. The four steps make up a logically necessary sequence if invention is viewed as the product of large numbers of individual insights and efforts. Any complete inventional process must satisfy this sequence, and an array of psychological and social processes is presumed to enforce it. In contrast, the local model emphasizes efficient causality, the processes necessary to drive the innovation from its present state to the next, without reference to a larger scheme of development. A local model corresponding to Usher's model might be the collaborative work model of scientific research. Final, formal, and efficient causes fit together, as pattern and mechanism which fills out the pattern.

In the global model, influences on development come from the larger surround and include factors such as economic trends, social needs, the legal system, the larger culture, and long-term changes in organizational structure. The local model focuses on influences on the immediate situation, including microlevel factors such as motivation level and group interaction processes and macrolevel influences like organizational structure, resource control, and the organization's culture.

Both global and local perspectives are necessary for an adequate theory of innovation because innovations are extended over long periods, yet driven through time by immediate action systems. The global theory incorporates the local theory of immediate action and explicates how immediate action unfolds in a larger temporal and social context. Indeed, some global theories describe how two or more local mechanisms combine.

Unfortunately, the distinction between global and local action theories has been blurred in most developmental work. In most global theories, immediate action is only implicit and remains vaguely described. As a result, global theories tend to be an overly simplistic view of local action. On the other hand, most local theories do not consider how immediate actions interact and aggregate into a larger developmental context. As a result, they tend to have overly simplistic views of the long run. Tushman and Romanelli's (1985) punctuated equilibrium theory best sets out both global and local components and the relations between them. On balance, however, the theory tends to focus on global periods of convergence punctuated by discontinuous ruptures without adequately articulating the local processes that go on in each period.

In part, previous theories have not spelled out both global and local elements because developmental theories are still in their infancy. As Nisbet (1970) notes, theories have depicted social and organizational development as an overly simplified progression through a single historically and logically necessary sequence of stages. Van den Daele (1969) termed this model a simple, unitary sequence. From the theoretician's point of view unitary sequences are elegant representations that set up a compelling case for progress building toward some end-state. However desirable unitary sequences might be at the abstract level, mounting evidence suggests that organizational innovation and change are not so simple. Some changes do follow unitary sequences, but many do not. At present we are somewhat at a loss to explain this complexity

49

because most of our theories can deal only with simple sequences and processes. More precise specification of a wider range of developmental models is needed. Only with a repertoire of long- and short-run models incorporating different explanatory mechanisms will we be in a position to tackle the complexity of innovation processes.

Two major global theories are well established in social theory: developmentalism or evolution and epigenesis or accumulation theories (Van de Ven and Poole 1988). The MIRP studies described in Chapters 4 through 19 outline a number of global and local models. We will pull them together in Chapter 20.

Testing Theories of Change Processes

The combinations of global and local models may produce several possible innovation process theories. It may come to pass that one of these theories is proven superior. It may also be the case that different theories hold for different circumstances or types of innovations. For example, one theory might be required to explain the development of radical, new-to-the-world innovations, and another for incremental innovations that consist of adapting existing knowledge to new contexts. So, also, an accumulation theory may be most appropriate in the initial stages of innovation, followed by a developmental theory to explain the innovation's diffusion and institutionalization. If this is the case, a meta-contingency theory of innovation processes is ultimately necessary that elaborates the comparative and temporal conditions for different explanations (see Poole and Van de Ven, Chapter 20). But at this stage it is not possible to say which innovation process theories are best.

These considerations suggest a need to be able to test the fit of several alternative developmental models to the data. This requires us to spell out for each theory criteria and pre-

dictions that would allow us to determine whether it fit observed sequences and processes. Several types of criteria or predictions are useful:

1. Each theory can predict the nature of the developmental path and its structural properties (for example, a logical necessity model would predict a unitary sequence).
2. Some theories imply causal or correlational relations among tracks (that is, that the idea track drives the innovation process).
3. Some theories specify the nature of relationships between successive elements in the tracks (for example, an accumulation model would imply additive or inclusive relations, with some events mediating others).
4. There is also often direct evidence that a given immediate action mechanism is operating (such as actors should be conscious of a politically motivated process if a political/negotiative/incremental/model holds).

For each theory a pattern or set of predictions can be identified and matched to observed data. This permits a narrowing of the range of acceptable models.

Mixing Models and Mediating Paradoxes in Developmental Theory

The quest for a single best theory or a "meta-contingency" theory of innovation may not yield fruit. Most likely, it will be necessary to combine different explanations in a larger synthesis. Van de Ven and Poole (1988) argue that the study of social change requires one to mediate seemingly contradictory explanations because both "halves" of the contradiction encompass key aspects of change. They discuss the tension between internal (developmental) and external (accumulation) explanations as one such "paradox" in need of mediation. They deduce four generic ways to resolve paradoxes and incorporate apparently conflicting explanations in a single theory:

Learn to Live With and Appreciate the Paradox. In this case opposing theories are kept separate and used as different lenses to view the same phenomenon. Theories are distinguished in terms of different and exclusive questions and problems that they address, and insights are gained through contrast.

Clarify Connections between Organizational Levels. Here one explanation, the local model, is presumed to operate at the microlevel of action, and the global model at the macrolevel. The contrast between micro and macro may be between individual, group organization, industry, or society. The key theoretical move comes in linking local to global explanations. Coleman's (1986) theory of social action is a good example of this. He uses mathematical models to show how individual actions aggregate to yield structural changes on the macrolevel, thus relating both action and structure in a single explanation. In terms of the models just discussed, one possibility is to use a political/negotiative local theory with a logical necessity global theory (hence having an accumulation model at one level and a developmental one on the other).

Use Time to Relate the Two Theories. A punctuated equilibrium model (as described by Tushman and Romanelli 1985) uses time as one avenue to reconcile and link both evolutionary and accumulation theories of change and both global and local models. In the punctuated equilibrium model, organizational development proceeds through occasional discontinuous reorientations in part or all of the system, which interrupt long periods of continuous convergence and gradual morphogenesis. Accumulation appears to be the basic process underlying discontinuous punctuations because the resulting transformation represents a metamorphic or radical change that no longer includes representations of the earlier organization. Evolution best describes morphogenic change—that is, where an organization converges over time toward increasing order, complexity, unity, or operational effectiveness.

Change in a given organization may be empirically observed to result from either evolutionary elaboration of existing functions or punctuated accumulations of totally new functions. In addition to incorporating different global models, the convergence and reorientation periods have different models of local action. The convergence model relies on incremental steps by which the system adjusts itself. The reorientation period seems to involve crisis reaction and political negotiations. Thus, as we argue elsewhere (Van de Ven and Poole 1988), time provides the vehicle for incorporating two different change processes in a punctuated equilibrium model.

Introduce New Concepts That Mediate the Two Theories. Sometimes it is possible to reconceptualize a situation in which two theories seem incompatible so as to "cut through" or dissolve the paradox. Doing so generally requires the introduction of new concepts and theoretical frames. Van de Ven and Poole (1988) note that the theory of structuration (Giddens 1979) is one such theory. This theory introduces the novel conception of *modalities of structuration,* which can be used to mediate developmental and accumulation explanations. Briefly put, a modality is the actor's appropriation of structure for use in a particular action context. A structure is a rule or resource people use in acting or interacting; it is usually embodied in some social institution or in the store of generally available social knowledge. Actors take these general structures and adapt them for their own purposes and, in so doing, reproduce the structures. This is a complex process, discussed at length by Giddens (1979, 1985) and Poole, Seibold, and McPhee (1985, 1986).

51

The following example illustrates how global and local models can be linked in structuration.

In the short run, actors might interact in a *political/negotiative system* guided by their long-run conception of a structure of necessary stages *(logical necessity)* in the innovation process. Insofar as the situation lets them follow this structure, the global developmental process follows the sequence implied by logical necessity. However, if conflict erupts or other unforeseen events in the negotiation intrude, the logical flow may be interrupted and the innovation process will depart from the unitary sequence. The structure of logical necessity that the innovators use to guide their behavior is appropriated from the general stock of social knowledge, which contains many such "ideal" sequences. Poole and Doelger (1986) explicate this model in detail for group decisionmaking, although it has yet to be spelled out for innovation.

It is important to struggle with the task of comparing and mediating existing theories. There are as yet no clear answers.

CONCLUSION

Each reader is likely to have different answers to a few basic normative questions: Where do we go with this process analysis and search for process models? Where do we want to end up? It seems to us, however, that we want to not only *describe* observed innovating processes but also *explain* why they occur. Scientifically valid explanation requires systematic procedures for observing and tabulating longitudinal data (as proposed with the first three requirements for process analysis) as well as the development and evaluation of theories or motors of the change process itself.

We must therefore ask, "What are the requirements for a good theory of change?" Hernes (1976), Dahrendorf (1959), and Van de Ven and Poole (1988) have proposed standards for a theory of change that is useful in answering this question. Adapting their criteria to innovation management suggests that our theories should attain the following four requirements for a good theory of innovation.

1. It should explain how structure and individual purposive action are linked at local and global levels of analysis. The dominant paradigm of social science rests on the firm belief that any macro theory of organizational or industrial change must be grounded in the purposive actions and ambitions of individuals (Coleman 1986).

2. It should explain how innovation and change is produced both by the internal functioning of the structure and by the external purposive actions of individuals. If one concludes that innovation is totally controlled by natural or structural forces imminent to the social system, no room is left for individual purpose, and no theory of action can result. If one concludes that organizational change is totally controlled by purposive individual action unconstrained by natural or structural forces, only a teleological or utopian theory can result (Van de Ven and Poole 1988).

3. The theory should explain both stability and instability. "In its fundamental structure a theory of organizational innovation or change should not be remarkably different from a theory of ordinary action" (March 1981: 564). Without this requirement, any theory of innovation would explode and be unable to explain the amazing persistence and fixity observed in common organizational life.

4. It should include time as the key historical metric. By definition, change is a difference that can be noted only over time in an entity. Chronos (or calendar time) tends to predominate in studies of structural change, while kiros (periods of peak experi-

ences—as in the planting and harvesting periods of a growing season) appears to be the most common metric of time in studies of individual creativity and purposive action. A theory of innovation and change that links structure and action must therefore link chronos and kiros time metrics (Van de Ven and Poole 1988).

REFERENCES

Abbott, A. 1984. "Event Sequence and Event Duration: Colligation and Measurement." *Historical Methods* 17: 192–204.

Aldrich, H., and E. Auster. 1986. "Even Dwarfs Started Small: Liabilities of Age and Size and Their Strategic Implications." In L. Cummings and B. Staw, eds., *Research in Organizational Behavior,* pp. 165–98. San Francisco: JAI.

Argyris, C. 1968. "Some Unintended Consequences of Rigorous Research." *Psychological Bulletin* 70 (3): 185–97.

———. 1985. *Strategy, Change, and Defensive Routines.* Marshfield, Mass.: Pitman.

Aristotle. 1941 *The Basic Works of Aristotle.* R. McKeon (ed.). New York: Random House.

Coleman, J.S. 1986. "Social Theory, Social Research, and a Theory of Action." *American Journal of Sociology* 16 (May): 1309–35.

Daft, R., and S. Becker. 1978. *Innovation in Organizations.* New York: Elsevier.

Dahrendorf, R. 1959. *Class and Class Conflict in Industrial Society,* Stanford, Calif.: Stanford University.

Etzioni, A. 1963. "The Epigenesis of Political Communities at the International Level." *American Journal of Sociology* 68: 407–421.

Flavell, J.H. 1972. "An Analysis of Cognitive-Developmental Sequences." *Genetic Psychology Monographs* 86: 279–350.

Folger, J.P., D.E. Hewes, and M.S. Poole. 1984. "Coding Social Interaction." In B. Dervin and M. Voight, eds., *Progress in Communication Sciences,* Vol. 5, pp. 115–65. Norwood, N.J.: Ablex.

Freeman, John. 1983. "Granger Causality and the Time Series Analysis of Political Relationships." *American Journal of Political Science* 27: 337–58.

Freeman, J., J. Williams, and T. Lin. 1987. "Modeling Macro Political Processes." Paper prepared for the Fourth Annual Meeting of the Political Methodology Society, Duke University, August 6–9.

Frey, Peter W. 1986. "A Bit-Mapped Classifier." *Byte* (November).

Giddens, A. 1979. *Central Problems in Social Theory.* Berkeley: University of California.

Glaser, B.G., and A.L. Strauss. 1967. *The Discovery of Grounded Theory: Strategies for Qualitative Research.* Chicago: Aldine.

Holland, John. 1986. "Escaping Brittleness: The Possibilities of General Purpose Learning Algorithms Applied to Parallel Rule-Based Systems." In R.S. Michalski, J.G. Carbonell, and T.M. Mitchell, eds., *Machine Learning II.* Los Altos, Calif.: Morgan Kaufmann.

Kaplan, A. 1964. *The Conduct of Inquiry.* New York: Intext.

Lazarsfeld, P. 1978. "Some Episodes in the History of Panel Analysis." In D.B. Kandel, ed., *Longitudinal Research on Drug Use,* pp. 249–65. London: Wiley.

March, J.G. 1981. "Footnotes to Organizational Change." *Administrative Science Quarterly* 26: 563–77.

Miles, M.B., and A.M. Huberman. 1984. *Qualitative Data Analysis: A Sourcebook of New Methods.* Beverly Hills, Calif.: Sage.

Mintzberg, H., D. Raisinghani, and A. Theoret. 1976. "The Structure of 'Unstructured' Decision Processes." *Administrative Science Quarterly* 21 (2) (June): 246–75.

Mohr, L.B. 1982. *Explaining Organizational Behavior: The Limits and Possibilities of Theory and Research.* San Francisco: Jossey-Bass.

Nisbet, R.A. 1970. "Developmentalism: A Critical Analysis." In J. McKinney and E. Tiryakin, eds., *Theoretical Sociology: Perspectives and Developments,* pp. 167–204. New York: Meredith.

Nutt, P. 1984. "Types of Organizational Decision Processes." *Administrative Science Quarterly* 29, no. 3: 414–450.

Pelz, D.C. 1985. "Innovation Complexity and the

Sequence of Innovating Stages." *Knowledge: Creation, Diffusion, Utilization* 6: 261–91.

Pettigrew, A. 1985. *The Awakening Giant: Continuity and Change in ICI.* Oxford: Basil Blackwell.

Poole, M.S. 1981. "Decision Development in Small Groups I: A Comparison of Two Models." *Communication Monographs* 48: 1–24.

———. 1983a. "Decision Development in Small Groups II: A Study of Multiple Sequences in Decision Making." *Communication Monographs* 50: 206–32.

———. 1983b. "Decision Development in Small Groups, III: A Multiple Sequence Model of Group Decision Development." *Communication Monographs* 50: 321–41.

Poole, M.S., and J.A. Doelger. 1986. "Developmental Processes in Group Decision-Making." In R.Y. Hirokawa and M.S. Poole, eds., *Communication and Group Decision-Making,* pp. 35–62. Beverly Hills, Calif.: Sage (1989).

Poole, M.S., and J. Roth. 1988. "Test of a Contingency Model of Decision Development." *Human Communication Research,* 15, 3, 323–356.

Poole, M.S., J.P. Folger, and D.E. Hewes. 1987. "Analyzing Interpersonal Interaction." In M.E. Roloff and G.R. Miller, eds., *Interpersonal Processes,* pp. 220–256. Beverly Hills, Calif.: Sage.

Poole, M.S., D.R. Seibold, and R.D. McPhee. 1985. "Group Decision-Making as a Structurational Process." *Quarterly Journal of Speech* 71: 74–102.

Poole, M.S., D.R. Seibold, R.D. McPhee. 1986. "A Structurational Approach to Theory-Building in Group Decision-Making Research." In R. Hirokawa and M.S. Poole, *Communication and Group Decision-Making.* Beverly Hills: Sage.

Riegel, K.F. 1969. "History as a Namothetic Science: Some Generalizations from Theories and Research in Developmental Psychology." *Journal of Social Issues* XXV: 99–127.

Ruttan, V.W., "Usher and Shumpeter on Invention, Innovation, and Technological Change, *Quarterly Journal of Economics,* 1959, LXXIII, 596–606.

Schrodt, Philip A. 1986. "Predicting International Events." *Byte* (November).

Schroeder, R., A. Van de Ven, G. Scudder, and D. Polley. 1986. "Managing Innovation and Change Processes: Findings from the Minnesota

Innovation Research Program." *Agribusiness* 2, 4: 501–523.

Tuma, N.B., and M.T. Hannan. 1984. *Social Dynamics,* Orlando, Fla.: Academic Press.

Tushman, M.L., and E. Romanelli. 1985. "Organizational Evolution: A Metamorphosis Model of Convergence and Reorientation." In B. Staw and L. Cummings (eds.), *Research in Organizational Behavior* 7, pp. 171–222.

Van de Ven, A.H. 1986. "Central Problems in the Management of Innovation," *Management Science,* 32, no. 5 (May): 590–607.

Van de Ven, A.H., and M.S. Poole. 1988. "Paradoxical Requirements for a Theory of Organizational Change." In R. Quinn and K. Cameron (eds.), *Paradox and Transformation: Toward a Theory of Change in Organization and Management.* Cambridge, Mass.: Ballinger.

Van de Ven, A.H., Y. Chu, T. Lin, M. Chinnappan, and R. Hudson. Revised 1987. *Methods Manual for Minnesota Innovation Research Program.* Minneapolis: University of Minnesota Strategic Management Research Center.

Van de Ven, A.H., P.E. Johnson, A.A. Marcus, M.S. Poole, and P.S. Ring. 1988. "Processes of Innovation and Organizational Change." Proposal to the National Science Foundation from the University of Minnesota Strategic Management Research Center (February).

Van de Ven, A.H., and Associates. 1984. "The Minnesota Innovation Research Program." University of Minnesota, Strategic Management Research Center, discussion paper #10 (April).

Van de Ven, A.H., and Associates. 1987. The Minnesota Innovation Research Program. Final report to the Office of Naval Research Program on Organizational Effectiveness.

Van den Daele, L.D. 1969. "Qualitative Models in Developmental Analysis." *Developmental Psychology* 1, no. 4: 303–310.

———. 1974. "Infrastructure and Transition in Developmental Analysis." *Human Development* 17: 1–23.

Welck, K. 1979. *The Social Psychology of Organizing.* 2nd ed. Reading, Mass.: Addison-Wesley.

Zimmermann, H.J. 1985. *Fuzzy Set Theory and Its Applications.* Boston: Kluwer-Nijhoff Publishing.

A PSYCHOMETRIC ASSESSMENT OF THE MINNESOTA INNOVATION SURVEY

Andrew H. Van de Ven

Yun-han Chu

This chapter evaluates the measurement properties of an innovation questionnaire developed and being used as one of several measurement instruments to observe the development of a wide variety of innovations over time by the Minnesota Innovation Research Program (MIRP). Evidence of the reliability and validity of the Minnesota Innovation Survey (MIS) is provided by applying standard psychometric

The authors are grateful for the cooperation of their colleagues who provided data from their innovation studies for this chapter, particularly John Bryson, Raghu Garud, Daniel Gilbert, Alfred Marcus, John Mauriel, Douglas Polley, Michael Rappa, Peter Ring, William Roering, Roger Schroeder, and Gary Scudder. This research was supported in part with a major grant to the Strategic Management Research Center to study the management of innovation from the Program on Organizational Effectiveness of the Office of Naval Research under contract No. N00014-84-K-0016. Additional support for the research program is being provided by ADC Telecommunications, Bemis, CENEX, Control Data, IBM, Dayton-Hudson, Dyco Petroleum, Farm Credit Services, First Bank System, Honeywell, and 3M Corporations and by the McKnight and Bush Foundations.

procedures for evaluating tests and measurement instruments. Such a psychometric assessment is a necessary prerequisite to developing or testing substantive theories based on the MIS data, and provides important information about the strengths and weaknesses of the MIS instrument to other potential users who wish to use the survey to conduct research on the management of innovation.

The core framework of the MIRP centers on five basic concepts: *ideas, people, transactions, context,* and *outcomes.* In a nutshell, the research examines the innovation process by tracking the development and implementation of new ideas that are carried by people who over time engage in transactions or relationships with others within a changing institutional context. Multiple methods are being used to conduct the longitudinal study of each innovation. This chapter evaluates the measurement properties of only the MIS being used in this research and draws on the other MIRP instruments only to evaluate parallel measures for some of the indices in the survey questionnaire. Definitions and measures of these other indices in the MIS are presented in the appendix to this chapter.

After a review of the conceptual basis of the dimensions included in the MIS, we present the methods and findings from this psychometric analysis.

CONCEPTUAL BASIS FOR MEASUREMENT MODEL IN MIS

Figure 3–1 outlines the dimensions that are examined in this evaluation of the measurement properties of the MIS. The dimensions are grouped into four clusters: (1) The internal innovation dimensions all relate to the processes

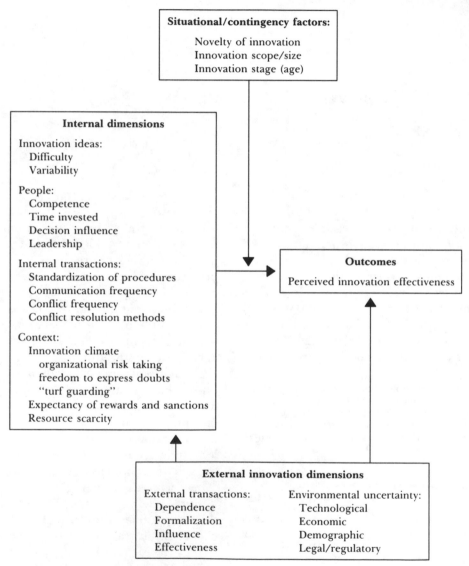

FIGURE 3–1. *Dimensions in Measurement Model of Minnesota Innovation Survey.*

and context within the innovation organizational unit and are evaluated in part 1 of this psychometric assessment. (2) The external innovation dimensions all pertain to the transactional and global environment of the innovation unit and are evaluated separately from the internal innovation dimensions because they pertain to a different level of analysis. (3) Perceived innovation effectiveness is used as the ultimate dependent criterion to assess the predictive and concurrent validities of the MIS internal and external dimensions. (4) Finally, the situational/contingency factors were measured with other instruments (not the MIS) and are used to examine the basic contingency theory that underlies the MIS measurement model.

In the overall MIRP framework, perceived innovation effectiveness is hypothesized to be a function of the internal innovation dimensions: (1) the uncertainty and difficulty of the innovation idea; (2) the leadership, influence in decisionmaking, time invested, and competence of the people involved in developing the innovation; (3) the standardization of procedures, and frequency of communications, conflict, and methods of conflict resolution in relationships or transactions among people engaged in developing the innovation; and (4) organizational context, in terms of organizational climate, rewards, and resource scarcity. Specific hypotheses for each of these dimensions will be developed.

The relationships between these internal innovation dimensions with perceived innovation outcomes are expected to be moderated by the nature of the innovation (that is, its novelty, scope or size, and its stage of development). Thus, an overall contingency theory underlies the measurement model. We expect the hypothesized relationships between the MIS dimensions with perceived innovation effectiveness to be substantially stronger when the innovation's novelty, size, and stage are taken into account because qualitatively different design patterns are expected to exist and influence outcomes for innovations that vary in novelty, size, and stage of development. Empirical evidence for this overall contingency-theory proposition has been provided by Munson and Pelz (1981) and Pelz and Munson (1982).

Factors external to an innovation unit are expected to influence internal processes within an innovation unit as well as its performance. In particular, an innovation unit is often highly dependent on external organizations and groups for technology, competence, financing, and raw materials or components in order to develop and implement its innovation over time. Therefore, the MIS includes several measures of the external transactions of an innovation unit with other groups: dependence, formalization, influence, and perceived effectiveness of external transactions. These dimensions are based on a theory of how and why interorganizational relationships emerge and are maintained over time discussed by Van de Ven (1976) and empirically evaluated by Van de Ven and Walker (1983).

Finally, the MIS includes measures of the perceived uncertainty of an innovation's general environment. As Van de Ven and Garud (1989) review, organization theorists and management of technology scholars have often emphasized the inherent environmental uncertainty associated with innovations. This uncertainty not only deals with the nature of the innovation idea itself, but also with the general technological, economic, regulatory, and demographic environments that innovations both construct and are constrained by. The more uncertain the general environment, the more one should observe many trials and errors in the search process of finding the relevant resources and market or user niches to develop and implement an innovation. Thus, the greater the environmental uncertainty, the less efficient the innovation development process, as reflected in lower perceived innovation effectiveness.

57

Specific definitions, measures, and hypotheses for these MIS dimensions will now be presented, followed by the psychometric evaluation procedures and measurement results.

Perceived Innovation Effectiveness

Innovation effectiveness in the MIS is the degree to which people perceive that an innovation attains their expectations about process and outcomes. Process expectations deal with making progress in developing the innovation and solving problems as they are encountered (measured by items 1 through 3 below). Outcome expectations deal with perceived ratings of the innovation's present effectiveness and the contributions it makes to attaining overall organizational goals (items 4 and 5 below). Thus, perceived innovation effectiveness was measured as the average response to the following questions in the MIS:

1. Progress satisfaction (MIS Q31);
2. Problem-solving effectiveness (MIS Q33);
3. Progress meeting expectations (MIS Q34);
4. Effectiveness rating (MIS Q32);
5. Innovation attains organizational goals (MIS Q35).

Kimberly (1981) points out that a positive bias pervades the study of innovation. Innovation is often viewed as a good thing because the new idea is expected to be useful—to be profitable or constructive or to solve a problem. New ideas that are not perceived as useful are not normally called innovations; they are usually called "mistakes" (Van de Ven 1986). Objectively, of course, the usefulness of an idea can be determined only after the innovation process is completed or implemented. In this sense, it is not possible to determine if work on new ideas will turn out to be "innovations" or "mistakes" until the very end of the longitudinal research. In the interim, our perceived effectiveness construct obtains periodic assessments of the degree to which people judge the developmental process to be achieving their expectations. This is a valid measurement because we assume that people will continue to invest their energies and make commitments to an undertaking they consider successful and withdraw their investments and commitments to ventures not considered successful.

MIS Internal Innovation Dimensions

Innovation Uncertainty is defined as the *difficulty* and *variability* of the innovative ideas being developed as perceived by the individuals involved in its development. The construct stems from March and Simon (1958), has been extensively applied to examine the work performed by organizational units (Perrow 1967; Van de Ven and Ferry 1980), and was further developed and applied to the study of innovation by Lawrence and Dyer (1983). *Innovation difficulty* (measured by items 1 and 2 below) refers to clarity of the innovative idea and the ease with which one can specify in advance what sequence of steps or tasks are needed to develop the innovation. *Innovation variability* (items 3 and 4 below) is the perceived variations or number of exceptions encountered in developing the innovative idea. Both innovation difficulty and variability indicate the extent to which the development processes are trivial and programmed at one extreme or must rely on chance and guesswork at the other extreme of the task uncertainty continuum. All other factors held constant, we hypothesize that the greater the uncertainty, the lower the perceived innovation effectiveness. Because uncertainty makes it difficult to connect means with ends, greater randomness is expected in relationships between innovation uncertainty and effectiveness.

Innovation uncertainty is computed as the average of the following MIS questions:

1. Difficulty know innovation steps (MIS Q2, reverse scale);
2. Predictability of innovation outcomes (MIS Q3, recode to 1–5 scale);
3. Frequency difficulty problems arise (MIS Q11);
4. Uniqueness of problems that arise (MIS Q12).

Resource Scarcity refers to the amount of work undertaken by innovation participants and the perceived degree of competitiveness for obtaining critical resources to develop and innovation. Although workload pressure (measured by items 1 and 2 below) and resource competition (items 3 through 6 below) are conceptually distinct constructs, they both indicate the scarcity of resources (including time, finances, materials, personnel, and management attention) needed to develop an innovation. Thus, overall resource scarcity is measured as the following items in the MIS:

1. Heaviness of workload (MIS Q6);
2. Lack of advance time (MIS Q7, reverse scale);
3. Competition for finances (MIS Q47a);
4. Competition for materials (MIS Q47b);
5. Competition for management attention (MIS Q47c);
6. Competition for personnel (MIS Q47d).

As Lawrence and Dyer (1983) suggest, high perceived resource scarcity implies that innovation participants have little slack to exercise discretion or to exercise control over the development of their innovation. Heavy workloads and high competition for resources represent pressing demands to accomplish a set of tasks and meager means with which to complete them. Resource scarcity creates a "brush fire fighting" orientation of reacting to short-term pressures and deadlines, as opposed to a more proactive and long-term orientation required to develop an innovation. At the other extreme, no or very little resource scarcity provides little stimulus for initiating action, thereby encouraging daydreaming or lethargy. Thus, we follow Lawrence and Dyer (1983) in hypothesizing that moderate levels of resource scarcity are positively related to innovation success.

Standardization of Procedures is defined as the degree to which work rules, policies, and standard operating procedures are formalized and followed to develop an innovation. Processes to develop an innovation are programmed when there are a large number of rules to follow (items 1 below) and when they are specified in detail (item 2 below). Thus, standardization is the average of the following two MIS questions:

1. Number of rules to follow (MIS Q4);
2. Detail of rules and procedures (MIS Q5).

There is a fine but important conceptual distinction between standardization and the variability component of innovation uncertainty. Innovation variability is concerned with the uniformity or number of exceptional problems encountered in developing an innovation, whereas standardization deals with the uniformity of methods for doing the work. It is important to make this distinction between innovation variability and standardization because the former is an aspect of the innovation idea, the latter of how the innovation unit organizes its work. Standardizing procedures for those tasks that can be programmed increases efficient performance of routine or repetitive tasks and permits people to focus a greater proportion of their attention on novel or exceptional tasks that require greater discretion and creativity in developing an innovation over time. Hence, other things held constant, the

59

greater the standardization of routine tasks, the greater the perceived innovation effectiveness.

Decision Influence refers to the amount of discretion or authority that innovation group members perceive they exercise in making decisions on the goals or directions of the innovation, what work needs to be done, obtaining resources for the innovation, and recruiting personnel to work on the innovation (see items 1 through 4 below). Decision influence is a key indicator of the perceived degree of control by innovation group members on the developmental process and is hypothesized to be strongly associated with innovation success. As Pressman and Wildavsky (1959) stated, a person is more likely to implement his or her own ideas than those of someone else.

Decision influence was measured as the average to the following four MIS questions:

1. Deciding on innovation goals (MIS Q10a);
2. Deciding on work to be done (MIS Q10b);
3. Deciding on funding (MIS Q10c);
4. Deciding on personnel recruitment (MIS Q10d).

Expectations of Rewards and Sanctions are two separate constructs that refer to the degree to which innovation participants, individually or as a group, anticipate that good work performance will result in some reward (whether formal promotion or informal recognition) and that poor work performance will result in some punishment (either informal reprimand or formal demotion). Therefore, incentives can take four extreme combinations: individual rewards and group sanctions, group rewards and individual sanctions, individual rewards and sanctions, and group rewards and sanctions. The particular combination of individual and group incentives relates not only to individual and group performance but also the degrees of cooperation and competition among group members. These constructs were developed by Van de Ven and Ferry (1980) and found to be highly

related to job satisfaction and work motivation. A tight link between job performance and expectancy of rewards or sanctions has been a central feature in expectancy theories of motivation (Hackman and Oldham 1975).

The MIS measures expectations on individual and group rewards with two questions:

1. Chance of group reward (MIS Q23a);
2. Chance of individual reward (MIS Q23b).

Expectations of individual and group sanctions are measured with

1. Chance of group reprimand (MIS Q24a);
2. Chance of individual reprimand (MIS Q24b).

Innovation Group Leadership refers to the degree to which leaders of an innovation are perceived by participants to encourage innovative behavior, which the literature suggests is achieved by the leader encouraging initiative, delegating clear responsibilities to members, providing clear feedback, placing trust in members, and maintaining a balanced emphasis on task accomplishment and human relationships (Filley, House and Kerr 1976). The stronger the leadership, the more participants will perceive the innovation to be effective.

Leadership was measured as the average response to the following questions in the MIS:

1. Initiative encouraged (MIS Q15);
2. Members clear of responsibilities (MIS Q16);
3. Clear feedback received (MIS Q30);
4. Emphasis on task (MIS Q18);
5. Emphasis on human relations (MIS Q19);
6. Leader puts trust in members (MIS Q21).

Freedom to Express Doubts refers to the degree to which innovation participants perceive pressures to conform to group and organization norms about not expressing their own beliefs and opinions about the innovative effort. The construct derives from extensive theory and research on the powerful influence of group norms on individual behavior and performance (see reviews in Shull, Delbecq, and Cummings 1970; McGrath 1984). Freedom to express doubts is measured as the average response to the following MIS questions:

1. Criticisms encouraged (MIS Q17);
2. Others speak their doubts (MIS Q20, reverse scale);
3. Freedom to "rock the boat" (MIS Q22, reverse scale).

The more participants perceive an open climate to express their opinions, the more they will perceive the innovation to be effective, both because a felt freedom to express doubts provides a conducive organizational climate for learning and because by expressing doubts problems will likely be detected and/or prevented from "snowballing" into serious obstacles to innovation (Van de Ven 1980).

Learning Encouragement is the degree to which members of an innovation group perceive the organization to place a high priority on learning, to value risk taking, and to minimize personal or career retributions from failure. These factors are indications of an organizational climate or culture often cited to be conducive to innovativeness (Peters and Waterman 1982; Schein 1985). The MIS measures this construct as the average of three questions:

1. Failure not a career blight (MIS Q44, reverse scale);
2. Organization values risk taking (MIS Q45);
3. Learning a high organizational priority (MIS Q46).

External Innovation Transactions

A basic proposition of the research framework is that innovation is not an isolated, individual activity; instead it is a network-building enterprise among a group of people who become sufficiently committed to the innovation to push it into good currency. These people include members of the innovation team as well as members of other organizations and groups who become involved in relationships with the innovation unit. Based on Van de Ven and Walker's (1983) study, the MIS includes measures of the following dimensions of external unit relationships.

Resource Dependence is defined as the extent to which parties in a relationship perceive they need resources (money, information, materials) from the other party in order to develop the innovation. It is measured with three items in Part II of the MIS.

1. Innovation unit needs this group (MIS IIQ5);
2. Other group needs innovation unit (MIS IIQ6);
3. Amount of work done for this group (MIS IIQ7);
4. Amount of work done for innovation unit (MIS IIQ8).

Formalization is the extent to which the terms of the relationship have been verbalized and codified. It is measured with two MIS questions:

1. Relationship verbalized (MIS IIQ3a);
2. Relationship written out (MIS IIQ3b).

Perceived Effectiveness is the extent to which the parties involved believe that (1) each carries out its commitments and responsibilities and (2) the relationship is equitable, worthwhile, and satisfying. It is measured with four MIS questions:

1. Satisfied with relationship (MIS IIQ10);
2. Other group maintained commitments (MIS IIQ11);

3. Innovation unit maintained its commitments (MIS IIQ12);
4. Equity in relationship (MIS IIQ19).

Influence is the perceived degree to which parties to a relationship have changed or affected each other. Since one or both parties in a relationship can be influenced, both sides of the relationship are examined in the MIS:

1. Innovation unit influenced other party (MIS IIQ14);
2. Other group influenced innovation unit (MIS IIQ15).

Perceived Environmental Uncertainty

Finally, the MIS also included scales adopted from Khandwalla (1977) to measure respondents' perceived levels of uncertainty of an innovation's general environment. Uncertainty is a broad concept that can stem from a number of factors, including complexity, predictability, stability, and restrictiveness of various environmental segments relevant to an innovation. Environmental complexity is the diversity or number of different external groups or issues in environmental segments that are perceived to potentially affect the development of an innovation by people within the innovation unit. Environmental predictability is the degree to which it is easy (or difficult) to predict or forecast the future state of affairs in each environmental segment. Environmental stability is the perceived amount and rapidity of change in each environmental segment. Environmental restrictiveness is the perceived degree to which an innovation group perceives it is restricted or constrained by external mandates and regulations. These indicators of environmental uncertainty were measured with the following items in each of four segments of an innovation general environment: economic, technological, demographic, and regulatory segments. The scales used for these measures of environmen-

tal uncertainty were developed and used with good results by Khandwalla (1977).

Economic Environment (market structure and competition):

1. Number of competitors (MIS Q49c);
2. Predict competitors in economic environment (MIS Q49b);
3. Stability of economic environment (MIS Q49a).

Technological Environment (other research and development units):

1. Number of other R&D efforts (MIS Q50c);
2. Predict technological developments (MIS Q50b);
3. Stability of technological environment (MIS Q50a).

Demographic Environment (social trends, population shifts, and educational levels):

1. Number of demographic factors (MIS Q51c);
2. Predict demographic trends (MIS Q51b);
3. Stability of demographic environment (MIS Q51a).

Legal Environment (government policies, regulations, incentives, and laws):

1. Predict legal/regulatory environment (MIS Q48b);
2. Restrictiveness of regulations (MIS Q48a);
3. Hostility of regulators (MIS Q48c).

Situational/Contingency Factors

As suggested in the introduction, the hypothesized relationships between the above MIS dimensions and perceived innovation effectiveness will vary depending on the settings and kinds of innovations being assessed. In particular, we expect these relationships to be conditioned by the novelty, scope, and stage of development of an innovation. We will now present definitions and measures of these situational or contingency factors.

Innovation Novelty refers to whether the work undertaken on an innovation represents an imitation, adaptation, or totally new-to-the-world developmental effort. Pelz and Munson (1982) found systematically different patterns of structure and process to exist between these three categories of innovation novelty. We borrowed Pelz and Munson's three categories to judgmentally code the novelty of the innovations examined in MIRP. Using the codes and descriptions below, research teams were asked to indicate the novelty of the innovations they were studying over time:

1. *Borrowing:* A well-packaged technology, product, process, or service exists; the organization copies it and applies it with little modification. This is sometimes called "imitation," "adoption," or "diffusion."
2. *Adaptation:* A few prototype solutions exist but are not well packaged. The innovation unit draws from these precedents to design and apply an adaptation. This has been called "reinvention," "revisions," or "redesign of a solution to fit local needs."
3. *Origination:* No solution to a problem is known to operate elsewhere. An invention may exist, but it has not been applied. A first-time solution or prototype is developed. This has been called "new to the world innovation."

Because the novelty of the work effort often varies between these categories as an innovation is developed over time (that is, once an original solution is worked out, the research team can often borrow or adapt ideas from elsewhere to work out various components of the solution), the researchers coded the novelty of

the innovation during each round of the MIS survey.

Innovation Scope/Size refers to the number of people and the amount of resources employed in developing an innovation. Two measures obtained from organizational records were used:

1. The number of people directly involved or employed in developing the innovation between each MIS data collection round;
2. The total budget (or dollars spent) on the innovation between each MIS data collection round. Where actual figures were not available, estimates were made.

Innovation Stage/Age. Two measures were used to capture the age and amount of progress toward completion of the innovations. Age was measured by the number of years since dedicated work began on the innovation by the organization. This required determining what was regarded as the innovation's startup birthday.

Innovation developmental stage is an approximation of the number of steps undertaken before an innovation is securely established. Our use of the term "stages" does not imply that the sequence is invariant; stages may recycle, reverse in sequence, and so forth. Instead, our intent is to obtain an overall approximation of the major functional steps that have been completed in the development of an innovation. For purposes of making empirical comparisons between innovations, the following stages and descriptions were adapted from Pelz and Munson's (1982) and coded by the research teams:

1 *Idea stage:* A problem is recognized, search for a solution is undertaken; alternatives are diagnosed; a prototype does not yet exist.
2 *Design stage:* An innovative solution or prototype is developed, adapted, or adopted

and detailed guidelines for action are established.

3 *Implementation stage:* The innovation is put into action; scaleup operations begin. The innovation may be evaluated to decide whether to expand, modify, or discontinue it.

4 *Incorporation, or diffusion, routinization, or institutionalization stage:* The innovation becomes accepted as part of standard operating procedures and is no longer viewed as "an innovation."

ASSESSMENT OF MEASUREMENT PROPERTIES

Psychometric procedures based on confirmatory factor analysis were used to evaluate the measurement properties of the MIS scales. Because the internal and external innovation dimensions belong to different levels of analysis, they will be treated separately in the psychometric analysis. Before presenting the psychometric findings, we will first describe the method, strategy, and data employed in this analysis. The specification of hypothesized measurement model evolves around the nine indices of the innovation process dimensions, and the same specification can be easily extended to other indices in the MIS scale.

Psychometric Evaluation Methodology

Convergent, discriminant, concurrent, and construct validity of the various indices in the MIS scale can be assessed with the following confirmatory factor analysis model:

$$\Gamma = Ax + z \qquad (3.1)$$

$$\Sigma = A\varphi A' + \chi \qquad (3.2)$$

where y is a vector of p items, x is a $k < p$ vector of hypothesized latent dimensions (indices), φ is the intercorrelation matrix among dimensions, A is a matrix of factor loadings relating y to x, z is a vector of unique scores (that is, random measurement errors), Σ is the variance/covariance matrix of item scores, and χ is a diagonal matrix of error variances. Equation (3.1) is commonly known as the factor equation and equation (3.2) as the covariance equation. We now elaborate on this basic measurement model before presenting the results.

Convergent Validity. Convergence is defined as the amount of agreement among multiple measures of the same latent variable. One indicator of convergent validity for an individual index is a formal test of the hypothesized factor structure. A second indicator is the internal consistency of items within an index. To assess the convergent validity of the nine indices in the MIS scale, we test the following null hypotheses:

1. The multiple measures of perceived innovation effectiveness, standardization of procedures, innovation group leadership, freedom to express doubts, and learning encouragement will each converge to indicate a unidimensional scale, and this holds for both time 1 and time 2 samples.

2. As defined above, we expect the multiple measures of the following indices to each converge on two subdimensions: innovation uncertainty (difficulty and variability), decision influence (task and resource decisions), and resource scarcity (workload pressure and resource competition). However, we expect these subindices to converge into their overall respective indices.

These two hypotheses assess the convergent validity of the MIS measures only at a given point in time. Because the MIS was administered over two different points of time with roughly a six-month interval, we can also test the hypothesis for convergence across time. Accordingly, we hypothesize

3. The number of underlying factors and the factor loading matrices of the MIS measures will be invariant across time 1 and time 2 surveys.

The computer program LISREL VI provides a formal means to test the above hypotheses. First, to test hypothesis 1 the LISREL program computes a chi-square goodness-of-fit statistic. This statistic, under the assumption of multivariate normality, allows us to test the null hypothesis that a given measurement model provides an acceptable fit of the observed data. Under this same assumption, which justifies maximum likelihood estimation, the LISREL program computes the variances of individual parameter estimates, which can be used to test the significance of individual items in the factor loading matrices.

To test the null hypothesis 2, we need to examine not only the overall goodness-of-fit and the significance level of individual factor loading scores, but also the intercorrelation matrix among indices and subindices. The convergent validity of a multidimensional index is established if the correlation between its component subindices is significantly greater than the correlations of its subindices with all other indices and subindices in the MIS.

To test the null hypothesis 3, one must analyze the factor structures for time 1 and time 2 samples simultaneously and constrain the factor loading matrices in the two samples to be equal. The detailed procedures for multiple group analyses are described and illustrated in Joreskog and Sorbom (1986).

The confirmatory factor analysis results that provide acceptable fit with a model will then be used to compute the reliability coefficient for individual items and the composite reliability of each index. Although an indication of internal consistency within an index can be gained with an examination of a more conventional reliability estimate, coefficient alpha, it is a conservative estimate of the composite reliability of a scale, as Zeller and Carmines (1979) point out, when the items are not essentially tau-equivalent measures—that is, when items making up a scale contribute differentially to the dimension. Building on the classical test score model of Lord and Novick (1968), Werts, Linn, and Joreskog (1974) developed the following formula to compute the reliability of an individual item:

$$P_i = (\lambda^2_i \, \text{var}A)/(\lambda^2_i \, \text{var}A + \chi_i) \quad (3.3)$$

where $\Sigma\iota$ is the reliability of individual item i, λ_i is the factor loading relating item i to its respective construct A, and χ_i is the error variance of item l.

Substantively, in equation (3.3) the reliability of an individual item is defined as the proportion of variance of the observed scores attributable to the true score—that is, the construct. In a congeneric factor model, when the correlation matrix, instead of the unstandardized covariance matrix, is used in the confirmatory factor analysis, the reliability of an individual item is equal to the square of the factor loading score on its construct. Similarly, the composite reliability of n items of dimension X can be calculated as

$$P_i = (\sum_{i=1}^{n} \lambda_i)^2 \text{var}A / [(\sum_{i=1}^{n} \lambda_i)^2 \text{var}A$$
$$+ \sum_{i=1}^{n} \varphi_i] \quad (3.4)$$

where Σ_i is the composite reliability.

Discriminant Validity. Discriminant validity refers to the degree to which measures of different constructs are unique from each other. The confirmation of convergent validity for individual latent variables is a precondition for examining the uniqueness between them. In the context of confirmatory factor analysis, an appropriate way to assess the degree of discrimination is to examine the intercorrelation matrix of factors in a model providing an acceptable fit, as suggested by Bagozzi (1981: 336). If the latent variables are unique, then all elements (that is, correlation coefficients) in the matrix must be significantly less than unity. Thus to establish the discriminant validity for all nine indices in the MIS in both time 1 and time 2 samples, we propose that

4. The absolute value of all the off-diagonal parameter estimates in the intercorrelation matrice of factors are significantly less than one, and this holds true for both time 1 and time 2 data.

This can be determined by the use of the standard errors of the indices to compute confidence interval for a one-tailed test.

Concurrent and Construct Validity. For each of the nine indices in the MIS, construct validity is evaluated on the basis of how well-estimated correlations among factors correspond with the theoretically expected pattern of relationships among latent variables. In particular, of theoretical importance are the correlations between the Perceived Innovation Effectiveness index and indices that measure various dimensions of innovation ideas, people, transactions, and context. Specifically, we propose that

5. The correlations of perceived innovation effectiveness with standardization of procedures, group leadership, freedom to express doubts, learning encouragement, and decision influence indices will be positive and significantly different from zero, whereas the correlations with innovation uncertainty and resource scarcity will be negative and also significantly different

from zero (as hypothesized in the previous section).

Although we expect that hypothesis 5 will hold for both time 1 and time 2 data, we do not anticipate that these correlations will be uniform in their pattern over time. On the contrary, we adopt a longitudinal research strategy because we expect the structural relationships between the characteristics of the innovation process and innovation outcomes will change over time, depending on the phases or stages of a particular innovation project. Therefore, no cross-sample equality constraint will be imposed on the two matrices as we analyze the factor structures of time 1 and time 2 data simultaneously. We therefore constructed the confidence intervals of these correlation matrices for time 1 sample and time 2 sample separately.

Of course, hypothesis 5 is a crude test of construct validity. A more thorough test must be performed in a structural equation framework that incorporates the consideration of contingent causality and time lag of causal impact because both are expected to attenuate the temporal bivariate correlations. The specification and test of the hypothesized structural model will be presented after the initial results are reported.

In assessing the measurement properties of the innovation environment dimensions, we will test the same five hypotheses. We will not repeat them here, since the extension is straightforward.

Finally, given its centrality in establishing construct validity for other indices in the MIS scale, a parallel measure of perceived innovation effectiveness was used in the interview schedule in order to assess the validity of this MIS scale. Interviews were conducted with a subset of the same individuals who completed the MIS questionnaire. In the interviews, respondents were asked, "What current criteria are you using to judge the success of this innovation?" (open-ended response) and "How do you grade the innovation on these criteria? (A = 5, B = 4, C = 3, D = 2, F = 1)." Although an indication of concurrent validity can be gained with an inspection of the raw intercorrelations between the effectiveness grade obtained in the interview and the MIS index items, such an approach does not take into account random measure errors in the innovation effectiveness scale explicitly. The confirmatory factor analysis will provide us with an estimate of the unattenuated correlation between the two parallel measures, and is therefore a stronger and more conclusive test for concurrent validity. Accordingly, we propose that

6. The unattenuated correlation between the MIS perceived innovation effectiveness index and its parallel measure will be positive and significantly high.

Hypothesis Testing Strategy

A seemingly straightforward way to test the above six hypotheses is by fitting a congeneric measurement model in a multiple group simultaneous analysis, in which (1) all items in the MIS are included and the correlation matrix of items of the two samples are analyzed simultaneously; (2) all items are restrained to have nonzero loadings on their respective hypothesized factor and zero loadings on all other factors; (3) factor loading matrices in time 1 and time 2 samples are constrained to be equal; and (4) all covariance estimates among unique factors within each sample are fixed to zero. However, this approach has some serious drawbacks. First, even if the above simple factor structure provides an acceptable overall fit to the data, there is no guarantee that any part of the model will also fit the corresponding subset of data reasonably well (Joreskog and Sorbom 1986). As the number of items increase, the overall goodness-of-fit statistic might conceal a poor fit in some part of the model when the rest fit the

data extremely well. On the other hand, when dealing with a large data set, it is possible that a robust model that is capable of explaining all variances of substantive importance is rejected on the basis of a chi-square test when the differences between the observed covariance matrix and the predicted matrix is trivial. This is what Bentler and Bonnet (1980: 591) refer to as the "minimally false model" problem.

As a practical matter, it should not be a reasonable expectation that, without relaxing any fixed parameters, a parsimonious model with very large degrees of freedom, such as ours, will provide a statistically acceptable fit to a data set with as many as thirty-seven items and composed of two medium-sized (N1 = 193; N2 = 179) samples. In all likelihood, we will have to allow a few pairs of unique factors to be correlated to reach a statistically acceptable model while keeping the essential factor structure intact and the overall model mathematically identified. The presence of correlated unique factors may be caused by some systematic measurement errors or simply by sampling variability. As items increase, some unique factors will become significantly correlated simply by chance. It should not concern us too much if only a small portion (no more than five percent) of the off-diagonal elements in the theta matrix need to be relaxed and, more critically, if they do not repeat themselves in another sample.

Our hypothesis-testing strategy, then, is to begin our evaluation of the convergent, discriminant, and construct validities of each index by assessing a series of two-factor models, which include the Perceived Innovation Effectiveness index and one of the other eight indices. Based on these preliminary results, we can screen out indices that do not converge on the hypothesized dimension, and then work gradually toward an overall analysis that includes only those indices passing the two-factor model convergence tests.

This approach has several advantages.

67

First, it provides a stronger test of the psychometric quality of an individual index, allowing our null hypotheses to be tested in both a partial and overall analysis and assessed in two samples separately as well as simultaneously providing additional assurance that the hypothesized measurement model is adequate and stable and that the estimation procedure is robust. Second, it is simply a more cost-effective way to test and reach an overall model that is both substantively and statistically acceptable. Fitting a series of two-factor models allows early detection of correlated unique factors. We can then begin the overall analysis by specifying a more realistic model that allows a few measurement errors to be correlated. This substantially avoids the costly "fishing expedition" of relaxing each fixed parameter, one at a time, to specify the model.

Finally, a few minor adjustments are needed to accommodate some practical restraints presented by our data. First, the resources competition subindex of resource scarcity was not included in our time 1 questionnaire, so it will not be available for cross-time assessment. Also, for the same reason, the indices of innovation environment dimensions will be assessed only with our time 2 sample. Next, hypothesis 6 will be tested separately because interviews were conducted with a subset of the same individuals who completed the questionnaire. Also, because the effectiveness scores obtained in the interviews were based on a single-item question, its own psychometric quality cannot be evaluated. We fixed its factor loading to 1 assuming that it is perfectly measured. This is not a heroic assumption because all of our interviewees were managers or project leaders, persons who are commonly judged to be the most informed to evaluate the effectiveness of their innovation.

Innovation Sample and Field Setting
To conduct this psychometric assessment of the MIS, two rounds of data collection were conducted with all key personnel involved in nine different innovations, each located in a different organization. Because the nature and the contextual setting of each innovation has been described in the Appendix of Chapter 1, we include here only a brief summary of the field setting of each of these innovations and the respondents to the MIS survey as follows.

1. *The cochlear implant study:* MIS survey data examined here were obtained from twenty respondents in February 1985 and seventeen respondents in August 1985. Among them, ten respondents who are the innovation's top managers of all the functional areas (such as R&D, clinical trials, FDA regulatory affairs, manufacturing, marketing, and sales) participated in both surveys.
2. *The therapeutic apheresis study:* MIS survey data were obtained from forty-nine respondents in February 1985 and twenty-three respondents in August 1985. Twenty-one respondents including the same eleven top functional managers of the innovation and who represent their respective organizations in the joint venture participated in both surveys.
3. *The naval systems study:* MIS survey data examined here were obtained from ten respondents in January 1985 and in August 1985 from thirty-five respondents, representing middle- and upper-management levels in all functional areas of the innovation. Ten respondents took part in both surveys.
4. *The site-based public school management study:* The MIS data examined here are from one school district that is adapting and implementing this innovation and consist of responses obtained in March from thirty-three respondents and in October 1985 from nineteen individuals who are members of the school's site council and represent teachers, parents, administration, and student constituents. Among them, sixteen individuals participated in both surveys.

5. *The strategic planning study:* The specific innovation included here for this evaluation of the MIS is a county department of nursing and health care services. Data were obtained in August 1985 from forty-eight respondents and in January 1986 from thirty-one respondents who represent different segments of the professional, civil service, and administrative personnel affected by the strategic planning system. Twenty-seven people completed both surveys.

6. *The university research program study:* The sixth innovation to be examined here consists of a university research program that was created to investigate an applied social science problem. The MIS data examined here consist of responses from sixteen investigators obtained in November 1984 and twenty-five of them in May 1985.

7. *The study of small business startup:* The first MIS survey was completed by all thirty-one personnel in the business startup in November 1984. However, the small business encountered major financial setbacks during 1985, when only two of the four remaining personnel completed the MIS. Because the number of cases in the second-wave survey are too few, they are not included in our data set.

8. *The advanced integrated circuits and commercialization of space studies:* Also included are two studies that began their surveys with the revised round 2 questionnaire (T2). The first study examines the development of a joint-endeavor agreement between NASA and a private firm to undertake experiments in low-gravity earth orbit. The innovation process by a gallium arsenide integrated circuits R&D team is being tracked in another study. This R&D team is located within a large computer manufacturing company, and six personnel completed the MIS in February 1985. The R&D team manager chose not to have the MIS survey administered in a subsequent round of interviews.

9. *The nuclear power safety regulations study.* A ninth study examines the implementation of a new set of safety regulations for nuclear power utilities issued by the Nuclear Regulatory Commission. The MIS was completed in 1986 by twenty-three safety review managers of different nuclear power plants throughout the U.S.

In summary, the database for this psychometric evaluation involves five technological and four administrative innovations. Two waves of panel data, approximately six months apart, were obtained with the MIS from a total of 193 respondents during the first-round data collection and 179 respondents during the second. There are a total of 100 respondents involved in both surveys. Chapter 4 provides a comparative qualitative description of the developmental processes of these innovations over time.

RESULTS FOR INTERNAL INNOVATION DIMENSIONS

All the following analyses were performed with LISREL VI. The correlation matrices, instead of the covariance matrices, were used in the analysis. The fits of the models to the data were tested with maximum likelihood ratio chi-square tests. Acceptable fits are based on the conventional .05 level of significance. Parameter estimates and their standard errors will be presented for each overall factor model. However, to avoid redundancy and conserve space, these estimates and standard errors will not be presented for the two-factor models (they are available from the authors).

Convergent Validity
Table 3–1 presents the chi-square test results based on the series of two-factor models. Among the nine MIS indices, innovation effec-

TABLE 3–1.   *Maximum Likelihood Estimation of Two-Factor Model Fitting for Eight MIS Indices with Effectiveness Index.*

Indices	Time 1		Time 2		Cross-time	
	χ^2 (DF)	p	χ^2(DF)	p	χ^2(DF)	p
1. Innovation uncertainty[a]	23.0(22)[b]	.400	Unreasonable estimates		Unreasonable estimates	
2. Workload pressure	7.0(13)	.901	Unreasonable estimates			
3. Standardization	18.4(13)	.142	16.4(11)[b]	.127	45.5(33)	.072
4. Decision influence[a]	24.9(24)	.413	29.9(23)[b]	.152	71.0(57)	.101
5. Reward and reprimand	34.6(24)	.075	Unreasonable estimates			
6. Group leadership	52.3(43)	.156	53.8(40)[b]	.072	115.7(94)	.064
7. Freedom to express doubt	15.2(19)	.719	27.5(19)	.093	45.6(46)	.489
8. Learning encouragement	24.2(19)	.189	20.9(16)[b]	.187	56.6(44)[b]	.096
9. Resources competition	(N/A)		30.8(24)[b]	.160	(N/A)	

a. A three-factor model is specified as the index is hypothesized to converge on two subdimensions.
b. Some error terms are allowed to correlate with each other.

tiveness, standardization, leadership, freedom to express doubts, and learning encouragement consistently converge to indicate a unidimensional construct in time 1 data, time 2 data, and cross-time analyses. Also, as hypothesized, the decision influence items consistently converge on two subdimensions. Further, for all six indices, one cannot reject the hypothesis of factor invariance across time.

Innovation uncertainty items converge on the hypothesized difficulty and variability subdimensions for time 1 data only. Although unidimensional specifications provide a good fit to workload pressure items for the time 1 data, the factor loading estimate of the item dealing with lack of advance time is too low (.176) and statistically not significantly different from zero. For the time 2 data, model fittings for both indices yield unreasonable estimates (either the standardized factor loadings or interfactor correlation coefficients exceed 1 in absolute value). This is an indication that the model might be seriously misspecified. Similarly, reward and reprimand items converge on a single dimension only for the time 1 data. Finally, the resource competition items yield a single structure, but it is not known whether this structure is stable over time, since data on this index were obtained only in the time 2 survey.

Overall, strong evidences exist for the convergent validity of many of the MIS indices. The psychometric quality of innovation uncertainty, resource scarcity, and reward and reprimand indices need further investigation.

In the overall analyses, for each sample a generic factor model was specified for the six MIS indices that passed the preliminary convergence tests (resource competition will be added to the time 2 factor analysis below). As can be seen in Table 3–2, the hypothesized factor structure provides a reasonably good fit to each data set. More significantly, the hypothesis of factor invariance across two samples yields a

TABLE 3–2. *Standardized Parameter Estimates for Overall Factor Models.*

Index: Component Items	Time 1		Time 2		Cross-Time	
	Factor Loading	Error Variance	Factor Loading	Error Variance	Factor Loading	Error Variance
Perceived effectiveness						
Progress satisfaction (MIS Q31)	.831	.309	.868	.246	.851	.302
Problem-solving effectiveness (MIS Q33)	.853	.292	.868	.223	.847	.305
Progress meeting expectations (MIS Q31)	.600	.639	.686	.529	.651	.637
Effectiveness rating (MIS Q32)	.650	.577	.691	.511	.662	.576
Attains organizational goals (MIS Q35)	.444	.803	.511	.757	.478	.800
Standardization of procedures						
Number of rules to follow (MIS Q4)	.823	.323	.985	.029	.966	.027
Detail of rules and procedures (MIS Q5)	.821	.344	.582	.657	.652	.550
Decision influence						
Deciding on innovation goals (MIS Q10a)	.766	.414	.932	.132	.829	.362
Deciding on work to be done (MIS Q10b)	.938	.153	.832	.308	.888	.225
Deciding on funding (MIS Q10c)	.987	.026	.738	.454	.853	.253
Deciding on personnel (MIS Q10d)	.502	.745	.701	.496	.620	.663
Leadership						
Initiative encouraged (MIS Q15)	.654	.550	.641	.576	.659	.532
Members clear of responsibility (MIS Q16)	.499	.737	.663	.560	.577	.723
Clear feedback received (MIS Q30)	.562	.684	.478	.772	.529	.686
Emphasis on task (MIS Q18)	.500	.745	.364	.868	.380	.831
Emphasis on human relations (MIS Q19)	.460	.795	.674	.546	.524	.779
Leader puts trust in members (MIS Q21)	.671	.533	.569	.676	.604	.613
Freedom to express doubt						
Criticisms encouraged (MIS Q17)	.429	.814	.630	.603	.552	.774
Others speak their doubts (MIS Q20)	.552	.698	.656	.556	.639	.630
Freedom to "rock the boat" (MIS Q22)	.695	.517	.703	.506	.674	.586
Learning encouragement						
Failure not a career blight (MIS Q44)	.394	.854	.604	.624	.563	.735

TABLE 3–2. (continued)

Index: Component Items	Time 1		Time 2		Cross-Time	
	Factor Loading	Error Variance	Factor Loading	Error Variance	Factor Loading	Error Variance
Organization values risk taking (MIS Q45)	.852	.270	.637	.609	.691	.470
Learning a high priority (MIS Q46)	.762	.419	.787	.380	.840	.222
Number of cases after deletion of missing values:	N = 135 $\chi^2 = 229.18$ df = 200 p = .077		N = 148 $\chi^2 = 220.29$ df = 198 p = .133		$\chi^2 = 429.32$ df = 416 p = .316	

relatively better fit than fitting the same model to each time period sample individually. Specifically, the chi-square statistics for the time 1 sample is 229.18 with 200 degree of freedom; for the time 2 data 220.29 with 198 degree of freedom, and for the cross-time analysis 429.36 with 412 degree of freedom. In each case, the chi-square to degree of freedom ratio is close to 1 (a result that is much better than the conventional criterion of 2).

Table 3–2 presents the factor loadings and error terms for each item. Generally, most of the factor loadings are moderately high to high, and error variances are low to moderate in absolute value. Also, evidence of convergent validity was achieved between the two subindices of decision influence. As can be seen in Table 3–3, in the cross-time analysis the interfactor correlation coefficient estimates between task decisions and resource decisions is 0.608

TABLE 3–3. Intercorrelation Matrix of MIS Indices for the Cross-Time Congeneric Model.

	Effective	Standard	Decide 1	Decide 2	Leader	Exp/Doubt	Learning
Time 1							
Effective	1.000						
Standard	.134	1.000					
Decide 1	.417[a]	−.054	1.000				
Decide 2	.109	−.157	.608[a]	1.000			
Leader	.625[a]	.062	.371[b]	.204	1.000		
Exp/doubt	.454[a]	.134	.088	.077	.631[a]	1.000	
Learning	.378[a]	.023	.108	.014	.416[a]	.353[b]	1.000
Time 2							
Effective	1.000						
Standard	.104	1.000					
Decide 1	.328[b]	−.146	1.000				
Decide 2	.449[a]	−.086	.623[a]	1.000			
Leader	.721[a]	.003	.457[a]	.494[a]	1.000		
Exp/doubt	.497[a]	.132	.393[a]	.398[b]	.852[a]	1.000	
Learning	.176	.037	.048	−.033	.268[b]	.334[b]	1.000

a. T-value significance level < .01
b. T-value significance level < .05

with a standard error of 0.079 for the time 1 data, and 0.623 with a standard error of 0.074 for the time 2. In both cases, at a .05 probability level, they are significantly greater than the correlations of the two component subindices with all other factors in the congeneric model.

Reliability

Table 3–4 illustrates the composite reliabilities for the each MIS index in the congeneric models. Although some of the factor loadings are moderate, the composite reliabilities are all quite high in magnitude. Except for the freedom to express doubts items, all reliability estimates are fairly stable among the three factor models. In most cases, a Werts, Linn, and Joreskog (1974) formula yields a greater composite reliability estimate than the lower bound estimate, alpha coefficient (as one should expect). In a few cases, our composite reliabilities are close to, or even a bit smaller, than alpha coefficients. This is because, in the model estimation, one or two error term correlations were relaxed to take nonzero values. Overall, then, the six indices all achieve good internal consistency.

Discriminant Validity

Table 3–3 presents the interfactor correlation estimates based on the cross-time factor invariance model. To assess discriminant validity among the six MIS indices, one can examine the correlations among them to see if these values are significantly less than unity. Inspection of the confidence interval for every coefficient in Table 3–3 shows that all these estimates are significantly less than unity at the .05 level. Even the very high correlation between the leadership and freedom to express doubts indices in time 2 marginally pass the criterion; its standard error is 0.071. Thus, as hypothesized, discriminant validity is achieved among different MIS indices.

Concurrent Validity of the Innovation Effectiveness Index

As stated earlier, hypothesis 6 is tested among a subset of thirty-one respondents. Based on a two-factor measure model, in which the single-item interview question is assumed to be perfectly measured and the five effectiveness items are specified to converge on one dimension, the estimate of the unattenuated correlation between the MIS perceived innovation effectiveness index and its parallel measure is .746, and the model itself provides a reasonably good fit to the data. With nine degrees of freedom, the chi-square statistic is 7.73 and probability level .562. Even the correlation between the average of these five MIS items and the effectiveness grade obtained in the interviews is impressive (.732). Together they provide a

TABLE 3–4. *Composite Reliabilities for MIS Indices.*

Index	Time 1		Time 2		Cross-time
1. Effective	0.813	(.817)	0.853	(.853)	0.823
2. Standard	0.802	(.783)	0.782	(.738)	0.819
3. Decide 1	0.837	(.808)	0.876	(.864)	0.834
4. Decide 2	0.742	(.708)	0.686	(.733)	0.703
5. Leadership	0.735	(.697)	0.742	(.741)	0.720
6. Exp/Doubt	0.518	(.560)	0.704	(.713)	0.636
7. Learning	0.723	(.651)	0.718	(.723)	0.754
8. Resource	—		0.849	(.841)	—

Note: For comparison, corresponding alpha coefficients are parenthesized.

strong footing for the criterion-related validity of the perceived effectiveness index.

Construct Validity

If the MIS indices are to be useful for diagnosing innovations, they should be able to detect systematic variations in the organizational arrangements of different kinds of innovations and explain variations in their effectiveness. Moreover, these variations should make logical sense; that is, they should be consistent with a theory that explains why different organizational arrangements are needed for different types of innovations if they are to be effective. Ideally, the explanation would also be predictive in time, so that one can tell in advance what performance outcomes will likely result from various organizational patterns at different stages of innovation development.

We will begin to address these construct validity considerations with a relatively straightforward test of the ability of the MIS dimensions to distinguish between the six different innovations from which data were collected for this psychometric evaluation of the MIS. Table 3–5 presents the mean scores of the MIS dimensions obtained on the eight different innovations as well as the F-ratios and probabilities of significant differences among the eight innovations obtained from a series of one-way analyses of variance conducted on the time 2 survey wave. Table 3–5 shows that significant differences beyond the .01 level are detected among the eight innovations by all ten MIS dimensions. Actually, the F statistic for most of them is beyond the .001 level. This provides solid evidence of the ability of the MIS dimensions to discriminate between different types of innovations. An explanation of how and why these differences exist between the innovations requires an examination of the theory underlying the MIS measurement model, which we turn to next.

As has been suggested, we expect the relationships between the MIS dimensions and perceived innovation effectiveness to be contingent on the innovation's context (that is, its novelty, size, and stage of development). In particular, we expect the MIS dimensions to explain a greater proportion of the variation in perceived innovation effectiveness when the innovation's novelty, size, and stage are taken into account. The reason for this expectation is that qualitatively different design patterns are expected to exist and influence outcomes for innovations that vary in novelty, size, and stage of development. Table 3–6 shows the simple correlations *(r)* and standardized beta coefficients obtained from multiple regression analyses of these innovation context and MIS dimensions on perceived effectiveness within time 1, within time 2, and lagged between times 1 and 2.

In addition to the overall regression results, the bottom of Table 3–6 reports the regression results when only the MIS dimensions are included as the independent variables in the equations. The results in Table 3–6 indicate two major findings. First, with only the MIS dimensions included in the regression equations, 31 percent and 48 percent of the variances are explained in perceived effectiveness within time 1 and time 2, or 22 percent and 42 percent, respectively, after adjustment for degree of freedom, but only 14 percent (and .2 percent after adjustment) of the variance in time 2 perceived effectiveness is explained by the time 1 MIS dimensions. Thus, the results show that although there is solid evidence to support the concurrent validity of the MIS dimensions, predictive validity is limited. Apparently, the temporal period required for the MIS dimensions to exert a causal influence on perceived effectiveness is shorter than expected, since the cross-sectional results are stronger than the six-month lagged regression results.

Second, although evidence to support a contingency theory is not significant at time 1, in time 2 innovation size and stage of development

TABLE 3–5. *One-way Analyses of Variance Between Eight Innovations on the Perceived Innovation Process Dimensions.*

MIS Dimensions	Social Research Group	Naval System Program	Government Planning Program	School Site Management	Gallium Arsenide Innovation	Nuclear Safety Program	Hearing Device Innovation	Blood Device Innovation	ANOVA Results F-ratio	P
Uncertainty	2.8	3.2	2.8	2.6	3.4	2.7	3.1	2.8	5.36	.00
Workload pressure	3.5	3.9	2.8	3.2	4.0	3.7	4.1	3.1	7.25	.00
Resource competition	2.4	3.0	3.1	2.3	2.4	2.9	3.9	3.4	6.43	.00
Standardization	2.4	2.1	1.7	1.8	1.8	2.4	2.1	2.0	2.82	.00
Decision influence	3.0	2.2	2.2	2.9	3.6	3.6	2.5	3.1	5.55	.00
Expect rewards	3.2	3.4	2.7	2.9	3.1	2.8	3.2	3.3	5.36	.00
Expect sanctions	2.9	3.2	2.0	1.7	3.2	3.1	3.1	2.6	19.06	.00
Leadership	3.7	3.7	3.5	3.9	3.9	3.9	3.2	4.2	5.09	.00
Freedom to express doubts	3.5	2.8	3.7	3.1	3.7	3.9	2.5	4.0	9.66	.00
Learning encouraged	3.8	3.4	4.0	3.8	4.2	2.9	3.7	3.7	4.24	.00
Perceived effectiveness	3.0	3.6	2.6	3.6	3.7	3.3	2.7	3.7	10.04	.00
Sample size:	N = 25	N = 35	N = 31	N = 19	N = 6	N = 23	N = 17	N = 23		

TABLE 3–6. *Correlations and Regressions of Contingency Factors and MIS Dimensions on Perceived Effectiveness within T1, T2, and between T1 and T2.*

	Time 1 on T1 Perceived Effectiveness			Time 2 on T2 Perceived Effectiveness			Lagged T1 on T2 Perceived Effectiveness		
	r	beta	p	r	beta	p	r	beta	p
MIS dimensions									
Uncertainty	−.09	−.06	.62	−.10	.11	.16	.02	−.04	.74
Workload pressure	.18[a]	.12	.98	.24[b]	.24	.01	.17	.03	.79
Resource competition	NA	NA	NA	−.18[a]	−.18	.02	NA	NA	NA
Standardization	−.04	−.04	.72	−.01	.06	.45	−.04	.02	.83
Decision influence	.18[a]	.14	.24	.40[b]	.06	.45	.19[a]	.04	.70
Expect rewards	.41[b]	.22	.06	.33[b]	.06	.48	.21[a]	−.08	.45
Expect sanctions	.02	−.04	.77	.10	.01	.91	.25[b]	.18	.18
Leadership	.40[b]	.24	.07	.61[b]	.33	.00	.15	−.06	.65
Freedom to express doubts	.22[a]	.11	.33	.27[b]	.25	.02	.08	.03	.79
Learning encouragement	.31[b]	.16	.14	.04	−.07	.36	.04	−.08	.45
Contingency factors									
Innovation novelty	−.10	−.25	.12	.07	−.16	.13	.09	−.15	.33
Innovation size	.04	.13	.37	.51[b]	.38	.00	.33[b]	.25	.07
Innovation stage	.08	−.01	.94	−.33[b]	−.12	.30	−.11	−.16	.18
Perc. innovation Effectiveness at T1	NA	NA	NA	NA	NA	NA	.53[b]	.58	.00
For overall regression:									
R square (adjusted R2)		.35	(.23)		.62	(.56)		.42	(.31)
F-ratio		2.92			9.65			3.69	
P-level		.002			.000			.000	
For regression with only MIS dimensions entered in equations:									
R square (adjusted R2)		.31	(.22)		.48	(.42)		.14	(.02)
F-ratio		3.47			7.40			1.21	
P-level		.001			.000			.300	

a. Correlations significant at .05 level.
b. Correlations significant at .01 level.

are significantly correlated with perceived effectiveness, and they along with innovation novelty substantially increase the explained variation in perceived effectiveness over and above what is accounted for by the MIS dimensions. In the lagged prediction of time 2 effectiveness from time 1 context, MIS dimensions and effec-

tiveness, it is clear that most of the variance in time 2 perceived effectiveness is explained by perceptions of effectiveness in time 1. This is not an unusual finding because past performance is often the best predictor of future performance. Overall, these findings suggest that innovation novelty, size, and stage operate as contingency factors on the relationships between MIS dimensions and perceived effectiveness in time time 2. In order to examine these contingency effects further, we will now report the findings for different levels of innovation novelty, size, and stage of development.

Tables 3–7 to 3–9 show the simple correlations and multiple regression analyses results on perceived effectiveness in time 2 when the sample of respondents is divided into innovations that differ by levels of innovation novelty (adaptations versus originations in Table 3–7), size (less than or more than fifty individuals in Table 3–8), and stage of development (idea versus implementation stage in Table 3–9). These results indicate the following major findings.

First, Table 3–7 shows that whereas team leadership is found to be the only signifi-

cant predictor of perceived effectiveness for highly novel, new-to-the-world originations, workload pressure, leadership, and resource competition (in that order) are the most important predictors of effectiveness for less novel adaptations of innovations developed elsewhere. In addition, whereas the MIS dimensions after adjustment for degree of freedom explain 56 percent of the variation in effectiveness of adaptations, they account for only 35 percent of the variation in effectiveness of originations. These findings indicate that the influence of the MIS dimensions on perceived effectiveness weaken as the innovations being studied are more novel. One explanation for this result is that with increasing novelty, it is more difficult for people to connect means with ends, and thus randomness increases in relationships between the innovation idea, people,

TABLE 3–7. *Correlations and Regressions of MIS Dimensions on Perceived Effectiveness for Adaptations versus Original Innovations in Time 2.*

MIS Dimensions	ADAPTATIONS N = 47 Perceived Innovation Effectiveness in Time 2			ORIGINATIONS N = 46 Perceived Innovation Effectiveness in Time 2		
	r	beta	p	r	beta	p
Uncertainty	.00	.06	.53	−.02	−.00	.99
Workload pressure	.50[a]	.40	.00	−.16	.06	.69
Resource competition	−.35[a]	−.32	.01	.02	.15	.29
Standardization	.12	.08	.51	−.20	−.12	.35
Decision influence	.49[a]	.16	.20	.24	−.10	.52
Expect rewards	.36[a]	.05	.73	.25	.08	.54
Expect sanctions	.18	−.09	.55	−.11	.02	.87
Leadership	.60[a]	.42	.01	.64[a]	.60	.00
Freedom to express doubts	.05	.10	.46	.50[a]	.26	.16
Learning encouraged	−.03	−.17	.14	.16	−.14	.32
For overall regression:						
R square (adjusted R2)		.67	(.56)		.50	(.35)
F-ratio		6.56			3.40	
P-level		.000			.004	

a. Correlations significant at .01 level.

TABLE 3–8. *Correlations and Regressions of MIS Dimensions on Perceived Effectiveness for Small (less than 50 people) versus Large (more than 50 people) Innovations in Time 2.*

MIS Dimensions	SMALL INNOVATIONS N = 41 Perceived Innovation Effectiveness in Time 2			LARGE INNOVATIONS N = 53 Perceived Innovation Effectiveness in Time 2		
	r	beta	p	r	beta	p
Uncertainty	.10	.15	.22	−.07	.02	.89
Workload pressure	.41[a]	.23	.08	−.12	.21	.15
Resource competition	−.31[b]	−.34	.00	−.26[b]	−.16	.27
Standardization	.20	.20	.15	−.14	−.10	.42
Decision influence	.45[a]	.27	.04	.31[b]	−.30	.06
Expect rewards	.42[a]	.04	.76	−.05	.03	.80
Expect sanctions	.34[a]	−.31	.03	−.24[a]	.01	.94
Leadership	.43[a]	.42	.00	.67[b]	.72	.00
Freedom to express doubts	.27[b]	.08	.54	.46[b]	.28	.13
Learning encouraged	.24[b]	−.04	.71	−.06	−.13	.34
For overall regression:						
R square (adjusted R2)		.58	(.47)		.62	(.48)
F-ratio		5.42			4.52	
P-level		.000			.001	

a. Correlations significant at .01 level.
b. Correlations significant at .05 level.

transactions, context, and perceived outcomes. Pelz and Munson (1982) obtained similar results to support this contingency-theory explanation.

Second, the MIS dimensions appear to be equally robust in explaining variations in effectiveness for both large and small innovations. However, Table 3–8 shows that different sets of MIS dimensions are the significant predictors of perceived effectiveness in large versus small innovations. Although leadership is the most important predictor in both groups, the perceived effectiveness of small innovations is significantly explained by less resource competition, smaller chances of receiving sanctions for doing poor work, greater influence in decisionmaking, and greater workload pressure; whereas in large size innovations, the most im-

portant influences on effectiveness are less decision influence and more freedom to express doubts. These findings suggest that as an innovation grows beyond fifty people in size, the pendulum of positive effects from increasing decision influence for participants in small innovations swings in reverse for large innovations. In addition, whereas competition for resources and workload pressure are prominent influences on perceived effectiveness for small innovations, these factors become subordinate to organizational contexts where participants feel free to express their doubts in large innovations.

Finally, Table 3–9 compares predictions of perceived effectiveness for respondents involved in the idea versus implementation stages of their innovations. A small

TABLE 3–9. *Correlations and Regressions of MIS Dimensions on Perceived Effectiveness for Innovations at Idea versus Implementation Stages in Time 2.*

MIS Dimensions	IDEA STAGE N = 20 Perceived Innovation Effectiveness in Time 2			IMPLEMENTATION STAGE N = 69 Perceived Innovation Effectiveness in Time 2		
	r	beta	p	r	beta	p
Uncertainty	−.41[a]	−.49	.16	.06	.12	.23
Workload pressure	.52[b]	.35	.30	.28[a]	.34	.00
Resource competition	.08	.32	.25	−.31[b]	−.22	.03
Standardization	−.07	−.08	.73	.00	.02	.88
Decision influence	.27	.12	.74	.39[b]	.09	.42
Expect rewards	.30	.02	.95	.28[b]	.08	.51
Expect sanctions	−.40[a]	−.03	.92	.10	−.11	.34
Leadership	.45[a]	.25	.44	.58[b]	.51	.00
Freedom to express doubts	.23	.11	.73	.18	.07	.58
Learning encouraged	−.01	−.06	.81	.07	−.08	.44
For overall regression:						
R square (adjusted R2)		.61	(.21)		.51	(.43)
F-ratio		1.54			6.14	
P-level		.255			.000	

a. Correlations significant at .05 level.
b. Correlations significant at .01 level.

sample of twenty respondents limits our ability to draw statistical inferences about regression analyses for innovations in the idea stage—although the beta coefficients suggest that uncertainty of the innovation idea is the most important predictor of perceived effectiveness of innovations at the idea stage. In the implementation stage, however, the most important predictors are leadership, workload pressure, and resource competition (in that order). Furthermore, the signs of the correlations flip from positive to negative for uncertainty, resource competition, and expectations of sanctions for doing poor work with perceived effectiveness for innovations at the idea versus implementation stage. These findings again suggest the need for developing situational or contingency statements about the factors that influence innovation success.

Indeed, the only factor that is consistently and significantly positively correlated with perceived effectiveness for originations versus adaptations, small or large, and idea versus implementation stages of innovations is leadership. Strong team leadership appears to be the only consistent "universal" predictor of perceived innovation effectiveness.

Of course, these substantive conclusions on this limited sample of innovation respondents are premature. Our objective here has not been to develop or test a substantive theory on innovation effectiveness. Further examination of these findings on a larger and different sample of innovations is required to draw any substantive conclusions on what factors influence innovation effectiveness. Instead, our objective has been to obtain an estimate of the construct validity of the MIS dimensions by evaluating them in relation to the measurement model underlying the MIS. We believe the results are sufficiently meaningful and significant to conclude there is substantial evidence for the construct validity of the MIS contingency-theory measurement model.

RESULTS FOR INNOVATION ENVIRONMENT DIMENSIONS

The assessment of the measurement properties of the MIS innovation environment dimensions was conducted on the time 2 survey data, which include eight innovations. The hypothesis-testing procedures are essentially a replication of those reported above for evaluating the MIS internal innovation dimensions with two differences. First, the hypothesis about cross-time factor invariance cannot be examined because data on many environmental dimensions were

TABLE 3–10. *Standardized Parameter Estimates for Overall Factor Model for External Transaction Items with Perceived Effectiveness.*

| | Time 2 | | |
Index: Component Items	Factor Loading	Error Variance	Composite Reliability
Perceived innovation effectiveness			.845
Progress satisfaction (MIS Q31)	.848	.281	
Problem-solving effectiveness (MIS Q33)	.879	.227	
Progress meeting expectations (MIS Q31)	.681	.536	
Effectiveness rating (MIS Q32)	.693	.520	
Attains organizational goals (MIS Q35)	.474	.775	
Influence			.427
Group influenced unit (MIS IIQ15)	.331	.890	
Unit influenced group (MIS IIQ14)	.691	.513	
Formalization			.824
Relation verbalized (MIS IIQ3a)	.978	.044	
Relation written (MIS IIQ3b)	.677	.541	
Effective relation			.866
Trust with group (MIS IIQ9)	.819	.330	
Satisfied with relation (MIS IIQ10)	.928	.138	
Group fulfilled commitments (MIS IIQ11)	.742	.449	
Balance in relation (MIS IIQ19)	.634	.599	
Dependency			.575
Need group to achieve goals (MIS IIQ5)	.393	.852	
Group needs innovation unit (MIS IIQ6)	.847	.283	
	N = 143		
	χ^2 = 107.30		
	df = 79		
	p = .019		

obtained in only one time period (time 2). Second, as a result, tests of predictive validities of the environmental indices are also not available. In addition, the environmental dimensions constitute the larger context in which an innovation unit finds itself. Their influence on innovation outcomes is mediated by other innovation process dimensions as well as situational/contingency factors. In structural equation terms, their theoretical significance can be evaluated only in a fully specified recursive model, an exercise well beyond the scope of this chapter.

Reliability, Convergent and Discriminant Validity

To assess the convergent, discriminant, and construct validities of the innovation environment dimensions, we fit two congeneric factor models to the data: one including the four indices of external transactions and perceived innovation effectiveness, and the other including the four indices of environmental uncertainty and innovation effectiveness. Tables 3–10 and 3–11 present the results in estimating the two models, respectively.

Overall, in both cases, the hypothesized

TABLE 3–11. *Standardized Parameter Estimates for Overall Factor Model for Environmental Uncertainty Items with Perceived Effectiveness.*

Index: Component Items	Time 2		Composite Reliability
	Factor Loading	Error Variance	
Perceived effectiveness			.864
Progress satisfaction (MIS Q31)	.860	.260	
Problem-solving effectiveness (MIS Q33)	.891	.207	
Progress meeting expectations (MIS Q31)	.684	.532	
Effectiveness rating (MIS Q32)	.728	.469	
Attains organizational goals (MIS Q35)	.551	.708	
Uncertainty of economic environment			.698
Predictability of economic environment (MIS Q49b)	.729	.459	
Stability of economic environment (MIS Q49a)	.832	.300	
Complexity of economic environment (MIS Q49c)	−.371	.860	
Uncertainty of technological environment			.732
Predictability of technological environment (MIS Q50b)	.598	.676	
Stability of technological environment (MIS Q50a)	.898	.193	
Complexity of technological environment (MIS Q50c)	−.563	.685	
Uncertainty of demographic environment			.754
Predictability of demographic environment (MIS Q51b)	.729	.435	
Stability of demographic environment (MIS Q51a)	.872	.240	
Complexity of demographic environment (MIS Q51c)	−.482	.739	
Restrictiveness of legal environment			.565
Restrictiveness of regulations (MIS Q48a)	.404	.837	
Hostility of regulators (MIS Q48c)	.820	.319	

$$N = 143$$
$$\chi^2 = 106.96$$
$$df = 87$$
$$p = .072$$

factor structures provide reasonably good fits to the data. The chi-square statistic for the first model is 107.3 with 79 degree of freedom, and the goodness-of-fit statistic for the second model is 106.96 with 87 degree of freedom. In the latter, the significance level is .072. In the former, its significance level fell slightly below the .05 benchmark, but the chi-square ratio is well above the widely used criterion of 2.

Evidence of convergent validity is also found in the factor loading of individual item. Generally, most of the factor loadings are either fairly high or moderately high. For environment uncertainty dimensions, only one item fell below .40, and for external transactions dimensions only two. On Table 3–11, the three items measuring the complexity aspect of environmental uncertainty carry negative factor loadings because of their reverse wording (as compared to the other two on the scale). This may be part of the reason that their factor loadings are consistently lower than others.

Tables 3–10 and 3–11 also report the composite reliabilities for each innovation environment dimension. Most of the reliability co-

efficients based on Werts, Linn, and Joreskog formula are quite high. The only exception is the one measuring Perceived influence of transactions. Table 3–12 presents the interfactor correlation estimates based on the two congeneric factor models. Evidence for discriminant validity is achieved as all of the correlation estimates are significantly less than unity.

Construct Validity

To assess construct validity, we begin with an examination of bivariate correlations among environment dimensions. As can be seen in Table 3–12, all four indices of external transactions maintain moderately weak positive correlations with Perceived Innovation Effectiveness while all environment uncertainty dimensions are either weakly correlated with innovativeness or not correlated at all. This does not suggest that environment uncertainty is less relevant a set of factors than transactions. Instead, it suggests only that the latter, dealing with the immediate transactional environment, have a stronger affect, as should be expected.

TABLE 3–12. Intercorrelation Matrix of MIS Innovation External Dimensions for Time 2.

Indices of External Unit Transactions:

	Influence	Formalization	Effective	Dependency	Innovative
Influence	1.000				
Formalization	.284[a]	1.000			
Effective	.315[a]	.490[b]	1.000		
Dependency	.344[a]	.423[b]	.640[b]	1.000	
Innovative	.379[b]	.359[b]	.385[b]	.464[b]	1.000

Indices of Environmental Uncertainty:

	Technology	Demography	Economic	Legal	Innovative
Technology	1.000				
Demography	.256[b]	1.000			
Economic	.480[b]	.395[b]	1.000		
Legal	−.009	−.120	−.631[b]	1.000	
Innovative	−.001	−.090	.328[b]	−.218[a]	1.000

a. T-value significance level < .05.
b. T-value significance level < .01.

TABLE 3–13. One-way Analyses of Variance between Eight Innovations on the Perceived External Innovation Dimensions.

	Social Research Group	Naval System Program	Government Planning Program	School Site Management	Gallium Arsenide Innovation	Nuclear Safety Program	Hearing Device Innovation	Blood Device Innovation	ANOVA Results F-ratio	P
I. External unit transactions:										
Influence	2.6	3.0	2.7	2.8	3.1	2.9	3.0	2.8	1.05	.40
Formalization	2.9	3.0	2.7	3.0	3.4	3.1	3.2	3.7	2.43	.02
Effective relation	3.8	3.5	3.3	3.5	3.5	3.2	3.3	3.6	1.42	.20
Dependency	3.3	4.0	3.6	3.8	4.0	3.0	3.4	4.0	5.19	.00
II. Environmental uncertainty:										
Economic	2.8	2.9	2.7	2.8	2.4	2.9	1.9	2.9	7.03	.00
Technological	2.9	2.9	2.7	2.8	2.7	3.0	2.5	3.0	2.35	.03
Demographic	2.8	3.2	3.2	3.1	2.9	3.1	3.1	3.1	1.55	.15
Legal	2.4	3.3	3.6	2.1	2.3	3.3	3.4	3.4	14.06	.00
Sample size:	N = 25	N = 35	N = 31	N = 19	N = 6	N = 23	N = 17	N = 23		

A series of one-way analysis of variance were computed to determine if the eight environment dimensions detect significant differences between the eight different innovations. The mean scores of MIS dimensions and the F-ratios and significance levels obtained from the ANOVA tests are reported in Table 3–13. Significant differences beyond the .05 level are detected among the eight innovations by five of the eight innovation environment dimensions. The exceptions are influence (which was found earlier to have low reliability), effective relationship, and demographic uncertainty.

Finally, in lieu of structural equations, we present simple correlations and standardized beta coefficients obtained from multiple regression analysis of MIS innovation environment dimensions on perceived effectiveness. The regression results when the situational/contingency factors are added to the equation are reported in the two columns on the right side on Table 3–14. The inclusion of innovation size, stage, and novelty increase the explained variation in perceived innovation effectiveness from 24% to 33%. The inclusion of contingency factors, however, either attenuate the prediction power of MIS environmental dimensions or introduce little noticeable

TABLE 3–14. *Correlations and Regressions of Contingency Factors and MIS Innovation External Dimensions on Perceived Effectiveness for T2 Sample.*

		For Regression with Only MIS Dimensions Entered in Equations:		For Regression with Contingency Factors Also Entered:	
	r	beta	p	beta	p
External unit transactions					
Influence	.25[a]	.11	.16	.10	.17
Formalization	.33[a]	.17	.05	.09	.28
Effective relation	.32[a]	.09	.33	.18	.05
Dependency	.42[a]	.27	.00	.15	.10
Environmental uncertainty					
Economic	.23[a]	.17	.04	.15	.07
Technological	.08	.03	.73	−.01	.91
Demographic	−.09	−.13	.10	−.15	.04
Legal	−.10	−.03	.66	−.07	.34
Contingency factors					
Innovation novelty	−.01	N.A.		−.13	.21
Innovation size	.36[a]	N.A.		.24	.00
Innovation stage	−.30[a]	N.A.		−.22	.02
For overall regression:					
adjusted R2		.24		.33	
F-ratio		6.36		7.19	
P-level		.00		.00	

a. Correlations significant at .01 level.

changes. The exception is effectiveness of transactions, which was expected to have the strongest direct relationship with overall innovation effectiveness. These tentative findings suggest that innovation size, stage, and novelty operate more as intermediate variables than contingency factors as far as the relationships between the MIS environmental dimensions and perceived innovation effectiveness is concerned. Overall, the evidence for the construct validity of the eight environmental dimensions, while far from being conclusive, are quite consistent with their respective theoretical locations in our conceptual framework.

CONCLUSION

This chapter has evaluated the measurement properties of the Minnesota Innovation Survey (MIS), which is one of several data collection instruments being used for longitudinal real-time study of innovations as they develop by the Minnesota Innovation Research Program. Substantial evidence supporting various estimates of the reliability and validity of most of the MIS dimensions was obtained by evaluating data obtained with the MIS in two time periods on a sample consisting of nine very different innovations. In addition, weaknesses in a few specific indices (innovation uncertainty, chances of rewards and sanctions, and influence of parties in transactions) were found, and they require revision before being used in subsequent empirical research. By examining how well the MIS dimensions explain perceived innovation effectiveness, strong empirical support was found for the concurrent (contemporaneous) validity of the MIS internal innovation dimensions, but predictive (time lagged) validity was found to be weak. As expected, the MIS external environmental dimensions were generally weakly related to overall innovation effectiveness, since they are only indirectly related to innovation outcomes. Finally, substantial evidence of construct validity was obtained by identifying the moderating or contingency effects of innovation novelty, size, and stage of development on the relationships between the MIS dimensions and perceived innovation effectiveness.

APPENDIX A

Other Indices in MIS

Several other dimensions are measured in the MIS but were not evaluated for their psychometric properties because they are single-item scales. Most of them deal with relatively straightforward biographical or behavioral descriptions.

Individual Differences include basic biographical information about the respondent's age, number of dependents, previous work experience, and educational background. They are measured with the following MIS items:

1. Age (MIS Q37);
2. Number of dependents (MIS Q38);
3. Years working experience in related field (MIS Q43);
4. Months of tenure in positions within organization (MIS Q39);
5. Months of tenure in positions outside organization (MIS Q40);
6. Years of education after high school (MIS Q41);
7. Highest educational degree (MIS Q42).

Time Allocation on Tasks refers to the different kinds of work that occupy the time of innovation unit members during a period of time. The measure is important for determining what occupies the working time of innovation group personnel and what they are paying attention to. Time allocation is measured in the MIS with the following questions:

1. Total hours/week worked on innovation (MIS Q8);
2. Hours/week supervising employees (MIS Q9a);
3. Hours/week technical work (MIS Q9b);
3. Hours/week potential user contact (MIS Q9c);
4. Hours/week obtaining resources (MIS Q9d);
5. Hours/week coordination among members (MIS Q9e);
6. Hours/week administrative review (MIS Q9f);
7. Hours/week administration (MIS Q9g);
8. Hours/week personal education (MIS Q9h).

In addition, an open-ended question was used to determine the qualitative nature of the work the respondent performed by asking "list the major tasks you performed the past six months to develop this innovation," and the percentage of time spent on each task.

Communication Frequency refers to how often unit members communicate with others within and outside of the innovation unit. The construct has been related to innovation success in many studies, and provides an indication of the local/cosmopolitanism of unit members, and the kind of network they maintain over time. It was measured with the following MIS questions about the frequency of communications with:

1. Other innovation group members (MIS Q26a);
2. People in other departments in the organization (MIS Q26b);
3. Managers at higher levels (MIS Q26c);
4. External consultants (MIS Q26d);
5. Existing or potential customers (MIS Q26e);

6. Existing or potential vendors (MIS Q26f);
7. Government or industry regulators (MIS Q26g).

Problems Encountered refers to obstacles or barriers experienced in the development of an innovation over time. The scale focuses on the incidence of experiencing six categories of problems, which were initially developed in a study of implementing innovations by Gross, Giacquinta, and Bernstein (1971) and found to be strongly associated with innovation success by Bass (1971) and Van de Ven (1980). Each of these six problem categories are measured with one item in the MIS as follows:

1. Personnel recruitment problems (MIS Q14a);
2. Lack of clear goals or plans (MIS Q14b);
3. Lack of clear implementation methods (MIS Q14c);
4. Lack of finances or resources (MIS Q14d);
5. Coordination problems (MIS Q14e);
6. Lack of support or resistance (MIS Q14f).

In addition, an open-ended question (MIS Q13) asks respondents to qualitatively describe the problems experienced in developing their innovation.

Conflict refers to two basic kinds of discord within an innovation unit. One kind is the frequency of disagreements or disputes among personnel involved in the development of an innovation (item 1 below). The second kind emerges from "segmentalism" (Kanter 1983) or "turf guarding" in which innovation group members withhold cooperation across disciplinary or functional lines (item 2 below). Only single item scales were used to measure the two kinds of conflict in the MIS.

1. Conflict frequency (MIS Q27);
2. "Turf guarding" (MIS Q25).

Conflict Resolution Processes refer to the methods by which disagreements and disputes are handled by the innovation group. Based on the previous work of Blake and Mouton (1964), Lawrence and Lorsch (1967), Burke (1970), and Van de Ven and Ferry (1980), four basic methods of conflict resolution are examined in the MIS. (The anecdotes below are based on Burke's (1970:394) description of the conflict resolution methods.)

1. *Ignoring or avoiding the issues* (MIS Q28a): easier to refrain than to retreat from an argument; silence is golden; see no evil, hear no evil, speak no evil;
2. *Smoothing over the issues* (MIS Q28b): play down the differences and emphasize common interests; issues that might cause divisions or hurt feelings are not discussed;
3. *Openly confronting the issues* (MIS Q28c): open exchange of information about the conflict or problem and a working through of differences to reach a mutually agreeable solution;
4. *Resorting to hierarchy* (MIS Q28d): having the supervisor or other person with power or authority over the contesting parties resolve the matter.

External Unit Relationships

Complementarity is the perceived level of mutual benefit or synergy between parties to a relationship, and is measured with two MIS questions, which do not combine as one factor:

1. Complementary objectives (MIS IIQ2);
2. Alternative uses of relationship work (MIS IIQ15).

Consensus/Conflict is the degree of agreement and conflict among the parties in the relationship, and the degree to which they trust each other. It is measured with three MIS items, which tap very different sources of conflict:

1. Frequency of conflict (MIS IIQ17);
2. Competition between parties (MIS IIQ4);
3. Trust among parties (MIS IIQ9).

Communication Frequency refers to how often parties to a relationship were in contact with each other during the past six months. It is measured with MIS IIQ16.

Duration of the relationship refers to the length of time it is expected to last. It is measured with MIS IIQ18.

APPENDIX B

Minnesota Innovation Survey

Cover Letter The Strategic Management Research Center at the University of Minnesota is studying the development of a wide variety of innovations in a broad cross section of organizations. The purpose of the research is to better understand how to manage the innovation process and to learn what factors influence the successful development of innovations over time.

One of the innovations being studied is the development of [innovation]. The study will track this innovation over time by obtaining periodic information through this survey, interviews, and from various documents.

We would appreciate your cooperation in this study by completing this survey. It should take less than forty minutes. Most questions can be answered by simply circling or writing a number that reflects your best judgment on an answer scale. All questions are straightforward, and there are no right or wrong answers.

Throughout the survey the term "innovation" refers to [innovation]. Similarly, unless used otherwise, the term "organization" or unit refers to the individuals who are directly involved in the development of the [innovation].

You will receive a feedback report on the findings of this survey. We promise that the information you provide will remain confidential. Data will be averaged across individuals and organizational units, and no individual will be identified in any of the study findings. Thank you.

_____

Project Director

_____

Today's Date _____

Name of Your Organization _____

Your Name: _____ Telephone: _____

Your Position: _____

Your Address: _____

Please describe your role or part in this innovation by listing below the major tasks you performed during the past six months to develop this innovation.	What percent of your total time on this project was spent on task?
_____	_____ %
_____	_____ %
_____	_____ %
_____	_____ %
_____	_____ %
_____	_____ %
_____	_____ %
_____	_____ %
_____	_____ %

Please circle a number on the answer scale that best reflects your answer

1. How big of an undertaking does this innovation represent?

   1 = very small undertaking
   2 = small undertaking
   3 = moderate undertaking
   4 = large undertaking
   5 = very large undertaking

2. How easy is it for you to know in advance what correct steps are needed to develop this innovation?

   1 = very easy
   2 = quite easy
   3 = somewhat easy
   4 = quite difficult
   5 = very difficult

3. What percent of the time are you generally sure of what the outcomes will be of your efforts to develop this innovation?

   _____ %

4. How many rules and procedures exist for doing your part of the work on this innovation?

   1 = very few if any
   2 = a small number
   3 = a moderate number
   4 = a large number
   5 = a great number

5. How precisely do these rules and procedures specify how your work is to be done?

   1 = very general
   2 = mostly general
   3 = somewhat specific
   4 = quite specific
   5 = very specific

6. How heavy was your work load during the past six months on this innovation?

   1 = often not enough to keep me busy
   2 = sometimes not enough to keep busy
   3 = just about the right amount
   4 = hard to keep up with
   5 = entirely too much to handle

7. How far in advance do you generally know what kind of work is required of you on this innovation?

   1 = about 1 hour in advance
   2 = about a day in advance
   3 = about a week in advance
   4 = about a month in advance
   5 = about 6 months in advance

8. On the average, how many hours per week did you work on matters related to the innovation in the past six months?

   _____ Hours/week

9. Of this time, about how many hours per week were spent on each of the following activities during the past six months:

a. Supervising individuals connected with the innovation? _____ hours

b. Working on technical aspects of the innovation? _____ hours

c. Discussing the innovation with potential customers or users? _____ hours

d. Obtaining funding and resources for the innovation? _____ hours

e. Coordinating the innovation with other organizational units _____ hours

f. Preparing for or conducting administrative reviews of the innovation? _____ hours

g. Administrative work (scheduling, planning, paperwork)? _____ hours

h. Personal education (reading & seminars to remain current)? _____ hours

10. How much influence did you have in each of the following decisions that may have been made during the past six months?

	Decision Not Made	None	Little	Some	Quite A Bit	Very Much
a. Setting goals and performance targets for the innovation.	0	1	2	3	4	5
b. Deciding what work activities are to be performed on the innovation?	0	1	2	3	4	5
c. Deciding on funding and resources for the innovation?	0	1	2	3	4	5
d. Recruiting individuals to work on the innovation?	0	1	2	3	4	5

11. During the past six months, how often did difficult problems arise in the development of this innovation?

1 = monthly or less
2 = about weekly
3 = about daily
4 = several time daily
5 = many times daily

12. How different were these problems each time they arose?

1 = very much the same
2 = mostly the same
3 = quite a bit different
4 = very much different
5 = completely different

13. Please describe any special problems or difficulties you are now experiencing in developing this innovation.

14. During the past six months, to what degree have you experienced each of the following difficulties?

	None	Little	Some	Much	Great
		Amount of Difficulty			
a. *Personnel recruitment problems:* Finding individuals who are properly qualified for jobs?	1	2	3	4	5
b. *Lack of clarity* about certain goals or plans for the innovation	1	2	3	4	5
c. Lack of understanding about *how to implement* certain goals or plans for the innovation?	1	2	3	4	5
d. *Lack of finances or resources* necessary for developing the innovation?	1	2	3	4	5
e. Problems in *linking or coordinating* aspects of the innovation with other organizational units?	1	2	3	4	5
f. *Lack of support or resistance* from key sponsors of the innovation?	1	2	3	4	5

**The next questions focus on how this innovation is organized.
Please circle a number that best describes your views.**

	Disagree Strongly	Disagree Somewhat	Neutral	Agree Somewhat	Agree Strongly
15. Leaders of this innovation encourage individuals to take the initiative.	1	2	3	4	5
16. Individuals connected with the innovation are clear about their individual responsibilities.	1	2	3	4	5

	Disagree Strongly	Disagree Somewhat	Neutral	Agree Somewhat	Agree Strongly
17. Criticising or providing information which challenges the feasibility of what is being done to develop the innovation is encouraged.	1	2	3	4	5
18. Leaders of this innovation place a strong emphasis on getting the work accomplished.	1	2	3	4	5
19. Leaders of the innovation place a strong emphasis on maintaining group relationships.	1	2	3	4	5
20. I sometimes get the feeling that others are not speaking up although they harbor serious doubts about the direction being taken.	1	2	3	4	5
21. A high level of trust is placed in individuals connected with this innovation by leaders.	1	2	3	4	5
22. Often I feel pressured *not* to "rock the boat" by speaking my mind about what's going on with this innovation.	1	2	3	4	5

23. When performance targets for this innovation *are attained,* how likely will the following happen:

	No Chance	Small Chance	50% Chance	Quite Likely	Almost a Certainty
a. All involved *as a group* are rewarded or recognized for their collective achievements?	1	2	3	4	5
b. Only *specific individuals* are rewarded or recognized for their individual achievements?	1	2	3	4	5

24. When performance targets for this innovation are *not attained,* how likely will the following happen:

	No Chance	Small Chance	50% Chance	Quite Likely	Almost a Certainty
a. All involved *as a group* are reprimanded or told to "shape up" to improve their performance.	1	2	3	4	5
b. Only *specific individuals* are reprimanded or told to "shape up" to improve their individual performance.	1	2	3	4	5

25. How much "turf-guarding" is there between different departments and professional groups connected with this innovation?

Not At All	A Little	Somewhat	Quite A Bit	Very Much
1	2	3	4	5

26. During the past six months, how frequently have you personally communicated on matters related to the innovation with:

	No Con-tact	Monthly Or Less	About Weekly	About Daily	Every ½ Day	Every Hour
a. Other individuals who are working on the innovation?	1	2	3	4	5	6
b. people in other departments in your organization?	1	2	3	4	5	6
c. managers at higher levels in your organization?	1	2	3	4	5	6
d. consultants from other organizations?	1	2	3	4	5	6
e. existing or potential customers?	1	2	3	4	5	6
f. existing or potential vendors?	1	2	3	4	5	6
g. people in government or industry regulatory agencies?	1	2	3	4	5	6

27. During the past six months, how often were there disagreements among people related to the innovation?   1   2   3   4   5   6

28. When disagreements or disputes occurred, how often were they handled in each of the following ways in the past six months?

	Almost Never	Seldom	Half the Time	Often	Always
a. by ignoring or avoiding the issues?	1	2	3	4	5
b. by smoothing over the issues?	1	2	3	4	5
c. by bringing the issues out in the open and working them out among the parties involved?	1	2	3	4	5
d. by having a higher-level supervisor resolve the issues among the parties involved?	1	2	3	4	5

	Not At All	A Little	Somewhat	Quite A Bit	Very Much
29. When problems occur, how often are the basic assumptions re-examined about the goals of this innovation?	1	2	3	4	5
30. How often are individuals involved in the innovation given constructive feedback on how to improve their work?	1	2	3	4	5
31. Overall, how satisfied are you with the progress made in developing this innovation during the past six months?	1	2	3	4	5

	Poor	Fair	Good	Very Good	Excel- lent
32. Overall, how would you rate the present effectiveness of this innovation?	1	2	3	4	5
33. How well do people connected with the innovation anticipate and solve problems?	1	2	3	4	5

	Far Below	A Little Below	Where Expected	A Little Above	Far Above
34. To what degree is your progress with the innovation below or above your initial expectations?	1	2	3	4	5

	None	Little	Some	Much	Very Much
35. How much does this innovation contribute to attaining the overall goals of your organization?	1	2	3	4	5

36. What specific suggestions do you have for improving this innovation effort?

The next questions ask about your background.

---

37. What is your age? _____ years old

38. How many individuals in your family are financially dependent on you for their livelihood?

Number of dependents _____

39. List below your last two positions or or job titles with this organization.

Indicate the months and years during which each position was held.
mo/year   mo/year

a. _____ From _____ To _____

b. _____ From _____ To _____

40. List below the last two positions or jobs you held with other organizations before you began with this organization.
Position   Organization

a. _____ From _____ To _____

b. _____ From _____ To _____

41. Years of academic or professional education you obtained after high school?

_____ Years after high school

42. What is the highest degree you obtained
in school? (check only one at right)

_____1 High School
_____2 1–3 yrs. College or
Technical School
_____3 Bachelor's degree
_____4 Master's degree
_____5 Doctoral degree

43. Please list the number of years of working experience in
fields related to the innovation?

_____ years

The next questions focus on the overall organization and environment in which your innovation is located.

	Disagree Strongly	Disagree Somewhat	Neutral	Agree Somewhat	Agree Strongly
44. When a person tries something new and fails at it, it will be considered a serious blight on the individual's career in this organization.	1	2	3	4	5
45. This organization seems to place a high value on taking risks, even if there are occasional mistakes.	1	2	3	4	5
46. In this organization, a high priority is placed on learning and experimenting with new ideas.	1	2	3	4	5

47. How much does your innovation have to compete with other organizational units for each
of the following:

	None	Little	Some	Much	Very Much
a. financial resources?	1	2	3	4	5
b. materials, space, and equipment?	1	2	3	4	5
c. management attention?	1	2	3	4	5
d. personnel?	1	2	3	4	5

48. How would you characterize the *legal/regulatory environment* of this innovation—including
government policies, regulations, incentives and laws?

a) Very lenient, Few Regulations	1	2	3	4	5	Very Restrictive, Many Regulations
		moderate				

b) Very unpredictable;   1   2   3   4   5   Very predictable,
   hard to anticipate   moderately   very easy to forecast
   the nature or   predictable   the future state of
   direction of changes   affairs in the
   in the environment   environment

c) Hostile, Adversarial   1   2   3   4   5   Friendly, supportive
                                 moderate

49. How would you characterize the *economic environment* of this innovation, including market
    structure and competition?

a) Very dynamic   1   2   3   4   5   Very stable,
   changing rapidly   moderate   virtually no change

b) Very unpredictable;   1   2   3   4   5   Very predictable,
   hard to anticipate   moderately   very easy to forecast
   the nature or   predictable   the future state of
   direction of changes   affairs in the
   in the environment   environment

c) Very simple,   1   2   3   4   5   Very complex,
   few competitors   moderate   many competitors

50. How would you characterize the *technological environment* of this innovation, including
    advances in research and development of new products, devices and processes?

a) Very dynamic   1   2   3   4   5   Very stable,
   changing rapidly   moderate   virtually no change

b) Very unpredictable;   1   2   3   4   5   Very predictable,
   hard to anticipate   moderately   very easy to forecast
   the nature or   predictable   the future state of
   direction of changes   affairs in the
   in the environment   environment

c) Very simple,   1   2   3   4   5   Very complex,
   few other R&D efforts   moderate   many other R&D efforts

51. How would you characterize the *demographic environment,* including social trends, population
    shifts, income and educational levels, that may affect this innovation?

a) Very dynamic   1   2   3   4   5   Very stable,
   changing rapidly   moderate   virtually no change

b) Very unpredictable;   1   2   3   4   5   Very predictable,
   hard to anticipate   moderately   very easy to forecast
   the nature or   predictable   the future state of
   direction of changes   affairs in the
   in the environment   environment

c) Very simple, few   1   2   3   4   5   Very complex, many
   demographic factors   moderate   demographic factors
   affect this innovation   affect this innovation

PART II: Relationships With Other Groups

This section concentrates on the most important "other groups" that your innovation unit has dealt with during the past six months in order to develop the innovation. Please answer the questions for each group identified below.

Names of Key "Other Groups"	What Activities Were Undertaken With This Other Group During the Past Six Months?
1. _____	
2. _____	
3. _____	

4. _____

_____

5. _____

_____

Please select the "other group" from the previous page that you have worked with most closely during the past six months. Write the name of this group in the space below, and answer the following questions about this group.

Name of group you have worked with most closely: _____

1. What type of relationship does your   1 = they are part of our organization
   unit have with this group?   2 = we have a joint venture with them
      (check one on right)   3 = we are a vendor-supplier for them
                                          4 = we are a customer-user of them
                                          5 = they are a government regulator

	Not At All	Little	Some	Quite A Bit	Very Much
2. How much do the objectives of this other group complement those of your innovation group?	1	2	3	4	5
3. How much have the terms of the relationship between your innovation group and this other group:					
a. Been explicitly verbalized or discussed?	1	2	3	4	5
b. Been written down in detail?	1	2	3	4	5

	Not At All	Little	Some	Quite A Bit	Very Much
4. To what degree does your group compete with this other group?	1	2	3	4	5
5. In order for your unit to accomplish its goals, how much do you need this group?	1	2	3	4	5
6. In order for this group to accomplish its goals, how much does it need your unit?	1	2	3	4	5
7. How much of your unit's work or services was done for this group in the past six months?	1	2	3	4	5
8. How much of this group's work or services was done for your unit in the past six months?	1	2	3	4	5
9. How much trust exists between people in your group and this group?	1	2	3	4	5

	To No Extent	Little Extent	Some Extent	Much Extent	Great Extent
10. Overall, to what extent are you satisfied with this relationship?	1	2	3	4	5
11. To what extent has this group carried out its commitments in regard to your group?	1	2	3	4	5
12. To what extent has your group carried out its commitments in regard to this group?	1	2	3	4	5
13. During the past six months, how much has this group changed or influenced your group?	1	2	3	4	5
14. How much has your group changed or influenced this other group?	1	2	3	4	5
15. How much of the work done with the group can be used for other purposes in your organization.	1	2	3	4	5

	No Contact	Monthly Or Less	About Weekly	About Daily	Every ½ Day	Every Hour
16. During the past six months, how frequently have people from your group communicated or been in contact with people in this group?	1	2	3	4	5	6
17. How often were there disagreements or conflicts with this group?	1	2	3	4	5	6

18. How long do you expect the relationship between your group and this group to continue in the future?

1 = terminate soon
2 = about 6 months
3 = about a year
4 = 2–3 years
5 = indefinitely

19. For the effort and resources you devote to dealing with this group how great a value does your group receive?

1 = we get much less than we ought
2 = we get somewhat less than we ought
3 = balanced
4 = we get somewhat more than we ought
5 = we get much more than we ought

20. What specific suggestions do you have for improving the relationship with this group?

REFERENCES

Bagozzi, Richard P. 1981. "An Examination of The Validity of Two Models of Attitude." *Multivariate Behavioral Research* 16: 323–59.

Bass, B. 1971. "When Planning for Others." *Journal of Applied Behavioral Science* 6 (April/June): 151–72.

Bentler, P.M., and D.G. Bonett. 1980. "Significance Tests and Goodness-of-Fit in the Analysis of Covariance Structures." *Psychological Bulletin* 88: 588–606.

Blake, Richard R., and Jane S. Mouton. 1964. *The Managerial Grid*. Houston, Tex.: Gulf.

Burke, Ronald J. 1970. "Method of Resolving Superior-Subordinate Conflict: The Constructive Use of Subordinate Differences and Disagreements." *Organizational Behavior and Human Performance* 5 (4): 393–411.

Filley, Alan C., Robert J. House, and Stephen Kerr. 1976. *Managerial Process and Organizational Behavior*, 2d ed. Glenview, Ill.: Scott Foresman.

Gross, N., J. Giaguinta, and M. Bernstein. 1971. *Implementating Organizational Innovations*. New York: Basic Books.

Hackman, J. Richard, and Greg R. Oldman. 1975. "Development of the Job Diagnostic Survey." *Journal of Applied Psychology* 60 (2): 159–70.

Hudson, Roger L., and Harold L. Angle. 1986. "The Minnesota Innovation Research Program (MIRP): Collective Action in an Individualist Culture." Paper presented at the annual conference of Academy of Management, Chicago, Ill. August.

Joreskog, Karl G., and Dag Sorbom. 1986. *LISREL VI*. Mooresville, Ind.: Scientific Software.

Khandwalla, P.N. 1977. *The Design of Organizations*. New York: Harcourt Brace Jovanovich.

Kantner, R.M. 1983. *The Change Masters*. New York: Simon and Schuster.

Kimberly, J. 1981. "Managerial Innovation." In P. Nystrom and W. Starbuck, eds., *Handbook of Organizational Design*, Vol. 1; Oxford: Oxford University Press. pp. 84–104.

Lawrence, P., and P. Dyer. 1983. *Renewing American Industry*. New York: Free Press.

Lawrence, P.R., and J.W. Lorsch. 1967. "Differentiation and Integration in Complex Organiza-

tions." *Administrative Science Quarterly* 12 (June): 1–47.

Long, J. Scott. 1983. *Confirmatory Factor Analysis*, Sage University series on Quantitative Application in the Social Sciences, series no. 07-033. Beverly Hills, Calif.: Sage.

Lord, F.M., and M.P. Novick. 1968. *Statistical Theories of Mental Test Scores*. Reading, Mass.: Addison-Wesley.

March, J.G., and H.A. Simon. 1958. *Organizations*. New York: Wiley.

McGrath, J.E. 1984. *Groups: Interaction and Performance*. Englewood Cliff, N.J.: Prentice-Hall.

Munson, Fred C., and Donald C. Pelz. 1981. *Innovating in Organizations: A Conceptual Framework*. Ann Arbor: School of Public Health and Institute for Social Research, University of Michigan.

Pelz, Donald C., and Fred C. Munson. 1982. "Originality Level and the Innovating Process in Organizations." *Human Systems Management* 3 (3): 173–87.

Perrow, C.B. 1967. "A Framework for the Comparative Analysis of Organizations." *American Sociological Review* 32 (April): 194–208.

Peters, T., and R. Waterman. 1982. *In Search for Excellence: Lessons from America's Best-Run Companies*. New York: Harper & Row.

Pressman, S., and H. Wildavsky. 1973. *Implementation*. Berkeley: University of California Press.

Ring, P.S., and G. Rands. 1986. "Outcomes of Programmatic Longitudinal Research: Some Preliminary Observations from the Minnesota Innovation Research Program (MIRP)." Paper presented at the annual meeting of the Academy of Management, Chicago, August.

Schein, E., 1985. *Organizational Culture*. San Francisco: Jossey-Bass.

Shull, F.A., A.L. Delbecq, and L.L. Cummings. 1970. *Organizational Decision Making*. New York: McGraw-Hill.

Van de Ven, A.H. 1976. "On the Nature, Formation, and Maintenance of Interorganizational Relationships." *Academy of Management Review* 1 (1):24–36.

———. 1980. "Problem Solving, Planning, and Innovation. Part I. Test of the Program Planning Model." *Human Relations* 33:711–40.

———. 1986. "Central Problems in the Management of Innovation." *Management Science* 32: 590–607.

Van de Ven, A.H., and D.L. Ferry. 1980. *Measuring and Assessing Organizations.* New York: Wiley.

Van de Ven, A.H., and R. Garud, 1989. "A Framework for Understanding the Emergence of New Industries." In R. Rosenbloom and R. Burgelman, eds., *Research on Technological Innovation, Management, and Policy,* Vol. 4. Greenwich: Conn.: JAI (forthcoming).

Van de Ven, A.H., and G. Walker, 1983. "The Dynamics of Interorganizational Coordination." *Administrative Science Quarterly* 29: 598–621.

Werts, C.E., R.L. Linn, and K.G. Joreskog. 1974. "Interclass Reliability Estimates: Testing Structural Assumptions." *Educational and Psychological Measurement* 34: 25–33.

Zeller, Richard A., and Edward G. Carmines. 1979. *Measurement in the Social Sciences.* New York: Wiley.

SECTION
II

The Minnesota Innovation
Research Program Framework

This section contains four chapters that, collectively, address the organizing framework (ideas, people, transactions, context, and outcomes) that has guided the collaborative, longitudinal research program known as the Minnesota Innovation Research Program (MIRP). In the same sense that these concepts have guided research, they also provide an organizing framework for this section of the book.

As discussed in Chapter 1, the innovation studies that comprise MIRP did not just "happen" but emerged from a series of meetings that explored how to leverage the efforts of several research teams and attain a degree of synergy from their parallel efforts. One view that emerged immediately from these meetings was that a common framework was needed to organize the teams' thinking and to integrate results across the separate studies. Indeed, without such a common guiding framework, our efforts might well be chaotic.

After many discussions, the program participants identified the five key innovation concepts listed above. An innovation is a new *idea,* which may be a recombination of previous ideas, a new scheme that challenges the present order, a formula for change, or a unique approach to an old problem that is perceived as new by the people involved. In Chapter 4, Schroeder, Van de Ven, Scudder, and Polley show that innovation ideas, as they develop over time, are less simple and unitary than either the conventional wisdom or earlier research on organizational innovation might lead one to believe.

Clearly, it is *people* who push, develop, modify, and/or drop innovation ideas. In Chapter 5, Angle discusses the role of people in the process of organizational innovation, indicating that innovation is largely a joint function of organization members' ability and motivation. *Transactions* are the deals, relationships, and exchanges that bind together people in an institutional framework or context, and it is through transactions that the structural arrangements that support (and resist) innovation ideas are created. In Chapter 6, Ring

105

and Van de Ven show that individual transactions cannot be considered in isolation; rather, a sort of serial interdependence occurs across a stream of agreements as transaction norms take shape.

Innovation *outcomes* are the value judgments about success or failure that various people make about the developmental process and end results of an innovation. In Chapter 7, Dornblaser, Lin, and Van de Ven point out that such value judgments are often nonconsensual and differ systematically at various stages of the innovation, depending on the specific stakeholder whose perspective is taken. Their chapter also points up the difficulty of connecting actions with desired outcomes, which leads to difficulties in organizational learning.

Finally, *context* is the setting or institutional order within which innovative ideas are developed. Four levels of context are examined in MIRP: the broad institutional/industry environment, organizational strategy and structure, organizational practices, and the characteristics of the innovation team. Although no individual chapter in this section focuses on context, per se, they all incorporate contextual issues into their fabric, because context is so pervasive that it interacts at all levels with the other four MIRP guiding constructs.

All in all, the four chapters in this section ground the reader in a framework that should help put in context the remaining chapters in the book. We recommend that these chapters be read before the book's other sections are explored.

THE DEVELOPMENT OF INNOVATION IDEAS

Roger G. Schroeder

Andrew H. Van de Ven

Gary D. Scudder

Douglas Polley

This chapter describes how innovation ideas develop over time based on findings emerging from seven innovations included in the Minnesota Innovation Research Program. These observations are very different from typical

We greatly appreciate the willingness of our colleagues involved in the Minnesota Innovation Research Program to make this chapter possible by giving permission to use their research cases and data on their innovations: Harold Angle, Jeanne Buckeye, Roger Hudson, Raghu Garud, Daniel Gilbert, Mary Knudson, Karen Lindquist, John Mauriel, William Roering, and Vernon Ruttan. This research was supported in part by a grant to the Strategic Management Research Center at the University of Minnesota by the Organization Effectiveness Research Programs, Office of Naval Research, under Contract No. N00014-84-K-0016. Additional research support for this research has been provided by the Underseas Systems Division of Honeywell Corporation. An earlier version of this chapter appeared in *Agribusiness Management Journal* 2, 4 (1986): 501–523.

models in the literature of how innovation ideas develop from concept to reality. The process we observed is fluid and includes an initial shock to propel the innovation idea into being, proliferation of the original idea, setbacks and surprises along the way that provide numerous opportunities for trial-and-error learning, and a blending of old and new ideas as the innovation is implemented and diffused. This chapter is one small step in developing descriptively more accurate and useful models of the innovation process based on longitudinal MIRP research studies.

As stated in Chapter 1, the MIRP framework focuses on five interdependent concepts: ideas, people, transactions, context, and outcomes. Its central research focus is to describe and explain how and why innovation ideas are developed and implemented by people who engage in transactions with others within an institutional context. This chapter focuses on the concept of innovation ideas and examines how the other MIRP concepts are related to the development of innovation ideas over time.

Comparing longitudinal patterns in the development of innovation ideas in seven of the innovations being studied by MIRP with the literature led us to find existing process models increasingly unsatisfactory in explaining observed developmental patterns. The seven innovations included four new product technologies (hybrid wheat, cochlear implants, therapeutic apheresis, and a naval weapons system) and three new administrative arrangements (a new business startup, site-based management of public schools, and strategic human resources management). The gaps between theory and practice led us to adopt a grounded theory approach (Glaser and Strauss 1967) to the research and to develop and present below a set of six observations that appear to capture significant process patterns in the development of innovation ideas across the seven cases. We also examine the conceptual links between the six observations and outline the rudimentary ideas for developing a new process model for describing the innovation process that differs in several fundamental respects from existing models in the literature.

CONCEPTUAL OVERVIEW

Although invention is the creation of a new idea, innovation is more encompassing and includes the process of developing and implementing a new idea—whether it is a new technology, product, organizational process or arrangement (Ruttan 1959; Rogers 1983; Van de Ven 1986a, 1986b). Usually this innovation idea involves a set of ideas that may represent

a novel recombination of old ideas, a scheme that challenges the present order, or a unique approach that is perceived as new by the individuals involved (Zaltman, Duncan, and Holbeck 1973). As long as the idea is perceived as new to the people involved, it is an "innovative idea," even though it may appear to others to be an "imitation" of something that exists elsewhere.

From this definition, it follows that the *process* of innovation centers on the temporal sequence of activities that occur over time in developing and implementing new ideas from concept to concrete reality. Table 4–1 summarizes the major process models available in the literature to describe how this process unfolds. Most of these models are derived or borrowed from process models of either: individual or group decision making (March and Simon 1958; Mintzberg et. al. 1976; Cohen, March, and Olsen 1972; Lindblom 1965; Quinn 1980), small group development (Lewin 1947; Bales and Strodtbeck 1951), organizational change and development (Lippitt, Watson, and Westley 1958; Dalton, Lawrence, and Greiner 1970; Hage and Aiken 1970), or organizational planning (Friedmann 1973; Ackoff 1974; Lorange 1980). To be sure, models have also been proposed for the innovation process itself. For example, there is Utterback and Abernathy's (1975) model of joint process and product innovation, Usher's (1954) cumulative synthesis theory, and a host of innovation diffusion process models reviewed by Rogers (1983). However, as Table 4–1 illustrates, there is a striking similarity between these innovation process models and those proposed for individual decision making and group and organizational development.

Two issues stand out when assessing the process models listed in Table 4–1. First, most of the models are normative and lack empirical research to substantiate their validity. With the exception of retrospective case histories of organizational innovations, very little longitudi-

TABLE 4–1. Comparisons of Developmental Phases in Process Models.

Authors and Summaries	Beginning ←——————— Activity Phases ———————→ End				
Group development models					
Lewin (1947): Social psychological description of change process in individuals and small groups	Unfreezing: Recognition of a need; willingness to give up	Moving: Activities undertaken to design and implement a change	Freezing: Institutionalizing new behavior		
Bales & Strodtbeck (1951): Observations of phases in group decision process (8 cases)	Orientation: Address problem in cognitive terms without committing	Evaluation: Speaking in evaluative terms	Control: Attempts to control by joint action		
Decision process models					
March and Simon (1958): Description of administrative problem solving with bounded rationality conditions	Parenthood of invention is necessity, opportunity, and moderate stress	Intelligence: Problem formulation	Design: Search, screen, and evaluate alternatives	Factor design elements into substantive programs	Choice: Decision to implement satisfactory solution
Mintzberg et al. (1976): Field study of 25 strategic, unstructured decision processes	Identification phase: decision recognition routine; diagnosis routine	Developmental phase: Search routine; Design routine		Selection phase: Screen routine; Evaluation – choice routine; Authorization routine	
Cohen, March & Olsen (1972): Model of organized anarchies based on universities study	Decisions are outcomes of relatively independent streams within organizations: Stream of choices; stream of problems; stream of energy from participants; rate of flow of solutions				

TABLE 4-1. (continued)

Authors and Summaries	Beginning ← ———— Activity Phases ———— → End					
Quinn (1980): Case studies of nine major corporations	Fourteen process stages beginning with need sensing and leading to commitment and control systems. Flow is generally in sequence but may not be orderly or descrete.					
	Sense need	Develop awareness and understanding	Develop partial solutions	Increase support	Build consensus	Formal commitment
Organizational planning models						
Friedmann (1973): Normative model of innovative planning for "retracking American" institutions and communities	Identify and describe problem as collective phenomenon	Identify and describe patterns of collective behavior	Identity and involve institutions responsible for behavior	Analyze and relate performance of institutions to the problems	Formulate proposal for structural innovation in response to problem	Implement strategy and make adjustments on an ongoing basis as necessary in the course of action
Ackoff (1974): Normative model of "interactive planning" for designing idealized futures of social systems	Ends planning: Design idealized future goals that are technologically feasible	Means planning: Invent alternative courses of action		Resource planning: Determine type, amount, and sources of resources needed	Organizational planning: Determine and design organizational arrangements needed	Implementation and control planning: Implement, control, evaluate, and improve plan under changing internal and external conditions

TABLE 1–1. (continued)

Authors and Summaries	Beginning ———— Activity Phases ————→ End					
Delbecq and Van de Ven (1971): Normative model of program planning and evaluation process for groups, organizations, and communities	Prerequisites: Identify complexity of problem or goal to be dealt with	Problem exploration: Involve users to identify needed priorities	Knowledge exploration: Involve experts to reconceptualize problems and identify solutions	Program design: Involve affected parties in developing new program proposal and implementation plan	Program activation and evaluation: Trial implementation and formative evaluation of new program	Program operation and diffusion: Institutionalize program as ongoing activity and/or transfer program to adopters
Lorange (1980): Normative model of corporate strategic planning	Objectives setting: Identify relevant strategic alternatives	Strategic programming: Development of programs for achieving chosen objectives	Budgeting: Establish detailed action program for near-term		Monitoring: Measure progress toward achieving strategies	Rewards: Establish incentives to motivate goal achievement
Organizational change and development models						
Lippet, Watson, and Westley (1958): Normative model from cases of original change	Develop need for change	Establish relation with change	Diagnose problem of client	Examine alternatives and establish goals and action routes	Transform intentions into change efforts	Generalize and stabilize change → Terminate relation with change agent
Dalton, Lawrence, and Greiner (1970): Normative model of change process developed from case studies of large-scale organizational changes	Pressure on top management	Intervention from outside and reorientation	Diagnoses and recognition of problems and their determinants by gathering information at all levels of organization	Invention of solution and obtain commitment to change through participation	Experiment with solution and search for results	Reinforce new practices with rewards and feedback

111

TABLE 4-1. *(continued)*

Authors and Summaries	Beginning ←			Activity Phases ← →		→ End
Hage and Aiken (1970): Description of program change process within	Evaluation: Organizational elite identify need for change and consider alternative means	Initiation: Elite choose solution and search for alternatives		Implementation: Start of new activity		Routinization: Stabilize program change
Innovation process models						
Usher (1954): Normative model of innovation emergence	Perception of the problem: Recognition of partial or incomplete need satisfaction	Setting the stage: Elements necessary for solution are brought together		Act of insight: Essential solution is found		Critical revision: New relations become understood and worked into context
Abernathy and Utterback (1975): Model for process and product innovation based on 120 firms in five industries	Product development; Performance maximizing: Rapid product change emphasizing unique product and performance; Process Development; Uncoordinated: Process largely unstandardized and manual			Sales-maximizing: Reduced uncertainty; some designs dominate; Segmental: Production systems become mechanistic and rigid		Cost-minimizing: Product variety reduced; product change coupled with process; Systemic: Large integrated process; tightly coupled process elements
Rogers (1983): On the development and diffusion of innovation	Need recognition: Problem or need	Research application of knowledge to problem	Development idea into useful form	Commercialization: Manufacturing, packaging, marketing, and distribution of product or process	Diffusion and adoption of innovation	Consequences of innovation

nal empirical study has been conducted to evaluate these process models.

Second, with the exception of Cohen, March, and Olsen (1972), the most common feature in these models in Table 4–1 is that they propose a simple unitary progression of phases or stages of development over time. For example, the innovation process has traditionally been viewed as a sequence of separable functional stages (such as design, production, and marketing) sequentially ordered in time and linked with transition routines to make adjustments between stages. These simple unitary stage-wise progression models are increasingly being discredited because of their lack of empirical validity or correctness. When researchers use a priori stages or phases to design their research and collect data, their results can quite easily become self-fulfilling prophesies (Poole 1983; Tushman and Romanelli 1985). Van den Daele (1969, 1974) argues that the unitary model of development is simply inadequate to deal with the complexities of many innovation processes because it assumes invariance between and within all units in following a prescribed order of developmental phases, one locked in step after another. As described in Chapter 2, Van den Daele (1969) proposes a typology of developmental models that go beyond simple unitary progression and includes multiple, cumulative, and conjunctive progressions of convergent, parallel, and divergent streams of activity sequences that may unfold as an innovation develops over time.

The tendencies to reduce complex innovation processes to simple unitary stages and their lack of empirical substantiation suggest that many of the process models in the literature are suspect or simply inadequate. The present research was undertaken with the aim of avoiding these problems and hopefully making a contribution that partially corrects these problems. Specifically, as will be discussed in the next section, efforts were made to avoid these problems by undertaking a longitudinal study that tracks the development of a variety of innovations over time and by using a grounded theory approach (Glaser and Strauss 1967) to identify the developmental processes actually followed in each innovation instead of entering into the research with a particular process model in mind. Such an inductive approach has recently been suggested by Lewin and Minton (1986) as useful and needed in order to begin to appreciate the complexity and "messiness" of organizational processes.

RESEARCH CASES

The seven research cases used in this study are part of the Minnesota Innovation Program summarized in the Appendix of Chapter 1 and described elsewhere in this book. It should be noted that the research cases selected represent a wide range of different types of innovations. Some of the innovations are very complex, involving hundreds of people, while others only involve a few people. Some of the innovations are administrative in nature, while others represent new products. Thus, a wide range of different innovations are used as the basis for this study. This makes the findings necessarily broad in nature but also provides the basis for quite generalizable conclusions.

The framework for the innovation studies examines innovations as ideas carried by people who engage in transactions or relationships with others within a changing institutional context. Key elements in understanding the process of innovation emergence are the significant changes that occur in these factors (ideas, people, transactions, and context). These changes represent events in the innovation process. Mapping these events over time is a central part of the innovation studies reported here.

Cases developed by the individual innovation studies, as well as event listings and supplemental notes from the studies, were the data on which this research was based. This material was carefully reviewed for the key events and their interrelations. Detailed mappings have been developed for the events. These mappings summarize the longitudinal development of the innovation activity. A brief synopsis of these cases is provided along with the event mappings.

Case 1. The hybrid wheat study examines the research and development, marketing, and diffusion of a new strain of wheat seed for increasing the yield of wheat. This study concerns the entire industry and follows the innovative activity from the early 1950s up to the present. This innovation has involved both private and academic organizations.

Early emphasis on development of hybrid wheat was due to interest in increased yields and disease resistance. The latter was due to an outbreak of stem rust blight that caused epidemic yield and quality losses in the 1950s. A variety of research occurred during the 1950s, and commercial activity was stimulated by research discoveries in the early 1960s. During this same period the conventional wheat yields were also being improved in competition with the hybrid effort.

By the early 1970s major changes had occurred in the innovation participants. Almost all public institutions had dropped out as well as some of the private companies. This was primarily due to the high risk attributed to this research.

The first release of hybrid varieties came from two firms in 1978. This release proved premature due to difficulties with stability, impure seed stock, and low seed availability. This release was withdrawn and one of the firms subsequently sold its hybrid program. Recent efforts in hybrid development have been more successful. Several firms have released hybrid varieties and some 1983 tests

have shown hybrid varieties to outperform varietal seeds.

Figure 4–1 is an "event map" for hybrid wheat. Note how stem rust blight is shown as an initial shock that started the development of new ideas—in this case, a search for hybrid varieties. Then several organizations started research and development activities as indicated by the branching of activities in Figure 4–1. The termination of some of these branches occurred when organizations stopped their research and development efforts. This particular innovation is therefore characterized by a shock that propels the idea into being, followed by multiple innovative activities, and some setbacks and termination of activities as the innovation progressed.

Case 2. The cochlear implant program involves the development of an artificial, or bionic ear, which provides some profoundly deaf people an ability to discriminate sound. This program involves joint efforts by researchers and commercial developers. Alternative competing technologies are being explored.

Early development of this product began in the 1950s and 1960s, but a concentrated effort did not begin until the early 1970s. Shortly after this new idea began to take hold, the chief researcher was censored by colleagues. They contended that it was improper to implant this technology when newer ones were expected to be developed soon. This controversy over choice of technology still remains with the current development.

Commercial effort by the company under study began in 1978 but did not gain momentum until the promotion of an early proponent. This individual asked that independent activities be pulled together into a coherent program under one manager.

Recent work has included two different technologies plus diagnostics. One product was

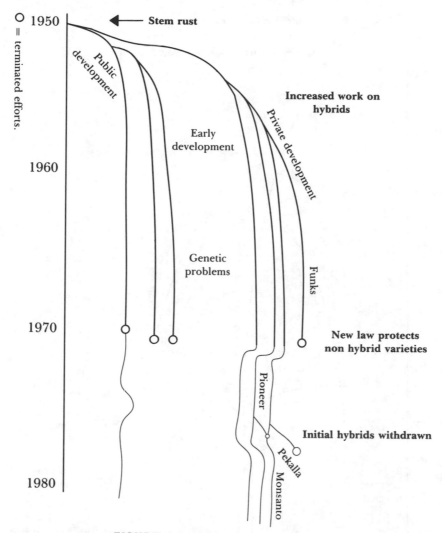

FIGURE 4-1. *Hybrid Wheat Development.*

approved for sale by the FDA in late 1984. This product was subsequently recalled. Active research continues on the other ideas. Pursuit of these different technologies is indicated by different branches in Figure 4-2.

The organizational context has also changed for this innovation. In late 1984 the reporting structure for this program was revised when it was placed with a group of other new product ventures. In late 1985 the program manager was changed and a new program manager was brought in from another project.

Case 3. Therapeutic apheresis is a new medical technology (the idea) designed to remove specific pathogenic components of blood from a patient in order to treat specific diseases. This program is a joint venture between three firms. The innovation involves development of the

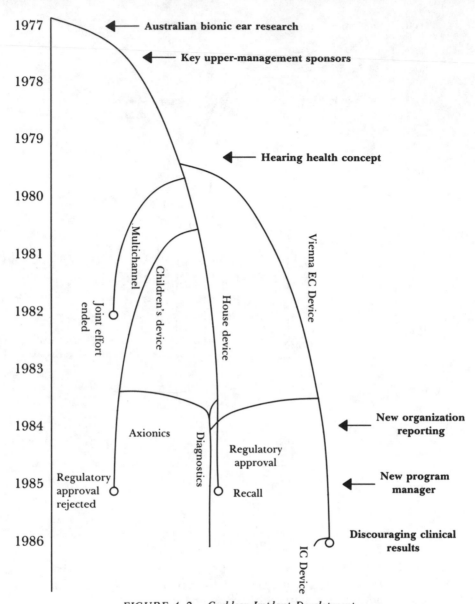

FIGURE 4–2. *Cochlear Implant Development.*

separation technology as well as the choices of the diseases for treatment.

The initial conception of the product idea has been elaborated into five products involving two phases of technological development. During the most recent year the first product has been undergoing extensive field trials. These took place after the initial FDA approval was received in early 1985. The second phase of the technology was under development during most of 1985. Initial prototypes of the second phase were created at the end of 1985 for use in clinical trials.

In late 1985 budgetary reductions

116

caused a change in the program planning. The development effort would not have funds for market entry as originally planned. This caused the development of a new program for the marketing of the device.

The organizational environment was also changing for two of the partners during recent times. In one case there was promotion of the vice president responsible for the innovation to a higher position. In the second case the division president resigned. This has led to other management changes including some changes in membership on the innovation management team. Despite these changes the program has been reviewed twice during 1985 and received good appraisals from the reviewers.

The event map for therapeutic apheresis, in Figure 4–3, shows several branches corresponding to the different technologies (ideas) attempted. It also indicates the terminations of some ideas and the management actions along the side of the figure. Note how therapeutic apheresis started with an opportunity, then evolved into five product variations. Setbacks included budget reduction and device failure. Recently, further proliferation has been considered in the form of a broader business charter. These changes are neatly shown by the event map.

Case 4. The naval systems program is a multimillion dollar developmental effort by a defense contractor aimed at the development of a new weapons system for the U.S. Navy. This system includes a variety of process and management ideas that began in the early 1970s. Specific innovations have been identified in the area of factory automation, material management, strategic planning, human resources, and design-to-production transitions.

The Naval Systems innovation started with a major product failure amounting to several million dollars when the company could not deliver a quality product on time. As a result several changes in management occurred, including a change in general manager of the division, and an extensive program of quality improvement was initiated. This resulted in a proliferation of innovative ideas in human resources, materials management, factory automation, and strategic planning in order to ensure a quality product. Figure 4–4 indicates extensive branching corresponding to these many different micro innovations undertaken.

The Naval Systems innovation had direct involvement of the general manager over an extended period of several years, as the general manager personally directed many of the innovations which occurred. These innovations, which started with a shock, were eventually successful as a quality product was produced and a follow-on program was initiated.

Case 5. The site-based management of public schools program is an administrative idea that decentralizes the decision making process in public school systems from the superintendent to school principals. This involves the shifting of responsibility from central administration to individual schools and the sharing of these responsibilities with representatives of the school's constituency.

This innovation started when the school district experienced budget cuts and a new superintendent was hired. The new superintendent brought along ideas of school-based management that could be used to operate the school district on a more decentralized basis. To implement the school-based management idea, committees were established in each of the individual schools in the district. This required a restructuring of the way in which each school was operated.

Throughout the entire implementation process the superintendent maintained a visible and direct role as leader of the innovation. Many ambiguities were encountered in implementing school-based management. These occurred because the idea was constantly being refined and defined. The idea was very new to

117

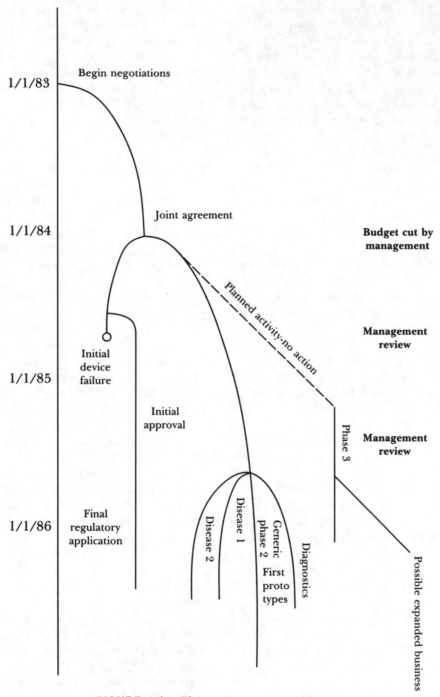

FIGURE 4–3.   Therapeutic Apheresis Development.

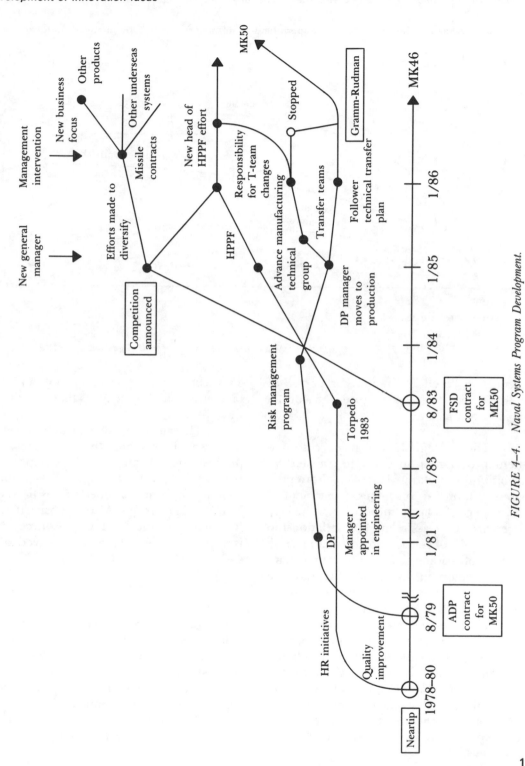

FIGURE 4–4. Naval Systems Program Development.

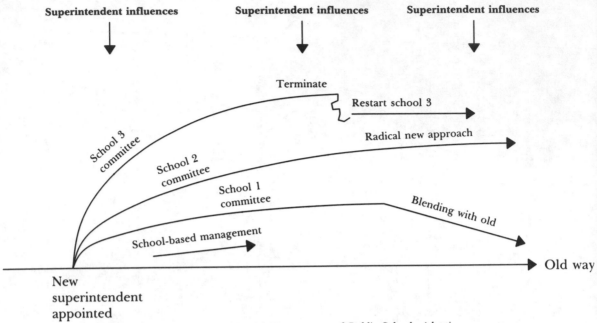

FIGURE 4–5. *Site-Based Management of Public Schools Adoption.*

the people involved and required a period of time for acceptance. See Figure 4–5 for an event map of this case.

Case 6. The strategic human resources management program is an effort to transfer responsibility for traditional personnel-oriented activities of a human resources department to line managers of a large corporation. The idea is to replace the classic line/staff relationship with a new spirit of cooperation. Programs have begun to professionalize human resource management, develop line managers, integrate line and human resource management, and reconfigure the human resource organization.

The human resource innovation started with a new vice president of human resources. He was not satisfied with the apparent isolation of the human resources area from the line operations of the company. As a result he began discussing the idea that human resources would take on more of a line orientation while the line would become more human resource oriented.

This evolved into several programs for implementation of these ideas that are still underway.

The vice president of human resources is personally leading this effort toward implementation throughout the company. Some restructuring will be required as human resource people are assigned to positions or projects within divisions. Also extensive training of line management in human resource concepts is needed. The event map is displayed in Figure 4–6.

Case 7. The process of creating a new organization is being examined in the startup of a computer software company. This company resulted from the merger of two new companies in 1983. These two companies had themselves been started in 1980. This innovation activity involves the development of new products in a rapidly evolving environment.

This case exemplifies a typical new startup company with continual problems

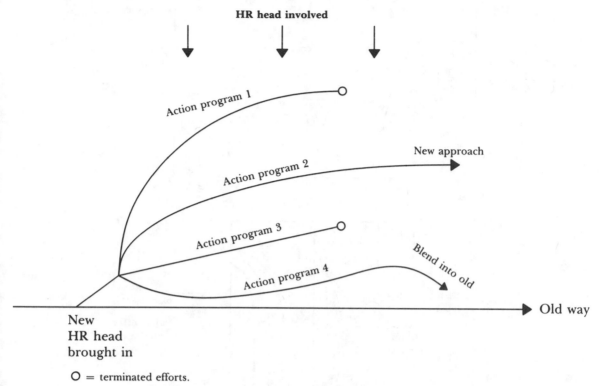

O = terminated efforts.

FIGURE 4–6.   Human Resource Management Development.

being encountered in financing the new product ideas and in generating sufficient sales to remain viable. The software company started when one of the entrepreneurs decided to leave a secure job with a large company in order to start up the new venture. Many new product ideas are being developed as the company continues to search for the right combination of new products that have market potential. This is shown by the many terminations of events in Figure 4–7.

Two companies with similar interests were joined together in an effort to restructure things to take advantage of joint competencies. The new company is now exploring a joint venture with several other companies to gain needed capital and expertise. This case clearly demonstrates the fluid and rapidly changing pace of product innovation and the need to remain extremely flexible and responsive to the market in the initial stages of innovation. It also illustrates the problems in attracting capital and maintaining positive cash flow in the small firm.

OBSERVATIONS ON THE INNOVATION PROCESS

Based on the seven cases described above, observations were formulated about the relevant changes observed in the cases. Although every observation was not noted in every case, overwhelming support is evident for these observations in most of these cases.

In late 1985 a preliminary version of the observations was reviewed with a number of the

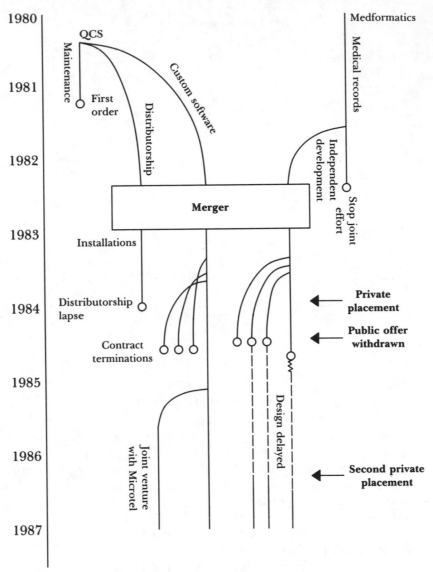

FIGURE 4–7. *New Organization Startup.*

case clients at a feedback conference. They were supportive of the observations and their comments were included in later revisions of the material.

The observations stated below are aimed at inductively generalizing what we have learned from the seven cases. After stating each of these observations, we will provide a gener-

alized event map that seems to encompass these seven cases and permits at least some level of generalization.

Six major observations were developed based on the cases already described. These observations are of a descriptive nature. They are reflections of what we observed in those situations, not what we think should have happened. It is hoped that these descriptive observations can later be expanded into prescriptive actions as the longitudinal research progress.

Table 4–2 summarizes the evidence observed across the seven case histories on the six observations, which will be discussed in detail below. Of the six observations, the most general evidence occurs for the propositions concerning shocks, setbacks, and restructuring. However, evidence for the remaining observations is also relatively strong, meaning that instances of the observations exist in at least six of the seven cases.

Observation 1. Innovation is stimulated by shocks, either internal or external to the organization

An overall observation from the seven case studies is that innovation was more prevalent when some major change occurred in the organization or its environment. Ideas were often generated but are not acted on in an organization until some form of shock occurred. Shocks included such things as new leadership, product failure, a budget crisis, and an impending loss of market share, although it is evident from the participants in our project that a shock can come in many different forms.

In the naval systems program, a major failure in a product improvement program and a resulting large financial loss caused the organization to expend considerable effort to uncover the underlying problems in the organization. This shock (product failure) resulted in more attention to manufacturing by management as well as an increased emphasis on human resource management.

In two of the cases, a new leader in the organization was the shock required to initiate innovation. In the strategic human resources management program, the personal vision of a new leader was needed to get line managers to think as human resource managers and vice versa. In a local school district, new leadership combined with a major budget crisis caused a total rethinking about managing schools in a more decentralized manner.

In the hybrid wheat industry, the impetus for developing the hybrid was a disease called stem rust blight. A hybrid variety of wheat was expected to resist this disease and provide better yields.

Shocks do not need to be viewed as negative. In the therapeutic apheresis case study the approach by another company with an offer to enter into a joint venture opportunity was seen as the shock necessary for the renewal of an abandoned effort.

Thus, in all seven cases, efforts to begin work on an innovation can be traced to some kind of shock that stimulated peoples' action thresholds to pay attention and initiate novel action. This process observation is consistent with the general belief that necessity, opportunity, or dissatisfaction are the major preconditions for stimulating people to act. March and Simon (1958) set forth the most widely accepted model by arguing that dissatisfaction with existing conditions stimulates people to search for improved conditions and that they will cease searching when a satisfactory result is found. A satisfactory result is a function of a person's aspiration level, which Lewin et al. (1944) indicated is a product of all past successes and failures that people have experienced.

This model assumes that when people reach a threshold of dissatisfaction with existing conditions, they will initiate action to resolve their dissatisfaction. However, although a given event may be perceived as a shock and stimulate action for some people, it may not do so for others. This is because individuals have widely varying and manipulable adaptation levels (Helson 1948, 1964). When exposed over time to a set of stimuli that change very gradually, people do not perceive the gradual changes—they unconsciously adapt to the changing conditions. Their threshold to tolerate pain, discomfort, or dissatisfaction may not

TABLE 4–2. *Evidence for Six Process Observations.*

Proposition	Start-up Company Software and Hardware	Hybrid Wheat Industry	Cochlear Implant	Naval Systems	Therapeutic Apheresis	School-Based Management	Human Resources Management
Shocks	Entrepreneur left a steady job to start the company	Hybrids needed to solve the wheat rust problem and to improve yields	News of a "bionic ear" being developed in Australia	Major product failure; multi-million dollar loss	Firm approached with opportunity for joint venture on new technology	State budget cuts caused a crisis; new superintendent hired	New head of H.R. brought in from outside
Proliferation	Many new products developed after the initial startup	Several systems of breeding were developed in the different companies	Expansion into hearing health business; many products	Major innovations in several areas were started including human resources, materials management	Five product variations were created	School committees were formed; each school had its own version	Strategic management split into four action programs
Setbacks and surprises	The need for venture capital caused continual problems and setbacks; had difficulty marketing	Firms dropped out of the program after spending great amounts of money	Bad quality; management changes	Material management suffered several reversals from the original idea	Cutbacks in the budget caused program revisions; changes in target diseases	Several attempts needed to develop staff leader role	Difficult to get all depts involved

124

TABLE 4–2. *(continued)*

Proposition	Start-up Company Software and Hardware	Hybrid Wheat Industry	Cochlear Implant	Naval Systems	Therapeutic Apheresis	School-Based Management	Human Resources Management
Degree of linking old and new	No old yet	New hybrid wheat comes from the old; new yield must exceed the old	Old—single; new—multi-channel	The line is learning to work as teams; old and new products in the same organization	Three phases of product being developed: current, new, and future	Teachers learning to administrate; administrators learning to delegate	The line learning staff roles; the staff learning line roles
Restructuring	Two companies were joined to form the new one; developed license arrangement	Created separate laboratories for hybrid wheat; mergers occurred	Joint venture used; also new program formed	Transistion teams formed; organization charts constantly changing	Joint venture formed; Spinoff from corporate division	The innovation itself is a restructuring of teams that cross traditional lines	Partial decentralization of the HR function
Hands-on top management	Two entrepreneurs were the driving force in all innovations	VP research involved at one company	Division VP involved on a frequent basis	Pervasive influence of the general manager; a blitz of memos	Constant attention by the division VP; some attention by the CEO	The superintendant pushed the idea constantly	Head of H.R. led the effort personally

have been reached. As a consequence, opportunities for innovative ideas are not recognized. In general, we would expect that direct personal confrontations with the sources of problems or opportunities are needed to trigger the threshold of concern and appreciation required to motivate most people to act (Van de Ven 1980). Shocks serve this function in stimulating innovation.

From a macro perspective, we have observed a shock that triggers an action threshold to start innovative activity in new directions. Actually, a shock may consist of several smaller shocks or changes, each of which is barely perceptible. According to this micro perspective of shock (explained in Chapter 8), innovations do not suddenly spring into action but rather are the result of prolonged activities. Whether a macro or micro perspective is used, the literature strongly supports the need for a threshold to be reached, which we have defined as a shock, before action is realized.

Observation 2. An initial idea tends to proliferate into several ideas during the innovation process

More generally, proliferation of ideas, people, and transactions over time is a pervasive but little understood characteristic of the innovation process, and with it comes complexity and interdependence—and the basic problem of managing part/whole relationships (Van de Ven 1986b).

As we observe the innovations develop over time, in all cases it was found that the initial ideas that served to stimulate the innovation projects proliferated over time into an increasing number of alternative paths, and it is often impossible to know which paths may yield fruit. Some ideas go on the "shelf" for a long time. Others lead to important innovation spin-offs. Still others converge at later times as central to making the innovation a reality. For example, in the case of hybrid wheat's development, three alternative ideas were followed simultaneously, and it took several years

of extensive investments in each idea to determine which one was appropriate in developing a hybrid strain of wheat.

Moreover, most innovations studied did not consist of a single new device, product, or procedure. Instead, families of related new products and procedures are being developed to create sufficient critical mass and penetration to become commercially or organizationally viable. This exponentially increases the complexity of managing innovation. For example, in the development of therapeutic apheresis, the initial product idea has already proliferated into five product ideas. The cochlear implant innovation has expanded from initial work on a single channel device to current developments of five new devices using three different technologies. Although these new ideas and products are at different stages of development, work on them tends to occur simultaneously by different and overlapping subgroups within the innovation programs.

As a consequence, the cases show that managing an innovative idea over time usually involves linking overlapping and parallel cycles of development efforts. Each cycle may require linking R&D, prototype development, testing, manufacturing scale-up, and marketing activities for a given product idea. However, subsequent cycles must be simultaneously integrated in order to create a related family of products, and yet differentiated enough to permit creation of a unique new product idea or component.

As these preliminary observations suggest, after an initial shock that stimulates a simple unitary progression of activity to develop an innovative idea, the innovation process soon proliferates into a multiple divergent progression of developmental activities (Van den Daele 1969). Some of the activities in this diverging multiple progression are conjunctive (or related by a division of labor among functions to develop a given alternative), while many activities and ideas are unrelated as disjunctive alternative paths pursued by different people or organizational units. As a consequence, after a short initial period of simple progression of a unitary set of activities, the management of innovation soon proliferates into an effort of trying to direct controlled chaos. As one manager stated, "The problem is like trying to grow an oak tree when there are inexorable pressures to grow a bramble bush."

Observation 3. In managing an innovation effort, unpredictable setbacks and surprises are inevitable; learning occurs whenever the innovation continues to develop There is no doubt that setbacks or surprises occur in any innovation process. Organizational learning can result from these setbacks as the innovative idea progresses. Understanding the learning process has, in fact, been prescribed as a better method of management than attempting to remove all setbacks and surprises (Jaikumar and Bohn 1984).

The foremost example of setbacks and the positive learning that can result is found in the naval systems case. In this case, a major product failure occurred, resulting in a major financial loss. However, from this failure, the company created a learning process and vowed never to have this happen again. The actual experience was documented in a case history that is used in divisional training. The product failure also resulted in a major thrust in the human resources management area.

In several of the other cases, multiple setbacks have also occurred to date. In hybrid wheat, the first commercial release of the seed failed due to contamination. Genetic problems also were uncovered in the primary breeding source. In school-based management, the high school staff rejected the new idea, requiring many amendments. The therapeutic apheresis project faced reduced funding levels, as well as an initial field failure of a unit due to a poor user interface. Finally, the cochlear implant innovation has had to deal with the recall of a newly introduced product idea in the market, a change in program management, along with failures to complete negotiations on joint agreements with competitors.

In general, learning tends to occur in three ways: (1) by imitation of something done elsewhere, (2) by extrapolation from the past into the future, and (3) by trial-and-error through error detection and correction (March and Simon, 1958). New-to-the-world innovative ideas must rely primarily on trial-and-error learning, along with a little extrapolation. Imitation is not possible for there is no precedent to follow or copy. The necessary (but not sufficient) conditions for trial-and-error learning are that setbacks and errors are detected and corrected.

Error detection requires (1) knowledge of desired or expected outcomes, (2) an ability to discriminate substantive issues from "noise" in the system, and (3) a felt freedom to communicate knowledge of errors or setbacks to appropriate individuals. As Van de Ven (1980) found in his study of planning new organizational startups, error detection was pervasive in the innovation cases. Indeed, in repeated interviews with innovation participants and observations of their meetings, the researchers obtained extensive lists of reported mistakes or "red flags" by participants in the development, schedule, or direction of their innovation over time.

However, although there are differences among the innovation cases, in general, relatively few attempts were observed to correct these reported mistakes. Different reasons appear to exist among the innovation cases for this observed lack of error correction when extensive errors are detected: (1) lack of recognition of who is responsible for an error, (2) lack of slack resources and tight project deadlines limit the time available to address mistakes, (3) the use of group problem solving processes that avoid or smooth over conflicts and that allow errors to fester and grow into larger mis-

127

takes, and (4) loss of memory for retaining past learning experiences through personnel turnover and inadequate documentation of developmental processes.

The fact that many detected errors were not corrected appears to have two consequences on the innovation process. First, as Van de Ven (1980) observed, when errors are not corrected, they "snowball" over time into vicious circles of even larger and more intractable complexes of problems resulting, as in the cochlear implant case in the replacement of the program manager. Alternatively, as Masuch (1985) suggests, the vicious circle may lead to organizational stagnation, apathy, or failure. Each failure to correct a detected error represents a lost opportunity for learning in the development of an innovation.

Observation 4. As an innovation develops, the old and the new exist concurrently, and over time they are linked together Whenever an innovation develops within an existing organization, the old and the new must exist concurrently, as parallel streams of activities. The new idea often represents a threat to the established order, and thus there is opportunity for establishing new organizations and linkages. These relationships between old and new may also exist between organizations and between products.

Where new product ideas have proliferated, as in therapeutic apheresis and cochlear implants, convergent relationships exist between current versions and future product versions. This is particularly salient where multiple products are under development simultaneously. In some situations product changes gave way to organizational changes as well.

In the naval systems innovation, a major linkage was observed that involved the redesign of the production facility to produce two different products simultaneously. Although the products are somewhat similar, the production processes required for the new product are

quite different. This effort is still underway and a special group was formed to develop this integrated facility. In this situation, the old and the new ideas are co-existing simultaneously for several years under the same roof.

In the human resources and school-based management cases, new administrative processes were introduced that differed greatly from the existing processes. In both cases, there was evidence of tension between the old and new ideas and in the human resources case, implementation has not yet occurred, possibly because sufficient linkages between old and new ideas have not been created. The school-based management innovation has required many linkages to approach implementation. These have been formed through the use of several meetings with the involved parties (the teachers, the union, and administrators) as well as cross-functional committees that address issues.

The implication of this observation is that innovations are often not simple additions to or replacements of existing organizational programs. Instead, if they are to be implemented and become institutionalized, the new innovative ideas must overlap with and become integrated into existing organizational arrangements. In other words, after a period of divergent and parallel streams of activities, we begin to observe convergent paths of activities that link the innovative ideas with ongoing organizational operations. The management of convergent integration processes appears to take several forms, as described in our next observation.

Observation 5. Restructuring of the organization often occurs during the innovation process; this restructuring can take many forms, including joint ventures, changes in organizational responsibilities, use of teams, and altered control systems We have already noted that the innovation process entails a period of divergent proliferation of activities, parallel streams of alternative innovation paths, and then convergent integration of the coexisting old and new ideas. One of the ways that the managers in the case studies handled these divergent, parallel, and convergent streams of progression was by changes in the organizational structure. Organizational changes observed included formal and informal as well as temporary and permanent changes.

In the naval systems innovation, a great amount of formal organizational restructuring has been observed. A materials management department was created by redefining responsibilities and needs. Prior to this reorganization, these functions were performed in many different segments of the organization. Another formal structure used in this company was the assignment of a design-to-production manager with a group concerned with the producibility of the product idea early in the design phase. As the product design became more established, this group was moved from engineering into the production department. Most of the group was absorbed into the production area, with only one or two key individuals maintaining a matrix relationship with engineering. Finally, this company also uses groups referred to as transition teams, cross-functional and ad hoc in nature, whose charge is to solve specific problems in the transition of the product into production.

In the therapeutic apheresis innovation, a strategic business unit (SBU) was established linking the major companies in the joint agreement. In addition several problem-oriented groups have also been formed as needed. For example, a so-called phoney group was established among different functional specialists between the organizations, who conducted telephone conference calls as needed when problems arose. In the human resources management innovation, an ad hoc task force was created to define "what was fixed and what was variable" about human resources in the company. Finally, the school-based management in-

novation required the creation of a district curriculum council and an overall leadership committee.

In each of these cases, many interdisciplinary functions needed to be combined to address issues regarding the innovation. Although the importance of integration and coordination mechanisms have long been recognized in the innovation literature (see Galbraith 1982), we were surprised to observe the number, fluidity, and variety of creative mechanisms used in the innovation cases to restructure, coordinate, and address problems. These mechanisms provided incremental ways to make continual transitions between divergent components of the innovations and between the new innovations and existing organizational operations throughout the innovation period.

Observation 6. Hands-on top management involvement occurs throughout the innovation period; several levels of management removed from the innovation itself are directly involved in all major decisions The literature has frequently mentioned sponsorship and product champions (Roberts 1980; Kanter 1983; Peters and Waterman 1982). However, we observed in the innovation cases much more than a management sponsor or product champion, which the literature has focused on. Top managers, meaning one to four levels above the innovation leader, were observed in many of the cases to be: (1) very knowledgeable of and keenly interested in the innovative ideas within their organizations, (2) directly involved in all major decisions about the innovations, and (3) performed different technical, managerial, and institutional roles that were critical to the innovation developmental effort but often not recognized by the immediate innovation unit managers. Having multiple levels of managers involved in the innovations appeared to provide balance among ever-present contradictory forces for expansions and contractions in program scope, resource allocation, time schedule, and performance targets. These levels of management also "ran interference" for the innovation, helped in controlling proliferation of ideas, assisted in restructuring, and helped get the necessary resources.

Thus, we observed in six of the seven cases a set of very active "hands-on" roles not only for one upper-level manager, but by hierarchical teams of top managers. (We did not have detailed knowledge of top management involvement in the seventh case, hybrid wheat.) These managerial roles appear to vary as different problems or opportunities unfold for an innovation, but managerial involvement does not appear to diminish over the life cycle of innovation development. The degree of time involvement of top managers appears to remain relatively constant over time in the longitudinal tracking of the innovation studies thus far.

In the cochlear and therapeutic apheresis projects, top management frequently monitors the progress of the innovations and became directly involved in operational decisions when significant problems arose, such as a product recall, replacement of a program manager, and in conflicts over the technological directions of the innovations. These activities involved the CEO, sector and group vice-presidents, and division general managers in which the innovations were housed. These executives were observed to take high levels of personal interest and ownership in the projects and provided the necessary clout to get things done.

The other innovation cases also exemplify the active "hands-on" roles of top managers. In the school-based management innovation, the superintendent pushed the idea throughout the district even though committees were set up in each school to decide on the details. In the case of naval systems the general manager of the division, two levels removed from most of the innovations, provided con-

stant direction and support. He ran interference, provided the budget, and was a key participant in all major decisions.

AN EMERGING INNOVATION PROCESS MODEL

We have described above six process observations about how a variety of technical and administrative innovative ideas evolve over time. These observations are grounded in data about seven innovations that are being intensively studied over time. The observations now need to be unified into a coherent process model about innovation.

Figure 4–8 is helpful in describing how all of these observations fit together. Imagine that the organization is proceeding in the general direction of point A as shown in Figure 4–8. At time zero a shock occurs that propels an idea or innovation in a new direction B. As indicated earlier, this shock may be a budget crisis, product failure, change of management, or environmental change that brings the new idea into being and provides a discontinuity in the old organization. So the innovation is now moving in direction B while the organization is still moving toward point A.

As the innovation begins to be implemented in the organization, we see that proliferation of the new idea occurs, and a divergent multiple progression process begins to emerge. This is represented in Figure 4–8 as a branching out of the original idea into several ideas. This proliferation may represent several new product variations spinning out of one original new product idea, or several related ideas that are required to get the original innovation implemented. As long as these proliferating ideas are, at least, loosely related, the innovation still proceeds toward point B, although it now has multiple divergent and parallel paths of progression.

130

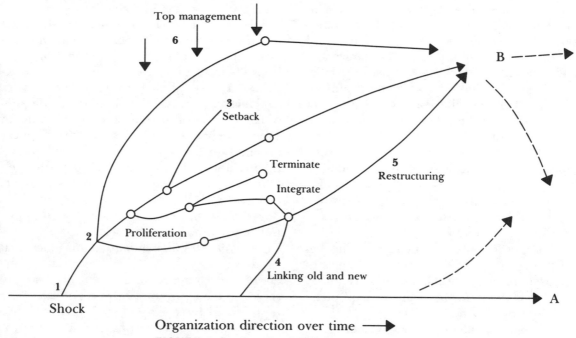

FIGURE 4–8. *Emerging Innovation Process Model.*

Source: Schroeder, Van de Ven, Scudder, and Polley, 1986.

As work proceeds on the ideas over time, setbacks and surprises are encountered. This is shown in Figure 4–8 by the lines that represent setbacks and terminations. Although some setbacks are treated as aborted dead ends to certain paths in the innovation process, more often they are terminated as incomplete or not-immediately useful ideas or components for progressing with the innovation at that point in time. These terminated ideas or components are often stored away in memory or placed on the "shelf" for possible subsequent use. For an ongoing organization that undertakes multiple innovations over time, these terminated ideas or components become a rich store-house of knowledge and materials for use in unforeseen ways in subsequent innovations. Indeed, in one of the organizations that houses one of the innovation cases, an internal study found that the average shelf-life of incomplete innovative

ideas or materials was nearly ten years, and most often those embryonic ideas were incorporated in totally unforeseen and unpredictable ways in subsequent commercially successful innovations. As Alexander Gray (1931) stated, "No point of view, once expressed, ever seems wholly to die . . . our ears are full of the whisperings of dead men."

As the innovation develops further, convergent linkages are established by integrating different component paths of the innovation, as well as by overlapping the old and the new. These linkages are represented in Figure 4–8 by lines that connect different paths of the innovation and by the converging lines between the operating organization moving toward point A and the innovation moving toward point B. This process of multiple convergent progression appears to take three forms: (1) The old organization can be moved toward

131

point B, as the entire organization fundamentally changes direction as a result of the innovation, or (2) the innovation can be moved toward point A and blended into the old organization, or (3) the old and new can coexist in parallel progression with linkages between the old and the new. All of these kinds of linkages appear to occur throughout the life cycle of the innovation cases in this study.

Also, the organization is constantly being restructured as it moves toward point B. This is shown in Figure 4–8 by the entire course of events. The initial shock may require restructuring, proliferation requires restructuring, setbacks and surprises may require restructuring, and linkages require restructuring. This restructuring may take the form of changes in organizational responsibilities, joint ventures, teams, new departments being formed or altered control systems.

Finally, hands-on top management takes place all along the route toward point B as shown by a continuous input by top management in Figure 4–8. Top management controls proliferations, deals with setbacks, helps to link the old and the new, and restructures the organization as needed, in addition to providing general goals and resources to support the innovation.

What we have been describing, through our process observations and Figure 4–8, is quite different from theories that have been proposed in the literature about innovative ideas. First, our process model is grounded in data and is empirical in nature. As such it is descriptive of how innovative ideas evolve over time. No attempt is made here to describe a normative theory of innovation. This would seem to require another step that relates descriptions of innovation processes to success and failure. But our descriptive observations are a good step toward developing and testing normative theories.

The process model we have described here does not resemble the unitary sequential stage theories found in the literature. As a matter of fact, we have observed innovation to unfold from a simple unitary process into multiple divergent, parallel, and convergent progressions of activities over time. Some of these activities in the multiple progression process are conjunctive (that is, related through a division of labor among functions and interdependent alternative paths of activities), and many appear to be disjunctive (that is, unrelated in any noticeable form of functional interdependence). Moreover, we observed that many components and paths that innovation participants perceived as being interdependent and conjunctive at one time were often reframed or rationalized as being independent and disjunctive at another time. In either case, it is clear that the messy and complex progression of ideas observed in the innovation cases is not accurately represented by a simple, sequential progression model of stages or phases as the vast majority of process models in the literature suggest.

Finally, the process model we have proposed here relates specifically to innovation. Most other process models found in the literature represent less discriminating and more general process models of individual and organizational decision making, development and change.

CONCLUSIONS

Based on an in-depth review of the longitudinal development of seven innovations, we have proposed a descriptive process model of how an innovative idea evolves over time. These innovations include four new product technologies and three new administrative arrangements.

Using the seven innovations, we have developed six observations about how innovation evolves over time. These six observations describe a rather fluid process where an idea seems to start off with a shock, then proliferates, is subject to setbacks and surprises, and then links with the old organization along the way. Restructuring of the organization occurs and top-management maintains hands-on involvement until the innovation finally becomes part of the accepted order or establishes a new order.

This process stands in contrast to the many sequential stage models that have been proposed in the literature. We conclude that a much more complicated multiple progression process of divergence and parallel and convergent streams of activities occurs in the development of innovations. Moreover, although some of these streams of activities are interdependent and conjunctive, many appear to be disjunctive and occurring independently of other streams of activities.

Of course, it should be recognized that these observations and the resulting process model that emerged inductively are preliminary. In the future this process model will be more extensively developed and modified, if necessary, as data collection and updates on these innovations continues over time in the Minnesota Innovation Research Program. Longitudinal tracking of these innovations is providing a rich database that will be suitable for more detailed understanding of the process of organizational and technological innovation.

REFERENCES

Ackoff, R.L. 1974. *Redesigning the Future: Systems Approach to Societal Problems.* New York: Wiley.

Angle, H.L., C.C. Manz, and A.H. Van de Ven. 1985. "Integrating Human Resource Management and Corporate Strategy: A Preview of the 3M Story." *Human Resource Management.* 24: 51–68.

Bales, R.F., and F.L. Strodtbeck. 1951. "Phases in Group Problem-Solving." *Journal of Abnormal and Social Psychology* 46: 485–95.

Cohen, M.D., J.G. March, and J.P. Olsen. 1972. "A Garbage Can Model of Organizational Choice." *Administrative Science Quarterly* 17: 1–25.

Dalton, G.W., P.R. Lawrence, and L.E. Greiner. 1970. *Organizational Change and Development.* Homewood, Ill.: Irwin-Dorsey.

Delbecq, A.L., and A.H. Van de Ven. 1971. "A Group Process Model for Problem Identification and Program Planning." *Journal of Applied Behavioral Science* 7: 466–92.

Friedmann, J. 1973. *Retracking America: A Theory of Transactive Planning.* Garden City, N.Y.: Doubleday.

Galbraith, J.R. 1982. "Designing the Innovative Organization." *Organizational Dynamics* (Winter): 3–24.

Glaser, B.G., and A.L. Strauss. 1967. *The Discovery of Grounded Theory: Strategies for Qualitative Research.* Chicago: Aldine.

Gray, A. 1931. *The Development of Economic Doctrine.* London: Longman's Green.

Hage, J., and M. Aiken. 1970. *Social Change in Complex Organizations.* New York: Random House.

Helson, H. 1948. "Adaptation-level as a Basis for a Quantitative Theory of Frames of Reference." *Psychological Review.* 55:294–313.

———. 1964. "Current Trends and Issues in Adaptation-Level Theory." *American Psychologist* 19: 23–68.

Jalkumar, R., and R. Bohn. 1984. "Production Management: A Dynamic Approach." Harvard Business School Working Paper 9-784-066.

Kanter, R.M. 1983. *The Change Masters.* New York: Simon and Schuster.

Lewin, K. 1947. "Frontiers in Group Dynamics." *Human Relations* 1:5–41.

Lewin, K., T. Dembo, L. Festinger, and P. Sears. 1944. "Level of Aspiration." In J. McV. Hunt, ed., *Personality and Behavior Disorders*, Vol. 1, Chapter 10. New York: Ronald Press.

Lewin, A., and J. Minton. 1986. "Determining Organizational Effectiveness: Another Look, and an Agenda for Research." *Management Science*, (May): 514–38.

Lindblom, C. *The Intelligence of Democracy.* New York: Free Press.

Lippitt, R., J. Watson, and B. Westley. 1958. *The*

Dynamics of Planned Change. New York: Harcourt, Brace and World.

Lorange, P. 1980. *Corporate Planning: An Executive Viewpoint.* Englewood Cliffs, N.J.: Prentice-Hall.

March, J.G., and H.A. Simon. 1958. *Organizations.* New York: Wiley.

Masuch, M. 1985. "Vicious Circles in Organizations." *Administrative Science Quarterly* 30 (1) (March): 14–33.

Mintzberg, H., D. Raisinghani, and A. Theoret. 1976. "The Structure of Unstructured Decision Processes." *Administrative Science Quarterly* 21: 246–75.

Mohr, L. 1982. *The Limits and Possibilities of Theory and Research.* San Francisco: Jossey-Bass.

Peters, T., and R. Waterman. 1982. *In Search of Excellence: Lessons from America's Best-Run Companies.* New York: Harper & Row.

Poole, M.S. 1983. "Decision Development in Small Groups, Ill: A Multiple Sequence Model of Group Decision Development." *Communications Monographs* 50: 321–41.

Quinn, J.B. 1980. *Strategies for Change: Logical Incrementalism. Homewood, Ill.: Irwin.*

Roberts, E.B. 1980. "New Ventures for Corporate Growth." *Harvard Business Review* (July-August): 134–42.

Rogers, E.M. 1983. *Diffusion of Innovations.* 3d ed. New York: Free Press.

Ruttan, V.W. 1959. "Usher and Schumpeter on Invention, Innovation, and Technological Change." *Quarterly Journal of Economics* 73: 596–606.

Tornatzky, L.G., J.D. Eveland, M.G. Boylan, W.A. Hetzner, E.C. Johnson, D. Roltman, and J. Schneider.1983.*The Process of Technological Innovation: Reviewing the Literature.* Washington, D.C.: National Science Foundation.

Tushman, M. and E. Romanelli, 1985. "Organizational Evolution: A Metamorphosis Model of Convergence and Reorientation." in B. Staw and L.L. Cummings (eds.), *Research in Organizational Behavior* 7, Greenwich, Connecticut: JAI Press, 171–222.

Usher, A.P. 1954. *A History of Mechanical Inventions.* Cambridge, Mass.: Harvard University Press.

Utterback, J.M., and W.J. Abernathy. 1975. "A Dynamic Model of Process and Product Innovation." *Omega* 3: 639–56.

Van de Ven, A.H. 1980. "Problem Solving, Planning, and Innovation, Part I: Test of the Program Planning Model." *Human Relations* 33: 711–40.

———. 1986. "An Update on the Minnesota Innovation Research Program Framework and Methods." Minneapolis: University of Minnesota Strategic Management Research Center, unpublished, May.

———. 1986b. "Central Problems in the Management of Innovation." *Management Science* (May): 590–607.

Van den Daele, L.D. 1969. "Qualitative Models in Developmental Analysis." *Developmental Psychology* 1 (4): 303–10.

———. 1974. "Infrastructure and Transition in Developmental Analysis." *Human Development* 17: 1–23.

Zaltman, G., R. Duncan, and J. Holbek 1973. *Innovations and Organizations,* New York: Wiley-Interscience.

5

PSYCHOLOGY AND ORGANIZATIONAL INNOVATION

Harold L. Angle

As an innovation idea moves from its inception through development and implementation, it is people who push, modify, or drop the innovation (Van de Ven 1986). This chapter is concerned with the ways in which personal attributes interact with activities, experiences, and social contexts to influence the development of innovative ideas over time. The chapter's primary purpose is to outline an agenda for investigation of the psychology of organizational innovation. Because people, their environments, and their behaviors have bidirectional impacts on one another (Bandura 1977), the discussion encompasses the reciprocal influence between people and innovations, which unfolds over time—that is,

The author greatly appreciates comments on an earlier draft of this chapter from Kenneth Craik (University of California, Berkeley), Andy Van de Ven (University of Minnesota), and numerous colleagues in the Minnesota Innovation Research Program, as well as the students in the spring 1987 doctoral seminar on the management of innovation at the University of Minnesota.

the ways in which people affect and are affected by the innovation process.

ORGANIZATION OF THE CHAPTER

The discussion begins by reviewing some widely accepted psychological principles that have particular relevance toward innovation, then turns to a projection of how future research can help fill the gaps in our present knowledge of the psychology of innovation, specifically as it occurs within complex organizations. Although the chapter's focus is on psychology, it is more specifically oriented toward *organizational* psychology. Accordingly, it is necessary to consider simultaneously both micro and macro issues (see Roberts, Hulin, and Rousseau 1978) in order to keep in mind the three factors that Porter and Miles (1973) suggested are necessary in order to understand organizational (and, by implication, innovative) behavior:

1. What the person *brings* to the situation;
2. What he or she *does* there;
3. What *happens* to him or her in the organizational setting.

The chapter first considers the role of motivation in organizational innovation. Following this, the discussion turns to limits on human information processing that may affect innovation, before turning to biographical characteristics that individuals bring to the situation and that impose both opportunities and constraints on the innovation process. As appropriate, socio-psychological issues are brought into the discussion. The chapter concludes with a discussion of change—that is, change that takes place in people as they engage in the process of innovation. At appropriate points throughout the chapter, evidence from MIRP studies is offered, and a number of researchable propositions are suggested, generally clustered together under the categories suggested above.

The data presented here are of two types: survey data from questionnaires, and qualitative data garnered from event mapping during longitudinal study of various innovation studies within the MIRP framework. The survey data are from two types of questionnaires: (1) the Minnesota Innovation Survey (MIS), which was administered to members of nine innovation teams, and (2) two organizational-assessment surveys conducted by MIRP researchers. The MIS is described in Chapter 3. The other two organizational-assessment surveys were administered in two settings: the central research laboratory of a *Fortune* 500 manufacturing company (N = 170) and one operating division of the same large organization (N = 265). Both surveys were conducted to identify factors that facilitate or impede innovation in those organizations.

SOME NOTES ON ABILITY, MOTIVATION, AND INNOVATIVE PERFORMANCE

Innovative behavior, like other organizational behavior, is motivated behavior. Although the principles discussed in this section are rather broad in their application, individual differences that may make a difference will be addressed, as appropriate.

Katz (1964) pointed out that organizations must motivate their members to perform three types of behaviors: (1) to join and stay, (2) to perform reliably in a prescribed manner, and (3) to perform such spontaneous, innovative behaviors as are necessary to fill in the gaps between what the organization can anticipate and what it cannot. It is readily apparent that the organization's task with respect to motivating its members becomes more complex as one proceeds from the first to the third requirement.

To attract people to join and stay may require nothing more sophisticated than adequate "system rewards" (Katz and Kahn 1978: 412), which are those rewards allocated to all persons solely on the basis of their membership in the system. Attaining reliable performance of prescribed behaviors requires the addition of what Katz and Kahn called "individual rewards" (1978: 412)—that is, rewards that are administered or withheld, contingent on delivery of the prescribed performance. The third motivational objective—a requirement that is obviously more difficult to manage than the first two—is where this discussion of motivation vis-a-vis innovation focuses, for this is clearly the domain of organizational innovation. Let us begin, then, by reexamining the relationship between motivation and innovation. This requires simultaneous consideration of three notions: (1) that ability and motivation are equally important to innovative performance, (2) that both motivation and ability come from within persons and from organizational contexts, and therefore (3) that both persons and contexts are important to organizational innovation.

INNOVATION: A MATTER OF PERSONS, SITUATIONS, EFFORT, AND ABILITY

Motivation is necessary but not sufficient to bring about a high level of performance. While the requisite motivating factors must be in place, enabling factors are equally important. According to expectancy theory (Vroom 1964), performance $= f$(ability $\times$ motivation). The functional relationship between ability/motivation and performance in Vroom's formulation is multiplicative. This indicates that *both* ability and motivation are necessary conditions for performance; if either is missing, performance equals zero in Vroom's equation (1964: 203):

> It makes little sense to ask which is the more important determinant of a performance—a person's ability or his level of motivation. Such a question would be analogous to asking whether the length or height of a rectangle is more important in determining its area.

Motivation and ability each appear in two forms—as properties of people and as properties of organizations or contexts. Campbell, Dunnette, Lawler, and Weick (1970), for example, drew a useful distinction between *content* and *process* theories of motivation. Content theories focus on what is inside the person, such as drives, needs or motives—that is, what the person brings to the situation. Process theories, on the other hand, more often focus on incentives and rewards—the mechanisms of motivated behavior. Here, it is what happens to the individual that is of primary interest. Accordingly, motivation (like ability) can be a trait variable or a state variable—something the person brings or encounters.

For instance, different people bring different amounts and kinds of internal motivation to their employing organizations. Whether such internal states are interpreted in terms of needs (McClelland 1961; Maslow 1954) or in terms of reinforcement history (Skinner 1938), people differ in their response tendencies and will respond to like situations in unlike ways. In a similar vein, people differ from one another with respect to the particular skills and abilities they bring to the situation. Some are better suited than others to specific organizational requirements, by virtue of their training and experience; others differ with respect to innate aptitudes, and so forth.

It is also true that people encounter different types and amounts of incentives in different situations. Organizations differ from one another in the inducements they offer for the contributions they seek (see March and Simon 1958), and these inducements may not be entirely consistent from time to time, within a given organization. According to Fritz Heider (1958), social systems affect the extent to which people "can" perform and on the extent to which they feel motivated to "try" to perform. In much more recent work, Amabile (1988) has developed a model of creativity (a construct having considerable domain overlap with innovation which is discussed in a subsequent section). Amabile's model contains two "can" factors (domain skills and creativity skills) and one "try" factor (internal motivation). It seems therefore incumbent on organizations that would be innovative to create two broad classes of conditions—facilitating conditions (so that people *can* innovate) and motivating conditions (so that they are willing to *try* to innovate).

Ability, then, like motivation, may reside partly in persons and partly in contexts; the ability to create and to innovate can result both from what the person brings to the situation and what he or she encounters there. People differ with respect to experience, knowledge, skills, and abilities, but there is ample evidence that creativity (as is brought out in a subsequent section) also is an important individual difference that enters into the "ability" component of the ability-motivation equation. Given the ability/motivation input from their human resources, organizations can establish and sustain situations that impede performance or that facilitate performance.

The Interaction Between Person and Environment

This brings us to consider a second maxim, often associated with Kurt Lewin: Behavior is a function of person and environment. The statement is, in essence, the interactionist manifesto, and there are two important points embedded in it, relevant to organizational innovation:

1. Personal characteristics and context have co-equal roles in bringing about innovative behavior.
2. The *interaction* between people and context will result in outcomes not fully accounted for by people and context taken separately.

Context can be a two-edged sword. At times it places constraints in the way of innovative behavior, while at other times it makes innovative behaviors more probable. Context also provides a basis for interpretation of experience: People make sense out of otherwise ambiguous stimuli in terms of the context in which they are perceived. Because, as discussed elsewhere in this chapter, the management of attention is such a vital component of the management of innovation, this information-shaping aspect of context is a vital consideration in the psychology of innovation.

It is not performance in the generic sense, but *innovative* performance that is of primary interest here. As a subset of general performance, innovation would seem amenable to understanding, prediction, and control, under the principles discussed above. We would therefore expect that organizational innovation, like other forms of performance, depends on what members bring to the situation and what they encounter there and that these factors affect both the ability to innovate and the motivation to innovate. People bring varying amounts of knowledge, skill, and internal motivation to the situation, but that situation also affects (1) what is attempted, (2) what is noticed, and (3) what is possible.

Let us focus for the moment, on the organization's role in the ability-motivation equation. Complex organizations comprise social milieux that shape and constrain the behavior of their members. This context contains incentives and disincentives—facilitating conditions and constraints. Organizations may create conditions that aid discovery and that facilitate the nurturing of new ideas into good currency—or they may present a context that obstructs the innovation process. These contextual conditions may well dominate other factors in influencing the level of innovation in both upward and downward directions. Amabile (1988), for example, has described three studies in which she interviewed R&D scientists and marketing personnel in a varied set of organizations. Interviewees mentioned qualities of environments more frequently than personal attributes such as creativity, in both high-creativity and low-creativity stories.

Organizations may provide incentives and rewards for members' innovative performance, or they may essentially ignore this aspect of performance. Even worse, they may unintentionally provide disincentives to innovation because of the actual (as opposed to intended) incentives and rewards they provide. Kerr (1975), for example, has elaborated this point, as he described several ways in which organizations have motivated their members to perform dysfunctionally.

The foregoing discussion leads to two specific, interrelated propositions.

Proposition 1: Organizational innovation is a joint function of members' personal attributes and the context for innovation in their organization.

Proposition 2: Organizational innovation occurs in organizations that provide a context that contains both enabling and motivating conditions for innovation; innovation does not occur where either factor is missing.

Motivating conditions are those incentives, rewards, and the like that lead organization members to want to innovate—in Heider's (1958) words, to *try*. In this respect, it is important to take both intrinsic and extrinsic motivation into account.

Intrinsic and Extrinsic Motivation

Incentives and rewards that are offered directly by the organization on a contingent basis are, by definition, extrinsic factors—that is, rewards that the system or agents of the system may offer or may withhold, at their option. Prescriptions offered managers by motivational psychologists, whether of the behaviorist camp (see Luthans and Kreitner 1985) or of a more cognitive bent (Lawler 1973) are more similar than not. Simply put (but in terminology that would violate the sensitivities of a radical behaviorist), the organization's task is to determine the outcomes that the individual values and then make these outcomes contingent on the individual's performing the desired behaviors.

Intrinsic rewards are also important; indeed they are probably much more important than extrinsic rewards for the specific purpose of attaining high levels of spontaneous, innovative behaviors from the organization's members. Amabile (1983), in her pursuit of a "social psychology of creativity," provided an impressive amount of evidence that intrinsically motivated people are more creative than their extrinsically motivated counterparts.

At risk of oversimplification, intrinsic outcomes are defined as those that are received automatically as a result of, or concomitant with, the relevant behavior rewarded. Although some of these rewards come from the behavior itself (such as having fun or doing interesting work), others come from one's accomplishments (enhanced self-esteem). What both types of intrinsic rewards have in common is that, rather than being mediated by the system, they are administered by the person and are accordingly much more reliably administered than are extrinsic rewards (Porter and Lawler 1968).

Despite the apparent straightforwardness of this comparison between intrinsic and extrinsic rewards, there are troubling areas of overlap. Even organizational psychologists have often disagreed on matters of classification (Dyer and Parker 1975). For instance, Herzberg, Mausner, and Snyderman (1959) categorized *recognition* as intrinsic. Yet recognition is externally mediated; others decide whether or not to provide it. How then can one argue that it is "intrinsic"? Perhaps the key is that it is an extrinsic reward that triggers intrinsic satisfaction. According to Festinger's (1954) social comparison theory, we often seek to learn about ourselves through others. If the organization *tells us* that we have achieved, we may then be ready to believe it.

Whether they are intrinsic, extrinsic or mixed, the following motivational principle summarizes the relationship between rewards and behavior: *Behavior is a function of its consequences.* This statement, represents a fundamental axiom of stimulus-response psychology. It stems from Thorndike's (1913) law of effect, which states that behavior that is followed by a valued outcome tends to be repeated while unrewarded behavior (or punished behavior) tends to disappear. In other words people do what pays off for them. The principle has its roots in the ancient Greek philosophy of hedonism.

As with more general performance, innovative performance appears to be influenced by its consequences, both intrinsic and extrinsic. This leads to our third proposition:

Proposition 3: Organizational innovation is a function of the intrinsic and extrinsic rewards associated with innovative behavior.

As this proposition implies, both intrinsic and extrinsic rewards are capable of eliciting innovative behavior. The use of extrinsic incentives and rewards to attain high levels of spontaneous, innovative behaviors from their members is easier, however, in principle than in practice. Not only does the organization have to provide incentives and rewards in a systematic and timely manner, but these must be *potent* incentives and rewards—that is, rewards that are valued by the individual. This is often a difficult requirement to meet because it is the person whose innovative behavior is at issue—not management—who is the final authority as to what constitutes a valued reward.

In discussions with managers in organizations participating in the Minnesota Innovation Research Program, our research teams have seen concern for finding powerful incentives and extrinsic rewards as a recurring theme. Managers have repeatedly expressed the need for their organizations to find ways to provide powerful incentives and rewards to elicit innovative behavior. The problem is that of finding the right "coin of exchange" (see Hackman 1976)—that is, the specific types of incentives and rewards that will appeal in particular to those members who have creative potential.

Money, the most widely recognized "generalized conditioned reinforcer" for employees (Opsahl and Dunnette 1966), may not always be the most effective approach. In one MIRP research project within a *Fortune* 500 manufacturing company, it became apparent that financial incentives can be problematic. A series of structured group discussions using the Nominal Group Technique (Van de Ven and Delbecq 1971) with groups drawn from the organization's research laboratories and from an operating division that enjoyed a reputation for innovativeness, encountered a recurring theme—money is *not* the major attraction for many innovators. Much more important than financial incentives was recognition—both from the organization and from peers. In a survey of the operating division, 52 percent of the respondents indicated that financial rewards are seldom related to innovative behavior. About one-quarter of persons surveyed expressed doubts that financial rewards *could* be used effectively as an incentive for innovation.

It may be, however, that for at least some persons money can be an effective motivator of innovative behavior—but not because of money's extrinsic nature. On the contrary, as has already been stated, the person receiving a reward is the final authority in defining the reward's properties. This recipient-specificity may extend to the fundamental issue regarding whether financial compensation is indeed an extrinsic outcome. Although Herzberg et al. (1959), in their well-known research on motivator-hygiene theory, categorized pay as a "hygiene factor" (that is, extrinsic), their data indicate that a number of their interviewees had what the researchers termed "second level" feelings about pay. Specifically, some of these individuals saw pay as an indicator of recognition (that some people do not work for money, per se, but use money to "keep score" on how they are doing). Indeed, it would not be difficult to argue that an organization that continually praises its members' performance without ever backing up that praise with financial rewards, as tangible evidence of its appreciation, might soon lose credibility.

The complex relationship between pay

and the various meanings different recipients are apt to assign to it makes it difficult to make a categorical statement about the efficacy of pay as an incentive for organizational innovation. Perhaps the following proposition, though equivocal, best expresses the status of pay as a motivator for organizational innovation:

> *Proposition 4:* Financial compensation, per se, is a relatively weak extrinsic reward for innovation. However, to the extent that it provides an effective means for organizations to express recognition, pay can help motivate future innovative behavior.

There are problems in contingency-based, extrinsic motivation schemes beyond those discussed thus far—particularly in systems based on individual incentives and rewards. In the first place, there are potential difficulties with reward equity (Adams 1965; Walster, Walster, and Berscheid 1978). According to equity theory, people evaluate the ratio between their inputs (contributions) and their outcomes (rewards). As long as this ratio is about the same as others' inputs/outcomes ratios, there is no problem. However, if people perceive that others are being rewarded (relatively) out of proportion to their actual contributions, a tension is established that motivates people to reduce the disparity. To remedy the perceived imbalance, people may (for example) begin to lower their own contributions, attempt to get others to alter their contributions, or perhaps even leave the system altogether.

Individualism and Collectivism

Western management is frequently chastised for being too individualistic in its orientation toward rewards (see Ouchi 1981). Equity problems may arise from attempts to reward individuals for performance that may have actually been the result of a group's collective efforts. Innovation, no matter how individually it may have been initiated, always becomes a collective undertaking at some point in its implementation. The unique contributions of various individuals may become blurred. Moreover, each stakeholder to an innovation may have a rather myopic perspective on the big picture, seeing his or her own contribution more clearly than the contributions of others. The most even-handed of individually based extrinsic reward systems may therefore leave at least some—perhaps all—of the contributors feeling underrewarded.

In the Minnesota Innovation Research Program, more than 170 members of nine innovation teams in various organizations were asked for their global assessments of their innovation projects' overall effectiveness. Although the correlation between innovation effectiveness and the perceived likelihood that *individuals* would be rewarded for goal attainment was not statistically significant, the sign of the correlation was negative ($r = -.09$; $p < .25$). By contrast, the correlation between *group* rewards and innovation effectiveness was positive and statistically significant ($r = .39$; $p < .001$).

It may be that difficulties in using individual rewards to motivate innovative behavior are more prevalent at the implementation end of the innovation process than in the earlier idea-generation phase. For example, one organization studied within the MIRP framework had a goal of creating new products at such a rapid rate that, in any time period, at least one-quarter of all items in the product mix will be less than five years old. In this organization, it appeared that the majority of the viable new product ideas generated over a three- or four-year period had been originated by one specific member of the research and development team: The organization could have virtually designated this one individual as their idea generator. Yet once these ideas were surfaced, the one-person show immediately had to become a team undertaking. At the point where interdependence emerges, collective action becomes perhaps a more essential aspect of innovating, than individual imagination.

It may also be that collectivist concerns are relatively more important where innovation is incremental, while individualism is better suited to radical innovation. Japan, for instance, which represents one of the world's most collectivist cultures, has a reputation for incremental innovation. Japanese industry has dominated many world markets by making relatively conservative adaptations of product ideas and then implementing, producing, and marketing the new products worldwide. Yet it is often said that nothing radically new ever seems to come from Japanese industry.

The relationship between culture (individualist versus collectivist) and innovation is likely rather complex. We offer the following propositions to guide further research on this issue:

> *Proposition 5:* The more individualized the reward system, the more effective is the idea-generation process; the more collectivist the reward system, the more effective is the innovation implementation process.
>
> *Proposition 6:* Individualized reward systems are better suited to motivation for radical innovation; group-based reward systems are more effective in motivating for incremental innovations.

There are still other problems with the use of extrinsic, individualized reward schemes. One such problem is the need for surveillance. In order to reward innovation, one must notice when it occurs and recognize it. Yet this may be a rather large order when judgment as to whether something is an innovation or a mistake (Van de Ven 1986) often depends on the test of time. By definition, an innovation is something new. By implication it is unexpected. Organizations' inability to provide ex-

trinsic rewards in a reliable, timely manner was what led Porter and Lawler (1968) to suggest that intrinsic rewards are more tightly linked than extrinsic rewards to future behaviors. In effect, the individual is much better able than the organization to monitor performance and apply appropriate rewards.

Another factor that might inhibit an organization's ability to elicit innovative behavior through extrinsic incentives and rewards is what has been called the "hidden costs of rewards." Deci (1971, 1972) has provided evidence that intrinsic motivation to perform a task decreases when an individual is provided monetary rewards on a contingent basis. Although Deci's theory has not met with universal acceptance, there is clearly food for thought here. Amabile's (1983) research has provided evidence that external constraints, demands, and rewards actually have a negative effect on creativity.

People may, indeed, formulate implicit theories about their relationship to their work through a process of self-observation and attribution (Bem 1972, 1978). They make judgments regarding the underlying causes of their own behaviors in the same way they attribute causation (see Jones and Davis 1965; Kelley 1973) for others' behavior—that is, they observe regularities in their own behavior and formulate personal theories. If the receipt of compensation is tied specifically and conspicuously to innovative acts, individuals may redefine themselves as employees "only doing it for the money." The organization may be hard-pressed to sustain a high level of innovative effort in the future as spontaneous creativity begins to give way to a simple economic exchange.

More to the point, if organization members begin to define their relationship toward their organization and their work as one of economic exchange per se, the member-organization relationship becomes what Etzioni (1975) termed calculative. There may develop a tendency toward bureaupathic behavior (Lawler and Rhode 1976). In this syndrome, the member considers it more important to meet the letter of the organization's demands than the spirit of those expectations. Behavior becomes focused on looking good on whatever performance indicators the organization employs and letting performance, in a deeper sense, go by the boards (also see Kerr 1975). All of this seems antithetical to the spirit of innovation.

In Chapter 16, Marcus and Weber illustrate a similar process in their description of rule-bound responses to externally imposed change. Such responses appear bureaupathic in that they meet the letter of enforcement agencies' requirements without attending unduly to their spirit. Marcus and Weber provide evidence that such responses are of limited efficacy and are associated with poor overall organizational performance.

As has been suggested, the exclusive use of extrinsic means for motivating spontaneous, innovative behaviors would probably be counterproductive. Intrinsic motivation seems to play a highly significant role. Some persons will be more inclined toward innovative behavior than others, even without obvious external incentives. In effect, people bring at least some of their own motivation with them and often behave as if innovation were its own reward. When such is the case, the organization's most effective strategy may be simply to create and sustain conditions so that the innovation process can occur—the so-called agricultural approach to management.

Although organizations cannot administer or withhold intrinsic rewards in the same finely tuned way that they can manipulate extrinsic consequences, they can establish (or fail to establish) general conditions that influence members' levels of intrinsic motivation. Typical prescriptions for enhancing intrinsic motivation are targeted toward the redesign of such organizational systems as the work itself (Hackman and Oldham 1980; Herzberg 1968),

which, inter alia, may raise the self-esteem of the employee, particularly where autonomy or personal control is enhanced (DeCharms 1968; Greenberger and Strasser 1986; Langer 1983). In the MIRP survey of the research laboratories of a major manufacturing organization, for example, perceived autonomy was significantly related to innovation effectiveness ($r = .46$; $p < .001$).

As elaborated by Amabile (1988) the relationships between intrinsic and extrinsic motivation and innovation may be both complex and nonlinear. Striking a balance seems necessary. If supervision is too tight, it stifles individual initiative and creativity, whereas if it is too loose the efforts of team members may become fragmented and unfocused (see Pelz and Andrews 1966). In a similar respect, it would seem counterproductive to rely on either intrinsic or extrinsic rewards to the exclusion of the other. As Amabile suggests, reward systems require a balancing act: Rewards must not be so tightly linked to output that people lose their internal motivation, yet sufficient generous and equitable rewards must be offered for innovation so that members know that the organization values their contributions.

The preceding discussion supports the view that both extrinsic and intrinsic rewards play a part in the innovation process. Indeed, despite the lack of research evidence to support Maslow's hierarchy of needs (Hall and Nougaim 1968; Lawler and Suttle 1972; Wahba and Bridwell 1976), there is evidence that at least some lower-level needs must be satisfied before higher-level needs become salient (Lawler and Suttle 1972). A complete motivational scheme, then, ought to incorporate contingency-based, extrinsic incentives and rewards, as well as intrinsic rewards.

Proposition 7: Extrinsic motivation is necessary but not sufficient for evoking spontaneous, innovative behaviors; it is at least equally necessary for organizations

143

to create and sustain conditions that facilitate intrinsic motivation. Above all, a suitable *balance* needs to be struck between intrinsic and extrinsic motivation.

Enabling Factors in the Organization

Although the distinction between micro and macro blurs somewhat in practice, the establishment of an innovation context that provides enriched opportunity for members' finding intrinsic rewards is perhaps more macro in its orientation than was the case for the earlier discussion of principles of individual motivation. Equally macro in its perspective is the need for organizations to establish and sustain conditions that *permit* innovation. In a preceding section, the discussion cited Heider's framework suggesting the equal importance of the words "can" and "try" in organizational innovation. This suggests that organizations need to provide conditions so that those who are willing to try to innovate also "can" innovate.

Even highly motivated persons find innovation difficult when they lack access to such basic resources as money, materials, information, and above all, slack time. In the same respect that organizations can create a demotivating climate, they can, through a number of mechanisms, create systems and structures, and an organizational climate, that impede the innovation process. On a more positive note, of course, it is also possible to establish ambient conditions—that is, *enabling factors*—that foster innovation.

One important enabling factor is information flows. The sharing of critical elements of information among people who have various roles in the innovation process is fundamental. This may be most crucial in the idea-generation phase. For purposes of reality-testing, as well as the cross-fertilization of ideas, people need to break "out of the boxes" to talk with people with different backgrounds and perceptual sets

than their own. Kanter (1988: 172) observed that innovation is more likely in organizations having cultures that emphasize diversity but at the same time have "connectedness."

One organization with which we are working in the Minnesota Innovation Research Program is designing a new headquarters and laboratory facility with the social psychology of information flows explicitly in mind. The design encourages people in different specialties to interact by location and arrangement of meeting rooms, eating facilities, traffic flows, and the like. This appears to be a sound, if commonsense, organizational application of social psychology to workspace design, as summarized by Becker (1981).

In the Minnesota Innovation Survey data, innovation effectiveness was found to be related both to communication frequency within the innovation teams ($r = .17; p < .03$) and communication frequency outside the teams ($r = .19; p < .02$). Innovation effectiveness also covaried with people's communication upward with their managers ($r = .22; p < .001$). Somewhat surprising was the lack of apparent relationship between innovation effectiveness and communication with customers ($r = .09; p < .24$) or communication with vendors ($r = .12; p < .12$). This corroborates data from a separate MIRP survey, conducted in a manufacturing organization, in which 69 percent of technical personnel indicated that they *never* get their new ideas from people outside their own company (about 35 percent indicated that they never get such ideas from others inside the company). Such results are clearly at variance with Utterbach's (1971) well-known finding that new ideas come largely from contacts outside the organization, as well as Peters and Waterman's (1982) notion that innovation results from being "close to the customer." Although our data may be sample-specific, it should be noted that they were obtained in organizations having solid track records in new-product innovation.

144

Despite this counterintuitive set of findings, nothing in our data would suggest that the cross-fertilization of ideas and the enhanced sensitivity to market needs that come from communication with clients would not have further enhanced the already high level of innovation in the organization. Indeed, proponents of group-based creativity techniques, whether using devices such as brainstorming (Osborn 1953) or simply interactive group processes, appear to agree that a high level of communication among persons *unlike* one another, so that new, creative ideas are generated, is necessary to create synergy (Maier 1970; Steiner 1972).

Proposition 8: Innovation effectiveness will be positively associated with frequency of communication among persons having dissimilar frames of reference.

Another way that organizations can free up members and create slack time to deal with innovation problems is through *uncertainty reduction*. Rather than minimizing uncertainty, however, it may be important to strike a reasonable balance in the amount of uncertainty in the system—that is, to optimize uncertainty. Where uncertainty is very low, people may be lulled into a state of complacency; where it is extremely high, they may become overanxious and may lose their focus on important goals.

Mechanistic, bureaucratic organizations (Burns and Stalker 1961) are typically characterized as totally lacking in innovative potential. Such organizations attempt to take uncertainty reduction to the limit by establishing formal procedures for nearly everything. It is often argued that this type of organization may stifle the spontaneity of its members by overroutinizing all aspects of organizational life.

On the other hand, it can be argued that the development of programs and procedures for certain repetitive activities can free up people to think about other things. If standardization were to be limited to the inherently routine

aspects of work (that is, those that lend themselves to programmed solutions), then people might be able to focus more attention on novel or exceptional tasks—those that require creativity. Uncertainty optimization, rather than uncertainty reduction, would then be the appropriate aim for organizations.

One key aspect of uncertainty in organizations is the ambiguity with which operative goals are established (Perrow 1967). In the Minnesota Innovation Survey, goal clarity was associated with perceived innovation effectiveness ($r = .43$; $p < .001$), as was an understanding of how goals and plans were to be implemented ($r = .41$; $p < .001$). Mixed-messages from management, in this respect, may be particularly dysfunctional. In the survey of the research laboratories, a disparity between stated and real goals was negatively associated with innovation effectiveness ($r = -.49$; $p < .001$).

On balance, it is suggested that the relationship between uncertainty and organizational innovation may be a curvilinear one. Too much certainty may be stifling, leading to rigid, stereotyped approaches to, and interpretations of, problems that arise. On the other hand, too much ambiguity may be disruptive, leading to an inordinate amount of energy expended on issues only peripherally related to innovation.

Proposition 9: Innovation effectiveness will be associated with a moderate amount of environmental uncertainty. At extremely high or low levels of uncertainty, innovativeness is reduced.

Obviously, physical and economic resources are high on any list of enabling factors for innovation. For many types of innovations, seed funding is essential. Such startup capital can be provided through official systems or by means of certain unofficial mechanisms. One of our participating organizations has a program whereby members with ideas can submit them to a research and development grants committee. Submissions are welcomed by members in all walks of life within the organization. The same organization has had for years a well-known informal policy that encourages innovators to "go around the system" and obtain "bootleg" funding for early project development.

As suggested by Lawrence and Dyer (1983), resource scarcity implies that people have little slack to exercise discretion or control over the development of their innovation. Heavy workloads, for example, or extreme competition for economic resources may result in a short-term "survival" orientation. On the other hand, there may be little stimulus to action where resource availability seems limitless. Accordingly, one would expect maximum facilitation for innovation where the level of resource availability is at least at a moderate level.

Relative resource scarcity was measured in the Minnesota Innovation Survey by items that asked about such factors as workload and competition for such scarce resources as finances, materials, people, and management attention. Perceived innovation effectiveness was positively related to workload pressures ($r = .19; p < .02$) but negatively related to resource competition for people ($r = -.23, p < .01$). Innovation effectiveness was also negatively related to competition for management's attention ($r = -.30.; p < .001$). The correlation between innovation effectiveness and competition for financial resources was not statistically significant but the sign of the correlation was negative ($r = -.12; p < .13$).

The positive relationship between workload and innovation effectiveness may simply indicate that a fast-paced operation is invigorating (or even that an innovation is effective *because* people are working hard). Overall, however, the data generally suggest that, as an enabling condition, the availability of needed resources appears to be positively related to

innovation performance. As competition with other uses of those resources reduces their availability, perceived innovation effectiveness declines.

Proposition 10: Competition with peer units in the organization for scarce resources is associated with reduced innovation effectiveness.

To summarize the foregoing, it has been argued that an organization's motivating and enabling conditions are importantly related to its level of innovation. Not only must organizations motivate their members to attempt to innovate, they must also manage the environment so as to reduce inhibitory influences and to increase the number of facilitating factors. Both the "try" and the "can" of innovation are important.

PEOPLE STRIVE TO MAKE SENSE OUT OF THEIR WORLD BUT ARE SOMEWHAT LIMITED IN THEIR CAPACITY TO DO SO.

What one individual experiences in an organizational setting can be substantially different from what another person experiences or what an outside observer would describe as "objective" reality. How a person perceives his or her environment is an active process. An individual's "perceptual readiness" (Bruner 1973) results in selectivity in noticing different aspects of the environment. What is perceived is based to a considerable extent on the person's past experiences, and the resulting perception is then evaluated in terms of personal needs and values. Because various people's needs and past experiences often differ markedly, it follows that people's enacted environments (Weick 1979) differ also. This section examines some of the implications, for organizational innovation, of people's apparent need to engage in a sensemaking process (Louis 1980).

As Van de Ven (1986) has pointed out, much of the research and folklore on the management of innovation seems to have ignored a large body of research by cognitive and social psychologists on the limited capacity of human beings to handle complexity and to maintain attention. Van de Ven (1986: 594) goes on to suggest that

> A more realistic view of innovation should begin with an appreciation of the physiological limitations of human beings to pay attention to non-routine issues, and their corresponding inertial forces in organizational life.

Limits on Human Information Processing

In his article titled "The Magical Number Seven, Plus or Minus Two," George Miller (1956) reviewed the evidence that people are so limited in their ability to process information in short-term memory that they are restricted to a total of about five to nine pieces of information at a time. (Indeed, this estimate might be over-optimistic at the upper end of the range.) In order for a person to process more data than that, it is necessary to exchange elements of information between short-term memory (STM) and long-term memory (LTM), which has a virtually inexhaustible "filling" capacity. This process of alternating storage and retrieval makes it necessary to process complex information serially, rather than simultaneously. The process of exchanging information between STM and LTM comes at considerable cost, in terms of time and effort. For each "chunk" the procedure takes from five to ten seconds (Simon 1974). Moreover, as chunks become more elaborate they take even longer to deal with.

A related phenomenon that limits people's information-processing capacity is *infor-mation input overload* (Miller 1960). People's perceptual defenses appear to cause them to reject stimuli that exceed the capacity of their "input gates." For this reason, and because organizational environments are inherently complex, the amount of information available at any one time is apt to exceed people's processing capacity. Thus, they do not perceive more than a fraction of the available environment. It is as if they look at their environment through an imperfect lens that filters out much of what tries to pass through it.

Perception is therefore a rather selective process. The individual is unable to notice many potentially relevant environmental aspects. Of course, stimuli that are highly distinctive are more likely to be noticed than those that do not "stick out" in some way. In an organization that generates dozens of memoranda every week, a written memo from the central office may not capture many people's attention. In that same organization, however, a video-taped message from the company president (which occurs rarely) will be distinctive and memorable to all who see it.

In addition to stimulus distinctiveness, people's prior experience plays an important part in determining what is noticed. In any organization, people soon learn to discriminate between those stimuli that must be attended to and those that can be safely ignored. Furthermore, people develop a "language" of sorts for processing familiar material, which perhaps leads to a bias in favor of easily managed, familiar stimuli over more difficult novel signals.

The psychological principle of *primacy* may also play a major role in determining the fate of stimuli that compete for people's attention. People seem loath to revise the judgments they have made on the basis of early input. Tversky and Kahneman (1974) have shown how people who have made estimates of one sort or another will not stray very far from those initial estimates, when revising, based on newer information.

A dramatic example of this tendency was demonstrated by Ross and his colleagues (Ross, Lepper, and Hubbard 1975). Subjects in a laboratory experiment were led to believe that they had been either accurate or inaccurate at intuiting whether some suicide notes were genuine or phony. In fact, the "successful" and "unsuccessful" conditions had been set up in advance and had no relationship to the subjects' ostensible insights (all the notes had been written by the experimenters). Even though the subjects were later debriefed and told that the level of their "success" had been a "set up," they still tended to estimate their own future performance (in tasks actually requiring such intuition) in accordance with the spurious performance feedback they had received in the early part of the experiment. Information that is first-in may be difficult, indeed, to replace with later data—despite improved relevance or accuracy provided by more recent data.

Perception is a highly active process. It is not simply a matter of novelty versus familiarity, or "first-in, stays in" that determines what is attended to and incorporated into one's cognitive map and what is excluded. As Ring and Rands show in Chapter 10, sensemaking—or the assignment of meaning—plays a major role. People exhibit a strong bias toward incorporating incoming information in such a way as to make sense against the backdrop of personal schema or scripts (Stotland and Canon 1972; Abelson 1981) that have been the result of prior experience and earlier cognitive processing. This means that even when a stimulus is noticed or attended to, there is no guarantee that it will be perceived accurately.

Because the filtering process results in an incomplete set of information, perceptual mechanisms are brought to bear in order to fill in gaps. As Bruner (1973) put it, people tend to go "beyond the information given." This process of confabulation is central to sensemaking. Individuals construct a system of categories or stereotypes and then assign to a novel situation all the attributes included in some specific stereotype, even though evidence may be completely lacking for some of these attributes. In a more dynamic sense, people seem to construct *scripts* of events that unfold over time (Abelson 1981). Acting on minimal information, they enact entire scripted behavioral sequences. This is reminiscent of March and Simon's (1958) concepts of *evoked sets* and *performance programs*—situations in which simple and incomplete stimulus conditions trigger an elaborate program of activity, which contains more detail than was explicitly signaled by present stimulus conditions. Paradoxically, then, human information processing results in both *less* information than was originally available and *more* information than was originally available!

Clearly, a major aspect of the sense-making process is rationalization. Aronson (1973) used the expression "the rationalizing animal" to indicate that people, in order to preserve logical consistency among all their cognitions (their beliefs, opinions, ideas, and so forth), frequently distort some cognitions in order to make them appear more consistent with others. Festinger (1957) theorized that such inconsistencies among cognitions motivate people toward resolution of the cognitive dissonance via various mechanisms such as post-hoc rationalization.

It is difficult to predict for any given individual how a particular stimulus will be interpreted. There are simply too many idiosyncratic factors involved, both in the event itself and in the psychological makeup of the person. It is possible, however, to be aware of conditions that tend to be associated with high levels of such cognitive activity—for example, where the person has a high emotional investment in the stimulus. This comes about, for instance, where the situation is relevant to important personal values or needs, where there appear to be opportunities for the achievement of important personal goals, or where self-esteem is challenged.

To summarize, people are rational but imperfectly so. Physiological limitations alone would seem to be sufficient to ensure that people's phenomenological worlds are less than perfect representations of objective conditions. Exacerbating this situation are the active sensemaking processes that people undertake both to preserve a sense of a consistent and logical world and to defend their egos. The process of defending themselves against complexity leads people to filter out information that may be important. As they settle into routines, people operate frequently on "automatic pilot," thinking very little about what they do.

This creates a particular problem for innovation—what Van de Ven (1986: 594) termed the problem of "management of attention." People tend not to notice gradual changes in their environments. Complacency and inattention result in opportunities for innovation being missed. In essence, people's action thresholds are not triggered by subtle changes in their environments. As situations deteriorate gradually, people adapt to the changes and are not motivated to act. New adaptation levels (Helson 1959) are established that "track" with gradually changing conditions, so that stimuli remain below attention thresholds. At the extreme, innovative response to threats or opportunities become forestalled until situations change to a fairly drastic extent.

Because the press of daily activities may cause people to become insensitive to gradual change or subtlety in their situations, they may miss many opportunities for innovating. In order to counter this, the organization might be well advised to create mechanisms to redirect and jostle the attention of its members so that subtle change and the opportunity that it may create are more widely noticed. Appointing people to positions whose purpose is environmental scanning and evaluation is possible, al-

though not always fully effective, solution to complacency. A better approach might be to establish internal information programs aimed at breaking up complacent frames of reference.

> *Proposition 11:* The level of innovation in an organization will be positively associated with the existence of mechanisms for focusing members' attention on changing conditions.

CREATIVITY IS NOT NECESSARILY INNOVATION

It is unlikely that any psychological construct could claim a more central position than creativity, in the study of innovation. This section outlines some key issues in the study of the psychology of creativity, in order to provide a general frame of reference, and then the chapter moves from creativity to a somewhat broader discussion of innovation, revisiting creativity as appropriate along the way.

At the outset, it seems fitting to quote Guilford's introduction to his (1950: 444) presidential address to the American Psychological Association: "I discuss the subject of creativity with considerable hesitation, for it represents an area in which psychologists generally, whether they be angels or not, have feared to tread." Essentially, Guilford's charge was that psychology had turned its back on creativity.

Much has changed, however, since 1950. There have been a number of major research programs on creativity, much of the research having at least some relevance to work organizations. Guilford was a central figure in systematizing the study of creativity in the years following his address (Guilford 1975). Possibly the most frequently cited stream of research on creativity—particularly on the characteristics of creative individuals—was begun a quarter-century ago by Donald MacKinnon and his colleagues at the Institute for Personality Assessment and Research (IPAR) at the University of California at Berkeley (MacKinnon 1960, 1975).

Despite the flowering of creativity research since Guilford's presidential address, until very recently few contributions on the topic were made within the management literature. A recent computer search of psychological abstracts in which "creativity" was combined with "management" or "business" resulted in fewer than two dozen relevant articles since about 1965. Much of the academic literature on the psychology of creativity has focused on creativity among pre-school or school children, but the preponderance of the published material on adults' creativity has appeared in the popular or applied literature rather than in academic journals.

This is not to say that there has been absolutely no recent work on adult creativity. Amabile (1983, 1988) has been conducting a series of studies on creativity, specifically within management settings in actual organizations. Another prominent exception to the trend toward neglect of adult creativity in the workplace appears to have been initiated at the University of Sheffield by Michael West and his colleagues (West, Farr, and King 1986). These research oases in what has been something of a wasteland are very welcome additions to the literature on the psychology of creativity and innovation.

What Is Creativity?

Schwab (1980) made a strong plea for more attention to construct validity in organization science. The casual use of various terms by writers has muddied the waters of understanding in virtually every area of inquiry in the study of organizations. With Schwab's caution in mind, a review of the creativity literature shows a wide range of underlying definitions. As pointed out by Taylor (1975), early definitions of creativity tended to be unifactory in nature (see Mednick 1962; Barron 1969). Examples of this type of definition would include the Gestaltist view that creativity involves an act of "insight" (Kohler 1929) or Mednick's (1962) suggestion that creativity is the forming of associative elements into new combinations.

By contrast, more recent approaches have often tended toward multifactor or multiprocess perspectives (Taylor 1975; MacKinnon 1975). For purposes of this chapter, MacKinnon's multiprocess definition of creativity will be adopted. MacKinnon (1975: 67–68) states essentially that creativity involves a process that is extended in time and characterized by (1) originality, (2) adaptiveness, and (3) realization—that is,

> novelty or originality, while a necessary part of creativity, is not sufficient . . . it must also be adaptive to reality. It must serve to solve a problem, fit a situation, or accomplish some recognizable goal. And thirdly, true creativeness involves a sustaining of the original insight, an evaluation and elaboration of it, a developing of it to the full.

Creative process, as depicted in the above definition, has been only one popular framework within which creativity has been studied. In addition, we have seen at least three other approaches. One prevalent approach has attempted to identify creative people (Guilford 1975; MacKinnon 1975). Another, somewhat related, orientation has emphasized the search for *ways to increase* people's level of creativity through training or other means (Osborn 1953; Parnes 1962; Gordon 1956). Finally, a few scholars have studied the contextual or organizational factors that tend to either inhibit or nurture creativity (e.g., Cummings, 1965).

This set of approaches to studying creativity fits well with the overall framework of this chapter—that is, the view that behavior is best understood in terms of people's personal attributes in interaction with certain contextual factors, including both ambient conditions and more discrete experiences. Consistent with this viewpoint, one would expect creativity in organizations to depend on all these factors, in interaction with one another.

How uncommon, then, might we expect creativity to be in work organizations? Is creativity a rare flower that blooms only on infrequent occasion? Worse, is it an attribute that one can expect to see "selected out" of most organizations, by virtue of their staffing and reward systems? Or are our organizations a natural habitat for creative activity?

We may take heart from Campbell (1960), who saw creativity not as a rare and precious phenomenon but one that is found quite widely distributed in nature. According to Campbell, creative thought is but one aspect of a general process by which people acquire knowledge concerning the world. In a logic strongly reminiscent of Darwin, he proposed that creative production comes from a combination of three conditions or mechanisms: (1) the production of variations, (2) a selection process, and finally (3) a means of preserving and reproducing those variations selected. Thus, creativity in Campbell's view seems largely to be a matter of capitalizing on chance. Quantity seems to be more important than quality in the pool of "thought trials" that eventually lead to the selection of a winning idea.

Campbell warned against our attributing special powers to "creative" people. Instead he asserted that the recognition of creativity is largely a matter of hindsight, which may lead us to make light of the amount of mundane and random activity that preceded the sudden insight. This fallacy may lead us further to attribute causality to what were essentially random events. Campbell, does not, of course, completely write off the role of individual differences in creativity. He holds, for example, that people will be more creative to the extent that they can produce a wider range and number of thought trials. Life experience may be an important factor, with more well-rounded individuals having more elements to juggle while seeking new combinations. Such a viewpoint suggests that creative potential may be much more widely distributed in organizations than a more elitist perspective would lead one to expect.

Others have proposed that individual differences play a much larger role in creativity than Campbell's point of view would suggest. The IPAR research (MacKinnon 1975) suggests, for example, that creative persons are intelligent (but intelligent people are not necessarily creative); are independent in thought and action; tend to be "loners"; are unusually open to their own experiences; tend to have strong aesthetic and theoretical values; and are strongly convinced of the worth of their own ideas. Contrary to the conventional wisdom of the time, creative persons studied at IPAR did not display symptoms of poor mental or physical health; on the contrary, they appeared to be healthier in both respects than the "normal" population.

Guilford (1975) offered a similar set of attributes to those identified in the IPAR research (although he indicated it would be rare to find all at the same time, even in a creative person): a high level of curiosity; effective work habits; an interest in reflective thinking ("thinking introversion"); unconstrained by reality; a high tolerance for ambiguity; and some preference for risk taking (among other attributes). Both Guilford and MacKinnon noted that creative persons are strongly individualistic— often to the point of appearing antisocial and even socially offensive.

Based on interviews with organizational scientists, Amabile (1988) has complied a list of personal qualities that appear to promote creativity or that seem to inhibit creativity. The personality traits listed had considerable overlap with the types of traits identified in the IPAR studies. Self-motivation appeared to be

151

an important personal attribute (recall our earlier discussion of the importance of intrinsic motivation). Other consequential factors included risk orientation and expertise in the relevant subject matter, as well as social or political skill. Perhaps the much shorter list of traits that Amabile's scientists associated with low creativity implicitly makes the most concise statement of what creativity is by defining what it is not. Uncreative people were seen as unmotivated, unskilled, inflexible, extrinsically motivated, and socially inept.

The foregoing listing of characteristics, incomplete as it is, portrays the sort of person one might expect to find in greater numbers in some organizations than in others. People tend to self-select into work environments. Furthermore, organizational systems tend to shape behavior through rewarding some behaviors, while either ignoring or actually punishing others. One suspects that the cultures of many organizations are antithetical to the type of individual described by Guilford, MacKinnon, or Amabile.

This poses a potential problem for organizations. If one of the key distinguishing features of creative individuals is their aversion to social conformity, it may be difficult for organizations to incorporate such individuals into their collective activities. Finding a way for highly creative people to interact effectively with others within cooperative systems may be one of the key problems in the management of innovation.

> *Proposition 12:* Innovation effectiveness will be positively related to the extent to which the organization is able to integrate creative personalities into the organizational mainstream.

Can Creativity Levels in Organizations Be Enhanced?

The foregoing description of creative persons suggests that people differ from one another in the extent to which they tend to display creative behaviors. If we assume that creative acts are the events that stimulate and redirect innovations, it would seem that innovative organizations need to have at least some creative members. How might organizations increase the prevalence of creativity in their ranks? Traditionally, there have been two approaches to increasing the number of the "right" kind of people in organizations—the selection approach and the development approach.

The selection approach is consistent with the assumption that people are essentially unchanging—that what you see is what you get. The appropriate strategy would be to develop a profile of the "creative" person and set about systematically to attract applicants fitting that profile and to screen them on the basis of characteristics assumed to indicate creativity. The development system is based on the premise that people are capable of learning, changing, and developing, given the proper opportunity. In order to apply this second set of assumptions to organizational creativity enhancement, one would bring to bear one's training and development system toward teaching creativity techniques to existing members.

According to Amabile (1988), two broadly defined types of skills are prerequisites to creativity in the workplace: domain-relevant skills and creativity-relevant skills. The former type of skills are the fundamental tools of the trade and vary somewhat from occupation to occupation. These skills appear to lie in the domain of traditional organizational training and development programs. The latter category seem to be less familiar territory to most organizations. The arsenal of creativity techniques available for training and development has grown fairly large, however, including brainstorming (Osborn 1953), synectics (Gordon 1961), eclectic creativity technique (DeBono 1972), and nominal group technique (NGT) (Van de Ven and Delbecq 1971), to name only the better-known approaches. Some of these (NGT and brainstorming) are aimed at enhancing the synergistic potential of problem solving groups, while others are aimed at individual creativity.

With the exception of brainstorming (which has not been well-supported) (Bouchard 1971; Taylor, Berry, and Block 1958) and NGT, most of these techniques have not been well represented in academic research. Indeed, some of these procedures lack any trace of academic respectability in their origins. Accordingly, organizations that want to enhance their members' creative potential are often forced to consider the use of untested, unvalidated procedures.

To further complicate the picture, there is little reason to expect even an effective creativity-enhancement program to lead to a permanent improvement in the amount of creativity, all by itself. Dunnette and Campbell (1968) illustrated the ineffectiveness of training (in this instance, T-groups) when trainees' posttraining experiences occur in a culture that does not support the new learned behaviors. The most effective creativity training may be for nought, unless the organization's climate and culture support the lessons learned (that is, unless both enabling and motivating conditions are in place).

What is a creative organizational climate? Torrance (1967) suggested that in a creative climate there will be respect for unusual questions and unusual ideas and opportunities for performance to occur without high levels of evaluation apprehension. Cummings (1965) characterized the creative organization in the following way: little formalization of relationships; flexibility in power-influence structures; large areas of discretion and autonomy; long time horizon for results-measurement; many open communication channels; and a managerial philosophy that projects the assumption that employees are capable and creative. Cummings' description was clearly congruent with Burns and Stalker's (1961) notion of the "organic" organization.

It is not intended to imply, however, that organizational structure is the cause, while creativity is the effect. A more realistic view would be that structure and creativity (or at least the need for creativity) share some of the same determinants. Proponents of environmental determinism (such as Burns and Stalker 1961) have long held, for instance, that organic organizational forms evolve as ways to cope with unpredictability and volatility in organizations' environments. These environmental conditions also make it necessary to innovate in order to deal with rapidly changing conditions. Whether one takes the view that necessity leads to new organizational forms or the more Darwinist perspective that organizations that are well adapted to conditions survive while others do not (Hannan and Freeman 1977; McKelvey 1982), one is left with the same result. In the long run, turbulent environments lead to organic organizational forms.

By definition, organic organizations are those in which information flows, particularly lateral flows, are facilitated, expertise replaces position power as the basis on which input is evaluated, and decision authority is decentralized. Through the diversity implicit in such organizational arrangements, it would appear that an organic form would create enhanced opportunities for innovation. A number of researchers on innovation, including Kanter (1983, 1988) and Zaltman, Duncan, and Holbek (1973), have noted that learning and adaptation to environmental shocks are more effective where organizations are structured so as to diffuse decision influence. On the other hand, even this may not be quite straightforward. Kimberly (1981), for instance, concluded that such effects may occur only where environments are relatively turbulent; in stable and predictable environments, some degree of formalization and centralization of decisionmaking may actually enhance the organization's ability to implement innovations. Hence, in suggesting that organic forms foster organizational innovation, we hedge a bit and specify that this occurs specifically in unstable environments (which is perhaps the "normal" state of most organizations these days).

> *Proposition 13:* Particularly under conditions of environmental change and uncertainty, the level of organizational innovation will be higher in organic organizations than in mechanistic organizations.

Thus far, the effects of human information processing on organizational innovation have been described essentially in terms of individual psychology. The situation becomes further complicated by the social nature of organizations and their members.

PEOPLE ARE SOCIAL ANIMALS AND ORGANIZATIONS ARE SOCIAL SYSTEMS

As has been indicated, the perceived organizational environment may be an incomplete and distorted representation of reality, because of limitations on individuals' ability to manage complexity. The issue is no less relevant for collectivities. The choices people make when deciding what to filter out of complex stimulus fields, and what meaning to impose on ambiguous events, are only rarely made in a social vacuum. People depend on others for ambiguity reduction, so that the process becomes largely one of the *social* construction of reality (Berger and Luckmann 1966).

Under circumstances of complexity or ambiguity, people in organizations often depend on others for the interpretation of information. This social construction of reality is one function of the informal work group, which is a ubiquitous phenomenon in organizations. Whether informal or officially constituted, cohesive groups tend to homogenize the frames of reference of their members. It has been ob-

served, for example, that no matter how heterogeneous a research and development team is on its formation, the team will become homogeneous by the time it has been together for as long as three years (Pelz and Andrews 1966).

The literature on creativity in problem-solving groups suggests that too much homogeneity is counterproductive to the generation of new ideas (Maier 1967). Perhaps innovation teams should not be allowed to remain intact for very long to avoid becoming stagnant through loss of individuality. Janis (1983) suggested, for example, that tendencies toward "groupthink" could be countered by changing the membership of long-standing, cohesive groups or at least inviting in outsiders from time to time.

Yet any task team has to go through a series of growth and development stages before becoming effective as an integrated system (Tuckman 1965). Too much disruption might lead to a group that never reaches the stage of development that permits effective unit performance. This suggests that there is a tradeoff to be made between stability of membership, on the one hand, and turnover in the group's membership for the re-infusion of novel approaches, on the other. As Staw (1980) and others have suggested, however, it is not so much the amount of turnover but the *quality* of turnover that determines its effects on the organization. Organizational renewal depends on the importation of new people with new (and appropriate) ideas not accessible through the continuing contributions of "old" members.

Taking all this into consideration at once, it appears that too little turnover leads to organizational stagnation, while too much leads to loss of continuity or direction. In addition, it is important that the replacement of members result in a net increase in the likelihood of new, appropriate ideas.

> *Proposition 14:* Assuming that newcomers bring new, useful ideas to the group, in-

154

novation effectiveness will be positively associated with a moderate amount of turnover in the innovation team. Too little turnover, like too much turnover, will be dysfunctional.

Social Influences on Behavior and Belief

Groups leave an indelible fingerprint on their members. At the very least, they affect members' level of motivation, their affective state, the information they possess, their skill development, and the approaches they use to perform tasks or solve problems (Hackman 1976). Moreover, groups can have a striking impact on members' very belief systems (see, for example, Asch 1951).

There are at least three ways in which a group can affect the attitudes and beliefs of its members (Salancik and Pfeffer 1978). Very directly, colleagues can render statements about the environment that persuade the recipient of the message. Especially where these messages appear to represent consensus within the group, they can be fairly powerful. A second way that social influence processes can affect reality construction is by calling an individual's attention to certain events (which results in the target person's disregarding other events because of short-term memory limits). Finally, the group can impose its interpretation of the "true" meaning of inherently ambiguous events on group members. This is most likely under conditions of stress or uncertainty or ambiguity. People appear particularly motivated to validate those opinions for which there are few objective criteria by comparing them with the opinions of others (Festinger 1954). Thus, much of what people know, or believe they know, is derived through processes of social influence.

There is yet another sense in which social processes underlie learning. Although people often learn what behaviors lead to what outcomes through their direct experiences, they also can learn through observing the experiences of others—that is, vicariously. According to social learning theory (Bandura 1977; Davis and Luthans 1980), vicarious learning can often more readily account for the acquisition of complex patterns of social behavior than can an explanation based on individual outcomes.

The mentoring process in organizations has come in for a great deal of psychologists' attention of late (Kram 1985). A major aspect of mentoring is the modeling of appropriate (that is, effective) behaviors by more experienced organization members, for the benefit of the less experienced. Modeling has been shown to be a particularly powerful educational device under conditions of ambiguity and complexity (Manz and Sims 1981; McGehee and Tullar 1978; Goldstein and Sorcher 1972). It follows that the availability of competent mentors may be particularly important to the process of innovation. In Chapter 8, Van de Ven, Venkataraman, Polley and Garud explicate the multiple roles enacted by managers at various levels of an innovation, not the least of which is the mentor role.

Effective vicarious learning may depend in part on the process of interpersonal identification (Kagan 1958; Sanford 1955). When positive or negative experiences are observed to occur to someone with whom one has identified, it is almost as if they have occurred to the observer. This can be a very efficient process, when compared to other means of changing people's complex behaviors. People learn from experience, but that experience need not be exclusively their own. Organizations are complex social networks and provide many possibilities for social learning.

This has strong implications for innovative behavior in organizations. The most effective behavior models are those persons with whom others are most apt to identify. Although there are several types of personal attributes that increase the likelihood of identification, personal prestige is clearly one of the major ones. If an organization's culture venerates heroes on the basis of their innovative behavior, then others may follow in their footsteps. When new innovators emerge in the organization as a result of having learned their innovative behaviors through modeling, there is an escalation of the general level of innovation in an organization. Whether or not this happens, however, may depend on how well people's attention is managed—that is, how salient the innovators' behavior is made so that others notice it. In particular, innovation role models are likely to engender identive imitation to the extent that organizations *visibly* value their contributions through recognition and reward.

> *Proposition 15:* The level of innovative activity in an organization is positively associated with the availability of innovation role models or mentors who enjoy high status in the organization and who are appropriately rewarded for their innovative contributions.

Another aspect of social motivation that has implications for innovation is role enactment. The role metaphor, borrowed by social science from the theater, conveys the notion that people's conduct tends to adhere to certain socially transmitted expectations for appropriate behavior. As social animals, people in organizations clearly take on particular roles simply because it is expected of them by others—that is, their "role set" (Katz and Kahn 1978). In an earlier section it was asserted that behavior is a function of its consequences. In view of people's apparent readiness to acquiesce to the role demands of others, a similar statement of functional relationship can now be added: *Behavior is a function of expectations.*

The implications for innovation are quite straightforward. If an organization expects its members to behave innovatively and makes those expectations clear, it is likely that this alone will exert some influence toward raising the level of innovative behavior in the organization. People may adopt innovative organizational roles if their role set so demands (Graen 1976; Katz and Kahn 1978).

The Psychological Contract and Self-Fulfilling Prophecies

Two additional concepts are also relevant to people behaving in accordance with their received expectations: (1) the so-called Pygmalion effect and (2) the psychological contract. The Pygmalion effect (Livingston 1969) derives indirectly from the work of Merton (1948) on self-fulfilling prophecies and more directly from the research of Rosenthal (1966), who discovered that classroom teachers' expectations regarding the intelligence of their students resulted in actual differences in students' "intelligent" behaviors. The term was adopted from Shaw's play *Pygmalion,* and suggests that if an organization expects its members to innovate, they are apt to do so in response to those expectations. On the other hand, if the organization's expectations regarding its members' innovative behavior are low, then that too will come to pass as a self-fulfilling prophecy. Recently, Kanter (1988) provided a number of examples of how organizations can make it clear to their members that they *expect* innovation.

The notion of psychological contracts (Argyris 1960; Kotter 1973; Levinson, Price, Munden, Mandl, and Solley 1962; Schein 1980) stems from a separate research stream but appears equally relevant. Ring and Van de Ven suggest in Chapter 6 that although people in organizations sometimes contract formally with one another, as in the case of labor-management contracts, much of the "contracting" that takes place between organizations and their members is implicit—a "series of mutual expectations of which the parties to the relationship may not themselves be even dimly aware but which govern their relationship to each other" (Levinson et al. 1962: 21). It has been argued that psychological contracts, being inherently vague, can be more powerful in shaping behavior than more formal exchange arrangements—that is, it is more difficult to make a clear decision that a given behavior is *outside* the agreement (Barnard 1938; Simon 1976, on zones of indifference or zones of acceptance). If the organization expects its members to exhibit innovative behaviors as part of the psychological contract, *and if members' psychological contracts agree,* the probability of spontaneous, innovative behaviors should rise substantially, as compared with situations in which innovation is not even an implicit part of the bargain.

Despite the tacit nature of these "contracts," it is important that the organization communicate such expectations in some reasonable manner. Relying on the imaginations of the members is not likely to produce consistent results.

Proposition 16: The level of innovative activity in an organization is higher in organizations where there is a consensus between organization and member that spontaneous, innovative behaviors are a legitimate part of the psychological contract.

The Effects of Multiple Commitments

Psychological contracts imply some level of commitment on the part of the parties (member and organization) to meet the other's expectation. In contracting parlance, the term "commitment" is often used to indicate that, at some point in the agreement process, the parties bind themselves to carry out the letter and spirit of the agreement (see Chapter 10). This term, "commitment," has become quite popular in the study of organizations, too, although it has been applied with rather different meanings.

Perhaps the earliest use of "commit-

ment" was to signify that a person, by her or his own behaviors, had become constrained to behave in certain ways in the future. Becker (1960), for example, argued that people commit to consistent future courses of action as a result of investments or "side bets" they have made. Similarly, a number of social psychologists have been concerned with the way in which public, explicit, irrevocable, and voluntary behaviors (Kiesler 1971; Salancik 1977) can lead to a locking-in process such that the person's future behaviors and beliefs become bound to earlier behaviors. As implied by this description, this is often referred to as "behavioral" commitment (Staw 1976). Recently, there has been a flurry of research on the escalation of behavioral commitment to a decision or to a chosen course of action (Bazerman, Giuliano, and Appleman 1984; Northcraft and Wolf 1984; Staw 1976, 1981; Staw and Ross 1987). Although behavioral commitment may have some implications for innovation (individuals' early innovative acts may commit them to perform future innovative acts), it is to another type of commitment altogether that the discussion now turns.

In the "organizational behavior" usage of the term (Staw 1976), commitment is really a form of psychological attachment a person forms with some social system, such as an employing organization (Buchanan 1974; Mowday, Porter, and Steers 1982). It is often described as a syndrome in which the individual identifies with the organization, develops a strong membership bond, and internalizes the organization's values and goals (Mowday et al. 1982). Implicit is the likelihood that the member will work extra hard to meet the organization's goals, even when there is no obvious quid pro quo for the member. In a manner of speaking, Thorndike's (1913) law of effect is overridden by organizational commitment; people behave in a way that is not economically rational (at least from the individual's point of view) and exert considerable effort on the organization's behalf, even without any expectation of a pay-off.

Recall Katz's (1964) motivational requirements for organizations. A committed member would be expected to join and stay—a strong attachment to membership is part of commitment, by definition. The member would also be expected to work hard to meet known goals and objectives—that is, to carry out one's role prescriptions diligently. But what about the "spontaneous, innovative behaviors" specified in Katz's third motivational requirement? Would a committed individual be expected to be more spontaneous or more innovative? The answer probably depends on the specific form the commitment takes.

Attachment to an organization, like other forms of work attachments, can either be expressive or instrumental. Angle and Perry (1981) provided evidence that people form at least two different types of attachments to their organizations. They designated these attachments "value commitment" (a sense of identification with the organization and internalization of its values) and "commitment to stay" (a resolve to remain a member of the organization). The research showed that commitment to remain a member of an organization does not necessarily imply a resolve to perform to a high level. If members are committed first and foremost to remaining in the system, rather than taking risks to innovate, they may instead become risk averters rather than innovators. Such persons would tend to behave in ways that don't "rock the boat," as a first priority. In Angle and Perry's study, members' value commitment was more strongly associated with organizations' economic performance and adaptability; commitment to stay was more strongly associated with organizations' low rates of turnover.

The foregoing discussion of implicit obligations and commitment raises a number of issues relevant to people and organizational innovation. Walton (1985) contrasted the "control strategy" that traditional organizations have used to manage their members' performance, with an alternative he termed the "commitment strategy." The creation of high-commitment systems was suggested as a way to get "stretch" performance from organization members (and, by implication, a greater likelihood of spontaneous, innovative behavior).

In discussions with executives that eventually led to the initiation of the Minnesota Innovation Research Program, a recurring concern was expressed: "How can we increase commitment and thereby enhance our organization's level of innovation?" There seemed to be a shared, implicit theory that high commitment and high innovation go hand in hand. As we review the foregoing discussion of types of commitment, however, the picture appears less intuitively clear. Might not the particular form that commitment takes be important in determining its effect on innovation?

> *Proposition 17:* The level of innovative activity in an organization is positively related to members' value commitment but is not related systematically to these members' commitment to stay.

Not only do people become committed in different ways to a given organization, they may also become committed to different anchor points altogether (Morrow 1983; Reichers 1985). For example, members may become committed (more accurately, psychologically attached) to the entire organization, or they may instead become committed to some subgroup. If the goals of the subgroup are completely congruent with those of the organization, it may not matter which attachment is formed. On the other hand, if the values and goals of the group are antithetical to those of the larger organization, then it matters indeed where loyalty lies. This observation leads the discussion to a more explicit treatment of individuals as members of groups—more specifically, as members of *innovation teams.* Thus, in organizational innovation, the small group—the innovation team—becomes an important focus for analysis.

What type of commitment in an innovation team would be most conducive to high levels of innovation within the team? On the one hand, members might become committed to the team, per se. On the other hand, they might be more committed to carrying forward the innovation that the team is working on. Although these are not mutually exclusive options, it is interesting to contrast two extreme situations. The former example would be one of attachment to the social system; the latter, an example of commitment to a course of action. In these two alternatives, the means and ends change places. If one's commitment is to the innovation itself, the team is merely a means to that end. If, on the other hand, one's primary attachment is to a cohesive innovation team, then the innovation is in effect only a justification for participation in the system. Furthermore, it is not difficult to imagine circumstances under which commitment to the team leads to conservative rather than innovative behaviors. Commitment is clearly a two-edged sword when applied to the dynamics of innovation teams.

It would be expected that the norms and climate of the innovation team will affect whether or not commitment to the group leads to greater innovativeness. Most obvious, perhaps, would be the ways in which conflict is dealt with. If conflict is either avoided or resolved through power plays, innovativeness will probably suffer. Having too little conflict is probably not indicative of a healthy climate for innovation. Some minimal level of disagreement is necessary to stimulate the search for new directions. Complacency can lead to premature closure and suboptimal problem solving (Maier 1970). A healthier group climate would be one in which a reasonable amount of conflict is present and one in which conflict is resolved by direct confrontation and negotiation.

Herein may lie one of the dilemmas of organizational innovation. An innovation team is often intentionally isolated from the rest of the organization in order to create a special environment to foster innovation, outside the bureaucratic milieu of the larger organization. This has been referred to as the "skunk works" approach (Peters and Waterman 1982). In some instances, the members of the innovation team are intentionally made to feel like organizational outsiders, in order to intensify their commitment to the group (Kidder 1981).

It is well known that nothing solidifies a small social system like a feeling of being isolated or under siege, yet this ethnocentric solidarity may come at great cost to the team members' relationship with the organization as a whole. As Schroeder et al. show in Chapter 4, as an innovation develops over time, the "old" and the "new" need to become linked in order for implementation to occur. Balanced commitments to the innovation unit and the organization should increase the likelihood of innovation success. Group cohesiveness can either foster or thwart organizational effectiveness, depending on how well integrated the goals of the group are with those of the larger organization. Indeed, resolving the potential clashes between commitments to the subunit and commitments to the organization constitutes one of the classic dilemmas facing management. Because of the intensity of people's engagement and the inherent discontinuity between the innovation and the established order, resolution of this dilemma assumes particular importance in the case of innovation.

Proposition 18: Innovation success is positively related to management's ability to balance innovation team members' commitment to the innovation and to the larger organization.

As mentioned in a preceding section, Janis (1971) described a potential dysfunction he called "groupthink" that often occurs in small, cohesive groups. Group members' creative efforts become subordinate to preserving that "cozy 'we' feeling" (1971: 71). Rather than becoming committed to innovative performance, the members may instead become committed to the status quo. Nystrom (1979) agreed in part but also indicated that members of cohesive groups may actually be more innovative because of the psychological safety they feel within the group. This is an intuitively appealing view that indicates that it is not the cohesiveness per se but the norms for dealing with disagreement within the group that influence the level of innovation.

An organization's climate for innovation is perhaps less tangible than its systems and structures but is no less vital. Organizational subsystems also have climates (Powell and Butterfield 1978). The climate within an innovation team may be the most salient aspect of organizational climate for the members. Siegel and Kaemmerer (1978) developed a measure of innovative climate that served as the basis for several of the questionnaire items included in the Minnesota Innovation Survey. The relationship of these innovation-climate measures to perceived innovation effectiveness provided few surprises. Trust was positively related to innovation effectiveness ($r = .28$; $p < .001$), as were leaders' encouragement of members to take the initiative ($r = .21$; $p < .01$) and to express doubts ($r = .24$; $p < .001$). Negatively related to perceived innovation effectiveness was pressure "not to rock the boat" by speaking one's mind ($r = -.31$; $p < .001$). Finally, innovation effectiveness was positively related to the likelihood that innovation teams reexamine their basic assumptions when they hit a snag ($r = .20$; $p < .01$).

Another important aspect of innovation teams' climates seemed to be the norms for dealing with the inevitable conflicts that arise in a collective undertaking. Innovation team members were asked how often various methods were used to handle disagreements or disputes which arose. There was a negative relationship between "ignoring or avoiding the issues" and innovation effectiveness ($r = -.30$; $p < .001$), while there was a positive relationship between innovation effectiveness and "bringing the issues into the open and working them out ($r = .32$; $p < .001$). Taken as a whole, these climate indicators suggest that an "enabling" climate is one in which there is a high level of trust among members of the innovation effort and in which important issues are subject to renegotiation in an open, assertive manner.

> *Proposition 19:* Innovation effectiveness is positively associated with group cohesiveness, provided that an open, confrontive climate for conflict resolution exists within the innovation team. Absent such a climate, cohesiveness is negatively related to the level of innovation in the team.

DYNAMIC INTERACTIONS AMONG PEOPLE

To review the foregoing, we have moved in our discussion from the psychological to the social-psychological. In the former context, beginning with the factors that bear on innovative performance we moved to issues of cognition and then to individual creativity. It was not possible to stop at the individual level, however, because of the inherently social-psychological nature of cooperative efforts in organizations. Accordingly, we have offered a set of propositions dealing with such issues as role enactment and conflict management.

In the same respect that it is necessary to break out of the limitations of individual psychology and consider the psychology of human interaction, it seems equally necessary to move from the static perspective to a more dynamic approach to the interactions among people, their behavior, and their contexts. As one innovation team leader informed us rather forcibly, creation may well be a static event, but innovation is clearly a dynamic process. Nicholson (1987) pointed out that the stable equilibria that seem to be observed in organizations are really only snapshots of a continual flow of transformation and transaction. Change, as he puts it, is a constant. We agree completely and conclude with a plea for a more process research perspective on the psychology of organizational innovation.

Mobilizing and directing an innovation team significantly complicates the innovation management task. Contrary to the view sometimes implicit in the literature that innovation consists of an entrepreneur who works with a fixed set of full-time people to develop a new idea, a more descriptive of the process appears to be Cohen, March, and Olson's (1972) garbage can model, where many stakeholders fluidly engage and disengage in the innovation process over time as their needs for inclusion, identity, and facticity dictate (Turner 1987).

As we continue our study of organizational innovations, it becomes increasingly clear that we need to devote more attention to the ways that participation in the innovation process changes people over time. Future psychological research on organizational innovation would be well advised to consider the differences in the reciprocal influences among persons, behaviors, and environments (Bandura 1977) that occur in at least three distinct temporal phases of an innovation—that is, the beginning, the middle, and the culmination. Our real-time observations of innovation teams suggest that future research should be sensitive to the following dynamics in these three time periods, as described by Van de Ven (1985):

1. *Startup.* Here we observe dynamics of individual recruitment and engagement in an innovation team and the problems of the "hung jury," "acquiescent team player," and "tolerance for closure and trust." Growing pains similar to Tuckman's (1965) sequence of "forming, storming, and norming" are encountered. The processes of interpersonal attraction, subgroup formation, role-taking and specialization, and leader emergence and legitimacy are critical processes during this period. In the MIRP we have often observed the startup period to include an emotional euphoria, great expectations, and confidence among participants in the success of the innovative undertaking. Startup is an exciting time and the emotions that are associated with getting an innovation off the ground could provide an interesting anchor-point for psychologists to use as a "handle."

2. *Middle period.* Here, "the bloom is off the rose" for many people on the innovation team. Euphoria wanes, problems surface and are not solved, expectations are not met. Emotions are again a potentially exciting area for study, but here the emotions experienced are likely to be stress and frustration. Reality shock may set in—that is, the reality of the difficulty, complexity, and high risk of success becomes apparent, and clear solutions are not yet in sight to provide closure to the innovation effort.

In this middle phase, symptoms of lack of trust or confidence in fellow workers and leadership are most likely to surface. Some people may desert the effort, which creates problems of continuity. New people come on board without the same "organizational memory" as the people they have replaced, and this discontinuity causes problems for organizational learning. MIRP research reveals that remarkably little learning takes place in the middle period.

161

Friction, dysfunctional subgrouping, and the like become potential problems. Team members "label" one another as someone who is core or peripheral to the group or who is to be listened to or dismissed for his or her contributions. Team progress can easily grind to a halt. One MIRP study observed a mutiny and rejection of the leader during this phase. Although somewhat more graceful, replacements of the innovation manager also were observed in other cases.

External interventions by management or investors to rectify the situation often occur but frequently contribute to an increasingly complex and worsening vicious cycle. As performance deteriorates, outside intervention further curtails freedom and produces even worse performance.

3. *Ending period.* The innovation or at least some major aspect of it terminates. This provides a set of cues that trigger evaluation of the experience. Unmet expectations may emerge as the endpoint comes into view for the group and team members develop some sense of closure.

If the intended outcome is not achieved, members concoct reasons for why their ordeal was not in vain in order to bolster their egos. Or participants may accept their failure by vowing to "make it the next time." Attributions of the causes of such failure are probably subject to some well-known biases (Mitchell and Green 1983), which probably cause the participants to externalize blame to outside, "uncontrollable factors."

These well-known dynamics cause more problems for organizational learning in cases of innovation failure. Real learning, as opposed to rationalization, probably will not take place before an extended period of time has elapsed. If, on the other hand, the intended outcomes

are achieved, or if the innovation has been dubbed a "success" by outsiders, team celebrations occur. As opposed to the "external" attributions of the causes of failure, biases in the attribution process tend toward overattributing to internal causes. Success is seen as the proximate result of the commitment, talent, and heroic efforts of the team; the leader is accorded an almost superhuman quality. Association with such a triumph, and the team that "won," brings about a reluctance to see it all end. A variety of efforts to prevent or forestall termination of the group may be seen at this time.

Although they have been alluded to in the literatures on policy implementation (Pressman and Wildavsky 1974) and organizational development (Schein 1969), adequate recognition or treatment of these social-psychological, and temporal, dynamics of group development has not often been found in prior innovation research. Perhaps this omission has resulted because these dynamics become apparent only in the context of real-time, longitudinal field research. Based on such longitudinal observations in the Minnesota Innovation Research Program, the following additional proposition is offered:

Proposition 20: As people engage in an innovation over time, prevalent affective states tend to change—that is, from *euphoria* at first to *pain* as the innovation develops. Post hoc evaluations of the experience are biased toward the positive, which interferes with organizational learning.

In the MIRP research that has been conducted thus far, these temporal dynamics have been observed to represent some of the most intense experiences for innovation participants and managers. They will continue to be carefully documented and examined as observations of the MIRP innovations continue.

It is readily apparent that a complete psychology of organizational innovation needs, inter alia, to address not only motivation, information processing, and creativity. Equally important are the implementation problems of obtaining commitment, acceptance, and compliance through persuasion, bargaining, incentives, and power relations with others both within and outside the innovation unit. Moreover, these issues need to be considered as processes, rather than simply as static relations among sets of variables. A good process theory of innovation management needs to incorporate a strong concern for the ways in which people interact with one another and their contexts, over time, and how both the people and contexts are changed as organizational innovations unfold.

CONCLUDING DISCUSSION

People who study communication processes in organizations are prone to remark that the reason there is so little known about communication is because the topic is so important. Communication is everywhere, and so it is sometimes difficult to "see" because of its very ubiquity (it probably wasn't a fish who first noticed water!). By analogy, much the same can be said about the importance of people in organizational innovation. This chapter has attempted to provide an organizing framework for our consideration of the "people" dimension of the Minnesota Innovation Research Framework. The chapter began with the premise that the management of innovation is the management of people. Using two compatible theoretical frameworks, interactionism and social learning theory, it was contended that innovative performance is a complex function of what people are, as well as what they do and what they experience. Further, it was asserted that organizations need to establish conditions that *enable* innovation, just as they must provide the proper *motivation* for innovation. Neither is sufficient in itself.

These premises, combined with a number of specific observations from research conducted within MIRP, suggested a series of propositions. Table 5–1 recapitulates those propositions, for purposes of review, highlighting the essentially social nature of innovation. Although some aspects of the process (such as discovery or invention) might be an individual undertaking, the complex, interdependent nature of organizations makes it certain that collective action will be required as well.

Many of the propositions deal with interactions between the personal attributes of people and the contexts within which they behave. Proposition 1 suggests that what people bring to the situation and what they encounter there are co-equally important to organizational innovation. Proposition 2 subdivides the contextual side of the preceding proposition into environmental factors that (1) motivate or demotivate or (2) enable or impede. Several propositions deal specifically with contextual influences on innovation. Proposition 13 holds that when the surrounding context is highly uncertain, organic organizational forms are most efficacious. Propositions 4 through 7, in combination, indicate that organizations need to strike careful balances between intrinsic and extrinsic motivation strategies and between individually based and group-based reward and incentive schemes.

Other propositions also speak to the need for balance between extremes, as well as the likelihood that functional relationships are not monotonic. Proposition 14 asserts that the impact of turnover on organizational innovation is probably nonlinear. Proposition 9 also raises the issue of curvilinear relations between innovation and its antecedents, suggesting that the functional relationship between environ-

TABLE 5–1.   *Research Propositions on Psychology and Innovation.*

Proposition 1: Organizational innovation is a joint function of members' personal attributes and the context for innovation in their organization.

Proposition 2: Organizational innovation occurs in organizations that provide a context that contains both *enabling* and *motivating* conditions for innovation; innovation will not occur where either factor is missing.

Proposition 3: Organizational innovation is a function of the intrinsic and extrinsic rewards associated with innovative behavior.

Proposition 4: Financial compensation, per se, is a relatively weak extrinsic reward for innovation. However, to the extent that it provides an effective means for organizations to express recognition, pay can help motivate future innovative behavior.

Proposition 5: The more individualized the reward system, the more effective will be the idea-generation process; the more collectivist the reward system, the more effective will be the innovation implementation process.

Proposition 6: Individualized reward systems are better suited to motivation for radical innovation; group-based reward systems are more effective in motivating for incremental innovations.

Proposition 7: Extrinsic motivation is necessary but not sufficient for evoking spontaneous, innovative behaviors; it is at least equally necessary for organizations to create and sustain conditions that facilitate intrinsic motivation. Above all, a suitable *balance* needs to be struck between intrinsic and extrinsic motivation.

Proposition 8: Innovation effectiveness is positively associated with frequency of communication among persons having dissimilar frames of reference.

Proposition 9: Innovation effectiveness is associated with a moderate amount of environmental uncertainty. At extremely high or low levels of uncertainty, innovativeness is reduced.

Proposition 10: Competition with peer units in the organization for scarce resources is associated with reduced innovation effectiveness.

Proposition 11: The level of innovation in an organization is positively associated with the existence of mechanisms for focusing members' attention on changing conditions.

Proposition 12: Innovation effectiveness is positively related to the extent to which the organization is able to integrate creative personalities into the organizational mainstream.

Proposition 13: Particularly under conditions of environmental change and uncertainty, the level of organizational innovation is higher in organic organizations than in mechanistic organizations.

Proposition 14: Assuming that newcomers bring new, useful ideas to the group, innovation effectiveness is positively associated with a moderate amount of turnover in the innovation team. Too little turnover, like too much turnover, is dysfunctional.

Proposition 15: The level of innovative activity in an organization is positively associated with the availability of innovation role models or mentors who enjoy high status in the organization and who are appropriately rewarded for their innovative contributions.

Proposition 16: The level of innovative activity in an organization is higher in organizations where there is a consensus between organization and member that spontaneous, innovative behaviors are a legitimate part of the psychological contract.

TABLE 5–1. (continued)

Proposition 17: The level of innovative activity in an organization is positively related to members' value commitment, but is not related systematically to these members' commitment to stay.

Proposition 18: Innovation success is positively related to management's ability to balance innovation team members' commitment to the innovation and to the larger organization.

Proposition 19: Innovation effectiveness is positively associated with group cohesiveness, provided that an open, confrontive climate for conflict resolution exists within the innovation team. Absent such a climate, cohesiveness is negatively related to the level of innovation in the team.

Proposition 20: As people engage in an innovation over time, prevalent affective states tend to change; i.e., from *euphoria* at first, to *pain* as the innovation develops. Post hoc evaluations of the experience are biased toward the positive, which interferes with organizational learning.

mental uncertainty and innovation may be shaped like an inverted U. Finally, Proposition 18 holds that a balance needs to be struck between commitment to the innovation or innovation team and commitment to the larger system.

A number of propositions highlight the importance of diversity. Proposition 8 emphasizes the need for interactions among unlike people. Proposition 12 recognizes that the integration of such diversity is a key task, as organizations need to integrate especially creative persons into the mainstream. Cohesiveness tends to counter diversity, and although cohesiveness can be a motivating force, it may be counterproductive unless conflicts are resolved in an open manner that legitimizes diversity (Proposition 19).

The management of attention and of expectations also appears critical. Mechanisms for focusing people's attention on change are needed (Proposition 11), as well as clear psychological contracts in which people are aware that innovation is part of their obligation toward the organization (Proposition 16).

As stated in the final proposition, perspectives and affective reactions do not remain stable over the entire course of the innovation. Moreover, the inherent complexity of the innovation process makes it difficult to reach unambiguous conclusions as to results. The sense

making that takes place may not always be beneficial and can actually degrade the learning process (see Chapter 7 on superstitious learning).

To summarize, what people bring to the situation in terms of their creative talents is not insignificant in setting the stage for the innovation process, but more than personal creativity is required to sustain a high level of organizational innovation. Specifically, organizations need to create and sustain conditions so that people *want to* innovate and so that people *can* innovate. Organizations that neglect either aspect place their innovative capacity at risk.

REFERENCES

Abelson, R.P. 1981. "Psychological Status of the Script Concept." *American Psychologist* 36: 715–29.

Adams, J.S. 1965. "Inequity in Social Exchange." In L. Berkowitz, ed., *Advances in Experimental Social Psychology*, vol. 2. New York: Academic.

Amabile, T.M. 1983. *The Social Psychology of Creativity*. New York: Springer-Verlag.

———. 1988. "A Model of Creativity and Innovation in Organizations." In B.M. Staw and L.L. Cummings, eds., *Research in Organizational Behavior*, Vol. 10. Greenwich, Conn.: JAI.

Angle, H.L., and J.L. Perry. 1981. "An Empirical

Assessment of Organization Commitment and Organizational Effectiveness." *Administrative Science Quarterly* 26: 1–14.

Aronson, E. 1973. "The Rationalizing Animal." *Psychology Today* 6 (12): 46–50, 119.

Argyris, C. 1960. *Understanding Organizational Behavior.* Homewood, Ill.: Dorsey.

Asch, S. 1951. "Effects of group pressure on the Modification and Distortion of Judgments." In H. Guetzkow, ed., *Groups, Leadership and Men.* Pittsburgh: Carnegie.

Bandura, A. 1977. *Social Learning Theory.* Englewood Cliffs, N.J.: Prentice-Hall.

Barnard, C.I. 1938. *The Functions of the Executive.* Cambridge, Mass.: Harvard University.

Barron, F. 1969. *Creative Person and Creative Process.* New York: Holt, Rinehart, and Winston.

Bazerman, M.H., T. Giuliano, and A. Appelman. 1984. "Escalation of Commitment in Individual and Group Decision Making." *Organizational Behavior and Human Performance* 33: 141–52.

Becker, F. 1981. *Workspace: Creating Environments in Organizations.* New York: Praeger.

Becker, H.S. 1960. "Notes on the Concept of Commitment." *American Journal of Sociology* 66: 32–40.

Bem, D.J. 1972. "Self-Perception Theory." In L. Berkowitz, ed., *Advances in Experimental Social Psychology,* vol. 6. New York: Academic.

———. 1978. "Self-Perception Theory." In L. Berkowitz, ed., *Cognitive Theories in Social Psychology.* New York: Academic.

Berger, P.L., and T. Luckmann. 1966. *The Social Construction of Reality.* Garden City, N.Y.: Anchor.

Bouchard, T.J. 1971. "What Ever Happened to Brainstorming?" *Industry Week* (Aug. 2): 26–27.

Bruner, J.S. 1973. *Beyond the Information Given: Studies in the Psychology of Knowing.* New York: Norton.

Buchanan, B., II. 1974. "Building Organizational Commitment: The Socialization of Managers in Work Organizations." *Administrative Science Quarterly* 19: 533–46.

Burns, T., and G.M. Stalker. 1961. *The Management of Innovation.* London: Tavistock.

Campbell, D.T. 1960. "Blind Variation and Selective Retention in Creative Thought Processes." *The Psychological Review* 67: 380–400.

Campbell, J.P., M.D. Dunnette, E.E. Lawler, and K. Weick. 1970. *Managerial Behavior, Performance and Effectiveness.* New York: McGraw-Hill.

Cohen, M.D., J.G. March, and J.P. Olson. 1972. "A Garbage Can Model of Organizational Choice." *Administrative Science Quarterly* 17: 1–25.

Cummings. L. 1965. "Organizational Climates for Creativity." *Academy of Management Journal* 8: 220–27.

Davis, T.R.V., and F. Luthans. 1980. "A Social Learning Approach to Organizational Behavior." *Academy of Management Review* 5, 281–90.

DeBono, E. 1972. *Po: A Device for Successful Thinking.* New York: Simon and Schuster.

deCharms, R. 1968. *Personal Causation: The Internal Affective Determinants of Behavior.* New York: Academic.

Deci, E.L. 1971. "Effects of Externally Mediated Rewards on Intrinsic Motivation." *Journal of Personality and Social Psychology* 18: 105–15.

———. 1972. "Intrinsic Motivation, Extrinsic Reinforcement and Inequity." *Journal of Personality and Social Psychology* 22: 113–20.

Dunnette, M.D., and J.P. Campbell. 1968. "Laboratory Education: Impact on People and Organizations and a Response to Argyris." *Industrial Relations* 8: 1–27.

Dyer, L., and D.F. Parker. 1975. "Classifying Outcomes in Work Motivation Research: An Examination of the Intrinsic-Extrinsic Dichotomy." *Journal of Applied Psychology* 29: 455–58.

Etzioni, A. 1975. *A Comparative Analysis of Complex Organizations.* New York: Free Press.

Festinger, L. 1954. "A Theory of Social Comparison Processes." *Human Relations* 7: 117–40.

———. 1957. *A Theory of Cognitive Dissonance.* Evanston, Ill.: Row, Peterson.

Goldstein, A.P., and M. Sorcher. 1972. "A Behavior Modeling Approach in Training." *Personnel Administration* 35 (2): 35–41.

Gordon, W.J. 1956. "Operational Approach to Creativity." *Harvard Business Review* 34 (6): 41–51.

———. 1961. *Synectics: The Development of Creative Capacity.* New York: Harper.

Graen, G. 1976. "Role-Making Processes within Complex Organizations." In M.D. Dunnette, ed., *Handbook of Industrial-Organizational Psychology.* Chicago: Rand-McNally

Greenberger, D.B., and S. Strasser. 1986. "Development and Application of a Model of Personal Control in Organizations." *Academy of Management Review* 11: 164–77.

Guilford, J.P. 1950. "Creativity." *American Psychologist* 5: 444–54.

———. 1975. "Creativity: A Quarter Century of Progress." In I.A. Taylor and J.W. Getzels, eds., *Perspectives on Creativity.* Chicago: Aldine.

Hackman, J.R. 1976. "Group Influences on Individuals." In M.D. Dunnette, ed., *Handbook of Industrial-Organizational Psychology.* Chicago: Rand-McNally.

Hackman, J.R., and G.R. Oldham. 1980. *Work Redesign.* Reading, Mass.: Addison-Wesley.

Hall, D.T., and K.E. Nougaim. 1968. "An Examination of Maslow's Need Hierarchy in the Organizational Setting." *Organizational Behavior and Human Performance* 3: 12–35.

Hannan, M.T., and J. Freeman. 1977. "The Population Ecology of Organizations" *American Journal of Sociology* 82: 929–64.

Heider, F. 1958. *The Psychology of Interpersonal Relations.* New York: Wiley.

Helson, H. 1959. "Adaptation Level Theory." In S. Koch, ed., *Psychology: A Study of a Science*, Vol. 1. New York: McGraw-Hill.

Herzberg, F. 1968. "One More Time: How Do You Motivate Employees?" *Harvard Business Review* 46 (1): 53–62.

Herzberg, F., B. Mausner, and B.B. Snyderman. 1959. *The Motivation to Work.* New York: Wiley.

Janis, I.L. 1971. "Groupthink." *Psychology Today* 5 (6): 43–46, 74–76.

———. 1983. *Groupthink: Psychological Studies of Policy Decisions and Fiascoes*, 2d ed. Boston: Houghton-Mifflin.

Jones, E.E., and K.E. Davis. 1965. From Acts to Dispositions: The Attribution Process in Person Perception." In L. Berkowitz, ed., *Advances in Experimental Social Psychology*, Vol. 2. New York: Academic.

Kagan, J. 1958. "The Concept of Identification." *The Psychological Review* 65: 296–305.

Kanter, R.M. 1983. *The Change Masters.* New York: Simon and Schuster.

———. 1988. "When a Thousand Flowers Bloom: Structural, Collective and Social Conditions for Innovation in Organization." In B.M. Staw and L.L. Cummings, eds., *Research in Organizational Behavior.* Greenwich, Conn.: JAI Press.

Katz, D. 1964. "The Motivational Basis of Organizational Behavior." *Behavioral Science* 9: 131–46.

Katz, D., and R.L. Kahn. 1978. *The Social Psychology of Organizations*, 2d ed. New York: Wiley.

Kelley, H.H. 1973. "The Process of Causal Attribution." *American Psychologist* 28: 107–28.

Kerr, S. 1975. "On the Folly of Rewarding A while Hoping for B." *Academy of Management Journal* 18: 769–83.

Kidder, T. 1981. *The Soul of a New Machine.* New York: Avon.

Kiesler, C.A. 1971. *The Psychology of Commitment: Experiments Linking Behavior to Belief.* New York: Academic.

Kimberly, J.R. 1981. "Managerial Innovation." In P. C. Nystrom and W.H. Starbuck, eds., *Handbook of Organizational Design.* Oxford: Oxford University.

Kohler, W. 1929. *Gestalt Psychology.* New York: Liveright.

Kotter, J. 1973. "The Psychological Contract: Managing the Joining-up Process." *California Management Review* 15 (3): 91–99.

Kram, K.E. 1985. *Mentoring at Work.* Glenview, Ill.: Scott, Foresman.

Langer, E. 1983. *The Psychology of Control.* Beverly Hills, Calif.: Sage.

Lawler, E.E., III. 1973. *Motivation in Work Organizations.* Monterey, Calif.: Brooks/Cole.

Lawler, E.E., III., and J.G. Rhode. 1976. *Information and Control in Organizations.* Santa Monica, Calif.: Goodyear.

Lawler, E.E., III., and J.L. Suttle. 1972. "A Causal Correlational Test of the Need Hierarchy Concept." *Organizational Behavior and Human Performance* 7 265–87.

Lawrence, P., and P. Dyer. 1983. *Renewing American Industry.* New York: Free Press.

Levinson, H., C.R. Price, K.J. Munden, H.J. Mandl, and C.M. Solley. 1962. *Men, Management and Mental Health.* Cambridge, Mass.: Harvard University.

Livingston, J.S. 1969. "Pygmalion in Management." *Harvard Business Review* (July-Aug.): 81–89.

Louis, M.R. 1980. "Surprise and Sense Making: What Newcomers Experience in Entering Unfamiliar Organizational Settings." *Administrative Science Quarterly* 24: 226–51.

Luthans, F., and R. Kreitner. 1985. *Organizational Behavior Modification and Beyond.* Glenview, Ill.: Scott, Foresman.

MacKinnon, D.W. 1960. "The Highly Effective Individual." *Teachers College Record* 61: 367–68.

———. 1975. "IPAR's Contribution to the Study of Creativity." In I. A. Taylor and J.W. Getzels, eds., *Perspectives on Creativity*. Chicago: Aldine.

Maier, N.R.F. 1967. "Assets and Liabilities in Group Problem Solving: The Need for an Integrative Function." *The Psychological Review* 74: 239–49.

———. 1970. *Problem Solving and Creativity in Individuals and Groups*. Belmont, Calif.: Wadsworth.

Manz, C.C., and H.P. Sims, Jr. 1981. "Vicarious Learning: The Influence of Modeling on Organizational Behavior." *Academy of Management Review* 6: 105–13.

March, J.G., and H.A. Simon. 1958. *Organizations*. New York: Wiley.

Maslow, A.H. 1954. *Motivation and Personality*. New York: Harper & Row.

McClelland, D.C. 1961. *The Achieving Society*. Princeton, N.J.: Van Nostrand.

McGehee, W., and W.L. Tullar. 1978. "A Note on Evaluating Behavior Modification and Behavior Modeling as Industrial Training Techniques." *Personnel Psychology* 31: 477–84.

McKelvey, B. 1982. *Organizational Systematics: Taxonomy, Evolution, Classification*. Berkeley: University of California.

Mednick, S.A. 1962. "The Associative Basis of the Creative Process." *The Psychological Review* 69: 220–32.

Merton, R.K. 1948. "The Self-Fulfilling Prophecy." *The Antioch Review* 8: 193–210.

Miller, G.A. 1956. "The Magical Number Seven, Plus or Minus Two: Some Limits on Our Capacity for Processing Information." *The Psychological Review* 63: 81–97.

Miller, J.G. 1960. "Information Input Overload and Psychopathology." *American Journal of Psychiatry* 116: 695–704.

Mitchell, T.R., and S.G. Green. 1983. "Leadership and Poor Performance: An Attributional Analysis." In R.M. Steers and L.W. Porter, eds., *Motivation and Work Behavior*, 3d ed. New York: McGraw-Hill.

Morrow, P.C. 1983. "Concept Redundancy in Organizational Research: The Case of Work Commitment." *Academy of Management Review* 8: 486–500.

Mowday, R.T., L.W. Porter, and R.M. Steers. 1982. *Employee-Organization Linkages: The Psychology of Commitment, Absenteeism, and Turnover*. New York: Academic.

Nicholson, N. 1987. "The Transition Cycle: A Conceptual Framework for the Analysis of Change and Human Resource Management." In J. Ferris and K.M. Rowland, eds., *Research in Personnel and Human Resource Management*. Greenwich, Conn.: JAI.

Northcraft, G.B., and G. Wolf. 1984. "Dollars, Sense and Sunk Costs: A Life Cycle Model of Resource Allocation Decisions." *Academy of Management Review* 9: 225–34.

Nystrom, H. 1979. *Creativity and Innovation*. London: Wiley.

Opsahl, R.L., and M.D. Dunnette. 1966. "The Role of Financial Compensation in Industrial Motivation." *Psychological Bulletin* 66: 94–118.

Osborn, A.F. 1953. *Applied Imagination*. New York: Scribner's.

Ouchi, W. 1981. *Theory Z: How American Business Can Meet the Japanese Challenge*. Reading, Mass.: Addison-Wesley.

Parnes, S.J. 1962. "Can Creativity Be Increased?" In S.J. Parnes and H.F. Harding, eds., *A Sourcebook for Creative Thinking*. New York: Scribner's.

Pelz, D.C., and F.M. Andrews. 1966. *Scientists in Organizations: Productive Climates for Research and Development*. New York: Wiley.

Perrow, C. 1967. "A Framework for the Comparative Analysis of Organizations." *American Sociological Review* 32: 194–208.

Peters, T.J., and R.H. Waterman, Jr. 1982. *In Search of Excellence: Lessons from America's Best-Run Companies*. New York: Harper & Row.

Porter, L.W., and E.E. Lawler. 1968. *Managerial Attitudes and Performance*. Homewood, Ill.: Irwin-Dorsey.

Porter, L.W., and R.E. Miles. 1973. Motivation and Management. In J.W. McGuire, ed., *Contemporary Management: Issues and Viewpoints*. Englewood Cliffs, N.J.: Prentice-Hall.

Powell, G.N., and D.A. Butterfield. 1978. "The Case for Subsystem Climates in Organizations." *Academy of Management Review* 3 (1): 151–57.

Pressman, J., and A. Wildavsky. 1974. *Implementation*. Berkeley: University of California.

Reichers, A.E. 1985. "A Review and Reconceptualization of Organizational Commitment." *Academy of Management Review* 10: 465–76.

Roberts, K.H., C.L. Hulin, and D.M. Rousseau. 1978. *Developing an Interdisciplinary Science of Organizations.* San Francisco: Jossey-Bass.

Rosenthal, R. 1966. *Experimenter Effects in Behavioral Research.* New York: Appleton-Century-Crofts.

Ross, L., M.R. Lepper, and M. Hubbard. 1975. "Perseverance in Self-Perception and Social Perception: Biased Attribution Processes in the Debriefing Paradigm." *Journal of Personality and Social Psychology* 32: 880–92.

Salancik, G.R. 1977. "Commitment and the Control of Organizational Behavior and Belief." In B.M. Staw and G.R. Salancik, eds., *New Directions in Organizational Behavior.* Chicago: St. Clair.

Salancik, G.R., and J. Pfeffer. 1978. "A Social Information Processing Approach to Job Attitudes and Task Design." *Administrative Science Quarterly* 23: 224–53.

Sanford, N. 1955. "The Dynamics of Identification." *The Psychological Review* 62: 106–18.

Schein, E.H. 1969. *Process Consultation: Its Role in Organizational Development.* Reading, Mass.: Addison-Wesley.

———. 1980. *Organizational Psychology,* 3d ed. Englewood Cliffs, N.J.: Prentice-Hall.

Schwab, D.P. 1980. "Construct Validity in Organizational Behavior." In L.L. Cummings and B.M. Staw, eds., *Research in Organizational Behavior.* Greenwich, Conn.: JAI.

Siegel, S.M., and W.F. Kaemmerer. 1978. "Measuring the Perceived Support for Innovation in Organizations." *Journal of Applied Psychology* 63: 553–62.

Simon, H.A. 1974. "How Big Is a Chunk?" *Science* 183: 428–88.

———. 1976. *Administrative Behavior,* 3d ed. New York: Free Press.

Skinner, B.F. 1938. *The Behavior of Organisms.* New York: Appleton-Century-Crofts.

Staw, B.M. 1976. "Knee-Deep in Big Muddy: A Study of Escalating Commitment to a Chosen Course of Action." *Organizational Behavior and Human Performance* 16: 27–44.

———. 1980. "The Consequences of Turnover." *Journal of Occupational Behavior* 1: 253–73.

———. 1981. "The Escalation of Commitment to a Course of Action." *Academy of Management Review* 6: 577–87.

Staw, B.M., and J. Ross. 1987. "Behavior in Escalation Situations: Antecedents, Prototypes and Solutions." In L.L. Cummings and B.M. Staw, eds., *Research in Organizational Behavior.* Greenwich, Conn.: JAI.

Steiner, I.D. 1972. *Group Process and Productivity.* New York: Academic.

Stotland, E., and L.K. Canon. 1972. *Social Psychology: A Cognitive Approach.* Philadelphia: Saunders.

Taylor, D.W., P.C. Berry, and C.H. Block. 1958. "Does Group Participation When Using Brainstorming Facilitate or Inhibit Creative Thinking?" *Administrative Science Quarterly* 3: 23–47.

Taylor, I.A. 1975. "A Retrospective View of Creativity Investigation." In I.A. Taylor and J.W. Getzels, eds., *Perspectives on Creativity.* Chicago: Aldine.

Thorndike, E.L. 1913. *The Psychology of Learning.* New York: Teachers College.

Torrance, E.P. 1967. "Give the "Devil" His Dues." In J.G. Gowan, G.P. Demos, and E.P. Torrance, eds., *Creativity: Its Educational Implications.* New York: Wiley.

Tuckman, B.W. 1965. "Developmental Sequences in Small Groups." *Psychological Bulletin* 63: 384–99.

Turner, J.H. 1987. "Toward a Sociological Theory of Motivation." *American Sociological Review* 52: 15–27.

Tversky, A., and D. Kahneman. 1974. "Judgment under Uncertainty: Heuristics and Biases. *Science* 185: 1124–31.

Utterbach, J. 1971. "The Process of Technological Innovation within the Firm." *Academy of Management Journal* 14: 75–88.

Van de Ven, A.H. 1985. "Spinning on Symbolism." *Journal of Management* 11: 101–02.

———. 1986. "Central Problems in the Management of Innovation." *Management Science* 32: 590–607.

Van de Ven, A.H., and A.L. Delbecq. 1971. "Nominal versus Interacting Group Process for Committee Decision-Making Effectiveness." *Academy of Management Journal* 14: 203–12.

Vroom, V.H. 1964. *Work and Motivation.* New York: Wiley.

Wahba, M.A., and L.G. Bridwell. 1976. "Maslow Reconsidered: A Review of Research on the Need Hierarchy Theory." *Organizational Behavior and Human Performance* 15: 212–40.

Walster, E.H., G.W. Walster, and E. Berscheid. 1978. *Equity: Theory and Research.* Boston: Allyn and Bacon.

Walton, R.E. 1985. "From Control to Commitment in the Workplace." *Harvard Business Review* 63 (2): 77–84.

Weick, K.E. 1979. *The Social Psychology of Organizing,* 2d ed. Reading, Mass.: Addison-Wesley.

West, M.A., J.L. Farr, and N. King. 1986. *Innovation at Work: Definitional and Theoretical Issues.* SAPU Memo No. 814. MRC/ESRC Social and Applied Psychology Unit, Department of Psychology, University of Sheffield, Sheffield, England.

Zaltman, G., R. Duncan, and J. Holbek. 1973. *Innovations and Organization.* New York: Wiley.

6

FORMAL AND INFORMAL DIMENSIONS OF TRANSACTIONS

Peter Smith Ring

Andrew H. Van de Ven

Innovation is not the isolated enterprise of a single entrepreneur. It is a collective enterprise that centers on a network of relationships that bind together people and their organizations in order to transform an abstract concept into reality. Thus, the management of innovation involves developing and maintaining a variety of cooperative relationships from inside and outside an organization.

The core concept used in the Minnesota Innovation Research Program to examine these relationships is a transaction. John R. Com-

We gratefully acknowledge useful comments and observations on propositions in this paper from Raghu Garud, Roger L. Hudson, Mary Knudson, Doug Polley, Gordon Rands, Michael Rappa, Roger Schroeder, Gary Scudder, and other colleagues involved in the Minnesota Innovation Research Program. This research program has been supported in part by a grant to the Strategic Management Research Center at the University of Minnesota from the Program on Organization Effectiveness, Office of Naval Research, under contract No. N00014-84-K- 0016, and by Grant No. 0350-3312-22 from the Graduate School of the University of Minnesota.

mons (1934, 1950), the originator of the concept, argued that transactions are the fundamental building block of social, economic, and legal relationships. Transactions are "deals" on the exchange of property rights between two or more parties within an institutional order. In managerial terms, these "deals" include a variety of legal and organizational arrangements such as joint ventures, licenses, franchises, or employment contracts. Because transactions entail a present commitment to achieve future expectations, they are a primary mechanism to actualize plans and forethought into action and performance. They also provide a conceptual framework for examining how relationships emerge and develop among parties within and between organizations. Understanding how parties negotiate, commit to, and execute transactions provides a tangible way to appreciate how people interact to develop and implement innovative ideas within an institutional context. Moreover, success and failure in

171

the creation and execution of transactions may help to explain, in part, the success and failure of an innovation.

RELATIONSHIPS AMONG TRANSACTION CONTEXTS, STRUCTURE, AND PROCESSES

This chapter is concerned with two basic questions about the design of these kinds of transactions: (1) What formal and informal processes are important for understanding the dynamics of how transactions develop, grow, and dissolve over time? (2) What legal and managerial considerations are important in structuring transactions in an equitable and efficient manner? These two questions surface three pervasive problems associated with the structure, conduct, and performance of transactions.

First, both questions must be answered before either can be considered answered because the two questions are interdependent; transaction structure and process are like opposite sides of a coin. Traditionally, however, the two questions have been addressed independently because their disciplinary bases have evolved independently. The structure of transactions has been the dominant focus in the law of contracts (Macneil 1974, 1978, 1980) and institutional economics (Coase 1937; Commons 1934, 1950; Williamson 1975, 1985). This legal and economics perspective provides a rigorous framework for understanding the institutional governance and structural safeguards of transactions. However, because of a number of simplifying assumptions, this perspective treats transactions as static and largely devoid of human processes. The human processes in transactions have been emphasized in the social psychology of interpersonal and group behavior (Moment and Zaleznik 1963; Hare 1962; Pfeffer and Salancik 1978; McGrath 1984) and the sociology of interorganizational relationships (Levine and White 1961; Warren 1971; Aldrich and Whetten 1981; Van de Ven and Walker 1984). But while psychologists and sociologists emphasize the interpersonal dynamics of transactions, their frameworks often lack economic and managerial components.

Second, understanding transaction structures and processes requires an appreciation of both formal and informal processes. Formal legal requirements impose constraints on transaction structure and process, yet transaction negotiation and execution involves informal, interpersonal interactions. These, in turn, affect formal processes. Clearly, answers to our two basic questions must account for both formal and informal effects.

Third, the two questions must deal with the tension produced by inherently contradictory roles enacted by transacting individuals. On the one hand, each individual represents his own personal interests in a transaction. At the same time, each individual represents an organization or institution. Consequently, the individual also attempts to secure the interests of the organization. Thus each individual must play dual (multiple) roles, those of principal and agent. The manner by which an individual resolves this role conflict affects both transaction structure and process.

Similarly, individuals assume different organization roles, entailing different expectations. For example, lawyers are usually responsible for establishing an acceptable legal structure for transactions, whereas managers are responsible for administering and executing the transaction. Although both are agents for the same organization, we frequently see tension between the two. Lawyers, due to professional (and individual) interests, may be most concerned with risk minimization, whereas the manager's professional and personal interests may produce incentives to complete the transaction with much less regard for risk.

Because role specialists, acting either as principals or agents, tend to perform unique

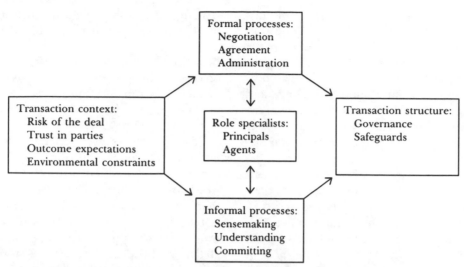

FIGURE 6-1.   *Relationships among Transaction Context, Structure, and Process.*

tasks and apply different performance criteria to a transaction, conflicts often arise within and between transacting units. These conflicts are addressed in different ways depending on how the role specialists stand in relation to each other. In particular, the balance of power between role specialists varies from symmetrical to asymmetrical. This holds at either intra-, or inter-, organizational levels.

Symmetrical power among agents, or principals, will facilitate bargaining relationships, ceteris paribus. *Bargaining transactions* are social relationships that occur between independent and equal parties, each with differing private interests and goals, and each with resources that can aid the other's realization of interests.

Asymmetrical power differences, on the other hand, tend to produce managerial relationships. *Managerial transactions* occur between parties who stand in a hierarchical relationship of supervisor and subordinate to each other. Here, one party's actions are carried out under the control of another and advance the other's interests. As Coleman (1986: 1325) notes, the associated institution is the formal organization or authority structure.

These distinctions between transaction structures, processes, and role relationships are summarized in Figure 6-1. These concepts provide the basic framework to address our two basic questions about the structure and processes of transactions in managing innovations.

First, we will examine the managerial and legal considerations in structuring transactions. Transaction structure refers to the legal forms of governance that apply to different kinds of transactions (ranging from markets to hierarchies), and the structural and procedural safeguards that parties negotiate into a transaction. We will argue that the efficiency and equity of alternative governance structures and safeguards are a function of the risk of the deal and trust among the parties. Specifically, we propose that *as risk of a deal increases and trust among parties decreases, an increasingly secure and costly governance structure will be required to motivate parties to enter into a transaction.*

Next we will examine how transactions develop, grow, and dissolve over time by distinguishing between formal and informal activities in each transaction. As Commons (1950) emphasized, transactions go through three formal stages (negotiation, agreement, and administration), and the courts have applied due process criteria to each of these stages to judge the enforceability of a contract. Corresponding to these formal stages are three informal processes of sense making, understanding, and committing that deal with the interpersonal dynamics between transacting parties. Our basic processual proposition is that *the greater the congruence between formal and informal processes, the more equitable and efficient the transaction.*

Finally, we will examine the relationships between transactions structure and processes. We will argue that *because transactions are socially contrived mechanisms for collective action, they are continually shaped and restructured by actions and symbolic interpretations of the parties involved.* Thus, just as a structure of governance and safeguards establishes a context for interparty action, so also do prior processes construct and embody future transaction structures. Given this interdependence, it is logical that neither transaction's structure nor process can be understood independently. Most transactions emerge incrementally and begin with small informal deals that initially require little trust because they involve little risk.[1] As these transactions are repeated through time and meet basic norms of equity and efficiency, the parties feel increasingly secure in committing more of their available resources and expectations to transactions with the other parties. Moreover, what may start as an one-time solution to a specific problem may eventually become a long-term web of interdependent commitments and relational contracts, if they are perceived by the parties to be equitable and efficient.

The Structuring of Transactions

We propose that the structure of a transaction is a function of the perceived risk of a deal and the degree of trust that exists between the parties. The key dimensions in this proposition for structuring transactions are illustrated in Figure 6–2 and will now be discussed.

Governance Structures and Safeguards of Transactions. The purpose of a transaction is to manage the transfer of property rights between two or more parties. In many situations the object of exchange and ownership rights to the goods or services are well understood by all parties. But innovation often involves the exchange of intellectual property, the rights to which are much more open to legal questioning than are rights associated with more material things. In addition, the intended outcomes of transactions in the innovation process are usually new "property" in whose "rights" many parties may have an interest.

In all but the simplest of transactions,[2] the transfer of property rights is conducted within a legally sanctioned *transaction structure.* It consists of a governance structure (the contracting mode) and a set of safeguards negotiated by the parties involved. A *governance structure* specifies the institutional guarantors. These institutional guarantors represent Common's (1950) concept of sovereignty, an overarching institutional framework that ensures contract compliance and recourse for nonperformance. In the United States the institutional guarantor includes state and federal laws and regulations that both prescribe and proscribe the behavior of transacting parties and establish enforceable procedures for resolving disputes. Although the vast majority of transactions are negotiated and conducted without institutional intervention, guarantors provide a "safety net" that "liberates, constrains, and protects" all lawful transacting parties (Commons 1950).

Using efficiency as the primary objective, Williamson (1975) classifies governance structures along a continuum ranging from market to hierarchically mediated transactions.[3] Included between these two ends of the continuum are contracts, licenses, franchises, joint ventures, partnerships, and equity ownership.

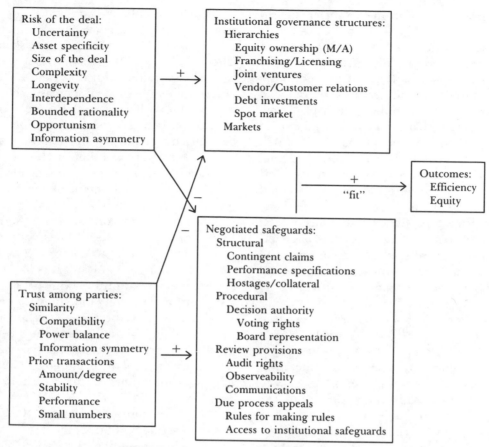

FIGURE 6–2. A Model of Transactions Structure.

These governance structures, in order, represent increasingly costly mechanisms for executing transactions but also provide the parties increasingly greater control over the exchange of property rights.

Although a particular governance structure establishes the general nature of relationships between parties, the specific terms of a transaction can be further modified and enhanced by *structural and procedural safeguards.* Transaction safeguards are used to elaborate rights and responsibilities and to allocate business risk among the parties. Examples of safeguards include performance specification, hostages, collateral, and board of director representation (structural safeguards); and

audit rights, arbitration procedures, and rules for making rules (procedural safeguards). These safeguards are endogenous to the transaction—that is, they result from independent bargaining by the parties.

In summary, the structure of a transaction consists of a mode of governance structure, an associated institutional guarantor, and a set of safeguards that are negotiated among the transacting parties. All three elements regulate an exchange.

Characteristics of Risk. All deals entail some risk, particularly those associated with the development of an innovation. One of the sources of risk is the uncertainty that pervades the processes of innovation. Can an idea such as hybrid wheat be transformed into a reality (see Chapter 14)? Can laboratory "samples" of gallium arsenide integrated circuits (Chapter 13) or apheresis filtration modules (Chapter 8) be mass produced without the loss of valuable properties? Are there commercially profitable organic materials that can be produced only in a microgravity environment (Chapter 10)? Will a new medical product be approved by regulatory agencies (Chapters 8, 15)? These are a few of many examples of uncertainties that abound in the processes of innovating under study by MIRP investigators. The perceptions of the parties to a deal regarding these types of uncertainties will be factors in their decisions as to the extensiveness of governance structures and safeguards they will demand to obtain security in any transaction.

Another aspect of deal risk is asset specificity (Williamson, 1975, 1985). Risk increases as the required assets are specific or unique to a deal and difficult to redirect to other uses without incurring significant costs. Asset specificity, according to Williamson (1981), can take any of four forms: physical, site, dedicated, and human. In the first form, assets are created for highly specialized uses (such as equipment designed to be used on board the space shuttle). An example of the site specificity occurs when a company builds a laboratory in close proximity to Cape Canaveral to facilitate integration of payloads into the space shuttle. In the third form, equipment may be acquired to suit the very specific needs of a major purchaser, such as when a software developer purchases specialized single-purpose computer equipment in order to create software for a large customer. Human asset specificity can arise from the need to move teams of workers (Williamson 1981) or because of the firm-specific knowledge that workers acquire (Teece 1985). The latter frequently occurs when program developers learn unique software languages that are not applicable in other settings or to other programs.

The size of the deal (in terms of the relative value of assets committed by the parties) may also be a factor affecting the risk involved. The complexity of the deal also is likely to alter the risk associated with it. As the relative size and complexity of a deal increase, ceteris paribus, so too will riskiness. In one innovation study, a party to a three-company joint venture committed a significant portion of its assets to the coventure. In another study, the defense contractor made major investment commitments to the development of the MK50 program in anticipation of receiving a single-source production contract from the navy. In both cases, if the "deal" goes sour, the two companies are likely to confront hard times.

The longevity of a deal is likely to have a curvilinear relationship with perceptions of risk. Very short or very long time frames within which deals must be completed are likely to increase their risk, all other things remaining equal. In short time frames the risk may result from uncertainties in developing an innovation and too little time during which to correct problems.

For example, in Chapter 10, Ring and Rands describe how under the initial 3M-NASA agreement the parties had only a ten-month time frame to design and fly an experiment. Although the time frame was met, it imposed significant costs on the parties. These same parties have also negotiated a ten-year agreement, during which time they will cooperatively engage in a series of experiments in low earth orbit. The risks inherent in such a relationship were all too graphically demonstrated with the explosion of the space shuttle *Challenger* on January 28, 1986. Other inherent risks in the relationship that are a function of the protracted time frame include the likelihood that a minimum of three presidents will be elected during its duration, as well as an equal number of CEOs at 3M. As Scudder et al. describe in Chapter 12, the defense contractor's fifteen-year development of the MK50 torpedo system entails extensive temporal risk of changing military defense priorities and government administrations. As a consequence, the contract has a complex governance structure and detailed set of contingent safeguards.

Characteristics of Trust. Although the concept of risk focuses on the characteristics of a deal, the concept of trust examines the relationships among the parties to a transaction in predicting the structure of a transaction. Transaction costs economics treats the identity of the transacting parties as relevant, but implies that opportunism is an ever present danger regardless of the parties involved. Yet there is a great deal of theoretical and practical support for including trust as a critical factor in the design of transaction structures. As Arrow (1973: 24) points out, "ethical elements enter in some measure into every contract; without them, no market could function. There is an element of trust in every transaction," and it varies with transacting parties.

As Figure 6–2 illustrates, an increase in trust leads to an increase in the likelihood that the parties will select a market-mediated contract mode and reduce their reliance on contingent claims and safeguards. This is because parties attempt to minimize transaction costs. As Fried (1981: 8) argues,

> When my confidence in your assistance derives from my conviction that you will do what is right (not just what is prudent), then I trust you, and trust becomes a powerful tool for our working our mutual wills in the world. So powerful a tool is trust that we pursue it for its own sake; we prefer doings things cooperatively when we might have relied on fear or interest or worked alone.

Two explanations are possible for the emergence of trust. The first is based on the norms of equity; the second is based on more utilitarian assumptions. Norms of equity refer to the degree to which contracting parties judge that another party will fulfill its commitments and that the relationship is equitable (Van de Ven and Walker 1984). The concept of equity is developed in exchange theory, which argues that participants in a relationship desire: (1) reciprocity, by which one is morally obligated to give something in return for something received (Gouldner 1959), (2) fair rates of exchange between costs and benefits (Blau 1964), and (3) distributive justice, through which all parties receive benefits that are proportional to their investments (Homans 1961).

Personal trust is more likely to be present in a deal when contracting parties have successfully completed transactions in the past and the negotiating individuals perceive one another as following the norms of equity. The more frequently the parties have successfully transacted, the greater the level of trust they bring to the deal. In one MIRP study, Cray Research obtained construction of a laboratory facility without undue concern for the quality of the work done or its timeliness simply because of a long working relationship with the construction contractor.

In a similar vein, an organization that engages in many transactions with many other firms is likely to attempt to reduce the numbers of firms with which it does deals over the long run. To the extent that it develops a high degree of trust in a smaller number of firms, it will tend to look first to those firms as transaction partners. Thus, trust, as well as asset specificity, may lead to the kind of small numbers conditions that Williamson and other transaction cost economists assert increases the likelihood of opportunistic behavior.

Organizational trust also arises when organizations earn a reputation in the marketplace for following the norms of equity. It permits parties to a transaction the luxury of using market-mediated transactions and less specificity in the design of their governance structures. Both NASA and 3M went out on limbs

177

during negotiations leading to a Two-Year Joint Endeavor Agreement, in large measure because each perceived the other as organizations with reputations for being true to their word. Indeed, 3M had come to NASA's attention because of its high standing in the *Fortune* magazine reputation survey of U.S. business. Conversely, several MIRP respondents at Control Data reported that transaction negotiations were aborted when they discovered that the other parties, in their opinions, could not be trusted.

An alternative explanation for the emergence of trust is based on more direct utilitarian reasoning. First, there are many non-legal sanctions which make it expedient for individuals and organizations to fulfill commitments (Macaulay 1963). At the personal level, repeated personal interaction across firms encourages some minimum level of courtesy and consideration, and the prospect of ostracism among peers attenuates individual aggressiveness. At the organizational level, the prospect of repeat business discourages attempts to seek a narrow, short-term advantage (Maitland, Bryson, and Van de Ven 1985). More important, organizational reputation is a valuable business asset. If dissipated through opportunistic behavior, loss of reputation leads to more costly and tightly drawn contracts and more safeguards in future transactions.

Second, when the parties to the deal stand relatively equal in terms of the power that they can exercise, trust is likely to be enhanced. This is especially so when the parties are organizations. The implicit threat is that opportunistic behavior by one party can be effectively matched by the other, if not in kind then certainly in degree. Finally, to the extent that the missions, objectives, or goals of the parties to the transaction are compatible, trust can be expected to rise, ceteris paribus. There is little to be gained by acting opportunistically. In a deal designed to produce samples of the new type of circuit for testing by a computer manufacturer, a firm is likely to be accorded greater trust by the circuit manufacturer if it does not have the capacity to produce the circuits than will a firm that has that capacity. In the former case, compatibility is likely to be high. The computer manufacturer may get the chips it needs, and the chip manufacturer may obtain a major new customer. Needless to say, if the two parties have pursued similar coventures in the past, and they have been trustworthy with each other, then the norms of equity previously discussed will reenforce compatibility factors.

In a subsequent section of this chapter dealing with relations among formal and informal transactions processes, we elaborate further on the affects that context has on governance structure choice. In the interim, it will suffice to say that to the extent that the relationships between trust and risk that we have set forth hold, the parties to a transaction perceive their governance structures to be efficient. To the extent that the relationships produce expected outcomes, the parties are also likely to view the governance structures as equitable.

Transactions Processes

Formal/Legal Transaction Processes. As Commons (1950) has indicated, transactions are dynamic and go through three temporal stages: negotiations, agreements, and administration. Although these three stages often overlap and recycle, it is analytically useful to distinguish these phases in order to examine the interpersonal and social-psychological processes that tend to predominate in each phase of a transaction.

The *negotiations stage* highlights the strategies and choice behavior of parties as they select, approach, and avoid alternative parties and as they persuade, argue, and haggle the terms, contingent safeguards, and governance structure involved in a transaction. Consequently, it is important to consider what kinds of factors might lead independent parties to enter into negotiations with one another. The next section addresses these issues in a consideration of the emergent interpersonal transaction process broadly described as sensemaking.

In the *agreement stage,* the "wills of the parties meet" by agreeing to the terms of the relationship and the working rules or procedures of action. It is here where governance structures and safeguards are set to organize the deal; whether that deal results in the establishment of a partnership relationship among peers, a hierarchical relationship between supervisors and subordinates, or a market relationship to contract, joint venture, license, franchise, borrow, lend, or otherwise undertake various activities needed to execute the transaction over time.

Finally, in the *administrative (or execution) stage,* the rules, procedures, and terms of the transaction are carried into effect. It is at this stage where misunderstandings, conflicts, and changing expectations of a relationship often occur—resulting in renegotiation, mutual adaptation, litigations, or termination of the transaction. Those transactions that do endure over time become institutionalized—meaning that the parties involved unconsciously begin to take the terms of the agreement for granted. Only when significant precedents occur do the parties involved reflect and reconstruct in memory or from files the initial, but now hazy, terms of the transaction they initially negotiated and agreed on.

To be legally enforceable, the above three stages of negotiation, agreement, and administration of contracts must adequately specify the nature of the parties' agreement on five critical dimensions of contracts: (1) *risk,* the level and nature of the risks accepted or imposed on transacting parties; (2) *returns,* the parties' expected rewards and outcomes; (3) *control,* the structures and procedures for allocating authority and responsibility between the parties; (4) *duration,* the length of time during which the parties are committed to a transaction (Klein 1982); and (5) *termination,* the events and procedures that allow parties the right of exit from a transaction.[4]

Although these criteria specify the "due diligence" that is necessary for a legally enforceable contract, they do not address many of the interpersonal dynamics we observed in the development of transactions in the innovation studies. These interpersonal dynamics deal with a more informal set of processes that are analogous to, but quite different from, the formal stages discussed above.

Emergent Interpersonal Transaction Processes.

It is often recognized that underlying formal contracts are a host of backstage interpersonal dynamics that mobilize and direct the formal contracting process but are seldom visible or explicitly written into the formal contract. Yet these informal processes often reflect more closely the "real" motivations for a transaction among parties than a formal contract. In other words, it is widely held that the formal contracting process discussed above does not address or reveal the basic underlying forces that mobilize, drive, and energize individuals to act, interact, and engage in transactions with others. The motivational basis for individuals to engage in transactions is taken for granted in law and economics and remains a controversial area in psychology, social psychology, and sociology. However, recently Turner (1987) made a significant contribution in reconciling this controversy by bringing motivational dynamics back into contemporary social interaction theory.

As Turner (1987: 16) indicates, the theory of motivation implicit in legal and economic treatments of transactions is that individuals engage in relationships to maximize gratifications, or utilities, and to avoid deprivations or punishments. These desires are mediated by rational calculation of probable payoffs and losses for pursuing alternative lines of conduct.

However, this utilitarian theory of motivation side-steps basic issues of how individuals order values, how they decide on alternative lines of behavior, and how the social context influences individual cognitive processes. It is an appreciation of these issues that permits one to go beyond an ideal-type rational actor to a more realistic description of a social human being.

Turner (1987) goes on to develop a composite model of motivation by synthesizing the key concepts in theories of motivation implicit in exchange, interactionist, phenomenology, and psychoanalytic social theories. Although his model is more complex than we need to address here, Turner suggests three fundamental forces motivate human thought and action: needs for identity, facticity, and inclusion. We will summarize these drives below because they ground an explanation for the emergence of collective transactions in the motivational predispositions within individuals. These micro-individual needs or drives appear to manifest themselves in three informal transactions processes that we have often observed in the innovation studies: sense making, understanding, and committing.[5]

Sensemaking. We view this as an enactment process in which organizational participants come to appreciate the nature and purpose of a transaction with others by reshaping or clarifying the identity of their own organization. By projecting itself onto its environment, an organization develops a self-referential appreciation of its own identity, which in turn, permits the organization to act in relation to its environment (Morgan 1986: 243). Although events may occur in an objective environment that modify the development of transactions between parties, the meaning of these transactions is always an "enacted" environment (Weick 1979). The parties to a transaction in the course of sense-making processes can be expected to be enacting their own version of the environment while, at the same time, integrating the enactments of other parties into their own enactments.

For example, the 3M-NASA coventure could not have emerged had it not been for an extended process of sensemaking in which 3M and NASA personnel interacted and came to appreciate a totally new possibility for 3M—space commercialization. So also, the navy's announcement of competitive procurement on the MK50 system led Honeywell's division to reassess its business and articulate a new diversification strategy and engage in exploratory discussions with new potential customers. Finally, the failure of Qnetics to market its medical records software after repeated efforts to enter into marketing relationships with a variety of distributors lead the principals to realize that their small company was a technology-, as opposed to a marketing-, driven company. In each of these examples, the processes of sensemaking saw organizational members significantly revised the identity of their own organizations (their cognitive maps) as a result of communications with potential or existing transacting parties.

Psychologically, Turner (1987) indicates that sensemaking processes derive from the need within individuals to have a sense of identity—that is, a general orientation to situations that maintains esteem and consistency of one's self-conceptions. Sensemaking processes have a strong influence on the manner by which individuals within organizations begin processes of transacting with others. If confirmation of one's own enacted "self" is not realized, however, sensemaking processes recur and a reenactment and rerepresentment of self follows. If the presented self is not reinforced by others, then the individual is likely to resort to the use of defense mechanisms. When the results of sensemaking are not confirmed by others, because of the need for cooperation, the individual will engage in another process of reenactment, and representation. However, if individuals repeatedly fail to confirm their enactments of objective reality, "then needs for cooperation (and indirectly, needs for identity) go unconsummated, thereby setting into motion defense mechanisms and/or the development of new self conceptions, such as a deviant identity" (Turner 1987: 18).

At a more macro level, the process of sensemaking may be initiated by "shocks," either internal or external to the organization as Schroeder, Van de Ven, Scudder, and Polley describe in Chapter 4. In the course of the MIRP studies, a number of these shocks have been observed, including new leadership, product failure, impending loss of market share, and budget crisis. Renewed sensemaking processes can also be a product of repetitious challenges to causal maps developed by prior enactment processes. In the NASA/3M coventure, the two organizations began to transact only after repeated urging by Astronaut Dr. Bonnie Dunbar that U.S. industry was in danger of having "its lunch eaten" by the Japanese and West Germans who were investing heavily in the commercial development of space.

Understanding. As we employ the concept, this involves a process by which interacting parties socially construct and agree to the terms of their relationship. The process of understanding emerges from the need of individuals to construct an external factual order "out there" or to recognize that there is an external reality in their social relationships. Interaction is constrained by actors' needs to feel that they share a common understanding of the world they jointly occupy. As Turner (1987:19) states,

> Actors are motivated to create a sense, even an illusionary sense, that they share a common universe. The basic motivational force behind interaction is, therefore, the use of implicit stocks

of knowledge to make indexical interpretations of gestures in order to generate feelings of a shared universe.

As applied to transactions, understanding processes provides this shared interpretation of a relationship. The shared facticity derived from the processes of understanding between parties on the terms of their transaction often emerges gradually and incrementally. For example, in the 3M/NASA transactions, we observed numerous instances in which a common lexicon appeared to be essential to decisions regarding contracting. In negotiating the Ten-Year Joint Endeavor Agreement, the 3M team frequently experienced breakoffs in discussions when either it or the NASA team had to reach agreement on the meaning of an issue such as the definition of "flight." In this particular case, the result was the creation of a totally new term in the lexicon of space commercialization, "flight experiment opportunity," which was defined in terms of physical space within the shuttle.

In addition, the facticity derived from a process of understanding often changes as a result of experiences in carrying out initial commitments. A mutual agreement that two individuals think they have developed and is operational may not be in place or may have evaporated. Within 3M we observed such an occurrence in the course of development of one of the early low-orbit experiments. After the experiment had been flown on the shuttle, some of the parties reflected on their experiences and decided not to have anything further to do with the project. It was apparent from interviews with all individuals involved in this project that understandings that had been reached with others had dissipated under the stress of preparing the experiment and that nothing had replaced the commitments that flowed from these understandings. In Turner's terms, the parties' needs for a sense of facticity,

181

while perhaps strong, were no longer congruent. They had lost a sense of a "common, shared, obdurate world."

Committing. In the transactions model, this process usually leads to the creation of psychological contracts among transacting parties (Argyris 1960; Levinson et. al 1962; Schein 1970; Lawless 1972; Kotter 1973; Thomas 1976). Psychological contracts, as opposed to legal contracts, consist of unwritten and largely unverbalized sets of expectations and assumptions held by transacting parties about each other's prerogatives and obligations. These expectations of what each party will give to and receive from the relationship vary in their degree of explicitness; the parties are often only marginally aware of their exact nature. Yet these expectations can exercise a powerful effect on the parties' relationship.

Many commitments are likely to be incorporated in a psychological contract, although at any given point in time an individual may be aware of only a limited number (Kotter 1973). Expectations address areas such as norms, work roles, the nature of the work itself, social relationships, or security needs. Individuals' expectations are shaped by such things as past experiences (many of which predate the individual-organizational relationship), personal values, technical specialization, and hierarchical level. The organization's expectations arise out of its history and current environment and typically are more specific for some individuals than for others (Levinson et. al 1962). Nicholson and Johns (1985) also suggest that psychological contracts vary significantly across occupational roles in the amount of trust that is involved in the contracts.

As a transaction evolves over time, new experiences can modify the psychological contract that was negotiated and agreed to on entry. Individuals who come to feel that their expectations are unmet or that there has been a breach of commitments to a transaction have

three basic options: attempt to renegotiate, continue the relationship in an alienated manner, or sever the relationship (Thomas 1976). In such circumstances, we would expect parties to choose the first option and adopt the third option only at last resort because of the psychological needs within people to cooperate with others and to avoid anxieties revolving around three dimensions of social relationships: (1) the need to feel included, (2) the need to sense predictability or trust the responses of others, and (3) the need to feel secure that things are as they appear. Turner (1987) suggests that the level of motivation and commitment to a transaction is an inverse function of the anxiety associated with needs for inclusion, trust, and security.

If exchange relations do not confirm self, then an individual's level of anxiety is increased, causing redoubled efforts to sustain self through intensified self-presentations. In other words, maintaining the substance, esteem, and coherence of self is dependent on an individual's capacity to confirm self directly through self-presentations and also to achieve a sense of inclusion, trust, and security, thereby avoiding more deep-seated sources of diffuse anxiety that cause self-doubt (Turner 1987: 25).

Relations among Informal Transactions Processes. Turner suggests that three fundamental forces underlie human action and thought. These three forces appear to provide the necessary motivational base for parties to commence sensemaking, understanding, and commitment processes. Because the motivating forces described by Turner are are highly interdependent, it is not surprising that we have found the informal processes of transactions to be interdependent as well. This may explain why transactions often emerge incrementally, grow with relationships that are perceived to be equitable, and often develop into a web of multiple commitments among transacting parties. Interdependence among the informal transactions processes may also explain high rates of transactions failure and why the vast majority of sense-making processes are not accompanied by understanding or committing.

The timing of sensemaking, understanding, and committing processes is a key factor in explaining the emergence, growth, and decline of transactions. The case study by O'Toole et al. (1972) is particularly insightful in its description of how transactions emerge as a result of a slow, flexible, developmental process, with many small thrusts of exchanges around specific problems, followed by periods to make sense of and re-understand new developments. O'Toole emphasizes that transactions grow and build on previous small but successful exchanges between organizations. By participating in small but escalating commitments, each party is able to adjust cognitive maps regarding the positive aspects of a transaction as well as its negative aspects. Further, commitments and formalized arrangements are not developed prematurely, when sense-making and understanding processes are still producing unclear products and the parties are ambivalent as to the benefits of a transaction for themselves or for their organizations.

Van de Ven (1985) reports that the problem of ambivalence was a common phenomenon observed in the development of relationships among members of many innovation units, where participants often send and receive all kinds of mixed messages about the group's identity, strategy, and direction. When pressed to get on with the task by an entrepreneur, groups got involved in frustrating periods of internal "navel-gazing" and emotionally torn group meetings that appeared to accomplish little. Indeed, on most issues of substance, groups appeared ambivalent or incapable of committing themselves to clear courses of action. It is clear that committing processes are critical to the overall process of innovation. We propose that in the absence of freely negotiated

psychological contracts, effective innovation teams will not emerge.

In conclusion, the development of a transaction is a dynamic process that is continually shaped and recreated by the sensemaking, understandings, and commitments of parties engaged in a transaction. The ambivalence problems explain why transactions are likely to emerge incrementally with small transactions that initially require little trust because they involve little risk. As these transactions are repeated through time and meet basic norms of equity, the participants feel increasingly secure in committing more of their resources to a transaction. Conversely, if parties cannot come to reconcile their own identity in, understanding of, or commitment to a transaction, they will withdraw from the transaction.

Relations among Formal and Informal Transactions Processes

Figure 6–3 diagrams the interdependencies envisioned among the formal and informal transactions processes discussed above. The informal processes of sensemaking, understanding, and committing are dynamic processes that go on in each of the formal stages of negotiating, agreement, and administration. Over time the formal and informal transaction processes are expected to interact in three basic ways to explain the emergence, maintenance, and demise of transactions.

First, many transactions are not formally specified because the informal processes serve as substitutes for formal transaction processes. This substitution effect is expected to occur in situations where high consensus and trust is achieved in during informal sensemaking, understanding, and committing processes. Under these conditions, transactions costs are minimized by dispensing with formal processes and structures for governing the relationship among the parties involved. Conversely, where informal processes yield only moderate levels

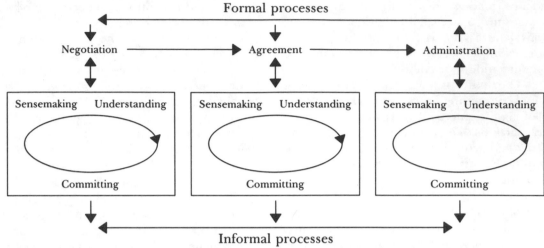

FIGURE 6–3. *Formal and Informal Transaction Processes.*

of consensus and trust, the parties may press for the establishment of extensive formal and costly procedures of negotiations, agreements, and administration to obtain the set of governance structures and safeguards with sufficient degrees of coupling and specificity they perceive are required to enter into a transaction with parties with whom they feel interpersonally insecure.

However, even where high levels of interpersonal trust and consensus emerge among transacting parties, two factors require the simultaneous development of formal transactions structures and processes. First, interorganizational transactions imply that organizations, not individuals, are the principal parties to a transaction, and that individuals act as representatives or agents for their respective organizations in negotiating a transaction. Second, transactions of long duration require formal institutionalization, which permits the transaction to remain in force beyond the time span of the individuals who negotiated the transaction. In both circumstances, the initial products of sensemaking, understanding, and committing processes arrived at between agents will be formally specified in each of the formal stages of

a transaction among principal organizations. This formalization provides structure and predictability in a transaction and thereby permits a transaction to operate efficiently. As a consequence, the causal direction of the paths in Figure 6–4 between informal and formal processes is reciprocal. Initially they may go from informal to formal transactions processes, but once established the causal paths reverse.

Formal and informal transactions stages often do not coincide in their temporal development. For example, often observed in the innovation studies were situations in which transacting parties reached mutual agreement facticity and developed psychological contracts prior to their organizations' negotiating a legal contract. The specific actions to be taken with respect to certain provisions of legal contracts were not yet spelled out. However, shared norms of acceptable behavior inherent in the psychological contract permitted the parties to negotiate, reach agreement, and administer a legal contract in an atmosphere of mutual trust.

Of course, in many transactions it is difficult to foresee all the possible situations that might arise in the course of the agreement that the parties to a negotiation seek to finalize.

Only those contingencies that are seen as possible, important, and open to significant disagreement between the organizations appear to be addressed in detail in legal contracts. The remaining issues, we contend, are worked out in informal understandings and commitments among parties. In general, we would expect that as informal commitments among parties are upheld, subsequent transactions among the parties will contain fewer formal safeguards. However, if commitments are violated, we should expect greater specificity in legal contracts among the parties.

Transactions institutionalize over time through the repetition of acts by the successors of the parties who established a transaction, meaning that they come to share the idea of "the ways things are done." As Berger and Luckmann (1966: 57) note, "Man is capable of producing a world that he then experiences as something other than a human product." As a consequence, what was once recognized as a socially constructed transaction takes on the form of an externally specified objective reality, where transacting parties play out preordained roles and "action routines" (Starbuck 1983) without understanding the initial and changing intentions of a transaction. As this drift between appreciation of the formal and informal processes of a transaction develops over time, we would expect conflicts to erupt among transacting parties. These conflicts signify either the termination of a transaction or the initiation of another cycle of sense making, understandings, and commitments to a transaction.

Relationships between Structure and Process in Transactions

To this point transactions structures and processes have been addressed in a somewhat independent manner. As suggested by Figure 6–1, however, we consider them to be interrelated. Moreover, in our discussion of the relations between formal and informal transaction processes, we foreshadowed structure and process relations. We will now make explicit a number of important relationships between structure and process observed in the MIRP studies.

We began our analysis with a conclusion drawn from the MIRP studies that the processes of innovating invariably require cooperative efforts between organizations. These efforts take the form of "deals" that are realized by transactions. Inherent in these transactions is risk, which the parties to the transaction seek to minimize. Absent other considerations, greater risk will lead to more hierarchical governance structures for the transaction and greater specificity of endogenous safeguards by the parties to it.

A second conclusion drawn from the MIRP studies, and to this point only implicitly made, is that parties to innovation engage in repeated transactions and that one of their objectives is to develop stable recurrent transactions. This leads to the question, "What aspects of structure and process lead to stable, recurrent transactions?"

The answer is explicit in the model in Figure 6–2. Increases in trust between the parties, coupled with outcome efficiency and equity, lead to stable, recurrent transactions, ceteris paribus. Greater trust, we argue, is associated with less hierarchical governance structures and less specific endogenous safeguards. By themselves, these factors increase the opportunities for managerial flexibility and thus enhance the probabilities that potentially effective transactions will not be inhibited by unnecessary legal barriers.

Trust emerges from the formal and informal processes of transacting. The formal process of negotiating enables, perhaps even forces, the parties to engage in sensemaking. If the sensemaking is effective, understanding and committing processes should evolve. These processes will facilitate reaching agreement. To an extent, these informal processes

185

can mitigate market and organizational failures frequently described in the literature. For instance, sensemaking processes may reduce the risk of a deal, while understanding and committing processes may increase trust among parties.

Time permitting, even in an initial transaction these informal processes may produce levels of trust that mitigate against the need for more hierarchical arrangements and greater specificity in safeguards. To the extent that the formal processes of negotiation and agreement permit transacting parties to develop psychological, as well as legal, contracts, successful administration of the transaction is enhanced. As these formal and informal processes are successfully repeated, trust is increased and governance structures and safeguards can be "relaxed" further, if not in the "letter of the law," then in its spirit.

Of course, transactions not only grow and stabilize; they also fail. In particular, we will discuss four reasons that many attempted transactions do not reach an agreement stage or are subsequently terminated: (1) market and organizational failures, (2) role specialization, (3) escalation of commitment, and (4) structure-process incompatibilities.

Market and Organizational Failures. These have been frequently discussed as reasons for transaction failure (see, e.g., Williamson 1975, especially Chapter 7). Bounded rationality and opportunism are classes of reasons that markets might fail due to human considerations, while uncertainty/complexity and smaller numbers define broad classes of environmental reasons underlying market failures (Williamson 1975). In the organization, Williamson attributes transaction "failure" to conditions such as bias toward internal procurement, internal expansion and program persistence, all supported by communication distortion and conflict between systems and subgroup rationality.

Role Specialization. As the discussion associated with Figure 6–1 implies, managerial transactions occur within organizations, and different role specialists within these organizations are often engaged in developing different parts of an interorganizational transaction. Not surprisingly, the lawyers with whom we discussed the issue in the context of MIRP studies invariably favored greater specificity of the structure and safeguard of a transaction than did managers. Aside from any consideration of training or sense of obligation to "protect" their "client," we believe an important underlying reason that lawyers generally are more conservative stems from the fact that professional norms create barriers to the development of informal commitments and/or psychological contracts during the formal processes of transacting. In the one MIRP case in which the lawyers spoke openly on this issue, they were quite explicit in describing behaviors that clearly prevented any development of informal understandings or commitments with the "other side." For example, they steadfastly refused to even have "a good natured drink" after negotiating sessions.

A second aspect of the effects of role specialization between lawyers and managers arises out of the formal process of negotiating. It appears that it matters who is "in charge" of developing the negotiating strategy. More effective transactions appear to evolve when managers design and have control over the strategy than is the case when lawyers are the architects. More efficacious negotiation appears to result when the counterpart architects of negotiating strategy between two organizations play the same roles (that is, managers and managers or lawyers and lawyers).

Escalation of Commitment. Invariably, there are occasions in the course of a transaction when it ought to be terminated in the interest of one or more of the parties. Yet transactions often persist beyond a point of producing any of the objectives desired by the parties. One reason that this may occur stems from the level of commitment that one or more of the parties brings to the transaction.

Self-justification is one reason that commitment may escalate over the course of a transaction (Staw 1976). In an innovation, the individual responsible for an idea is likely to become more committed to the underlying objectives than are those who come to the project at a later date. If the originator of the idea has significant control over resources necessary to pursue the objective, a transaction designed to achieve the objective is likely to be continued, even when an effective outcome is clearly in doubt. Fox and Staw (1979) conclude that when an individual's job was at stake, or a pet policy was at issue, escalation of commitment occurs, even in the face of evidence that the course of action being pursued was a losing one. Face saving is another source of escalating commitment.

In organizational contexts, transactions may persist in the face of evidence of failure simply because the organization has "invested" so heavily in a project. The commitment to the project may have been escalated because of political alliances and coalitions that were developed over the evolution of the project. As a result, decisions to discontinue transactions require agreements from others in the organization, and they may have developed independent commitments to the transaction.

Structure-Process Incompatibilities. We have previously proposed the need for "fit" between structure and process if transactions are to be efficacious. The reason for this proposition is that the seeds for disintegration of a transaction are contained in the very governance structures, safeguards, and behaviors that lead to the formation and growth of transactions (Van de Ven and Walker 1984). Excessive formalization of governance structures and monitoring of safeguards leads to conflict, dissensus, and

distrust among parties, who can be expected to desire to maintain their expectations regarding autonomy in the face of growing interdependence, which inevitably results from cooperative transactions. Where the design of the structure largely has been in the hands of lawyers, the chances of this occurring appear to be enhanced. Legal considerations in many cases inherently conflict with managerial needs for autonomy and flexibility.

From a purely managerial perspective, the increased transfer of proprietary resources among transacting parties over time implies that their identities and unique domains will gradually shift from being complementary to being undistinguished, which increases the likelihood of territorial disputes, conflict, and competition. To a degree, these dysfunctional consequences of structure may be offset by emergent interpersonal transaction processes such as sensemaking, understanding, and committing. When the fit between structure and process is altogether absent however, territorial disputes will begin to dominate among parties, leading to the termination of the transaction.

CONCLUSIONS

In the course of innovating, an entrepreneurial unit must often cooperate with others and hence engages in transactions processes. Innovation processes are inherently uncertain. Parties frequently do not fully appreciate or understand their needs. They may not even know who can provide for those needs. In all likelihood, those providers will turn out to be few in number.[6]

Under such circumstances, the unit's normal scanning and search routines (Cyert and March 1963) are not likely to provide answers to the questions of what the units needs are or what other organizations can satisfy

187

those needs. The behaviors of a number of the innovating groups under investigation by MIRP researchers, however, suggests that sense-making and understanding processes will help to identify one or more potential parties with whom the organization can effectively transact. Those same informal transaction processes may produce agreement among parties that they indeed to have complementary objectives or slack resources that they can usefully exchange. Assuming that these parties are managers, they can be expected to use all their wits to find a transaction structure that permits them to operate with as much flexibility as levels of trust and risk permit.

The analysis undertaken in this chapter suggests a number of courses of action for managers who are seeking or need partners to develop their innovations. First, from a purely managerial perspective, trust and risk appear to be paramount concerns in structuring efficient and equitable transactions. Low risk and high trust permit parties to exercise a broader range of potential governance structures in designing their transactions. Even when managers may not be able to substantially reduce levels of risk associated with the transaction,[7] they may be able to significantly elevate levels of trust in the course of the emergent interpersonal transactions processes that we have described. Sensemaking and understanding processes provide opportunities for enhancing trust during negotiation stages such as reciprocal sharing of know-how. These same processes also permit parties to a transaction to evaluate the norms and beliefs of others and to individually reach conclusions about how much trust they might be willing to accord others.

The somewhat differing norms of managers and lawyers suggests that care be given to decisions regarding lead roles in transaction processes. Experience in the 3M/NASA case suggests that when managers determine the objectives and strategies behind a transaction and thereafter "determine" the conduct of

their attorneys, the outcomes appear to be more efficacious than when attorneys act as "deal-makers" from start to finish.[8] As reliance on house counsel increases, it may be necessary to reconsider the specialized roles of negotiating teams.

The general presence in the United States of reliable exogenous safeguards, such as the courts, relieves the parties to a transaction of the legal necessity of providing a full range of endogenous safeguards for a transaction. The presence of generally accepted business norms in many cases further reduces the need to rely on legally oriented endogenous safeguards. Too frequently, however, parties to transactions tend to lose sight of these realities and to engage in extensive negotiation over safeguards. Since resolution of these kinds of issues tends to require zero-sum solutions, positional bargaining overtakes the negotiations (Fisher and Ury 1981). A consequence may be that levels of trust built up during sensemaking and understanding processes evaporate.

This observation leads us to a second tentative conclusion that we draw from the MIRP studies, a conclusion that also provides an answer to the first question we set forth in the beginning of this chapter. The evidence from the 3M/NASA case indicates that early in the negotiations period, issues related to safeguards were accorded more importance than the overall structure of the transaction itself. As the negotiations progressed, however, and the sensemaking and understanding processes in which the parties were engaged informed their decisions, they came to a conclusion that the key to an ongoing relationship lies in finding a transaction structure that was adaptable to the uncertainties of the future that both parties confronted. Thus, it may be that a governance structure is more critical to success than endogenous safeguards and that, commensurate with considerations of risk and trust, interpersonal processes of sensemaking, understanding, and committing may obviate the need for many common safeguards. If the informal processes facilitate the emergence of trust, or reenforce its prior existence, then the chances of an appropriate governance structure being designed in the course of negotiation is enhanced. Sensemaking and understanding may lead to a better appreciation of the risk involved in the transaction that, again, should result in the design of a more appropriate governance structure.

Our analysis also suggests the need to more fully consider the role of fit between formal and informal processes and structure. The model in Figure 6–2 implies that governance structure issues focus intensively on the allocation of deal risk (or its flip side, opportunity). The formal processes of negotiation, agreement and administration tend to drive the parties to design a transaction structure that minimizes deal risk (or, conversely, optimizes deal opportunity). Party risk—the safeguard issue—on the other hand, is different and should be viewed as a function of levels of trust that exist or can be created during the informal processes of sensemaking, understanding, and committing.

Our review of the transactions literature cited throughout this chapter suggests that too little attention has been paid to the design of governance structures that facilitate an effort to more effectively cope with both deal and party risk. There is a clear need to identify a broader range of the kinds of intermediate governance structures between markets and hierarchies that cope with both forms of risk in equitable and efficient ways.

Many of the observations and propositions about transactions structures and processes presented in this chapter were developed inductively from observations of the innovations included in MIRP. As such, it is important to conclude with the caveat that these propositions require rigorous testing. For example, our propositions about the actions of parties have been developed independently of the environment in which a firm is operating at the time a transaction is undertaken. To complete our understanding of the model, we would need to consider its operation in the specific contexts in which managers of a business firm find themselves as parties to a transaction seeking to minimize total ex ante and ex post costs associated with the transaction. These contexts might include the strategy that a firm was pursuing, its strategic predisposition, or the stage in industry or product life cycle that the firm confronts. These aspects of the model set forth in Figure 6–1, which we defined as environmental constraints, while critically important, are beyond the scope of the chapter.[9] Innovation provides a requisite degree of context for our purposes.

By their very nature, tests of the propositions set forth in the chapter will require intensive, real-time longitudinal studies. They dictate use of multidisciplinary concepts principally from law, economics, psychology, sociology, and organizational theory and behavior. The continuing longitudinal nature of MIRP will permit testing some of the propositions. But rigorous investigation of these observations and propositions are required in other organizational settings in order to establish their generality. Only through such efforts will we begin to develop a framework regarding transaction structure and process that managers will find helpful in the design, maintenance, or dissolution of relationships that facilitate the processes of innovation.

NOTES

1. We appreciate that some transactions may not appear, at first glance, to fit this mold. Large mergers or acquisitions between seemingly unrelated firms were frequently offered as counterexamples to this proposition. However, in all of the MIRP studies these major transac-

tions were, in fact, preceded by smaller transactions, sometimes involving only individuals. Similar examples can be found elsewhere. Reginald Jones, former CEO of General Electric, describes that company's acquisition of Utah Mining in terms entirely consistent with our view (HBS Video Case No. 883-004).

2. In the simple case of barter, property rights can often be transferred without a formal transaction structure. For example, in the startup of Qnetics, the entrepreneurial managers agreed to the exchange of using one's office space in return for the use of the other's computers without any formal contract. Transactions that consist of more than simple barter typically require a more formal agreement that more fully outlines the nature of the expected relationship between the parties.

3. These two ends of the continuum correspond exactly to Commons' (1950) bargaining and managerial transactions. In both cases a distinction is made between relationships involving only principals and those relationships that involve principals and agents. Some legal theorists classify these same types of relationships along a continuum that ranges from discrete to relational contracts (see Macneil 1974). Perfectly discrete transactions are those that stand apart from all past, current, and future transactions, with all provisions explicitly drawn so that no gaps in the agreement need ever be resolved. In general practice, discrete contracts are more analogous to bargaining and market mediated transactions. At the other extreme, relational contracts are those in which past, present, and future interactions impinge on the agreement between parties. Gaps may exist in relational contracts because complete presentiation, the act of bringing the future into the present, is not possible. When uncertainty does not allow complete planning, relational contracts are used in governing relationships, which may change as subsequent events unfold before the parties to the transaction. Relational contracts are more analogous to Commons' managerial transactions and Williamson's hierarchical transactions, in which the future responsibilities and obligations of the parties are subject to modification. However, it is possible for some bargaining transactions to appear to be much more relational than discrete, and for some managerial transactions to be discrete.

4. We do not treat in detail here the basic ingredients of legal contract: (1) an offer and acceptance that constitutes an agreement among the parties, (2) consideration that is legally sufficient and bargained for by the parties, (3) capacity to enter into the contract by all the parties, (4) reality of assent in which the parties genuinely agree to the terms and conditions of the contract, (5) form, or that the contract meets legal requirements, and (6) legality, or that the subject of the contract does not violate public policy.

5. This model of emergent interpersonal transaction processes is outlined in greater depth in Chapter 10 by Ring and Rands.

6. The authors wish to acknowledge the helpful comments of Gerald Salancik that led us to this insight.

7. It may be possible, however, through more careful attention to sense-making and understanding processes to reduce adverse effects of bounded rationality and information asymmetry.

8. Hambrick (1987) in a study of top management teams associated with failed organizations has found some evidence suggesting that these teams are overly populated with individuals with legal backgrounds.

9. See Hudson and Ring (1986) for a discussion of these kinds of issues in relation to the model presented in this chapter.

REFERENCES

Aldrich, Howard E. 1979 *Organizations and Environments.* Englewood Cliffs, N.J.: Prentice-Hall.

Aldrich, Howard E., and David Whetten. 1981. "Organization Sets, Action Sets, and Networks: Making the Most of Simplicity." In P.C. Nystrom and W.H. Starbuck, eds., *Handbook of Organizational Design.* New York: Oxford University.

Argyris, Chris. 1960. *Understanding Organizational Behavior.* Homewood, Ill.: Dorsey.

Arrow, Kenneth. 1973. *Information and Economic Behavior.* Stockholm: Federation of Swedish Industries.

Berg, S.V., J. Duncan, and P. Friedman. 1982. *Joint Venture Strategies and Corporate Innovation.* Cambridge, Mass.: Oelgeschlager, Gunn & Hain.

Berger, P.L., and T. Luckmann. 1966. *The Social Construction of Reality.* Garden City, N.Y.: Doubleday.

Blau, Peter M. 1964. *Exchange and Power in Social Life.* New York: Wiley.

Coase, Ronald. 1937. "The Nature of the Firm." *Economica N.S.* 4 (1937); reprinted in G.J. Stigler and K.E. Boulding, eds., 1952. *Readings* in Price Theory. *Homewood, Ill.: Irwin.*

Coleman, James S. 1986. "Social Theory, Research, and Theory of Action." *American Journal of Sociology* 91: 1309–35.

Child, John. 1972. "Organizational Structures, Environment and Performance: the Role of Strategic Choice." *Sociology* 6: 1–22.

Commons, John R. 1934. *Institutional Economics,* Madison: University of Wisconsin.

———. 1950. *The Economics of Collective Actions.* Madison: University of Wisconsin.

Cyert, Richard, and James March. 1963. *A Behavioral Theory of the Firm.* Englewood Cliffs, N.J.: Prentice-Hall.

Fisher, Roger, and William Ury. 1981. *Getting to Yes.* New York: Penguin.

Fox, Frederick, and Barry Staw. 1979. "The Trapped Administrator: The Effects of Job Insecurity and Policy Resistance Upon Commitment to a Course of Action." *Administrative Science Quarterly* 24: 449–71.

Fried, Charles. 1981. *Contract as Promise.* Cambridge, Mass.: Harvard University.

Gouldner, Alvin. 1959. "Reciprocity and Autonomy in Functional Theory." In L. Gross ed., *Symposium on Sociological Theory;* pp. 241–70. New York: Harper & Row.

Hambrick, D. 1987. "Top Management Team Composition and Firm Failure." SMRC Colloquium, University of Minnesota, February.

Hare, A.P. 1961. *Handbook of Small Group Research.* New York: Free Press.

Homans, George. 1961. *Social Behavior: Its Elementary Forms.* New York: Harcourt.

Hudson, Roger L., and Peter S. Ring. 1986. "Design-

ing Low Cost Governance Structures" Strategic Management Research Center, University of Minnesota, mimeo.

Klein, William. 1982. "The Modern Business Organization: Bargaining Under Constraints." *Yale Law Journal* 91(8): 1521–64.

Kotter, John P. 1973. "The Psychological Contract: Managing the Joining Up Process." *California Management Review* 15: 91–99.

Lawless, David J. 1972. *Effective Management: A Social Psychological Approach.* Englewood Cliffs, N.J.: Prentice-Hall.

Levine, L., and Paul E. White. 1961. "Exchange as a Conceptual Framework for the Study of Interorganizational Relationships." *Administrative Science Quarterly* 5: 583–601.

Levinson, Harry, Charlton R. Price, Kenneth Munden, Harold J. Mandl, and Charles M. Solley. 1962. *Men, Management, and Mental Health.* Cambridge, Mass.: Harvard University.

Macaulay. S. 1963. "Non-contractual Relations in Business." *American Sociological Review* 28: 55–70.

Macneil, I.R. 1974. "The Many Futures of Contract." *Southern California Law Review* 47 (1974): 691–816.

———. 1978. "Contracts: Adjustments of Long-Term Economic Relationship under Classical, Neoclassical, and Relational Contract Law." *Northwestern University Law Review* 72: 854–906.

———. 1980. *The New Social Contract.* New Haven, Conn.: Yale University.

Maitland, Ian, John Bryson, and Andrew Van de Ven. 1985. "Sociologists, Economists, and Opportunism." *Academy of Management Review* 10: 59–65.

McGrath, Joseph E. 1984. *Groups: Interaction and Performance.* Englewood Cliffs, N.J.: Prentice-Hall.

McMillan, Ian C., and Patricia E. Jones. 1986. *Strategy Formulation: Power and Politics,* 2d ed. St. Paul, Minn.: West.

Moment, D., and A. Zaleznik. 1963. *Role Development and Interpersonal Competence.* Cambridge, Mass.: Harvard University.

Morgan, G. 1986. *Images of Organization.* Beverley Hills, Calif.: Sage.

Morely, Ian, and Geoffrey Stephenson. 1977. *The Social Psychology of Bargaining.* London: George Allen & Unwin.

Nicholson, N., and G. Johns. 1985. "The Absence Culture and the Psychological Contract—Who's in Control of Absence?" *Academy of Management Review* 10 (3): 397–407.

O'Toole, R., A.W. O'Toole, R. McMillen, and M. Lefton. 1972. *The Cleveland Rehabilitation Complex: A Study of Interagency Coordination,* Cleveland, Oh.: Vocational Guidance and Rehabilitation Services.

Pfeffer, J., and G.R. Salancik. 1978. *The External Control of Organizations: A Resource Dependence Perspective.* New York: Harper & Row.

Ring, Peter S., and Gordon Rands. 1987. "Sensemaking, Understanding, and Committing: Emergent Transaction Processes in the Evolution of 3M's Microgravity Research Program." In A. Van de Ven, H. Angle, and M. Poole, eds., *Research on the Management of Innovation.* Cambridge, Mass.: Ballinger.

Schein, Edgar. 1970. *Organizational Psychology.* New York: Prentice-Hall.

Schroeder, Roger, Andrew Van de Ven, Gary Scudder, Douglas Polley. "Managing Innovations and Change Processes: Findings from the Minnesota Innovation Research Program." *Agribusiness* 2: 501–23.

Starbuck, W. 1983. "Organizations as Action Generators." *American Sociological Review* 48 (1): 91–102.

Staw, Barry M. "Knee Deep in the Big Muddy: A Study of Escalating Commitment to a Course of Action." *Organizational Behavior and Human Performance* 16: 27–44.

Teece, David J. 1985. "Applying Concepts of Economic Analysis to Strategic Management." In J.M. Pennings, ed., *Organizational Strategy and Change.* San Francisco: Jossey-Bass.

Thomas, R. 1976. "Managing the Psychological Contract." In P.R. Lawrence, L.B. Barnes, and J.W. Lorsch, eds., *Organizational Behavior and Administration: Cases and Readings,* 3rd ed. Homewood, Ill.: Irwin.

Thompson, James D., and Arthur Tuden. 1959. "Strategies, Structures and Processes of Organizational Decision." In James D. Thompson, et al., eds, *Comparative Studies in Administration.* Pittsburgh, Pa.: University of Pittsburgh. pp. 195–216.

Turner, J.H. 1987. "Toward a Sociological Theory of Motivation." *American Sociological Review* 52: 15–27.

Van de Ven, A.H. 1985. "Spinning on Symbolism: The Problem of Ambivalence." *Journal of Management* 11: 101–2.

Van de Ven, A.H., and G. Walker. 1984. "The Dynamics of Interorganizational Coordination." *Administrative Science Quarterly* 29: 598–621.

Warren, R. 1971. *Truth, Love, and Social Change.* Chicago: Rand-McNally.

Weick, K. 1979. *The Social Psychology of Organizing.* Reading, Mass.: Addison-Wesley.

Williamson, Oliver. 1975. *Markets and Hierarchies.* New York: Free Press.

———. 1981. "The Economics of Organization: The Transaction Cost Approach." *American Journal of Sociology* 87: 548–77.

———. 1985. *The Economic Institutions of Capitalism.* New York: Free Press.

7

INNOVATION OUTCOMES, LEARNING, AND ACTION LOOPS

Bright M. Dornblaser

Tse-min Lin

Andrew H. Van de Ven

A core concept in the Minnesota Innovation Research Program (MIRP) is innovation outcomes, defined as the value judgments about success or failure that various people make about the developmental process and end results of an innovation. Kimberly (1981) rightly points out that a positive bias pervades the study and management of innovation. Innovation is often viewed as a good thing because the new ideas being developed are expected to be useful—profitable, constructive, or solve a problem. New ideas that are not perceived as useful are not normally called innovations; they are usually called "mistakes" (Van de Ven 1986). Objectively, of course, the usefulness of an idea can be only partially determined as results become apparent, which is typically after the innovation process is completed and implemented. It is only then when new ideas can be characterized as "innovations" or "mistakes."

However, several years of intensive investment and effort are often required to develop an innovation to the point where its end-results can be determined. As a consequence, a central problem in managing innovation is determining the success or failure of a developmental process in the absence of objective or concrete information about the end-results of an innovation. Yet it is clear that subjective judgments of innovation success are made throughout the developmental process. These

We are grateful to helpful suggestions on an earlier draft of this chapter from Harold Angle, Scott Poole, as well as our colleagues in the Minnesota Innovation Research Program for providing and verifying data on the innovations examined here. Support for this research has been provided in part by a grant to the Strategic Management Research Center at the University of Minnesota from the Program on Organization Effectiveness, Office of Naval Research, under contract No. N00014-84-K-0016.

outcome assessments provide the bases on which innovation participants, managers, investors, and policymakers repeatedly evaluate the need for and invest in an innovation, guide its direction, and allocate their limited resources. Moreover, these in-process assessments largely predict the future course of events in the development of an innovation. People act on the basis of their assessments, and tend do more of what they think leads to success and less of what is perceived to lead to failure.

A basic assumption in the MIRP framework is that value judgments about innovation success are socially constructed in situ and change as an innovation develops over time. It leads one to examine the following questions, which are addressed in this chapter: What criteria are used to evaluate success while innovations develop? How and why do these criteria change over time? What are the consequences of these changes?

After a description of the methodology used to examine innovation outcomes in the MIRP studies, this chapter addresses the first question by providing a chronological description of changes in outcome criteria used by resource controllers and managers to assess their innovations over time. In response to the second question, it will be shown that some of these changes can be explained by concurrent changes in the innovation ideas, the people involved, their transactions with other units, and contextual events as the innovations developed over time. However, as the third question suggests, in-process innovation outcomes should not be simply viewed as dependent variables, as they traditionally have been treated in past research; they are also important independent predictors of subsequent courses of action in the development of innovations. We will conclude by proposing a dynamic model of action loops that explains the temporal antecedents and consequences of in-process assessments on innovation development. Achieving this will

hopefully provide a more robust and practically useful understanding of innovation success or failure than has been available in the past.

METHODOLOGY

The data used for the temporal analysis of innovation outcomes in this chapter come from repeated observations, surveys, and interviews obtained from eight innovations being studied by MIRP investigators. The eight innovations include a diverse set of product, process, and administrative innovations in public and private organizations. They are described in depth in other chapters: apheresis, cochlear implants, and a computer software company (see Chapter 8), a naval weapons system (Chapter 12), site-based management in two school districts (Chapter 17), and strategic planning in two governmental organizations (Chapter 18). Qualitative studies of these innovations were supplemented with quantitative survey data obtained from other MIRP innovations reported in Chapter 3.

As described in greater detail in Chapters 2 and 3, longitudinal data on innovation outcomes were obtained with questionnaires and interviews conducted approximately every six months with key innovation participants by the MIRP researchers. The questionnaire included five items designed to measure the degree to which respondents perceive that an innovation attains their expectations about process (progress in developing the innovation and solving problems as they are encountered) and outcomes (assessments of the innovation's present effectiveness and ability to contribute to attaining overall organizational goals). These five items were averaged to obtain a composite perceived effectiveness scale and was found to achieve good estimates of reliability and validity (see Chapter 3).

In addition, a parallel measure of perceived innovation effectiveness was used in the interview schedule: "How do you grade the innovation on these criteria? (A = 5, B = 4, C = 3, D = 2, F = 1)." As Chapter 3 reported, this parallel measure correlated .73 with the perceived effectiveness scale in the survey and provides a strong footing for the criterion-related validity of the perceived effectiveness measure.

The interview schedule also included three questions that were designed to provide a qualitative understanding of in-process innovation outcomes: "What current criteria are you using to judge the success of this innovation? Have changes occurred in the criteria used to judge the innovation in the past six months? How and when did the changes occur?" Repeated interview responses to these questions were content analyzed to identify the in-process outcome criteria used and how they changed over time.

For each innovation, responses were classified into two groups: (1) the innovation's internal management team, usually consisting of the innovation general manager and functional managers, and (2) external resource controllers, consisting either of venture capitalists and investors in the innovation or of top corporate managers in the organizations that housed the innovation units. This classification was made in order to determine the congruency of outcome criteria and ratings between internal innovation managers and their external resource controllers.

QUALITATIVE RESULTS

Temporal Patterns in Innovation Outcomes

The outcome criteria reported by the internal managers and external resource controllers for each innovation over time are summarized in Table 7–1.[1] This table presents the core qualitative data from which temporal patterns in the development of outcome criteria will be drawn in this section. However, it should be recognized that the table represents only a small distilled snapshot of the evidence necessary to empirically substantiate the patterns. We are confronted with the typical dilemma of having to tersely tabulate and summarize thousands of pages of longitudinal qualitative data obtained on the innovations. We will provide only examples that support and refute temporal patterns. In so doing, we emphasize that our objective is not theory testing; Instead it is to present and interpret empirical evidence that may lead to the discovery of a grounded-theory about temporal patterns in the development of innovation outcome criteria.

An examination of Table 7–1 shows that the specific outcome criteria used by resource controllers and innovation managers tended to be particularistic and differed substantially from innovation to innovation, even when the innovations are of the same type. It appears that managers and resource allocators select a variety of outcome criteria that are unique to their innovation in their particular organizations at specific times, and subjectively aggregate them to develop an overall judgment of the success of the innovation. This finding of managers using multiple, diverse, and idiosyncratic performance criteria is consistent with other effectiveness studies by Price (1968), Steers (1975), Campbell (1977), and Cameron (1980), and lead Van de Ven and Ferry (1980:-25–26) to draw three conclusions:

(1) It makes little sense to search for "objective" and universal measures of a concept that is inherently subjective—and is generalizable only to the unique set of decision makers who make the same value judgments in choosing effectiveness criteria. (2) Organization theory and logic of the mind is of little help in defining a concept that reflects the basic values, or simply, the "gut" feelings of people on "what they really

TABLE 7–1. *Outcome Criteria of Resource Controllers and Innovation Managers over Time.*

Time	Resource Controllers' Criteria	Innovation Managers' Criteria
Cochlear Implant Innovation		
6/84	Build new business in new market with new technology. Become self-sustaining business in ten years.	Create new product for deaf. Establish strong relationships with clinical research centers. Be first in market with FDA approval of first cochlear device.
2/85	Achieve operating income, expense, and profit objectives. Become number one player in new market in 10 years. Accomplish technical objectives.	Take advantage of opportunity window with FDA-approved device in market. Meet sales targets. Develop otological distribution and training centers.
7/85	Implement marketing plan. Take advantage of window of market opportunity. Meet sales targets.	Correct internal organizational problems. Develop production capacity for sales. Improve quality control.
12/85	Competence of program managers. Can program make it?	Solve image problem of product recall. Market and sell alternative products. Improve morale of program members.
8/86	Adequacy of market size for cochlear implant business. Shift from business to lab project—meet technical milestones and cost controls.	Reorganize program. Develop children's cochlear device. Leap-frog competitors with new device. Expand program into hearing aids.
12/86	Competence of program management. Over investment in single-channel device. Misreading of market demand. Resources drain of product recall.	Build credibility for program. Manage new acquisitions. Start work on hearing aids device.
9/87	Loss of market share. Technically inferior device. Failure in cochlear devices, but hope in hearing aids. Lack of matching people to jobs they are competent in.	Transition from cochlear to hearing aids business quickly. Get cost credibility to buy time. Become fastest growing division in three years.
Apheresis innovation		
11/84	Potential to build new business with new technology in 5–10 yrs. Build strong relationship with corporate partner and explore other opportunities with them. Undertake innovation until program manager decides its time to quit.	Develop an effective and marketable filtration device. Develop technological understanding. Meet developmental schedule.
4/85	Accomplish technical objectives. Show reasonable progress. Meet timelines and budget.	Accomplish product development schedule. Build strategic business unit group.

TABLE 7–1. *(continued)*

Time	Resource Controllers' Criteria	Innovation Managers' Criteria
9/85	Is the market there for the apheresis?	Meet budget and technical schedules. Solve engineering/production problems. Solve coordination problems among co-venturing firms.
4/86	Competence and productivity of innovation personnel. Meet financial and technical targets.	Survival: keep program alive. Plan execution and implementation. Show financial and technical performance.
11/86	Demonstrate "bottom line" in market entry of Phase I product.	Get "bugs" out of apheresis system. Prediction accuracy of technology and market competition.
6/87	Two-year schedule slippage. Investment attractiveness of program relative to others. Find another investor.	Maintain program credibility. Solve device quality problems. Improve communications among firms. Demonstrate program success relative to others.
12/87	Resource drain of program. Competence of program managers. Renegotiate program with partners.	Maintain commitment to program. Demonstrate clinical need in market. Relocate program personnel.

New company startup

Time	Resource Controllers' Criteria	Innovation Managers' Criteria
6/83	Attractive startup investment with public offering in a year (venture capitalist).	Start up and run own company. Develop and market custom design and medical records software. Distribute and service computer hardware.
12/83	Provide public financial offering during favorable market climate (venture capitalist).	Build and scale up organization. Search for additional capital Create marketing/distribution function.
7/84	(Planned public offering cancelled because market is "too soft.")	Narrow organizational focus on medical records software design and marketing; hardware distribution and custom software become intermediate revenue generators. Build marketing/distributor arrangements.
12/84	Ability of firm to follow planned financial milestones (banker).	Manage financial crisis and organizational retrenchment. Generate revenues from product sales and custom software. Develop joint venture to develop utilities load management device.
7/85	(Bank closes line of credit.)	Live out financial crisis.
3/86	Establish financial controls and marketing capabilities in small company. Bring in "hard-nosed" CEO (venture capitalist).	Financially restructure company with a second private placement or close the business.

197

TABLE 7–1. (continued)

Time	Resource Controllers' Criteria	Innovation Managers' Criteria
12/86	Conduct second private placement. Sell medical products division; future of company rests with future royalties from utilities software contract (board member).	Complete creditors' agreement with funds from second private placement. Keep medical software viable for sale. Obtain additional funding and cut costs.

Naval systems development

9/84	Meet objectives of navy full-scale development contract.	Incorporate navy's design-to-production templates into program. Organize and realign functional areas to develop program.
3/85	Achieve profit objectives. Dominate selected defense department markets Get corporate capital for production facility.	Meet customer expectations for technical quality, productivity, and cost criteria. Achieve high DOD award fees. Make revisions in materials management, human resources, and manufacturing.
12/85	Build management competence in program.	Solve software development problems. Achieve management coordination. Develop integrated and automated factory.
5/86	Division is to become integrated player in corporate group. Respond to DOD "second sourcing" proposal.	Resolve production problems. Develop transition teams.
12/86	Renegotiate contract with navy. Develop strategic plan to diversify business.	Respond to resource reductions. Meet productivity and cost expectations Maintain program continuity in new competitive environment.

River School District site-based management

9/85	Obtain stakeholders' contributions to school problem solving. Improve program quality. Individual school goal setting.	Increase power at the school level. District level management support. Improve upward communication channels. Improve competence of school councils.
8/86	School cost-effective competition. Obtain union support.	Clarity of SBM goals, structure, and roles. Obtain district superintendent attention. Reduce cost to match the benefit.

Metro School District site-based management

9/85	Improve participatory decisionmaking process.	Better school decisions with broad stakeholder inputs.
8/86	Competence of school management council and school leadership. Operate SBM within available fiscal resources.	Improve school management council's membership, leadership, and success Obtain possible additional resources.

Strategic planning—City E

2/85	Contribute to city council's performance.	Improve strategic planning capabilities. Bring key decisionmakers together. Address important performance issues. Develop issue resolution strategy.

TABLE 7–1. (continued)

Time	Resource Controllers' Criteria	Innovation Managers' Criteria
10/85	Minimize psychological/social/political cost of a change in city council's planning roles	Reduce costs to match reduction in slack resources for strategic plan
6/86	Implement strategies.	Leadership needed from champion. Support from city council.

Strategic planning—Nursing Service

3/86	Contribute to county board's performance.	Improve strategic planning capabilities. Increase organizational power.
11/85	Satisfy grant foundation requirements.	Minimize risk of participation
9/86	Match strategic planning's workload costs and benefits.	Build members' competence to resolve conflicts and problems.

want" and "what is important to them." (3) The *processes* by which people in organizations do, don't, and can articulate answers to these questions have been ignored.

Indeed, as we content analyzed the qualitative data underlying Table 7–1 for *processes* in the development of outcome criteria over time, three basic patterns became apparent: (1) Outcome criteria shifted over time and reflected the dominant concerns of respondents at the point in time they were being interviewed. (2) The outcome criteria of innovation managers and resource controllers were different at the beginning, converged during the developmental process, and diverged in conflicting opposite directions as innovation implementation problems arose. (3) Given conflicting criteria between internal innovation managers and external resource controllers, the outcome assessments that mattered in subsequently shaping an innovation's direction was that of the more powerful resource controllers. Each of these temporal patterns will now be discussed and elaborated.

Emergent and Shifting Outcome Criteria. Table 7–1 shows that outcome criteria (however idiosyncratic) changed over time for each innovation. Not only did initially nebulous targets crystallize later into more operational criteria,

but the targets themselves were reconstructed to redirect the innovations. For example, the success criteria used by the therapeutic apheresis innovation manager shifted over time from: whether an apheresis product could be developed at time 1, to accomplishing a product development schedule at time 2, to meeting innovation budget projections at time 3, and to keeping the innovation program alive at time 4. As this example suggests, outcome criteria can be viewed largely as indicators of the substantive concerns capturing the attention of innovation participants at different points in time along the innovation journey.

These changes in outcome criteria appeared to coincide with three kinds of events. First, the innovations frequently encountered setbacks, mistakes, and surprises (see Chapter 4), and as these unanticipated developmental problems arose, outcome criteria were added or revised concomitantly to evaluate their resolution. For example, difficulties of the cochlear implant program to scale up production capacity to support sales demand lead to establishing outcome criteria to do so. Second, unexpected events in the organization housing the innovation also influenced outcome criteria. For example, a corporatewide asset management decision by the defense contractor not to build new buildings and to refurbish existing production facilities required an unanticipated major factory automation and redesign effort. So also, work load in preparing the county's annual budget diverted time away from strategic planning and thus threatened the continuation of that innovation. Hence, a criterion was added concerning continuation of the strategic planning process. Finally, external environmental events that occurred independent of the innovations also had "spillover" effects on outcome criteria. For example, technological developments in competing innovations or substitute technologies created moving targets against which the hybrid wheat and cochlear implant innovations were judged.

As different types and degrees of criteria were used and emphasized at different points in time, innovation outcome evaluations became increasingly difficult to judge. This often created contradictory assessments of the overall performance of an innovation. For example and as described in Chapter 12, success evaluations of the naval weapons system decreased during a period when many important administrative "mini-innovations" were completed successfully. During the same period other technical problems in developing the torpedo had surfaced, and their salience may have dominated the evaluations. So also, ratings of innovation effectiveness by these innovation participants decreased during a period when the naval customer awarded the defense contractor an all-time high award fee for outstanding contract performance.

In other instances, judgments of innovation success were apparently influenced more by its "spin-off" effects than by its intended achievements. For example, Bryson and Roering report in Chapter 18 that while the strategic planning objectives in City E were not achieved, positive outcome assessments were made because the innovation stimulated city managers to initiate other activities that had good potential for improving other organizational outcomes. Although perceived effectiveness of the apheresis innovation dropped

sharply in recent periods as the innovation was struggling for survival, top managers indicated the innovation was successful in achieving the intended "side-effect" of using the innovation as a vehicle to explore other business opportunities with the co-venturing firm.

In other innovations studied, the spin-offs and side effects were often so tightly linked with preexisting orders that it was not possible to isolate and attribute outcomes from a specific innovation. Yet they may have far greater effect than any direct outcome from an innovation, as might be reflected in any single measure of ultimate innovation success or failure. These observations suggest that it may be more productive to view innovation outcome assessments as "remnants of a journey," rather than as objective benchmarks of developmental success.

Finally, in some cases outcome assessments appeared to be absolutely defined; in others success was a matter of degree. For example, Chapter 8 describes how failures in a few key components of developing initial products in the hearing health, apheresis, and company startup cases, derailed the entire first-product development efforts. However, in the development of school reform legislation in Chapter 9, compromise and partial goal attainment was considered progress or partial success. For example, although the Freedom of Choice bill failed for kindergarten through sixth graders, it passed and was adopted for seventh and eighth graders. Thus, perhaps due to the decomposibility or interdependence of innovation components, it appeared that an all-or-nothing view of success was taken in these private-sector innovations, while an incremental view of success was taken in the public-sector case.

Divergent and Convergent Criteria between Internal and External Groups. A second basic pattern is that outcome criteria reported by internal innovation managers and external resource controllers were different in the beginning, converged during the development process, and then diverged in opposite directions when innovation implementation problems were encountered. This pattern is illustrated in Table 7–2, which lists the typical outcome criteria cited by innovation managers and resource controllers for those innovations on which data are available over three or more time periods in Table 7–1. Although not mutually exclusive, the performance criteria emphasized by resource controllers shifted over time from long-run output goals to process concerns and then to more immediate input criteria, whereas those of the innovation managers tended to go in the opposite direction—that is, from input to process to output criteria.

At the outset, external resource controllers commonly justified their initial innovation investments and commitments with the guarded optimism that their innovations had the potential of contributing to long-run organizational goals, programs, profit, or business growth. For example, top managers of the cochlear implant and therapeutic apheresis in-ternal corporate ventures emphasized that although these innovations entailed high risks, (1) such risk taking was an inevitable part of creating the next generation of technologies and businesses necessary to sustain the long-term viability and growth of the corporation, and (2) that a long-term (five- to ten-year) corporate commitment to the development of these innovations could achieve this potential. In contrast to these long-run outcome criteria, the initial success criteria of internal innovation managers or entrepreneurs focused much more on short-run hurdles of obtaining resources, recruiting personnel, team building, planning, and mobilizing technical activities needed to initiate innovation development. Although the innovation managers often acknowledged a longer-term vision that complemented that of the resource controllers, it was clear in the beginning that their central and immediate attention was devoted to launching startup activities as quickly as possible.

As developmental work progressed, however, each innovation encountered unanticipated setbacks and problems, causing ini-

TABLE 7–2. *Divergent-Convergent Pattern of Outcome Criteria Held by Internal Innovation Managers and External Resource Controllers.*

Period	Resource Controllers	Innovation Managers
Beginning	*Outcome criteria* Create a new self-sustaining business/program in 5–10 years. Contribute to organization's goals. Achieve profit objectives.	*Input criteria* Obtain resources needed to launch innovation. Build innovation team. Develop technical design of innovation idea.
Middle	*Process criteria* Meet targets and schedules in the plan. Organization of team and implementation of plans. Compare progress relative to competing others.	*Process criteria* Achieve technical milestones. Debug the system. Meet deadlines and budget. Maintain credibility.
Ending	*Input criteria* Competency of managers. Cost and resource drain. Effort and commitment of innovation people.	*Outcome criteria* Demonstrate market success and potential ROI. Contribution to organizational goals. Survival and/or growth.

tially set targets and schedules to slip. As stated before, these problems led innovation managers to emphasize meeting process criteria of solving these problems, achieving technical milestones, and meeting deadlines and budgets in order to maintain credibility for their innovation development efforts. Information about these problems were reported to resource controllers during occasional meetings with innovation managers and in periodic administrative review sessions. However, in attending many of these administrative review sessions, researchers observed that innovation managers tended to reconstruct negative information in a positive frame with assurances that they were in control of problems and with action plans for addressing the problems. Because resource controllers were not involved in daily innovation activities, they relied on this indirect and more ambiguous information to assess innovation progress. Thus, during this "middle period" resource controllers tended to mimic success criteria in the areas they were told the innovations were experiencing process problems.

More objective performance information did not become available until a later time (often several years) when attempts were made to introduce the innovation, either through pilot testing, customer inspections, manufacturing scaleup, market introductions, or initial organizational implementation. When these initial "acid tests" were judged successful, the process problems identified earlier were either dismissed as unimportant or overcome. In response to probes by the researchers, no efforts were reported by innovation managers to seriously study or record the correctness of inferences drawn from these experiences.

However, when innovation implementation efforts went awry, outcome criteria of resource controllers and innovation managers diverged. Disappointed with results, the external resource controllers tended to focus on input criteria by expressing concerns about the competency and efforts of innovation managers, the innovation's developmental costs, and its resource "drain" relative to other investment opportunities and priorities that had surfaced during the interim period. In contrast and in efforts to "salvage" their innovations, internal managers appealed to long-run potential contributions their innovations could make if resource controllers looked beyond the immediate "temporary" setbacks and maintained a long-term commitment to their innovations. Thus, as Ross and Staw (1986) observed elsewhere, innovation managers interpreted failures less as a symptom that their innovation was incorrect than as an indication that it had not been pursued vigorously enough. As a result, disagreements arose over the meaning of innovation implementation setbacks as resource controllers and innovation managers developed alternative stories that interpreted the same experience quite differently.

Whose Assessments Matter? This question arises naturally, given the conflicting criteria between resource controllers and innovation managers. Although the outcome assessments of innovation managers and participants appeared to trigger changes in innovation developmental activities, those of the more powerful resource controllers influenced the very survival and form of the innovations. Three interaction effects appeared to be present between evaluations of resource controllers and innovation managers: (1) Actions taken in response to resource controllers' evaluations were delayed when these evaluations conflicted with those of innovation managers, (2) resource controllers' pronouncements of innovation success or failure "spilled over" on innovation managers' definitions of success or failure, and (3) in the end if outcome assessments of resource controllers and innovation managers did not converge, the assessments that mattered in determining the fate of an innovation were those of the resource controllers. These observations on the relative power of evaluators were observed in both positive (continuation) and negative (termination) directions on innovation development.

An example of the positive case is the strategic planning innovation in the nursing services organization described in Chapter 18, which continues perhaps because the outcome evaluations by its resource controllers are substantially more positive than those of the innovation participants. The latters' more negative assessments have stimulated a number of procedures to minimize risk of participation and improve conflict resolution among strategic planning participants.

When the resource controller is also the innovation champion, as in the case of the River School District superintendent (see Chapter 17), his or her judgments of innovation outcomes may result in a self-fulfilling prophesy. If he or she judges the innovation to be successful, resource commitments increase, which provide slack to mask or correct mistakes and increase the probability of successfully overcoming obstacles to complete or implement the innovation.

However, in other cases the innovation champion was typically not the sole resource controller. For example, Chapter 8 describes how corporate innovation champions continued to infuse resources into their hearing health and apheresis programs for a year or two despite mounting technological setbacks and budget overruns. But even the corporate innovation champions' mediating or buffering role between innovation managers' escalating commitments and other resource controllers' rising skepticism could be sustained for only so long. Growing concerns among peer corporate resource controllers combined with normal periodic structural reorganizations of corporate offices (resulting in both cases in innovations that no longer reported directly to their corporate champions) resulted in significant budget reductions in both innovations. These budget reductions decreased the probabilities of correcting mistakes or surmounting implementation hurdles and thereby reinforced other resource controllers' conclusion of not "throwing good money after bad money."

In the wake of the drastic budget cut, the hearing health innovation managers resurrected the idea to shift the innovation from cochlear implants to hearing aids development—an idea playfully explored two years earlier by innovation participants when slack resources existed. In a meeting attended by the researchers, they successfully persuaded a corporate investment committee to commit to the idea. One resource controller stated "the idea provides a way to partially recoup our investment" because technological competencies in cochlear implants could be productively redeployed on hearing aids development. As a result, the program obtained a "new lease on life" by embarking on a new innovation.

In the wake of its July 1987 announcement of a budget cut effective January 1988, despite several trials the apheresis managers proposed no convincing alternative courses of action to change resource controllers assessments that the program had "failed." This judgment was subsequently found to "spill over" on the innovation manager. Over the course of four years of bimonthly meetings with the researchers, December 1987 was the first time the apheresis innovation manager labeled his innovation a "failure," as he and other participants were in the process of searching for other jobs in their organizations or elsewhere.

RATIONAL AND SUPERSTITIOUS LEARNING

A learning process involves the feedback of outcomes to actions. Basic to principles of rational learning is that outcomes are a function of actions that are believed to lead to those outcomes and are not a result of spurious unknown factors. Here, as before, "outcomes" refer to desired results or ends of a developmental process, and "actions" refer to the means or activities undertaken to achieve these ends. Based on this causal means-ends logic, rational learning increases by decreasing the ambiguity of knowing what actions determine what outcomes. As illustrated in Figure 7–1, the rational learning model admits to the possibility that outcomes may be influenced by a variety of spurious unknown factors (such as cognitive limitations, shifting frames of references, or extraneous environmental events). However, these spurious factors are assumed to be exogenous to the model and to exert little influence on outcomes because they are normally distributed, and therefore they are treated as "noise" or error terms in most rational models of learning.

However, the troubling finding in the foregoing qualitative analysis is that numerous instances were observed where actions appeared loosely connected to outcomes and where judgments about these connections were incorrect—that is, desired outcomes appeared to be less influenced by actions that were thought to lead to these outcomes than by extraneous factors unknown to the participants.

As a consequence, relatively little rational learning appeared to occur.

Superstitious learning occurs when the subjective experience of learning is compelling but the connections between actions and outcomes are loose or misspecified (Skinner 1953; Levitt and March 1988). Numerous instances were shown to exist for such misunderstandings in learning from experience as the innovations developed. First, desired innovation outcomes changed from target A to B to C over time as evaluators' reference points, experiences, and aspirations changed. As a result, actions initially undertaken to achieve target A became irrelevant and difficult to redirect to achieve targets B or C. Second, in-process outcome assessments appeared to rely less on an innovation's course of action and more on changing external factors that emphasized the innovation's success relative to shifting organizational opportunities, technological knowledge, and environmental contexts. What was learned appeared to be less influenced by the history of events and activities accomplished by a given innovation than by the shifting contextual frames of reference that were applied to that history (as observed elsewhere by Fischoff 1975; Kahneman, Slovic, and Tversky 1982; Pettigrew 1985). What happened was not always obvious, causality between actions, outcomes, and exogenous events was difficult to untangle, and thus the difference between success and failure appeared to be largely a product of "enactment" (Weick 1979).

Yet the qualitative data provide clear evidence that innovation participants and investors formed interpretations of events, classified outcomes as good or bad, and acted in manners largely consistent with their assessments. Innovation units continued to pursue or expand their courses of action when outcomes were judged to be successful and changed their courses of action when outcomes were judged to be failures. This pattern of outcomes influencing subsequent actions was apparent

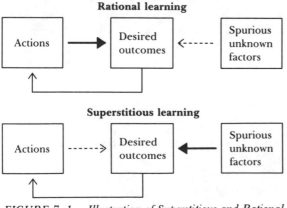

FIGURE 7–1. Illustration of Superstitious and Rational Learning Models.

irrespective of whether the learning process appeared to be rational or superstitious.

Resource controllers and innovation managers often had different targets and evaluated the same events differently. As a result, the ambiguous process by which these interpretations developed made it relatively easy for conflicts of interest to spawn conflicting interpretations. Under these conditions, Levitt and March (1988) suggest that the relative power of evaluators in a nested organizational hierarchy makes it possible to largely predict which evaluations will govern the developmental progressions of innovations. During very good times—that is, when the more powerful resource controllers interpret events positively—only exceptionally inappropriate courses of action will lead to judgments of innovation failure. In like manner, during very bad times, or when pessimistic resource controllers reinterpret outcomes negatively, no course of action will lead to outcomes judged to be successful. Thus, what may appear to be a rational learning process to powerful resource controllers may be a superstitious learning process for innovation participants. But even this micro-macro levels comparison is tentative because rational learning denies tautologies. As observed before, success or failure can become a self-fulfilling prophesy when evaluators have the power to allocate resources that reinforce their assessments.

Causal Antecedents and Consequences of Innovation Outcomes

The foregoing qualitative analysis can be transformed into a relatively straightforward crucial proposition (Stinchcombe 1968) that may discriminate superstitious learning from rational learning in the development of innovations. As illustrated in Figure 7–1, superstitious learning is evident when the connections between actions and outcomes are either loose or misspecified. Rational learning presumes that out-

comes and actions are connected, that changes in outcomes are caused by actions, and that the action-outcome links are correctly identified. If outcomes are unconnected to actions, or if the causation is incorrect (that is, spurious unknown factors change outcomes that, in turn, cause actions), then superstitious learning prevails because whatever outcomes people think they are learning from their actions, they have incorrectly specified the connections. This line of reasoning suggests the following crucial propositions:

> Superstitious learning is evident when actions and outcomes are either unconnected or misspecified—that is, a substantial proportion of variation in perceived outcomes cannot be explained by actions, and outcome assessments determine actions more than actions determine subsequent outcomes.

The alternative and prevailing proposition is that

> Rational learning is evident when actions and outcomes are tightly coupled and correctly specified—that is, when a substantial proportion of variation in perceived outcomes can be explained by actions.

This section will present one approach to test these propositions by empirically examining the temporal causal ordering of innovation outcome assessments with core MIRP dimensions of innovation ideas, people, transactions, and organizational context. Figure 7–2 transforms the propositions into a path analytic model of key dimensions that were selected from the MIRP framework to test the propositions. Because operational definitions and measures of these dimensions are provided in Chapter 3, we only briefly outline why these dimensions were included in the causal path model.[2]

Innovation outcome is the core construct in

205

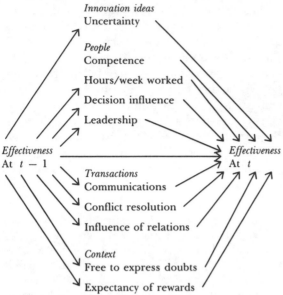

FIGURE 7–2. Path Model for Testing Rational-Superstitious Learning Propositions.

the path model. We examine its temporal antecedents and consequences in terms of the other dimensions outlined below. In-process outcome assessments are operationalized in terms of perceived innovation effectiveness, which is a composite scale consisting five items designed to measure the degree to which respondents perceive that the innovation attains their expectations about process (progress in developing the innovation and solving problems as they are encountered) and outcomes (assessments of the innovation's present effectiveness and ability to contribute to attaining overall organizational goals).

Innovation uncertainty is the analyzability and predictability of the innovation idea being developed as perceived by the individuals involved in its development. Because idea uncertainty makes it difficult for an individual to connect means with ends, we hypothesize that the greater the uncertainty the lower the associations between between dimensions of innovation action and outcome.

Four key dimensions of *people* involved in innovation development are included in the path model: *competence* (highest educational degree), *effort* (hours per week worked on the innovation), *decision influence* (perceived authority in making decisions about innovation goals, tasks, resources, and personnel recruitment), and *leadership* (the degree to which innovation managers are perceived by participants to encourage innovative behavior). Whereas competence is a key individual difference characteristic of what people bring to an innovation (and hence treated as an exogenous variable in the model), effort and decision influence capture how much energy people exert as an innovation develops, and leadership taps a critical aspect what participants perceive is done to them to develop an innovation. The standard expectation in the literature (reviewed by Angle in Chapter 5) is that innovation effectiveness (outcomes) is produced by the following actions: highly competent people exert high levels of effort and are given wide latitude to influence decisions with a group leadership style that encourages innovative behavior.

Three dimensions of *transactions* among people within the innovation unit and with other organizational units are examined in the path model: *communication frequency* among innovation participants, *conflict resolution* by openly confronting disagreements or disputes among innovation participants, and *relationship influence* in dealings with other organizational units. As discussed in Chapter 6, management scholars and practitioners typically believe that innovation effectiveness increases with frequent communications and open methods of conflict resolution among innovation participants, as well as strong ties with other organizational units relevant to the development of an innovation (Kanter 1983; Peters and Waterman 1983).

Finally, two dimensions of *organizational context* were included in the path model. *Freedom to express doubts* is the degree to which the orga-

206

nizational norms encourage people to express their own opinions and criticisms about the innovation. *Expectancy of rewards* is the extent to which innovation participants, individually and as a group, anticipate that good work performance will result in some rewards (whether formal promotion or informal recognition). Freedom to express doubts and expectancy of rewards are two key enabling and motivating conditions that organizations often promote to facilitate innovative behavior (see Chapter 5, as well as Peters and Waterman 1983 and Kanter 1983).

Evidence to support the rational learning proposition would be shown if the paths leading from these action dimensions to outcomes (perceived innovation effectiveness) in Figure 7–2 are strong and larger in magnitude than the paths leading from the exogenous lagged effectiveness variable. Reverse evidence would show support for the superstitious learning proposition.

The data used to test the path model in Figure 7–2 consist of three waves of the Minnesota Innovation Survey (MIS), which was completed approximately every six months by innovation participants and managers. The innovations include hearing health, therapeutic apheresis, naval system development, government strategic planning, and an innovation research program in all three data collection waves. However, survey data are available on school site-based management in only times 1 and 2, while surveys on space commercialization are available in times 2 and 3 only. Data obtained in each period from the same respondents were matched and merged in order to permit analysis of scores of the same respondent over time. This longitudinal data file produced 100 matched respondents across times 1 and 2, and 94 across times 2 and 3 for analyzing the path model in Figure 7–2.

The temporal ordering of variables in the path model is based on the findings reported in Chapter 3 on the predictive validity of the MIS dimensions. There it was reported that perceived effectiveness is largely associated with *concurrent* (not lagged) dimensions of ideas, people, transactions, and context. The LISREL model found that these latter MIS dimensions at time 1 had very low explanatory power on perceived effectiveness at time 2. (In a subsequent analysis it was found that while the MIS dimensions lagged at time 2 were moderate predictors of perceived effectiveness at time 3, concurrent predictors at time 3 were still stronger). The fact that perceived effectiveness responds to concurrent rather than lagged predictors suggests that the underlying causal patterns work themselves out far more quickly than six months. Therefore, the path model was specified to examine the concurrent effects of the MIS action dimensions on perceived effectiveness, whereas the effects of the latter on the former were lagged one time period.

Table 7–3 presents simple descriptive statistics obtained in the three survey waves on the eleven dimensions of innovation actions and outcomes. The left columns show the mean responses for each dimension in times 1, 2, and 3. The middle columns show the mean square differences between consecutive time periods (T1–T2 and T2–T3) for each dimension, as averaged across the respondents. Finally the right columns show the autocorrelations of each dimension between consecutive time periods (that is, the lagged self-correlation of each dimension between T1 and T2 and between T2 and T3).

Evident in Table 7–3 is a time-dependent pattern of *stabilization,* which is highly relevant to understanding the connections between actions and outcomes as innovations develop over time. Stabilization is indicated both by a decrease in the rate of change, and an increasing homogeneity in the direction of change. The middle columns in Table 7–3 show that the absolute changes or differences in each dimension are consistently smaller between T2–T3

TABLE 7-3. *Descriptive Statistics on MIRP Dimensions in Three Periods.*

	Mean Responses in			Mean Square Differences:		Autocorrelations (Same variables)	
	T1	T2	T3	T1–T2	T2–T3	T1&T2	T2&T3
Outcomes							
Effectiveness	3.21	3.30	3.23	.58	.50	.51	.66
Innovation ideas							
Uncertainty	2.82	3.09	2.96	.64	.47	.28	.59
People							
Competence	3.60	3.73	3.76	.11	.01	.93	.99
Hours/week worked	20.89	25.46	25.30	66	246	.92	.75
Decision influence	2.32	2.73	2.61	1.13	.79	.53	.69
Leadership	3.82	3.78	3.70	.49	.26	.29	.73
Transactions							
Communications frequent	3.62	3.78	3.79	.94	.89	.72	.75
Conflict resolution	3.36	3.55	3.52	1.37	.71	.28	.63
Influence of relations	2.81	2.84	2.86	1.62	.57	−.13	.52
Context							
Free to express doubts	3.39	3.39	3.32	1.02	.69	.32	.67
Expectancy of rewards	3.57	3.11	3.21	.83	.40	.23	.25

than they are between T1–T2. They indicate a slowdown in the rate of change as the innovations are maturing. On the other hand, the right columns of Table 7–3 show that the autocorrelations of each dimension are higher between T2 and T3 than they are between T1 and T2 (except for hours worked per week). This suggests an increasing homogeneity of changes in that more innovation participants are responding in the same direction on each dimension over time. This finding is consistent with that of Pelz and Andrews (1966) that with frequent interactions over time, a group of initially heterogeneous research scientists becomes homogeneous in perspectives and approaches to problems. As the rate of change in a developmental progression decreases, homogeneity of responses from its participants increases.

This time-dependent pattern of stabilization in the development of innovations suggests that, over time, as participants think more alike and are more stable in their perceptions about actions and effectiveness, the causal connections between actions and outcomes should increase. If they do, we should, as a consequence, expect greater evidence to support the superstitious learning proposition in the initial volatile period of innovation development and greater support for the rational learning proposition in later periods as innovation development progressions stabilize. We will therefore examine the path analytic model in Figure 7–2 over two lagged periods; first between T1 and T2 and second between T2 and T3.

Figure 7–3 presents the test results of the path model in the two lagged time periods. Written on the paths are the standardized beta coefficients (and in parentheses their probabilities of statistical significance) obtained from regression analyses. The results clearly show that the fit of the model is better for the later T2–T3 period than the earlier T1–T2 period. This is true for both parts of the model: In the T2–T3 versus T1–T2 period, lagged outcome effec-

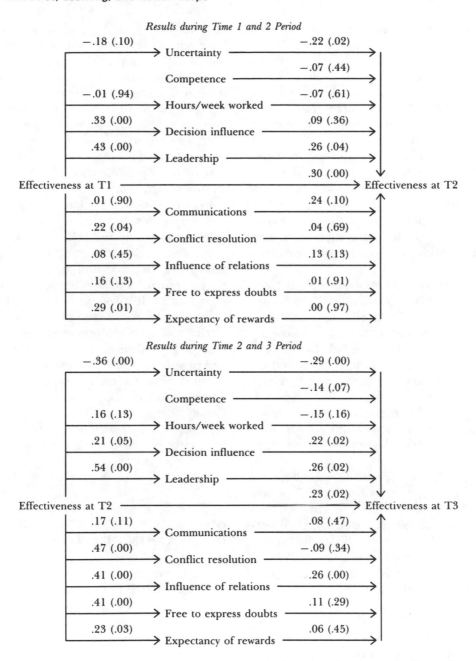

On paths are standardized beta coefficients (with probabilities of statistical significance in parentheses).

FIGURE 7–3. *Path Analytic Results for Testing Rational-Superstitious Learning Propositions.*

tiveness has more significant impact on the ideas, people, transactions, and context dimensions; and these action dimensions, in turn, explain significantly greater proportions of variations in concurrent perceived effectiveness (in T1–T2, adjusted R2 is 47 percent, at $p < .00$ level; while in T2–T3, adjusted R2 is 67 percent at $p < .00$ level of significance.)

As the innovations stabilize in their developmental progressions over time, the connections between actions and outcomes become stronger. Thus, the first elements in our propositions pertaining to the tightness of means-ends relationships are supported over time in favor of the rational learning proposition.

In addition, four relationships that are statistically significant (beyond a .05 level) between the action and outcome dimensions in times 2 and 3 are consistent with typical assumptions of rational learning held by management scholars and practitioners. Perceived innovation effectiveness decreases when the innovation idea becomes more uncertain, and increases as people exercise more influence over decisions and their relationships with other units, and have a leader who encourages innovative behavior.

However, a comparison of the beta coefficients on the paths in the T2–T3 period show that these four relationships between actions and outcomes are reciprocal; lagged perceived effectiveness is an equally strong predictor of idea uncertainty and decision influence, and clearly a stronger predictor of leadership and the influence of interunit relations, than the latter are of perceived effectiveness in the concurrent period. Moreover, perceived effectiveness at T2 significantly predicts conflict resolution, freedom to express doubts, and expectancy of rewards at T3; while these dimensions exert no influence on perceived effectiveness at T3, as was hypothesized according to the rational learning proposition.[3]

Thus, partial support and disconfirma-

tion has been found for both the rational and superstitious learning propositions in this sample of innovations. Although the action-outcome connections were found to increase over time, and some were consistent with rational learning principles about how actions predict outcomes, the data show that prior outcomes exert a significant reciprocal temporal relationship on four of these action dimensions. In particular, prior judgments of the effectiveness of an innovation significantly predicted decreases in the subsequent uncertainty of the innovation idea being developed, and increases in influence on decisions and relationships with other units, and having leaders who encouraged innovative behaviors; and these action dimensions, in turn, were significant predictors of innovation effectiveness. In addition, as innovations were judged to be increasingly effective at T2, more open methods followed for resolving conflicts among the parties involved, and people more freely expressed their doubts about the innovation and expected rewards for doing good work. Contrary to commonly held views, these methods of conflict resolution and organizational "enabling conditions" were found to be insignificant predictors of perceived effectiveness.

DISCUSSION: ACTION LOOPS OF INNOVATION OUTCOMES

A clear inference from the above qualitative and quantitative findings is that there are elements of both rational and superstitious learning occurring as innovations develop over time. Judgments about in-process innovation outcomes are partially a consequence of action, partially a predictor of future actions, but often unconnected to actions that occur in the development of innovations. The predictors and consequences of outcome assessments become clear only with time as developmental progressions stabilize. Even then, while perceived effectiveness judgments may provide a rational basis for choosing subsequent actions to develop an innovation, unspecified and conflicting targets or changing frames of reference often produce superstitious learning and struggles between innovation managers and resource controllers on the developmental progressions of their innovation.

One model that may explain these complex dynamics of in-process innovation outcomes has been provided by Van de Ven, Venkataraman, Polley and Garud in Chapter 8. Based on an accumulation theory of change, they point out that during the startup of an innovation the relative power of external resource controllers must shift to internal managers if the innovation is to successfully takeoff and sustain itself when implemented. They propose a model of action loops (illustrated in Figure 7–4) to explain the processes of transition in power from external to internal innovation groups during the innovation development process.

The model argues that if the course of action being pursued by an innovation unit is perceived to be successful, this increases the external resource controllers' confidence in and willingness to delegate greater control to the entrepreneurial unit, which in turn permits the innovation unit greater discretion to pursue and expand its chosen course of action. However, perceptions of failure produced either by negative assessments of an innovation's course of action or from external environmental events increases the resource controllers' uncertainties about the innovation, triggers their external intervention, and often results in a struggle between internal and external groups on the appropriate course of action to follow. When a new or revised course of action is selected, the struggle subsides and this renews the action loops. Resolutions to power struggles explained by Figure 7–4 may entail confusing, contradictory, and abrupt shifts in innova-

tion development. As a consequence, innovation startups may often reflect periods of rupture and discontinuous change as described earlier.

Thus, the model explains transitions in power between outside to inside groups on the basis of the perceived outcomes of the actions taken by the innovation unit. However, the model also acknowledges that outcome attributions may be influenced by contextual events (such as shifting priorities by external groups, progress of competitors, and other environmental factors) as well as learning and disabilities in correctly judging innovation outcomes. Thus, as observed in several MIRP cases, pursuit of an innovation idea may be terminated for reasons exogenous to successful accomplishing an agreed on course of action. Alternatively, an innovation unit may continue to escalate in the left loop of Figure 7–4 because either resource controllers or innovation managers erroneously perceive the course of action to be successful, while mistakes and failures are not detected or dismissed, and accumulate into vicious cycles.

In particular, Van de Ven et al. report in Chapter 8 that three kinds of learning disabilities were frequently observed that made success judgments error prone. First, it was often difficult for innovation managers to discriminate substantive issues from "noise" because positive and negative messages about performance outcomes were mixed in random order over time. As a consequence, at any particular point in time innovation outcome evaluations are highly subject to recency effects and over time are highly volatile as positive and negative information is received in random order. Second, hypervigilance often triggered fast learning, but fast learners tend to specialize in the inferior alternative, leaving the superior alternative to their somewhat slower competitors (Herriott, Levinthal, and March 1985). This suggests that moderate (not fast nor slow) rates of learning are related to success. Third, ratio-

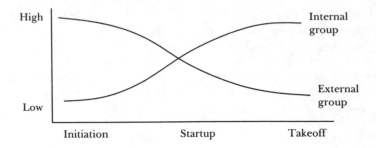

Action Loops in Transitions of Power
between Internal and External Business Groups

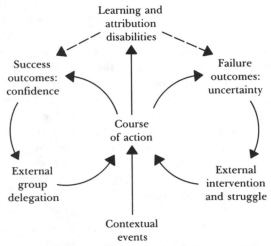

FIGURE 7–4. *Relative Power of Internal and External Groups in Accumulation Process Model.*

nal logic implies that modifications in courses of action are taken as failures become apparent. However, Brunsson (1982) points out that an entrepreneurial logic of escalating commitments to a course of action is the most reasonable logic of action to withstand the ever-present "nay sayers" and organizational inertia to an innovation. This entrepreneurial logic of action, of course, questions the rationality of much trial-and-error learning behavior.

As stated before, resource controllers and innovation managers were found to hold different criteria of innovation success, and each group imparted its own particular direction to the developmental process when occa-

sions arose. Innovation managers viewed their innovations up close and personal, whereas resource controllers were more detached and regarded the innovations as but one of a larger set of alternative investments. The innovation managers governed the immediate developmental work, while the resource controllers set the basic resource and strategy parameters for the innovations. In "good times" the two groups appeared in synch, but when failures arose it was clear they were not, and sudden shifts in the developmental sequences were observed.

As the model suggests, when developmental activities were perceived to proceed successfully, transitions in power from resource controllers to innovation managers were uneventful because the external groups were satisfied with progress and had little reason to intervene and impose their differing views on the internal group. In other words, in such "good times" the lack of outside intervention permitted the internal innovation units increased discretion to pursue and expand on their course of action.

However, when mistakes or failures were perceived, uncertainties arose about the appropriateness of the course of action being taken by the innovation unit, and this stimulated the resource controllers to intervene and explore alternative courses of action. These instances were occasions when power struggles erupted between the resource controllers and innovation managers, with the former usually imposing their preferred course of action on the latter.

Of course, many resource controllers and innovation managers were not oblivious to these dynamics and adopted a variety of suboptimization strategies to cope with the success and failure action loops—many of which only reinforced or exacerbated vicious cycles in the action loops. Three strategies in particular warrant discussion: "sugarcoating," "laying low," and "snipe control."

First, as discussed in Chapter 8, initial plans and proposals to undertake innovations were overly optimistic because they were used more as vehicles to obtain commitments from resource controllers than they were to develop realistic alternative scenarios of innovation development. Although innovation managers acknowledged that their innovation targets and costs were "sugar coated," none were willing to document likely pitfalls or a more extended development timetable because they perceived that would decrease their likelihood of obtaining resources. The resource controllers, in turn, held the innovation managers accountable for achieving these targets and projec-

tions, even when they recognized and discounted certain projections in the plans as "fluff." Thus, in the fear of not obtaining startup capital, innovation managers committed themselves to courses of action and outcome criteria that had low probabilities of being achieved, while resource controllers invested in innovation proposals and entrepreneurs that they hoped would produce long-run returns or benefits even though "operational details are uncertain."

As developmental work progressed, unforeseen setbacks and mistakes arose, and slipping schedules and cost overruns became apparent. Innovation managers recognized that such evidence could shake the confidence of resource controllers, resulting in their external interventions with directives that could misdirect innovation development and in decreased commitments to support the innovation. As a consequence, prior to periodic innovation administrative reviews several innovation units were observed to make extensive preparations and rehearsals of their presentations with fancy visuals of innovation progress. As stated earlier, in the administrative reviews negative information was "sugarcoated" into a positive frame with assurances that innovation managers were in control of problems. For resource controllers, these "slick" presentations appeared to reinforce their earlier (but private) recognition that initial plans contained some unattainable targets, and management's proposed positive action steps to correct setbacks promoted confidence in the innovation unit's abilities to cope with its problems. Innovation managers thereby extended their "grace period" to pursue and expand their course of action.

Several innovation managers were observed to adopt a strategy of "laying low" during these "grace periods" to address unforseen problems and setbacks in innovation development. "Laying low" included a variety of tactics, such as restructuring the innovation unit

213

so that innovation managers could report to lower-level organizational managers or to "fair-haired" managers who were viewed positively by resource controllers. It also included efforts to develop different methods of financial accounting that "pooled" the innovation unit's costs with other organizational operations and thereby become less identifiable.

But perhaps the most inventive strategy adopted by one innovation manager was what he called "snipe control." Reasoning that resource controllers' time and attention were scarce and sequentially focused on addressing only their most pressing problems, the innovation manager stated his objective was not to be among the top, highly visible "showcase" innovations. Instead it was to "stay one above the bottom of the pack, because it is the one at the bottom that gets all the scrutiny and top management attention."

CONCLUSION

Four basic patterns became apparent as we examined how innovation outcome criteria developed over time. First, the specific outcome criteria used by management evaluators tended to be particularistic, even when the innovations were of the same type. Managers and resource controllers selected a variety of outcome criteria that were unique to their innovation in their particular organizations at specific times and subjectively aggregated them to develop their overall judgments of success or failure.

Second, these outcome criteria shifted over time and appeared to reflect the dominant concerns of managers and investors at different points in time along the innovation journey. Not only did initially nebulous targets crystallize later into more operational criteria, but the targets themselves were reconstructed to redirect the innovations. These changes in outcome criteria coincided with unanticipated setbacks and problems in the development of the innovations, shifting priorities and events in the organizations housing the innovations, and external environmental events that occurred independently but had "spillover" effects on the innovations.

Third, the outcome criteria of internal innovation managers and external resource controllers were different in the beginning, converged during the developmental process, but then diverged in opposite and conflicting directions as innovation implementation problems arose. Whereas the performance criteria emphasized by resource controllers shifted over time from long-run output goals to process concerns and then to more immediate input criteria, those of the innovation managers tended to go in the opposite direction—that is, from input to process to output criteria over time.

Given these conflicting criteria between resource controllers and innovation mangers, we examined whose assessments mattered. Although the outcome assessments of innovation managers appeared to trigger incremental changes in developmental activities, those of the more powerful resource controllers not only influenced the very survival and form of the innovations, they affected innovation managers' definitions of innovation success or failure. In some cases resource controllers' judgments of innovation success appeared to be self-fulfilling prophesies when they allocated resources that reinforced their assessments.

In exploring these four qualitative themes about how innovation outcome criteria develop over time, numerous instances were observed where actions appeared loosely or incorrectly connected to outcomes, producing situations of superstitious learning. Rational learning presumes not only that actions and outcomes are connected but also that actions cause outcomes. If outcomes are unconnected to actions or the causation is incorrect (that is, spurious unknown factors determine outcomes), then superstitious learning prevails because whatever outcomes people think they are learning from their actions, they have incorrectly specified the connections.

A quantitative test was conducted of two crucial propositions to discriminate between the prevalence of rational or superstitious learning in the development of innovations. Based on three waves of MIRP survey data in which a time-dependent pattern of stabilization existed in the development of innovations, partial support and disconfirmation was found for both the rational and superstitious learning propositions in this sample of innovations. Some of the action-outcome relationships were found to be consistent with rational learning principles, but it also was found that prior outcomes exerted a significant reciprocal temporal relationship on these and other action dimensions—thereby establishing the conditions for superstitious learning.

The overall inference drawn from the these qualitative and quantitative findings is that there are elements of both rational and superstitious learning occurring as innovations develop over time. In-process innovation outcomes are partially a consequence of action, partially a predictor of future actions, but often incomplete explanations of actions that occur in the development of innovations. These predictors and consequences of outcomes assessments only become clear with time as developmental progressions stabilize. Even then perceived effectiveness judgments may provide a rational basis for choosing subsequent actions to develop an innovation, but unspecified and conflicting targets or changing frames of reference often produce superstitious learning and struggles between innovation managers and resource controllers on the developmental progressions of their innovation.

A model of success and failure action loops was proposed to explain these contradictory dynamics as innovations develop over time. The model argues that when the course of action being pursued by an innovation unit is judged successful, this increases the external resource controllers' confidence in and willingness to delegate greater control to the entrepreneurial unit, which in turn permits the innovation unit greater discretion to pursue and expand its chosen course of action. However, when failure is perceived, uncertainties arise, and they trigger external resource controllers to intervene and engage in a struggle with innovation managers on the appropriateness of the innovation's course of action. When this struggle subsides (often by the imposition of a new or modified action plan), the loop is completed and recycles in either positive or negative directions.

Thus, the model not only explains how innovation outcomes are both a cause and consequence of action, it also acknowledges that outcome attributions can be produced by spurious unknown factors. In addition to an assessment of an innovation's course of action, outcome assessments are influenced by environmental events, shifting organizational priorities, as well as three observed learning disabilities in judging innovation outcomes correctly: mixed messages randomly ordered over time, hypervigilance, and entrepreneurial logic.

These research findings and inferences call into question a firmly held assumption about the management of innovation. Applied managerial research has typically searched for the actions that lead to success and treated success or failure as the "bottom-line" dependent variable. The research reported here questions the wisdom of this practice, for it may contribute to superstitious learning. Innovation success might be more usefully viewed as "by-products along the journey" than as final end results. Although perceived effectiveness judgments during innovation development can provide useful rationales for choices of subsequent actions, unspecified and conflicting targets and changing frames of reference produce compelling opportunities for superstitious learning, which can dominate current outcome evaluations by innovation participants and resource controllers. These practices account for many of the difficulties experienced in both the management of innovations and our abilities to predict the outcomes of the innovation journey.

Finally, most applied theories of successful management practice assume that consistency or congruence is necessary between the effectiveness criteria of different groups; agreed-on goals provide the basis for unity of direction. However, the prevalence of contradictory success criteria found here between resource controllers and innovation managers directly contradicts this assumption. It suggests that contradiction and nonliniearity may be inherent in most innovative undertakings. As a consequence, a central problem in the management of innovation may be management of paradox (Van de Ven and Poole 1988). As Cameron (1980) found, highly effective organizations were paradoxical in that they performed in contradictory ways to satisfy contradictory expectations. In addition, Quinn and Cameron (1983) found that effective organizations did not overemphasize one set of criteria over others; they maintained some balance or capacity to respond to multiple and conflicting criteria of effectiveness.

NOTES

1. The entries in Table 7–1 may appear to represent the "current squeaky wheels" of respondents more than their "real" ultimate outcome criteria. However, they are the responses obtained in repeated interviews over time to the question, "What criteria are you using to judge the success of this innovation?" Instead of second-guessing the respondents' answers, we view them as representing the substantive concerns that captured respondents' attention and frames

of reference at different points in time along the innovation journey.

2. It should be recognized that this will represent a humble preliminary test of the propositions. The MIRP survey instrument was not initially designed to directly test the propositions. As a result, we have selected variables in the MIS that "cover the bases" of the model, but do so imperfectly. Nevertheless, the data used for this quantitative test of the propositions are far better than no data at all. But as a result of these limitations, a longer than desirable chain of inference is necessary to draw conclusions about this evaluation of the propositions.

3. Remember that in Chapter 3 these dimensions were found to have even weaker predictive power on perceived effectiveness in a subsequent time period.

REFERENCES

Brunsson, N. 1982. "The Irrationality of Action and Action Rationality: Decisions, Ideologies, and Organizational Actions." *Journal of Management Studies* 19:29–34.

Cameron, K. 1980. "Critical Questions in Assessing Organizational Effectiveness." *Organizational Dynamics* 9: 66–80.

Campbell, J.P. 1977. "On the Nature of Organizational Effectiveness." In P. Goodman, J. Pennings, and associates, *New Perspectives on Organizational Effectiveness.* San Francisco: Jossey-Bass: 13–55.

Fischoff, B. 1975. "Hindsight or Foresight: The Effect of Outcome Knowledge on Judgment under Uncertainty." *Journal of Experimental Psychology: Human Perception and Performance* 1: 288–99.

Herriott, S.R., D. Leventhal, and J.G. March. 1985. "Learning from Experience in Organizations." *American Economic Review* 75: 298–302.

Kahneman, D., P. Slovic, and A. Tversky, eds. 1982. *Judgment under Uncertainty: Heuristics and Biases.* Cambridge: Cambridge University.

Kanter, R. 1983. *The Change Masters.* New York: Simon and Schuster.

Kimberly, J.R. 1981, "Managing Innovation." In P.

Nystrom and W. Starbuck, eds., *Handbook of Organizational Design,* Vol. 1, pp. 84–104. Oxford: Oxford University.

Levitt, B., and J.G. March. 1988. "Organizational Learning." *Annual Review of Sociology* 14 (forthcoming).

Levinthal, D.A., and J.G. March. 1981. "A Model of Adaptive Organization Search." *Journal of Economic Behavior and Organization* 2: 307–33.

Pelz, D., and F. Andrews. 1966. *Scientists in Organizations.* New York: Wiley.

Peters, T., and R. Waterman. 1983. *In Search of Excellence: Lessons from America's Best-Run Companies.* New York: Harper & Row.

Pettigrew, A.M. 1985. *The Awakening Giant: Continuity and Change in Imperial Chemical Industries.* Oxford: Basil Blackwell.

Price, J.L. 1968. *Organizational Effectiveness.* Homewood, Ill.: Irwin.

Quinn, R.E., and K.S. Cameron. 1983. "Organizational Life Cycles and Shifting Criteria of Effectiveness." *Management Science* 9: 33–51.

Ross, J., and B.M. Staw, 1986. "Expo 86: An Escalation Prototype." *Administrative Science Quarterly* 31: 274–97.

Skinner, B.F. 1953. *Science and Human Behavior.* New York: Macmillan.

Steers, R.M. 1977. *Organizational Effectiveness: A Behavioral View.* Santa Monica, Calif.: Goodyear.

Stinchcombe, A.L. 1968. *Constructing Social Theories.* New York: Harcourt, Brace & World.

Van de Ven, A.H. 1986, "Central Problems in the Management of Innovation." *Management Science* 32 (5) (May): 590–607.

Van de Ven, A.H., and D. Ferry. 1980. *Measuring and Assessing Organizations,* New York: Wiley.

Van de Ven, A.H., and M.S. Poole. 1988. "Paradoxical Requirements for a Theory of Organizational Change." In R.E. Quinn and K.S. Cameron, eds. *Paradox and Transformation: Toward a Theory of Change in Organization and Management,* pp. 19–63. Cambridge, Mass.: Ballinger.

Weick, K. 1979. *The Social Psychology of Organizing,* 2d ed. Reading, Mass.: Addison-Wesley.

SECTION
III

Studies of Business Creation

8

PROCESSES OF NEW BUSINESS CREATION IN DIFFERENT ORGANIZATIONAL SETTINGS

Andrew H. Van de Ven

S. Venkataraman

Douglas Polley

Raghu Garud

Organizational structures for business creation are different for small company startups and joint interorganizational ventures and within the large diversified corporation. It is widely

We greatly appreciate comments on earlier drafts of this chapter from M. Scott Poole (University of Minnesota), Robert Burgelman (Stanford University), John Hake and Robert Schoenecker (Qnetics), William Coyne, Sheldon Klasky, Robert Oliveira, and Keith Wilson (3M), Michael Mirvis (Sarns), and James Bray (Millipore), as well as other colleagues involved in the Minnesota Innovation Research Program. In addition, we greatly appreciate the intensive and ongoing access provided to us by the 3M, Sarns, Millipore, and Qnetics corporations to conduct the real-time longitudinal studies of business creation reported here. Funding for this research was provided by the Program on Organization Effectiveness in the Office of Naval Research, under contract N00014-84-K-0016.

held that the *processes* of business creation in these organizational settings are also different. We question this conventional belief for two reasons.

First, although different organizational arrangements have been associated with different levels of innovativeness, we argue that these findings provide little theoretical basis for concluding that the process of creating a new business in these different organizational settings also must be different. An equally plausible conclusion is that creating a business entails fundamentally the same process regardless of organizational setting. If empirical evidence is obtained to support this conclusion from the research to be reported here, then significant benefits and efficiencies can be obtained by applying principles for business creation from new company startups to internal corporate venturing and interorganizational joint ventures, and vice versa.

A second basic reason for questioning commonly held beliefs about relationships between organizational structures and processes of business creation is that very little comparative longitudinal research has been conducted on processes of business creation within the large corporation, in new company startups, and in joint interorganizational ventures. There is very little empirical evidence to substantiate any conclusions about if and how new business creation processes are similar or different across organizational settings.

To address these issues, this chapter reports preliminary findings of three intensive real-time studies begun four years ago. The studies address business creation processes in three different organizational settings:

1. One new business creation effort is housed within a large diversified corporation (3M) and is the Hearing Health Program (HHP), which is creating a totally new business by developing a line of new products, including cochlear implants, hearing aids, and otological diagnostics instruments for the hearing health industry.
2. The second new business, a Therapeutic Apheresis Program (TAP), is a joint interorganizational venture between 3M, Sarns, and Millipore Corporations. The TAP is developing a new line of medical products and diagnostics instruments for treating a variety of diseases by separating pathogenic substances from blood and returning the beneficial blood components to the patient. Both HHP and TAP represent new-to-the-world technologies and products, and both are major long-term investments and commitments to create totally new

businesses that are expected in ten to fifteen years to generate significant revenues for the corporations involved.

3. The third new business venture occupies the exclusive attention and total resources of a new company startup, called Qnetics, Inc. In efforts to become a financially viable company, Qnetics has pursued a variety of new business creation efforts over its five-year history. They include a computer distributor and maintenance business, a custom-design computer software business, a line of medical software products on patient and financial records for hospitals and third-party payors, and an electrical load management hardware and software business for the power utilities industry.

CONCEPTUAL FRAMEWORK

Organizational Structures for Business Creation

Conventional wisdom attributes a significantly higher rate of innovation per technical employee to the smaller firm than to the larger firm. A study of innovation by the U.S. Department of Commerce (Charpie 1967) called attention to this disparity. More recent research by Gellman (1976) confirmed this by examining 635 innovations that reached the marketplace. The study found that small firms (less than 500 employees) produce 2.5 times as many innovations as large firms per employee and that small firms bring their innovations to market 27 percent more rapidly than large firms. (No study has been found that includes interorganizational joint ventures in these comparisons.)

Goldman (1985) notes that the recent proliferation of successful high technology companies nurtured by an explosive venture capital market widens the disparity between large and small firms in innovation rate. The irony in this trend is that in the early postwar years, it was generally assumed that the innovation proclivity—particularly in the U.S. economy—was fed by the industrial R&D establishments and that these tended to be highly concentrated in large firms.

Why are small firms more innovative than large firms? The typical answer is that small firms are more flexible and quick to adapt to changing environmental opportunities and threats than larger organizations, which experience greater inertia due to their "liabilities of aging and bigness." However, this explanation is not adequate because it ignores a more imposing set of "liabilities of newness and small size" that small businesses must surmount in order to survive and be innovative.

Liabilities of newness include both internal and external obstacles that make survival difficult (Stinchcombe 1965). Aldrich and Auster (1986: 177–79) and Schoonhoven and Eisenhardt (1985) point out that barriers to entry into a new domain are the most frequently cited external obstacles for new firms, including (1) product differentiation, (2) technological barriers, (3) licensing and regulatory barriers, (4) problems of vertical integration, (5) illegitimate acts of competitors, and (6) lack of experience. New organizations also face internal obstacles, which primarily revolve around the creation and clarification of roles and structures consistent with external constraints and the ability to attract qualified employees.

In addition to these liabilities of newness, many new organizations also face liabilities of smallness, including (1) problems of raising capital, (2) tax laws pertaining to normal income versus capital gains work against the survival of small organizations, (3) government regulations placing a proportionately higher overhead cost on small than large firms, (4) major disadvantages in competing for labor with larger organizations, and (5) limited ability

to obtain benefits from specialization and economies of scale.

These liabilities of newness and small size tend to be more difficult to overcome than the liabilities of aging and bigness. As a result, one should expect the success probabilities of new business creation to be lower in new small firms than in larger and more mature firms. Aldrich and Auster (1986) and Singh, Tucker, and House (1986) review the empirical evidence showing that older and larger organizations have a substantial advantage over younger and smaller organizations in terms of not being decimated and replaced. New, small organizations fail at disproportionately higher rates than do older and larger organizations. Moreover, of those that survive and grow old, the vast majority of firms (75 percent according to Reynolds and West 1985) remain small "mom and pop shops" throughout their existence.

As suggested above, older and larger organizations must deal with a different set of "liabilities of aging and bigness," which lead to organizational inertia and make them increasingly less fit for changing environments. As Aldrich and Auster (1986: 183) insightfully point out,

> The obstacles faced by new, small organizations can be easily overcome by larger, more established organizations, whereas the constraints faced by larger, more established organizations can often be easily surmounted by new, small organizations.

In a similar vein, Williamson (1975) argues that although new company startups may be especially good at developing new innovations, large corporations are better at introducing and marketing new products in the market.

The joint interorganizational venture, which is a form of organization intermediate to new small and old large firms, provides an or-

223

ganizational setting that should provide the most conducive environment for new business creation, for in theory it can overcome the liabilities of newness, small size, aging, and bigness. In other words, the newness and small size of a joint venture promotes flexibility and minimizes forces of inertia, and the competence, resource base, and institutional legitimacy of the large mature parent organizations help the joint venture overcome the liabilities of newness and smallness. By this logic, the joint interorganizational venture should be the most conducive organizational context for creating new businesses.

However, this conclusion is premature, for it does not consider the problems endemic to joint interorganizational ventures, which we will label "the liabilities of double parenting and conflict." As Harrigan (1985) and Killing (1982) describe, the problems in managing joint ventures largely stem from having two or more parent organizations, which "often disagree on just about anything: How fast should the joint venture grow? Which products and markets should it encompass? How should it be organized? What constitutes good and bad management?" (Killing 1982: 121). Harrigan (1985: 36) adds to this list the problems of antitrust, sovereign conflicts, loss of autonomy and control, and loss of competitive advantage through strategic inflexibility. As a consequence, joint ventures have a high overall failure rate, and many of the failures are very costly for the partner companies.

Both Killing and Harrigan conclude that the existence of these problems implies that joint interorganizational ventures should be managed better, not avoided. A similar conclusion could be applied to better managing the problems inherent to new business creation efforts in new company startups and within the large corporation. But how? The purpose of this research is to examine the management process of new business creation in these different organizational settings in order to ad-

dress this question. This brief review suggests *that no clear a priori hypotheses can or should be drawn about the comparative superiority of different organizational settings for new business creation.* Each organizational setting has its potential benefits, but each also is exposed to unique liabilities. The critical issue requiring systematic longitudinal observation becomes one of understanding the similarities and differences in the problems that arise and how they are managed as new businesses are created in new company startups, within the large corporation, and in joint interorganizational ventures. To begin this task requires an initial appreciation of the nature of new business creation.

The Nature of New Business Creation

Creating a new *business* is more encompassing than *product* and more proprietary than *technological* innovation. Creating a business usually entails developing and organizing all the functions necessary to exploit a technological invention into a family of related proprietary products, which if successful, constitute a self-sustaining ongoing economic enterprise. Seldom can a business achieve commercial viability with a single product in the marketplace. An ongoing business requires the creation of synergy and economies of scale across functions, which are obtained from developing and applying functional competencies (that is, R&D, clinical trials, manufacturing, marketing, and service) in creating a family of related products and services over time. Thus, a description of the business creation process entails making statements about (1) how a business idea (or strategy) emerges over time, (2) when and how different functional competencies are created to develop and market the first proprietary product, (3) when and how these functional competencies are redeployed to develop subsequent new products in a family of products believed to result in a sustaining business, and (4) how these business development efforts both influence and are constrained by organizational and industry contexts.

Timely orchestration of resources among interdependent functions to develop each product and redeployment of functional resources to subsequent generations of products become two key challenges for business managers. The first challenge is typical of any single product innovation effort. If it is true that functions tend to be established sequentially for the first generation of products, then each preceding functional step becomes a "bottleneck" for succeeding steps in the product development sequence. Errors not detected in preceding steps are passed along to the next functional step. The consequence is that these errors may compound developmental work for subsequent functions (or at the extreme result in market recalls, if undetected) and may entail costly revisions in all functional steps of the product development cycle.

The new business creation challenge is more complicated because it involves developing a family of product innovations. Here another set of "bottlenecks" arises in redeploying resources and attention from one product to the next generation of products. When a functional step is completed on a preceding product, those specialized resources are freed up and must be redeployed or they remain idle. These specialized resources are usually reassigned to begin similar functional work on the next generation of products.

Each product generation involves its own unique mix of required skills, interdependent processes, and development timetable. However, the business usually does not have that mix of skills; it has the mix of skills and resources that it learned and used for the last product development effort. Thus, without significant retraining and learning, simple redeployment of specialized resources to the next product when the prior one is completed invariably results in replicating the previous product development effort and compromising the design specifications and timetable for the new generation product. Alternatively, the excitement and timetable for a new generation product may result in a "flee" from the predecessor product without adequate completion or market exploitation.

Accumulation Model of New Business Creation

An *accumulation* or *epigenetic* model of change (as described by Etzioni 1963) will be used to examine the process of business creation. This model views change as springing from the business ideas and purposeful behavior of entrepreneurs. Over time, these entrepreneurs accumulate the external resources and technology necessary to transform their ideas into a concrete reality by constructing a new business unit that produces outputs (a marketable family of products and services) that, if successful, eventually sustain the business as an ongoing economic enterprise.

Initiation, startup, and takeoff are key temporal periods in this accumulation process. *Initiation* is the time when entrepreneurs decide to form a business venture (and if successfully launched becomes the birthday of the business), and *takeoff* is the time when the unit can do without the external support of its initiators and continue growing "on its own." The period between initiation and takeoff could be called *startup*, where the new unit must draw its resources, competencies, and technology from the founding leaders and external sources in order to develop the proprietary products, create a market niche, and meet the institutional standards established to legitimate the new business as an ongoing economic enterprise.

The process of accumulation is like that of an airplane that first starts its engines and begins rolling, still supported by the runway, until it accumulates enough momentum to "take off" and continue in motion "on its own" energy to carry it to higher altitudes and

225

speeds. Although the business relies initially on external support, the necessary conditions for autonomous action are produced through a process of accumulation (Etzioni 1963).

As applied to business creation, one should expect to observe cycles of initiation, startup, and takeoff for each of the multiple components and products that may constitute the business. As a consequence, the overall business creation process may be nested with multiple, overlapping, and iterative progressions of initiation, startup, and takeoff activities of various business components. Some of these components or products that are initiated may not start up, and some startups may not take off; and takeoffs in one component may facilitate or inhibit the initiation, startup, or takeoff of other components. Because a given business component or product may never be able to sustain itself economically (and hence be unable to "take off" on its own), an understanding of overall business takeoff requires study of the interactions among business components and their relationships with external sources of support and resistance.

The accumulation model directs attention to three basic research questions to study these business creation processes.

Where are the resources and power that control the business creation process located? From the initiation to startup to takeoff periods, the relative power (resources and control) shifts from external to internal sources in the accumulation model. For example, whereas outside investors or corporate managers may largely control the resources and set the direction to initiate a new business, the business unit entrepreneurs/ managers should increasingly gain power and control of the process during the startup period, culminating in the takeoff period with all the resources and power necessary for the new business to operate on its own. This implies that both internal and external groups govern the business startup period, and that a central

dynamic of this period is the transition in power from external to internal groups. When the business creation process proceeds smoothly and according to plan, this transition may be uneventful. However, because business creation is a highly uncertain process, mistakes and setbacks usually occur, and power struggles often erupt between external and internal groups on the appropriate actions to take to further develop the business. Resolutions to these struggles may entail confusing, contradictory, and abrupt shifts in business development.

As a consequence, the process of new business startup often tends to reflect the period of rupture and discontinuous change in a punctuated equilibrium model of change, as discussed by Tushman and Romanelli (1985). If the new business surmounts these business startup hurdles by the takeoff period, activities in the accumulation model begin to reflect the convergent period of continuous change and increasing business refinements in the punctuated equilibrium model.

What is the temporal sequence in the development of various business components and products?
This question examines how and which business ideas, functional competencies, and technological products were developed and introduced first, second, and so on? How did they affect subsequent developments of business components? These questions deal with the progression of paths followed in the development of a new business. As discussed in Chapter 2, van den Daele (1968, 1974) provides a useful vocabulary for examining whether unitary or multiple paths of divergent, parallel, and convergent progressions were followed, and if these paths of activities were conjunctive (related), iterative (repetitive), and structured (hierarchically nested in time and overlapping or sequential over time). Examining these paths should lead to an appreciation of the timing, momentum, and resource allocation condi-

tions where the takeoff of one component or product facilitates, preempts, or exhausts the initiation of other new business components and products.

What outcomes are associated with different paths and progressions?

Given that the business creation process is inherently uncertain, one should expect many trials and errors, as witnessed by multiple divergent paths leading to dead-ends, stalemates, and terminations over time. Moreover, an examination of in-process outcomes leads one to examine if and how learning occurs from previous paths that provide guidance as to the next paths taken to create the business. By examining the outcomes of alternative paths, we hope to identify the feasible sets of paths available in business creation over time.

Timing, momentum, and *focus* of these business development paths are key concepts for examining the success of these paths (Etzioni 1963: 490). The longer and more gradual the events in each path, and the more dispersed the resources among business development paths during the startup period, the lower the overall market awareness and interest in the new business venture, and the more difficult it is to mobilize commitments from investors and potential key suppliers and customers. On the other hand, if the startup period for one or more business components is too fast and concentrated, there may be insufficient time during startup to learn and build requisite competencies, develop quality products, or build the momentum and critical mass needed for market penetration to sustain the new venture at takeoff. As the principle of "moderation in all things" would suggest, we hypothesize that the probability of successful business takeoff increases with intermediate amounts of time, momentum, and focus that are invested in business initiation and startup activities.

These three basic process questions, which serve as the starting point for our study

of business creation, are related in terms of a standard input-transformation-output model from systems theory. The first question examines system inputs in terms of the resources and power that a new business draws from its external environment in the initiation period, and how the power center that orchestrates the new business shifts from external to internal groups during the startup period. At any given time, the ideas, competencies, and products produced by the new business can be viewed as system outcomes (question 3). The processes of transforming inputs into outputs are addressed by the second question, in terms of how the ideas/competencies/products developed, and the forms of progression in the accumulation process.

LONGITUDINAL RESEARCH METHODOLOGY

Data Collection

Multiple methods were used to observe how the key managers created their new businesses over time. In the case of Qnetics, the key managers were two entrepreneurs who jointly formed the company, as well as the key personnel they hired and who became the top functional managers of their business. In the case of the Hearing Health Program (HHP), we focused on thirteen key managers who were in charge of various program functions and who were members of a Program Steering Committee that met for day-long meetings each month to set program policies, priorities, and directions. In the case of the Therapeutic Apheresis Program (TAP), there were seven key managers who represented their respective corporations in the joint venture and who met as a strategic business unit every two months for two-day meetings to set program policies, priorities, and directions. The researchers regu-

227

larly attended and took detailed notes of discussions at the HHP and TAP strategic group meetings over three years. Minutes of all Qnetics' board meetings were obtained, and beginning in the third year the researchers attended and observed regular meetings of the board of directors.

Following the procedures described in Chapter 2, the researchers also conducted repeated interviews and administered the MIRP questionnaires with all entrepreneurs and general managers of HHP, TAP, and Qnetics every six months. In the HHP and TAP cases, yearly interviews were conducted with top-level managers in the direct reporting hierarchy from the new businesses to the corporate CEOs. Finally, archival and published information were regularly collected on all the cases over the years.

Listing and Coding of Chronological Events

Using these multiple data sources, chronological listings of events in the development of HHP, TAP, and Qnetics were developed. The decision rule used for designating an event was when changes were observed in the business idea (or strategy), in the key people involved in the business creation process, or in the transactions or relationships engaged in with other parties to develop the business; and when changes were made in the form or structure of the encompassing organization or industry of the business. Events so designated were arrived at by a consensus among the researchers. Where differences of opinions existed among the researchers on the designation of an event, they were discussed and resolved through consensus.

In order to organize these events into a format conducive to process analysis, they were next coded by "key words" to identify the substantive kinds of issues involved in each event. Because new business creation is fundamentally concerned with the development of personnel competencies and resources needed to develop and transform business ideas into a line of products, key words pertaining to different kinds of (1) business ideas, (2) resources, (3) functional competencies, and (4) products were used to identify the substantive nature of each event. Because these issues are unique to the business being examined, a computer software program (R-Base V) was used to search for key words used to describe each event. These key words permit events to be connected and how substantive issues unfolded over time to be tracked. A simple test of interrater reliability was conducted, and it was found there was an 87 percent consistency among the researchers in coding events into the categories. Again, discussions and consensus decisions were used to resolve inconsistent event codings among the researchers.

The objective with this coding scheme was to identify the different paths that were taken in business creation. Charts were also drawn to identify when and how alternative routes that were undertaken, and what outcomes were associated with each path.

Using the conceptual framework and methodology described here, the next three sections describe the business creation processes observed over time in a new company startup, within a large corporation, and in an interorganizational joint venture, respectively. We follow the three new business cases with a discussion of empirical similarities and differences in business creation processes among the three cases.

QNETICS NEW BUSINESS CREATION CASE
S. VENKATARAMAN AND ANDREW H. VAN DE VEN

This case describes an ongoing effort to create a new computer software business, called Qnetics, which was established in September 1983 as the result of the merger of two new startup companies called Quality Computing Systems (QCS) and Medformatics. Located in Minneapolis, Minnesota, QCS was incorporated in April 1980, and Medformatics was incorporated in December 1980. Qnetics grew to thirty-one employees in 1983 and then dropped to four employees in 1985; currently it has eight employees and 1986 sales revenues of about $1 million. In efforts to become a financially viable company, Qnetics and its predecessor firms pursued a variety of new business creation efforts in its short history, including distributing and maintaining computers, custom-designing computer software, designing medical software products for patient records in hospitals and for third-party payors, and developing software for a load management system for electric power utilities industry.

Longitudinal study of Qnetics began in May 1983 when the QCS and Medformatics founders were involved in discussions to merge their firms. Data collection involved quarterly interviews with company founders and top managers, and two questionnaire surveys of all company personnel in 1983 and 1984. Because Qnetics had only between two and five employees between July 1985 and October 1986, no questionnaire surveys were administered. During this period, data collection focused on repeated quarterly interviews with company principals, as well as information obtained from company records and minutes of management meetings and board meetings. From the data collected, 100 events were identified as having occurred in the development of Qnetics. To reduce the complexity of this chronological list to manageable proportions, the events were content analyzed and grouped into four temporal periods. Events within each of these periods are graphically illustrated in Figure 8–1 and presented in the appendix to this case.

The first period, labeled *initiation,* occurred from January 1980 to November 1983 and includes the founding of QCS and Medformatics and the events leading to their merger to create Qnetics effective September 1983. The second period, *startup,* details the growth in the products, competencies, and resources within the firm, and occurred from November 1983 to September 1984. A third period of *survival and retrenchment* occurred from about October 1984 to February 1986 and includes a tragic series of failed takeoff events that led the company to the brink of collapse. Finally, a *restartup* period, begun March 1986 and now in progress, includes a financial restructuring of the company, new management, and efforts to position the firm as a viable growing business.

Table 8–1 provides a yearly summary of the financial and personnel status of Qnetics. As the table shows, Qnetics reached peak sales revenue and net income in 1983, the year of merger between QCS and Medformatics. Since then revenues and profits have eroded. Much of this erosion came about due to an aborted computer hardware sale and an inability to market medical records software products. Although operating costs steadily increased from 1983 to 1985, operating expenses were drastically cut in 1986 through large-scale personnel layoffs and retrenchment of operations. Based on these data, we now describe the progression of events in each of the four periods of Qnetics' development.

Initiation Period

1. *Quality Computing Systems* (QCS) was incorporated in April 1980 by three investors who made personal investments and obtained a $25,000 line of credit from a local bank. One of the investors, who became one of the two full-time principals of the firm, reported he enjoyed the entrepreneurial atmosphere at his former employer but stated he foresaw limited advancement opportunities there, and had "a

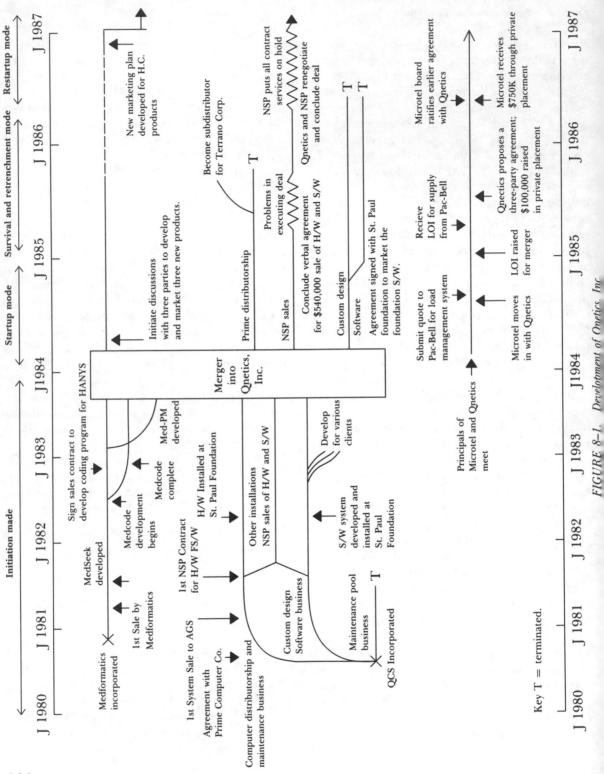

FIGURE 8-1. *Development of Qnetics, Inc.*

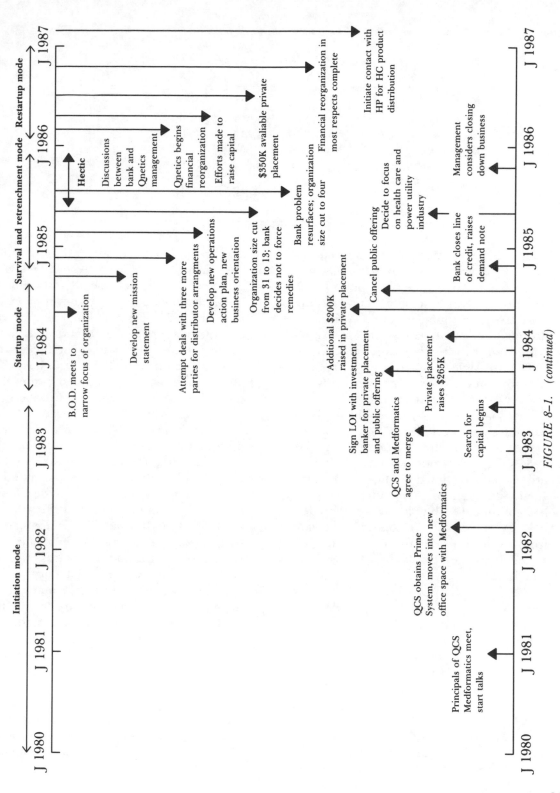

FIGURE 8–1. (continued)

TABLE 8–1. *Financial Summary of Qnetics for Fiscal Years 1983 to 1986 (All figures in $ 000s).*

	1983	1984	1985	1986
Revenues	$2,369	$1,727	$ 990	$ 938
Direct expenses	1866	1,517	302	560
Operating costs	484	862	1,399	676
Total expenses	2,350	2,379	1,701	1,236
Other income	3	2	1	0
Net income	22	(650)	(710)	(298)
Number of personnel	$ 31	$ 31	$ 13	$ 5

strong urge to start something on his own." QCS was begun with three business ideas: computer maintenance pool, Prime Computer distributorship, and custom software development. The principals dropped the maintenance pool idea within a few months when it proved to be financially unfeasible. The idea for a Prime distributorship came from the principal's fellow investor, who used his position in Prime Computer Company to obtain Prime hardware distribution rights for QCS. The first Prime computer sale occurred in May 1981. The first major sale was to a power utility company based in Minnesota, in fall 1981. Since then the power utility company has continued to buy hardware upgrades, new systems, and customized software from QCS and later from Qnetics. QCS added three key employees during 1981 and 1982.

2. *Medformatics* was incorporated in December 1980 by two principals, one who devoted his energies full time, without pay, to starting the firm from his home office. The two principals made personal investments totaling $1,000 and obtained a $37,000 line of credit from a local bank. The principal could not persuade his former employer to develop his business idea of creating hospital medical records and started Medformatics with the idea of developing and marketing a system of computerized medical records that were required by hospitals to

manage and maintain medical information in their medical records departments. The principal initially contracted with another medical software expert to develop an automatic coding system, but dropped the contractor's services in December 1982 because of different opinions about how this service should be provided to hospitals. Medformatics then proceeded to develop the following line of medical products.

The first medical product developed in summer 1981 was MEDSEEK, which was designed as the database manager for the entire medical software line. MEDCODE, the second medical product, was an automatic coding system that was developed between July and December 1982 with the help of QCS. It used the data obtained from the MEDSEEK database and automatically assigned the codes for diseases in accordance with international coding systems. These codes are then submitted by hospitals to agencies controlling reimbursements under Medicare and Medicaid. In 1983 QCS developed a third system in the medical products software—MEDPM—which aimed at improving the efficiency of hospital maintenance operations.

In addition to the principal and his part-time partner, one employee experienced in medical information areas joined Medformatics in 1981 as a marketing manager. Medformatics made its first sale to a group of hospitals in the Twin Cities in

April 1981 and also won a contract for $60,000 to develop a coding program for a hospital association in New York in December 1982.

3. *Qnetics.* The principals of QCS and Medformatics met each other through a common acquaintance in January 1981. They met several times in the next few months to discuss their common problems in business startup. One year after startup both companies faced resource and competency constraints in developing their businesses into viable entities. Medformatics needed computer resources to develop their programs, some programmers, and capital, whereas QCS needed proprietary products in the area of software programming. In April 1982 QCS obtained its Prime Computer and moved into new office space, and Medformatics moved in with QCS. The requirements of the two companies complemented their resource endowments, which prompted them to work closely together. After working together for more than a year, QCS formally acquired Medformatics (although personnel in the two firms considered it a merger), effective September 1, 1983. The combined companies were renamed Qnetics and continued to pursue their three combined businesses in healthcare software, Prime equipment sales, and custom software development. QCS's principal became chairman and chief executive officer of Qnetics. In July 1984 Medformatic's principal became president and chief operating officer. The combined companies continued to recruit new employees in the areas of programming, design, and software services.

Startup Period

From August to November 1983 the principals of Qnetics undertook a strategic planning process to create a formal mission statement for the newly merged organization. This new mission statement was reviewed and approved by Qnetics' board of directors in June 1984 and indicated that Qnetics would focus on delivering information systems to the health care industry. Qnetics management stated that clients not in health care would continue to be supported and would be revenue generators for development work in the health care areas.

Products. During the startup period Qnetics' management proposed developing six new products in the healthcare area, three of which were to be jointly developed and marketed with three other parties during 1984. The company was successful in developing only one of these products during 1984.

The first product was a hospital cost accounting system to be developed jointly with a hospital consulting firm. This product was completed in early 1985. The second product, a claims processor for third-party payor situations, was to be completed jointly with a health care management firm specializing in claims processing and cost containment systems for third-party administrators. Although development work began in October 1984, Qnetics shelved the project in August 1985, when the health care management firm ceased investing in the project. The third product was software for a memory key device in which demographic and medical information about the carrier of the key can be stored. Qnetics entered into an agreement with another firm to develop software for a health records memory key. They also entered into a distributor agreement with this firm to market the key, the peripherals, and the software. Distribution costs turned out to be prohibitive, and Qnetics did not proceed to develop this product.

No new products were developed outside of the healthcare area during the startup period. However, Qnetics concluded two key agreements, one in April 1984 and the other in November 1984. Qnetics reached an agreement with a foundation to market a software product Qnetics had developed for them ear-

233

lier and concluded a verbal agreement with the power utility for a $540,000 sale of hardware and software as a follow on to earlier contracts.

In October 1984 the principals of Qnetics and Micro-Tel, Inc., a Minnesota-based company, met. Micro-Tel had designed an intelligent remote terminal for use in the home and business premises that could address load management problems for electric utilities. Micro-Tel approached Qnetics to manage and raise capital for them. Qnetics' management saw this as a significant opportunity for their utilities business and signed a letter of intent with Micro-Tel in February 1985 for merger between the two companies.

Organization. By June 1984 Qnetics employed thirty-one people involved in management, software development, marketing, technical support, finance, accounting, and staff support. Qnetics management reported organizational growth problems of unclear divisions of labor and job assignments among its employees. In July 1984 Qnetics management undertook a reorganization. The new organization grouped technical functions together under the president's direction. Salary studies, compensation changes, and job descriptions for all employees were completed by the end of 1984. In addition, management also attempted to improve personnel skills and communications by conducting companywide meetings and occasionally featuring guest speakers who lectured on various technical and management topics. According to Qnetics management, these activities helped reinforce the organizational changes.

In July 1984 the vice president of finance established the first operating and capital budgets for the firm. In addition, Qnetics transferred its computerized accounting system to run on Altos. The newer system allowed more integrated financial reporting than the old Prime system. The transfer to the new system was completed by January 1985.

The CEO concentrated on raising external financing and marketing activities. Developing a solid marketing program was a major concern to Qnetics management. The vice president of sales was judged not to be generating sufficient sales growth for the firm. In September a vice president of marketing was hired, and by November 1984 he created a marketing plan to sell the firm's healthcare software products by obtaining leads through mass media magazine advertising that would be followed up by four regional sales representatives.

External Financing. Obtaining additional external financing to support most of the firm's employees engaged in R&D work was another pressing problem for Qnetics management immediately after the merger. Qnetics reached an agreement with an investment banking firm specializing in small companies to arrange for a private placement followed by a public offering. A private placement drive for $400,000 began in October 1983 and raised $265,000 by February 1984. The public offering that was to follow was planned to raise $1.8 million. The private placement drive, however, soon stalled, and so the period of the drive was extended and the amount to be raised was increased to $600,000.

By late August 1984 Qnetics obtained an additional $200,000 of finances through the private placement, which now totaled $465,000. However, it became apparent that the investment banker was having great difficulty raising the remainder of the capital under the prevailing market conditions. Qnetics management and the investment banker canceled the planned $1.8 million public offering scheduled for March 1984. The investment banker stated that the financial market had become "very soft."

Survival and Retrenchment Period
In October 1984, after the aborted public offering, Qnetics' bankers expressed concern

about the financial position of the company and, after a review of the situation, stated that without changes in the organization the bank would not renew Qnetics' line of credit. In fact, on December 31, 1984, the bank closed the line of credit held by Qnetics, which denied Qnetics the use of $30,000 in remaining borrowings on the original working capital line of credit of $350,000. It also forced Qnetics to use cash to pay the $17,000 issued in checks prior to the closing of the line of credit. Further, the bank raised a demand requiring full repayment on all Qnetics' notes including interest. However, the bank did not exercise its rights to take control of Qnetics' assets at that time by enforcing the remedies.

Qnetics' CEO and president developed a plan in January 1985 in an attempt to overcome Qnetics' serious financial crisis and took action on the basis of this plan. Qnetics' bankers were satisfied with the actions and changes taken by the management and agreed not to exercise the remedies at that time, on the condition that each of the seven owners make a joint personal note to cover a new loan in the amount of $20,000. However, the bank did not renew its line of credit to Qnetics at that point in time.

In the fiscal year ending June 30, 1984, Qnetics realized a loss of about $694,000 on sales of $1.7 million. The firm carried nearly $430,000 in long-term debt and $1 million in short-term debt and had a net worth of $498,-000. Qnetics management stated that the implementation of its operating plan helped reduce operating costs by $75,000 per month. Qnetics also realized $100,000 in revenues from sales in December 1984 and $100,000 in January 1985. In December 1984 the firm employed thirty-one full-time staff members, as illustrated by the organization chart in Figure 8–2.

Unrealized Marketing Efforts. By early 1985 management observed that the cost of direct marketing and distribution of healthcare products to hospitals were prohibitively expensive for Qnetics because it lacked capital and competencies in the area. In early 1984 management tried to enter into joint venture agreements with some exclusive or even nonexclusive distributors for distribution of healthcare products to hospitals. They approached six different parties between February 1984 and May 1985 but were not successful in coming to an agreement with any one of them.

During 1985, faced with a serious financial crisis and a lack of distribution resources, outlets, or capital to develop its own distribution outlets, Qnetics discontinued further development of medical software and discontinued selling or promoting most of its medical software. According to the president, the company dropped all of its medical products except MEDCODE and MEDSEEK in 1986, and MEDCODE continued to remain its "only window of opportunity" existing in the healthcare field in 1987.

The power utility agreement for $540,000—which was verbally committed by utility officials in October 1984 and was awaiting the signatures of its managers—did not materialize. On December 31, 1984, Qnetics had to withdraw as an authorized distributor for Prime Computer Company and was relegated the role of subdistributor because Qnetics' bank canceled its letter of credit worth $250,-000 to Prime Computer. At the end of 1984 the local office of Prime Computer approached the power utility and attempted to sell directly to the utility a Prime Computer system that was being negotiated by Qnetics. However, the power utility accepted Qnetics' proposal and thwarted the bid of Prime Computer to take business away from Qnetics. Prime Computer did not have the software required to deliver the services to NSP. In the process of renegotiating the deal, Qnetics' revenue margin from Prime Computer for sale of the hardware was reduced from $180,000 to $60,000.

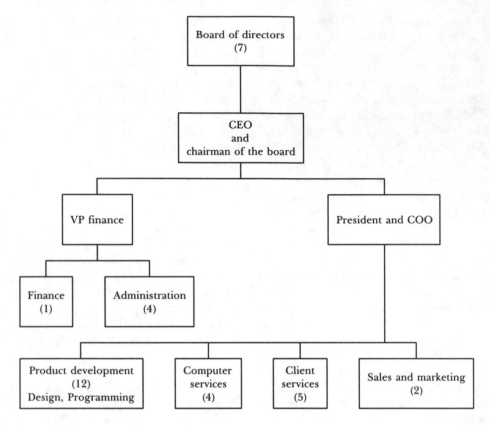

Note: Figures in parentheses indicate number of employees.

FIGURE 8–2.   Organization Chart for Qnetics, Inc., December 1984.

Strained financial circumstances led Qnetics to discontinue work on the foundation software in May 1985, and the agreement to market the software remains unexecuted. Along with the foundation software, Qnetics dropped servicing two other contract software clients. The November 1984 proposal to develop a long-term management system for Associates was declined by the power utility because Associates was fighting for its own survival.

In July 1986 the power utility became involved in a scandal by hiring an employee from a regulatory body and put all contract services in the area of research activities on hold until new management was able to develop new budgets. This step was taken due to a change in management at the power utility. Several of Qnetics' software development contracts were put on hold.

During the period May 1985 to February 1986 the business was managed entirely on a cash basis. Much of the revenue accrued from medical software products and the power utility sale. According to the CEO, Qnetics management succeeded in reducing costs and the company's cashflow breakeven point by about $60,000 per month, a two-thirds reduction in cash requirements.

The only promising prospect for new business with significant revenue potential during 1985 occurred in May, when Pacific Bell issued a letter of intent to supply a load management system to Qnetics and Micro-Tel. The potential contract was worth $56 million. However, at this time, Micro-Tel had difficulty in getting the product commercialized. Micro-Tel had a cross-licensing joint venture agreement with AST, which gave AST joint proprietary rights to the technology. The CEO of AST refused to enter into an agreement with Micro-Tel that would permit Micro-Tel to have access to the technology in order to develop the software required to make the hardware run. Without an integrated product Micro-Tel was experiencing difficulty raising the required capital for further development of the project because no investor was willing to make an investment under the current arrangement.

On December 30, 1985, Micro-Tel made one more attempt to reach an agreement with AST that would give it access to the technology. This time, a new CEO of AST expressed willingness to enter into a relationship with Micro-Tel. A letter of intent was drawn on February 12, 1986, that gave Micro-Tel access to the technology in order to develop the software for the hardware product. Qnetics' CEO stated that on March 14, 1986, a letter of intent between Qnetics and Micro-Tel was signed that permitted Qnetics to proceed with the software development.

Qnetics management considered creating a new company which would acquire Micro-Tel. Qnetics would then be a contractor to integrate the hardware and software. This approach was considered so that capital-raising measures by Micro-Tel would not be hampered by Qnetics' poor financial situation. No progress was made on this alternative.

In August 1985 Qnetics proposed a three-party agreement to private investors for funding and operating Micro-Tel. Under this agreement Qnetics assigned the services of its CEO and COO, a facilities contract, and software development to Micro-Tel. This three-party agreement resulted in a commitment of $600,000 from private investors, of which $100,000 was funded at that point for Micro-Tel.

Bank Confrontation. In July 1985 the bank again informed Qnetics' management of its intention to terminate its relationship with Qnetics, by exercising its remedies. The ensuing confrontation between the bank and the owners of Qnetics resulted in the bank's giving Qnetics' owners until September 3, 1985, to restructure the business. Failure to do so would lead the bank to exercise its remedies, which would, in turn, lead to the closure of the business.

According to the CEO, after long and agonizing discussions, the owners of Qnetics decided not to be intimidated by the bank and not to satisfy the bank at the expense of the other stakeholders. The owners of Qnetics returned to their bankers on August 27, 1985, with two alternatives: The bank should either support the plan proposed by Qnetics to the bank in January 1985 or exercise its remedies, which would lead to closing the business. If the bank chose the latter alternative, the owners would each file for personal bankruptcy and the company would file for Chapter 7 bankruptcy. These moves would not leave the bank with sufficient assets to fully cover its loan. On September 3, 1985, the bank agreed not to exercise the remedies. Qnetics' owners felt that this was the "first major battle" they had won in saving the business.

Revised Company Strategy. The strained financial circumstances and emerging opportunity in the load management system prompted Qnetics management to revise its business strategy, which they articulated late in February 1985:

1. Qnetics is a product development company and not a marketing company;
2. Qnetics' business is information and com-

munication and is not restricted to medical industry;

3. Qnetics would enter large markets (load management systems for power companies) rather than concentrate solely on markets with relatively few clients (healthcare industry).

Organizational Retrenchment. To deal with Qnetics' serious financial crisis, management undertook several steps to reduce the size of the organization. First, the salaries of management employees were reduced. The owners went without pay for the last quarter of 1984. Second, the organization size was reduced from thirty-one to twenty employees (which included the owners). The eleven employees who left the firm were mainly from the areas of marketing, administration, support, finance, and technical writing. During summer 1985 the size of the company was further reduced. Four of the owners were laid off, including the CEO and president. However, the CEO and president continued to work without pay for several months. That left only two employee owners, who were producing enough direct revenue to pay for themselves.

According to the CEO, 1985 and early 1986 had been a big disappointment for the owners, but he stated that the knowledge, skill, and experience they had gained in weathering the various crises could be acquired only by experiencing it, not in classrooms or textbooks. He stated that "we have become good street fighters and learned survival." He also strongly felt that "somewhere down the line Qnetics will make a lot of money." When asked to sum up the principal reason for this survival, the CEO stated that tenacity carried them through all the crises and gave the will to draw the line beyond which they would not go (as in the case of the confrontation with the bank).

In the fiscal year 1985, Qnetics had sales revenue of $990,000 and a loss of $709,000. Although sales remained stagnant in 1986 at $938,000 Qnetics managed to reduce the loss to $245,000 through extensive cost cutting. From a high of thirty-one employees in December 1984 the company size was reduced to just four employees in mid 1986.

Restartup Period

The focus of Qnetics' management in early 1986 was to formulate a strategy to financially reorganize the business. An experienced investment banker and a consultant assisted Qnetics' management with reorganization. Further, the investment banker extended $20,000 on a no commitment basis to provide Qnetics with "a little breathing room" during the months of February and March. This, in addition to averting another closure threat from the bank, helped Qnetics' management concentrate on the task of restructuring rather than firefighting. Because the Micro-Tel project was moving ahead, investors were again interested in investing in Qnetics, according to the CEO. Efforts were directed toward raising $350,000 through private placement. A number of existing debenture holders converted their debenture holdings in order to fund the proposed private placement drive. This capital became available in July 1986.

With a new infusion of funds from the second private placement, Qnetics' financial problems eased somewhat by mid-1986. The CEO and president were able to be reinstated on the payroll, and Qnetics was able to hire three people to work on custom contract basis for software work in health care and power utility areas. By October 1986 the company also hired a general manager for healthcare software marketing in order to renew marketing efforts. By January 1987 Qnetics had eight employees.

During February 1987 the board was reconstituted and four new members were voted in. The experienced businessperson who helped in reorganizing the business became the new president and CEO of the company, and the previous CEO became the chairman of the board.

No new products had been developed by the company since the financial restructuring. The major focus of the company in late 1986 and early 1987 has been to renew marketing of the medical products and to continue contract software development for the Micro-Tel load management device. The board of Micro-Tel ratified the agreement with Qnetics in February 1987. Under this ratification Qnetics will develop software for the load management system and receive a 50 percent royalty split for past services and future services. This royalty was expected to yield $3 million.

In the Health care area, the general manager reported that he began making contacts with Minnesota's hospitals and has been following up on leads in efforts to generate revenues from the medical software. Management also reported that that it had initiated contact with Hewlett-Packard corporation and had received interest from them in the medical software business. HP has a vertical market niche in health care. Qnetics' strategy is to build a strong partner with an organization such as HP for distribution of its products. Based on demonstration provided to a HP representative in early February 1987, Qnetics' management hopes it will receive an invitation for further demonstration. If it receives this invitation, the chairman stated that there is a 90 percent chance of entering negotiations for a possible alliance with HP.

Conclusion

In this section we try to interpret the different forms of progression that the overall business creation took. As Figure 8–1 shows, the overall business creation process has always followed multiple paths. Within each business product line there is first an elaboration of the business ideas leading to proliferation in products.

Three paths have been discernible since the inception of QCS and Medformatics—the medical software business, the custom software business, and the Prime Computer equipment sales business. To these three business paths Qnetics added a fourth path—the load management software business.

Although the overall business has followed multiple paths in the brief history of Qnetics, one can see sequences of proliferation or elaboration followed by pruning of products within each business product line. Thus the healthcare line began with two products in MEDSEEK and MEDCODE and grew to eight products. However, by 1985 the healthcare line was reduced to two, and the COO sees MEDCODE as the only "window of opportunity" in 1987 as far as healthcare business is concerned. Similarly, in the custom software business, the clients grew from one to six by 1984. But all these clients were discontinued by the end of 1985 in order to focus management's attention. Finally, as a result of the company's financial problems, it lost its Prime Computer distributorship, which in turn significantly reduced its expected profit margin from the power utility sale. Even though some of these business lines may appear independent, each affected the fortunes of others through its cash generating or absorbing capacity.

A central feature of the business development process at Qnetics has been a cycle of crises. The firm has generally been characterized by lack of slack resources, which forces it to operate on a cash basis. Further, the management of Qnetics decided to use the custom software business and the Prime distributor business as revenue generators for development work in the healthcare business. This made the three businesses highly interdependent, at least in terms of cash requirements. Hence a crisis in one business affected the operations of another, often resulting in a domino effect as far as the overall firm was concerned.

In order to market its healthcare prod-

239

ucts, the company pursued the "make and buy" strategies of creating its own marketing and sales branches and contracting with independent dealers and distributors. However, Qnetics was unsuccessful in implementing both strategies. Qnetics had to abort the planned public offering because of poor market conditions. This paucity in finances restricted the company's ability to develop its own marketing and distribution outlets. Further, Qnetics could not consummate any contract with third-party distributors even though it explored relationships with as many as six parties. The poor capital situation led the bank to withdraw its line of credit, thus worsening an already resource-poor situation. Elimination of the bank's line of credit also led to Qnetics' losing its Prime Computer Co. distributor status. Loss of this distributorship in turn, stimulated Prime district sales representatives to take control of the large computer sale that Qnetics had negotiated with the power utility over an extended time period. Loss of this NSP sale would have occurred had it not been for the intercession by the customer in favor of Qnetics. Nevertheless, Qnetics received a substantially lower margin on the large sale than it would have if it had not lost its Prime distributorship.

Faced with lack of development capital and a marketing capability, the company dropped several products in the healthcare line, dropped current services in the custom software line, stopped all development work on new products in all business areas, and laid off or terminated twenty-seven employees. Lack of new products and discontinued services further hampered revenue generation, thus creating a further cash crisis in the almost never ending loop.

The major source of power to break this cycle came from external sources and never from within the organization. The external support to survive the critical period of 1985 and 1986 came from several sources. The first was in the form of financial support from their bankers, their investment banker, and the experienced businessperson. The second was the unexpected new business opportunity with Micro-Tel that emerged in early 1986.

With the help of a survival loan from its investment banker, temporary working arrangements with the bank, and the help of the expert, the company was able to stave off bankruptcy. The new opportunity that emerged in the electric power utility business enabled Qnetics to raise a fresh round capital through private placement and helped the company rearticulate its business orientation and initiate its present second effort at company startup.

Qnetics Case Appendix.
Chronological Events in Qnetics History.

Initiation Period

1979	Principal of Medformatics tries to interest his employers in developing business out of abstracting and automatic coding services for hospitals.
Oct. 1979	Principal of QCS begins search for business opportunities in discussions with two other acquaintances.
Jan. 1980	Principals of QCS draw up an informal business plan with three business ideas.
	One of the fellow investors recognizes opportunity to undercut Prime's maintenance prices. Principal and fellow investor prepare plan for a Prime maintenance pool.
April 1980	QCS is incorporated.
Oct. 1980	Maintenance pool idea is abandoned.
	Prime system builder agreement is signed by QCS with Prime Computer Co.
Nov. 1980	FHCE repeatedly refuses to fund Medformatics principal's coding project, and he leaves employment with the foundation.
Dec. 1980	Medformatics is incorporated.

Jan. 1981	Principals of Medformatics and QCS meet and begin monthly discussions to explore joint venture opportunities.
March 1981	QCS obtains its first contract for a Prime system.
April 1981	QCS hires VP of marketing, who also becomes company director.
	Principal of Medformatics completes data entry system for the automatic coding system. Medformatics makes first sale of coding system.
Jun. 1981	A VP and company director joins QCS.
Summer 1981	Medformatics develops medical records information system.
Fall 1981	QCS wins first contract for a Prime hardware and software system.
April 1982	A Power Utility buys stock in QCS and funds QCS for purchase of a Prime computer.
April 1982	QCS obtains its Prime computer and moves into new office. Medformatics moves into the same space.
July 1982	Medformatics begins development of MEDCODE.
Aug. 1982	QCS hires VP of finance, who also becomes company director and treasurer.
Fall 1982	QCS begins joint venture discussions with a power utility company.
Dec. 1982	Medformatics completes development of MEDCODE.
	Medformatics accepts $60,000 contract to develop coding program for a New York Hospital Association.
Spring 1983	A case mix analysis system is developed for a group of three hospitals in the twin cities.
April 1983	A Power Utility offers to acquire 80 percent of QCS for $1.5 million.
	Acquisition offer denied by owners of QCS.
	QCS develops and installs a hardware and software system at a local foundation.
May 1983	QCS and Medformatics agree, tentatively, to merge.
June 1983	Companies begin search for capital.
Oct. 1983	Companies sign letters of intent with an investment banking firm for private placement and public offering.
Nov. 1983	QCS acquires Medformatics effective September 1, 1983.
	Combined companies renamed Qnetics.

Startup Period

Nov. 1983	A director of client services joins Qnetics.
Feb. 1984	Private placement drive by investment banker raises $265,000.
March 1984	Qnetics moves into expanded office space.
April 1984	Qnetics establishes an agreement with a health care consulting firm to jointly develop and market a software module for hospital cost analysis.
	Qnetics initiates negotiations with a Health Care Group to develop and market medical software for third-party payors.
	Qnetics initiates development of a patient information system, and negotiates an agreement with a firm to distribute product.
	Qnetics reaches agreement with a foundation to market the foundation package. Also begins a consulting project with another foundation on a specialized data processing job.
May 1984	Qnetics advertises medical software products nationwide and receives 500 responses and requests for 200 demonstrations.
June 1984	Board of directors meets and narrows the focus of the organization.
	Investment banker proposes that private placement be raised to $600,000 and Qnetics board agrees.
July 1984	Qnetics establishes first operating budget.
	Qnetics makes internal organizational changes include grouping technical and operating responsibilities under principal of Medformatics, who becomes the COO. Principal of QCS, as CEO, is responsible for sales, external and financial management activities.

241

Aug. 1984	Qnetics obtains $200,000 in private placement. Private placement drive raises $465,000.
Sept. 1984	Qnetics hires new VP of marketing and sales.

Survival and Retrenchment Period

Oct. 1984	Entered joint venture agreement with a Health Management Group (to develop and sell a third-party payor product). Qnetics obtains a verbal commitment from power utility for $540,000 sales of hardware and software as a follow-on to earlier contracts. Qnetics negotiates to be facilities manager and to market software for Associates, Inc. Venture capitalist cancels the $1.8 million public offering scheduled for fall 1984, due to refusal of underwriter to proceed with the offering. Micro-Tel contacts Qnetics to explore mutual working arrangement.
Dec. 1984	Qnetics discontinues Prime Computer distributorship and becomes a subdistributor. Qnetics develops new mission statement and marketing plans. Qnetics' bankers close the line of credit and demand repayment on Qnetics' notes for full payment including interest on the loan.
Jan. 1985	CEO and COO develop an operating action plan to overcome the financial crises. Also develop a new business orientation. Micro-Tel moves into same office space as Qnetics. Qnetics Commences discussions with CDC, Honeywell, and CPT for sale of clinical software.
Jan.-March 1985	The organization size is cut from 31 to 13 employees. Employee owners operate without pay during January and February, and pay to other management employees is reduced. Bank decides not to force remedies on their demand, which allowed Qnetics to continue operations.
Feb. 1985	Qnetics and Micro-Tel submit a quotation to Pacific-Bell for a load management system.
April 1985	A letter of intent is developed to merge Micro-Tel with Qnetics. Principals of Qnetics and Micro-Tel verbally present a load management system proposal to Pac-Bell. Qnetics' management enters into negotiations with Computing Systems to supply integrated systems to health care industry. Qnetics discontinues custom-design software development to four customers. Qnetics' proposal to develop upgrades and supply computer system for Associates is turned down.
May 1985	Pacific Bell awards a letter of intent to Qnetics to supply load management system. Qnetics' management decides to focus on healthcare and electric power utility industries. Qnetics discontinues support of Associates software contract. Also discontinues third-party distributing of general products.
July 1985	Qnetics relocates to a smaller office space. Bank again states its intention to terminate relationship with Qnetics. The bank gives Qnetics' management until September 3, 1985, to provide the restructuring of the business or it will exercise the remedies that lead to the closure of the business.
July 1985	Organization size is further reduced, leaving only four full-time employees and some owners working without pay.
Aug. 1985	Qnetics stops development work on joint-venture with health care management group. Qnetics proposes a three-party agreement to private investors for funding and operating Micro-Tel, which results in $600,000 commitment, with $100,000 funded at that point. Qnetics signs a contract for facilities, services of its CEO and COO, and software development to Micro-Tel.

Aug. 27, 1985	The owners present the bank with two alternatives: either support the plan proposed by Qnetics or exercise the remedies.
Sept. 3, 1985	Bank agrees to Qnetics plan and not to exercise the remedies.
Oct. 1985	Maintenance services VP leaves Qnetics, and Qnetics sells the business to another firm.
Nov. 1985	Qnetics management considers alternatives of closing down the business or selling the business under best possible conditions.
	Qnetics reaches agreement with power utility company to provide $540,000 worth of hardware.
Feb. 1986	Financial reorganization of Qnetics begins with the help of an experienced businessperson.

Restartup Period

May 1986	Qnetics moves to a new location.
June–Nov. 1986	Financial reorganization of Qnetics is completed in all respects except for settling creditors' agreement. Proposal is sent to creditors committee requesting them to settle with Qnetics.
Feb.–June 1986	Efforts are made to raise $350,000 for Qnetics. This capital became available in July, 1986.
June 1986	Micro-Tel secures $750,000 investment from private placements.
July 1986	Power utility customer places all contract services in the area of research activities on hold until new budget is drawn up.
Oct. 1986	Qnetics hires a general manager for marketing of medical products.
	A letter is sent to creditors committee informing them that if no action was taken by the creditors on the earlier proposal, Qnetics will file for Chapter 11 bankruptcy.
	COO of Qnetics carries out his long-term plans of relocating to Hawaii.
	A new marketing plan is being developed for the medical products business.
Nov. 1986	A manager of client services joins Micro-Tel.
Dec. 1986	Marketing plan for medical products developed.
Jan. 1987	The experienced businessperson becomes CEO and president of Qnetics. The ex-CEO becomes a director on the board of Micro-Tel and also takes over as general manager of Micro-Tel.
	The former-COO ceases permanent employment with Qnetics but continues on the board and to serve as a consultant in the healthcare business.
Feb. 1987	Qnetics initiates negotiations with Hewlett-Packard resulting in a demonstration of the healthcare software line.
	The board size is increased from two to seven. Three outside board members are elected to the board: a banker, a physician, and a professor.
Feb. 1987	Micro-Tel board ratifies agreement with Qnetics whereby Qnetics will develop software for the load management device and receive 50 percent royalty split.

THE DEVELOPMENT OF 3M'S HEARING HEALTH PROGRAM
RAGHU GARUD AND
ANDREW H. VAN DE VEN

This case describes an ongoing effort to develop a new business within the 3M corporation consisting of a set of related hearing health products. Begun in 1977, the program presently consists of a family of cochlear devices and plans to manufacture and market hearing aids. Cochlear implants are surgically implanted electronic devices that enable the profoundly deaf a sensation of sound. Figure 8–3 illustrates the different parts of a cochlear implant as fitted in a human ear. Hearing aids are

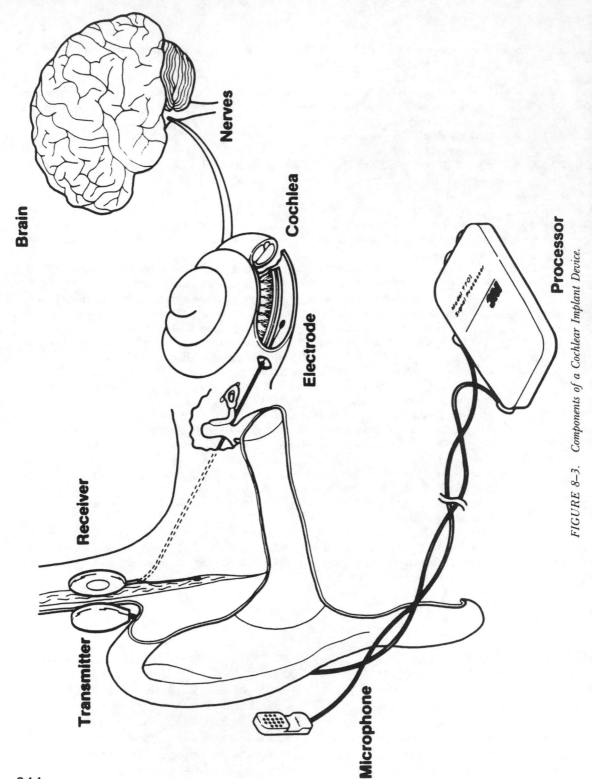

Brain

Nerves

Cochlea

Receiver

Transmitter

Electrode

Microphone

Processor

Model 770L
Signal Processor

3M

FIGURE 8–3. *Components of a Cochlear Implant Device.*

244

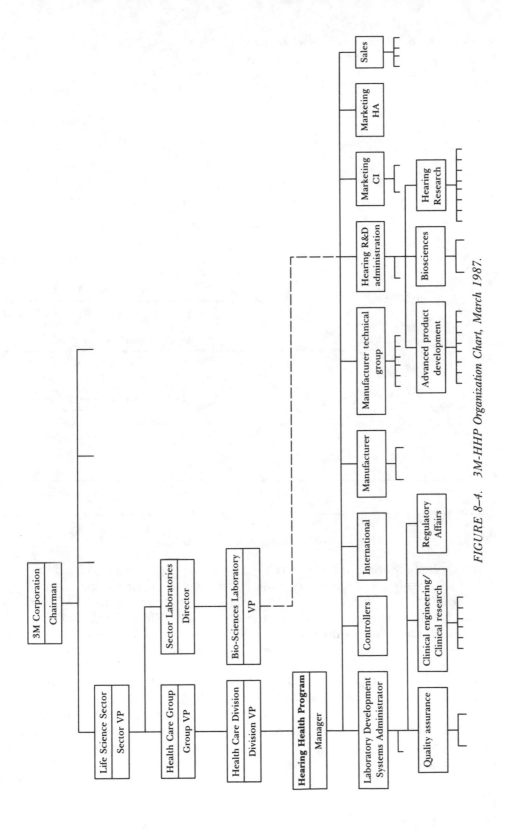

FIGURE 8–4. *3M-HHP Organization Chart, March 1987.*

245

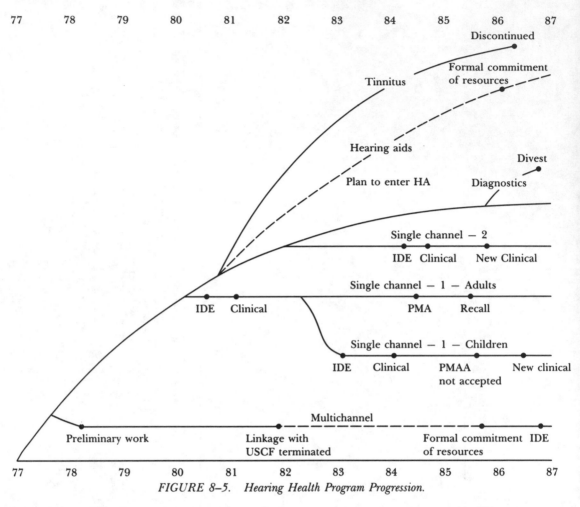

FIGURE 8–5. *Hearing Health Program Progression.*

products that facilitate hearing in less severely impaired deaf patients by the principle of amplification of sound.

Organizationally, the Hearing Health Program (HHP) is a part of the health care specialties division, which in turn is a part of the health care group of 3M's life sciences sector (see organization chart in Figure 8–4). Sales revenues of the life sciences sector was $1.4 billion in 1985. The program, consisting of about fifty full-time people in 1986, is managed by a steering committee comprising of nine functional managers and a program director. The steering committee meets once every

month to review and decide about the strategic direction of the program.

Real-time study of 3M's HHP began in December 1983. Data collection involved observing monthly meetings of the HHP management steering committee, semi-annual interviews with all HHP managers and questionnaires with all key HHP personnel, annual interviews with 3M top managers, as well as information obtained from company records, published materials on industry developments, and attendance at major otological trade conferences. From these data sources, ninety-seven chronological events were identified as having occurred in HHP's development from its inception in 1977 until the end of 1986. To reduce the complexity of this list to manageable proportions, the events were grouped into four segments: program initiation, the development of both single and multichannel cochlear implant devices, and efforts by the program to diversify from cochlear implants. Events classified by these segments will be described in this case, are graphically illustrated in Figure 8–5, and are presented in the appendix to this case.

Table 8–2 shows the number and turnover of key HHP personnel. As this table indicates, by December, 1986, most of the founding members of the program were replaced by new members, with some positions now being occupied by a third generation of people. Figure 8–6 summarizes the perceptions of key HHP members to various characteristics of the program that were obtained from four repeated MIS surveys completed in February 1985, June 1985, January 1986, and August 1986. As this table indicates, program members' perception about their own effectiveness was least favorable during a survey conducted in January 1986. Interviews conducted with program members during this period indicated that their morale was low because the program had suffered major setbacks; this is described later in the case. However, during this survey, program members also reported that they had demonstrated great resilience in their efforts to overcome these setbacks. The overall consensus of program members was, and has been throughout the development of this program, that the HHP represents one of the few real long-term opportunities for 3M and that consequently 3M would continue to fund the program in spite of setbacks in its implementation.

In contrast to program members' per-

TABLE 8–2. *Date of Entry/Exit of Key People, 3M–Hearing Health Program.*

Program Member	Joined in:	Left in:
First program manager	December 1980	July 1985
Second program manager	September 1985	
Quality assurance manager	NA	
First finance manager	NA	July 1986
Second finance manager	January 1987	
Bio-Sciences Laboratory manager	1979	
Product development manager	January 1984	
First manufacturing manager	1978	Fall 1984
Second manufacturing manager	March 1985	October 1986
Third manufacturing manager	October 1986	
Marketing manager–cochlear implants	April 1982	
Marketing manager–hearing aids	October 1985	
Sales manager	NA	March 1986
International marketing manager	NA	
First regulatory manager	NA	December 1985
Second regulatory manager	NA	December 1986
Clinical research specialist	NA	February 1986
Senior specialist	NA	October 1985
Design specialist	NA	
Senior clinical engineer	1980	
Clinical research specialist	NA	

NA=means information not available.

247

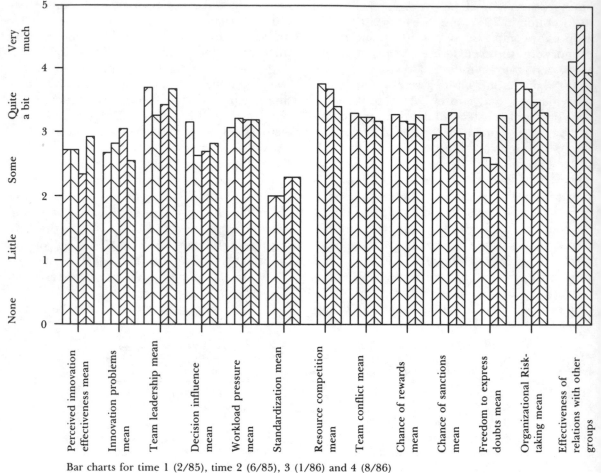

Bar charts for time 1 (2/85), time 2 (6/85), 3 (1/86) and 4 (8/86)

FIGURE 8–6. Overall Survey Results for Hearing Health Program

ceptions, members constituting top management have expressed conflicting opinions about their support of the program. One top manager expressed ambivalence about HHP in 1985 and again in 1986. He observed that the HHP is one of the most risky programs among the thirty-five new ventures in his organization. On the other hand, another top manager, who was the original sponsor of the business idea, has consistently expressed strong support for the program throughout its development. He stated that cochlear implants, and hearing health in general, offer 3M a significant long-term business opportunity. He added that 3M is less concerned about immediate financial payoffs from an innovative program than venture capitalists would be. Instead, 3M is more concerned that its innovative programs grow to become self-sustaining division-sized businesses in ten years. On this criterion, about three of ten programs such as HHP succeed within 3M. The third member constituting top management has not taken either of the extreme views of unequivocal support or of reservation as expressed by the other two top managers.

We will now describe the four segments in the development of the HHP.

Initiation of the Hearing Health Program

In 1977, the technical director of 3M's surgical laboratories was informed that 3M's subsidiary company in Australia had been approached by the University of Melbourne for a possible cooperative agreement to develop a "bionic ear." The technical director was intrigued by the concept and decided to explore it further by visiting various research centers and clinics in the United States to learn of developments in otology. By 1978 the idea of an electronic ear had gained sufficient credibility to persuade the technical director and his supervisor to commit 3M to it.

The idea to explore development of a cochlear implant was assigned to an unrelated products group in 3M. About this time, two other groups within 3M were also initiating preliminary work on cochlear implants and other related products such as hearing aids. One group was the sector biosciences laboratories (see Figure 8–3), which hired a full-time scientist in 1979 to startup a research program in hearing and speech in 1979. The other group involved in hearing health products was an acquisitions group within 3M, which in 1980 looked into the commercial viability of entering the hearing healthcare product by acquiring a firm manufacturing hearing health products.

In 1978 the technical director was transferred to California for two years to direct McGhan Medical Products, a 3M subsidiary. His replacement at the surgical laboratories, along with the head of the unrelated products group, established a relationship with a researcher at the University of California at San Francisco (UCSF) in 1978. Through this 3M-UCSF relationship, a cochlear device was developed and implanted in two or three individuals during 1980 and 1981. This effort was gener-ally considered successful but not pursued further.

In September 1978, while the technical director was at McGhan, a vendor relationship was established between McGhan and the House Ear Institute, whereby McGhan supplied parts of cochlear devices to the House Ear Institute (HEI). According to a manager involved in this relationship, McGhan was successful in solving HEI's problems with the device. This later proved to be the basis for developing a stronger relationship with HEI.

With the promotion of the technical director's supervisor to group vice president in fall 1980, the technical director returned from California to direct the surgical products division in St. Paul as the division VP. So as to increase the speed of development of the program, the division VP asked a 3M employee to direct and integrate three independent 3M activities into a single unit in December 1980. These three activities included (1) a manufacturing relationship between McGhan and the House Ear Institute, (2) a developmental relationship with the University of California at San Francisco, and (3) an ongoing 3M laboratory otological research effort.

The newly appointed program director, although initially skeptical about the value of cochlear implants, decided to seriously examine the idea because of his great respect for his supervisor. Two events motivated his commitment to the program. First, a hearing-impaired 3M colleague strongly encouraged him to invest his energies to developing a cochlear implant program for both social and business reasons. Second, the program manager was influenced by an anecdotal story of a profoundly hearing-impaired woman who was intentionally bumped by shoppers with carts in grocery stores when she failed to move after being asked to.

Having become committed to cochlear implants in 1981, the program manager began building his team and understanding prior ac-

249

tivities in the three areas that were integrated into the program. Team discussions resulted in a simple yet important philosophy, which the program manager characterized as follows—"You have to crawl—walk—then you run." By taking this position, members of the program were acknowledging the complexity of the phenomenon of hearing, and thus the foolhardiness of expecting that actual hearing could be accomplished by a cochlear device with the technology then available. In the words of the head of the biosciences laboratories, the formal initiation of the cochlear implant program (later to be known as the Hearing Health Program) was crucial

> because if this action had not taken place at the time, I think I would have cut the research projects because there is no reason for us to do research if there is no business unit doing anything.

Several events after the formal birth of the program are relevant to its initiation. In 1981 various people working on the cochlear implant effort realized that they were involved in something of a much bigger magnitude than just cochlear implants—the whole area of hearing health. This major insight was an outcome of the different efforts underway at the HHP and the biosciences laboratories as well as a study carried out by an acquisitions group in 1980 on the commercial viability of acquiring a firm in the hearing health business. Consistent with this broader vision, program members redefined their strategy: "As AT&T is to communication, and IBM is to computers, 3M will be to hearing health." The program managers developed the vision of planting a seed that one day would grow into an oak tree consisting of a family of hearing health products. As one of the founding members of the program explained, it was difficult to justify resource commitments for cochlear implants alone; a

broader vision of a family of products was needed.

In December 1982 the program manager arranged a visit for the 3M chairman and other top company officials to the House Ear Institute. According to the program manager, this trip proved important a short time later when additional support was requested for the program in the midst of a "belt-tightening" period at 3M. This request for funds was approved by top management because of their first-hand awareness of the program. In May 1983, to effect an acceleration of the program, the surgical products division was permitted to reduce its profit target, so as to avoid the need for continual justification of a lower income statement. Obtaining corporate support in May 1983 was significant because it allowed for substantial personnel additions.

Single-Channel Cochlear Devices

In September 1978 3M's subsidiary McGhan Medical Products became a vendor to the House Ear Institute by supplying custom-designed parts for the single-channel cochlear device. McGhan's manufacturing manager reported that McGhan's success in solving HEI's problems with the device led to a joint development agreement with HEI in December 1981. He further stated that the HEI vendor relationship also led 3M to believe that the single-channel cochlear implant was the most appropriate technology to launch the program with because it was believed to be safer than multiple channel devices. This conclusion however was not without controversy because many in the scientific community were urging otologists to wait for further developments in the multichannel implant technology instead of embracing the single-channel technology advocated by Dr. William House of the House Ear Institute.

Thus, even prior to the formal agreement with HEI in December 1981, 3M had begun developmental activities with HEI in 1979. In December 1980 the House Ear Institute had already obtained an Investigational Device Exemption from the FDA to conduct clinical studies with the device on humans. Immediately after the formal agreement with the House Ear Institute, clinical trials of the device were undertaken. On the basis of the results obtained from these clinical tests, in October 1983 3M submitted an application to the FDA seeking approval to market this device in the United States on a commercial basis. One year later, in November 1984 the FDA granted its approval for commercial sale of the device in the U.S. While granting its approval, the FDA announced that this was the first time that one of the five human senses was being replaced by an electronic device.

While the process of seeking regulatory approvals was underway, other functional competencies were being added by 3M to manufacture and market the device. In April 1982 a marketing manager was added to the program. One of his first tasks was to convince various insurance carriers to extend coverage to cochlear devices even before the FDA had granted its approval for commercial sale of the device. The marketing manager's efforts were fruitful when in early 1983, insurance carriers agreed to cover the device. After November 1985, when the FDA had granted 3M's competitor approval to market its device, 3M along with other firms in the industry undertook efforts to seek Medicare coverage for cochlear devices. By the end of 1986, firms in the industry had successfully convinced the Prospective Payment Group of the U.S. Public Health Service to extend Medicare coverage to cochlear implants.

In December 1983 3M sponsored its first course to train physicians to implant cochlear devices. In 1984, the biosciences laboratory manager, who until then had been in charge of research and development, began to devote his efforts to conduct research alone, and another person was appointed to assume development activities. Immediately after FDA's approval of Device 1 in November 1984, manufacturing operations were consolidated into a manufacturing plant in Sarns, a 3M subsidiary, and a pilot plant in St. Paul.

In April 1985 program members reported that the sales of the House device were above forecasts for the first quarter of 1985 and that the HHP was one of the few innovative programs in the life sciences sector to have met its sales forecasts. Program members used this achievement to requisition for five additional salespersons to the program.

In March 1985 the group and sector VPs recommended that the HHP should not invest any more resources to further develop Device 1, and that it should start developing the next generation of products. However, two incidents of product failure occurred in March and November 1985 that consumed considerable resources and prevented significant development of the next generation of devices. First, a product failure occurred in March 1985 when program members uncovered problems with the epoxy coating of Device 1. This led to a technical audit of the program in August 1985, which the quality assurance manager characterized as an "inquisition" that consumed considerable resources of the program. The auditors found that the program was complying with "good manufacturing practices" set down by the FDA.

This incident was symptomatic of growing internal conflict among HHP managers and resulted in increasing upper-management dissatisfaction with the progress of the program. In July 1985 3M's division and group VPs intervened. The founding program manager was replaced by another manager transferred from another 3M division.

A second product failure incident took place in November 1985. In April 1985, program members had sought and received an approval from the FDA to make modifications in the electrodes of the device. Essentially, the modification called for using a hermetically sealed titanium electrode rather than an epoxy-coated one. In November 1985 three of the newly approved titanium electrodes failed in quick succession in the market. The HHP technical manager stated that 3M had decided to voluntarily recall the device from the market as this was an unusually high failure rate for a biomedical device. During the next six months HHP members were involved in rectifying defects in Device 1 and in obtaining a reapproval of the device from the FDA.

All program members interviewed in February 1986 stated that the product recall created extremely low morale and consumed most of the program's resources and attention for six months until the FDA granted its approval to release the redesigned device in May 1986. The recall also stimulated a reassessment of the HHP organization. In February 1986 a separate "manufacturing technologies group" was established to interface between the laboratory and the manufacturing group to ensure better product quality. In addition, the HHP manager reported a shift in strategy from that of getting the product out to the market as fast as possible, to that of developing better product quality, reliability, and performance.

In October 1985 it was announced that a competitor's device had received FDA approval for commercial release. Test results circulated in late 1985 by the University of Iowa, a newly established and self-appointed independent cochlear implant testing center, indicated that the competitor's device was superior in performance in comparison to the performance of HHP's Device 1. At the same time, three-month sales figures of Device 1 prior to its recall strongly suggested that this device would not enable 3M to develop or capture the market for cochlear implants. Both these indications stimulated the HHP in February 1986

to reduce its emphasis on Device 1 and to renegotiate 3M's contract with the House Ear Institute. This included significantly reduced consultancy fee to Dr. House for services rendered to the program. In an interview held at this time, the program manager reported that HHP's relationship with HEI was cordial but should no longer be looked on as a "marriage." In November 1986 sector and group VPs again confirmed their belief that it was necessary for HHP to reduce its emphasis on Device 1 and to move on to other products.

In addition to the above events, which pertain to the developments of Device 1 for adults—patients over age eighteen, another set of events was undertaken to develop an extension of this single channel device for profoundly deaf children. In August 1985 3M submitted a PMA application to the FDA for a version of the device suitable for children. The FDA did not accept the application because of the manner in which the HEI clinical data had been collected and presented. In June 1986 two HHP regulatory affairs personnel were assigned full time for six months to collect and compile additional data on the children's device in order to revise and resubmit a PMA application to the FDA in early 1987.

While efforts were underway to develop and commercialize Device 1, activities were being initiated to develop a second single-channel cochlear implant device for adults. 3M entered into an agreement with the Hochmiars in Vienna, Austria, in June 1981. Program members reported having begun intensive developmental work by mid-1983. In March 1984 the FDA granted an IDE for this device, and clinical studies were begun soon thereafter. By February 1985 the program manager reported that outstanding success had been documented with five U.S. patients. On this basis, the device was promoted at the Thirteenth International Otoghinolaryngology Conference as a multifrequency device with superior performance and safety features as compared to competitors' multichannel devices. Later, in August 1985, salespersons reported that the promotion of Device 2 at the conference was adversely affecting sales of Device 1. Audiologists apparently delayed ordering cochlear implants until the second single-channel device became available.

In October 1985 clinical evidence on a larger sample of U.S. patients implanted with Device 2 failed to corroborate the same encouraging results that had been stated for the five patients earlier in February. Consequently, program members decided to initiate new clinical studies to test the efficacy of Device 2 with a different, more typical, electrode orientation. In November 1985 3M sought and obtained another IDE from the FDA to begin clinical trials with the different surgical orientation of the electrode. As of December 1986, 3M is in the process of collecting clinical data on this device for a possible PMA application in 1988.

Multichannel Cochlear Devices

3M's original contact with the multichannel technology also begins with the inception of the program in 1977 when the University of Melbourne in Australia approached 3M's Australian subsidiary company to jointly manufacture and market a twenty-two-channel cochlear device that they had been developing since 1975. After several months of negotiations, discussions with the University of Melbourne were terminated for financial and Australian nationalistic reasons. Subsequently, in 1979 the University of Melbourne linked up with an Australian firm, Nucleus Corporation, to commercialize its multichannel device. By 1986 Nucleus became 3M's major competitor in the cochlear implant area, offering a multichannel device.

As stated before, in 1978 3M established a relationship with a researcher at the University of California San Francisco to develop a multichannel device. Through this 3M-UCSF relationship a multichannel cochlear implant was developed and implanted in two or three

individuals during 1980 and 1981. This effort was generally considered successful but not exceptional. According to 3M's research person involved in this relationship, 3M and UCSF had somewhat different orientations. UCSF had an academic orientation that focused on answering the "why" behind the basic questions related to electrical stimulation and hearing, whereas 3M had an industrial orientation that concentrated on the utility of electrical stimulation and hearing. Because of this, the relationship with UCSF was terminated in 1982. Subsequently, UCSF continued its development of cochlear devices and entered into a relationship with Storz in 1983 to develop and commercialize a multichannel cochlear implant device, thereby establishing another 3M competitor.

In May 1985, 3M promoted its second single-channel device at the thirteenth international otorhinolaryngology conference. Test results of this single-channel device displayed by 3M showed that the device performed better than competitors' multichannel devices. The basic theme adopted by 3M was, "It is not important whether the device is multi- or single-channel; it is important that the device be able to transmit multiple frequencies of sound impulses." 3M claimed that its device could transmit such multiple frequencies more effectively than competitors' multichannel devices. Other firms contacted who exhibited their cochlear devices in this conference were unanimous in their opinion that the multichannel device would prove to be superior to the single-channel device in the long run and that 3M was pursuing the wrong technological path. 3M informed other firms that they were making exaggerated claims about the efficacy of their devices.

In May 1985 an HHP member reported that FDA personnel were also convinced of the superiority of multichannel device over single-channel devices and that the FDA was promoting this in its publicity releases about cochlear implants. HHP members protested to FDA personnel that the single-channel device was indeed more desirable from an overall safety/efficacy point of view than were multichannel devices.

In April 1985 an HHP member conducting research on multichannel devices stated that he had not been able to make progress in developing this device because of resource constraints. Other program members too began voicing their concerns about the program's lack of progress in developing a true multichannel device. In September 1985 the new program manager stated that "we should not end up believing our own bull-shit" about the single-channel device's being superior to multichannel devices. In September 1985 the research manager stated that HHP personnel were continuously defending the single-channel device and had paid "lip service" to developing a multichannel device. He reported that little actual progress had been made to develop a multichannel device. At this time, another technical person working on cochlear implants, a recipient of the 3M citation for technical excellence, also stated that the multichannel device would eventually prove to be superior in performance to the single-channel device. In January 1986 a security analyst's report referenced tests conducted by the University of Iowa that clearly showed that the multichannel device performed better than the single-channel device.

As a consequence of all these indications, in January 1986 the biosciences laboratory submitted a plan and committed research personnel and funds to develop an advanced multichannel device. In July 1986 the program was granted an IDE by the FDA for this device, and clinical trials were begun thereafter. While explaining the basic approach in designing the advanced multichannel device, the program manager stated that the upgradability of the device to incorporate future technological advances was a key consideration. The biosciences laboratory manager was confident that the advanced multichannel device would be capable of this kind of flexibility.

In its efforts to accumulate the basic competencies required to speed the development of the multichannel device, the program also explored linkages with other firms. In April 1986 Storz, which had a relationship with UCSF to develop their multichannel device, was reported to be searching for an organizational sponsor to continue funding its cochlear implant program. 3M reviewed a possible relationship with Storz in order to gain technical knowledge about multielectrodes. This relationship was terminated when Storz was acquired by another company that did not want to pursue work on cochlear implants. Similarly, in June 1986 another firm in the industry, Symbion, attempted to seek a partnership with 3M by offering its technical knowledge in multichannel devices. 3M decided not pursue this offer because Symbion's technology was not entirely compatible with 3M's and did not offer the desired future technological flexibility. In July 1986 Storz was acquired by another company, American Cyanimide, which prompted 3M's group VP to ask at the August 1986 sector review meeting of the program why HHP had not considered acquiring Storz.

In October 1986 the 3M Biosciences and the HHP laboratories were merged into one laboratory. According to the program manager, this was done to increase the effectiveness and efficiency of operations of the two laboratories.

In December 1986 the U.S. Office of Health Technologies Assessment published a report on cochlear implants required by the Prospective Payment System of the U.S. Public Health System as a necessary procedure for granting Medicare coverage to cochlear implant recipients. The report clearly stated that the multichannel technology was superior in performance to the single-channel technology.

As of December 1986, the HHP has a clear commitment to developing a multichannel device. Program members expect to complete the basic developmental work by the end of 1988 and to be able to submit an application to the FDA sometime in 1989.

Diversification of HHP Business

An earlier section described the events that led the HHP manager to expand program scope from cochlear implants to hearing health. The hearing health idea was influenced by an early appreciation of a possible application of cochlear implants to solve tinnitus—a debilitating ringing sound in the ear that afflicts many deaf people. Little is known about progress made on tinnitus except that research began in 1981 and continued till 1986. In June 1986 the HHP research head stated that further work on tinnitus was being discontinued as it was found that neither cochlear implants nor any other research approach that 3M was pursuing could solve this ringing problem.

In February 1985 the first program manager reported that the division VP had encouraged him to develop a broader vision of the hearing health industry. To explore this, in April 1985 the program manager arranged for a special discussion to examine whether HHP should diversify into the less profoundly impaired market with cochlear implants and hearing aids or to continue serving the profoundly deaf market with cochlear implants alone. The program manager stated that HHP did not have adequate resources to pursue both efforts and that he preferred to follow the former route. Other program members unanimously voted to continue with their cochlear implant efforts for the profoundly deaf patients and rejected the program manager's proposal to pursue placing the cochlear implant in the residual hearing market. HHP members wanted to win in cochlear implants first. With the reassignment of the program manager in August 1985, the idea was tabled.

In March 1985, as a part of another ac-

quisition, 3M acquired a firm that manufactured devices that measured patients' hearing loss. Because this device appeared to fit well with cochlear implant activities, 3M offered HHP an opportunity to sell the product as a part of the business. After evaluating the opportunity, HHP decided to sell diagnostic devices in May 1985. In the same month HHP promoted the diagnostic device at the thirteenth International Otorhinolaryngology Conference. Immediately after this, the international marketing manager reported that efforts to promote the diagnostic device at the conference had been premature and abortive as HHP was unable to supply the device to customers immediately on demand.

The October 1985 realization that Device 2 was not demonstrating acceptable performance results and the November 1985 market recall of Device 1 implied that HHP had no cochlear devices to sell in the market for the first half of 1986. As a result, at year-end 1985, HHP management decided to focus full sales efforts to promote and sell diagnostics devices. The reported objective was to generate revenues from diagnostic devices to partially sustain the program, both immediately and in the long run, and add products to the line.

Later, in June 1986, the HHP sales manager reported that the resources required to sell and service the diagnostic devices far exceeded the revenues that they could generate. As a result, in July 1986 both top management and program members decided to divest from diagnostic activities. By January 1987 HHP had divested from the diagnostics business.

As stated before, in October 1985 Nucleus Corporation was granted a PMA by the FDA for commercial release of its twenty-two-channel device. In February 1986 the HHP finance manager stated that this would enable 3M to determine how big the market for cochlear implants really was. By mid-1986, the HHP marketing manager stated that evidence suggested that the cochlear implant market was not growing fast enough for HHP to sustain a program solely on cochlear implants. He observed that Nucleus Corporation's sales failed to grow as expected. In addition, as reported earlier, two other competing firms, Storz and Symbion, were searching for alliances, reportedly because they expected the cochlear implant market to grow at a much slower pace than they had earlier anticipated.

Consequently, in February 1986 a special programwide meeting was held to discuss possible product ideas for diversification that could sustain the program while the cochlear implant market emerged over time. One idea evaluated was hearing aids (as was suggested by the founding HHP personnel in 1980). After this meeting, program members went about gathering information about the hearing aids industry and about possible organizations that HHP could enter into an agreement with to develop and manufacture technically advanced hearing aids.

In May 1986 a plan presented to the sector committee by program members to continue work on cochlear implants and expansion into other unspecified products was tabled. Sector committee members asked the group to present more details of the alternate products that the HHP could pursue. One of the sector committee's main concern was that the cochlear implant market was not increasing fast enough and therefore, Cochlear implant activities should not be the immediate thrust of the program. Later, in July 1986, the program presented a revised plan in which HHP proposed entering the hearing aids industry by acquiring a company manufacturing advanced hearing aids. This time, the plan was approved by the sector committee. In October 1986 the sector VP granted permission to acquire a hearing aids company. In January 1987, HHP acquired Diaphone of Sweden to manufacture hearing aids; HHP research personnel had been following the development of the key idea by the Diaphone personnel since 1980.

Analysis of HHP's Developmental Progression

Having described the major events in the four periods of HHP's development, we will now examine the patterns of progression between and within periods, as illustrated in Figure 8–5.

1. *Unitary paths.* An examination of the chronological list of events indicates that between 1981 (when HHP was formally initiated) to November 1984 (when FDA approved device 1), 63 percent of events were associated with the development of Device 1. This suggests that the development of the program essentially followed an "unitary path" between the period beginning 1981 to the end of 1984. Some other pieces of qualitative data support this inference.

First, Figure 8–5 indicates that although HHP entered into formal agreements with Hochmiars and HEI at about the same time in 1981, clinical engineering studies of Device 2 on humans was initiated only in 1984. According to the marketing manager, this was because HHP did not utilize its resources well and because a lack of focus plagued the program. A related and more interesting reason provided by the HHP research manager was that HHP made a tactical error in working toward an FDA premarketing approval application for commercial sale of Device 1; considerable resources were expended in creating this "window of opportunity." He also stated that by the time the FDA had granted its approval, Device 1 had become technologically obsolete in comparison to other devices available on an IDE basis. He suggested that it might have been wiser to have only progressed to the investigational stage for Device 1 and then to have moved on to develop the next generation of devices.

Second, in March 1985 the group VP suggested that HHP should not expend any more resources to further develop Device 1 and that they should instead move ahead to commercialize Device 2 and initiate work on multichannel devices. However, little progress was made in either direction. For instance a member of the program reported in 1985 that the lack of resources was severely hindering progress on the multichannel device and that failure to develop a multichannel device was a critical omission for the program. In October 1986 the program manager stated that HHP "had taken four years dragging their feet" with Device 2 and that now they may decide to remain at an IDE stage only. In November 1986 the group VP again expressed frustration that the program had not shifted focus from single-channel technology as he had suggested earlier.

One reason for this unitary development of the program between the period 1981 to 1984 can be linked to a key objective laid down by the program manager in November 1984. This objective was to create and exploit "a window of opportunity" by being the first in the market with a cochlear device. Because the cochlear implant industry is socially regulated by the Food and Drug Administration, any device that is first to obtain FDA approval for commercial sale enjoys such a window of opportunity till such time that a competing firm's product also gets FDA approval.

2. *Product transitions.* The case illustrates one of the problems associated with product transition management. According to the sales manager, the promotion of Device 2 in May of 1985 at the Thirteenth International Otorhinolaryngology Conference resulted in a situation where audiologists and patients preferred to wait the release of the Device 2 rather than pursue the implantation procedure with Device 1. This was one reason attributed for the poor sales of Device 1 in the second quarter of 1985.

3. *Sequencing of activities.* The HHP has de-signed a sequential schedule to use resources and functional competences for different generations and families of products over time. Essentially, the plan is to complete R&D and clinical work on Device 2 while the biosciences laboratory completes its work on the multichannel device. This in turn will be taken over by the hearing health laboratory, while the biosciences lab will take on research of advanced hearing aids and other hearing health care products. Eventually, the advanced hearing aids will be scaled up by Diaphone and the hearing health personnel for commercialization. There is a consensus among HHP personnel that such a sequential plan will effectively handle the increased scope of program activities without necessarily increasing resources.

4. *Divergent paths.* After 1984 there was a proliferation of products, both in the cochlear implant stream, as well as the wider concept of the business idea. For instance, in the cochlear implant area, the development of a device for children and Device 2 were taken up during the end of 1984. In 1986 development of the multichannel device was taken up. The overall concept of the business of hearing has evolved from tinnitus and diagnostics to hearing aids. These were all "disjunctive" changes from each other as each required different specific skills and knowledge, although utilizing the same stock of basic knowledge on the science of hearing.

5. *Convergent paths.* One convergent effort between cochlear implants and hearing aids is emerging. Cochlear implants have been reserved for the profoundly deaf, who cannot avail of hearing aids. Hearing aids, on the other hand, have been prescribed for patients with some residual hearing. In 1985, so as to increase the potential market base, program members first articulated a strategy to use cochlear implants to alleviate the

257

deafness of patients with some residual hearing as well. This convergence appears to be emerging as cochlear implants, and hearing aids were combined into one hearing health program in 1987.

6. *Sequential development of functional competencies for products.* An examination of Figure 8–5 shows a sequential development occurred in functional competences for each of the cochlear implants. For instance, in the case of the development of Device 1, basic R&D was followed by clinical engineering, which was followed by regulatory approvals, which in turn was followed by manufacturing and sales. One reason for this sequential development of functional competences is the prescription of the order of creation of functional competences by the FDA.

However, two instances of opportunistic "leap-frogging" were observed in the development of cochlear implants by HHP. At the 1985 otolaryngology conference 3M promoted the introduction of the second single-channel device even though it had not yet (1) completed clinical trials, (2) obtained FDA regulatory approval, and (3) scaled up manufacturing of the device. In other words, the marketing function had begun much prior to the completion of three other preceding value-creation functions. The second instance of leap-frogging occurred when 3M prematurely introduced its newly acquired diagnostic device. However, after having generated customer interest in the diagnostic devices, 3M found that it was unable to service the demand.

7. *Iterative paths.* One set of iterative events is the process of seeking FDA approval. Each product has to obtain an "investigational device exemption" status from the FDA after having been clinically tested on animals. Next, each of the clinical sites has to obtain an "institutional review board" clearance to certify its capability to conduct clinical engineering on humans. After test results show that a minimum level of safety and effectiveness has been achieved, the device has to be submitted to the FDA panel for a premarket approval. If the FDA finds the device to be safe and effective, it grants its approval for commercial sale after having ascertained the prevalence of "good manufacturing practices." The entire procedure of obtaining FDA approval from the time of initiation of clinical trials on animals can take anywhere from three to five years.

Hearing Health Program Case Appendix.
List of Events in the Emergence of 3M Hearing Health Program.

Initiation Period

Jan. 1, 1977	University of Melbourne approaches 3M for a joint venture to manufacture cochlear implants. 3M decides to pursue the idea separately.
Jan. 1, 1978	Idea of cochlear implants assigned to Unrelated Products Group.
Jan. 1, 1978	Cochlear Corporation links up with the University of Melbourne.
Jan. 1, 1978	3M initiates relationship with UCSF to work on multichannel devices.
Sept. 1, 1978	3M becomes vendor for House Ear Institute Device 1.
Jan. 1, 1980	Formation of cochlear implant program at 3M.

Single-Channel Cochlear Implant Technology

Jan. 1, 1979	3M takes up initial research on Device 1.
Feb. 1, 1980	HEI obtains IDE approval from the FDA for Device 1.
Jan. 1, 1981	Clinical trials initiated by 3M for Device 1.
June 1, 1981	Agreement between Hochmier and 3M established to develop Device 2.
Dec. 1, 1981	Agreement between 3M and HEI to develop Device 1.
Jan. 1, 1982	Marketing activities initiated by 3M for Device 1.
Jan. 1, 1983	3M promotes Device 2 at a training program.
	3M convinces third-party payors to extend coverage to Device 1.
Jan. 2, 1983	HEI gets IDE approval from the FDA for Device 3 (children).
June 1, 1983	3M begins to work intensively on Device 2.
Oct. 1, 1983	3M submits first PMAA for Device 1 to FDA.
Dec. 1, 1983	3M sponsors its first courses to train physicians on cochlear implants.
March 1, 1984	IDE for Device 2 granted by the FDA.
June 1, 1984	Clinical trials commenced for Device 2.
Oct. 1, 1984	Major strategic decision to take up children's Device 3 toward FDA review.
Nov. 1, 1984	FDA grants PMA approval for 3M-House Device 1.
Dec. 1, 1984	3M establishes pilot manufacturing facilities at St. Paul and manufacturing facilities at Sarns.
Jan. 1, 1985	3M establishes titanium vendor for Device 1.
Feb. 5, 1985	3M enters into a relationship with a local hospital and forms CISM to conduct rehabilitation.
	Outstanding success documented with Device 2 with five U.S. patients.
March 6, 1985	Lack of sales becomes HHP credibility problem in 3M.
April 10, 1985	HHP obtains program status in 3M.
	HHP actual sales for the first quarter was 5 percent ahead of its forecast. HHP use this sales performance to requisition for five additional salespersons.
April 26, 1985	HHP is one of the few in the life sciences sector to achieve its forecasts.
May 1, 1985	3M promotes diagnostic device at the Thirteenth International Otorhinolaryngology Conference.
	3M promotes Device 2 at the Thirteenth International Otorhinolaryngology Conference. Each firm in the cochlear implant industry claims that others are making exaggerated claims and expresses the need for standards at conference.
May 7, 1985	Plan to file a pre-market approval application for Device 3 (children) in three months' time reported to 3M sector review committee.
May 10, 1985	The FDA states that the multichannel device is better than the single-channel device.
May 23, 1985	The HHP gets an A grade as compared to the other programs in the 3M sector by top management. Sector committee criticizes HHP for its cost expenditures.
June 1, 1985	Epoxy failure found in Device 1.
June 21, 1985	Change in HHP program manager.
July 1, 1985	Unfavorable 3M internal survey findings.
	3M evaluates linkages with Richards and decides not to enter into relationship.
Aug. 1, 1985	Program members find that the premature promotion of Device 2 cramps sales of Device 1.
	3M corporation profits reported as stagnant for the quarter.
Aug. 21, 1985	Program member complains to 3M chairman about the quality of Device 1. Audit reveals many hidden skeletons, although HHP is found to be complying with good manufacturing practices.

Sept. 1, 1985	HHP steering committee discusses that HHP is no longer the industry technological leaders for the first time.
	It is found that the Hochmaier Vienna device does not perform well in U.S. clinical trials.
	Change in thrust in Device 2—extracochlear instead of intracochlear.
Sept. 11, 1985	HHP steering committee discusses that the realistic time schedule for filing the PMAA for Device 2 is fall 1987 and not fall 1985.
	Program manager informs HHP members that they have credibility problems with 3M top management about sales of the Device 1.
Oct. 1, 1985	Nucleus Corporation obtains FDA approval for a multichannel device.
	FDA grants 3M IDE for Device 2 different electrode orientation.
Nov. 1, 1985	3M starts servicing Device 1.
	Pressure on HHP to achieve sales target.
	3M voluntarily recalls Device 1 from the market due to product failures.
Dec. 1, 1985	Decision to divest from Cochlear Implant System of Minnesota.
	Formation of Cochlear Implant Industry Council consisting of 3M, Nucleus Corporation, Stortz, and Symbion to attempt to get cochlear implants covered under Medicare.
	3M alerts competing firms about recall and arranges to have a joint meeting to reduce negative impact of the recall.
	FDA does not accept 3M's children's Device 3 pre-market approval application.
Dec. 17, 1985	HHP reorganizes its laboratories.
Jan. 1, 1986	FDA spot audit of HHP manufacturing facilities reported very successful.
Feb. 1, 1986	Separate manufacturing technologies group set up for HHP at 3M.
Feb. 6, 1986	Reduction in emphasis of Device 1 and renegotiation of contract with HEI.
	HHP makes shift in emphasis from speed to quality and reliability.
April 16, 1986	HHP members meet to share with each other what they have learned from the recall.
May 21, 1986	Device 2 failures reported.
May 27, 1986	FDA grants its reapproval for the redesigned Device 1.
June 18, 1986	HHP commits personnel to complete the children's Device 3 PMAA by the end of the 1986.
	HHP decides to do further work on rehabilitation services.
July 16, 1986	3M sector committee reports that resources have been squandered and states intention to focus on costs incurred by HHP.

Multichannel Cochlear Implant Technology

Jan. 1, 1982	UCSF and 3M relationship terminated.
Feb. 15, 1985	Emergence of the multichannel device is reported as threat.
March 1, 1985	Some work initiated on multichannel devices.
Oct. 16, 1985	HHP once again starts research on the multichannel device.
Jan. 1, 1986	Biosciences commits resources formally to begin work on their advanced multichannel device.
	Storz approaches 3M for a possible joint venture.
	Security analyst report uses University of Iowa's test results that shows multichannel device performs better than the single-channel device.
June 18, 1986	Symbion approaches 3M for a possible joint venture.
July 1, 1986	FDA grants 3M an IDE for advanced multichannel device.
Oct. 6, 1986	Biosciences merges with HHP labs for increasing effectiveness and efficiency.
Oct. 15, 1986	Nucleus Corporation obtains an IDE for a children's device from the FDA.
	3M technical audit of HHP multichannel lab program reported very successful.
	OHTA report suggests that multichannel is superior to single channel.
Dec. 1, 1986	Medicare coverage extended to cochlear implants.

Diversification Events

March 1, 1985	Significant progress reported in ways of doing diagnostics without doing surgery.
April 26, 1985	Program members discuss the possibility of getting into hearing aids. Program manager suggests that HHP does not have the resources to pursue multiple hearing aids technological paths.
Sept. 1, 1985	Change in the constitution, charter, and meeting frequency of steering committee.
Sept. 11, 1985	HHP begins pruning products/program that do not add to the bottom-line figures.
Oct. 16, 1985	HHP members agree to acquire Axonics.
Jan. 1, 1986	Significant reduction in head count (eleven people) and operating budget for the program.
Jan. 15, 1986	Report that HHP efforts to promote the diagnostic devices at the Miami conference was premature and abortive.
Feb. 10, 1986	Project Voyager meeting to identify alternate avenues of resource generation.
	Bethovan 2000 is adopted as the programwide objective.
March 19, 1986	HHP members change the charter of the steering committee to examine strategy issues in more detail.
May 21, 1986	Sector does not accept HHP plan to continue efforts to first win in cochlear implants.
June 18, 1986	HHP decides to drop further work on tinnitus.
July 3, 1986	Second HHP plan to enter into HA (hearing aids) accepted by sector committee.
July 16, 1986	3M decides to divest from Diagnostics/Axonics.
Oct. 15, 1986	HHP receives permission to acquire Diaphone to enter into HA.
Nov. 19, 1986	3M sector review of HHP. Program is once again given the go ahead.

Note: This list of events does not include entry or exit of personnel.

THERAPEUTIC APHERESIS PROGRAM JOINT-VENTURE CASE
DOUGLAS POLLEY AND
ANDREW H. VAN DE VEN

This case describes an ongoing joint venture by three corporations to create a new business based on a new medical technology called therapeutic apheresis. The technology includes a blood separator that removes noxious blood components and returns the remaining blood to the patient. This involves removal, separation, and return of a patient's blood during a continuous treatment period. The medical device itself consists of hardware that pumps the blood, filtration modules, and tubing that connects the filtration module to the patient. The process of removal of blood components and return of the remainder to the patient is called apheresis.

The idea of using blood removal to treat diseases dates back to the use of leeches to remove blood. In the time of George Washington bloodletting was still popular. The idea of removing specific parts of the blood originated in 1914 but did not gain popularity until World War II when it was used for obtaining needed plasma. The first therapeutic uses of apheresis (specific blood component removal) appeared in the late 1950s. The first continuous flow separator was introduced by IBM (now Cobe) in 1966 and was followed by other continuous flow machines. The dominant technology used by these manufacturers is a centrifuge device, which separates blood components by centrifugal force in a spinning cylinder. Three companies dominate the domestic marketplace for this technology: Baxter-Travenol, Haemonetics, and Cobe.

More recently filtration and adsorbtion technologies have been introduced. The filtration devices (first introduced in 1978) pass the

261

blood through a membrane that rejects certain blood components. The adsorbtion technology is more recent still. It utilizes the capabilities of certain materials to bind blood components and thus remove them from the system. This case examines the process in which one of the first commercial apheresis filtration technologies is being developed. The initial development of this therapeutic apheresis program (TAP) began in late 1983 with a joint venture agreement between 3M, Millipore, and a 3M subsidiary, Sarns, Inc. Millipore, headquartered in Bedford, Massachusetts, develops, manufactures, and markets products for analysis and purification of fluids with total 1985 sales of $367 million and 4,450 employees. The TAP activity at 3M is a part of the health care products and services group of 3M's life sciences sector. The Sarns subsidiary is also a part of the health care products and services group.

The management of TAP is organized as a strategic business unit (SBU) and is comprised of managers from Millipore, Sarns, and 3M. The operation of the venture is managed jointly, with each SBU member responsible to his own organization. This results in a set of activities located in three different geographic areas with several reporting chains for the SBU members. Figure 8–7 illustrates the respective organization charts in an abbreviated fashion. TAP is shown with indirect connections to these organizations based on the current SBU membership. No individual has sole responsibility for the entire TAP activity.

Longitudinal study of the development of the TAP began in December 1983, just as the joint venture agreement was established. Data collection involves attendance at bimonthly meetings of the TAP SBU operating committee; semi-annual interviews with TAP SBU members and questionnaire surveys of all key TAP personnel in 3M, Sarns, and Millipore; annual interviews with top managers of the co-venturing firms; as well as information obtained from company records and industry trade pub-

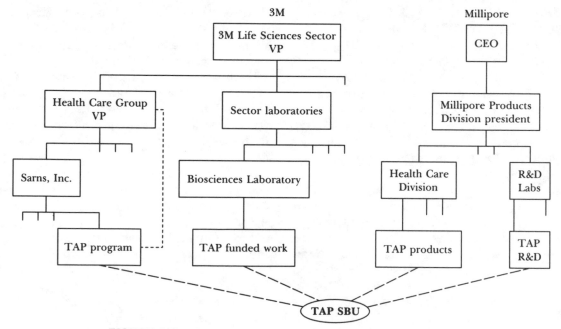

FIGURE 8–7. *Organization Chart of Therapeutic Apheresis Program.*

lications. From these data sources a chronological list of sixty-nine events were identified as having occurred in the development of TAP. To reduce the complexity of this list to manageable proportions, the events were content analyzed and grouped into four segments: initiation and organization of the joint venture, the development of two phases of apheresis products, and explorations of related businesses for the TAP. Events within each of these segments are graphically illustrated in Figure 8–8 and presented in the appendix to this case. The progression of events within each of these segments is described below.

Initiation and Organization of the Joint Venture

Events occurring from early 1980 until November 1983 mark the initiation period of the program. At the beginning of this period 3M, Millipore and Sarns were independently exploring technological developments related to apheresis. From 1980 to 1982 research was undertaken in 3M laboratories to develop a new technology for the removal of noxious blood components. Although this effort was discontinued because of technical and marketing difficulties, interest in apheresis remained.

Millipore was interested in medical applications of its filtration expertise and had developed a prototype blood filtration module. In 1980 the Millipore CEO approached the president of Sarns (located in Ann Arbor, Michigan) to discuss the possibility of a joint venture. Being a leading developer and manufacturer of heart/lung equipment, Sarns was perceived as having the expertise to develop the blood pumps and tube sets necessary for the filtration module.

Soon after the initial approach by Millipore, Sarns became engaged in discussions with 3M regarding a possible merger, and Sarns discontinued its discussions with Millipore. Negotiations to execute an acquisition by 3M of Sarns occurred throughout the first half of 1981 and were completed in June 1981.

Subsequent to the merger it was learned that personnel at 3M and Sarns were still interested in developing apheresis business opportunities, and a new round of negotiations began in March 1983. These negotiations resulted in the announcement of a joint agreement between Millipore, 3M, and its Sarns subsidiary in November 1983. This marked the formal beginning of the blood treatment program (TAP) examined here. The basic idea for the TAP was treatment of specific diseases by removal of specific blood components. As initially conceived, this involved the development of blood filters, hardware for blood pumping and monitoring, clinical testing of the devices, and protocols designed to instruct physicians how to effectively utilize the devices in disease treatment.

The following division of labor among the firms was also specified in the formal agreement. Millipore's filtration technology would be combined with Sarns' manufacturing expertise and 3M's research and worldwide marketing capabilities. Millipore was responsible for the development of the filtration modules, and 3M-Sarns was responsible for conducting clinical trials, FDA regulatory review procedures, manufacturing the hardware, and marketing the devices worldwide. 3M and Millipore agreed to cover their respective developmental costs and share revenue from unit sales.

The joint venture agreement also specified three phases of product development, labeled Phases I, II, and III. The Phase I product was designed to compete with current apheresis products on the market, which were based on a centrifuge technology rather than filtration. The apheresis market currently focuses on plasma removal and does not involve extensive treatment of diseases. Because Millipore had already developed a Phase I filtration device, development of a Phase I product was initially expected to require relatively minor resources.

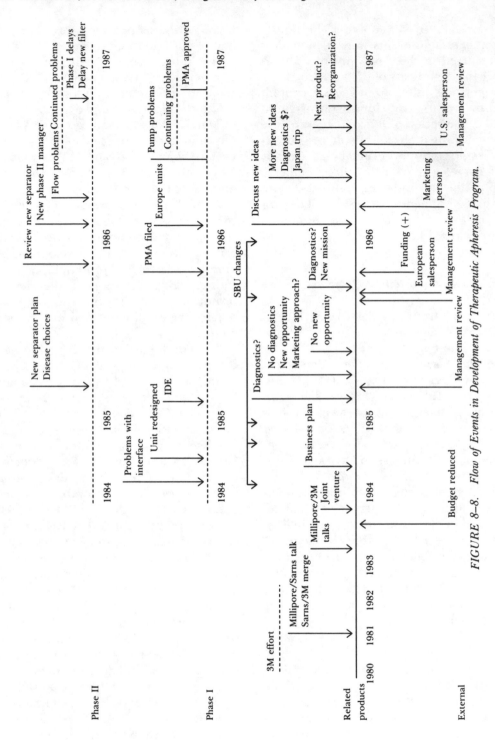

FIGURE 8–8. Flow of Events in Development of Therapeutic Apheresis Program.

Although the Phase I product was not perceived by TAP managers to be a significant product innovation, its commercial introduction was judged to be important for gaining a "market presence."

Phase II was perceived to represent the significant technological innovation by the TAP managers. As initially conceived, the Phase II product line would be targeted to treat specific diseases with technologically advanced filtration modules that could separate particular molecules in blood. Thus, the Phase II plan called for targeting specific diseases and obtaining FDA approval for therapeutic claims. These claims would involve extensive clinical field trials to establish effectiveness. The Phase II device was also planned as a unit that could be attached to and upgrade a Phase I device. This would utilize much of the Phase I development hardware and would allow early buyers to upgrade their units with the Phase II filtration modules.

Phase III in the TAP agreement stated an intent by the firms to jointly develop future apheresis technologies. The intention was to structure an agreement sufficiently broad to incorporate a wide array of potential technological developments and related business opportunities as they arose.

Organizationally, the TAP is managed by a strategic business unit (SBU) operations committee that meets bimonthly and is responsible for the strategic direction of the joint venture. Table 8–3 summarizes the key events in the internal organization and management composition of this SBU and other key management personnel related to TAP since its formation in 1983 when the joint agreement was reached. Charter members of the SBU involved three managers from Millipore and three from 3M. They included a program manager from 3M, a Sarns development director, and an R&D manager from the 3M biosciences laboratory. The Millipore members included the marketing manager, a research scientist, and the director of apheresis development activities. Initially, the TAP SBU operated without permanent staff. Individuals were added to the effort from other units in 3M, Millipore, and Sarns as the need arose.

The two-day SBU meetings alternate between Minnesota and Massachusetts. They are always held at locations away from the corporate facilities and include formal meetings as well as evening social activities that provide an informal opportunity to maintain communications. The meetings typically include status reports on program activities and strategic issues but avoid operational decisions whenever possible.

Within the SBU, members have been observed to play dual roles—as representatives of their respective host organizations and as TAP members. In some cases they represent decisions as already made at their respective organizations and at other times they look first for a TAP perspective. Resolving these two approaches has been a source of conflict and debate within the SBU. One manifestation of this dual situation has been a need for leadership felt by both groups. The 3M program manager has been a key leader from that organization, but the nominal leader from Millipore has changed several times over the history of the program, which has led to concerns on some occasions about whether Millipore was getting adequate representation.

As Table 8–5 indicates, the composition of the SBU team changed periodically over time with the departure of the original Millipore marketing manager and the addition of a new 3M TAP marketing manager. By mid-1984 the team consisted of seven members. Further changes occurred in late 1984 when two of the Millipore members were replaced by new representatives. Further Millipore changes occurred in September when a new marketing VP replaced a previous Millipore marketing representative. This individual left in December 1985 and was replaced on the SBU by the VP

TABLE 8–3. *Internal Organizational Events in TAP SBU.*

Date	TAP Organization and Managerial Turnover Events
1983	SBU formed with three members from 3M and three from Millipore
1984	Millipore marketing manager leaves and is replaced
4/84	3M hires marketing manager for TAP—added to SBU
12/84	Two new SBU members from Millipore
12/84	Marketing staff person from 3M transferred to another program
12/84	TAP reports to SARNS after reorganization of 3M Health Care
9/85	New VP of marketing at Millipore replaces Millipore SBU member
10/85	European sales representative added for TAP
12/85	President of Sarns retires and is replaced by 3M executive
12/85	Millipore VP of R&D returns to SBU to replace VP of marketing
12/85	Millipore SBU member named VP and general manager of health care
2/86	Marketing staff person added to Sarns TAP staff
5/86	Domestic sales representative added at Sarns
10/86	Marketing manager for TAP promoted to Sarns marketing manager

of R&D from Millipore who had previously been an SBU member. In late 1985 there were senior management changes at Millipore. As a result the development director was promoted to vice president of Millipore's health care products. This promotion increased his responsibility and made him the most senior Millipore person in the SBU. This individual had been an active leader inside Millipore and thus became the leader of the SBU representation as well. During 1986 this helped to balance the leadership within the SBU.

The TAP SBU has used a variety of techniques to organize and control the management process. Team efforts were utilized to assist in the process of disease choice and competitive assessment. Technical staff met in January 1985 and 1986 to draw up detailed PERT type charts for the development activity for phase II. These diagrams indicated the complexity of the development process and were dubbed "spider webs" by program participants.

In 1985 the SBU's initial program risk analysis meeting was held with an outside consultant in May 1985, and additional risk analysis meetings were held. These meetings provided a foundation of risk analysis for the various parts of the program that was used by the SBU in discussions regarding program priorities and direction. Some SBU members indicated that the analysis was beneficial, but others questioned its value as a management tool.

Figure 8–9 summarizes the perceptions of the SBU members and key participants in the TAP on a set of composite attitudinal dimensions measured with the innovation study questionnaire in November 1984, June 1985, January 1986, and August 1986. As the table indicates, perceptions of TAP participants remained highly favorable over time, but there was a measurable decline in the most recent survey. That survey occurred during a time when development activities were encountering difficulties in the development of Phase I and Phase II products, which are described in the next sections.

Phase I Product Development Efforts
As noted above, prior to the joint agreement, Millipore had independently developed a prototype blood filtration device. However, initial customer focus sessions identified difficulties with the user interface components of the device. To correct these, an external vendor was

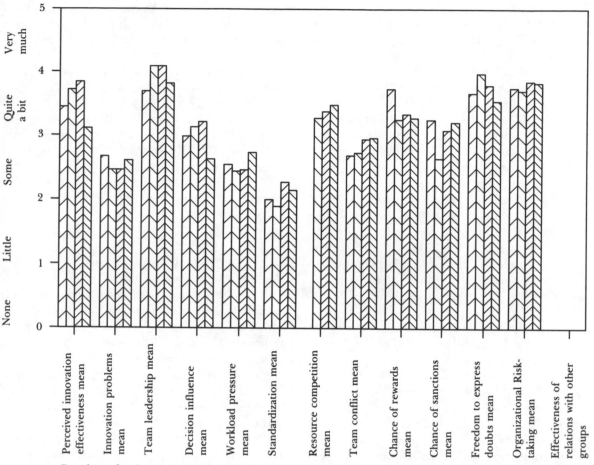

Bar charts for time 1 (1/85), times 2, (6/85), 3 (1/86), and 4 (8/86).

FIGURE 8–9. *Overall Survey Results for Therapeutic Apheresis Program.*

contracted to redesign the unit in early 1984. This redesigned unit went on to become the basic Phase I TAP product.

The use of an outside contractor to redesign the unit was consistent with plans for resource use in the program. The program manager reported that TAP would minimize employing permanent staff until it became clear that the venture would be successful. The assembly of the Phase I filtration module and tubes for the device was also contracted outside. Most of the development and testing of TAP devices were internally contracted to 3M and Millipore research laboratories.

Having completed the initial development of a Phase I system, in June 1984 the TAP initiated tests on animals to determine its safety and efficacy. Results from the animal tests provided the basis for submitting an application to the FDA for an IDE (investigational device exemption) that would permit conducting clinical trials on humans. The FDA approved the IDE application for the Phase I device in February 1985. The clinical trials on the Phase I device occurred at several hospitals throughout the

United States and were monitored to ensure compliance with FDA regulatory procedures by a full-time TAP staff person with experience in these areas. These clinical trials produced good results, and in December 1985 the TAP filed its PMAA (premarket approval application) with the FDA. This event occurred as initially scheduled by TAP management at the beginning of the trial period. These clinical trials represented the major Phase I technical activity of the TAP during 1985.

Throughout 1986 the regulatory approval process was largely administrative as the FDA review panel evaluated the proposal. The TAP individual monitoring this process maintained contact with the FDA to answer questions and provide additional information upon request from the FDA. No problems were detected by the FDA review panel, and in November 1986 the FDA approved the PMA application, permitting the commercial release of the TAP Phase I device throughout the United States.

With the submission of the PMA application to the FDA on December 1985, TAP management believed it was possible to begin sales of the Phase I system in Europe where different regulatory procedures apply. A European salesperson was hired in October 1985, and initial shipments of Phase I products were made to Europe in January 1986. However, installations of the Phase I units sent to Europe were delayed three to six months (depending on the specific country) because it was found that regulatory approval in European countries did not occur as quickly or simply as anticipated. The TAP management enlisted assistance from 3M European subsidiaries and outside consultants to obtain approvals from European regulatory bodies for the Phase I systems. As a result, installation and revenues of the Phase I systems in Europe did not begin until second quarter 1986.

At the May 1986 SBU meeting there were reports of a variety of technical problems involving systems in Europe and systems built for testing in the United States. In the August SBU meeting there were further reports of problems in the European systems. One individual at Sarns was then assigned full time to field service installation and repair of the installed Phase I units. As a result of these problems, at the October 1986 SBU meeting it was reported that projected 1986 revenues would be about only 10 percent of what had been projected at the beginning of the year. A reliability review subsequently resulted in sales being stopped until problems could be corrected.

Phase II Product Development

The Phase II product was conceived as being a technologically more sophisticated blood filtration module that would be connected to a Phase I device. In an apheresis treatment, a patient's blood would first flow through the Phase I separator and then to the Phase II filtration module in order to treat specific diseases. R&D efforts on the Phase II technology began in 1984 when Millipore designed a new filtration module with capabilities beyond that of Phase I. By late 1984 efforts were underway by a TAP task force to determine which diseases were the most likely targets for treatment with the new filtration technology. The strategy was to develop and test a Phase II device that could claim demonstrated effectiveness in treating diseases.

However, these plans were redirected in late 1984 by news of a 1985 budget reduction in the 3M portion of the TAP operation. The 3M group VP reported this decision was made in order to change the TAP strategy of making therapeutic claims that the Phase II device will treat specific diseases, to that of more simply providing protocols on the technical filtering capabilities of the Phase II device. This change in strategy would avoid major clinical testing costs that the FDA would require in proving that the device could treat specified diseases and thereby substantially shorten the time for market entry. Because the group VP's strategy for the TAP was simpler to implement and entailed lower developmental costs than the one on which the proposed 1985 TAP budget was based, the budget was accordingly reduced.

However, during the latter part of 1984 TAP management had begun program expansion anticipating that its proposed 1985 budget would be approved. As a result, the 1985 budget that the group VP approved was begrudgingly perceived by TAP management as a "budget cut." The scaled-down approved budget resulted in reassigning a full-time TAP marketing coordinator to another area of 3M. After extensive discussion at the February 1985 SBU meeting, the budget reduction also resulted in dropping plans of extensive clinical testing to support claims of the therapeutic effectiveness of the Phase II module.

The Phase II task force made its recommendations of a set of treatable diseases at the February 1985 SBU meeting. This list of diseases was reduced to a recommended set of two classes of diseases in a report by the 3M laboratory manager at the April SBU meeting. The revised plans for the Phase II products, then, consisted of developing a generic product that would be combined with protocols for treatment of chosen diseases and targeted to appropriate users. The division of labor to implement these plans during 1985 was for Millipore to develop the Phase II filtration module, while 3M was responsible for developing clinical protocols for disease treatment, developing an instrument, and a strategy to market the products.

The marketing manager indicated that 85 percent of the TAP marketing effort during 1985 was spent in support of Phase I. A marketing strategy was debated in several SBU meetings beginning in April 1985. At issue was whether market entry of apheresis products should be directed toward key leaders in the targeted disease areas or toward the larger population of prescribing physicians. The marketing manager favored the former while technical members of the SBU were uncertain of the preferred market-entry approach. The conclusion reached from discussions of these alternatives at several SBU meetings in 1985 was to diffuse the innovation through key leaders.

Key aspects of the program involved the clinical, development, and marketing portions of the program. This was carefully articulated by the Millipore development manager at the April meeting. The development included module and hardware design. Clinical effort included protocol development and publication of scientific results. Marketing included the development of the market and reimbursement procedures—efforts that were seen as separate but tightly connected aspects of the program whose creative combination was seen as one of the major program strengths.

In the April 1985 SBU meeting a decision was made to develop additional advanced separation technology beyond that being developed for the first Phase II product. This involved development of another filtration module to remove additional blood components. However, this decision stirred controversy among SBU members because the laboratory manager could not guarantee that available tests would demonstrate the effectiveness of the new module. Development of the advanced separator resurfaced in a controversial discussion in the September SBU meeting. Again the decision was to move ahead with the development despite reservations about how to test the efficacy of the device.

Difficulties were encountered by Millipore in building Phase II filtration prototypes. This came to the attention of the SBU at its June 1985 meeting, and the difficulties were reported as resolved by the September meeting. The Millipore development director requested that he be replaced as Phase II program manager because testing of the module

269

now largely became the responsibility of 3M and Sarns.

Clinical efforts on the Phase II product during 1985 focused on the development of protocols for treatment of the diseases targeted. Plans presented at the January 1985 SBU called for development of a protocol for one of the chosen diseases. This effort was noted as successfully completed in the year-end summary by the TAP program manager.

In March 1986 it was reported to the SBU that the projected blood flow rate on the Phase II modules had not been met. At the May SBU meeting further delays were reported in manufacturing modules for clinical testing. These difficulties were reported to stem from efforts to move from prototype development to pilot production of the Phase II filtration modules.

At the August and October 1986 SBU meetings Millipore reported further difficulties in manufacturing the Phase II modules. The manufacturing process was reestablished, but other technical problems had become apparent in module operation. These difficulties were reported to further risk delays in the schedule. By this time the availability of reliable Phase I machines also affected the further Phase II testing. (As noted above the Phase II unit consisted of a Phase I unit with additional filtration apparatus.)

These delays resulted in revision of the planned schedule for testing and seeking regulatory approval for the Phase II product. In addition, given the technical difficulties that delayed pilot production of the Phase II modules and full production of the Phase I devices, it was decided at the October 1986 SBU meeting that the Phase II advanced separation module would take a lower priority than initially set. This effectively postponed development of the advanced filtration module until these manufacturing difficulties were resolved.

The October 1986 SBU meeting also included a report by the TAP program manager that the Phase II portion of the program had slipped by as much as one year from estimates at the beginning of the year.

Related Opportunities in the Joint-Venture Agreement

The joint-venture agreement indicated that additional unspecified products would be developed beyond the Phases I and II products discussed above. Discussions among TAP management on new business opportunities emerged quite regularly beginning in 1985 and occurred parallel with Phases I and II product development efforts. Although these discussions have to date not resulted in any clear strategy or resource commitment to a new line of products, they add an important segment to TAP's early history of strategy development and renegotiation of the joint agreement. During 1986 most of the program's resources were devoted to resolving unanticipated Phases I and II regulatory, production, and marketing problems.

One of the first opportunities to expand the TAP business arose in February 1985. At this SBU meeting it was noted by the 3M laboratory manager that some of the diagnostic work done in support of the Phase II product might have commercial potential. Because development of diagnostic tests was necessary to test the Phase II filtration module and these tests were judged to be superior to any currently available in the market, the potential existed to sell them commercially. Discussions among Millipore and 3M SBU members led to the conclusion that diagnostic activity was outside the domain of the TAP joint agreement. This was reported at the April 1985 SBU meeting, and both 3M and Millipore agreed to proceed independently. Diagnostic work by 3M labs would continue on the Phase II module, but its commercialization would not become a part of the TAP.

Millipore reported in April 1985 an opportunity to participate in bidding for blood treatment equipment to be purchased by a government agency. This equipment was not disease related but did use technology available in the TAP program. The opportunity was discussed outside of a regular SBU meeting by 3M, Sarns, and Millipore personnel. Sarns and 3M staff concluded that Sarns did not have sufficient resources to pursue this effort. It was reported at the June 1985 SBU meeting that Sarns would not participate in this venture as either a vendor or a partner. Millipore subsequently proceeded independently to bid for the government business.

Although decided against at the April SBU meeting, discussions of commercializing diagnostics tests developed at 3M reappeared in subsequent SBU meetings. In September 1985, for example, further progress was reported in the development of diagnostics tests. Millipore SBU members again asked if the diagnostics could be part of the TAP program. At this time the 3M program manager referred to the April decision not to proceed but proposed that diagnostics be considered at a later time.

In September 1985 there was a report by the Millipore research director on possible new business opportunities for TAP. At this time the list included extensions into further blood purification and other medical uses of filtration technology. The SBU discussed additional apheresis technology beyond Phase II. Discussions at the January 1986 SBU meeting expanded the list of possibilities for new business. The possibility of adding diagnostics as a part of the business was made a part of the list of future possibilities.

In March 1986 the Millipore program manager noted that several specific business opportunities were being considered that used technology similar to that of the TAP. He noted

that TAP was being given the right of first refusal to join in these efforts. It was reported that if the TAP did not decide to participate, outside partners would be sought to assist in the development of these ideas. Further discussion of these opportunities was scheduled between SBU meetings.

There was a discussion of Phase III developments at the May 1986 SBU meeting. This was noted as more than just a product addition to the Phases I and II efforts. One of the differences in opinions within the SBU was how to proceed in determining new opportunities. One development manager (3M) favored examining available technology for potential use (technology push innovation), while another (Millipore) favored examining the marketplace to identify opportunities (demand pull innovation).

In the March 1986 meeting there was a report from the marketing manager of recent travel to Japan, where new competitors were uncovered. The SBU discussed the need to examine the alternative technologies being developed by competitors. Further discussions of competitor technologies and opportunities took place in August 1986. Additional competitors were identified. A report was made of combined therapies of apheresis with drugs by the 3M program manager. It was not clear that drug companies would want to work with TAP to develop joint therapy.

The result of these discussions was that the TAP program continued with little change from its initial program plan. Outside opportunities were not exercised and no new products were added to the Phases I and II products.

As the above discussion indicates, the joint agreement was sometimes used as the basis for SBU discussions regarding appropriateness of new activities. However, interviews of upper management indicated much less concern regarding the terms specified in the joint agreement than was reflected in the discussion among SBU members. In repeated interviews the senior executives of Millipore and 3M indicated that they looked on the effort as a partnership and stated that they were not familiar with the detailed terms of the agreement. They also indicated that the contractual terms in the joint agreement were far less important than meeting operational necessities as determined by the SBU.

The SBU members initially viewed top management as only weakly committed to TAP, as indicated by concerns among SBU members about 3M's budget reduction in late 1984. The top management review in April 1985 was basically favorable, but 3M executives continued to express concern regarding the market potential for TAP products. Top management continued to be supportive of the program in its September 1985 review. In the June 1986 review management continued to be supportive despite some delays in schedules. This support was reflected in the decision to move to annual reviews instead of semi-annual reviews.

Analysis of the TAP Business Development Process

As noted earlier the TAP case has unique structural properties that helped to create its character. It is a joint venture of geographically remote and culturally diverse companies. It is organized under a committee governance structure that has multiple links to the respective parent organizations. In addition, it has faced the full set of problems and opportunities that other new product innovations face.

During the period to date it has been able to build and test a Phase I machine and receive FDA approval of this unit as a result of successful field tests. It has also successfully developed prototypes of Phase II modules, although additional work is proving necessary in this area. The TAP team has gained the respect of the top management of the respective organizations even though the program has suffered some setbacks.

These prototypes were the result of overlapping research efforts in product development and new business creation. The case also illustrates a number of events particular to business creation in a joint interorganizational venture.

Despite the joint nature of the TAP activity much discussion centered on interests of the individual organizations as opposed to actions beneficial to the joint effort. This may have resulted from the organizational form of TAP as a joint venture as well as from the reliance on part-time and contract-type employment. Separate leaders also emerged from each organization rather than a single joint leader.

The SBU placed greater reliance on the nature of the joint venture agreement than might have been expected given the opinions expressed by senior management. This reliance might have been an outgrowth of the lack of role clarity that members who were both SBU managers and organizational representatives experienced. It might also have been related to the considerable uncertainty that surrounds any innovative activity and the need to find criteria that can be used to reduce this uncertainty.

Within the scope of providing a device for the treatment of specific diseases there was considerable convergence. All of the major TAP activities were related to this goal. Even the apparent opportunity to engage in a diagnostics business was closely related to the TAP innovative idea. This and more divergent ideas for expansion were discussed but remained largely speculative as the TAP effort focused on the primary objective. The diagnostic effort continues as a key element in the TAP program but still has not been accepted as a business opportunity independent of its role in apheresis.

The case demonstrates that many components of the new TAP business were designed to be loosely coupled, which permitted much slack and flexibility in developing various TAP components. However, as problems and slipped deadlines accumulated over the years, the components became tightly coupled and adversely affected total performance of the new business creation effort. As a result, by December 1986 the Millipore program manager emphasized that "TAP has become a complex system, and failures in one or two components derail the entire effort, even though other parts are proceeding effectively."

In 1983 Phases I and II products were planned to be developed in parallel. Although the Phase II filtration module was to be hooked on the Phase I hardware, it was not necessary for Phase I to complete market entry to begin Phase II development. The division of labor among coventurers also permitted parallel activities: Millipore was largely involved in developing the Phase II filtration module, while 3M clinical staff performed field tests on the Phase I product, 3M labs developed Phase II protocols, marketing developed its strategy, and Sarns scaled up production operations. Throughout 1984 and 1985 progress and problems in these areas were reported at SBU meetings as they occurred, and solutions to problems were developed largely independent of problems in other areas. Finally, discussions of new TAP business opportunities proceeded independently of Phases I and II product development efforts.

Using Van den Daele's (1969) distinction of conjunctive and disjunctive progressions, the activities illustrated in Figure 8–8 of the different phases can be seen as disjunctive, and activities within each phase can be seen as conjunctive. For example, from the beginning of the observation period until late in 1986 the development activities of Phases I and II remained largely separated and proceeded independently of each other (disjunctive). Within each phase the efforts were quite closely connected although they continued to be worked on in parallel in multiple organizations (conjunctive).

Within each phase the efforts to develop modules, hardware, clinical protocols, and marketing were pursued independently but required close coordination. With the exception of marketing the separate portions of the product were all required before moving from development to clinical trials or to sales. There is less connection among the activities of Phase II. The efforts on Phase II were primarily involved in the early research and development phases, and these could proceed without much of the apparatus of the Phase I product. Portions of the development activities were sequentially interdependent as well. For example, Phase I effort entailed the development of a prototype filtration module first, then pilot production of a few custom designed devices for testing, and finally scaleup of full production capabilities. In addition, the FDA regulatory process imposed sequential execution of animal tests to obtain an IDE approval, followed by clinical trials on humans to obtain a FDA approval to manufacture and distribute the product in the U.S. market. Marketing was permitted to begin European sales before the FDA PMA approval, but required prior approval from European regulatory bodies.

The case shows that all but three of these sequential activities in the development of the Phase I occurred smoothly. Difficulties with early tests on the Phase I prototype were corrected quickly by an outside contractor. The initially uninformed efforts to gain initial product approval from European regulatory bodies was effectively solved with the help of 3M European subsidiaries and consultants, although the problem delayed sales by six months. The third and only major problem encountered thus far is the scaleup of hardware manufacturing at Sarns for the Phase I device. Technical manufacturing problems emerged incrementally, and the case shows they were reported in piecemeal fashion in SBU meetings over time. However, their cumulative effect did not become apparent until the latter part of 1986.

A similar pattern was observed in the development of the Phase II product. A prototype filtration module was successfully developed, and a few modules were produced in pilot production to permit testing to begin. However, problems were encountered in shifting from pilot to full-scale production of the Phase II filtration module at Millipore. During most of 1986, development work and problem resolution of the two product phases continued in parallel and largely independent of each other. It was not until the latter part of 1986 that the SBU began to appreciate the lagged interdependencies among the two product development efforts.

By year end 1986, the two phases of the program had ceased to be entirely parallel. The key underlying assumption that made the parallel activity feasible was that the Phase I devices were ahead of the Phase II devices in the development cycle. Thus the Phase I units could be used in the continued development for Phase II. By 1986 year end this assumption was no longer true. Had it not been for the difficulties encountered in Phase II module design, the Phase I impact on Phase II would have been even more severe. Even so there were delays in clinical tests of Phase II modules because of the inability of manufacturing to produce working Phase I machines. This indicates how parallel activities that appear disjunctive can become conjunctive when lagged interdependencies become activated.

Budgeting and marketing provide related examples of difficulties in anticipating the timing of events necessary to build momentum in the startup of a new business. Anticipating approval of the 1985 budget, TAP management began program expansion in late 1984. When the group VP approved a reduced budget in line with a revised strategy, TAP management begrudgingly downsized its program. Similarly, marketing acquired resources in late 1985 and 1986 in anticipation of Phase I product market entry. This entry was effectively delayed by the manufacturing problems encountered.

Although efforts were made to design a largely nonsequential new business creation process, the case described here has turned out to be much like that of the baton passage at a relay race. However, it is a "just-in-time" relay race, where one runner starts before the others arrive at the track. At other times the next runner begins running before the baton arrives in order to gain sufficient momentum to match the prior runner's pace. Although several hand-offs occurred successfully within and between Phase I and Phase II component parts, it is clear that the race stops when the next component does not arrive on time for the baton hand-off.

Finally, discussions of new TAP business opportunities appears to be following an iterative and preoccupied gestation process. For example, diagnostics business opportunities were repetitively proposed, rejected, and reproposed for TAP business expansion in the 1985 and 1986 SBU meetings. Outside opportunities for TAP business expansion were asynchronous to the program's progress, but systematic consideration of alternatives was not. Proposals began in mid-1985 when efforts were successful in both phases of the product development. The 3M lab manager reported that this lead to concerns about keeping research and development resources actively employed in TAP after current assignments would be completed. However, by the end of 1986 these resources were actively involved in problem resolution and the discussion of long-run alternative activities gave way to short-run problem resolution activities.

Therapeutic Apheresis Joint-Venture Case Appendix.
Events in the Development of Therapeutic Apheresis Program.

Initiation and Organization of TAP

Jan. 1980	3M works on blood treatment systems from 1980 to 1982, when work discontinued but interest remains.
Jan. 1981	Millipore initiates talks with Sarns regarding plasmapheresis.
	Millipore contacts Sarns about treatment device. Discussion delayed by merger.
June 1981	Sarns is acquired by 3M.
March 1, 1983	Negotiations begun between Millipore and 3M regarding joint venture.
	After Sarns merger it is discovered that 3M also has interest in apheresis.
Oct. 1983	Clinical development manager hired for TAP.
Nov. 16, 1983	Signing of 3M-Millipore agreement to pursue apheresis, including specific disease treatment.
Jan. 1, 1984	Initial business plan completed in January 1984.
April 1, 1984	Marketing manager hired by TAP.
Oct. 1, 1984	Sector planning committee supports TAP 1985 budget request, but at a reduced level.
Dec. 1, 1984	Health care division reorganized, with TAP now reporting to Sarns.
Dec. 31, 1984	Marketing person leaves the program as result of budget reductions.
Jan. 30, 1985	Millipore marketing manager leaves the SBU. New Millipore individuals become permanent members.
Feb. 13, 1985	3M program manager presented a business plan for 1985 that includes several goals for 1985.
April 15, 1985	Need recognized for additional resources in marketing. 3M offers money for research but no staff.
May 1, 1985	Management review of TAP by top management of 3M, Sarns, and Millipore.
May 1985	Initial risk analysis meeting.
July 23, 1985	New Millipore marketing VP joins the SBU, replacing Millipore marketing person.
Sept. 30, 1985	Management review of TAP by top management of 3M, Sarns, and Millipore.
Nov. 13, 1985	Millipore reorganizes along lines of business.
	TAP budget for Millipore for 1986.
Dec. 8, 1985	Sarns general manager resigns and is replaced by 3M executive.
Dec. 10, 1985	TAP meets five of its six 1985 goals.
	3M budget for TAP approved.
	Millipore development director appointed VP of Millipore products and general manager of health care products.
Jan. 21, 1986	Millipore VP of R&D returns to SBU.
Feb. 1, 1986	Marketing person joins TAP marketing group. European sales rep added in October 1985.
April 1986	Risk analysis meeting.
May 9, 1986	Effective May 12 domestic sales representative appointed.
May 28, 1986	Report of new Millipore organization chart. New Millipore products president appointed following death of previous president of corporation.
May 29, 1986	New Sarns lab director replaces previous director on SBU.
June 26, 1986	3M and Millipore joint top management review of TAP.
Aug. 6, 1986	A new organization for TAP activities is proposed based on Detroit meeting.
Oct. 1, 1986	Considerable debate and discussion regarding planned reorganization.
	TAP marketing manager has been promoted to marketing manager at Sarns.
Oct. 2, 1986	Report on 1986 goals. Sales only at 10 percent of projections. Credibility is in question.
Dec. 10, 1986	Report of reliability review results in hold on sales until September 1987.

Phase I Product Developments

Oct. 1983	Focus panel shows need for changes in initial prototype system.
June 1, 1984	A redesign of the apheresis system is performed by an outside consultant. Animal tests began in June 1984.
Feb. 6, 1985	Phase I IDE approved by FDA.
April 15, 1985	Initial deliveries of units are delayed due to technical manufacturing difficulties.
Dec. 17, 1985	Phase I PMA filed with FDA.
Jan. 13, 1986	Five instruments shipped to 3M France. In service now planned for February 10, 1986, a delay from January 6, 1986.
May 28, 1986	Difficulties encountered in pump motors. Distribution of other units are being held up.
Aug. 5, 1986	Three problems noted with European machines.
Nov. 18, 1986	Final PMA approval for phase I device obtained from FDA. No postmarket information required.

Phase II Product Developments

Feb. 13, 1985	Discussion of diseases to treat. Short list developed.
April 15, 1985	Discussion on the next system to be developed. This activity will go on despite some difficulties in the lab.
	Discussion of marketing strategy to leaders or a broad base of prescribers.
	Lab manager reports on research into diseases. He recommends that the list be reduced to two diseases.
June 12, 1985	Some difficulties encountered in Phase II module construction.
Sept. 5, 1985	Prototype advanced separator to be built despite concern over ability to match with disease states.
Jan. 20, 1986	Millipore development manager replaced by 3M regulatory manager for Phase II program manager.
March 27, 1986	Phase II module encounters additional technical difficulties.
March 28, 1986	Diagnostics will be ready for clinical trials in November 1986. More resources requested.
May 29, 1986	Delays reported in Phase II modules due to performance.
Aug. 5, 1986	Phase II modules still have some problems. Schedule slips by four months.
Aug. 6, 1986	Priorities will be to finish current products. Advanced separations priority reduced.
Oct. 1, 1986	Production process for modules reestablished. Continued problems may cause schedule delay.
	Problems with Phase I instrument; conflicts over Phase II testing.

Related TAP Business Opportunities

Feb. 13, 1985	Issue of whether diagnostics belongs in the basic agreement is raised in SBU meeting.
April 15, 1985	It is decided not to fold diagnostics into the current joint agreement.
	Millipore offers TAP the right of first refusal for spinoff product.
May 13, 1985	Sarns is not to participate with Millipore on spinoff due to resource limits.
Sept. 5, 1985	A list of possible new mission statements for use beyond Phase III is presented at SBU meeting.
	Progress reported in diagnostic area. Diagnostics not to be included in TAP, but can be brought in later.
Jan. 13, 1986	Additional discussion is held on new products in SBU meeting.
March 28, 1986	Millipore VP reports on new business opportunities. TAP has right of first refusal.
	Marketing manager and 3M program manager visit Japan and assess competition.
May 29, 1986	SBU meeting continues discussion on the next big product idea for TAP.

PROCESS COMPARISONS ACROSS BUSINESS CREATION CASES

We now compare the three business creation cases in order to identify (1) the process similarities that are inherent to new business creation irrespective of organizational context, as well as (2) the process differences in new business creation that can be attributed to organizational settings. Because the new businesses are still in the startup period, it is premature to draw conclusions about the factors that may lead to success or failure. Indeed, none of the new businesses are presently self-sustaining and no successful business takeoff has yet occurred. Qnetics and HHP have each failed in product market introductions thus far, and TAP has as yet been unable to introduce its product into the market because of manufacturing problems. Continued longitudinal research will determine the ultimate fate of the new businesses.

Table 8–4 outlines the common processes of business creation and variations in these core processes that were observed in the three cases. These core processes and diverse overlays were identified by empirically comparing the three cases in terms of the inputs (resources), outputs (business ideas, products, functions developed), and transformation processes (developmental progressions) during the initiation, startup, and takeoff periods of the accumulation process model. This model was described in the introduction and used as the basic conceptual framework to guide this research.

We draw two overall preliminary conclusions from Table 8–4: (1) Business creation entails many common core processes irrespective of organizational and industry settings; and (2) diverse overlays on these core processes can be attributed to the different organizational and industry settings in which businesses are created. Although obvious structural differences exist in the technol-

ogy, organizational arrangements, and industry regulations among the three cases, there are similarities in core business creation processes across the three cases, as well as variations evident in these core processes that may be attributable to the different settings in the three cases. These common core processes and diverse overlays in the initiation, startup, and takeoff periods of the three business creation cases are discussed below.

Business Initiation

1. An Extended Gestation Period with Multiple Stimulants What precipitates the initiation of a new business? According to Louis Pasteur, "Chance favors the prepared mind." Event histories show that the new businesses were not initiated on the spur of the moment, by a single dramatic event, or by a single individual.

Each of the three cases experienced an extended gestation period lasting about four years in which entrepreneurs or organizations engaged in a set of activities that set the stage for initiating new businesses. Although these activities were not necessarily undertaken for the specific purpose of initiating a new business, they provided the founders with (1) knowledge and experience in the core technology on which the new business was ultimately initiated and (2) access to the financial, personnel, and technological resources needed to mobilize action on a new business idea when it became apparent. Moreover, none of the cases support the proposition that the initiation of efforts to create a new business was precipitated by a single dramatic incident or inspiration. Instead, the events that ultimately led to initiating each new business came from multiple and seemingly coincidental sources, and they had the common cumulative effect of triggering the recognition of and

277

TABLE 8–4.   *Core Processes and Diverse Overlays of Business Creation.*

Common Core Processes Observed	Diverse Overlays Observed
Initiation period	
Extended gestation process with multiple independent events, some intersecting by chance to set business creation stage.	Potential sources for innovation increase with organizational complexity and boundary crossing.
Business and financial plan to develop a family of related products in a loosely coupled parallel or lagged sequential order.	Matching idea scope and novelty with resource capabilities of the organization.
Resources to start new business come from outside the entrepreneurial unit.	Market venture capital is more short term, difficult, and risky to obtain than corporate capital.
Startup period	
Developmental sequence deviates from plans and proliferates into multiple, interdependent paths as unforeseen events arise, producing trial-and-error adaptation process.	None observed.
While most activities proceed successfully, failures in a few components derail entire first product development effort.	None observed.
First product errors spill over to subsequent startup efforts, accumulate with others, and snowball into vicious cycles.	None observed.
Little learning occurs in "noisy" systems overloaded with positive, negative, and mixed messages randomly ordered over time.	None observed.
Resource and product timelines diverge.	None observed.
Vicious cycles broken by external intervention.	Top management, directors, and investors perform different roles.
Frequent revisions of business idea, with convergent search in good times, divergent search in bad times.	Search boundaries broadest in new company, narrowest in joint venture, intermediate in internal corporate venture.
Fragile and unanticipated consequences of business transactions encountered.	Institutionally derived legitimacy from parent corporations.
Takeoff period	
Success attributions made to management.	Market, administrative, and consensus tests

commitment to a commercially feasible new business idea.

The hearing health case dramatizes the multiple, quasi-independent, and coincidental events leading up to the initiation of HHP within a complex organization. News from Australia in 1977 of the development of a "bionic ear" intrigued a 3M technical director, who then visited a variety of U.S. otological research centers and clinics and persuaded his division manager to explore development of a cochlear implant. The division manager could have rejected the proposal (and thereby close off one of many stimulants for innovation apparent in the case), but he assigned the idea to an "unrelated products" group. To take advantage of a normal career advancement opportunity, the technical manager accepted reassignment to a 3M manufacturing subsidiary in California, which happened to have a vendor relationship with the House Ear Institute. Meanwhile, his successor at 3M established a relationship with UCSF that developed and implanted a cochlear device in several patients in 1980, after which the relationship terminated. But with the termination of this one source for the innovation, two others were being cultivated independent of these activities. Research on hearing aids was underway within a 3M research laboratory, and in another part of the organization a 3M acquisitions group was exploring the acquisition of a hearing health company. All these parallel events clearly set the stage for initiating the program, but few were orchestrated by a central actor and none appeared to be individually sufficient to cause program initiation. It was not until his return from California and promotion that another stimulus occurred when the new division VP expressed disappointment at the lack of progress in developing a cochlear implant, combined the independent groups, and appointed a manager to initiate the program in fall 1980.

A similar extended gestation period of multiple coincidental events occurred before the initiation of TAP. 3M labs undertook research on blood treatment systems in 1980 but discontinued it in 1982 because no commercially feasible products were evident from the work. Independently, by 1981 Millipore had developed an apheresis filtration prototype and contacted Sarns as a potential vendor because of its recognized leadership in manufacturing heart blood pumps. But again, for unrelated reasons negotiations were terminated in 1981 because Sarns had entered into negotiations to be acquired by 3M. In March 1983 Millipore approached 3M about a possible joint venture with its new Sarns subsidiary, at which time it discovered that 3M also was interested in apheresis. Recognition of the complementary competencies of Millipore, 3M, and Sarns precipitated negotiating an informal joint venture and initiating TAP in November 1983.

Finally, the initiation of Qnetics includes two independent gestation periods. The first includes the independent parallel events that led two entrepreneurs to leave their employing organizations for different reasons in 1979, start up their own companies, and recognize limitations in making their independent companies commercially viable businesses. The second gestation period begins with the coincidental meeting of the two entrepreneurs through a common acquaintance, and their subsequent interactions, which led them to recognize potential opportunities to obtain venture capital support by merging their fledgling companies in November 1983.

In all three cases, it is clear that these gestation events were not directed toward the initiation of a new business—at least in the form that it subsequently unfolded. Instead, it is more reasonable to conclude that the events undertaken by the entrepreneurs and their organizations embarked them on courses of action, which often *by chance* intersected with the independent actions of others. These intersections provided occasions for interaction, which led the actors to recognize and access new opportunities and potential resources. Where these occasions were exploited, the actors modified and adapted their independent courses of action into interdependent joint actions and agreements to initiate their new business.

A comparison of the cases suggests that an important variation in this core gestation process, which may be attributable to organizational setting, is the number of potential sources from which stimulants for innovation can arise. The HHP within 3M exhibits the largest number of stimulants for innovation, the Qnetics new company startup the least, and the TAP joint venture an intermediate number of precipitating events. This observation is partially consistent with Hage and Aiken's (1970) empirically supported proposition that the greater the structural differentiation, the greater the innovativeness. However, the TAP joint venture between 3M, Millipore, and Sarns emerged from the most structurally differentiated organizational arrangement, and yet a fewer number of precipitating events were observed in the gestation of TAP than of HHP. It is possible that this empirical finding is idiosyncratic to the cases examined here, but it is not likely to be due to technology (both HHP and TAP are new-to-the-world biomedical innovations), industry (market entry for both HHP and TAP are regulated by the FDA), or organization (both HHP and TAP involve 3M).

We think that boundary-crossing difficulties may explain this empirical finding. The probability of intersecting stimulants for innovation increases with the permeability of organizational boundaries between the stimulant sources. We observed TAP to experience more difficulties crossing structural boundaries between organizations than HHP experienced crossing departmental/division boundaries within 3M. Organizational boundaries that are permeable through only limited and prescribed modes (such as through TAP's SBU) limits the probability that stimulants for innovation generated within boundaries will intersect between the boundaries. This line of reasoning suggests a need to modify Hage and Aiken's proposition with an important qualifier, "the greater the differentiation in structures the greater the innovativeness," given the existence of permeable boundaries.

2. Initial Business Idea   Each case clearly began with entrepreneurs who had a business idea that envisioned a family of two or three cumulatively related product initiatives that were to be developed initially with external capital and subsequently funded by expected revenues from initial product market introductions. This is consistent with our starting observation that a new business requires a family of products to

be viable in the marketplace. The initial business ideas also clearly reflect a temporal plan for startup of business components in either a parallel or lagged sequential order, with loose coupling planned between these components so that they could be developed in a quasi-independent parallel fashion of each other. But during the startup period, unforeseen difficulties were encountered and the business ideas shifted and adapted on a seemingly trial-and-error basis.

To be sure, operational steps for implementing the business ideas were unclear to the entrepreneurs in the initiation period. The clarity of these steps appeared to vary with the technological novelty and scope of the business idea. Thus, although years of prior R&D had been undertaken to suggest the commercial feasibility of cochlear implants or therapeutic apheresis devices, the most appropriate technological trajectory (Nelson and Winter 1982) and particular steps to develop, manufacture, test, and market devices within that trajectory were highly uncertain. Moreover, it was clear from the outset that commercialization of HHP and TAP products would require creating new markets and entail extensive and costly steps in obtaining FDA regulatory review and approval of the medical devices. In contrast, the technology to develop custom-designed medical records software and distribute computer hardware systems by Qnetics was far less uncertain and novel. As a consequence, the resource commitments and time projected to commercialize the HHP and TAP business ideas were far greater than those of Qnetics. Indeed, it is very unlikely that a small new company like Qnetics, with its limited capitalization, personnel competencies, and technological research base, would be able to initiate either the HHP or TAP business ideas. As Hauptman and Roberts (1985) found, the probability is very low that small new biomedical and pharmaceutical firms can successfully develop and commercialize a new biomedical product because the resources required for technology development and obtaining FDA product approval are prohibitive. Thus, an important variation on the core process of idea development is matching the scope and novelty of an innovation with the resource capabilities of the organization.

3. Resources and Exposure An infrastructure is required for new business creation. In each case, financial, personnel, and technological resources to start the new businesses initially came from outside the new business unit, as the accumulation model suggests. However, significant differences exist between the cases in the sources and ways external resources were obtained. In relation to HHP or TAP, Qnetics experienced greater difficulties obtaining external resources and exposed its entrepreneurs to greater personal risks. Its failure to establish an adequate capital base at the outset exacerbated subsequent problems of business startup and takeoff. The Qnetics case exemplifies the liabilities of newness and small size of a company startup.

HHP obtained its funding from 3M corporate venture capital, personnel with needed competencies were reassigned from other corporate units, and technology came from 3M R&D labs as well as relationships established with university research centers and otological clinics. TAP obtained its funding, personnel, and technology through investments, part-time use of professional staff, and lab R&D from its co-venturing parent firms. Thus, at the outset, significant corporate resources were committed to HHP and TAP for an extended time period. Top managers interviewed stated that although they perceived them to be highly risky, both HHP and TAP were understood to represent major long-term investments and commitments to create totally new businesses that could position their companies in lucrative new markets in ten to fifteen years.

Unlike HHP and TAP, Qnetics' entrepreneurs assumed personal financial risks by making personal investments in their new company. Instead of relying on corporate resources, Qnetics used the market to obtain the vast majority of its resources with a private placement arranged by a venture capitalist and by hiring personnel with the needed competencies. In making his sales pitch to potential investors in Qnetics' during winter 1983, we observed the venture capitalist to emphasize that an investment entails high risk but also offers investors a lucrative short-term return when the anticipated public placement of Qnetics was to be undertaken in fall 1984. As it turned out, because of a "soft market" the public placement was canceled in October 1984. This event not only led Qnetics into its financial crisis, but it eliminated any hopes of short-term returns that initial private investors were led to expect. Indeed, the value of their combined investment of $465,000 in Qnetics in August 1984 had reduced to near zero six months later. As these events suggest, liabilities in generating external resources for new, small companies influence the fortunes of entrepreneurs and investors alike. Thus, private venture capital was found to be more difficult and risky to obtain and had a shorter time horizon that corporate venture capital.

Startup Processes

Unlike the seemingly random intersections of multiple independent events leading up to the initiation of the new businesses, the startup period in the three cases began with noticeably more calculated and orchestrated business plans. In each case, these plans set forth a strategy, timetable, and budget for developing a family of loosely coupled products and functional components (some to be implemented simultaneously, others sequenced over time), which in years to come would generate increasing yearly revenues and returns on investment during business takeoff. But in each case, event histories show that over time the developmental sequence increasingly deviated from initial plans and proliferated into multiple, divergent paths as unforeseen problems and events arose, resulting in a trial-and-error process of business startup and adaptation. Analysis of the data suggests that three common patterns may explain why these plans went awry during the startup period.

1. Business Plans Were a Vehicle for Generating Investments First, the plans, colored as they were with optimism, were used more as a vehicle to obtain resource commitments from investors or corporate sponsors than they were to develop realistic alternative scenarios of business creation. Although all the entrepreneurs acknowledged in interviews that parts of their business plans (particularly product development timetables and projected revenues) were overly optimistic, none stated a willingness to document uncertainties in their plans or to propose a more extended timetable for business startup because that would decrease their chances of obtaining startup funding. The external investors and corporate sponsors, in turn, used the targets, schedules, and forecasts specified in the plans to evaluate business creation progress. Those interviewed admitted that they discounted certain projections in the plans "as fluff" and that business creation is an inherently uncertain process that commonly entails setbacks beyond the control of entrepreneurs. However, they steadfastly held to their convictions that the entrepreneurs be held accountable for business creation success, as evidenced by achieving the financial and performance targets in their business plans.

Thus, in the fear of not obtaining startup capital, the new business entrepreneurs committed themselves to a course of action and a set of optimistic expectations by investors and corporate sponsors that were difficult to achieve. When these expectations were not achieved for the reasons described below, the confidence of key sponsors and investors was shaken, resulting at first in external interventions that often misdirected business startup activities and later in decreased commitments to support the entrepreneurial ventures.

Ironically, if the business plans were a primary vehicle for obtaining startup capital, it is clear that the requested duration of funding for business takeoff was underestimated. As illustrated below, the duration of capital committed to business startup was significantly shorter than the time required for business develop-

ment and takeoff. In each of the cases initial timelines for product development and business takeoff at $t2$ were underestimated and slipped to $t3$; whereas estimates of the length of time that initial capital investments would be needed to support business startup ($t2$) were overestimated and ran out by $t1$.

Although Qnetics provides the most dramatic example, both HHP and TAP encountered similar gaps between the durations of capital investment (C) and business startup (S). Early anticipations of this $C < S$ gap were addressed by expecting that revenues from initial product market introductions would support later product development efforts. But in each case this strategy failed as setbacks and errors developed in commercializing the first products. Subsequent strategies to reduce the expanding $C < S$ gap included differentiation efforts that introduced interim revenue-producing products that would supplement C to reach S (custom software by Qnetics and hearing diagnostic devices by HHP), but in each case these interim products were insufficient to close the $C < S$ gap. Finally, in each case appeals were made to obtain additional investment capital, and each of these requests resulted in significant adjustments in the scope and strategy of the new business creation efforts to fit the resources available (Qnetics and TAP undertook significant budget cuts and program reductions, while HHP redirected its priorities from cochlear implants to hearing aids).

Thus, one reason that business startup plans went awry is that the plans themselves were awry in setting forth overly optimistic performance targets in order to attract investment capital—targets that ironically were grossly underestimated in amount and time required for business creation. Of course, the presence or absence of a gap between capital investment and business startup timelines is only a symptom (not a necessary or sufficient conditions) of business startup failure or success. Moreover, as the processes described below indicate, it is not at all clear that if additional resources had been made available to the cases, business creation processes would have lead to greater success. Under conditions of high technological and market uncertainties, additional or excess resources often mask underlying problems and delay subjecting product innovations to "acid" tests of the market (Burgleman and Sayles 1986).

2. A Few Key Failures in Developing the First Product

Event histories show that most activities in the development of the first business product proceeded as planned during the first year or two of business startup and that failures occurred in only a few critical components, resulting in slipped schedules, budget overruns, and prevented successful takeoff of that first product in the market. Although investors, corporate sponsors, and business managers attributed these failures *after the fact* to "management problems of implementation," it is not at all clear a priori that these errors can be attributed to mismanagement. Indeed, investors, sponsors, and business managers collectively agreed to the strategies and tactics for developing the first products in each business. Given

the high technological and market uncertainties in launching the innovations, it is surprising that so few setbacks were actually encountered in developing the first products. Indeed, laws of probability would have predicted a greater number of errors and setbacks. Yet the few setbacks that did occur were sufficient to produce failures in the entire first product development efforts.

For example, efforts to develop the Phase I product in the TAP case have been largely successful. Prototype filtration modules and equipment were built, clinical trials were successful, the unit was introduced in several foreign countries, and FDA approved the unit for commercial release in the U.S. market. The only critical problem encountered so far has been in manufacturing defects and scaleup production of the unit, which, in turn, resulted in slipped schedules for product market introduction, deferred sales revenues, as well as delayed development of the Phase II device. As a manager stated, "TAP has become a complex system, and failures in one or two components derail the entire effort, even though other parts are proceeding effectively."

So also, the major thrust in Qnetics startup has been to design, program, and market medical records software. Design and programming tasks proceeded very well—so well, in fact, that eight products were developed. But the company has thus far been equally unsuccessful in marketing its medical software products, in spite of multiple trials in building internal marketing competence and in establishing marketing alliances with distributors. These marketing failures, coupled with cancellation of the public offering, drove the company into a financial crisis.

3. Errors Accumulated Into Vicious Cycles

The failures in startup and takeoff of the first product had important spillover effects on the strategy, timing, cost, and confidence in starting up subsequent product development efforts.

Moreover, they accumulated, made the business venture more vulnerable to subsequent unforeseen events and problems, and appeared to trigger vicious cycles that jeopardized the survival of the overall business creation effort. As Masuch (1985) describes, a vicious circle is a complex action loop in which a set of activities entail a chain of other activities that, in turn, ultimately recreate and worsen the original situation.

Initially, these spillover effects were not recognized, perhaps because *logical* independence of activities masked their *temporal* interrelatedness. Business plans appeared to reflect an "illusion" that business startup could consist of undertaking parallel, independent streams in developing products and functions. This was reflected in plans made to simultaneously undertake various functional activities to create the new business, including research and development, manufacturing, marketing, and financial activities. A manager stated these parallel functional activities were undertaken both to increase the speed of business startup and to build functional competencies needed to run the business "from the ground floor up."

Parallel developments of business components continued for a time until setbacks and problems occurred in developing first products. Then it became evident that what were conceived of as parallel and independent activities were indeed sequential and highly interdependent activities. For example, with manufacturing defects in the first TAP product, the Phases I and II parallel product development paths became sequential, even though there was a clear division of labor between Millipore, Sarns, and 3M. In the HHP case, the House and Vienna devices developed in parallel until the premature marketing announcement of the second product release before the first product had completed market entry. In the case of Qnetics domino effects occurred among all product lines (medical and custom software and hardware distribution) as a result of their pooled financial interdependence and the cancellation of the public offering.

As problems were encountered, an elaboration or proliferation of business activities was observed in the cases, either by including more detailed products in one line or by expanding into new areas (that is, in depth versus breadth). TAP maintained its focus by exploring new applications of its filtration technology, while HHP and Qnetics ventured more broadly into alternative technologies and industries. This proliferation of developmental activities, appeared to be a risk reduction strategy of avoiding having "all eggs in one basket." However, as one manager stated metaphorically, "The problem is like trying to grow an oak tree when there are inexorable pressures to grow a bramble bush."

Ironically, although multiple parallel development efforts were initiated, management's attention was sequential and appeared to escalate and be capable of dealing with only one major effort at a time. For example, HHP managers were unable to cut their losses and leave the single channel device even when repeatedly told to do so and move on to other technologies and products by upper management and corporate sponsors. So also, Qnetics continued to make big investments in developing up to eight medical products, even though none of the products were successfully tested and entered into the market. Finally, TAP could not get beyond Phases I and II products to explore strategic business opportunities in the joint agreement.

Cyclical patterns were observed when crises occurred in each case and surprisingly little evidence of learning was observed. The new businesses began with the idea of a family of products, which led to a proliferation of product initiatives and the adding of new differentiated functional activities. When significant problems emerged, cuts were made in program functions leading to program retrenchment, which in turn exacerbated the initial product development problems. In HHP, for example, poor performance led to some program reductions including QA (quality assurance), and when problems subsequently were encountered with the first product, QA was not available to correct the problem. These cuts probably should not have been made when problems arose. Alternatively, problems may result from poor initial strategies or premature overinvestment in business creation activities, where additional resources exacerbate and mask core problems.

4. Learning Disabilities   All three cases amplify the need for learning but also its difficulties. In general, learning tends to occur in three ways: (1) by imitation of something done elsewhere, (2) by extrapolation from the past into the future, and (3) by trial-and-error through error detection and correction. New-to-the-world innovations must rely primarily on trial-and-error learning because there are few if any precedents to follow or copy and because no past history has developed from which to project extrapolations. Learning by trial and error requires that problems or errors are detected and corrected.

HHP's repeated exposure to information about the limitations of the single channel and superiority of multiple channel technologies, in particular, dramatically illustrate that warnings, "red flags," and questions about the program's direction were repeatedly raised. However, relatively few attempts were observed to (1) detect these warnings as errors and (2) correct the detected errors. Indeed, they were stubbornly and repeatedly not perceived as errors with HHP's single-channel technology; instead they were dismissed as exaggerated and unsubstantiated claims by competitors, the FDA, surgeons—and a mounting majority of industry stakeholders. So also, recorded observations of Qnetics' board meetings and TAP's SBU meetings produced extensive lists of questions, mistakes, or "red flags"

that were raised by participants regarding the development, schedule, directions, or financing of their new businesses. Although some of these warnings were detected as errors, which often generated considerable discussion and debate, most were simply aired without response or dismissed as irrelevant to the discussion at hand.

Although Schroeder et al. (Chapter 4) provide a number of reasons for this observed lack of error detection and correction when extensive warnings were raised, *the ability to discriminate substantive issues from "noise" in the system* surfaces as a particularly salient explanation. The "noise" consists of many mixed messages received by decisionmakers in a seemingly random order over time: Some bear good news, some bad, but most are contradictory; some issues are formulated well, some poorly, but most equivocally; some come from outside, some inside, but most from "nowhere"; some are expressed in meetings, some on expressways, but most are not expressed; some are credible, some incredible, but most are uncredible; some appear once, some disappear, but most reappear; some are stated emphatically, some are whispered, but most are indistinguishable from a din of grumbles, rumors, banter, vendettas, hidden agendas, small talk, and gossip.

In such a noisy social din where positive, negative, and mixed signals are randomly ordered over time, it is easy to understand why so few messages were detected as errors and why few attempts were observed to correct detected errors. The stimuli are far beyond the information processing capacity of individuals. Moreover, cognitive psychologists have found that individuals have widely varying and manipulable adaptation levels (Helson 1948, 1964). When exposed over time to such a "noisy" set of stimuli that change very gradually, people can not recognize the gradual changes—people unconsciously adapt to the slowly changing conditions. Their thresholds to detect errors

285

and tolerate discomfort are not reached. As a consequence, they do not move into action to correct their situation, which over time may become deplorable. Opportunities for correcting errors are not recognized, problems fester and swell into metaproblems, and at the extreme, crises are sometimes necessary to reach the action threshold (Van de Ven 1980).

The fact that many detected errors were not corrected appeared to have two consequences on the new businesses. First, because many errors were not corrected, they "snowballed" over time into vicious cycles of even larger and more intractable complexes of problems. When situations deteriorate to the point of reaching a crisis, peoples' action thresholds are triggered but often produce deleterious responses. As Janis and Mann (1977) describe, crisis decision processes are dominated by defense mechanisms of isolation, stereotyping, and attribution of problems to others to avoid negative evaluations. As a result, error-prone solutions tend to emerge from crisis situations. For example, in the HHP case errors and complaints accumulated into crisis proportions and resulted in the replacement of the founding program director, a product recall, and four years of lost time and misinvestment in the emerging technologically superior trajectory.

A second consequence is that errors and mistakes, by definition, are the major sources of learning by trial-and-error. Each failure to correct a detected error represents a lost opportunity for learning about unforeseen or misdirected courses of action during the business startup process. Given the evidence presented above about the limited utility of business plans for guiding the business startup process, trial-and-error learning assumes increasing prominence as a core monitoring process to redirect and adapt business startup in a purposeful way. Without this learning occurring, business startup activities begin to drift, in which "a set of beliefs, values and guiding principles may emerge that are counterproductive to the organization's mission or distinctive competence" (Selznick 1957: 139).

Of course, trial-and-error learning implies that errors are embraced rather than sanctioned (Michael 1973). Legitimate error stems from the uncertainty inherent in the nature of a situation. As Burgelman (1983) notes, the major problem in dealing with uncertainty is maintaining a balance on organizational diversity and order over time. In new business creation, diversity results primarily from autonomous initiatives of different stakeholders and technical personnel. Order results from establishing standards and a concept of strategy on the new business. Managing this diversity requires framing ideas and problems so that they can be approached through experimentation and selection. Burgelman suggests that the learning process is facilitated by probing into various dimensions of problematic situations and of promoting constructive conflict and debate between advocates of competing perspectives. Competing action strategies lead to reconsideration of the business idea and perhaps a reformulation of that strategy.

The above discussion of learning should be tempered with two caveats. First, James March and his associates through a series of experiments are finding that fast learners often do better than slow learners but there are many plausible situations in which slow learners do better than fast learners. In particular, when goals are diffuse and technologies highly uncertain (as in many instances of new business creation), fast learners tend to specialize, but often in the inferior alternative, leaving the superior alternative to their somewhat slower competitors (Herriott, Levinthal, and March 1985). These findings suggest that moderate (not fast nor slow) rates of learning are related to success and that not all opportunities for error detection and correction should be acted on. Moreover, although insights and "learning experiences" may be obtained by participants in the course of business creation, many cannot be acted on either because of the lag time involved in recommitting "chunky" resources to different courses of action or because many situations are unique or seldom represent themselves to apply the learned behavior.

Second, rational management of trial-and-error learning implies frequent modifications and adaptations in chosen directions as errors become apparent. Ironically, however, Brunsson (1982) points out that such behavior may be irrational for managing innovations. Because the mobilization of action toward a novel and highly uncertain goal requires inordinate commitments and investments of resources to overcome organizational inertia and to withstand the ever present criticisms of "naysayers," Brunsson argues that the most reasonable logic for innovation leaders is to "Damn the torpedoes, proceed full speed ahead." This logic of action, of course, questions the rationality of much trial-and-error learning behavior.

5. Business Idea Search and Redefinition Processes.
The new businesses were observed to frequently revise their business ideas or strategies during startup, sometimes in seemingly confusing and contradictory ways. Differences among the cases were observed (1) in search boundaries, (2) in search criteria, and (3) with good or bad times. A cycle of power struggles between internal and external groups, as suggested by the accumulation process model, may explain these dynamics.

First, in terms of search boundaries, Qnetics' search appeared to be unbounded by technology or lines of business. Whatever it took to become a financially viable entity appeared to be the open territory, as long as it related to the broad area of computer hardware and software systems. HHP and TAP, however, were seen to follow a more bounded proliferation of their strategy, perhaps because their domains were more clearly specified by their respective organizational sponsors.

Second, these search processes appeared to be governed in different degrees by *market, administrative,* and *consensus* criteria. Qnetics is the clearest example of following a sequential trial-and-error process of product developments that were largely governed by the "acid" test of the market. A large part of HHP's development, on the other hand, was observed to be governed by by 3M's administrative hierarchy, and the market test did not govern until a product was introduced into the market. Finally, explorations of new business ideas in the TAP SBU meetings required consensus, but consensus could seldom be achieved among pluralistic SBU members. Problems of trust and a divided house (hung jury) existed in selecting new diversified business ideas for development in the SBU. As a result, the SBU's strategy has thus far remained confined largely to the terms in its initial joint agreement.

Third, these search processes appeared to differ in good and bad times. When things were going well, search occurred within the program's existing business strategy. When things went bad, search focused outside of the program's business idea. As March (1981) described, disjunctive searches appear to be a way to get a new lease on life. For example, when everything was going well in 1985, TAP's SBU was concerned about keeping scientists and conducted its slack search to redeploy its assets to TAP-related activities. However, when many technical problems emerged in January 1987, the SBU began looking for other things to do because of problems apparent in what it was doing.

An action cycle produced by shifts in power exerted by inside and outside groups in the accumulation process may explain these three differences in business search processes. A model of this action cycle is illustrated in Figure 8–10. The model argues that a chosen course of action by the business unit that is perceived to be successful increases the exter-

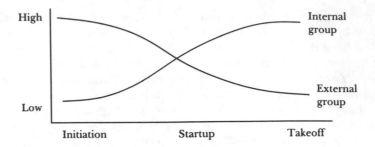

Action Loops in Transitions of Power
between Internal and External Business Groups

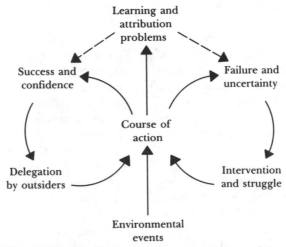

FIGURE 8–10. *Relative Power of Internal and External Groups in
Accumulation Process Model.*

nal group's confidence in and willingness to delegate greater control to the business unit, which in turn permits the internal business unit greater discretion to pursue and expand its chosen course of action. However, failures or setbacks encountered with a course of action increase the external group's uncertainties of the business unit's course of action, trigger their external intervention, and often result in struggles between internal and external groups on the appropriate course of action to follow. When a new or revised course of action is selected, the struggle subsides and this renews the action loops.

As stated in the introduction, the ac-cumulation process model posits that from the initiation to startup to takeoff periods the relative power (resources and control) of the new business shifts from external to internal groups if the new business is to successfully takeoff with the power necessary to sustain itself as a viable entity. This implies that during startup (the period between initiation and takeoff), a transition in power and control must occur between internal (business unit managers or entrepreneurs) and external (investors or corporate management) groups in the governance of the new business. For example, in both HHP and TAP the program managers and their committees were the internal power groups, while

288

the top managers and their administrative review committees constituted the external power groups. In the case of Qnetics, most of the time there was really only one active power center (the internal entrepreneurs), since the venture capitalist seldom became involved, and the banks and creditors were somewhat disconnected from the business creation process.

Internal and external groups were observed to hold somewhat different perspectives on the business development process, and each imparted its own particular direction to the developmental process when occasions arose. The internal power groups saw the business creation process up close and tried to develop a range of technical functions and products to support the business. The external power groups were more detached, regarded the business as but one of a larger set of alternative businesses to develop, and regarded many of the functions considered important by the internal groups to be unnecessary because they could be performed by other parts of the organization. The internal business unit governed the immediate work of business creation, while the external groups set the basic resource and strategy parameters for the business. Most of the time the two power centers appeared in synch, but when failures arose it was clear they were not, and sudden shifts in the developmental sequences were observed.

As the model suggests, when business creation activities proceeded smoothly and according to plan, transitions in power from external to internal groups were uneventful because the external groups were satisfied with progress and had little reason to intervene and impose their differing views on the internal group. In other words, in such "good times" the external group's nonintervention permitted the internal business unit increased discretion to pursue and expand its chosen course of action.

However as we have observed, when mistakes or setbacks occurred, uncertainties arose about the appropriateness of the course of action being taken by the business unit, and this stimulated the external group to intervene and explore alternative course of action. These instances, such as the TAP "budget cut" and the HHP product recall, were occasions when power struggles erupted between the internal and external groups, with the latter usually imposing its course of action on the former.

Additional insights to these action loops in the transition of power in the accumulation process model will become evident as we examine the differing roles and activities of the external power groups in relation to the business units.

6. Roles of External Power Groups As the model in Figure 8–10 would suggest, in each case vicious cycles were broken by external interventions of business investors or top managers in the corporate hierarchies. In no instances were significant problems found to be resolved internally by the new business management teams. Perhaps this is because of the learning disabilities observed within the businesses units where people may have been on "automatic pilot" because they were "locked" into a unitary course of action that required completion before they were open to consider alternatives (as Gersick 1988 observed is the common development of small groups). In such situations, interventions by external groups were necessary to trigger the attention and action thresholds of the business managers.

Each round of external interventions resulted in seemingly more complex, contradictory, and abrupt shifts in the developments of the new businesses. But on closer examination, the abrupt shifts represented nothing more than substitutions of one simple developmental sequence or formula for another. For example, when the internal TAP group decided to address specific diseases with its device, it initiated the simple sequence of tasks necessary to obtain FDA approval for this device. However,

3M's external top manager believed that FDA regulatory licensing steps for such a device would be too costly and time consuming, and instead proposed simply marketing the TAP device as a blood filtration product. This simple market penetration logic overruled the other simple regulatory licensing sequence. So what appeared to look like a complex and chaotic process is one simple sequence supplanting another. Although the business managers did not often agree initially with the alternative courses of action proposed by external groups, they participated in their formulation and implementation and tended to grow and accept them in two to four months after the interventions.

Very different perceptions of HHP and TAP were found in yearly interviews between managers in the corporate reporting hierarchies from the businesses to CEOs. Top managers from one to four levels removed from the new businesses were observed to (1) be more knowledgeable and interested in HHP or TAP than the researchers expected, (2) be directly involved and apply different criteria among each other in major decisions about the new businesses, and (3) perform different managerial and institutional roles that were critical to legitimizing the new businesses in ways that were often unknown to or unrecognized by the immediate business entrepreneurs or managers.

The involvement and roles of managers at multiple levels in an innovation have not been adequately recognized in the literature, most of which focuses only on a management sponsor who "runs corporate interference" for an innovation (Peters and Waterman 1982; Pinchot 1985). Multiple levels of management involvement appeared to provide a balance of cross-checks between contradictory forces among top managers pushing for expansions and contractions in the scope, resource allocation, time schedules, and performance expectations of business innovations. For example, the ongoing enthusiastic support of HHP's man-

agement sponsor, a group vice president, was counterbalanced with "hard-nosed" business skepticism by the sector vice president, while a division vice president viewed his role as being a "mentor" or "tutor" to the HHP program manager. In addition, critical functions of institutional support and endorsement assume increasing degrees of legitimacy the higher the organizational level at which they are performed. Thus, for example, a casual agreement made to explore an apheresis joint venture during a golf match between the CEOs of Millipore and 3M, when communicated to lower management levels, brought immediate credibility to the formation of TAP. Similarly, a brief visit by the 3M CEO to the House Ear institute and the election of the 3M group vice president to the board of directors of HEI solidified the 3M-House relationship to develop cochlear implants.

Many of these differing roles and activities across management levels were not apparent to the new business unit managers. However, they were signaled to program managers in indirect ways during occasional meetings with individual top managers and in biannual administrative reviews of the HHP and TAP businesses. In particular, they were reflected in the differing approaches taken to problems and solutions by business managers and upper management in these administrative review meetings. Although unit managers often stated they did not have enough resources, upper management reported a need for program redirection and resource conservation. So also in Qnetics, the venture capitalist interviewed stated that Qnetics management overinvested in health products development when it should have invested in building its marketing capabilities. In addition, although the venture capitalist and top managers viewed the new business as one of several ventures in which they were investing for future growth and profits, to the unit managers the new business was their only venture.

The multiple levels of top management

involvement in HHP and TAP obviously were not present in the Qnetics case. Analogous roles were performed by the venture capitalist and the board of directors for the new company startup. However, their roles differed in two critical respects from that of top management. First, the Qnetics' board of directors consisted only of the inside owners/entrepreneurs from 1983 to February 1987, at which time the board was reconstituted and three outside directors were elected: a physician, a banker, and a professor. Thus, throughout the period examined here, the new business entrepreneurs/owners/ and board members were "talking to themselves." The internal board composition prevented Qnetics' entrepreneurs from being exposed to the different perspectives and criteria that external board members could have provided (like that of HHP and TAP top managers). Indeed, Qnetics's entrepreneurs reported (and sometimes complained) that since February 1987 they have had to spend much greater time preparing for board meetings to respond to questions from external board members.

Second, as a new company startup, Qnetics does not enjoy the institutional legitimacy provided to HHP and TAP by their established and highly regarded parent corporations. Qnetics can draw on neither the infrastructure of functional competencies, resources, and systems available to HHP and TAP, nor the institutionally derived legitimacy of a parent corporation to launch and conduct its business. Nowhere are these liabilities of newness and small size for Qnetics better illustrated than in its repeated failures to secure marketing relationships with other firms and in the highly interdependent and risky set of transactions it has structured to conduct business.

7. Fragile Business Transactions. Although all cases exemplify the fragile, interdependent, and unanticipated consequences of engaging in interorganizational transactions, Qnetics is the

most dramatic. Qnetics' event history shows that even after numerous attempts to do so, it has experienced repeated failures in establishing marketing relationships with distributors or marketing distribution outlets for its medical records software products. Qnetics' principal reported that although highly enthusiastic responses have been consistently received from potential customers and distributors after product demonstrations, various reasons have been given for not closing deals. The two most frequent reasons were that (1) while Qnetics' product may be technically superior, its capabilities to provide product maintenance and upgrades are not as extensive as competitors like Unlsys, Hewlett Packard, and Texas Instruments, and (2) there is a six- to twelve-month decisionmaking cycle in large hospitals or medical software distribution houses (when Qnetics' time horizon of financial solvency is measured in weeks).

Qnetics liabilities of newness and small size are also reflected in the interdependencies that are structured in its set of seemingly independent transactions. Failures in refinancing the company by the venture capitalist led to problems with the bank, and the closing of the line of credit by the bank led Prime Computer Co. to terminate Qnetics as one of its distributors, which in turn led to almost losing and then settling for a significantly reduced profit margin on the large sale to a major customer. These domino effects, when combined with failures to establish medical product marketing or distribution relationships accumulated to push Qnetics to the brink of bankruptcy.

While Qnetics was experiencing these liabilities of newness and small size, the TAP joint venture was exposed to the liabilities of double parenting and conflict. Event histories show that the TAP SBU experienced the inflexibilities and rigidities of a "hung jury" in exploring new strategic directions for the joint venture. By repeatedly rejecting new business opportunities that came to the SBU's attention over time, the basic business strategy for TAP has not grown or developed much since the joint-venture agreement in November 1983.

Finally, HHP exemplifies some of the unanticipated temporal dynamics of interorganizational business relationships over time. First, two aborted attempts at establishing relationships with other firms with complementary resources turned out a few years later to be the major industry competitors of HHP (the University of Melbourne subsequently linked up with Nucleus, and UCSF linked up with Symbion). Second, highly "successful" and tight relationships with the House Ear Institute may have lead the HHP into a "Group Think" (Janis and Mann 1977) about the superiority of the single channel technology and to repeatedly reject claims of the emerging superiority of the multiple channel technology. Normatively, these observations suggest two bits of practical advice. "Watch out who you choose not to cooperate with for they may turn out to become your competitors, but don't get too cozy, for you may jointly become 'tunnel-visioned' and get 'blind-sided.' "

Takeoff Processes
Thus far, the three businesses being studied over time have not had successful takeoffs, so no comparisons can be drawn among how the new businesses sustained themselves as economically viable entities. However, HHP's unsuccessful attempt at developing and commercializing its 3M-House single-channel cochlear implant provides insights into the business takeoff process and the ability of management to control it.

From 1981 (when the HHP was formally initiated) to November 1984 (when FDA approved the first device for commercial market introduction), most work focused intensively on achieving the strategy of "being the first into the market with a safe cochlear implant device." Of course, in keeping with Maslow's (1965) observation that "all organiza-

tions grumble," there were some ongoing complaints and problems in the HHP organization, but none was sufficient to prevent accomplishing all the major steps in achieving the strategy. These included establishing the 3M-HEI relationship, prototype design and manufacturing, clinical trials, FDA approvals, and marketing, education, and distribution activities. The strategy was thought to be successfully accomplished in November 1984, with FDA's approval of the device and widely publicized announcement that this was the first time that one of the five human senses was replaced by an electronic device.

Unfortunately HHP's celebration was short lived as three critical setbacks were experienced in the ensuing year and closed its "window of opportunity" in being the first in the market: (1) Market demand for the device fell far short of expectations, (2) a competitor obtained FDA approval in October 1985 for a multichannel device claimed to be superior to the 3M-House device, and (3) 3M encountered product failures and a recall of its device from the market in November 1985. One could argue (as several HHP members and sponsors in 1986 did) that HHP should have conducted better market research, but with a revolutionary device that must create a new market, how can one truly test the market without introducing the device? After all, clinical trials and training centers stimulated much enthusiasm in the otological community. Critics also claimed that HHP overinvested in single-channel technology and should have invested more heavily in the multiple-channel technology. But limited resources demand commitment to and focus on only one technological path. In addition, the technical superiority of competing technological paths is difficult if not impossible to determine a priorl. HHP gambled on what is now believe to be an inferior path. It could just as likely have pursued the technologically superior path. Finally, design failures are typical, perhaps inevitable, in revolutionary product introductions and become apparent only with customer use (Von Hippel 1978).

Many HHP participants and corporate top management attributed errors to mismanagement by the program manager, who in turn was replaced by another manager. Although attributions for failures to mismanagement were frequently observed, the evidence suggests the attributions were misplaced and made managers scapegoats for events beyond their reasonable and fallible control (Pfeffer and Salancik 1978). Such scapegoating misdirects one's attention from understanding the more fundamental processes and risks involved in new business startup. By replacing the program manager, the attributions reinforced the myth that managing innovation is fundamentally a control problem, as opposed to one of orchestrating a highly complex, uncertain, and probabilistic process of collective action. Moreover, those business entrepreneurs who come to appreciate this orchestral role through the "school of hard knocks" were sanctioned with replacement and deselected from applying their learning experiences.

CONCLUSION

This chapter examined the creation of new businesses in a new company startup, within a large corporation, and in an interorganizational joint venture. The research was undertaken in order to determine (1) what business creation processes are the same across different organizational settings, and perhaps inherent to creating a new business regardless of organizational setting, and (2) what processes are unique across the cases and attributable to organizational setting. Because there are different but compensating benefits and liabilities associated with new company startups, internal corporate innovation, and joint interorganizational ventures, it was hypothesized that nei-

ther organizational setting is superior to the others for new business creation. Using an accumulation theory of change to guide the investigation, we examined processes of initiation, startup, and takeoff of new business in terms of (1) inputs—the sources from which resources and power were obtained to create the new business, (2) outputs—the ideas, competencies, and products produced by the new business, and (3) transformation of inputs into outputs—the progression of paths taken to develop the new business over time.

Based on historical and real-time longitudinal studies of new business creation processes in the Qnetics new company startup, the 3M hearing health program, and the 3M-Millipore-Sarns joint therapeutic apheresis program, we drew two overall preliminary conclusions: (1) *Business creation entails many common core processes irrespective of organizational, technological, and industry settings.* (2) *Several diverse overlays on these core processes can be attributed to the different settings in which the businesses exist.*

Specifically, the core processes commonly observed in the three cases during the initiation, startup, and takeoff periods of the new businesses are outlined below. The diverse overlays on these common core processes are indented below.

Business Initiation Processes

1. The new businesses were not initiated on the spur of the moment, by any single dramatic incident, or by a single entrepreneur. There was an extended gestation period lasting about four years in which entrepreneurs and their organizations engaged in a set of activities that set the stage for business initiation. Most events during the gestation period were not planfully directed toward starting a new business. Instead they embarked entrepreneurs on courses of action, which often by chance

293

intersected with the independent actions of others. These intersections provided occasions for interactions, which led the actors to recognize and access new opportunities and potential resources. Where these occasions were exploited, the actors modified and adapted their independent courses of action into interdependent collective actions to initiate their new businesses.

- Organizational arrangements influence the number of potential sources for innovation and new business ideas. The greater the differentiation in organizational structures with permeable boundaries, the greater the potential sources from which innovative ideas can spring to initiate new businesses.

2. The initial business idea envisioned a family of cumulatively related products that were to be developed initially with external capital and subsequently funded by expected revenues from initial product market introductions.

- An important variation on this core process of idea development is matching the scope and novelty of an innovation with the resource capabilities of the organization. The clarity of operational steps for initiating and starting up the business varied with the technological novelty and scope of the business idea.

3. An infrastructure is required for new business creation. Financial, personnel, and technological resources to start the new business came from outside the new business unit, as the accumulation model suggests.

- The new company startup experienced greater difficulties obtaining external resources and exposed its entrepreneurs to greater personal

risks than did the internal corporate venture or the joint interorganizational venture.

Business Startup Processes

1. The developmental sequence increasingly deviated from initial plans and proliferated into multiple, divergent paths as unforeseen problems and events arose, resulting in a trial-and-error process of business startup and adaptation.
2. Although business plans were used primarily as a "sales" vehicle for obtaining startup capital, requested amounts and duration of capital were grossly underestimated and speed to product market introduction and business takeoff were grossly overestimated.
3. Although most activities in the development of the first business product largely proceeded as planned, failures occurred in a few critical components, resulting in slipped schedules, budget overruns, and failure of the entire first product development effort.

- Failure to successfully enter the market with its first major product by the new company startup led it to the brink of financial collapse, whereas a similar failure by the internal corporate venture led it to revise its business strategy with significant additional corporate investment.

4. Failures with the first product had important spillover effects on subsequent product development efforts, accumulated and made the new business more vulnerable to subsequent unforeseen events, and triggered vicious cycles that jeopardized the survival of the overall business creation effort. These spillover effects were not initially recognized, perhaps because logical

independence of activities masked their temporal interrelatedness.

5. Significant trial-and-error learning disabilities were observed because actors could not discriminate substantive issues from "noise" in systems overloaded with positive, negative, and mixed signals randomly ordered over time. Extensive errors were detected, but very few were corrected. Instead they festered and "snowballed" to crisis proportions before they were often addressed. Crisis decisions are highly error-prone.

6. Vicious cycles were only broken with external interventions and resulted in frequent revisions in the business idea and startup strategy.

   · Search boundaries and criteria for revising business ideas varied with organizational setting. Search by the new company startup appeared to be unbounded by technology or lines of business, while the internal corporate and interorganizational joint ventures followed a more bounded proliferation of their business strategy. Whereas the new company startup was largely governed by the "acid" test of the market, in the internal corporate venture the administrative hierarchy prevailed, while consensus among corporate partners largely governed development of the joint interorganizational venture.

7. Multiple levels of top management involvement were observed to provide checks and balances between contradictory forces in the parent corporate hierarchies pushing for expansions and contractions in the scope, resource allocation, time schedules, and performance expectations of the new businesses. In addition, they provided institutional support and legitimacy to their new businesses.

· The new company startup did not enjoy the institutionally derived legitimacy and credibility of an established parent corporation, which significantly hindered its abilities to conduct business transactions with large customers and distributors. In addition, its board initially consisted of only inside directors, which limited the exposure of company principals to the kinds of divergent perspectives provided by top managers in the corporate settings.

8. Business startup and takeoff entails entering into many transactions with other firms and customers, and these transactions were found to be highly fragile and produce unintended consequences.

   a. Due to its scarce resources, the new company startup engaged in a leveraged set of highly interdependent transactions, and when one of the transactions failed the entire set collapsed in domino fashion.

   b. The liabilities of double parenting and conflict were particularly evident in the repeated inabilities of the joint interorganizational venture parties to reach consensus on future strategic directions for the new business.

   c. Unanticipated temporal dynamics of business transactions for the internal corporate venture included (1) aborted attempts at cooperative relationships a few years later became competitive relationships and (2) close and "successful" transactions with another organization may have led to "group think."

Business Takeoff Processes

   · It is premature to draw comparisons between the cases on business takeoff processes, since none have thus far completed takeoff. However, the factors explaining the unsuccessful business takeoff of the internal corporate venture leads one to question how much of business takeoff failure (or success) can be attributed to management. Although attributions for failures to mismanagement were frequently made, the evidence suggests many were misplaced and made managers scapegoats for events beyond their reasonable control. Such attributions reinforce the myth that managing innovation is fundamentally a control problem, when it should be one of orchestrating a highly complex, uncertain, and probabilistic process of collective action.

The above common core processes of business creation, along with diverse overlays, were found to occur in three very different organizational settings. They are preliminary findings, since the three businesses are still being studied in real time as they develop. However, the preliminary findings are rewarding because they indicate that many business creation processes are fundamentally the same regardless of organizational setting. If this conclusion continues to be supported in subsequent study of the three cases and is replicated in other studies, then we may call attention to the significant benefits and efficiencies that can be obtained from cross-fertilizing and applying principles for business creation from new company startups, internal corporate venturing, and interorganizational joint ventures.

Further conclusions are perhaps premature at this point in the longitudinal research. However, before the research is completed we plan to derive a basic set of propositions about managing the process of new business creation. In particular, three kinds of propositions are envisioned: (1) propositions about the sources and amounts of resources that lead to success-

ful business initiation, startup, and takeoff; (2) propositions about the relative effectiveness of alternative progressions in the chronological development of business ideas, proprietary products, and functional competencies; (3) propositions about the nature, temporal order, and interdependence among paths in the creation of new businesses. If possible, we hope to identify the feasible sets of paths in new business creation and examine if business creation outcomes can be explained and predicted from the prior paths undertaken by the business unit and the organizational contexts housing the unit. As these anticipated propositions suggest, our work has just begun in building a scientific body of knowledge of the processes of new business creation.

REFERENCES

Aldrich, H., and E. Auster. 1986. "Even Dwarfs Started Small: Liabilities of Age and Size and Their Strategic Implications." In L. Cummings and B. Staw, eds., pp. 165–198. *Research in Organizational Behavior.* San Francisco: JAI.

Brunsson, N. 1982. "The Irrationality of Action and Action Rationality: Decisions, Ideologies, and Organizational Actions." *Journal of Management Studies* 19: 29–44.

Burgelman, R.A. 1983. "Corporate Entrepreneurship and Strategic Management: Insights from a Process Study." *Management Science* 29 (12): 245–73.

Burgelman, R.A., and L.R. Sayles. 1986. *Inside Corporate Innovation: Strategy, Structure, and Managerial Skills.* New York: Free Press.

Charpie. 1967. "Technological Innovation: Its Environment and Management." U.S. Department of Commerce Report, 0-242-736.

Etzioni, A. 1963. "The Epigenesis of Political Communities at the International Level." *American Journal of Sociology* 68: 407–21.

Gellman Research Associates. 1976. *Indicators of International Trends in Technological Innovation.* Washington D.C.

Gersick, C.J.G. 1988. "Time and Transition in Work Teams: Toward a New Model of Group Development." *Academy of Management Journal* 31, no. 1: 9–41.

Goldman, J.E. 1985. "Innovation in Large Firms." *Research on Technological Innovation, Management and Policy* 2: 1–10.

Hage, J., and M. Aiken. 1970. *Social Change in Complex Organizations.* New York: Random House.

Harrigan, K.R. 1985. *Strategies for Joint Ventures.* Lexington, Mass.: Heath.

Hauptman, O., and E.B. Roberts. 1985. "The Impact of Regulatory Constraints on Formation and Growth of Biometical and Pharmaceutical Start-ups." MIT Sloan School of Management working paper WP/1651–85.

Helson, H. 1948. "Adaptation-Level as a Basis for a Quantitative Theory of Frames of Reference." *Psychological Review* 55: 294–313.

Helson, H. 1964. "Current Trends and Issues in Adaptation-Level Theory." *American Psychologist* 19: 23–68.

Herriott, S.R., D. Levinthal, and J.G. March. 1985. "Learning from Experience in Organizations." *Theory of Economic Organizations* 75 (2): 298–302.

Janis, I., and L. Mann. 1977. *Decision Making: A Psychological Analysis of Conflict, Choice, and Commitment.* New York: Free Press.

Killing, J.P. 1982. "How to Make a Global Joint Venture Work." *Harvard Business Review* 60 (3): 120–27.

March, J. 1981. "Decisions in Organizations and Theories of Choice." In A. Van de Ven and W. F. Joyce (eds.), pp. 205–244. *Perspectives on Organizational Design and Behavior.* New York: Wiley.

Maslow, A.H. 1965. *Eupsychian Management.* Homewood, Ill.: Irwin.

Masuch, M. 1985. "Vicious Circles in Organizations." *Administrative Science Quarterly* 30 (1): 14–33.

Michael, D. 1973. *On Learning to Plan and Planning to Learn.* San Francisco: Jossey Bass.

Nelson, R. R. and S. G. Winter. 1982. *An Evolutionary Theory of Economic Change.* Cambridge, Mass.: Belknap Press of the Harvard University Press.

Peters, P., and R. Waterman. 1982. *In Search of Excellence: Lessons fromn America's Best-Run Companies.* New York: Harper & Row.

Pfeffer, J., and G. Salancik. 1978. *The External Control of Organizations.* New York: Prentice-Hall.

Pichot III, G. 1985. *Intrapreneuring.* New York: Harper & Row.

Reynolds, P., and S. West. 1985. "New Firms in Minnesota: Explorations in Economic Change." *CURA Reporter.* Center for Urban and Regional Affairs, University of Minnesota, presented at Academy of Management Conference, San Diego (August).

Schoonhoven, C.B., and K. Eisenhardt. 1985. "Influence of Organizational, Entrepreneurial, and Environmental Factors on the Growth and Development of Technology-Based Startup Firms: A Research Proposal." Submitted to the Economic Development Administration, U.S. Department of Commerce.

Seiznick, P. 1957. *Leadership in Administration.* New York: Harper & Row.

Singh, J.V., D.J. Tucker, and R.J. House. 1986. "Organizational Legitimacy and Liability of Newness." *Administrative Science Quarterly* 31: 171–93.

Stinchcombe, A. L. 1965. *Constructing Social Theories.* New York: Harcourt, Brace & World.

Turner, J.H. 1987. "Towards a Sociological Theory of Motivation." *American Sociological Review* 52: 15–27.

Tushman, M.L., and E. Romaneill. 1985. "Organizational Evolution: A Metamorphosis Model of Convergence and Reorientation." In B. Staw and L. Cummings, eds., *Research in Organizational Behavior* 7: 171–222.

Van de Ven, A.H. 1986. "Central Problems in the Management of Innovation." *Management Science* 32 (5): 590–607.

———. 1980. "Problem Solving, Planning, and Innovation: Part I. Test of the Program Planning Model." *Human Relations* 33 (10): 711–40.

van den Daele, L.D. 1968. "Qualitative Models in Developmental Analysis." *Developmental Psychology* 1 (4): 303–10.

———. 1974. "Infrastructure and Transition in Developmental Analysis." *Human Development* 17: 1–23.

Von Hippel, E. 1978. "Successful Industrial Products from Customer Ideas." *Journal of Marketing* (January): 39–40.

Williamson, O. 1975. *Markets and Hierarchies.* New York: Free Press.

297

SECTION IV

Studies of Administrative Innovations

Whereas the previous section examined the processes of new business creation in the private sector, this section focuses on administrative innovations in both public- and private-sector organizations. Administrative innovations tend to be less tangible than new business creation; there are no hard devices, products, or prototypes in the case of administrative innovations. The "products" of administrative innovations are verbal or symbolic in nature—a law, a contract, a new organizational form, or a set of new reporting relationships. Administrative innovations are rather fluid, because—prior to final passage of a law, the signature of a contract, or introduction of a new organizational arrangement—there is only verbiage, subject to change and renegotiation. For this reason, administrative innovations create special challenges in managing attention to ideas and orchestrating people. All innovations require political management and coalition-building. But these are limit cases because they have no existence beyond the agreement and acceptance of the people involved. As a result, the study of administrative innovations brings into stark light several processes that show up less clearly in technological, product, or process innovations.

In Chapter 9 Roberts and King study the process of educational policy innovation via legislative initiative. Their case is a wideranging educational reform adopted by the state of Minnesota that some say may produce a second "Minnesota Miracle" in education. They conceive of the policy innovation process as a series of hurdles the innovation must clear. To pass each hurdle, innovators must transform the idea through an infusion of energy (money, time, effort). As each hurdle is passed, the innovation becomes more complex until it is sufficient to move into law. The driving forces behind this transformative process are key individuals who collectively advance, develop, and adapt the policy. Policy entrepreneurs initiate and advocate new ideas; policy champions lead political debate; policy administrators advocate and fine tune the policy during the implementation period; and policy evaluators protect policies during evaluation periods. Roberts and King focus particularly on policy entrepreneurs, who develop and push ideas that challenge received wisdom. Policy

299

entrepreneurs are the source of new ideas and give the innovation its basic shape. Interestingly, they tend to come from outside the system, and it is through the work of policy champions, administrators, and evaluators that policy entrepreneurs can bring about real changes in the status quo.

Ring and Rands examine how transactions developed in the evolution of 3M's microgravity research program with NASA. 3M and NASA were trying to work out contracts for unprecedented commercial projects, the sort of "gleam in the eye" projects that promise huge benefits at the cost of great risks. Ring and Rands focus on how governance structures regulating this unprecedented relationship evolve. They argue that these structures emerge through a transaction process that moves through three stages: negotiations, agreement, and administration. This global movement is mobilized and directed by emergent interpersonal transaction processes of sensemaking, understanding, and committing. Depending on the circumstances, the emphasis placed on each of the phases and the mix of informal interpersonal transactions in the phases vary. Ring and Rands recount a number of negotiations and derive some contingency propositions specifying that nature of global transaction processes under various conditions.

In Chapter 11, Bastien deals with mergers and acquisitions, perhaps the leading business phenomenon transforming the U.S. corporate landscape in the past five years. He starts with the premise that mergers and acquisitions may be traumatic for organizational members and outlines a common syndrome that threatens successful business combinations. The nature of this syndrome varies according to the type of merger and acquisition; Bastien distinguishes four types of business combinations and details how the syndrome may emerge in each case. The key to understanding the success or failure of combinations, Bastien argues, is to study the process by which mergers and acquisitions occur. He develops a multiple sequence tracking scheme to follow the processes of mergers and acquisitions in the acquired company. Based on his study, Bastien concludes that individual and organizational learning processes are central to the firms' adaptation to each other. He extends the traditional learning model to include cognitive, affective, and behavioral learning. Bastien argues that conditions conducive to effective learning result in fewer problems in mergers and acquisitions and documents cases in which blockages to learning led to multiplying problem cycles.

The cases in this section differ in terms of the institutional contexts under which the administrative innovations were undertaken. Roberts and King studied a situation in which there were many partisan interest groups and an extremely complex political issue and in which debate was carried out largely in public. By contrast, Ring and Rands have a case with only two organizational parties, but one that clearly shows the difficulties of crossing cultural boundaries of an entrepreneurial private firm and a large bureaucratic public government agency. In Bastien's mergers and acquisitions there were many parties, but they were arrayed on only two or three different sides of the issue, and transactions occurred largely in private. Despite these differences, there are communalities in findings across the three cases: all observed that many people and groups with different and partisan perspectives had an impact during the innovation process; that there was fluid participation among the major actors who changed often over time; all found that a lot of energy was used to maintain the continuity of transactions grounded in a fluid, intangible idea; and all found that transaction processes were complex and multifaceted, rather than a simple progression through preset phases. These and other differences and similarities show up clearly in the following chapters and provide interesting comparisons with other parts of this book.

9

THE PROCESS OF PUBLIC POLICY INNOVATION

Nancy C. Roberts

Paula J. King

Analyses of public policy innovation span a wide-ranging set of policy issues, and identify multiple antecedents and consequences of those innovative policies (Polsby 1984; Kingdon 1984: Schon 1971). Despite the range and diversity of the studies, however, there seems to be a fundamental point on which researchers agree. Central to their research is the acknowledgment of a group of individuals who, as Schon (1971: 55–56) described them,

Research for this chapter was supported from 1983 to 1985 from a grant to the Minnesota Research Program from the Organizational Effectiveness Research Program, Office of Naval Research (Code 4420E), under Contract No. N00014-84-K-0016. From 1986 through 1987 the research was supported by a grant from the Research Foundation, Naval Postgraduate School, Monterey, California. We would like to extend our thanks to Andy Van de Ven, Scott Poole, and Ray Bradley for their invaluable help on earlier versions of this chapter, and to all of those who so generously gave their time and insights to make this study a reality.

challenged the system, were irrationally committed to the inventions they championed, operated informally and subversively, exploited informal networks and mobilized outside pressures, engaged in life-long combat, and became heroes or martyrs to their cause.

Such individuals have been variously referred to as "champion," "guerrilla," "public entrepreneur," "revolutionary," and "missionary" (Schon 1971), as "inventor," "adapter," "policy entrepreneur," "broker," and "incubator" (Polsby 1984), or as "advocate," "broker," and "policy entrepreneur" (Kingdom 1984). Researchers have found them to be strategically involved with the whole process of policy innovation—as important sources of innovative ideas, as critical linchpins to galvanize a network of support for the ideas, and as essential advocates to push the ideas into the arena of debate for legitimation, adoption, and implementation.

We have chosen to refer to these individuals as policy entrepreneurs. These people, like their analogs in business, passionately stake their time, energy, reputation, and resources on an idea in which they deeply believe. Like business entrepreneurs, they risk "capital" for a return in seeing a desired policy enacted; they derive satisfaction from involvement in the policy process and seeing their values shape policy; they often experience such personal gain as enhanced credibility, legitimation of ideas, increased public visibility, or career advancement (Kingdon 1984). Unlike business entrepreneurs, however, they labor in the public arena—where profit is not a motive but where civic duty and interest are—forging new policies to redirect activity in the public domain.

Our purpose in this chapter is to describe how a constellation[1] of policy entrepreneurs initiated innovative public policy. Although most observers see them as "central figures in the (policy) drama" (Kingdon 1984: 189) and recognize their contributions to the policy process (Eyestone 1978; Kingdon 1984; Polsby 1984; Salisbury 1969; Walker 1974), we uncovered no in-depth studies of the policy entrepreneurs themselves—their activities, their strategies, their tactics. How specifically did they advocate ideas and serve as brokers in the policy process? How did they take advantage of propitious events or help to open up the "window of opportunity" in order to advocate their proposals and to push their new ideas? We believe our close scrutiny of these policy entrepreneurs over the past four years has provided us with a unique vantage point with which to examine their pivotal position in the innovative process.

Public policy innovation is a complex process, weaving together multiple parties and interests, all of whom compete for a hearing and acceptance of their ideas on the govern-

ment's decision agenda. As a whole, we view policy innovation as an attempt to mobilize an *innovative system*—a set of *ideas* connecting *people* in multiple *transactions,* the thrust of which is to forge a new policy or procedure to guide public action. This system is embedded in an *institutional context* and *environment*[2] that is constantly changing, providing new opportunities for some social actors, and setting up constraints for others.

This chapter provides a rich empirical description of how many people (policy entrepreneurs, policy champions, policy administrators, and others) developed and pushed a new set of ideas about educational reform into public policy and law. For this to happen, the ideas had to go through a set of logically and institutionally required hurdles as illustrated in Figure 9–1. These hurdles, which become increasingly complex, consist of idea generation and mobilization of support, a proposal to and endorsement by a powerful elite (governor), the drafting of a legislative bill, the transformation of the bill into law, and the administration and revision of a program in compliance with the new law.

Our analysis of this innovation process will be divided into four parts. The first section describes the methodology used in this longitudinal study, the second outlines the major activities performed by the various groups in the policy-making process, and the third examines the catalytic role played by policy entrepreneurs in the innovative policy process and poses questions for further research. We conclude in the fourth section with a discussion of the implications of this study for the future of public policy innovation.

METHODOLOGY

We began this study in 1983 with the goal of analyzing the leadership role in the innova-

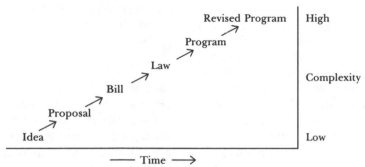

FIGURE 9–1. *Hurdles in Public Policy Innovation.*

tion process. The first author was involved in a longitudinal study of a leader who, as superintendent of education of a large, rapidly growing suburban school district, was the sponsor and champion of various innovations (Roberts 1985). Her appointment as state commissioner of education in late 1983 provided us the opportunity to observe the extent to which the commissioner could and would be able to continue her support of innovations at the state level. We envisioned a comparative study between the innovative processes at the state and district level, with a special emphasis on "institutional leadership" (Van de Ven 1986), in order to take advantage of the unique opportunity to hold the leader constant.

As events unfolded through 1984 and 1985, a group of individuals, eventually identified as policy entrepreneurs, surfaced as critical participants in the debate on state educational innovations. Our preliminary field work suggested it was they, not the commissioner, who were the initiators of innovation at the state level. Although it became clear the commissioner played an important role in the innovation process (Roberts and Bradley 1987; Roberts 1987), further observation reported below revealed strong evidence pointing to the policy entrepreneurs as the initiators of change.

This shift in emphasis highlights the difficulty a researcher in the field faces in identi-

fying key elements of the innovation process—particularly in the absence of a theory from which to derive and operationalize the major conceptual units (Blalock 1982; Lieberson 1985). We were more than a year into our field work before the important role of policy entrepreneurs in this system became apparent. Although initial field work pointed to the superintendent as pivotal for innovation at the district level, subsequent work revealed that this was less true at the state level. Indeed, the evidence made it clear that the initiators and mobilizers of policy innovation tended to hold *informal* positions outside the government, rather than formal leadership positions.

This lack of a guiding substantive theory and a viable operational model at the outset of a study makes a flexible methodological strategy imperative—one that can facilitate the discovery of useful conceptual units and appropriate operationalizations. Such an approach needs to be open-ended, aimed at uncovering and exploring the critical linkages in a social system. In the absence of clearly knowing what these might be a priori, more formalized methods, such as survey research, are inappropriate at this initial stage.[3] Consequently, we chose to pursue "naturalistic inquiry" (Lincoln and Guba 1985; Schatzman and Strauss 1973) to immerse ourselves in our research phenomenon in order to discover the parameters and properties of the policy innovation process.

Setting

The setting for our analysis is the Minnesota state capital. Our observations cover the time period from 1983 to 1987, while our archival research predates this period by at least fifteen years. Despite the difficulty in defining exact boundaries in a fluid system like state government, we decided our research focus would be the initiation, enactment, and administration of an innovative policy espoused by the policy entrepreneurs.

In defining this system, we do not use the terms *policy* and *policy innovation* lightly. We agree with Polsby's (1984:6–8) assessment that there is ambiguity about what counts as a policy and disagreement on what counts as innovation. There is "no standard that presently commands universal acceptance by which policies can be distinguished from non-policies or innovations from noninnovations" (1984:7). Given this current state of conceptual uncertainty, we will take our lead from Polsby, and recognizing the limitations of these definitions, define policies as "a course of action or a set of decisions leading to a particular course of action" (1984:7). Policy innovation will be defined as "relatively large-scale phenomena, highly visible to political actors and observers" that embodies "a break with preceding governmental responses to the range of problems to which they are addressed," and "unlike major 'crises,' with which they share the preceding traits, 'innovations' have institutional or societal effects that are in a sense 'lasting'" (1984:8). In what follows, policy entrepreneurs' ideas are shown to represent a large-scale, highly visible course of public action entailing a break from preceding governmental solutions. Moreover, following these ideas as they moved through the policy process (initiation, enactment, and administration) reveals that Polsby's three criteria for innovation are empirically satisfied. The successfully administered policy entrepreneurs' ideas can thus be regarded as innovative public policy.

Sample

The identification of policy entrepreneurs for this analysis was an emergent process. Without much guidance from the literature in helping us define policy entrepreneurs, we began with the definition of *policy entrepreneur* as one who espoused the "restructuring of education," especially those who advocated parent-student choice of schools and school districts. Further elaboration will show that this controversial idea was a dramatic departure from present educational policy.

The sample we drew with this very broad definition of policy entrepreneurs was very large. It included those working inside and outside of government, at many different levels, and in elected as well as appointed positions. In order for the concept to have meaning, however, we needed to make a finer-grained distinction among the restructuring advocates. The first distinction we made was between those who were the originators and developers of the innovative idea and those who mobilized support for it. This distinction helped somewhat; we were able to identify three individuals who were the primary formulators of the educational restructuring concept as it was applied to Minnesota.

We still faced a dilemma, however, because the list of those who mobilized support for the innovative idea began to grow longer and longer, especially by the time the restructuring bill was introduced to the legislature and the mobilization process extended throughout the state. Again, we needed a finer-grained distinction or we would have to eliminate the concept of support mobilization from our definition of policy entrepreneurs. We felt it was important to retain this concept, however, because in every case, the idea formulators also worked to mobilize support for their new ideas. The mobilization process appeared to be a very important aspect of their activities, especially since many of their creative ideas were more finely honed in the give and take of the mobilization process.

We resolved this dilemma by identifying two groups of individuals. The individuals in the first group were *outside* the formal apparatus of government and worked closely with the idea formulators to mobilize support. We added these six individuals to our original list of three and called them all *policy entrepreneurs*—social actors frequently outside the formal channels of government who seek the adoption of innovative policy into the public domain by formulating new ideas and mobilizing support for them.

All nine policy entrepreneurs espoused educational restructuring. All worked to mobilize support for this idea in their different capacities (policy analyst/former director of a citizens' group, educator/author, economist/professor/legislator, corporate executive/former director of a citizens' group, vice president/treasurer of a manufacturing firm, attorney/head of an interest group, director of a citizens' group, former governor, lobbyist for an interest group/former state senator). With the exception of the legislator, all were "outside" the formal positions of government. (The legislator was included because he was a source of innovative ideas).

For the purpose of this analysis, and to maintain confidentiality, we will treat these nine policy entrepreneurs as a *single entity* unless otherwise indicated in the text. Although there is variation among them in terms of their ideas, their activities, and their functions, space will not permit us an in-depth exploration of these differences at this point in time. Further research (King 1988) will investigate the variation among the nine policy entrepreneurs and their internal dynamics when they met as a group.

We called those who worked closely with the policy entrepreneurs to mobilize support *inside* the government *policy champions*. These individuals had formal governmental positions within the legislative and executive branches and were the focus of the policy entrepreneurs' influence attempts. Initially, in 1985 there were three policy champions—the governor and two key legislators. Convinced that the innovative idea had merit and was political feasible, they were in a position to serve as idea champions to give legitimacy and support to the ideas during the legislative process.

We made the distinction between policy entrepreneurs and policy champions because we found it necessary to differentiate between two categories of social actors who played a different, although complementary, role in the innovation process. Policy entrepreneurs, persons with expertise, competence, and special interest in educational policy, primarily concerned themselves with questions of policy substance and policy consequences. Policy champions (elected officials, their appointees, and/or administrators) concerned themselves with more immediate political considerations—enactment of the laws—especially those whose intention it was to to stay in office. Although policy champions did have some technical "expertise" in educational policy, they also had interests and other projects that competed for their time and attention. Policy entrepreneurs were more single-minded in their pursuits of educational policy changes; and in order to position their ideas and solutions on the decision agenda, their influence attempts were geared toward the policy champions, rather than the other way around.

Polsby (1984) also makes this distinction between the two groups in the policy process. Describing them as "specialists" and "nonspecialists," he refers to their relationship as a symbiotic one (1984:171–72). The specialist (policy entrepreneur) yields public credit of his ideas in exchange for the nonspecialist's (politician's) support during the enactment process. For his efforts, the nonspecialist gets new ideas and issues to champion in an election-dependent world. Thus, our sample of policy entrepreneurs, similar to Polsby's analysis, excludes policy champions.

Data Collection

Data for this research were drawn from multiple sources: interviews, archival records, and participant observations. One hundred four interviews were conducted with the policy entrepreneurs, the governor's staff, commissioner, representatives from various executive departments, legislators and legislative staff members, lobbyists, educators representing teacher unions and various associations for principals, superintendents, parents, and school board members, and members of various grassroots organizations and interest groups. These interviews spanned a time period from August 1983 to September 1987. Table 9–1 presents a time period/role matrix of the interviews conducted.

Archival material consisted of official state documents from the executive and legislative branches of government, interest group reports, and newspaper accounts. Research re-

TABLE 9–1.   *Time Period/Role Matrix.*

	Number of Individuals	Number of Interviews
August 1983–December 1985		
Executive agency official	5	26
Executive agency staff	4	4
Policy entrepreneur	6	6
Interest group member	6	6
Interest group lobbyist	1	1
Legislator or staff	6	6
Informed other	2	2
January–December 1986		
Executive agency official	1	2
Executive agency staff	2	2
Policy entrepreneur	6	6
Interest group member	3	3
Informed other	6	6
Media	2	2
January–September 1987		
Executive agency official	2	2
Executive agency staff	3	3
Policy entrepreneur	8	9
Interest group member	11	11
Interest group lobbyist	1	1
Legislator or staff	2	2
Informed other	2	2
Media	2	2
	Individuals = 58	Interviews = 104

ports from colleagues, doctoral dissertations, and studies from various research organizations also were available for analysis.

Both authors were involved in extensive observations of legislative hearings, formal and informal meetings of the Department of Education, interest group meetings, press conferences, and the governor's discussion group, an eighteen-month task force chaired by the commissioner of education. The task force, which brought together all the major educational interest groups in the state, including the policy entrepreneurs, was charged by the governor to create a "visionary proposal for state education."

ACTIVITIES IN PUBLIC POLICY INNOVATION

As a complex process, this public policy innovation involved a wide-ranging set of activities and included the collective efforts of many individuals. Although it is not possible to document all the activities in detail here, we have attempted to categorize them into a logically and institutionally required set of actions that are part of the lawmaking process: idea generation and problem framing to demonstrate how new ideas solve current problems; strategic thinking to evolve action plans in order to mobilize support for the innovative ideas; dissemination activities to reach the broadest audience possible; collaborative activities with high profile organizations to gain their endorsement; experimental activities to demonstrate that innovative ideas can work; initial legislative activities to float "trial balloons"; insurgence and intelligence activities to develop supporters within government; activities to enlist support of policy champions who will introduce ideas and carry them through the enactment process; outsiders' activities to gain grass-roots support for the purpose of lobbying the legislature; ac-

tivities to engage media attention and support; activities to administrate and evaluate the law; and finally, national-level networking activities to link state groups with others advocating reform.

These activities as they are presented do not follow an absolute chronology because many activities are concurrent and some are repeated. However, Table 9–2 outlines the major events as they flow through time.[4]

Idea Gestation Activities

In their capacity as writers, authors, analysts, researchers, and teachers, policy entrepreneurs traded in ideas, either of their own creation or as brokers of ideas from other sources and policy domains. They firmly believed in the power of ideas to shape the direction of history. One policy entrepreneur, an economist by training, quoted from John Maynard Keynes in *The General Theory of Employment, Interest, and Money* to illustrate how ideas can make a difference (policy entrepreneur 1985):

> The ideas of economists and political philosophers, both when they are right and when they are wrong, are more powerful than is commonly understood. Indeed, the world is ruled by little else. . . . I am sure the power of vested interests is vastly exaggerated compared with the gradual encroachment of ideas.

The central idea espoused by the policy entrepreneurs was a "restructured" or "redesigned" educational system. Rather than call for improvements in the current system, which they referred to as "tinkering around the edges," they advocated a total revamping of the present structure and a reexamination of its underlying assumptions or "givens" (policy entrepreneur 1985). One policy entrepreneur said the central issue was "whether education can be substantially better without the school system having to be significantly different" (Public Ser-

309

TABLE 9–2.   Timeline of Events in State Education Reform.

1980–82	Development of the innovative concept, restructuring education.
1980–81	Creation of Design Shop (public schools initiatives) to test out specific innovative ideas as part of the restructuring process. Financial support obtained from foundations to fund experimentation.
1982	"Levi law" passed that permitted school districts to enter into agreements with postsecondary institutions to allow students to enroll in postsecondary courses.
1982	Publication of the Citizens League Report "Rebuilding Education to Make It Work," which supports and develops the concept of educational restructuring and recommends parent-student choice as a mechanism to drive the restructuring process.
1984	Introduction of legislation to test the interest in choice, as a mechanism to restructure education.
1984	Publication of the Berman-Weiler study for the Minnesota Business Partnership, which advocates restructuring education and the concept of choice.
1984	Policy entrepreneurs appeal to governor to champion educational restructuring and the concept of choice.
1985	Governor unveils proposal in January that calls for educational innovation and parent-student choice in selecting and attending any public primary and secondary school in the state.
1985	Postsecondary Enrollment Options Act passes in June. Gives option (choice) to all junior and senior high school students to attend any postsecondary school in the state, receiving both college and high school credit at state expense.
1985	Governor sets up an eighteen-month Task Force on Education to develop a "visionary proposal for state education" in June. The task force includes every major educational group in the state as well as the policy entrepreneurs. The goal is to expand the efforts at innovation working with the "educational establishment."
1986	Preliminary evaluation of Postsecondary Enrollment Options Act completed in February.
1987	Final evaluation of Postsecondary Enrollment Options Act. Program approved and continued in February.

vices Redesign Project 1984:1). The policy entrepreneurs firmly believed that it could not be. This stand represented a challenge to the conventional wisdom in education and necessitated reframing educational problems to allow for a new set of solutions.

School system redesign actually represented a cluster of ideas that revolved around four themes: *empowerment* of parents and students to choose the school district or program of their choice based on the unique learning needs of the student (Nathan 1984); *deregulation* of state mandates in order to reduce red tape, permit more local control over the schools, and encourage creativity, innovation, and change; *decentralization* of decisionmaking authority from district-level administrators to local school site principals, teachers, and parents in order to empower those most intimately involved with student learning; and increased *accountability* in the educational system through expansion of outcome-based testing aimed at determining what students learn to ensure that quality education is being delivered.

The mechanism to drive this restructuring was "parent-student choice of school district." "Choice is the mechanism that drives all other changes. . . . It provides the incentive to change," said one policy entrepreneur (1985). By giving parents and students the options of *exit* (Hirschman 1970) from one district and/or school and *entry* into another, a market for educational services would be created. "Market forces" would then encourage educators to introduce innovative, high-quality programs to attract and retain students. If a school and its teachers were not able to deliver on their educational promises, over time they would lose students and the foundation aid that would follow student enrollment. The more effective schools and teachers would benefit as students would be drawn to their programs and classes. Less effective teachers and schools would be compelled to improve or risk loss of financial resources, decline in student enrollment, and state receivership. "Consumer sovereignty" in education would then provide a self-correcting lever to introduce innovations, improve performance, and ensure accountability (policy entrepreneurs 1984–86).

Sources for the policy entrepreneurs' ideas were many and varied. As brokers of ideas they "scanned the environment" in search of ideas, patterns, or models that could be applied to education from other policy areas. For example, several policy entrepreneurs frequently cited cases from the private sector, where market forces, competition, deregulation, and decentralization were common themes. In their writing, they emphasized consumer benefits derived from deregulation of health care, airlines, telecommunications, and financial services. Applying the same logic to education, they viewed education as a "regulated monopoly" and called for an end to the monopolistic hold school districts had on students within their boundaries.

Another borrowed analogy was the "ailing industry" theme. A Department of Education official said he was further convinced of the correctness of the policy entrepreneurs' analysis when one entrepreneur used the following example (policy entrepreneur 1985):

> Schools are like the steel industry or any other ailing industry. . . . If they are ailing, they ask for protection . . . they want import quotas, they want tariffs, they want anything to reduce the amount of competition on the system so that

only they can sell their product in the marketplace.

Other sources of ideas were their previous experiences in initiating large-scale system change and their own creativity in that process. We found evidence that not only did policy entrepreneurs broker ideas, but they also invented them. One entrepreneur was described as "visionary," a person "ahead of his time" for the new and provocative ideas he introduced (informed other 1986). Long active in policy analysis, he was described in the following way (informed other 1986):

> . . . just phenomenal. He is very creative. . . . He can come up with a comparison or critical question that jars your sense of the 'givens.' . . . Most of us assume the presence of boxes without even seeing the walls. . . . He pushes the walls down and looks beyond. . . .

The "givens" called into question in this instance were district boundaries, teacher training and tenure, certification procedures, the traditional governance structure, the state's core curriculum, the length of the school day and year, the use of paraprofessional help in classrooms, the loose system of accountability and the reward structure (policy entrepreneurs 1985).

Problem Framing Activities

It was not enough for the policy entrepreneurs to challenge the givens and trade in ideas in the abstract. To ground their ideas and make them more compelling, they needed to present their ideas as solutions to particular "problems" in the political context. Public problems, however, as we are reminded by policy theorists, are socially constructed. Multiple problem definitions are feasible. The issue is not which problem definition is correct but which is "most credible and politically acceptable at any partic-

311

ular time" (Cobb and Elder 1983:173). Thus the challenge for the policy entrepreneurs was to shape the definition of the problem in such a way so as to make it viable in the current context and do so in a way to establish a clear link between "the problem" and their ideas or proposed solutions.

They accomplished this problem framing in several ways. Drawing on educational theory and research (Boyer 1983; Goodlad 1984; Sizer 1984) and citing one theorist, Anthony Downs (1956), they drew parallels between education and similar "ills" typical of other "mature bureaucracies." Education, they claimed, suffered from structural inertia, lack of accountability between money spent and outcomes produced, excessive rules and regulations, attempts to stake out, defend, and expand a certain "territory of policy," goal inconsistency, inaccurate information about performance, protective self-interest, and reduced flexibility and responsiveness to those served. To correct for these systemic problems in education, nothing less than total system "redesign" was called for. It would not do to engage in "piece-meal tinkering" with system parts. Current strategies to "improve" education were "well-meaning but misguided and costly." A system that had been designed for an earlier industrial age had been pushed to its limits and needed a complete overhaul (policy entrepreneurs 1985). By cataloging the ills of mature bureaucracies, they then justified the need for outside pressure, claiming that self-interested bureaucrats would not voluntarily disrupt the status quo. As one policy entrepreneur said, citing Goodlad, "the cards are stacked against educational innovation" (policy entrepreneur 1985).

To further frame the problem and illustrate the limits to the current educational system, the policy entrepreneurs would reference data from a variety of reports and documents on the status of the nation's schools. For example, citing *A Nation at Risk* (1983), they would point out how serious the deficiencies of the current national system were. Again, they maintained, only a "reconfigured" system with a different set of operating assumptions could correct these deep-seated problems and address the current and future needs of the state and country (policy entrepreneurs 1985).

The policy entrepreneurs also argued that the current economic situation challenged reformers to find methods to change the educational system without adding excessive costs to the taxpayers and business community. In their opinion, those who advocated improvements in the current structure could offer only two choices for policymakers—either ante up more money for services rendered or suffer a continuing reduction in the quality and quantity of educational services delivered. From the policy entrepreneurs' perspective, neither alternative was deemed acceptable in the current political and economic climate. The policy entrepreneurs' position was that much more could be done with the present level of expenditures by ensuring greater accountability in the system, and creating new levers of change—choice for parents and greater control and participation by teachers and local schools. Thus, "the problem," as they defined it, should be shifted from "how education can be improved" to "how it should be restructured and redesigned to raise quality and accountability" (policy entrepreneurs 1985).

In order to expand public awareness and to mobilize support for their ideas, managing their ideas into "good currency" (Schon 1971), policy entrepreneurs engaged in a number of additional activities.

Strategic Activities

The policy entrepreneurs tended to organize themselves into a group of individuals for the purpose of brainstorming strategies and preparing an operational roadmap for change: *"long-term strategy"* to guide their overall efforts and *"short-term tactics"* to cope with day-to-day changing political realities. As one policy entrepreneur noted, long-term strategy provided the overall direction of the change effort; it answered the *how* part of the question of "what ought to be" (policy entrepreneur 1986). Short-term tactics often centered on the legislature and the shifting political mood. The tactical strategy was frequently modified to fit the evolving political context, and to clarify what was attainable given that mood. As one policy entrepreneur summed up the group's focus (policy entrepreneur 1985, emphasis added):

> What we're doing . . . is both identifying short-term tactics and long-term strategies . . . or how we make this happen. *Nothing like this happens that isn't orchestrated.* This doesn't fall into place. You've got to decide where to push, when you push, and what your position is.

The policy entrepreneurs usually met once a week during the legislative session in 1985 and regularly kept in contact by telephone when the legislature was not in session. Attendance at the meetings was fluid depending on the agenda, and any one of the policy entrepreneurs could call a meeting. Strategy meetings would attract the core group of policy entrepreneurs, normally six or seven individuals.[5] It was this core group that convinced the governor's executive team, who in turn, encouraged the governor to champion choice in the legislature (see below).

Other sessions were of an informational nature. "Informed others," such as strategically placed midlevel Department of Education managers, legislative staffers, education consultants, foundation staff members, representatives of business firms, lobbyists, or sympathetic members of the education establishment would be invited to sit in. They would be asked to fill any "information gaps" that existed and provide up-to-date feedback on the inclinations of key legislators, the governor, and education groups.

Policy entrepreneurs stepped up their pace when they had to gear up for the 1985 legislative session. Said one observer, "[They meet] much too often. . . . [Name] stops by here at least twice a week to fill me in" (Executive Agency Staff 1985). Yet this more frequent contact was necessary to coordinate all groups with which they were involved (see below) and to integrate their actions with groups who supported their cause (Public School Incentives, Minnesota PTA, Minnesota Association of Secondary School Principals, and League of Women Voters). Building greater coalitional support and broader appeal also was essential in this case because of the resistance of the "6M" group, a coalition of six education "establishment" groups formed in late 1984 to respond to the report issued by the Minnesota Business Partnership (see below). This "6M" group waged an intense and often bitter campaign in the media and in the legislature in 1985, attacking the governor and those who supported his innovative ideas. As the debate intensified, so did the need to meet and plan strategy.

Dissemination Activities

As part of their strategic efforts, policy entrepreneurs engaged in a continual effort to disseminate their ideas to as wide an audience as possible. They used a variety of media such as books, articles, newsletters, newspaper reports, and editorials to circulate their ideas. Forums to discuss their ideas ranged from informal discussion groups, to more formal ones such as public speaking engagements, university courses, leadership development programs and television debates. Our respondents also cited the policy entrepreneurs' telephone calls, personal contacts, "FYI" articles and documents sent for their assessment, and "special meetings" that brought together national education

313

"experts" with state and local political and educational leaders. One government official called one of them "phenomenal—the most effective organizer I've ever run into" (Department of Education member 1985):

> He just makes a couple of phone calls to people and says: Gee, I'm having a little gathering. . . . Would you like to come? The governor shows up. Everybody shows up.

Collaborative Activities with High-Profile Groups

In order to gain greater credibility for their ideas, the policy entrepreneurs often worked with other organizations and groups, especially those that had high visibility and good standing within the larger community. The objective was to encourage others to sponsor studies on the status of education and to propose recommendations for change. Two notable organizations were the Citizens League[6] and the Minnesota Business Partnership, both of which issued reports recommending changes that were compatible with ideas espoused by the policy entrepreneurs.

For example, the Citizens League, an "independent, nonpartisan, nonprofit, educational corporation dedicated to understanding and helping to solve complex public problems" (Citizens League 1976), issued a report in 1982 challenging the conventional wisdom of education by stating, "it will not suffice to merely pump more money into the same old system even if there were a willingness to do so. Instead, the system itself must be rebuilt" (Citizens League 1982:1). The report recommended a new structure that must

> Give parents—who should be the key decision-makers in buying education—more choice in what to buy. To put it another way, public

educational dollars should follow parents' choices about which schools or educational services to use.

Place more authority for shaping education at the place where it happens—the individual school.

Remove artificial barriers to excellence and encourage innovation, competition and entrepreneurship. Somehow, people in education must have the chance to break out of their stifling constraints, and others with new techniques and new technologies must have the chance to apply them to education.

A second organization was the Minnesota Business Partnership. Founded in 1977 by a group of chief executive officers of major corporations, its mission is "to identify and analyze the state's longer-range economic issues and help set priorities and plans for action" (Minnesota Business Partnership 1984:1). Its specific goal was to help create a political consensus on major economic issues among business, government, and other groups in the state. The "touchstone" was to increase economic prosperity for all citizens measured in jobs and growth in personal income, both of which depended on the quality of education in Minnesota. As they saw it, a strong educational system would maintain and even attract more business and industry and thus keep the state economically viable.

The Partnership selected Berman-Weiler Associates from California in 1983 to conduct an eighteen-month, $250,000 study of Minnesota schools with the following aims:

1. An assessment of how well students are prepared for college, work, and citizenship, and a description of the costs of the K through 12 system.
2. A diagnosis of the problems and needs of elementary and secondary schools, particularly in areas amenable to public policy or private assistance.

3. Formulation of a plan for strengthening education in Minnesota.

Berman-Weller issued its report, entitled "The Minnesota Plan: The Design of a New Educational System," in November 1984. The thrust of the report was that Minnesota had "a good system that can do much better" (1984:iv). Citing structural impediments in the current system that prevented high student performance, the report became an immediate standard bearer for the restructuring cause. Its summary was direct and to the point: "Despite its strengths, Minnesota education, like American education, has reached its limit of effectiveness. Adding more money will not help. Nor will tinkering. It is time for major restructuring" (1984:18). The improvement course of action that relied on such strategies such as raising state requirements, policing the teaching profession, and increasing the funding for K through 12 education would not solve Minnesota's shortcomings in education: Youth were learning well below their potential and not mastering "higher-order" skills needed for the increasingly complex world. Instead, the report recommended restructuring efforts or "deep reform" in (1) the structure of schooling, (2) the organization of instruction, and (3) the system's responsiveness to change.

"The Minnesota Plan," as the report was eventually called, framed educational problems in much the same way as had the policy entrepreneurs. In fact, one policy entrepreneur, who worked for the Partnership, had key input into the design of the study. He described his involvement in the following way (policy entrepreneur 1985):

> We [MBP] spent a lot of time negotiating with them [Berman and Weiler] about our expectations that were exactly—almost completely contrary to their previous work. . . . We felt, in the California work, that the data assessment part of it was excellent. We wanted to replicate that

here, that is, to really come in and critique and analyze what was happening here. That part of it of course was going to be identical to California. But the strategies that were appropriate here are *dramatically different* than California. . . . [We asked,] How do you spend money more effectively and how do you manage a more effective school system at a time of conserving resources? We explicitly said to them in our negotiations that spending more money was off the list.

It was not surprising, then, that the report included some policy entrepreneurs' recommendations for change: give eleventh- and twelfth-grade students *choice* that allowed them to use state-funded "stipends" for educational programs offered by either public or private vendors; decentralize authority for school governance and management; and measure student performance on statewide tests to assess how well students had mastered learning in core courses based on individualized instruction.

Experimental Activities

In the early 1980s, two policy entrepreneurs and some change-oriented individuals with strong ties to the Citizens League, began to discuss "dramatic" strategies to improve education. The group eventually formed *Public School Incentives* in 1981, the purpose of which was "to translate innovative, high potential ideas into demonstration projects to test their efficacy and potential for success in Minnesota schools" (PSI 1985). Results of these projects, if "successful," could be used to argue for widespread dissemination of the innovations, or at the very least, understand how and why they "failed" in order to avoid problems. This effort was timely as the national spotlight in 1983 was beginning to turn toward education, and members of the group sensed an opportunity to air their ideas (PSI member 1986).

315

Because resources were needed to translate ideas into tested programs, PSI also served as a fiscal agent for foundations willing to support innovative projects in education (PSI member 1986). A total of $1.2 million was eventually funneled through the organization, much of it studying and testing policy entrepreneurs' innovative ideas such as school-based management and parent-student choice. For example, PSI served as a fiscal agent when it received $135,000 from the McKnight Foundation over a two-year period; the purpose, in part, was to study various choice options and identify details of plans that would benefit students who had not succeeded in schools.

One of the policy entrepreneurs, an educator on leave, joined PSI as a part of the McKnight grant. Within a year he completed a book on educational reform that was widely disseminated at the state and national level (Minnesota Senator David Durenburger gave a copy to each of his Senate colleagues). The book summarized parent-student choice models and contained examples of choice (magnet schools, choice within and between districts, and private school voucher systems) (Nathan 1984).

PSI actively worked for expanding choice in Minnesota. The 1983 "voucher bill" (see below), advocating a limited voucher system for low-income families, was moved along by testimony of sympathetic superintendents, minority, and parent groups and coordinated by a policy entrepreneur and PSI member. Expanding choice, in their view, had the potential to break the hold of the regulated monopoly of schools by helping educators and disadvantaged students seek out learning opportunities that best met student needs and goals (PSI member 1986).

Initial Legislative Activities

In 1982 Representative Connie Levi authored a bill that eventually was enacted and came to

be known as the Levi law. It permitted school districts to enter into agreements with postsecondary institutions to allow students to enroll in postsecondary courses. Although some districts took advantage of this option, most did not. Disappointed with this response to her legislation, she authored revisions to the bill in 1984 but again was frustrated with the lack of action. By 1985, viewing the postsecondary option of the governor's bill as an extension of her earlier efforts and one that had a better chance for success because it gave parents and students, rather than the districts, the choice for postsecondary enrollment, she agreed to carry the governor's bill in the House (see below).

The policy entrepreneurs' also were actively involved in educational reform in the legislature in 1983–84. The Citizens League's report, "Rebuilding Education to Make It Work," had initially met with resistance from policymakers who refused to endorse its "voucher" plan—a controversial proposal that opened up the issue of public aid to private education. However, a state representative (and policy entrepreneur) ultimately introduced a modified version of the report that recommended a voucher option for students of low- and moderate-income families. These vouchers, according to the bill, could be redeemed in any private or public school of choice, as long as the school met certain selection criteria set up by the state.

The bill gave the topic of vouchers visibility and focused discussion on an issue that previously did not have enough legitimacy to warrant a hearing. According to one analyst, it was a "door opener" (Wilhelm 1984). Although it received support from a broad coalition of people of color, low-income whites, and others who worked with low-income families, and testimony was taken in the House hearing, neither proponents nor opponents of the bill called for a vote. "Nobody ever expected the bill to pass but everyone expected it to stir up a thoroughly interesting conversation about how you were going to change things," said one insider (Wilhelm 1984:257).

According to some observers, however, the bill did make an impact. Although it did not sell the idea of vouchers, it was able to enhance the legislature's receptivity to work for major educational change by softening up the system (Wilhelm 1984:266). It also created a loose, quasi-coalition of minority people, the disadvantaged, and discontented educators who had wanted alternatives to the present system and who had the potential to become activated in the future.

Insurgence and Intelligence Activities

The policy entrepreneurs were aware that the major structural changes they advocated would be difficult to initiate "outside" state government, especially given the power of the teacher organizations in the state. Although one policy entrepreneur was in the legislature, the rest were outside the formal governmental system. The support of strategically placed insiders would be important to help them develop and advance their ideas. Two examples are particularly important in their efforts to develop a governmental network of like-minded reformists.

In the first case, a policy entrepreneur put forward a district superintendent's name as a possible speaker for a presentation to state legislators. Her "Horizon speech," as it was called, attracted the governor's attention, and with some additional encouragement from policy entrepreneurs and others, the governor nominated the superintendent to be his commissioner of education. As a self-proclaimed change agent who shared some of the policy entrepreneurs' restructuring ideas, the commissioner's very visible and proactive stance toward educational change helped disseminate the concept of system redesign even further (Roberts 1987).

In another case, a member of the Department of Education (1985) reported his ties with the policy entrepreneurs since 1980. He had been a participant in a university-sponsored leadership development program in which several policy entrepreneurs taught. The initial contact exposed the staff member to the policy entrepreneurs' ideas (which also were compatible with his personal convictions about education and the need to change) and convinced him of the "logic of their arguments." The policy entrepreneurs stayed in touch with the staff member throughout his tenure in office through telephone calls, meetings, and discussion groups, often sending him articles and readings to keep him abreast of the latest developments in education. He, in turn, became a source of inside information on Department activities, keeping the policy entrepreneurs up-to-date on the critical developments (see below). Committed to restructuring, he also was invaluable in getting legislation championed by the policy entrepreneurs passed (see below). His efforts, most policy entrepreneurs agreed, were "crucial" (Policy Entrepreneurs 1987).

Enlisting Policy Champions in the Executive Office

Enlisting the support of executive and legislative leaders to champion the policy entrepreneurs' ideas in the legislature was deemed vital to their cause. As one policy entrepreneur commented (Mazzoni 1986:33):

> It was very clear . . . with the strength education groups—the organized groups—enjoyed in this state, we weren't going to get anywhere without a major, popular political leader making it his policy. It was absolutely of basic importance to find out if the Governor would undertake this cause.

Thus, securing the governor's advocacy became a very important political strategy: "First . . . getting education to be a top priority; second . . . getting them [the governor's executive team] to buy this plan" (Mazzoni 1986:33). The opportunity came along in the fall of 1984.

During the planning process that began in the summer of 1984, Department of Education representatives had proposed numerous ideas that could not pass the close scrutiny of the Department of Finance and the governor's executive team. The ideas were rejected as "too costly" in a period of budget constraint. They also were criticized as "line items" lacking "coherence" when an integrated educational "package" was needed. The finance commissioner, who also had close ties with the policy entrepreneurs, wanted something new. He opposed "throwing" more money at education, which was already the largest single budget in the state government. As one member of the Education Department expressed it, there was a "fundamental communication problem." Department of Education representatives did not understand the executive team's overall budget philosophy, and the team did not respond favorably to the policy recommendations coming from Education (Department of Education member 1985).

The stalemate wore on through the fall of 1984 and was exacerbated by the department's lack of access to the governor's policy team. One Department of Education member voiced frustration and concern: "We can't get in. The executive team just blocked us out. We can't get at them" (Department of Education member 1985).

Then on December 6, 1984, a very important call took place. A policy entrepreneur "tracked down" an Education Department representative who was out of town on agency business. He asked how things were going: Was the department going to have any new proposals? The representative explained the department's difficulties and why it was impossible to get its ideas before the executive team. He remembered his words to the policy entrepreneur this way (Department of Education member 1985):

We're not going anywhere—we're just dead, finished. The budget's got to be done and Finance makes the decisions and we can't get to the executive team.

When asked if the policy entrepreneur should try to intervene, the response was, "Yes, if you can get into that [meeting] you could do something." The department's representative recalled his advice to the policy entrepreneur (Department of Education member 1985):

If you want to change education policy in this state at this moment in time, you have to get to the executive team. You have to get to their meetings and you have to come with a fair amount of force and bring people with you. . . . You'll have to represent us [Education] because we can't be there. . . . We're dead—just finished—can't move. The executive team blocked us off. In fact, they not only blocked us off, they're ridiculing us. . . . We can't get anywhere. You'll have to break it open or something.

On the following day a select group of policy entrepreneurs had a meeting with the governor's executive policy team. The Education Department representative recalled the results of the meeting as they were later recounted to him (Department of Education member 1985):

For whatever he [policy entrepreneur] did after we talked, he arranged to get a select group of people and they had an audience with the executive team. And [he] sold them on choice. I mean, to the point where . . . [member of Finance Department] loved it so much he couldn't stop talking about it. So now all of a sudden, the door's open for us.

The meeting continued for about two to three hours. As one insider said, it "altered Minnesota's education agenda" (Department of Finance member 1985). Said another, "that

318

meeting probably turned around the willing-ness of [name] to introduce something. . . . I think [it] probably pretty well excited most of the group present" (Department of Finance member 1985).

By December 10 the Education Depart-ment formally heard news from the executive team that "we have found life." By December 17 they had pulled together a package to brief the governor. He was so enthusiastic about their proposal that they were given the go-ahead to draft a formal proposal that he would introduce to the legislature (Department of Ed-ucation member 1985):

> We got the green light on all of the ideas—learner outcomes, testing—we got the green light on everything. And off we went. And I mean, that was the quickest four days you'll ever see. . . . This was not planned change. This was serendipity. It just happened to be at some given time all the right things came together and it was sold entirely on an informal network. There's no formal procedures in that whatsoever. . . .

The governor surprised his advisors in his response to the new educational proposal presented in the briefing from the Finance and Education Departments. Said one member of the Finance Department (1985),

> When we went in—in fact we talked about how we would approach him on this issue—and we went in in a very quiet, soft, easy [way]. 'Here's the kind of steps you've been taking. We're interested in this kind of thing.' And frankly, *he* jumped on the bandwagon. The bandwagon started moving quicker than we thought possible. He liked it so much, he started dragging us. Honest to God, he started dragging us.

The governor unveiled his innovative proposal, "Access to Excellence," on January 4, 1985, several days before his State of the State address. He wanted a separate announcement

to dramatize the beginning of what the news media would describe as his "crusade" for edu-cation.

The basic ideas in the governor's pro-posal were almost identical to those advocated by the policy entrepreneurs, particularly the "open enrollment" provision that called on parents to assume greater responsibility for the education of their children. Beginning in the 1986–87 school year, students in all eleventh and twelfth grades would be allowed to select the school of their choice—either a public school in another district or a postsecondary institution. State aid would follow that choice. By the 1988–89 school year, students in any grade would be able to choose the public school district they wanted to attend, and state financial aid would follow (governor 1985). This and other aspects of "Access to Excel-lence," such as greater state financing, evalua-tion of student outcomes, and reduction of state mandates were virtually identical to the policy entrepreneurs' ideas on restructuring education.

Enlisting Policy Champions in the Legislature

The governor met some resistance from key legislators on his proposal. Some objected to the "choice" provisions, others resented his in-trusions into the issue domain of educational policymaking. Historically, *education policy was made by a few legislators on key education* commit-tees, their staff, and experts from the Depart-ment of Education, and educational lobbying groups. By initiating a proposal without con-sulting these traditional actors, the governor challenged legislative power and violated norms for educational policymaking (legislator 1985).

Partisan politics was also an issue. A for-mer chair of a key education committee refused to sponsor the bill because the opposition party was asked to introduce it first. He did not want to be on the "second team." He was especially incensed that the governor did not consult with legislators in the development of his proposal, nor did he seek support from his own party before he went to the opposition, even though his party was a minority in the House at that time (legislator 1985).

For a while it looked as if the governor's proposal would founder for lack of a policy champion in the legislature. Policy champions (whom we define as those who introduce and sponsor legislation) were very necessary to shepherd a bill through to enactment. If they were not institutional leaders, they at least had to be able to negotiate for the support of these leaders to leverage and broker power effectively. For as one astute observer of the legislative process in Minnesota commented (Mazzoni 1986:63),

> You can set the agenda, and you can move an issue forward, but you can't really exercise any political muscle unless you have the Speaker and Majority Leader as a strong proponent willing to help you out. That's where the trading is going to help you out. That's where the trading is going to take place in terms of getting things in the bills, and advancing things that legislators might want during the session.

After some negotiation, the Independent Republican House majority leader agreed to be the chief House sponsor of the governor's bill. In the Senate, the principal author became the chair of the Subcommittee on Education Aids and a member of the governor's party. These two legislators joined the governor as policy champions and took up the banner for "choice" in the schools.

From the governor's initial introduction to "Access to Excellence" on January 4 until the conference committee report approving parts of his bill during a special session in early June, the governor labored with enthusiasm and high energy. "I haven't worked for any-

thing this hard," he was reported as saying (Mazzoni 1986:91).

There were multiple centers for the choice campaign emanating from the governor's and the House majority leader's offices. They brought in, on a weekly basis, the advocates of choice—representatives from the Minnesota Business Partnership, the Citizens League, the Department of Education, the policy entrepreneurs, educators and special interest groups (see below) that supported open enrollment. The goals were to coordinate the actions of this growing coalition, plan some common strategy, build support among the public and the legislature, and counter the opposition of the 6M group.

The governor was personally involved in the campaign. Often accompanied by the commissioner, he visited some twenty high schools in over two and one-half months to champion his program. His media campaign also included a "flyaround" five-city state tour in April with the 3M CEO, the head of the Minnesota Business Partnership's task force on K through 12 educational reform. It too received excellent press. A good "media fight" with the opponents of "choice" also made headlines as the debate between the governor and the "education establishment" heated up. Media coverage from all of these activities was extensive both in the Twin Cities and throughout the state.

Besides these attempts to galvanize public support for open enrollment, the governor waged an "insider's" campaign to lobby key power brokers within the legislature, especially those members who would review his bill in the various House and Senate committees. At least one of his personal lobbying efforts made a difference as it was reported in the press (Sturdevant March 26, 1985: 1A, 9A):

> [The Governor] thought he was one vote short of the six he needed to get the proposal through its crucial first test. So he went to work. One by one, [the Governor] called six of the ten mem-

bers of the Education Aids Subdivision to his office for a little personal lobbying. One of these was DFL Representative [name], a critic of the idea. . . . Hours later, [name] a leading member of the Committee's DFL minority, told the Subdivision that he changed his mind. He said afterward the conversation with [the Governor] made a difference.

The two other policy champions, the House majority leader and the chair of the Education Aids Subcommittee in the Senate were also hard at work. The Senator was described as "trusted and popular," "one of the best," "very fair," and "always accessible" by his colleagues and lobbyists with whom he had contact (Sturdevant April 1, 1985: 9C, 10C). Yet he faced the distinct disadvantage of introducing a bill that did not have the backing of his party's Senate leader, nor its most senior member and well-respected former chair of the Education Committee. In fact, the former chair of the Education Committee became a chief sponsor, and the majority and minority leaders in the Senate became cosponsors of an alternative to the governor's Access to Excellence bill that eliminated the controversial "choice" options for students (Wilson March 23, 1985:7B).

Despite resistance to open enrollment among the leaders in his party, the Senator demonstrated his leadership and was successful in the first couple of rounds. The most controversial element of the governor's bill, open enrollment, passed the Senate Tax Committee, the Senate Education Aids subcommittee, and the Senate Education Committee. Yet this element of the bill was dealt a "mortal blow" when it was defeated in the Senate Finance Committee. It was estimated that it lacked four votes necessary to keep it alive (Sturdevant April 25, 1985:1B).

The majority leader, and policy champion in the House, had several advantages in maneuvering the bill through the legislature that her colleague in the Senate did not have.

She had a commitment to education and a history of success with legislation similar to the governor's (see above). Also, as House majority leader of the Independent Republican party, she was in a good leadership position to bargain with the governor and others for her sponsorship of their bills. She was known as a very effective leader and shrewd negotiator. The DFL governor in particular needed her support and cooperation in order to pass his legislative agenda in the IR-controlled House (legislator 1985).

The House majority leader achieved her first victory when the House Education Finance Division got the open enrollment proposal accepted by a close voice vote. However, she was quick to credit the governor and the Minnesota Business Partnership for their lobbying efforts: "I'd say the governor's office and the Minnesota Business Partnership made the difference today," she is quoted as saying (Sturdevant March 26, 1985: 9A). That feeling was shared by other observers as well. The ties between the Independent Republican legislators and the business leaders gave the plan more credibility than it might have had coming from a DFL governor (Sturdevant March 26, 1985:9A).

The next go-round in the full House Education Committee was not as successful. Despite vigorous lobbying, the open enrollment proposal was defeated by a 14 to 13 vote. The House majority leader's assessment was that "open enrollment may have died for this session with . . . [this] vote" (Sturdevant April 4, 1985: 11A). She went on to add that she doubted that any effort to put the option back into the House bill would succeed. Even if open enrollment stayed in the Senate's version, "the chances of bringing it back from conference committee are very slender" because the most likely House conferees oppose it (Sturdevant April 4, 1985:11A). "We gave this a good shot," she said. But the proponents of open enrollment were up against "probably the most

321

powerful lobby in the state. . . . I'm pleased that we did as well as we did" (Sturdevant April 4, 1985: 11A).

What the House majority leader did not mention, however, was that although open enrollment for grades one through ten failed, the postsecondary option provision that was "her baby," as others called it, passed. Not only were eleventh- and twelfth-grade students able to take courses at state postsecondary institutions at state expense, for both high school and college credit, but the option was expanded to include postsecondary schools in Wisconsin, North Dakota, and South Dakota, states that had reciprocity agreements with Minnesota. Little attention was given to its passage in the media, however.

The final deliberations on the governor's education bill took place in what some analysts have referred to as the "Third House," the conference committees that reconcile differences between the House and Senate versions of a bill. Bills coming out of conference committees are negotiated settlements among caucus, committee, issue leaders, and the governor in the form of a nonamendable report (Hanson 1985).

In this case, however, agreement was not forthcoming. The DFL-controlled Senate and IR-controlled House were deadlocked in conflicts in May and June over school funding, distributional questions, and the governor's demand for an arts school. The House majority leader's earlier assessment had been correct. There were too many opponents of open enrollment on both committees to reinstate open enrollment in the conference committee.

When discussions had completely broken down between the two groups, however, the House majority leader stepped in as a mediator in the disputes. She teamed up with the DFL leader and co-chair, the senator who had sponsored the legislation in the Senate, and together the two policy champions who had originally sponsored the governor's bill worked

out a compromise. Postsecondary enrollment, originally excluded in the Senate and passed in the House, would remain. In addition, a state high school for the arts, and testing to assess student learning, part of the governor's original proposal would be included in the legislation.

Except for the open enrollment proposal that was defeated, these policy champions were able to strike a compromise that looked very similar to the governor's initial bill, and that, in turn, was virtually identical to the policy entrepreneurs' proposal for educational system redesign. Thus, the governor, the majority leader, and senator acting as policy champions, were successful in moving the innovative ideas through the decision agenda of the legislature. Although they were not the originators of the ideas, they certainly were key in mobilizing and legitimating support for those ideas in the larger societal context and bringing pressure to bear on the legislature. Without their active involvement and influence, moving the ideas out of committee through to enactment would have been problematic.

Outsiders' Campaign Activities

While the governor and policy champions were engaged in campaign within the legislature, the entrepreneurs were engaged in an "outsider campaign" to build support for the governor's bill, lobby legislators who opposed it, and counter the resistance from many organized education groups who had added their voices to the opposition. Toward these ends, the policy entrepreneurs saw the value of attracting well-known figures in the public eye. They reasoned that political clout and name recognition would capture media attention (see below) and provoke greater interest in their ideas (policy entrepreneurs 1985).

For example, the Minnesota Business Partnership helped form "The Brainpower

Compact," which pulled together business leaders, four past governors of both political parties (one a nationally recognized leader in education), educational leaders, and the mayor of St. Paul to promote choice, educational excellence, and accountability. This group had direct ties with the Minnesota Business Partnership and received its funding from its member firms. During the 1985 legislative session, The Brainpower Compact employed a lobbyist and a public relations specialist (also a policy entrepreneur) to support the choice proposals and to advocate their enactment into law.

The Minnesota Business Partnership, an association of chief executive officers of the state's largest corporations, also was active in the reform debate (Mazzoni and Clugston 1987). Through fall 1984 a policy entrepreneur, who worked for the Partnership, attempted to influence the direction of the governor's education proposals. Later the Partnership conducted a letter-writing campaign, sponsored personal appearances of Lewis Lehr (then chief executive officer of 3M) with the governor, and under the guidance of its lobbyist/policy entrepreneur, was reportedly persuasive in creating initial legislative momentum for the governor's initiatives (Sturdevant March 26, 1985:1A, 9A).

In February 1985 the first meeting of what was eventually to be known as "People for Better Schools" was called by another policy entrepreneur. The group, formed to assist the governor, who wanted to mobilize grassroots support, consisted of the governor, some of his key advisors, and some thirty-five teachers, superintendents, principals, and school board members, all of whom were expected to be the nucleus of a growing statewide network to support choice. The goal was to begin a massive campaign to influence legislators and blanket newspaper editorial columns with favorable letters. Up against the teachers' unions, some of the state's most effective lobbyists, they sought to build a real grassroots movement to check

"the influence of the education establishment." By March 1986 they had a group of 780 members, many of whom lobbied for the governor's legislation. Orchestrated by the "tireless" leadership of two policy entrepreneurs, who were said to be at the legislature "night and day," this group sought to shape testimony with the skillful selection of credible parents and educational supporters, including principals, superintendents, school board members, and PTA members. They frequently were present at the capitol to disseminate research information on choice to key lawmakers and their staffs, and create a visible presence in the hearings. One legislative analyst noted their impact (Mazzoni 1986:67):

> This has been one of the few issues in all the years I've been here that I've seen develop that way.... This developed almost in textbook fashion. The testimony was convincing! People came in support of that idea that you never would have expected—the principals' organization for example. . . .

Activities to Gain Media Attention and Support

Throughout the legislative session every effort was made to get the media to cover the deliberations and testimony. Reasoning that it was important to help push the reform ideas forward and to keep the issues visible, policy entrepreneurs worked "very hard" to get and keep press coverage (policy entrepreneurs 1987). According to one, "We called, we begged" to get the press involved. They anticipated that media attention would increase public awareness of and support for the issues and would help to convince recalcitrant legislators of the merits of the governor's bill. It also was expected to counter the strength of the 6M group's opposition.

This strategy did not go unnoticed by members of the educational establishment. An educational insider criticized a major paper for its "lack of objectivity" in reporting on the educational issues and described one policy entrepreneur's media interventions in the following way (interest group member 1987):

> He has excellent access to the media. . . . In fact the [paper] has lost its perspective in news coverage. . . . They like to focus on change and controversy. 6M doesn't use the media. . . .

While convincing the press to cover the major events did not always come easily, the policy entrepreneurs did have important support on the editorial pages of the major metropolitan newspapers. Editorial writers reported their ties and associations with two policy entrepreneurs in particular (media 1986, 1987). One called a policy entrepreneur a "close friend," and another credited him with providing perspective and "helpful" information for his editorials.

Administration and Evaluation Activities

The Postsecondary Enrollment Options (PSEO) Act was signed into law by the governor in June 1985 as part of the $2.53 billion funding package for K through 12 education. The Minnesota Department of Education, in collaboration with the Higher Education Coordinating Board (HECB), was responsible for creating the mechanisms and processes to implement the statewide choice program. Specifically, the department's role was to provide guidance to the districts, information to students and parents, and direction to postsecondary institutions. Implementation efforts immediately began with the passage of the law.

It also was part of the department's responsibility to develop and coordinate the program evaluation study mandated by the legislature. To this end, the commissioner appointed a task force to provide direction to this initial evaluation, and several policy entrepreneurs were appointed as members. From their perspective and that of the governor's staff and the House majority leader, the primary objective during this initial evaluation phase was to "hold" the program, maintain its integrity, and allow for a meaningful full-year test (Department of Education member 1987). A team of evaluators from Washington, subcontractors to the federal government, were sent to assist in the evaluation.

By fall 1985 attention was focused on the Postsecondary Enrollment Options Act and the effect it was having in the schools. The House and Senate education committees scheduled hearings to discuss the bill's implementation. The hearings drew already familiar "battle lines." Traditional education groups (the 6M group), which originally opposed the legislation, described abuses and misuses of the program. The policy entrepreneurs and their supporters countered with accounts of "glowing success" and "renewed motivation" on the part of students allowed to attend postsecondary institutions (Department of Education member 1987).

Given these competing assessments of the law's efficacy, the legislature called for a "preliminary" evaluation report to be issued in February 1986, to supplement its mandated final report that was to be issued in February 1987. The planning began in October 1985 for the preliminary report.

Testimony and deliberations in the House and Senate for the preliminary report resulted in "fine tuning" and additional evaluation requirements, but no substantive changes in the original Postsecondary Enrollment Options Act (Department of Education member 1987). Some credited policy entrepreneurs' efforts to prevent the program from being gutted: Policy entrepreneurs had provided secretarial assistance to the Department of Education when the mass mailings were done; policy entrepreneurs coordinated pro-postsecondary enrollment testimony in the legislative hearings (Department of Education member 1987). Ultimately, however, observers cited the testimony of "the kids" participating in the postsecondary options program as providing the compelling evidence that the program was worth continuing (policy entrepreneurs 1987).

After the legislative session ended in June 1986, the commissioner convened a second task force to assist the Department of Education in completing its long-range evaluation study of the Postsecondary Enrollment Options Act. This time the task force consisted of representatives from all the major education groups. However, the policy entrepreneurs were excluded as a gesture to "mend fences" with the education establishment (Department of Education member 1987).

The commissioner of education, based on the group's evaluation, issued her final report to the legislature in February 1987. She hailed the program for providing "greater academic opportunities for high school students" (Doyle 1987:1A). In its first year of operation, the program attracted 3,668 students, and 95 percent of the 1,000 students surveyed reported that they were either satisfied or very satisfied with the program; only 1 percent said they were dissatisfied with the program. It also was noted that 6 percent of those participating had previously dropped out of the public schools.

The headlines in the press proclaimed: "Fears about the college program groundless" (Smith 1987:1B). An official of the state's largest teacher's union who had been a member of the 1986 evaluation task force, said, "It seems to be working and the bugs are being taken out of it" (Smith 1987:1B). He also commented that the mass migration predicted did not materialize, "partly because many school districts have strengthened their curricula in the past two years. . . . More and more college-level courses are available in high school buildings, some provided by colleges and others by high school teachers with

special credentials" (Smith 1987:1B). It would appear, based on this union official's observations, that the "market mechanism" of choice envisioned by the policy entrepreneurs seemed to be having its intended impact in at least some of the schools.

National-Level Networking Activities

During the legislative debates on the "choice" legislation, it became evident that the policy entrepreneurs had links to other associations and groups beyond the state level. These links were more than occasions in which to disseminate and exchange information among those interested in educational policy, which has been mentioned previously. These particular contacts appeared to be more far-reaching in scope. For example, one policy entrepreneur was credited with having some of his ideas incorporated into President Reagan's speech on education (policy entrepreneurs 1986). This same policy entrepreneur accepted a job with the highly visible and active National Governors' Association, which had, as one of its priorities, the investigation of the current state of education. This position was viewed as an opportunity to spread the policy entrepreneurs' ideas into a national audience, using the same patterns as had been employed at the state level: appealing to leaders who could champion the "cause"; building support among "visible types" who could help mobilize interest and attention of innovative policy and so on. Although this policy entrepreneur continued to participate at the state level, even joining another educational task force created by the governor in 1985 to fashion a "visionary proposal for the future of education in the State," his new obligations drew him more and more into the national educational arena. Thus, we see a network and linkages developing beyond the confines of local and state government. And with strategies that looked remarkably similar to some earlier activities, we may be witnessing the beginnings of future policy innovation at the national level.[7]

DISCUSSION: CATALYTIC ROLE OF POLICY ENTREPRENEURS

In the previous section we have described a wide-ranging set of activities that occurred during attempts to introduce innovation to a state educational system. The question is, How should we interpret these activities? Were they just of collection of diverse and idiosyncratic activities that happened by chance in Minnesota? We think not. We now analyze the innovation process and attempt to identify the critical role performed by those who pushed and rode parent-student choice into good currency, showing that policy entrepreneurs filled a key function in the innovation process. They are, we aim to show in this section, a critical *catalyst* in the innovative system—an agent who provokes or precipitates an action or reaction in others, without necessarily undergoing a transformation in itself (Webster 1971).

We begin with several assumptions about the innovation process. First of all, we view innovation as an energy conversion process in which energy in one form (ideas), through a series of transactions and infusions of additional energy (in the form of people's time, money, commitment), is transformed and becomes more and more complex over time. In this case, the ideas must pass through necessary procedural steps to become a legislative reality. The idea of educational restructuring eventually took shape as a proposal for the governor. The proposal then had to be transformed into a bill, and the bill transformed into a law. The law eventually had to be transformed through administration and evaluation. At each step, more and more energy was pulled in and attached to the idea in order to move it through the various stages of the process. And at each

juncture, the idea took on greater complexity: The proposal was more complex than the idea; the bill more complex than the proposal and so on (Figure 9–1).

Thus, as we see it, innovation as an energy conversion/accumulation process is an expensive proposition. Others have pointed out that just managing people's attention—attracting them to the new idea, and keeping them focused on it, as opposed to other options— becomes a major challenge (Van de Ven 1986). Beyond that, there are many other energy-demanding activities such as getting people's commitment, finding resources to fund experimentation, mobilizing the support of "visible types," to name a few.

Also, if one assumes there is a finite amount of energy in any system, then energy locked into the current structure needs to be "destructured" before it can be channeled into a new direction (Bradley 1987). It is this principle on which revolution is based. Energy trapped in current structures—the status quo, standard operating procedures—all build a resistance to change (Kanter 1984). Energy has to be "liberated" from social structure before it can be redirected. People have to be pulled away from the old system and competing ideas, and attracted toward the new, in order to release the requisite energy to support new ideas and to nourish them into innovation. It is this destructive as well as constructive force that some have warned is part of the innovative process (Schumpeter 1942).

We also assume that innovation as an energy conversion process needs something to touch it off. Without provocation of some sort, it would be difficult to move any system beyond the status quo. Some define this provocation as a "jolt" or a "shock" (Schroeder et al. Chapter 4), others a "disruptive event" (Schon 1971). Our analysis of policy entrepreneurs in public policy innovation leads us to understand the provocation in terms of individuals who initiate action against the current system. In questioning the current educational structure, by warning others of the dangerous trends in education, by offering alternative courses of action, and by "framing the problem" to fit their solutions, they served as *agents provocateurs*. They pushed, probed, questioned, and challenged the givens of the current system. They did not react to a crisis or a disruptive event as much as they helped create the perception that one would appear if no policy change was made, thereby prompting the system to act. Their function was that of catalyst.

We see catalytic agents, then, as particularly important for innovative policy. Defined as a recognized departure from a preceding governmental response to a problem, innovative policy is not a modification or an extension of current policy. Rather, it represents a significant change or disjuncture that offers a new approach and/or solution to the problem at hand. It is our contention that the catalytic agent is most critical when innovative policy is needed and is not occurring. Policy entrepreneurs in their function of catalyst help others break out of established patterns to overcome traditional bias in "policy subsystems" (Cobb and Elder 1983:184). Their commitment, energy, and action help redefine the problem, shift the frame of reference from incremental to innovative decisions, and galvanize public opinion around an issue. Through their activities, which we outlined in the previous section, we view the policy entrepreneurs as performing a very important function as the stimulus or catalyst for system-level change, the necessary but not sufficient condition for innovation to occur.

Our characterization of policy entrepreneurs as catalytic agents for policy innovation differs from that of Kingdon, who has the most developed view of policy entrepreneurs and their function in the policy literature. According to Kingdon (1984), policy entrepreneurs are valuable to the policy process because of their "coupling" function. They are responsible for linking solutions to problems and for coupling both problems and solutions to the politics of the moment.

At the macro level of analysis Kingdon's model had much to offer. The streams of problems, solutions, choice opportunities, and the coupling function of policy entrepreneurs comprise a process that is fluid, chaotic, and governed by chance, more in line with how Kingdon's modified "garbage can model of decision making" characterizes policy-making.

However, at the micro level, in an analysis of their activities over time, we found the policy entrepreneurs to operate more strategically, more deliberately, with greater conscious planning and orchestration than Kingdon's model would anticipate. There seemed to be a logic and an intentionality that guided action and its consequences, a view more in concert with a "boundedly rational" social actor in pursuit of self-interest. In this domain, the concepts of leadership, strategy, power, management, choice, and coordination had utility.

For example, several themes or assumptions surfaced during strategy deliberations. One thirty-five-year veteran of numerous policy "battles" summed up his "rules of thumb" that were guiding principles for some of the policy entrepreneurs (policy entrepreneur 1986, paraphrased):

> Know where you want to end up and don't lose sight of where you are headed.
> Don't play the "Washington game" by trading away the fundamental elements of the plan. Compromise may yield bad policy. Say no rather than give up the fundamentals of what you really want.
> Wait for the "background conditions" (political context) to change, thus necessitating the kind of change you want.
> Mature bureaucracies, like education, rarely initiate meaningful change from within, therefore outside pressure is needed to force them to respond.

327

Change never comes through consensus. Get the key leadership to back your idea and the "pack will rush to follow."

Money is needed to make change. . . . Get the elites involved.

Stay with issues where you have the advantage.

Keep the educational establishment talking about change and structural issues, and you'll change some minds.

Destabilize the opposition by co-opting one of the educational establishment groups.

Be willing to be bold.

Thus, rather than seeing policy entrepreneurs as taking advantage of a fortuitous set of circumstances, we saw them create opportunity, as in the case of their contact with the governor and his executive team. Rather than viewing them as only reacting to political climate, we observed them creating it by stirring up interest and debate for the questions of decentralization and deregulation. Rather than defining them as reactors to a set of circumstances beyond their control, we watched them proactively develop strategies and implement them, based on their assessment of the current weaknesses of the system and guided by their vision of what education could be in the future. Rather than waiting for a chance event or critical juncture to join the independent streams in the policy process, they created "choice opportunities," by employing successful change strategies developed in other policy arenas to provoke "the establishment" into responding to their ideas and challenges: They sponsored experimental educational programs; they convinced high status organizations to take up the cause of education innovation; and they encouraged others to introduce innovative legislation. Having studied earlier reform movements in education, they were convinced that successful efforts to reform the educational system would come only from outside the system, providing of course that some mechanism could be found to force this renewal. That mechanism was choice, and once set in place, they intended for it to unleash the self-renewing process of change without their direct intervention. In other words, they had a sense of some important levers of change that could precipitate system response. Their actions were deliberate, strategic, and rational—rational in Simon's sense of maximizing their valued outcomes in so far as their knowledge permitted (Simon 1957:76–77).

In sum, through our experience with the innovative process, we came to view it as both chaotic and rational, serendipitous and planned, random and structured—a system with many different levels of reality. One set of conceptual tools sheds some light at the global level; another is more appropriate for the micro level of analysis. The major challenge for us has been to understand this complexity as a coevolutionary process (Jantz 1980) uniting both micro and macro perspectives, enfolding parts into the whole and the whole into parts (Bohm 1980). Although no resolution or grand synthesis of the two has been offered, we feel at least we have been successful in outlining some of the broad contours for research in the future.

If we differ from Kingdon on the strategic nature of policy entrepreneurship and its catalytic function, we also differ from his description of where policy entrepreneurs are located in the policy system. According to Kingdon (1984), policy entrepreneurs can be found throughout the policy community. They surface both in and out of government, at executive levels as well as midlevels in the bureaucracy, in elected and appointed positions, interest groups and research institutions, and formal and informal positions (1984:129).

Contrary to Kingdon, we found policy entrepreneurs located in marginal positions in the governmental system—marginal in the sense that all except one were outside the formal positions of government at the time they acted as policy entrepreneurs, located either in research institutions, policy analysis organizations, or operating as free agents independent of any organization. The one policy entrepreneur who held a seat in the state legislature saw himself as someone who owed his allegiance to ideas rather than a constituency. He was unwilling to engage in political horsetrading; his values and principles were more important to him. Despite his formal position, he saw himself (as did others) as a philosopher who willingly challenged conventional wisdom and operated at the periphery of the legislative system rather than at its core (policy entrepreneur 1985).

The policy entrepreneurs derived some benefits from their marginality in the governmental system. At the periphery of the system they had several advantages: They had slack—energy not "locked up" in the system they were trying to reform; and they were structurally positioned to facilitate the free flow of ideas from many different domains leveraging their knowledge to create even more networks in other systems and subsystems. Much like the organizational boundary spanner, they "borrowed" ideas from other policy arenas and called on contacts in other policy subsystems for support and assistance.

The divergence between Kingdon's findings and ours in terms of the location of the policy entrepreneurs can be accounted for, to a great extent, in our distinction between policy entrepreneurs and policy champions. As we have defined them, policy champions are those who tend to be in positions of formal authority in the executive, legislative, and administrative branches; their function is to introduce, support, and defend policy initiatives. In the case of educational restructuring, the policy champions were the governor, the senator, and the House majority leader. They worked in tandem with the policy entrepreneurs to move the bill through the legislative process. Based on our reading of Kingdon, he would most likely include the policy champions in the category of policy entrepreneurs.

We also found it important to create another category for individuals in the policy innovation process—*policy administrators,* those who provided the administrative support to the policy champions and entrepreneurs through the adoption process. Originally we thought of the commissioner as a policy champion, appointed as she was to be the governor's "change agent" in the Department of Education. But with her "mandate" for change, we were puzzled why there was an impasse in getting innovative policy approved before the governor's executive committee and why other policy champions took the initiative to become the vanguard for change. As we watched the commissioner become more embroiled in the details of running a large state bureaucracy, especially in the first two years in office when internal problems were serious to the point they were aired in the public press, and as we saw her attempt to moderate the different factions that were lining up for and against the question of choice in the schools, we began to understand why it was important to make the distinction between policy entrepreneurs, policy champions, and policy administrators.

The commissioner had a very complex stakeholder map (Roberts and King 1989) that spanned multiple groups at the local, state, and national levels. In addition to her mandate for change, she had to administer a very large state bureaucracy, prepare and defend the state's largest single budget item, and at the same time, administer to the needs of her educational constituents in 435 state districts. The competition for her time and attention was enormous, exacting eighteen-hour days in her first years in office (Roberts and King 1989). The demands of restructuring education while administering the Department of Education, which by most accounts was very "traditional and conservative" (interest group members 1985; policy entrepreneurs 1985), placed her in the difficult and challenging role of defending an institution and its constituents while she was

329

trying to change them. These competing priorities eventually worked their way out over time by having the commissioner become more and more concerned with administrative issues. For example, in preparation for the legislative session, she formed a task force to consider how the governor's proposal could be implemented in the schools. Her concern focused on the question of how to implement the innovative ideas rather than how to initially form, structure, and enact them (commissioner 1985).

What we have learned from the examples of the commissioner and the policy champions is that given the constraints of elected and appointed political positions, it is difficult for the occupant to be the single-minded change agent whose major interest is the invention and development of innovative policy ideas. For this reason, we would argue that policy entrepreneurs are more likely to be found in organizations or groups that are free to be controversial and less inclined to question their loyalty or set up conflicting expectations. Instead, they give people the intellectual freedom and "the permission" to probe, question, and challenge the existing order. Competing loyalties force choices in what one attends to and how one allocates time. To initiate innovation in the policy arena more than likely means that one is relatively free of organizational constraints and able to think creatively independent of organizational maintenance activities. Thus, we view policy entrepreneurs as individuals who tend to be on the margin of the formal structure of government working to interest and attract those who have the formal authority to give their ideas legitimacy. In much the way that Schon (1971) describes learning at the periphery of a system, and as (Kuhn 1962) predicts paradigm shifts at the margins, so do we find policy entrepreneurs on the "edge" of the formal governmental process initiating innovative change.

We should be careful to point out, however, that we do not exclude the possibility that policy entrepreneurs can be in elected or appointed positions in government. We have, after all, one example of an elected official in our sample. Rather, the question to be addressed at some future point in time is: Under what conditions are we likely to find policy entrepreneurs in the locus of formal power rather than on its periphery?

IMPLICATIONS FOR FUTURE RESEARCH AND PRACTICE

Our research on policy entrepreneurs is just beginning, but already we see the broad outlines of some key areas that need greater exploration beyond what we have been able to develop in this initial effort.

The first question that needs greater scrutiny is, Who are these policy entrepreneurs as individuals? What motivates them to work at the periphery of the governmental system? What drives them to devote a large percentage of their time and energy to policy innovation, often times without formal recognition or tangible rewards? What light can their education, backgrounds, even personalities shed on their persistent and dedicated interests in policy innovation? As a related question, what parallels can be drawn between entrepreneurs in business and industry and entrepreneurs in the public domain?

An even more intriguing issue is to understand how the policy entrepreneurs pool their collective talents to work as a team. We have not uncovered other references to policy entrepreneurs working together as closely as this collection of policy entrepreneurs did. Was this collective experience unique to Minnesota or to the domain of educational policy? The question is an important one because we suspect that their success in educational policy could be attributed, in part, to their ability to work together. In the complicated terrain of educational policy-making, it helped to be able to pool the talents, time, and resources that all these individuals represented.

Perhaps given the number of them working together, it was not surprising to see them specializing. Evidence thus far suggests that three policy entrepreneurs initially focused more on problem definition and solution generation, creating and inventing new ideas and proposals; others saw themselves more as "strategists" planning the long-term and short-term strategies of change; and others viewed themselves as the "activists" who labored "in the trenches" working with interest groups to lobby the legislature (policy entrepreneurs 1985, 1986). These and other activities need to be explored in greater depth to understand how these individuals worked together, and with the policy champions, to affect change (see King 1988).

We should also add that working with others, such as the commissioner in her role as policy administrator, and those from the Department of Education and from Washington in their role as policy evaluators, further reinforces the finding from this case that *policy innovation was a collective effort*. We postulate, in fact, that policy innovation is successful to the extent that it brings together people who represent the various functions in the policy process: *policy entrepreneurs* as initiators or catalysts of change; *policy champions* as leaders of the political debate during policy enactment; *policy administrators* as advocates during policy implementation; and *policy evaluators* as protectors during the critical evaluation period. The close contact among the various individuals who performed these functions in this case suggests that coordination and integration of their activities will positively impact on the success of the innovation.

If greater coordination and integration of the various aspects of the innovation process are important for the success of the innovation,

we are left with a paradox that needs to be addressed. On the one hand, policy entrepreneurs seem to work most effectively on the periphery of government independent from those who would distract them from their function as *agents provocateurs*. On the other hand, we are suggesting that the degree to which policy entrepreneurs coordinate and integrate their efforts with champions, administrators, and evaluators, those very much part of the governmental structure, the more likely their success at initiating change will be. Yet how can policy entrepreneurs be both independent of the governmental system and of the system at the same time?

Parallel Organization

Our speculation is that the policy entrepreneurs who are successful are able to create a parallel organization that blends the two worlds. Instead of being part of the formal governmental system, they create their own organization and attract members of the governmental bureaus and interest groups. Although their parallel organization does not exist on paper, as far as we are aware, it does have characteristics of a formal organization: broadly defined goals and a vision of the future (restructuring education around the theme of choice); some degree of specialization (some policy entrepreneurs focus on developing ideas, others map long-range strategies, while still others concentrate on the day-to-day activities; some source of funding to support activities (foundation grant money encourages experimentation); some degree of coordination and control (the core group integrates member and nonmember activities). Drawing energy from within the formal governmental system, or from those who were not previously aligned with it, the policy entrepreneurs were able to build a constellation of individuals and groups that could serve as an alternative to the formal governmental process. Future research must examine the extent to which other successful policy entrepreneurs rely on parallel organizations to champion their causes.

Some practical implications for policy-making also need to be addressed. This case suggests that people "on the margin" are critically important to the innovative process. Their loyalty to ideas rather than to an organization or to others is a source of strength and creativity for the system. They need "safe havens" (temporary or institutionalized) that acknowledge their valuable function as system-level change agents whose purpose is to challenge current conceptual frameworks and their tidy assumptions. Academic institutions, think tanks, and foundations, in theory, can provide some protection for independent policy thinkers. Two policy entrepreneurs, in fact, were connected to a university institute, although they had other bases of support as well. Academic institutions, however, can pose their own dilemmas—such as questions of funding, quality and quantity of publications, appropriateness of research are raised, and demands for participation in institutional activities. One of our major concerns is that the current press for accountability in universities will result in policy entrepreneurs who have few options for support in the future.

An even more fundamental question about the phenomenon of policy entrepreneurs concerns the parallel organization as a mechanism for promoting and coordinating the innovation process. The contours of the parallel organization were barely visible to participants in and challengers of the process in 1985. What checks and balances exist to hold it accountable? In this case, because the policy entrepreneurs directed their efforts in the parallel organization back into the legislative and administrative apparatus of government, there was public accountability. In fact, through constant association with the media and opposition

groups, there was always accountability in one form or another.

But we can envision a case in which a group of policy entrepreneurs could operate their parallel organization beyond the reach of public scrutiny. In their desire to fight the "inertia" of bureaucracy and develop "innovative" responses to what they define as critical problems, they could position themselves above the law and beyond the checks and balances of the formal system. The challenge, especially for policy entrepreneurs, is to answer some important ethical questions about the innovation process: For whom and for what end is this innovation designed? What kind of process will be employed to reach the goal of implementation? We can attribute much of the success of the policy entrepreneurs in Minnesota to collective action that was promoted by those in the parallel organization. But the same process and mechanisms that led to their success also can be used by others for very different purposes. As the business entrepreneur's ideas, often hatched in isolated "greenhouses" and "skunkworks," must be integrated back into the mainline business organization, policy innovations also must be given reality checks through legislative and administrative processes to avoid destroying the systems that they are designed to innovate.

POSTSCRIPT

By June 1987 two additional initiatives to support choice in the public schools passed into law: The High School Graduation Incentives Program extended choice to "at risk" students who had dropped out of school or were in danger of dropping out; and the Voluntary Open Enrollment Options Program established voluntary enrollment for K through 12 students in districts that had formal school board approval of the program.

By June 1988 Minnesota had passed the nation's first mandatory enrollment options program that extended choice to all K through 12 students; the plan would be fully implemented by 1990–91 throughout the state. Policy entrepreneurs continued their involvement in choice legislation during this time period.

6. Three policy entrepreneurs had close ties with the Citizens League. Two were past executive directors and one was the present executive director.
7. According to Joe Nathan of the Spring Hill Center in Minnesota, expanding choice among the public schools is a growing national issue. As of May 1988, eleven states had taken action on choice in public schools and four others had pending state action.

NOTES

1. We use the term *constellation* here to describe numerous overlapping networks of groups to which the policy entrepreneurs belonged. Although acting in tandem to affect their desired changes, the policy entrepreneurs also represented various independent groups and constituencies.
2. The context of this study is a very complex one. National concerns over education provide the backdrop to this Minnesota case, as do the political and economic climate of the nation and state. We recognize these important contextual elements to the innovation process, but we unfortunately are unable to deal with them in any depth in this chapter. For a more indepth treatment of this issue, readers should consult Mazzoni (1974) and King (1988).
3. The latter presuppose a highly specified model from which to derive standardized instrumentation for systematic data gathering (Lieberson 1985). As we have already mentioned, this level of theoretical sophistication had yet to be attained in this research.
4. The timeline does not include the results of the 1987 and 1988 legislative sessions, both of which were notable for their achievements on educational choice. The postscript summarizes the highlights of these sessions.
5. The number of policy entrepreneurs varied from year to year depending on the time they had to devote to the innovation.

REFERENCES

Berman, P. and Weiler, D. 1984. *The Minnesota Plan The Design of a New Educational System.* Berkeley, Calif.: BW Associates.

Blalock, H.M. 1982. *Conceptionalization and Measurement in the Social Sciences.* Beverly Hills, Calif.: Sage.

Bohm, D. 1980. *Wholeness and the Implicate Order.* London: Routledge & Kegan Paul.

Boyer, E. 1983. *High School: A Report on Secondary Education in America.* New York: Harper & Row.

Bradley, R.T. 1987. *The Social Structure of Charisma: Love and Power, Wholeness and Transformation.* New York: Paragon.

Citizens League. 1976. *The Citizens League Itself.* Minneapolis: Citizens League.

———. 1982. *Rebuilding Education to Make It Work.* Minneapolis: Citizens League.

Cobb, R.W., and C.D. Elder. 1983. *Participation in American Politics: The Dynamics of Agenda-Building.* 2nd ed. Baltimore, Maryland: The Johns Hopkins University Press.

Downs, A. 1956. *An Economic Theory of Democracy.* New York: Harper.

Doyle, P. 1987. "High School Students in College Classes got Mostly As, Bs, State Says." *Minneapolis Star and Tribune,* Feb. 7. 1A–7A.

Eyestone, R. 1978. *From Social Issues to Public Policy.* New York: John Wiley & Sons.

Goodlad, J. 1984. *A Place Called School: Prospects for the Future.* New York: McGraw-Hill.

Hanson, R. 1985. "Legislative Stalemate Poses Basic

Questions," *Minnesota Journal* 11, no. 15 (June 25): 1–4.

Hirschman, A.O. 1970. *Exit, Voice, and Loyalty.* Cambridge, Mass.: Harvard University.

Lieberson, S. 1985. *Making It Count.* Berkeley: University of California.

Lincoln, Y. and Guba. E. 1985. *Naturalistic Inquiry.* Beverly Hills, Calif.: Sage.

Jantz, E. 1980. *The Self-Organizing Universe.* Oxford: Pergamon.

Kanter, R.M. 1984. *The Changemasters.* New York: Random House.

Keynes, J.M. 1961. *The General Theory of Employment, Interests and Money.* New York: St. Martin's.

King, P. "Policy Entrepreneurs: Catalysts in Policy Innovation." Ph. D. dissertation, University of Minnesota. 1988.

Kingdon, J.W. 1984. *Agendas, Alternatives, and Public Policies.* Boston: Little, Brown.

Kuhn, T.S. 1962. *The Structure of Scientific Revolutions.* Chicago: University of Chicago.

Mazzoni, T.L. 1986. "Educational Choice and State Politics: A Minnesota Case Study (1983–1985)." Unpublished manuscript, University of Minnesota, Department of Educational Administration, School of Education.

Mazzoni, T.L. 1974. *State Policy Making for the Public Schools.* Minneapolis: Minnesota Department of Educational Administration.

Mazzoni, T.L. and R.M. Clugston. 1987. "Big Business as a Policy Innovator in State School Reform: A Minnesota Case Study." *Educational Evaluation and Policy Analysis* 9(4) (Winter):312–24.

Minnesota Business Partnership. 1984. "Educating Students for the 21st Century." Minneapolis: Minnesota Business Partnership.

Minnesota Department of Education. 1986a. "Planning Document #5, 1986." St. Paul: Minnesota Department of Education.

———. 1986b. "Postsecondary Enrollment Options Program Preliminary Report". St. Paul: Minnesota Department of Education, 1986.

———. 1987. "Postsecondary Enrollment Options Program Final Report." St. Paul: Minnesota Department of Education.

Nathan, J. 1984. *Free to Teach.* Minneapolis: Winston.

National Commission on Excellence in Education. 1983. *A Nation at Risk: The Imperative for Educa-tional Reform.* Washington, D.C.: U.S. Governmental Printing Office.

Perpich, R. 1985. *Access to Excellence.* St. Paul: Governor's Office.

Polsby, N.W. 1984. *Political Innovation In America: The Politics of Policy Initiation.* New Haven: Yale University.

Public School Incentives. 1985. *School Based Management.* Minneapolis.

Public Services Redesign Project. 1984. "Two Alternative Routes to the Improvement of Education: Part I." Minneapolis: Humphrey Institute, University of Minnesota.

Roberts, N. 1985. "Transforming Leadership: A Process of Collective Action". *Human Relations* 38(11): 1023–46.

Roberts, N. 1987. "Transforming Leadership Part II: A Process of Collective Action." Unpublished manuscript, Naval Postgraduate School, Monterey, California.

Roberts, N., and R.T. Bradley. 1988. "Limits to Charisma," pp. 253–275. In Jay Conger and Rabindra Kanungo, eds., *Charismatic Leadership in Management.* San Francisco: Jossey-Bass.

Roberts, N., and P. King, 1989. "The Stakeholder Audit Goes Public." *Organizational Dynamics,* Winter, pp. 63–79.

Salisbury, H. 1969. "An Exchange Theory of Interest Groups." *Midwest Journal of Political Science,* 13 February, pp. 1–32.

Schatzman, L., and A.L. Strauss. 1973. *Field Research: Strategies for a Natural Sociology.* Englewood Cliffs, N.J.: Prentice-Hall.

Schon, D.A. 1971. *Beyond the Stable State.* New York: Norton.

Schumpeter, J. 1942. *Capitalism, Socialism, and Democracy.* New York: Harper.

Simon, H. 1957. *Administrative Behavior.* New York: Free Press.

Sizer, R. 1984. *Horace's Compromise: The Dilemma of the American High School.* Boston: Houghton Mifflin.

Smith, D. 1987. "Fears about College Program Groundless, State Agency Says." *Minneapolis Star and Tribune,* Feb. 7:1B.

Sturdevant, L. 1985. "Nelson, Olsen Hold Keys to School Bills," *Minneapolis Star and Tribune,* April 1, pp. 9c, 10c.

———. 1985. "Open Enrollment Cut from Bill,"

Minneapolis Star and Tribune, April 4, pp. 1A, 11A.

———. 1985. "Open Enrollment Plan Is Hanging By a Thread," *Minneapolis Star and Tribune,* April 25, pp. 1B, 4B.

———. 1985. "Perpich School Plan Clears House Hurdle." *Minneapolis Star and Tribune,* March 26:1A, 9A.

Van de Ven, A. 1986. "Central Problems in the Management of Innovation." *Management Science* 32(5):590–607.

Walker, J.L. 1974. "Performance Gaps, Policy Research, and Political Entrepreneurs," *Policy Studies,* Journal 3, Autumn, pp. 112–116.

Webster's Third New International Dictionary. 1971. Springfield, Mass.: Merriam.

Wilhelm, P.M. 1984. "The Involvement and Perceived Impact of the Citizens League on Minnesota State School Policymaking, 1969–1984." Unpublished Ph. D. dissertation, University of Minnesota.

Wilson, B. 1985. "Senate Alternative to Perpich's School Plan is Presented." *Minneapolis Star and Tribune,* March 23, p. 7B.

10

SENSEMAKING, UNDERSTANDING, AND COMMITTING: EMERGENT INTERPERSONAL TRANSACTION PROCESSES IN THE EVOLUTION OF 3M'S MICROGRAVITY RESEARCH PROGRAM

Peter Smith Ring

Gordon P. Rands

On February 8, 1984, Lewis W. Lehr, chairman and chief executive officer of Minnesota Mining and Manufacturing (3M), in St. Paul, Minnesota, announced at a joint press conference that his company would enter into a joint endeavor agreement (JEA) with the National Aeronautics and Space Administration (NASA) to conduct an initial series of three basic research experiments in the microgravity environment provided by the space shuttle. The announcement signaled 3M's commitment to work with NASA as its first non-aerospace *Fortune* 50 partner in the U.S. government's effort to further promote and develop the commercial use of space. Events associated with the development of this co-venture[1] are described in this chapter, and a theory of transaction processes that has evolved from the analysis is presented.

The chapter begins with an introduction

We gratefully acknowledge useful comments and observations from Andrew Van de Ven, Roger Hudson, Michael Rappa, and other colleagues involved in the Minnesota Innovation Research Program. This research was supported by Grant No. 0350-3312-22 from the Graduate Division of the University of Minnesota.

to the principal research setting, including a brief history of 3M and of NASA's Space Commercialization Program. The discussion then turns to the organizational setting of the project within 3M: its Science Research Laboratory (SRL).

The model derived from the initial research is then introduced. Its key variables are defined and their conceptual bases are established and exemplified in the 3M-NASA co-venture. The chapter concludes with a discussion of the research and managerial implications of the model.

THE RESEARCH SETTINGS

The study of emergent interpersonal transaction processes grew out of an investigation of the creation and management of a project by the 3M Corporations to conduct basic research in the microgravity environment of low orbit. 3M was able to actively consider such a project as a result of efforts of the Reagan administration to foster the commercial development of space. The genesis of that policy and the 3M microgravity research project is outlined below.

The Space Commercialization Effort[2]
A major goal underlying U.S. space policy is the maintenance of a national leadership position. This objective is set forth in a National Space Policy articulated by President Reagan on July 4, 1982, at Edwards Air Force Base (home of NASA's Dryden Space Flight Center) on the occasion of the landing of the space shuttle *Columbia*. Among objectives outlined in the policy were:

> obtaining economic and scientific benefits through the exploitation of space; and expanding U.S. private sector investment and involvement in civil space and space related activities.

NASA's Space Transportation System (STS) was declared to be a vital element in the space program and would serve as the "primary launch system" for civil space programs. NASA would "assure the shuttle's utility to the civil users." Among the policies that would govern the conduct of the civil space program was the following: "conduct appropriate research and experimentation in advanced technology and systems to provide a basis for future civil applications." In the course of the civil space program the U.S. government would "provide a climate conducive to expanded private sector investment and involvement in space activities, with due regard to public safety and national security."

On January 25, 1984, in his State of the Union message, President Reagan further enunciated key elements of the government's space program. He urged U.S. industry to engage in commercial ventures in space, noting that a primary governmental role in this effort would be to remove barriers to private-sector activities.

That role was further clarified by James M. Beggs, then NASA administrator, in testimony before the House Subcommittee on Space Science and Applications on June 19, 1984. Mr. Beggs stated that:

> The President's program is designed to encourage private enterprise in space and requires government and industry to work together to ease regulatory constraints and, with NASA's help, promote private-sector investment in space. . . . [W]e must ensure that the private sector has the freedom to organize and operate for profit in space. We must proceed thoughtfully to establish precedents that will best serve the commercial use of space and the civil space program for many years to come. I believe NASA's role is, in a word, to "facilitate."

Only with the capacity to carry into space and return large payloads, which the shuttle provided, did the commercialization of space begin to receive serious attention. This is not to say that commercial efforts in space were unheard of prior to the 1980s. To the contrary, an extremely profitable communications industry had been developed on a worldwide basis. This sector of the economy is valued in excess of ten billion dollars.

Once the shuttle program became a reality, NASA began its efforts to interest the private sector in commercial opportunities in space. In the last four years of the 1970s it commissioned five major investigations regarding the shuttle and its potential. Among those conducting studies for NASA was Rockwell International, which concluded that U.S. industry was generally unaware of the potential of the shuttle, but was prepared to engage in commercial adventures in space if their competitors did so. With the inauguration of the shuttle as an operational space vehicle, the drive to foster the commercial use of space shifted into high gear. NASA began to court U.S. industry, spurred on in large measure by increased efforts in the commercial use of space by the French, Germans, and Japanese.

The Internal 3M R&D Environment

In its current form, 3M has operations in forty-two states and operates subsidiaries in fifty-one countries outside the United States. It employs approximately 86,000 people, including 5,500 scientists. It is organized around four business sectors—Electronic and Information Technologies, Graphic Technologies, Industrial and Consumer, and Life Sciences—that produce forty-five major product lines. 3M's success has been built on a foundation of worldwide leadership in specialty chemistry and materials science. Today 3M ranks forty-sixth in sales in the *Fortune* 500 list of U.S. industrial companies.

The microgravity research project at 3M has been developed and conducted by the Science Research Laboratory (SRL) and by a newly established (January 1985) unit within SRL, the Space Research and Applications Laboratory (SRAL). SRL is one of four laboratories that together comprise 3M's Corporate Research Laboratories (CRL).[3] The mission of SRL is to conduct long-term, basic research science that can lead to new technologies and businesses for 3M. It focuses on the design and construction of materials that perform in unique and predictable ways. The focus of the research is on organic, inorganic, and polymer chemistry, physics of surfaces, molecular structures, and molecular interactions.

SRL is expected to transfer this knowledge and materials to sectors and divisions for the further development of their business needs. It is also to provide leadership within 3M in its interactions with universities, government labs, and other 3M-sponsored research and to enhance the 3M research image.

In the eyes of SRL management, the distinguishing characteristic of SRL's research, compared with that done elsewhere in 3M, is that it pushes the state of the art and is of a long-term nature. In short, the level at which the research is being done provides the uniqueness to SRL's research, and these characteristics provided a window of opportunity through which a series of transactions with NASA were initiated.

Space Commercialization in the MIRP Framework

As discussed in Chapter 1, the MIRP framework consists of five key concepts: ideas, people, transactions, contexts, and outcomes. The focus of this chapter is the processes of emergent interpersonal transactions. However, we briefly reintroduce the MIRP concepts at this point to describe how they appear to relate to the transaction processes investigated in this chapter. We also point out that in our analysis of the transactions within 3M and between 3M and NASA, we will, when appropriate, provide additional detail about the relationships between the MIRP concepts. For example, the contexts within which we investigated transactions varied substantially. Some emergent interpersonal transaction processes that we observed took place (1) among individuals acting as principals in interorganizational and intraorganization settings; (2) between individuals acting in their multiple roles as agents within each organization; and, finally, (3) between individuals acting as agents for their organizations in interorganizational transactions.

Contexts. In one respect, the evolution of space as an arena for human activity provided the larger context for the innovation processes undertaken by 3M and NASA. In more micro terms, at about the same time that NASA began to seek private-sector partners to commercialize space, 3M also had embarked on a program of developing new approaches to long-term basic research. That these two organizations came to co-venture might seem inevitable. Both organizations were known as being well managed and innovative. Both were said to pursue excellence.[4] As we shall demonstrate, however, the co-venture came into being only as a result of a series of extremely complex interpersonal transaction processes that were embedded in a more formal, and more familiar, structure of transactions.

Ideas. Two very different, but equally simple, sets of ideas were largely responsible for the initial 3M-NASA co-venture. President Reagan had committed his administration to using the national resource embodied in the space transportation system for commercial purposes. The private sector, however, was much less enthusiastic about the prospects of profits from space than was the administration. NASA, quite simply, was in the market for customers interested in using the space shuttle for commercial purposes.

At 3M, on the other hand, managers in the scientific community were looking for ways to better reward and motivate its key researchers—staff and corporate scientists. The concept

of doing research in space surfaced as one approach. Essentially, 3M's original interest in space boiled down to the simple idea of looking for new ways to reward a group of its employees.

These initial idea "sets" provided a basis for initiating a series of transactions, within and between the two organizations. However, the processes of transacting that we describe contributed to the generation of many new and diverse "sets" of ideas. By themselves, the new sets of ideas redefined contexts, roles, and outcome expectations. They led, as well, to additional transacting as the parties moved through the more formal structural phase of administration of existing agreements.

People. Transactions are inherently people oriented. Although who the people are may be important to the outcome of a transaction, the multiple and changing roles that they play (which we describe in detail in the next section) in the course of transacting are likely to be more critical. In the 3M-NASA co-venture, the people involved invariably fit within four roles, and most played more than one. These roles included, champions, lawyers, managers, and scientists.

At 3M, champion roles were played at all levels of the organization; at NASA they were relatively fewer and tended to be played at lower levels of the organization. In NASA, the lawyers tended to be the makers of policy. At 3M the role of lawyers was to craft management's policies into legally binding agreements. Most 3M managers worked with reasonably well-defined objectives, whereas at NASA they were still engaged in defining them. Finally, the scientists at 3M were internal competitors—that is, they were competing with other 3M scientists for resources. At NASA, on the other hand, the scientists were external competitors—they had to compete against outsiders for the use of the space shuttle. As we shall see, with the exception of the lawyers, many of the respective role occupants found counterparts in the other organization with whom they developed relationships that appear to have deepened beyond a purely professional level.

Outcomes. Both organizations were able to clearly articulate the outcomes desired from the co-venture. In 3M's case these appear to have been more diverse. The company clearly sought an "imaging" outcome that 3M was a "high tech" player in the business community. In an age in which high tech tends to be synonymous with success, it can be a valuable image for a company. 3M also sought a laboratory for learning. It wished to learn how to do transactions with the government again. It wished to learn about the effects of near-zero gravity. It sought successfully completed experiments and, in the longer term, new commercial products and processes. Finally, they also sought new transactions as the co-venture evolved.

For its part, NASA initially sought "commitment" from industry to commercialize space. It also sought new political constituents for its programs. Finally, it sought to reduce its dependence on the Congress for revenues.

Transactions clearly provide a vehicle by which individuals, functioning in a variety of roles, constrained by context, generate ideas, and seek to achieve desired outcomes. For example, conventional contract wisdom indicates that the critical ingredients to a successful "deal," whether they relate to people, ideas, or outcomes, will be contained within the "four corners" of the legal documents.[5]

On the other hand, interorganizational theory points to a number of variables as determinants of outcomes, including the accuracy of perceptions of the environment, degrees of decisionmaking uncertainty, past strategic choices, the degree of resource dependency experienced by the organization, and the strategic predisposition of the organization (see Miles and Snow 1978; Pfeffer and Salancik 1978;

Miles 1980). Many of these variables cannot be directly controlled for within a legal document. In the space commercialization case evidence suggests that the ordering of the emergent processes of interpersonal transactions, per se, may be a more important determinant of outcomes, however defined, than the variables typically defined in the organizational theory literature or than the ingredients of formal contract law.

A NEW TRANSACTION MODEL

In Chapter 6 a model of formal and informal transactions processes was described. That model was largely derived from the study of the 3M-NASA co-venture. In the following section, the model is briefly revisited.

Formal/Legal Transaction Processes

Commons (1950) posits that transactions—which we argue provide an appropriate unit of analysis in co-ventures—involve a dynamic process. He asserts that this process involves three temporal stages: negotiations, agreements, and administration. This model provided a basis for the theoretical departure of our analysis. Although these three stages often overlap and recycle, as was demonstrated in Chapter 6, it is analytically useful to distinguish these phases in order to examine the interpersonal and social-psychological processes that tend to predominate in each phase of a transaction. The *negotiations stage* highlights the strategies and choice behavior of parties as they select, approach, and avoid alternative parties, and as they persuade, argue, and haggle the terms, contingent safeguards, and governance structure involved in a transaction. In the *agreement stage,* the "wills of the parties meet" by agreeing to the terms of the relationship and the working rules or procedures of action. Thus, Commons views agreement as both process and outcome. It is here where governance structures and safeguards outlined in Chapter 6 are set. Finally, in the *administrative (or execution) stage,* the rules, procedures, and terms of the transaction are carried into effect. During this stage conflicts, misunderstandings, and changing expectations often occur—resulting in renegotiation, mutual adaptation, litigation, or termination of the transaction.

Emergent Interpersonal Transaction Processes

Underlying contracts one frequently finds backstage interpersonal dynamics that mobilize and direct formal contracting processes. These informal processes are seldom visible or explicitly written into the formal contract but often more closely reflect the "real" motivations of the parties to a transaction than does the formal language of the contract.

In Chapter 6, the micro-individual needs or drives that shape the form of contracts were said to manifest themselves in three informal transactions processes observed in the 3M/

NASA transaction: *sensemaking, understanding,* and *committing.*

Sensemaking was described as a process in which individuals develop cognitive maps of their environments. The processes of understanding, we argued, find individuals engaged in reciprocal sensemaking, seeking shared cognitive maps, or facticity. Finally, in processes of committing, individuals, seeking inclusion as principals or agents, bind themselves and/or their organizations to act. The model outlining these processes and their relationships to the formal processes of transactions is provided in Figure 10–1.

Because the interactions of these informal processes and their products are complex, the ensuing discussion is best understood by keeping in mind the kinds of products associated with each of the three emergent transaction processes. Table 10–1 illustrates these distinctions.

In the course of the studying the 3M-NASA transacting processes, we observed that the level of intensity of the processes of sensemaking, understanding, and committing varied within each of the three structural stages of

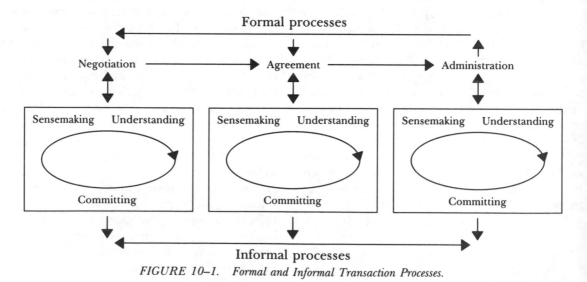

FIGURE 10–1. *Formal and Informal Transaction Processes.*

TABLE 10–1. *Distinctions between EITP Processes and Their Products.*

Process	Products
Sensemaking	Sense of identity-cognitive maps
	Enacted environments
	Revised expectations
Understanding	Facticity
	Mutually agreed on cognitive maps
	Organizational learning
Committing	Psychological contracts
	Legal contracts
	Inclusion

a transaction. Context, role, idea, and outcome variables also appear to affect the level of intensity of each of the informal processes observed. As we describe sequences of events in the following sections of the chapter, we will graphically represent the EITP model in a design similar to that illustrated in Figure 10–2. For each sequence of events, the model's bounda-

ries will be altered to reflect changes in the levels of intensity of the emergent interpersonal transaction processes taking place in the transaction(s) associated with the event.

METHODOLOGY

The data employed in developing the model presented in this chapter were obtained in three ways. Archival sources primarily were employed in developing background on 3M and the space program, although 3M personnel provided us with copies of relevant correspondence, internal memos, the MOU, the JEAs, and so forth. Interviews and observation provided the balance of the data. Two persons conducted most of the interviews, all of which were taped. Notes were also taken during the interviews, which proceeded generally according to a question schedule developed in advance by the interviewers.

As we reviewed written documents and

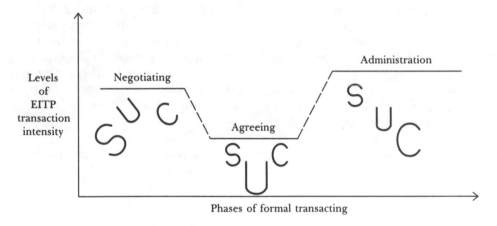

S = Sensemaking processes.
U = Understanding processes.
C = Committing processes.

The size of the letter reflects the relative importance of the process within each formal phase.

FIGURE 10–2. *Model Used to Depict EITP.*

listened to tapes, we sought to distinguish between negotiation, agreement, and administration stages of the formal transaction process. The grey area fell between negotiation and agreement. As a decision rule, we determined that the process moved from negotiation to agreement when the parties had concluded that they needed only to convert their work product to legally binding language. For example, in negotiating the two-year JEA, the parties spent some time discussing when research results had to be made available to the scientific community. Once they had settled on one year from the flight of the experiment, the process moved from negotiation to agreement under our decision rules. Once an agreement was signed, the deal moved into the administration stage. Only when a problem causes the parties to conclude that the legal document has to be amended do the formal processes cycle back to negotiation under our decision rules.

When the outline of the EITP model began to take shape conceptually, we developed decision rules for thinking about the three processes. It was relatively easy to distinguish the committing process from those of sensemaking and understanding. However, defining what was evidence of sensemaking and distinguishing it from evidence of understanding was much more difficult.

We decided that whenever the written material or responses from individuals reflected an intention on their part to simply enhance their own perspective on a subject, then such actions were indicative of sensemaking processes. Thus, a statement such as, "I asked the question because I needed the answer to improve my own appreciation of his circumstances," was treated as evidence of a sensemaking process at work, as were visits and research on a topic.

On the other hand, when these kinds of activities were pursued in activities that reflected reciprocity, we classified them as understanding processes. This is, of course, the grey area. The same activity may reflect, at once, sensemaking and understanding processes. I ask a question to enhance my own cognitive map, but at the same time, I immediately translate the results into a response, or a question, to a perons with whom I am attempting to reach agreement about the meaning of an event.

When the data reflected expectations of the parties, and efforts to reach mutually agreed on expectations, then we classified these as evidence of committing processes. Those that resulted in formal contracts were, of course, much easier to identify. Evidence of psychological contracts was much more problematic. In the final analysis, we listened for terms such as "informal commitments" and "contracts" used in contexts suggested by the literature on psychological contracts.

THE 3M-NASA CO-VENTURE

Table 10–2 reflects a chronology of the major events sequences in the evolution of the co-venture between 3M and NASA. In each of these sequences emergent interpersonal transaction processes of the type outlined in Figure 10–2 were observed. In the following sections we review a number of these event sequences in greater detail, outlining the evidence used in developing the EITP model.

A Memorandum of Understanding Is Negotiated and Signed

Table 10–3 summarizes the major events in the development of the memorandum of understanding between 3M and NASA. The memorandum provided a temporary legal basis from which 3M and NASA were able to reach agreement on the terms of a more permanent set of formal relationships.

In mid-1982 the idea of providing funding for a 3M scientist to place an experiment on

344

Event sequence

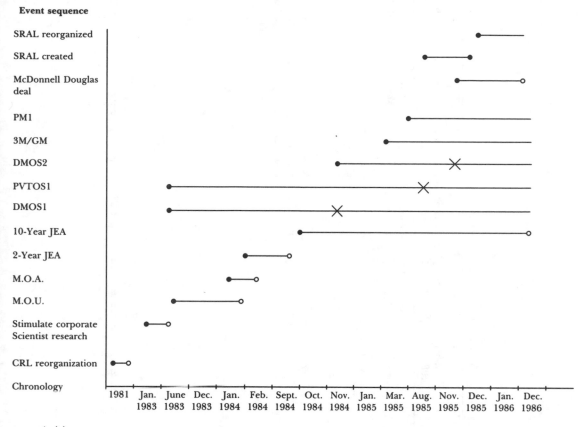

TABLE 10–2. Chronology of Major Event Sequences in the 3M/NASA Coventure.

one of the space shuttle's "Get Away Special" units ("gas cans") was first mentioned in 3M's Technology Assessment and Policy Committee (TAPC), as a possible incentive to encourage more innovative research. Interest began to grow in having someone speak to the TAPC about microgravity research in general. Other events were also taking place that helped to facilitate an agreement between NASA and 3M.

In February 1983 Gerry Kottong of 3M's Government Contracts Office, which had been created in January 1983, was informed of a chance meeting at a ceramics association convention between individuals from 3M's Indus-

trial Consumer Sector and astronaut Dr. Bonnie Dunbar. She had complained that U.S. industry failed to understand the importance of and opportunities inherent in commercial utilization of the space shuttle. Kottong, who admits to always having been something of a space buff, checked to see if the staff of various operating divisions that employed technologies relating to crystal growth had any interest in microgravity research, but none was expressed.

At a conference two months later Kottong encountered a professor who knew Dunbar well. The professor recounted Dunbar's feelings about the strides the Japanese were

345

TABLE 10–3. *Major Events in the Formal Processes of Transacting the Memorandum of Understanding.*

January 1983	February–April 1983	June 1983	August 1983	September 1983	October 1983	November 1983	December 1983	January 1984	February 1984
Government contract office created at 3M	Bonnie Dunbar haunts 3M	Coopers Lybrand visit 3M	Mike Smith of NASA visits 3M	Experiment proposal sent to NASA by 3M	NASA team visits 3M	3M team visits NASA	3M Top management reviews and gives OK to project	MOU signed by 3M	Beggs gives final OK
		Podsiadly gets involved at 3M				3M's counsel drafts MOU	Halpern at NASA revises MOU		3M/NASA announce MOU to press and public
		Debe and Egbert get involved at 3M							
		SRI visit to 3M							
		3M/NASA contact made							

making in preparing for commercially oriented space research. Kottong again checked for interest at 3M, with similar results as before. In June, Kottong was contacted by two marketing experts from the Stanford Research Institute who wanted to talk about space research during their visit to 3M. They informed Kottong that NASA had recently assigned two individuals (Bud Evans and Mike Smith) to promote space commercialization and that Kottong should talk to them. After a number of attempts, Kottong finally made contact with Smith in late June.

Also in late June, a team from Coopers & Lybrand and Grummann Aerospace visited 3M's TAPC. Cooper's was 3M's external accountant. In April they had contacted the company regarding a presentation on the space shuttle program, which they were making under a consulting and marketing contract with NASA. The presentation was scheduled for June 29, 1983.

3M Vice-President of Research and Development Dr. Lester Krogh, a TAPC member, invited Dr. Chris Podsiadly, the director of 3M's Science Research Laboratory to attend the presentation. Following the meeting, Krogh and Podsiadly discussed research opportunities that might exist aboard the shuttle, and Podsiadly subsequently discussed these with his six project leaders in SRL. Two of these individuals, physicists Dr. Mark Debe and Dr. William Egbert, indicated an interest and began developing proposals for microgravity experiments with members of their project teams.

Following the Coopers & Lybrand presentation, Dr. Krogh asked Gerry Kottong to find someone from NASA to speak to 3M's Technical Forum about microgravity research. Kottong and Smith engaged in a series of telephone calls in July that resulted in a plan to have a team of NASA people visit 3M in September. Kottong experienced difficulty in getting subsequent information from Smith and

NASA, however, and became so frustrated that in late August he told Smith's secretary that if Smith did not come to Minnesota that very evening to straighten things out, 3M would cancel the visit. Smith arrived in St. Paul that evening.

Meanwhile, Drs. Debe and Egbert continued work on the development of proposals for six microgravity experiments, and in September a report outlining these proposed experiments was sent to NASA. On October 3, 1983, 3M was visited by a delegation from NASA including Associate Administrator for Space Flight Lt. General James Abrahamson, Dr. Robert Naumann of the Marshall Space Flight Center (MSFC), astronaut Dr. Storey Musgrave, and Captain Chet Lee, director of the Office of Space Flight. This visit was a product of the collaboration of Kottong and Smith. A tremendous amount of interest was shown by 3M employees in the public activities associated with these visits, and Dr. Naumann spent a lot of time discussing the proposed experiments with Debe and Egbert.

The October visit of NASA personnel to 3M was followed in mid-November with visits to NASA headquarters in Washington, D.C., and to the MSFC in Huntsville, Alabama, by Drs. Podsiadly, Debe, and Egbert. Podsiadly returned to St. Paul with a letter from Lt. Gen. Abrahamson to 3M CEO Lew Lehr that promised that space would be made available on a shuttle flight as soon as 3M was ready to fly its experiments. Following the visits, Dr. Naumann wrote a helpful critique of 3M's proposed experiments, focusing on what 3M would need to do to fly the experiments.

On December 14, 1983, Dr. Podsiadly presented a review of the Science Research Lab's proposals to 3M's top management. A great deal of information regarding microgravity research opportunities had already been presented to top managers by Dr. Krogh, and members of the top management recognized that a series of experiments had a reasonably good chance of generating benefits. In what one participant described as "formalizing what had been done in the halls ahead of time," Lehr gave the go-ahead on the proposal at that meeting. Following the meeting, work began in earnest to develop a memorandum of understanding that would provide a legal basis for the relationship between 3M and NASA.

3M's associate counsel, Merritt Marquardt, had prepared a draft memorandum of understanding in November 1983, using information provided by Chris Podsiadly. This draft was revised in December by Richard Halpern, NASA's acting manager of materials processing in space. The two-page memorandum of understanding was in essence an agreement to agree. It called for 3M and NASA to work together to reach an agreement to conduct at least two experiments on the shuttle under terms and conditions that were mutually satisfactory to both parties. Although the memorandum of understanding was signed by NASA and 3M officials on January 17 and 18, respectively, final approval by NASA did not come until February 6, 1984, just two days before the public announcement of the agreement by Lehr and NASA Administrator James Beggs was scheduled to occur in Washington.

Our data regarding these events is not very extensive, but it appears that opposition to the memorandum of understanding was strong within many of the units at NASA. This opposition appears in large measure to have been a reflection of the longstanding existence within NASA of two competing and very different sets of views regarding the agency's mission, objectives, and strategies. A strong divergence of opinion has existed for some time regarding the emphasis that should be placed on manned spaceflight as opposed to unmanned interplanetary exploration; strong competition for budgetary resources between the groups has also existed. A collaborative relationship with a nonaerospace company, with the goal of commercially oriented research and development activities, represented an important new di-

mension of the manned spaceflight program. It is therefore not surprising that the 3M-NASA memorandum of understanding became a focus for conflict between those with strongly divergent views. Opponents attempted to convince Beggs that such an agreement would be a mistake for the agency. Smith's point of view was able to carry the day, however.

Emergent Interpersonal Transaction Processes and the MOU Negotiations

Study of the formal negotiation and agreement phases in the evolution of the memorandum of understanding led to an initial set of observations regarding informal interpersonal transaction processes. The entire string of transactions in the co-venture between 3M and NASA evolved from an effort by 3M scientists to develop a better picture of microgravity research opportunities.

Participants in the TAPC meeting seeking ways of motivating scientists discoved that they knew very little about the microgravity environment. They initiated a series of inquiries designed to provide them with information. Somewhat serendipitously, Gerry Kottong was being told, at the same time, that 3M ought to be doing basic research in microgravity. In addition, Coopers & Lybrand independently was asking 3M's management to consider the opportunities of space, and visiting scientists from SRI were also raising the issue.

In combination, these events caused a small number of individuals within 3M to challenge assumptions about space-based research. Initially, the entreaties by Bonnie Dunbar had little effect. However, as additional challenges to existing 3M views about space were made, individuals such as Kottong began to challenge their own assumptions. In the terms of social psychologists, individuals at 3M appeared to be redefining cognitive maps dealing with space-based research.

After the Coopers & Lybrand presentation, views began to change more rapidly. Our conversations with all of the key participants reveal that each had begun to think much more seriously about space research, each had begun to ask questions, to gather additional data from various sources. As important, Drs. Krogh and Podsiadly now began to think of space as a real research opportunity. Their involvement provided a focus for those who were now actively seeking more information about research opportunities in the microgravity environment, including Kottong. Active sharing of information occurred much more frequently after the Coopers presentation. Krogh, Podsiadly, Kottong, Debe, Egbert, and others sought common ground. The process also caused each to continue to rethink his or her views about space.

To this point, individuals within 3M had not yet made direct contact with NASA. For this reason, we argue that the activities in which they were engaged related to space can best be described as processes of sensemaking, although their discussions about space research with each other had begun processes of understanding within CRL. None of the activities associated with events described so far suggest that any committing processes were underway. There were, however, psychological contracts in place between Krogh and Podsiadly, Podsiadly and Debe and Egbert, among others, that would provide the foundation for committing processes.

The first evidence of committing processes between the organizations appears in the August meeting between Kottong and Smith. Their telephone conversations had permitted each to find out more about the other's organization, but had had limited effect. Kottong informed us that he had been promising Krogh and others that NASA was interested in 3M. Smith indicated in an interview that he was not sure about the strength of 3M's interest. This suggests to us that processes of understanding between Kottong and Smith were not very effective to this point. Kottong's ultimatum to Smith changed that, however.

Both parties described their evening meeting in St. Paul as one in which they both "laid their cards on the table." The setting was informal, conducive to "telling it like it is." Kottong explained to Smith that as they learned more about it, people within 3M were getting very interested in the possibilities of doing research in space. Kottong told Smith that he felt that the 3M people needed to hear directly from some top people in NASA and that he could arrange for NASA people to meet with top-level people at 3M. Smith suggested that bringing NASA people to 3M might indicate the seriousness of NASA's interest. Kottong agreed. During this meeting the two also exchanged views on the role of space in U.S. national security issues, foreign competition, and the state of U.S. industry in general. They found their views to be very similar on most of these issues.

In terms of the model, the understanding process that we believe is reflected in this meeting appears to have produced facticity between Smith and Kottong. This may have been facilitated by the similarity of their cognitive maps about subjects such as national security and foreign competition. The significant overlap in their "shared, obdurate world" also apparently facilitated processes of committing between the two, not only as agents for their organizations but also as principals. The latter conclusion manifests itself in friendship that has been sustained since the meeting.

We believe that the October 3 meeting at 3M is further evidence of sensemaking, understanding, and committing processes at work. NASA's top officials described to 3M employees NASA's vision for the commercial development of space. Experts in materials sciences such as Bob Naumann talked frankly with Debe and Egbert about the feasibility of the experiments they had proposed in September. Naumann also got a first hand look at 3M's labs and was impressed with the capabilities they reflected. NASA officials reported to us that they left St. Paul feeling that 3M was a company that they "could do business with."

The November visits to Washington and Huntsville by Podsiadly, Debe, and Egbert reflect the same interpersonal processes at work. The 3M scientists needed to see first hand what NASA could do for them, especially in terms of providing hardware. The trip enabled them to meet with material scientists at Marshall and to get first-hand information and feedback on the experiments they were considering. The give and take at the scientific level that the 3M scientists described as the core of the Marshall meetings suggests to us that at the science level sensemaking and understanding processes dominated the trip. However, the face-to-face contact between the parties afforded by the trip also appears to have been an important factor in the evolution of committing processes between 3M and NASA scientists.

At a managerial level, this trip provides evidence of an escalation in the level and intensity of committing processes. General Abrahamson's letter to Lew Lehr is evidence that the arena of committing within NASA had moved from the midlevels at which Smith operated to the very highest policymaking levels.

By this time the cumulative effect of the products of sensemaking and understanding processes within 3M apparently was sufficiently strong for top management to formally approve a microgravity project in December 1983. The work involved in drafting and revising the memorandum of understanding that was done in November and December may be viewed as evidence that understanding and committing processes between the organizations had been fruitful at some levels in NASA. This was not the case throughout NASA, however.

Smith and Kottong's descriptions, in great detail, of the last-ditch effort to scuttle the memorandum of understanding can be construed, at the very least, as evidence that understanding processes between 3M and NASA had not produced a unified goal across all levels at NASA. Whether is was due to failures in sensemaking or understanding processes at NASA, a simple desire not to commit to an agreement between 3M and NASA, or other reasons is still not clear to us.

In the formal processes of negotiating and reaching agreement on an memorandum of understanding, informal processes that we have described as sensemaking and understanding dominated. The fact that an agreement was signed reflects that some processes of committing were also taking place. However, we believe that the fact that the memorandum of understanding was simply an agreement to agree reflects the tentative nature of the level of commitment between 3M and NASA. In addition, the kinds of decisions made by 3M, and especially the ease with which they were made by top management, we believe are reflective of ongoing committing processes within 3M. At NASA, on the other hand, the little data we have suggests that the efficacy of a space commercialization program and relationships with 3M lacked strong organization-wide commitment. In Figure 10–3 we model the emergent interpersonal transaction processes associated with negotiating and reaching agreement on the memorandum of understanding.

THE TWO-YEAR JOINT ENDEAVOR AGREEMENT

The major events associated with the formal processes of negotiating and reaching agreement on a two-year joint endeavor agreement between 3M and NASA are identified in Table 10–4. In a JEA, NASA agrees to provide a private-sector firm with a free flight on board the shuttle. It may also provide the firm with the hardware necessary to house an experiment or

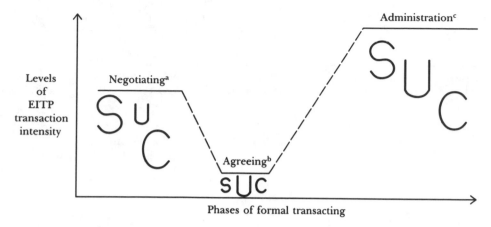

S = Sensemaking processes.
U = Understanding processes.
C = Committing processes.

The size of the letter reflects the relative importance of the process within each formal phase.

a. In terms of the transactions model outlined in Chapter 5, negotiating of the Memorandum of Understanding began when Kottong started looking for someone in NASA with whom 3M could discuss microgravity research.
b. Relative to negotiating the Memorandum of Understanding, the agreement processes were easy going. They occurred by phone and by mail.
c. The administration of the Memorandum of Understanding, and agreement to agree, is manifested in reaching agreement on a Two-Year Joint Endeavor Agreement and in designing and flying the DMOS1 experiment.

FIGURE 10–3. A Model of Emergent Interpersonal Transaction Processes Associated with the 3M/NASA Memorandum of Understanding.

in which to conduct an experiment (such as a furnace). It usually asks that the firm reciprocate (by providing NASA and its scientists with access to the data derived from the experiment). These provisions are all subject to negotiation.

Negotiating the Two-Year Joint JEA
Once the memorandum of understanding was signed, both organizations realized that they needed to formalize their relationships. In mid-February Mike Smith of NASA was designated as 3M's contact person at NASA. At about the same time, 3M CEO Lew Lehr designated Dr. Chris Podsiadly as the official contact at 3M. Podsiadly wished to improve his knowledge of

NASA, to meet people with whom he and his team would have to work if they were to conduct experiments on-board the shuttle, and to see, first hand, NASA operations.

Consequently, between March 11–15, 1984, Podsiadly and a group of 3M scientists toured five of NASA's research centers. Mike Smith accompanied them on this trip, which was done on 3M's corporate jet. The trip provided the first opportunity for Podsiadly and Smith to spend time with each other, and they used the time flying from one center to another and during the evenings in getting to know more about each other, their jobs, their organizations' capabilities, and their views on space research and the role of the private sector and NASA in commercializing space.

351

TABLE 10–4. *Major Events in the Formal Processes of Negotiating the Two-Year Joint Endeavor Agreement.*

February 1984	March 1984	April 1984	May 1984	September 1984
Contact persons named by 3M and NASA	3M people tour NASA sites	3M shifts focus to drafting a MOA	NASA says no to MOA	JEA signed by 3M and NASA
	Drafts started on JEA		Smith of NASA helps 3M in drafting JEA	
	Manning joins SRL			
	3M gets model JEAs from NASA			

Podsiadly also was able to use the time on the trip to provide Smith with a clear picture of the kinds of resources that 3M's SRL actually had to work with in a microgravity research program. Smith had assumed that media reports that had estimated that 3M was committing upwards of $5 to $10 million to the project were accurate. In fact, at that time Podsiadly had no formal budget for the proposed research. He had already begun to "bootleg," however, in the best of 3M's tradition.[6]

Back in St. Paul, work on the drafting of the initial joint endeavor agreement (JEA) began around March 1, 1984. On that date, Jan Manning joined the staff of SRL. An attorney, she had been employed at 3M for approximately twelve months, although not in a legal capacity. When she joined the staff at SRL, she became the first legal assistant permanently assigned to a project in 3M history.

3M's legal staff had no model agreements of its own to work from in drafting the JEA. Manning sought outside help. Mike Smith at NASA, and Peter Wood and Peter Stark with Booz-Allen-Hamilton in Washington, provided samples of JEAs that NASA had entered into with McDonnell Douglas, Microgravity Research, Inc., and Fairchild. Dick Halpern's of-

fice also provided them with a generic JEA model that had been developed earlier. The lawyers and the management team at 3M had initially assumed that a draft JEA would provide the basis for negotiating an agreement with NASA. The initial draft of a JEA was prepared by Jan Manning. She sent her draft to the patent counsel at 3M, Bill Ewert, and to Associate Counsel Merritt Marquardt for their review.

Midstream, however, Podsiadly suggested a change of course. In contrast to the MOU, the JEA on which the lawyers were working was a much more complex legal document. Podsiadly knew that NASA also used a document called a memorandum of agreement, which was different than a memorandum of understanding but was about as short and was much less complicated than a JEA. Podsiadly also believed that NASA was firmly committed to helping 3M fly with as little red tape as possible and based his conclusions on the efforts that had been made in 3M's behalf during the trips to NASA facilities in November and in March. Accordingly, the 3M legal team devoted two months to drafting and internally reviewing a two-page MOA.

The MOA was forwarded to NASA on May 10, 1984, having been signed by Les

Krogh. In a letter to NASA's Dick Halpern, dated April 19, 1984, Podsiadly outlined 3M's reasons for seeking the MOA rather than dealing with NASA through a JEA: Getting into space as quickly as and as efficiently as possible. NASA, however, had never used just as MOA as the basis for this kind of co-venture. Under the terms of MOA's, NASA worked closely with another organization in defining experiments, explaining NASA procedures in general, and those associated with doing experiments on board the shuttle in particular, and perhaps in conducting a joint experiment. In contrast to the autonomy granted an organization under a JEA, NASA remained firmly in control of all matters under a MOA.

It took NASA about a month to decide that they could not let 3M fly under the kinds of conditions that 3M had proposed in their draft of a MOA. 3M's lawyers then set about drafting a new JEA that they sent to NASA. Their contact point at NASA, Mike Smith, concluded that the draft was longer and more complicated than was needed, and so he helped them draft a much shorter version of a JEA. His redrafting was designed to minimize the chances that the 3M JEA would get caught up in NASA's internecine warfare. At this point in time it was almost September, and 3M was still planning on flying in November. Time was of the essence.

In drafting the JEA, both 3M and NASA had separate agendas of items that were critical from their respective points of view. They included issues related to patent indemnity, insurance, what each party was going to do, the duration of the agreement, and exact specifications of what 3M would fly. NASA had vetoed use of an MOA because that kind of legal document did not address many of these items.

As was the case with the MOU, most of the negotiating connected with drafting and revising the JEA (and the abortive MOA) took place over the telephone or by mail. Only as the deadline of being able to legally integrate the experiment on board the shuttle in preparation for the November flight drew near did face-to-face negotiations occur.

Emergent Interpersonal Transaction Processes in Negotiating the JEA

The trip that Podsiadly, his colleagues, and Mike Smith took in March reflects processes of sensemaking but primarily understanding at work. Smith thought he knew what kinds of resources 3M was committing to the project. The availability to Podsiadly of the 3M corporate jet initially gave Smith a false sense of the level of commitment, a sense that reenforced views that Smith had developed from media accounts of 3M's commitment. Both Smith and Podsiadly described an ongoing effort in the early stages of the trip by Podsiadly to convince Smith that 3M's commitment did not even begin to approach the lowest of the reported figures ($3 million). Podsiadly's efforts continued during dinner and in casual conversations in their hotels. Not until they got to California did Smith begin to realize that Podsiadly was being frank with him regarding resources. By the end of the trip, however, Smith had a more realistic view of what Podsiadly had to work with.

For Podsiadly, the trip provided first-hand evidence that NASA people, especially at the centers directly involved in the shuttle program, were, as he described them, "can do" people. They were able to answer his questions and to assure him that if his people had an experiment ready to fly by November, the NASA people would be able to fly it. The intense questioning and briefing sessions that took place throughout the trip refined Podsiadly's cognitive maps about NASA and the face-to-face nature of these sessions permitted both sides to begin to see that a November flight was possible. The NASA group, however, remained somewhat more skeptical: No one had ever prepared a complicated experiment from scratch and gotten it flight ready in seven months.

The five days of constant associations with each other during the trip had one other by-product. As Smith pushed Podsiadly on what kinds of resources were at his disposal, and refined his expectations about what resources the people in Science Research Laboratory actually had to work with, he realized the extent to which he was going to have to provide Podsiadly with help within NASA. The ability to observe Podsiadly over the five-day period, Smith says, lead him to conclude that they were "two of a kind." Podsiadly also described reaching the conclusion that he could trust Smith to get the job done within NASA. The evidence available suggests that a psychological contract between the two had developed during this trip. Their expectations of each other, and their respective organizations, were quite similar, and although unwritten, were becoming more explicit.

Within 3M, the formal processes of negotiating with NASA were accompanied by interpersonal transaction processes at two levels: within the legal team and between 3M's legal team and NASA.[7] Although no one at SRL, or in the general counsel's office, recognized it at the time, Manning's assignment to SRL was a key element in facilitating sensemaking and understanding processes within 3M on legal matters related to the co-venture with NASA. Her immediate access to Podsiadly and the research scientists provided her with a first-hand picture of their concerns and objectives. As she attempted to translate operational necessities into legal possibilities, she was provided immediate feedback by those who would have to live within the four corners of the legal document. She was also able to tell Podsiadly and the research team what the contractual constraints on them were likely to be.

For Manning especially, access to so-called model drafts of the JEA, accelerated sensemaking on legal options open to 3M in dealing with NASA. The results of the initial efforts to attempt to reach an agreement with NASA on a memorandum of agreement, rather than a JEA, suggest to us that both the sensemaking and understanding processes at work within 3M were flawed, however. First, as they were described by Dick Halpern at NASA, we believe that MOA's normally had been employed to facilitate sensemaking and understanding processes between NASA and other organizations. Second, what the lawyers (and managers) at 3M did not fully appreciate was the extent to which multiple camps within NASA used the JEA process as a forum for airing internal grievances and gaining power and resources.

Similar results were taking place within NASA. For example, Jan Manning pointed out that 3M was the first commercial firm to approach NASA and say, "All we want to do is basic research." Other firms had sought arrangements by which they would manufacture in space. Apparently, it took some time for managers at NASA to operationalize the differences and to translate their findings to its lawyers.

That the processes of understanding, at times, appeared laborious in the development of the JEA, was due, perhaps, to the degree to which 3M's requests diverged from the cognitive maps developed by the NASA negotiators in reaching agreement on earlier JEAs. On the other hand, the sensemaking, understanding and committing that had occurred during the negotiation of the memorandum of understanding appears to have facilitated comparable transacting processes in both negotiation and agreement phases of the JEA because most of the work was accomplished long-distance, via telephone or mail.

In reviewing the events leading up to the signing of the two-year JEA on September 25, 1984, it seems clear that both NASA and 3M sought only a formal contract which would allow 3M to fly by November 1984. Both sides made significant concessions to the other during the negotiations.[8] The legal formalities were reasonably well understood by the lawyers. Their application in the context at hand was less clear, in some measure because neither the managers nor the scientists of the two organizations had ever dealt with this kind of coventure. In addition to the constraints of time, the psychological contract between Smith and Podsiadly, and the trust between the two that it reflected, also appears to have provided a basis for overcoming short-term uncertainties associated with efforts to reach agreement on a two-year JEA. A model of the interpersonal transaction processes just described is depicted in Figure 10–4 below.

THE GENERAL MOTORS CO-VENTURE

In late 1984 General Motors contacted NASA about the possibility of conducting research in space. Mike Smith suggested that GM contact 3M. By now he was impressed with 3M's ability to get things done, and he believed that it might be easier for GM to accomplish its objectives by doing business with 3M than by going through the NASA bureaucracy. After some exploratory discussions that were described by Posdiadly as very routine, very much the way 3M did business with other firms all the time, and very much different from the ongoing negotiations with NASA, 3M and GM decided to engage in a collaborative venture, and a contract to do so was announced in March 1985. Table 10–5 indicates some of the major events in the development of the co-venture between 3M and General Motors.

Under terms of the agreement reached by 3M and GM, scientific experiments would be designed and conducted to "determine the effect of the structure of the crystal phase on the mechanical properties of Nylon which are important to automotive components." 3M would provide access to its staff and equipment and one of its flights on the shuttle, while GM would

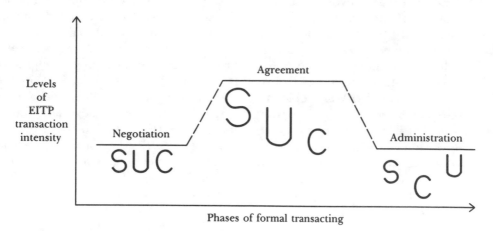

S = Sensemaking processes.
U = Understanding processes.
C = Committing processes.

The size of the letter reflects the relative importance of the process within each formal phase.
FIGURE 10–4. A Model of Emergent Interpersonal Transaction Processes Associated with the 3M-NASA Joint Endeavor Agreement.

TABLE 10–5. Major Events in 3M-GM Coventure.

November–December 1984	March 1985	July 1985	November 1985	January 1986
GM contacts NASA and then 3M	3M and GM announce contract	3M-GM experiment separated from DMOS2 Polymer morphology lab initiated	DMOS2 flies; more staff resources flow to PM1	Explosion of the space shuttle *Challenger*

provide the material. In typical joint-venture terms, the agreement provided for a 50-50 division of results. 3M's lawyers told us that there was absolutely nothing unusual about the terms of the deal or the manner in which it was negotiated.

The General Motors experiments were to use flight time already allotted to 3M under the terms of the two-year JEA; 3M initially intended to incorporate the experiment into the DMOS2 flight in November 1985. However, as the design of the experiment evolved, an increased appreciation of the dynamics of heat transfer demonstrated that the experiment would be incompatible with DMOS2. Plans were made to devote an entire flight to polymerization processes in microgravity. This flight, Polymer Morphology 1 (PM1), would incorporate the GM experiments. The loss of the *Challenger* put the PM1 flight on hold, however.

Smith's decision to put GM in touch with 3M suggests to us that his cognitive maps of his role in promoting the commercialization of space had been modified by his experiences with 3M. Rather than assuming that he needed to promote commercialization of space by securing companies as direct co-venturers with NASA, Smith said he now believed that, at least in certain situations, the commercialization of space might well be served by having interested companies co-venture with firms that had proven their ability to work effectively with

NASA. The processes of sensemaking, understanding, and committing in which Smith was engaged during the evolution of the co-venture between NASA and 3M are likely factors in the change in view which Smith described.

In Figure 10–5 the model of emergent interpersonal transactions processes associated with the 3M-GM joint venture is set out. At 3M, the decision to engage in this co-venture was guided, it appears, by processes in which 3M managers primarily resorted to cognitive maps describing customer-supplier relationships. GM was a 3M customer, and the co-venture was initially interpreted largely as an opportunity to engage in research that might generate new products that would further strengthen that relationship. Thinking about the co-venture in these terms permitted 3M's managers and lawyers to pursue formal negotiation and agreement processes for this co-venture in a manner that they described as business as usual. Podsiadly, Chris Chow, and Mike Runge all said this co-venture was like any other between two firms. Very few probing questions were asked about how the GM experiment would "fit" in the 3M research program. Neither side spent much time exploring the capabilities of the other. The dealings between the lawyers were described as routine.

Only during the administration phase of the co-venture, as 3M began to allocate resources to a polymer program, did problems

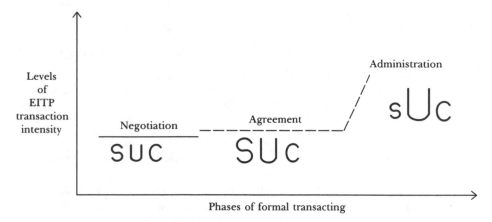

S = Sensemaking processes.
U = Understanding processes.
C = Committing processes.

The size of the letter reflects the relative importance of the process within each formal phase.
FIGURE 10–5.   A Model of Emergent Interpersonal Transaction Processes Associated with the 3M-GM Joint Venture Agreement.

begin to surface. The probable incompatability of the GM experiment with the DMOS2 might have been identified during negotiations had sensemaking and understanding processes been designed to explore the fit of the GM proposal with 3M's microgravity research program. They were not, however. Finally, only when they managers at SRAL perceived a need to develop alternative sets of resources, well after committing to the General Motors co-venture, did they begin to engage in sensemaking and understanding processes—that is, conducting microgravity research by actively seeking other co-ventures designed to determine if 3M could capitalize on its experience in opportunities.

THE 3M-MCDONNELL DOUGLAS ASTRONAUTICS COMPANY TRANSACTIONS

On November 18, 1985, 3M and McDonnell Douglas announced that they had tentatively agreed to establish a joint venture between 3M's Riker Laboratories and McDonnell Douglas Astronautics Company (McDAC). The co-venture would pursue the manufacturing, testing, and marketing of erythropoietin produced in a microgravity environment. McDonnell Douglas had been engaged in the production of pure erythropoietin, using an electrophoresis technology, in a microgravity environment since the mid-1970s. The major events associated with this transaction are outlined in Table 10–6.

McDonnell Douglas had been joint venturing with Johnson and Johnson's Ortho Labs on this project. However, Ortho decided to withdraw. Because of their mutual interests in the commercial uses of space, Chris Podsiadly and McDonnell Douglas's Jim Rose, project director for their Electrophoresis Operations in Space (EOS) program, had been communicating with each other for some time.

Both Rose and Podsiadly were active supporters of space programs, inside and outside their respective companies. Each had made

357

TABLE 10–6.   *Major Events in the 3M-McDAC Coventure.*

1984	April–June 1985	September 1985	October 1985	November 1985	January–February 1986
Podsiadly and Rose appear on space panels together	Podsiadly and Rose make first attempt at 3M-McDAC co-venture	Rose alerts Podsiadly to Ortho withdrawal	Podsiadly raises McDAC co-venture possibility at 3M	3M and McDAC announce intent to co-venture	3M withdraws before signing contract

trips to the other's organization, and they had appeared on a number of space commercialization panels together. As early as spring 1985 the two had discussed the possibility of working together. They had brokered a meeting of scientists from their respective organizations who were working on a related matter. There seemed to be an opportunity for collaboration. During a series of meetings between 3M scientists and their McDonnel Douglas counterparts in St. Louis, it became apparent to both sides that the 3M people could not help the McDonnell Douglas team. The lack of success here, however, apparently did not affect the relationship between Podsiadly and Rose. They continued to keep their organizations in contact with each other.

Consequently, very soon after Ortho broke off their agreement with McDonnell Douglas, Podsiadly learned that the latter organization was looking for a new co-venturer. Rose also informed Podsiadly that McDAC had evoked very little interest among U.S. manufacturers and was contemplating discussions with a number of Japanese firms.

Podsiadly raised the possibility of a co-venture with McDAC during a Technical Audit of SRAL. Krogh and others were immediately interested (it should be recalled that 3M began to give serious consideration to commercial uses of space only after learning of substantial Japanese activity in the area.) They passed the word along to Jerry E. Robertson, executive vice president of 3M's Life Sciences Sector.

Robertson contacted Ronald O. Baukol, vice president and general manager of Riker Laboratories, who initially saw that "Riker's program for the delivery of drugs through alternate means could benefit greatly from [a] joint project."

In terms of the emergent interpersonal transaction process model, the evidence suggests that Podsiadly and Rose had been engaged with each other in sensemaking and understanding processes regarding commercial opportunities in space for some time prior to the November 1985 announcements. The fact they had been seeking ways to bring their own organizations together for a co-venture suggests that a psychological contract may have also developed between the two individuals. Certainly there were informal commitments.

When Rose also informed Podsiadly that McDAC had evoked very little interest among U.S. manufacturers and was contemplating discussions with a number of Japanese firms, it may have initiated a new process of sensemaking by Podsiadly, in part involving his cognitive map regarding Japanese competition. This map had been developed by many within 3M as a result of their experience with the Japanese, especially in the video film market. Podsiadly's cognitive map on this subject also very likely reflected the admonishments of Bonnie Dunbar regarding Japanese commitment to the commercial use of space. Thus, when he heard that McDAC was contemplating doing business with the Japanese, he was spurred to act.

This episode, which is modeled in Figure 10–6, provides an interesting twist on the relationships that exist between the sensemaking, understanding, and committing processes. Because both organizations were involved with NASA, members such as Podsiadly and Rose had the opportunity to engage in sensemaking activities vis-à-vis each other and their respective organizations. As they sustained their contacts and acquired more information regarding the other, both gradually developed a commitment to find ways by which their respective organizations might co-venture. Their activities were largely responsible for the decisions to co-venture. However, neither Rose nor Podsiadly would be directly involved in the proposed activities, and this precipitated new sensemaking and understanding processes within 3M and McDonnell Douglas, by an entirely new cast of characters. One outcome of these processes was a decision by the 3M people that no benefits would come to 3M by engaging in the co-venture. As a consequence, no legal contract was ever concluded between the two organizations.

MANAGERIAL AND RESEARCH IMPLICATIONS OF EITP

As with much of the MIRP research, the findings outlined in this chapter reflect work in progress. Nonetheless, we feel confident that the data we have accumulated over the past three and one-half years is substantial enough to permit us to draw some tentative conclusions about emergent transaction processes that managers and researchers may find useful.

Managerial Implications

3M's involvement with NASA proceeded in an incremental, disjointed, fragile, and somewhat serendipitous manner, so much so that we believe managers need to refine their antennae to better pick up subtle signals from the external

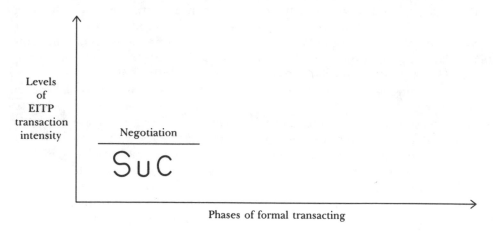

S = Sensemaking processes.
U = Understanding processes.
C = Committing processes.

The size of the letter reflects the relative importance of the process within each formal phase.
FIGURE 10–6. A Model of Emergent Interpersonal Transaction Processes Associated with the 3M-McDonnel Douglas Transactions.

environment during their sensemaking processes. Not all new ideas initially take the form of shocks or jolts.

In this case, both 3M managers and scientists "missed" some early signals, especially those from Bonnie Dunbar. In hindsight, this omission does not appear to have been fatal. However, we believe that there are many occasions when individuals inside organizations "miss the boat" altogether. The general response of U.S. industry to the opportunities of commercial development in space appears to reflect this kind of incomplete sensemaking and/or understanding process. The failure of the U.S. automobile industry to appreciate and respond to the competitive challenge from Japan and Western Europe is a clear and dramatic example of inadequate sensemaking.

The more general MIRP framework, and the "fireworks" process model discussed by Van de Ven, et.al. in Chapter 1, suggests that managers might wish to more actively involve people and their ideas in day-to-day operations as an adjunct to innovation processes. The disparate cognitive maps that these individuals will bring to the discussion may generate better processes of understanding. There is the danger, however, that attempts to reach and sustain facticity may lead to pathologies such as "groupthink" (Janis 1972), especially if the need for inclusion among many of these individuals is high.

We also suspect that another outcome of disjointed, incremental, sometimes serendipitous sensemaking and understanding processes is that innovations frequently begin without the formal commitment of resources. For example, because no budget for microgravity research existed, the program's initial development within 3M entailed substantial "bootlegging" of resources. At NASA, the program's managers frequently worked "ahead of the paperwork," making commitments to 3M before receiving formal authorization to do so.

This kind of outcome is not unique to the 3M/NASA co-venture. While it is an empirical question, we believe that the relative degree of success that 3M has had with the project to date is partly a function of its ability to adapt rapidly to new opportunities. This suggests the importance of processes that lead to the creation of slack (Cyert and March 1963; Bourgeois 1978). In addition to the kinds of prescriptions that the slack literature offers (such as retaining centralized control over portions of budgets, overstating expenses, and understating revenues), the EITP model outlined here implies a number of ways by which slack can be created.

First, greater reliance on informal, rather than formal, commitments is likely to increase slack, measured in terms of discretionary action. As trust increases, and looser coupling between organizations occurs, slack is increased. More comprehensive sensemaking and understanding processes may lead to psychological contracts in which intrinsic rewards substitute for extrinsic rewards, thereby increasing a manager's supply of "rewards" slack. Similarly, time expended on more comprehensive sensemaking and understanding processes early in a program of innovation may create slack, measured in units of time, down the road as the innovation moves from concept to commercial application.

Finally, the sensemaking and understanding and committing processes that we have described, if managed in effective ways, can create slack within organizational information and communications systems as individuals act on the products of these processes rather than formal communications, rules, and policies. The danger, of course, is that the pathologies to which we referred in Chapter 6 can consume this slack, and much more.

Somewhat related to slack is the concept of loose-coupling. Its value, especially in the context of innovation, has long been recognized (see March and Simon 1958; Singh 1983). In the 3M/NASA co-venture, the project was initially loosely coupled to SRL. This gave Podsiadly the flexibility of treating 3M—the organization—as a market from which he could "buy" or "borrow" resources on a short-term basis. His ability to do so, it appears, was contingent, first, on his ability to achieve facticity with his managerial counterparts elsewhere in CRL and/or 3M; and second, to establish psychological contracts between himself and their subordinates.

Research Implications

The model that has emerged from our investigation of the NASA/3M co-venture will require a substantial amount of additional research before we feel confident that it reflects a generalizable model of emergent interpersonal transaction processes. Discussions with our MIRP colleagues suggest that these processes have occurred in other sites, and our preliminary analysis of their research findings indicates to us that processes that resemble those described here are occurring elsewhere. In addition to further investigating whether the processes that we have observed are an inherent part of the processes of transacting, our research suggests a number of other issues that will also require further analysis.

Patterns of EITP.   The different, yet sometimes recurring, patterns of EITP observed in the course of the 3M-NASA co-venture, and reported on to a limited extent in this chapter, provide a basis for offering some propositions about the contexts under which certain interpersonal transaction processes might emerge. Needless to say, these propositions should be considered as extremely tentative. We note, however, that the patterns have been observed in a limited number of similar contexts outside the 3M-NASA study.

Figures 10–7a through 10–7d reflect four general propositions regarding the emergence of EITP's across differing types of trans-

actions. Figure 10–7a depicts a transaction that involves old parties but a new "deal." The data gathered during the 3M-NASA study provides a basis for arguing that the most intense EITP will occur in the agreement stage, since in many important respects the parties are likely to have established, in earlier deals, EITP processes governing their conduct in negotiation and administration phases. Trust has been established, the same lawyers are likely to be involved (although if they are new, then new sensemaking and understanding processes are likely), the same language is spoken. Hence, the need for extensive sensemaking and understanding processes is reduced.

A new deal, on the other hand, is likely to require significant new sensemaking and understanding processes related to its objectives and its impacts on the respective organizations. The new deal also may alter relationships established after the previous deals to such an extent that new committing processes also are required. These issues, among old parties, will be resolved in the agreement stage, not during administration. The parties will recognize the problems in advance and seek to resolve them in the context of the deal. Thus, when the deal is implemented, sensemaking and understanding processes, although ongoing, will be less intense than those that occurred during the agreement stage, at least until problems arise. Committing processes, to the extent that the new deal changes relationships, however, are likely to intensify during administration.

In the course of the evolving relationship between 3M and NASA, three formal agreements were reached (only two of which were discussed in this chapter). By the time the parties got around to attempting to reach agreement on their ten-year joint endeavor agreement, they knew each other and how each viewed the world, but the terms and objectives of the proposed ten-year JEA were significantly different from those of the MOU and the two-year JEA, as were the stakes for both 3M and

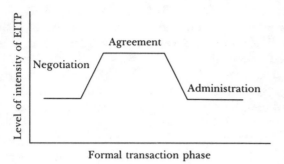

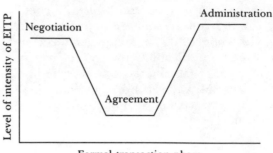

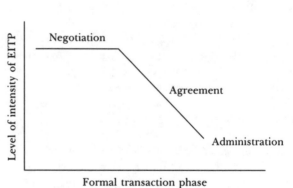

FIGURE 10–7.   *Illustrative EITP Transaction Processes*

NASA. The sensemaking and understanding processes that we observed in connection with the ten-year were much more intense than those described for the MOU and the two-year JEA.

Restated, sensemaking and understanding processes will be relatively more intense during agreement than other formal transaction phases when old parties negotiate new deals. If the new deal alters existing relationships, committing processes will intensify during administration.

In contrast, in Figure 10–7b, negotiation and administration stages of the transaction are more intense than the agreement stage. We argue that emergent interpersonal transaction processes of this type will be associated with deals in which parties from two very different cultures are engaged or in cases in which "sovereignty" or "institutional guarantors" are absent. In such circumstances, the "legal" document that is the usual product of the agreement stage may be difficult, if not impossible, to fashion. Consequently, we believe that much more sensemaking, understanding, and committing has to occur in negotiation and administration stages than during the agreement stage.

To an extent, the different cultures of the public and private sectors may have produced the patterns that emerged in the transactions associated with the MOU and two-year JEA. 3M had not done business with the government for a number of years, and there were few people in the organization, and virtually none in SRL, who had extensive experience in dealing with government agencies. At NASA, among those in the office of commercial programs only one had an M.B.A. degree and none had a background in business. Employees dealing with the space shuttle were accustomed to dealing with firms in the aerospace business but not with firms such as 3M driven by competitive market dynamics. Thus, in addition to technical issues related to an agreement, both sides had to engage in extensive sensemaking and understanding processes focused on the broader culture of the other's sector of the economy.

From this evidence we draw the following tentative conclusion: Deals between parties from very different cultures will reflect greater intensity of emergent interpersonal processes in negotiation and administration stages than during the agreement stage of formal transaction processes.

For routine or repetitive transactions between parties (Figure 10–7c), we believe that EITPs, in the main, will not vary much between formal stages. If the transaction is routine or repetitive, little sensemaking, understanding, or committing is required in the first instance. If the parties are old parties doing old deals, interpersonal transaction processes, in fact, probably will look like, and be treated as, "routines" as March and Simon (1958) employed the term. To an extent, this kind of behavior was reflected in the 3M-GM co-venture. 3M initially viewed the deal as just another buyer-supplier transaction. All with whom we discussed the deal at 3M saw it as routine. Consequently, we would argue that emergent interpersonal transaction processes will not vary substantially across formal transaction stages and will evidence a relatively low level of intensity throughout.

There are also transactions that involve either low uncertainty per se or are managed in such a way as to minimize uncertainty. We refer to the latter as "Japanese" style transactions because, in many important respects, they appear to reflect an approach to decisionmaking and transactions thought to be characteristic of the style favored by the Japanese. As reflected in Figure 10–7d the most intense EITP occurs during the negotiation stage, whereas in the agreement and administration stages of the transaction the intensity of EITP falls off dramatically. We have not observed such a pattern in the 3M-NASA transactions to date.

Interactions among the Processes. In light of the key role that the products of individual sensemaking appear to play in the processes of understanding and committing, additional research is needed on (1) the antecedents of comprehensive sensemaking and (2) whether the accuracy of cognitive maps, measured against some objective reality, matters to other emergent interpersonal transaction processes. Frederickson's (1983, 1984) work on comprehensive decision processes provides a basis for exploration of the first question. The research of Bourgeois (1978), Dess and Keats (1987), and Ring (1985) regarding the effects of accuracy of perceptions among top management teams on organizational performance serves as a starting point for the second issue.

Second, a number of issues regarding transitions from one EIT process to another require additional study. For example, the data from the case regarding the role that the lawyers at 3M played in the JEA negotiations raises an interesting question: Can sensemaking and committing occur without understanding processes also occurring? When two parties are involved, it seems to be inherent in the committing process as we have described it that understanding processes either preceded committing or occurred in sequence with committing. Similarly, when parties from two different organizations are involved, does moving from sensemaking processes to those of understanding imply that some committing processes will also be initiated?

The Mike Smith–Jerry Kottong meeting in August 1983 referred to in this chapter's discussion of the evolution of the MOU raises an interesting point regarding facilitating processes of understanding. Prior to this meeting, their discussion had been remote, by telephone or mail. Each had been developing cognitive maps of a relationship between 3M and NASA, and there was, as it turned out, a good deal of overlap between the maps. However, there were several issues, such as the intensity of 3M's interest in the program, that Smith had failed to pick up on. The Smith-Podsiadly conversations during the March 1984 trip to NASA research centers, and the facticity that they achieved regarding each other's organization's capabilities, provides additional evidence that face-to-face meetings may be more appropriate to understanding processes. In propositional terms, achieving facticity is facilitated by face-to-face communication.

There also appears to be a definitional question related to sensemaking and understanding processes: What is the relationship between one-way and two-way communication processes and sensemaking and understanding? Clearly, sensemaking can involve one-way communication links. A person tells me something, and it aids in the development of my cognitive map of some phenomenon. I need

not respond, but if I do, is the response associated with processes of sensemaking, understanding, or both?

Can understanding processes logically operate in the absence of two-way communications? The answer would seem to be No. On the other hand, two individuals could witness an event, and without ever communicating directly with each other achieve complete facticity as to its meaning. The issue clearly requires further analysis.

Another definitional question worthy of further consideration relates to the comprehensiveness and completeness of sensemaking and understanding processes. What does it mean for these processes to be comprehensive or complete? We have data that suggest that less than comprehensive sensemaking and understanding may be accompanied by committing processes that, in retrospect, are seen to have been premature and have had dysfunctional consequences. Yet as discussions of satisficing and optimizing (see March and Simon 1958; Simon 1976) make clear, attempts to ensure comprehensiveness in these processes generally entail dysfunctional consequences as well. Can the adequacy or comprehensiveness of the processes we have described be assessed by any measure other than functional results from actions based on their products?

We also note that in those cases in which time creates the luxury, initial sensemaking processes regarding innovation ideas, environmental shocks or jolts, and so forth may become quite comprehensive. However, as the cognitive maps that are the products of these processes develop, selective perception and other pathologies may begin to occur, thereafter reducing the efficacy of sensemaking. Moreover, in those instances when sensemaking is accompanied by a search for common ground, understanding processes may lead to groupthink (Janis 1972) or similar phenomena that diminish the effectiveness of understanding processes. Finally, if committing processes flow from those of sensemaking and understanding, the groupthink pathology may also lead to or be accompanied by an escalation of commitment (see Staw 1976; Staw and Ross 1987) to a course of action doomed to failure. In short, if the emergent interpersonal transaction processes we have described are interwoven and self-reenforcing, then poorly managed EITP may accelerate and amplify organizational crisis.

Finally, we note that this study has produced a number of additional conclusions regarding the legal and managerial dimensions of transactions in general that have been addressed in Chapter 6. Those interested in the larger issues are invited to explore our analysis therein. In addition, that analysis discusses a number of issues related to research design and methods that apply with equal force in the case of further testing of the EITP model outlined here.

NOTES

1. As we employ the term, *co-venture* refers to any cooperative arrangement between two or more organizations. As such, it has a broader meaning than *joint venture,* which usually requires some form of equity position by the parties. We have chosen the term to allow for the possibility that the range of cooperative arrangements that evolved out of this project might vary greatly.

2. More detail on the commercial uses of space can be obtained from Osborne (1985: 45–58); the June 25, 1984, issue of *Aviation Week and Space Technology,* which is entirely devoted to the subject; U.S. House of Representatives (1981: 800 *et seq.*); Grey (1983); Business-Higher Education Forum (1986).

3. At the time the research began, CRL was known as the Central Research Laboratories.

4. 3M was frequently cited in Peters and Waterman (1982), and NASA historically was viewed as among the best managed of federal agencies (see Sayles and Chandler 1971).

5. Legally enforceable contracts must include the following basic ingredients within the "four corners" of the contract document: (1) an offer and acceptance that constitutes an agreement among the parties; (2) consideration that is legally sufficient and bargained for by the parties; (3) capacity to enter into the contract by all the parties; (4) reality of assent—that is, the parties genuinely agree to the terms and conditions of the contract; (5) form—that is, the contract meets legal requirements; and (6) legality—that is, the subject of the contract does not violate public policy.

6. Bootlegging within 3M is a formally recognized informal process by which managers can, within reasonable constraints, "find" resources to use on innovative projects. For example, individuals are also free to devote up to 15 percent of their time on projects outside their normal work responsibilities. When a manager puts together a team of such individuals, much can happen.

7. As lawyers, Manning, Marquandt, and Ewert were dealing with a new form of legal relationship and clearly were attempting to define its legal and operational parameters from 3M's perspective—that is, to develop new cognitive maps. In addition, because Jan Manning did not work directly for either of the other two lawyers and held a position that was unique to their experience, all three were defining their expectations regarding this relationship. From the outset, it was made clear to Manning that all final legal judgments would be made by either Marquardt or Ewert, a condition that created no problems for her. Conversely, Marquardt and Ewert realized that Manning's close working relationship with those who would have to live with a legal agreement meant that her judgment's on 3M's objectives in a legal contract with NASA had to be given serious consideration. As explicit expectations regarding the roles each would play in the working relationship between the three lawyers evolved, psychological contracts undoubtedly evolved.

8. Many of these were concessions that they would not be willing to live with as they sat down to negotiate a ten-year joint endeavor agreement.

REFERENCES

Bourgeois, Lionel J., III. 1978. "Strategy Making, Environment, and Economic Performance." Ph. D. dissertation, University of Washington.

Business–Higher Education Forum. 1986. *Space—America's New Competitive Frontier.* Washington, D.C.

Commons, John R. 1950. *The Economics of Collective Action.* New York: Macmillan.

Cyert, Richard M., and James G. March. 1963. *A Behavioral Theory of the Firm.* Englewood Cliffs, N.J.: Prentice-Hall.

Dess, Gregory G., and Barbara W. Keats. 1987. "Environmental Assessment and Organizational Performance: An Exploratory Field Study," pp. 21–25. In F. Hoy (ed.), *Best Paper Proceedings, Academy of Management.* New Orleans: Academy of Management.

Frederickson, James W. 1983. "Strategic Process Research: Questions and Recommendations." *Academy of Management Review* 8: 565–75.

———. 1984. "The Comprehensiveness of Strategic Decision Processes: Extension, Observations, Future Directions." *Academy of Management Journal* 27: 445–66.

Grey, Jerry. 1983. *Beachheads in Space.* New York: Macmillan.

Janis, Irving L. 1972. *Victims of Groupthink.* Boston: Houghton Mifflin.

March, James G., and Herbert A. Simon. 1958. *Organizations.* New York: Wiley.

Miles, Robert H. 1980. *Macro Organizational Behavior.* Santa Monica, Calif.: Goodyear.

Miles, Raymond E., and Charles C. Snow. 1978. *Organizational Strategy, Structure, and Process.* New York: McGraw-Hill.

Osborne, David. 1985. "Business in Space." *Atlantic Monthly* (May): 45–58.

Peters, Thomas J., and Robert H. Waterman, Jr. 1982. *In Search of Excellence: Lessons from America's Best Run Companies.* New York: Harper & Row.

Pfeffer, Jeffrey, and Gerald R. Salancik. 1978. *The External Control of Organizations: A Resource Dependence Perspective.* New York: Harper & Row.

Ring, Peter Smith. 1985. "Objective and Perceived Environments, Environmental Predictability, and Strategic Choice: An Empirical Investigation of California Commercial Banking." Ph. D. dissertation, University of California–Irvine.

Sayles, Leonard R., and Margaret K. Chandler. 1971. *Managing Large Systems: Organizations for the Future.* New York: Harper & Row.

Simon, Herbert A. 1976. *Administrative Behavior,* 3d ed. New York: Free Press.

Singh, Jitendra V. 1983. "Performance, Slack, and Risk Taking in Strategic Decisions: Test of a Structural Equation Model." Ph. D. dissertation, Stanford University.

Staw, Barry M. 1976. "Knee-Deep in the Big Muddy: A Study of Escalating Commitment to a Chosen Course of Action." *Organizational Behavior and Human Performance* 16: 27–44.

Staw, Barry M., and Jerry Ross. 1987. "Behavior in Escalation Situations: Antecedents, Prototypes, and Solutions." *Research in Organizational Behavior* 9: 39–78.

U.S. House of Representatives. 1981. *U.S. Civil Space Programs, 1958–1978.* Subcommittee on Space Science and Applications of the Committee on Science and Technology, 97th Congress, 1st Session.

11

COMMUNICATION, CONFLICT, AND LEARNING IN MERGERS AND ACQUISITIONS

David T. Bastien

Although mergers and acquisitions (M/As) are common ways to leverage organizational growth and expansion, many, if not most, have failed to meet their strategic, financial, and organizational goals (Kitching 1967). Davidson (1985) argues that, as a group, M/As have been unsuccessful over the past century, and Magnet (1984) reports that well over two-thirds of recent corporate M/As examined in a study by McKinsey and Co. never earned as much as the acquirer would have made by investing the money in certificates of deposit issued by a bank. Williams and Feldman (1986) report that, in a large sample study in the healthcare industry, virtually all of the M/As showed sharp downturns in earnings, productivity, and profitability (in for-profit M/As) and a rise in unwanted employee turnover.

Managerial and organizational problems involved in executing an acquisition are often cited as the principal reasons for this unsatisfactory performance, yet most research on M/As has focused on the strategic, financial, or legal aspects of corporate M/As (Brunner, et al. 1985; Davidson 1985; Steiner 1975; Salter and Weinhold 1979). These works are driven by microeconomic and financial theory, advising managers to capture earnings and profit through synergies of various sorts. Unfortunately, these synergies have proven to be very difficult to realize (Davidson 1985). This chapter examines the issues of these difficulties, the processes through which postacquisition downturns and performance problems emerge, and the social and communicative processes of withdrawal (such as employee turnover) and post-M/A changes in task.

THE LITERATURE

Only recently has a coherent body of literature emerged to discuss the human organizational

TABLE 11–1.   *A Syndrome of Linked Organizational and Individual Behavioral Common Responses to a Merger or an Acquisition.*

Individual level

High levels of personal uncertainty about expected performance and behavior

Increased needs for receiving communication and information combined with a decreased willingness to give information

Heightened attentiveness to the congruence of communication from higher organizational levels (especially acquiring company management)

Simultaneous fight-flight response by managers

Resistance to change

Culture shock (cf. Harris and Moran)

Focus on personal security rather than organizational goals

Cultural differences reported as a source of hostility and conflict, including resentment of acquiring company managers

Job searches by acquired company staff

Organizational level

Tendency toward rumor generation in the absense of hard information, with the rumors carrying worst-case prophesies, especially at times of inadequate congruence or quantity of information

Levels and units isolate themselves from communication from the rest of the organization, especially top management

High rates of unwanted turnover of managers and professionals, especially on the acquired side

Generalized decrease in organizational performance as shown by earnings, profits, and productivity

Factors mediating the intensity of the syndrome

The level of personal uncertainty is inversely related to the quantity of communication from (acquiring company) management

High levels of formal communication reduce hostility and resistance to subsequent collegial/interpersonal communication

Collegial communication (communication of professional and social acceptance) must follow formal communication if either is to be effective; collegial communication serves to reduce uncertainty through increase of interpersonal trust between individuals from the two organizations

Congruence of communication between content and actual behavior is inversely related to personal uncertainty

---

behaviors and processes that occur after a change in ownership has been announced. First, on the individual psychological level, several studies have noted a universal syndrome of communication and behavior (Bastien 1987; Bastien and Van de Ven 1986; Marks 1982; Marks and Mirvis 1985; Marks and Mirvis 1986). As shown in Table 11–1, this syndrome includes both individual and organizational behavioral symptoms. On an individual level it includes (1) high levels of personal uncertainty about expected performance and behavior; (2) increased attention to the behavior of top man-

agement; (3) increased needs for communication and information with a tendency toward worst-case rumor generation in the absence of hard information; (4) strong resentment of the behavior of the acquiring organization's managers. On an organizational level, these same studies noted common phenomena: (1) tendency to not pass news of problems either up or down; (2) tendency among top management on both sides not to communicate with their respective organizations about the M/A; (3) tendency toward high rates of unwanted turnover; (4) conflict, often quite intense, within the ac-

Institutional Affinity

	Synergistic	Exploitive
Single hierarchy	Merger: Metaphor: union, marriage Assumption: co-equals Culture: joint (either synthetic or third) Process: attempts at mutual negotiation and teambuilding within functions Problems: culture clash, turf guarding, conflict resolution can overload management, very time consuming, decline in performance	Absorbtive acquisition: Metaphor: digestion Assumption: acquirer can buy capacity easier than build it Culture: acquirer's, but with substructure of old culture remaining in the acquired organization Process: complete imposition of acquirer's systems, structures, and management Problems: culture conversion, transfer of acquirer's systems and practices, retention of key acquired side employees
Two hierarchies	Additive acquisition: Metaphor: alliance, annexation Assumption: acquiree is fully competent in its business Culture: fully separate Process: gifts of tangible resources flow to acquiree, but no efforts at integration Problems: acquiree's perceived devaluation, misdiagnosis, intrusion of acquirer's values and practices, strategic errors	Conformative acquisition: Metaphor: conquest Assumption: acquirer knows acquiree's business better Culture: acquiree's remains, but with superstructure of acquirer Process: management and system infusion, training, but not pervasive in the acquired side Problems: turn-around crisis, requires lots of acquirer's management talent, acquired staff loses self-worth and assertiveness

Number of postacquisition hierarchies

FIGURE 11–1.   A Typology of Mergers and Acquisitions

quired organization or between the participating organizations; (5) tendency toward a decrease in earnings and productivity.

This pattern of uncertainty and performance problems is related to the change that is inherent in M/As. The minimum possible change is in who ultimately must be pleased. Even if what pleases the new owners is the same as what pleased the old owners, it takes time and testing just to learn that one simple thing. If new decision rules, task processes, the actual task itself, and new conventions of communication must be learned and new managers and standards must be figured out, then the learning task confronting both individuals and the organization will be powerful.

Bastien and Van de Ven (1986) identified a four-quadrant typology of M/As that specifies four different kinds and levels of organizational change. As shown in Figure 11–1, they each specify the characteristics and typical problems of each type, but, most important, they also specify who must change, and how they must learn to behave in the new organizational context. The minimum changes and learning are as follows:

1. *Additive acquisition,* where there should be no externally imposed change from the acquirer to the acquiree, and therefore the only learning task ought to be to learn that no rules have changed;

2. *Conformative acquisition,* where the acquirer imposes some changes in task, systems, processes, and personnel on the acquiree, but where separate hierarchies are maintained. Here, members of the acquired organization must learn to accommodate to some changes, and the new managers coming from the acquiring side also must learn to work within a framework of the acquiree's basic culture and personnel;

3. *Absorbtive acquisition,* where the acquirer imposes all of its systems, structures, and processes; often there is also a complete change of management as well. Here the entire retained acquired company staff must accommodate to all new practices;

4. *Merger,* where two roughly equal organizations attempt to form a single hierarchy and set of practices without a formal power differential between them. They attempt to develop practices, structures, and so forth that are synthetic (half-and-half) or develop all new practices and structures that are foreign to both organizations. Here all members of both organizations must learn (at least potentially) all new practices and relationships.

This set of issues presses a review of concepts from the various research on learning (both individual and organizational). Although there is considerable controversy over the legitimacy of the application of a learning perspective to organizations, several authors, especially those associated with the sociotechnical systems school of organizational theory (Schon 1983; Argyris and Schon 1974), have found it useful. These authors have conventionally defined organizational learning in terms of knowledge shared within an organization. Their general concern is with the development of reflective practices that will improve the ability of useful knowledge to be generated and shared within the organization.

This chapter adopts the organizational learning perspective and extends Argyris and Schon's (1974) formulation. Argyris and Schon's (1974) formulation approaches organizational learning from a prescriptive viewpoint and is adopted here because it affords considerable explanatory power as well as parsimony and specificity. Parsimony is achieved by this perspective because it allows the application of substantial amounts of developed and accepted pedagogical theory to the issue of the management of change in M/As. Specificity is achieved through the same vehicle. Pedagogical theories are currently specific about the types of learning outcomes, learning problems, appropriate teaching methods, and so forth. Finally, this perspective affords considerable explanatory power, in that it provides an explanation for the process of organizational change that links both individual and organizational level behavior in M/As.

Argyris and Schon (1974) and Schon (1983) also distinguish between single-loop (conventional) learning and double-loop learning. Conventional learning is directed toward error correction in a stable situation in which standards and organizational practices such as norms and roles are known and appropriate.

Argyris and Schon (1974) define double-loop organizational learning as where performance anomalies or errors are linked to both task processes and the normative structure of the organization, changing the organization's picture of its universe. In other words, double-loop learning occurs when performance errors emerge that (1) are the result of following organizational norms or (2) cannot be resolved through existing organizational norms. Because of the high probability of new norms, there is at least the potential for double-loop organizational learning in M/As. The quality of learning in the individual context can be examined through the following questions (Argyris and Schon 1974):

1. Are assumptions, especially attributions, treated as publicly testable or are they self-sealing?
2. Do individuals seek out disconfirmation as well as confirmation of the propositions they forward?
3. Do individuals recognize uncertainty, and do they respond to it only by problem setting?
4. Are individuals sensitive to incompatibility and inconsistency in their own theories-in-use and espoused theories?
5. Do individuals try to bring to the surface their own maps of the problematic situations in which they find themselves, and do they try to share the task of completing and coordinating incomplete and divergent maps?

These questions take the following form on the organizational level (Schon 1983):

1. Do members of the organization treat organizational assumptions as testable and search for disconfirming data?
2. Are members of the organization able to integrate their pictures of the production process with those of the acquiring company's management?
3. Do the members of the organization share memories of the past that provide them with a context for interpretation of the present?
4. Are members able to respond to uncertainty by reflection and by efforts at restructuring their perception of the problem?
5. Do members of the organization test for congruence of organizational espoused theories with theories in use, and do they test compatibility with norms?
6. Do the individual members oppose one another without awareness that this represents a conflict of organizational values?
7. Do members couple advocacy of their own positions with inquiry into the position of others?

The literature in pedagogy contributes the important concept that different kinds or domains of learning are learned in different ways and that each domain varies from the others in difficulty of the learning task. It is conventionally asserted that cognitive learning (learning of information and facts) is of a substantially different nature than is learning of skills (purposive motor and intellectual coordinated routines), which are both different again from the learning of affect (values, attitudes, feelings, and preferences) (Fogel 1980). Not only are the contents of each domain different, but they are learned differently. As shown in Figure 11–2, cognitive learning comes through observation, being told, reading, and so forth, whereas skills are learned through practice; affect is learned differently still. Obviously, learning of skills is much more time- and energy-consuming than learning of facts.

Returning to the discussion of organizational learning, it can be seen that, although distinctions are not made between domains of

371

Domain	Cognitive	Skill	Affective
Content	Information (learning about something)	Coordinated routines involving psychomotor activity	Emotions, values, beliefs, preferences, morality
How learning is accomplished	Observation, listening, reading, etc.	Practice	From modeling of others, social compliance, etc.

FIGURE 11–2. Processes of Rational Learning

organizational learning, Argyris and Schon are principally interested in the cognitive issue of informing future action rather than immediate coordination. The general action research and organizational development-oriented prescriptions they discuss are appropriate for cognitive organizational growth, but they do not discuss the skill domain either descriptively or prescriptively. In this chapter, then, the concept of domains of learning is applied to the concept of organizational learning, and the issues of organizational level skill learning and organizational level affective learning are explored. It is interesting to note that, as with individual skill learning, the organizational learning of interactive routines involves the intersection of cognitive learning (factual knowledge) and the actual effective practice of the routine. In other words, there are aspects of any skill that are cognitive (what the routine is, who is involved, when to use it, and so forth), but the central aspects involve the actual operation of the routine. Finally, there are aspects of both skill and cognitive learning that are linked to values and attitudes of the learner. The practice of any given skill must be linked to affective questions such as whether the learner wants to practice the skill, believes the skill is learnable, and believes the practice of the skill is useful.

This chapter, then, examines the pro-cesses of organizational change in the face of an M/A. This examination addresses the processes of accommodative learning on both individual and organizational levels as the organizations and their constituent individuals cope with learning to manage the changes.

THE STUDY

Data for this chapter were drawn from an overall sample of twelve M/A cases currently under study. Three cases demonstrate the issues of post-M/A accommodative learning and change most clearly. First, the contrast among them in areas of employee turnover, organizational change, and organizational problems is clear. Second, in all cases, the acquirer thought very highly of the acquired management, and in two of the cases, the acquired company management held the acquirer in high regard. Finally, the method of data gathering allows for close, fine-grained study of the dynamics of these three cases.

Data for these cases were collected through focused interviews (Merton, Fiske, and Kendall 1952) with individual managers and employees in three separate acquisitions: an additive type, a conformative type, and a merger. There were some differences in the data-gathering schedule. In two cases data were collected over a several-month period but without a concrete schedule. In the third, data were gathered in two rounds—one just after the announcement of the acquisition and the second about six months later. It was not possible to randomly select respondents from each organization, although all respondents except for one in case 1 were midlevel management staff. All individual respondents were either line managers (identified as managers by their organizations, having functional responsibility for a unit, and having people report to them) or professionals (identified by their organizations as such, having a functional specialty, and salaried).

On completion of the interviews in each organization, the data from the separate interviews were aggregated to construct the cases following Yin's principles of data collection in case studies (Yin 1984):

1. Use multiple sources of evidence (multiple interviews, in this situation);
2. Maintain a database (interview notes and tape recordings);
3. Maintain a chain of evidence (the case histories presented).

From these case histories, event maps were drawn that represent a chronology of critical incidents, coded to maintain continuity of causally related events (See Figure 11–3). As shown in Appendixes A through C, events that are causally linked all bear the same letter prefix and their sequence is shown by the following number. For example, all events coded with the letter X are causally linked; therefore, an event coded X–8 would be the eighth event in the

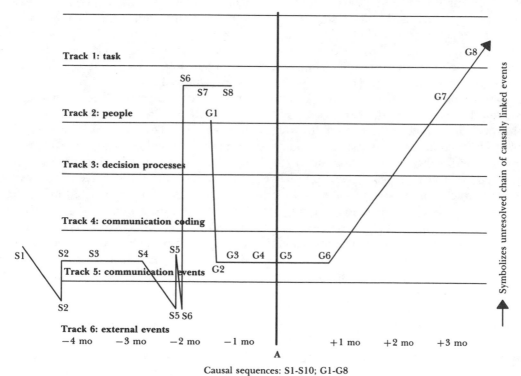

Causal sequences: S1-S10; G1-G8

FIGURE 11–3. Multiple Sequence Track for Case 1.

causally linked sequence. The events (and codes) of the event map are then recoded and arrayed on multiple sequence tracks.

Poole, in studying small groups, developed the multiple sequence model to relate differing aspects of group task processes (Poole 1983). This model suggests portraying the group processes as a set of parallel strands or tracks of activity. Each track represents a different aspect of the process and relates to a different level of data. Among the strengths of this approach to the data is that it allows for analysis of relationships between and across levels. Poole's original tracks (task processes, relationships, and topical focus) were developed in the context of laboratory research on small groups. As such, the groups were short-term, had single and explicit tasks, and had no extra-group constraints. Real organizations (and units within organizations) are long-term, have multiple and complex tasks without much explicit specification, and have substantial extra-group constraints. Van de Ven, in studying the broad topic of processes of organizational innovation, has altered and elaborated Poole's three tracks into five: (1) people, (2) ideas, (3) transactions, (4) context, and (5) outcomes (Van de Ven and Poole, Chapter 2). Six crucial tracks have been elaborated for this study: (1) changes in task activity; (2) changes in the people in the organization/unit; (3) changes in decision processes; (4) communication coding; (5) communication events; and (6) events in the other organization. Items 1 through 4, it should be noted, involve the use of interactive routines (organization level skills).

Data and analysis for this chapter adhere to the following sequence: The case histories are presented; each is followed by a multiple sequence tracking chart; then an analysis is presented; and finally conclusions are drawn from the cases and multiple sequence tracks. Appendixes 11–A through 11–C follow the analysis and present the sequential event map for each case.

374

THE CASES

Case 1 (1JCo acquired by 1SCo)

This was an acquisition of a large national consumer finance institution (1JCo) by a larger international consumer finance institution (1SCo). Prior to the acquisition, 1JCo had been privately held by a family corporation. 1JCo is headquartered in a large midwestern city, while 1SCo is headquartered in New York City. Prior to the acquisition, 1JCo had been marked by a rather low-key culture in which the management thought of itself as very competent but "unflashy." Performance had been somewhat off for the year prior to the acquisition, but that had been attributed principally to a slightly depressed market, undercapitalization by the family holding corporation, and several poor investments by that corporation. 1JCo thought of itself as a friendly, casual, consentual, and personal organization. 1SCo, on the other hand, is a large organization that had grown prior to the acquisition through several large related industry acquisitions. Its internal operations were more "state of the art," and its culture considerably more formal and less personal.

All five respondents in 1JCo reported both the actual events in the acquisition process and their individual reactions with remarkable consistency. Multiple interviews were conducted with three of the five, resulting in some longitudinal information. Although there was an early period when several of the respondents reported being ready to leave, later interviews had these same respondents change their minds completely. In later interviews, all indicated that they were very pleased with the acquisition, the new owners, and the new top management and that they intended to stay.

All of the respondents found out about the acquisition within a few hours of each other from an article in the *Wall Street Journal*. There had been no previous official communication to the employees, even though there had been rumors that the company might be up for sale.

Although uncertainty was expressed, no interviewees reported strong negative reaction to the news itself but all did report feeling angry and uncomfortable about the way that they heard about it.

About a month after this news in the *Wall Street Journal,* a large group of people from the 1SCo arrived to do an on-site assessment. Except for the CEO and a few top executives, most of the group were technical people—accountants, systems analysts, and so forth. The evening before the executives were to return home, a dinner meeting was held with the executives from 1SCo and some of the management of 1JCo. The 1SCo people were described by 1JCo management as being extremely rude and uncivil at this event. Some of the 1JCo managers, including one of the respondents, were so offended that they walked out of the dinner. This incident was discussed by all of the five respondents. Two of them reported considering resigning after that event and the others reported being upset by it.

About three weeks after the dinner, the *Wall Street Journal* ran a story indicating that 1SCo was no longer interested in the acquisition. The reason cited was that the price was too high because the management was not as good as had been thought. Again, 1JCo's employees greeted the medium of the message with resentment and the message itself was greeted with both anger and relief. The respondents did not want to be sold to 1SCo, and they were relieved that it no longer seemed to be likely. They were, however, offended by the perceived slap from the published comments about their not having strong management.

Six weeks later, an official announcement was made to 1JCo staff that negotiations were back on with the same company. Although the internal formal communication system beat the *Wall Street Journal,* it did not beat the rumor mill. The news was greeted with resignation but not as much anger. Upper management in 1JCo was informed of the negotiations several days before the announcement was made but was cautioned against mentioning this to other employees. Shortly after the negotiations were announced, the announcement of a deal came through.

Two weeks after the deal was agreed on, the CEO of 1JCo went to the offices of 1SCo for several days. He was well liked by 1JCo employees, and when he came back with the news that he was to be replaced as the CEO, three of the respondents reported that this caused or confirmed their intentions to leave. All, however, decided to wait for a while before searching for other jobs.

Six weeks before the actual exchange of stock and money, a new CEO arrived from 1SCo with a small team that included only one other executive. A few weeks after his arrival, the new CEO had dinner with one of the respondents. This was a critical incident in the course of the accommodation for this manager. The dinner was very successful in the sense that during it, the 1JCo manager completely changed his opinion of 1SCo and the acquisition, especially the CEO. Prior to this dinner, the respondent had referred to the new CEO by his last name only; following it he referred to him by his first name only. In an interview conducted the following day, the respondent stated that he was reconsidering all of his previously held opinions (although he maintained reservations). He reported being treated in an extremely civil and collegial way.

The office holiday party season started the following week. The new CEO made a point of appearing at the parties. At the one the respondents attended, the new CEO introduced himself to the individuals attending and, when it was appropriate, answered questions from the group at the party. His communication with the group was reported to be very friendly, personable, and open. This contrasted so strongly with earlier experiences with 1SCo that the respondents left the party feeling very happy about their new CEO and questioning their earlier opinions about 1SCo.

During this same eventful week, new

combined logos and corporate names were announced along with a major national advertising campaign. All respondents referred to this with some appreciation and pride. The lower-level managers indicated that this was more important to them than it was to the higher-level managers. Three lower-level managers pointed to this with considerable pride and claimed that it not only allowed for much greater public recognition of 1JCo but also accounted for an immediate increase in sales. On the day of the official change in ownership, a cocktail party was held for all 1JCo employees. This, again, was an important event for the managers. Several of the respondents commented that the previous ownership probably would not have had an all-staff event in the first place, but if it had, it would have been with cookies and coffee rather than cocktails and canapes. As one of the respondents said: "They treat us like adults." Again, the new CEO circulated at the reception, talking freely with 1JCo employees, particularly lower-level employees.

One month following the exchange of ownership, the new CEO initiated a program in which he weekly invited several mid- and lower-level employees to lunch. During these lunches the CEO set a social and conversational tone, encouraging employees to talk. All of the respondents commented on the lunches and the improvement in morale that accompanied the program. That the new CEO carefully listened to the employees was mentioned by all of the respondents, and a high degree of commitment to the new owners was attributed by them largely to the CEO's ability to pay respectful attention to the employees.

During this early postacquisition period, overall economic performance continued to improve, although both the CEO and the staff saw further opportunities for improvement. Two-and-a-half months after the exchange of ownership, a highly regarded manager was promoted to a new position responsible for quality improvement, especially in sales and strategic planning. This manager was charged with developing and overseeing highly participative programs in the two areas. The results of these two programs was an increased focusing of resources on the sales function of the organization, but it was a focusing that was directed from inside rather than from outside. These programs not only encountered no internal resistance, but instead had strong support even from those employees working outside of the focused areas.

Because of the continuing strong performance at the time of these later changes, the organization grew in numbers of employees during this time and none of the employees outside of the focus area lost their jobs. Although their jobs may have changed, the new positions seemed better than the old. There was almost no turnover during the postacquisition period, including the period of changing the organizational focus.

Several respondents continued to "wait for the other shoe to drop" for quite a while after the acquisition, but organizational performance has continued to improve in the three years since the acquisition. There has been almost complete retention of employees, especially management. The CEO continues to be highly thought of by the staff of 1JCo. Figure 11–3 illustrates the change patterns for the pre- and early post-M/A period in case 1. Appendix A contains the event map.

Case 2 (2JCo acquired by 2SCo)
This was an acquisition of a small (less than fifty employees) rural equipment leasing firm in a small town (2JCo) by a large regional financial institution (2SCo). The leasing firm was a separate business owned by an equipment dealer, but there was an association between the two operations in that they were housed together and, several years ago, shared some employees. There were eight respondents in 2JCo. The ini-

tial interviews were conducted about ten days after the final approval of the deal by the 2SCo board. Because it was a "white knight" acquisition, these early interviews indicated that the 2JCo staff felt excitement and optimism about the acquisition, and none indicated thinking about leaving.

2JCo had started as a sideline to a family equipment dealership during the 1950s but had grown tremendously during the 1960s and 1970s under the management of three professional managers (two of whom are still vice presidents with the organization). The dealership side of the operation had always been directly run by the family owners. Over the past several years the family owner had encountered such personal problems that he had been unable to effectively manage the business. The result was that both businesses suffered, although the leasing business had been somewhat more insulated from the problems than the dealership was.

All of the respondents knew by February 1985 of the intent of the owner to sell 2JCo. All indicated that they knew of the owner's problems long before and had been concerned that if something were not done, the leasing business would eventually lose its insulation from the owner's problems. All indicated that they saw this intention to sell as hopeful. They learned of it, in all cases, through interpersonal sources. The first official confirmation from the owner came in the announcement that a company was interested in acquiring 2JCo and that this company would be looking through the books. This first "suitor" lost interest within several weeks, but another appeared soon. It also lost interest, but several of the managers (including two vice presidents) became concerned that these first two suitors were not the type of company that would be best as a senior partner in 2JCo's operations. Their management therefore decided to go out and find a more suitable acquirer.

No surprises were expressed by any of the 2JCo managers and professionals interviewed. They indicated that they had always been kept well informed of everything going on in the company. This was due, in some measure, to the small size of the organization and the community in which it resides, but all respondents indicated that the two vice presidents had been very open about all management issues including both the acquisition and the preacquisition problems. The interviews revealed an organization that had a genuine family atmosphere in which 2JCo employees socialized together, went to each other's cabins for the weekends, recreated together, frequently ate together, knew each other's families well, and participated in civic functions together. All 2JCo respondents referred to all other 2JCo employees by first name only, and the receptionist called the vice presidents by their first names during telephone calls. There were few formal staff meetings, but there were also few management secrets. Both vice presidents had shared all information through casual communication channels (at meetings of community organizations, at lunch, at golf, having a beer after work, and so forth), and information got quickly to all organization members. The freedom of the information flow through the informal organization was further enabled by the generally long tenure of the employees. Only one respondent had worked for the company for less than two years, and he had been known professionally to the employees for two years before his hiring. The others had been employed by 2JCo for between five and seventeen years. The respondents were aware that 2JCo was a very close social system, all mentioning it and several saying that the closeness was one of the real benefits of working for 2JCo. The two vice presidents were credited with maintaining the closeness over the years.

The 2JCo VPs chose their next suitor on the basis of knowing both the company and the individuals who would be associated with the acquisition from the acquiring side. Their

377

choice was 2SCo, an institution that had done quite a bit of business with 2JCo over the years. Additionally, 2SCo had acquired a leasing business years before and had done well with it. Several of the 2JCo managers had known a top manager in the 2SCo leasing operation for some time and both liked and respected him. By early June, 2SCo was serious about the acquisition, and all of the respondents knew about the impending deal with 2SCo. No efforts were made to keep the impending deal secret, so all employees knew of it. The 2SCo manager had visited the 2JCo offices frequently and made no attempts to hide the acquisition. In fact, he talked freely of various stages of the 2SCo decision process with several 2JCo managers and employees on his visits. Additionally, two ex-2SCo employees in 2JCo were able to provide corroboration and added detail, and a "mole"—a 2SCo employee who was an ex-2JCo employee—provided inside information to both sides.

The organizational contrasts between 2JCo and 2SCo were striking. 2JCo, prior to the acquisition, had been very casual, with almost no formal, written communication. The staff members were all personally close to each other—attending the same churches, belonging to the same community organizations, even vacationing together. The hierarchy was flat, with minimal requirements for high-level approval of business decisions. These decisions also happened very quickly, usually within a few minutes of the request for a decision. Relationships with customers were casual and relaxed, partly because of the style of 2JCo and partly because the base of customers was stable and long-term. 2SCo, on the other hand, was a large and cautious urban institution. Employees tended not to be personally close to each other, although the norms of internal communication certainly were friendly and cordial. The hierarchy was steep, with many levels of approval for even small business decisions. Decisions were slow and cautious, and the organizational style was formal.

Somewhat later in the summer, more formal preacquisition activity began with an audit team sent in from 2SCo to 2JCo. Again, this team included people previously known to some of the 2JCo staff, and communication was open and personal. Also during this time the tentative decision was made to put the current 2SCo leasing manager in charge of both his own and the 2JCo operations. This was based both on his knowledge of the industry and on his personal relations with the 2JCo staff.

Between July and October there were several delays in the acquisition. The nature of 2SCo was that decisions of this sort are not hurried at all, and the employees of 2SCo involved with the acquisition knew that decision-making delays would happen. The 2JCo managers and professionals were not used to months of delay and started to become anxious. During this time, the respondents reported, there was both an increase in anxiety and a drop in productivity. Several reported feeling that they did not know what to do, so they just waited and maintained what ongoing work they could. The presence of the ex-2SCo employees and the mole were of considerable influence here because they kept reassuring the 2JCo people that slow formal decisions were normal and should not cause much concern. Even with this assurance, however, high levels of apprehension were reported.

It was also during this time that several of the respondents indicated that they started thinking about what might happen after the acquisition. As one of the respondents said: "You read that they [business acquirers] say that there won't be any personnel changes, and then they fire everybody." All respondents reported that morale in 2JCo went down each month when the 2SCo board did not approve the acquisition, even though several 2SCo managers (including the leasing manager) and the ex-2JCo employees tried to assure 2JCo that these delays were normal and that approval would eventually come.

During this period, the respondents reported high levels of emotional energy being expended especially by the two vice presidents. Emotions were strong enough that there were threats to call off the negotiations after both the August and September 2SCo board meetings, but the respondents indicated that they knew the threats were empty and were really an expression of the frustration of having to wait another month. This was reported to be a period of emotional ups and downs for all members of the organization interviewed. One of the respondents connected the organization's moods directly with the moods of the two vice presidents, noting that within a few minutes of their receiving bad news (news of the delay), the moods of all of the employees seemed to go bad, but when the vice presidents were happy, the rest of the employees were also happy.

During the delay time, 2SCo management continued negotiations as if there was no question of the acquisition. The audit team continued to work in 2JCo offices, and there was regular presence of 2SCo managers on site. In September a new benefits package was announced as well as a minor reorganization (two field offices were combined and one was added). All of the respondents were anxious to get the acquisition consummated for a number of reasons: (1) The benefits package was vastly superior to 2JCo's; (2) 2SCo was perceived as a comfortable and very stable senior partner; (3) working for 2SCo was perceived as allowing greater opportunity for upward mobility or for moving to the headquarters in the Twin Cities; and (4) the situation with the family owner of 2JCo was worsening.

At the time of the acquisition, the family owners of the business had been completely inactive for some ten years. With the sale of the business, 2SCo appointed a new, active CEO to the organization. This new CEO was a long-term employee of 2SCo but had several years of experience in the leasing industry. Because he was both professionally and socially known to the top two active managers of 2JCo, he was readily acceptable to them as CEO.

The initial business changes imposed on 2JCo by the acquirer were limited but important and were imposed at the point of the acquisition. The changes were as follows: (1) The new CEO must approve all leases above \$5,000; (2) all leases above \$15,000 must be approved by the new CEO's boss in 2SCo headquarters; (3) lease application packages would now include two documents not formerly required and must be put into 2SCo's loan format; and (4) payroll and benefits must be centralized in 2SCo's headquarters.

These initial changes were expected, even wanted, by 2JCo staff, but some of their side effects were not. Initially, the new layers of approval caused a slowdown of two or three days in approvals by themselves. Sometimes the new CEO and his boss were not in their offices for several days at a time, so there were additional delays. Even without these delays, the new people were anxious to learn the business and needed to take time to understand the data and documents in the packages. In addition to these delays, the administrative staff, knowing that the document packages had to be complete and in format, refused to forward applications for approval until the customer provided certain documents. This resulted in delays of several additional days. The customers and sales staff were both unprepared for this, and several long-time customers went to the competition rather than take the time to provide the documents and wait for approval, even though the new CEO kept assuring both the sales staff and the customers that the delays did not mean any "real change" in their approval process and that the leases would be approved. When the second long-time customer went to a competing leasing company, the sales manager started challenging the controller and harassing the administrative staff to speed up the process. This conflict was foreign to 2JCo and was extremely unpleasant for both the controller and the sales manager. From these initial incidents, this conflict became regular at least until the most recent data gathering.

379

The situation was complicated because the controller had to take considerable extra time to put the package in format. Although he had had sufficient time and assistance to do the job prior to the acquisition, following it he started falling progressively behind, causing additional delays. When he called this to the attention of the new CEO, a person was transferred from headquarters to help. This person clung tenaciously to the culture of 2SCo, including wearing business suits and refusing to work other than normal Monday through Friday business hours. In the following weeks, a rumor that was believed by all of the staff of 2JCo except the new CEO (who was never included in the rumor mill) was circulated that the new person was a spy from the top management of 2SCo. Because no one, including the controller, was willing to ask him to change his work habits to conform with 2JCo's work patterns, his presence did not materially relieve the work overload or relieve the issue of the conflict.

About two months after the acquisition, top management and headquarters staff decided that the long-range strategic plan should be accelerated, with expansion to start immediately rather than in two years. In doing this, the three top managers became more functionally differentiated in the approval process. It also meant that these three would have to spend much more time away from the office. Both of these changes meant further delays in the approval process, with the controller being the one stuck in the middle between the demands of the top management and the hostility of the sales manager at the increased delays. Conflict accelerated again.

The top three managers were only minimally aware of the problems at the point of T2 data gathering. During those interviews, it is interesting to note that the top three managers, the auditor, and the controller discussed their need to satisfy 2SCo for long-term career expectations. In other words, each of them indicated explicitly that they wanted to rise in 2SCo, eventually leaving 2JCo, the leasing industry, and the small town. The people on the sales side did not discuss this at all, although the sales manager did suggest that if he wanted to rise, it would be in the leasing industry. The clerical and lower-level administrative staff members were almost all married, self-described second-income earners. Some mentioned perhaps working for a local branch of the acquirer, but none were interested in leaving the area or in rising in the organization more than one step.

During the year following the M/A, it should be noted, the economic performance of 2JCo remained on target despite the problems and conflict. This can be attributed to three things: (1) This was not the first M/A for 2SCo, and its expectations were somewhat lower than they would be for a first attempt; (2) the accelerated strategy generated enough new business to counterbalance any losses of old business; and (3) the conflict and problems never got to extreme proportions. Figure 11–4 illustrates the events for case 2 arrayed along the six tracks, and Appendix 11–B contains the event map.

Case 3 (3ACo merger with 3BCo)

This was a merger between two large regional healthcare organizations based in the same city that had been each other's direct competition prior to the merger. There were eight respondents—five in 3ACo and three in 3BCo. Although little outright hostility was expressed, there was considerable turnover during the post-M/A period, including one respondent who resigned shortly after the second interview with her.

Although the negotiation discussions between the CEOs of the two organizations started in March 1983, it was not until July that the official announcement of the negotiations

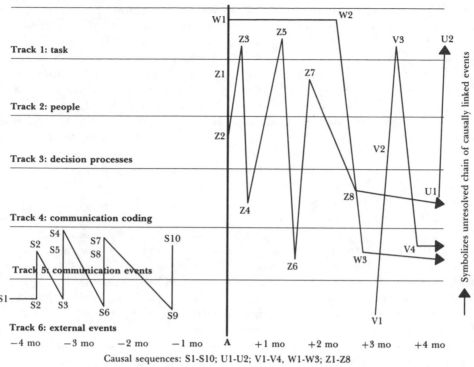

Causal sequences: S1-S10; U1-U2; V1-V4, W1-W3; Z1-Z8

FIGURE 11–4. Multiple Sequence Track for Case 2.

was made. A few managers in both organizations were informed of the negotiations (including three of the respondents) in March but were instructed not to mention them. The five respondents who were not informed knew that the discussions were taking place through active grapevines. Those five expressed resentment at not being informed by their organizations, and the three who knew also expressed discomfort at not being able to communicate with their peers and employees. During this period, seven of the eight respondents indicated that they knew of several other 3ACo and 3BCo employees who started actively seeking other employment as a direct result of the impending merger. As both organizations had enjoyed good economic and organizational health prior to the negotiations, anxiety and resentment were not as great as they could have been.

By the time of the actual announcement,

the respondents indicated that they felt that they already knew most of the details of the merger, so there were no immediate surprises. The actual announcement clearly stated that this was not an acquisition of one company by the other, but a merger in which one hierarchy would be created from two without a power differential between them, even though 3BCo was somewhat larger than 3ACo. The announcement also included a "no fire" statement that said that layoffs and firings would be suspended for two years following the merger. A new organization chart specified managers for each new combined operating unit. Both organizations had used acquisitions to grow, but neither had participated in a co-equal merger before.

The two CEOs had developed a respectful, cordial, and cooperative relationship, but in the necessary reorganization into a single hierarchy, the CEO of 3BCo became the new CEO of the merged operation while the CEO of 3ACo became the chief financial officer (subordinate to the CEO). This pattern held in the subsequent reorganization of the functional units of the merged organization, and some staff from both sides started to view it as a "polite acquisition" of 3ACo by 3BCo. The pattern also held in that relationships between 3ACo and 3BCo managers and professionals in the merging functional units were also cooperative, respectful, and cordial, although most respondents expressed a sense of resignation rather than the sense of excitement evident in the first two cases.

Although unit managers were designated by top management and announced with the organization chart, organizational integration was approached functional unit by functional unit, using group dynamics organizational development approaches to intraunit integration planning (organizational and technical). In addition to its normal task and its own integration planning and activity, the personnel department was also charged with helping to facilitate the intraunit planning processes throughout the rest of the organization. This phase of the merger proved to be a discouraging and upsetting time for five of the respondents. The two older respondents and the most highly placed respondent simply did not become personally invested in the process, so they were not upset by it, but the other five had strong reactions to it. Not only were there serious technical problems (such as financial systems that kept track of different things using different procedures, different hardware, and different computer programs), but in the effort to gain agreement among the staff members about both technical and organizational problems, turf guarding emerged. Two years after the merger these problems remained unresolved in several units, with the prognosis that they would take another three years to

work out. A number of the respondents said that they found the time frustrating and would "just like to get back to regular business." Seven of the respondents indicated that the work of integrating the functions had taken much of their regular work time and their normal task functions suffered.

Informants from both sides commented at length about differences in the cultures of the two organizations, claiming that those differences were in part responsible for the delays. 3ACo had been a consensus-oriented culture in which members had grown to expect consultation before decisions were made, while 3BCo's culture was built around a more traditional chain of command. Because 3BCo's decision-making style dominated, top management announced orders for various aspects of the integration (such as the organizational chart, moving schedules, and deadlines). Professionals and managers from both sides indicated that they learned to resist commands by claiming technical problems. The older respondents said, in fact, that they and their peers from the other side had simply agreed to disagree until retirement, keeping the two operations separate. Another aspect of the cultural difference was the question of accessibility of the top management. In 3ACo, management members were in frequent personal contact with employees prior to the merger, but the 3BCo CEO was not personally well known by the employees. His style was to make a decision, communicate it to the appropriate manager, then assume that it would be done. His style prevailed in the merger, but his assumptions were not always warranted. This style was picked up by the CFO and others in top management, sealing them from news of what was happening below them in the organization.

As the efforts at integration ground on, most of the respondents indicated that they gained respect for their peers from the other side but increasingly lost commitment to the

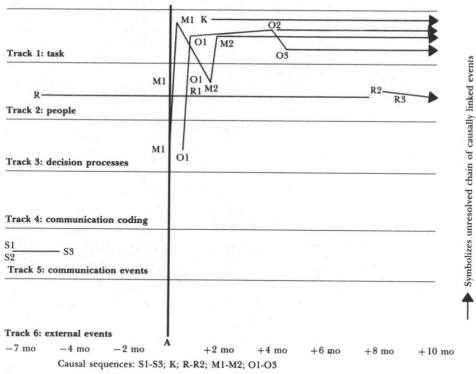

FIGURE 11–5. *Multiple Sequence Track for Case 3.*

ANALYSIS

new organization. Only a few months after the M/A, all but three of the respondents indicated that they were either actively searching for new jobs or at least ready to accept offers from other employers (one respondent accepted another offer and resigned immediately after her second interview). Overall, the turnover rate has been high with midlevel managers and professionals from both sides. Shortly after the interviews the CEO accepted an offer from an other firm, and his departure was followed by an orderly but nearly complete turnover of the upper management of the combined organization. This, of course, includes all of the managers who were involved in planning and executing the merger. Figure 11–5 illustrates the events in this case on multiple sequence tracks, and Appendix 11–C contains the event map.

Although many differences exist among the three cases, all respondents found the situation frought with uncertainty and unpleasantness at times. These times of uncertainty were not restricted to only one individual at a time but were pervasive throughout the organization. These were also times of task difficulty and relatively low productivity. By comparing the multiple sequence track charts (Figure 11–6), it can be seen that actual changes in people, task, and decision processes are associated with uncertainty. Uncertainty is in turn associated with turnover (Marks and Mirvis 1986).

We also saw formal communication diminish uncertainty initially through clarifying the interorganizational relationship, but interpersonal and collegial communication later

383

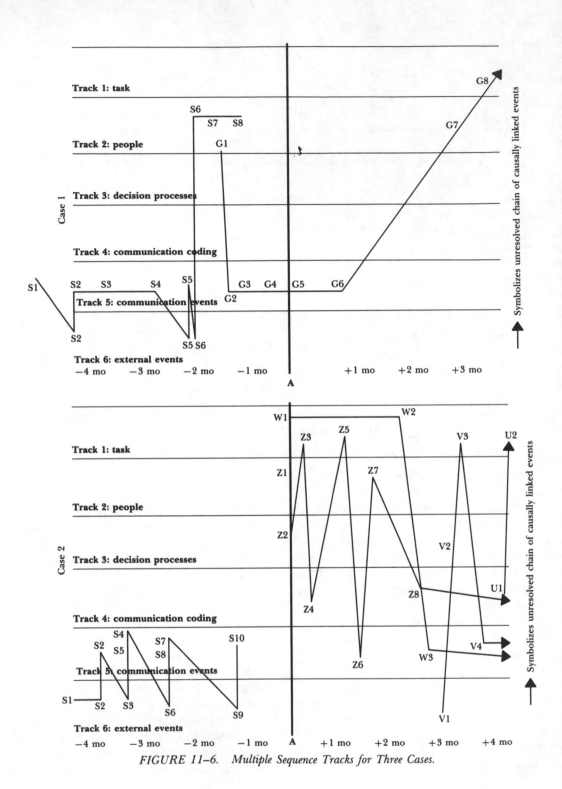

FIGURE 11–6. Multiple Sequence Tracks for Three Cases.

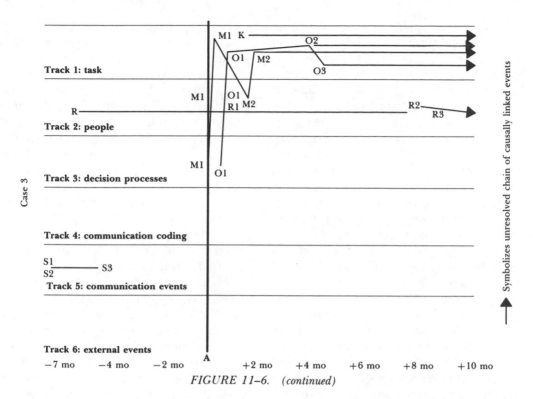

FIGURE 11–6. (continued)

defused the volatility of the situation (Bastien 1987). Communication of both sorts informed uncertain staff members of new relationships and expectations and provided feedback to them regarding their places in the new scheme of things.

As can be seen in Figure 11–6, causally linked chains of organizational changes are shown, with arrows representing unresolved chains as of the most recent data gathering. This linking of causally connected changes allows the change process to be visualized and shows that until resolution is reached in each chain, uncertainty about its resolution is felt by the members of the organization. In case 1, introducing one change at a time and allowing it to become fully resolved before the next change was introduced generated considerably fewer problems than introducing many changes in a short period of time (case 3). Even in case 2, where rather small changes were introduced

on a number of tracks, conflict and uncertainty were present. Certainly in case 3, all tracks in the organization were subject to change, and conflict and uncertainty abounded. Organizational change involves new knowledge and skills that *must be learned*. Organization members must know what is expected in the change (in terms of both actual activity and standards) and how to do new things (or do old things differently), but they also must learn to want to accomplish the change. This leads to the generalization that the greater the number of unresolved chains of change at any given time, the greater the amount of uncertainty within the system and, therefore, the greater the likelihood of evocation of the syndrome (Table 11–1). This also suggests that the longer multiple chains remain unresolved, the greater the likelihood of evocation of the syndrome.

Case 1 saw no early change in task or decision processes. The only changes were the gift of the ad campaign and the new CEO. The CEO waited until he had become well known to the entire staff before allowing any other changes to occur and worked hard at understanding and becoming well connected with the organization before any other changes were initiated. When changes were made, they were made through the promotion of a popular insider to manage the change decision process. Although a great deal of hostility was expressed before the new CEO became known, there was no conflict in case 1. There was also no turnover and no drop in work output, productivity, earnings, or profit. Furthermore, the pace of changes was leisurely, with the CEO taking several months to make sure the organization was used to him and that he understood the organization. This mutual understanding was achieved through the new CEO's visibility and his high level of collegial communication.

In a study of unrehearsed, improvised group jazz, Bastien and Hostager (1987) found that, although musical changes (in time, tempo, chordal embellishment, progressions, and so forth) can be introduced potentially anywhere, successful practitioners of the art typically resort to introducing one change at a time, with the group working through the implications of that change before moving on to the next. They also saw a great deal of communicative behavior (typically nonverbal) in signaling both that a change was going to happen and what the change would be. We saw something like this strategy applied in case 1.

In case 2, there were limited changes, but those changes were in people, task, and decision processes and they were simultaneous (see Figure 11–4). The change in timing of expansion strategy happened later, but it was initiated before any of the earlier changes had been resolved. It is interesting to note that all of the initial changes were anticipated by all of the staff of 2JCo and were agreed to before they

were made. They seemed to be minor changes when they were talked about, but the actual implementation of the redefined processes and routines was somewhat slow and painful. The side effects of these changes were unforeseen by all parties, and these unforseen effects caused the conflict that subsequently emerged. Because the top level of management (the CEO and the VPs) had become insulated from the rest of the organization because of concentration on their own new tasks, the conflicts and problems below them were compounded and grew to high levels.

This escalation of problems can be best understood in terms of learning, especially skill learning. Individuals had to simultaneously learn to manage a new set of decision processes, learn what counted with a new CEO, and learn how to manage a slightly redefined task in a context where norms and standards were not certainly known. It may seem obvious that learning takes time, but none of the actors in 2JCo were prepared for the time actually needed. In fact, none of the respondents recognized their situation in learning terms but instead saw it as one of immediate performance. It seemed to the employees of the organization that they were being blocked from good performance by each other (and by the acquiring company that was mandating the changes). Because many of the employees of 2JCo were anxious to please 2SCo (many had hopes of moving up in the hierarchy of 2SCo), each individual's extra effort to do things right (accomplish a new individual task with new formats and standards) was interpreted by others as blocking or delaying their own individual efforts. Not only were individuals presented with revised tasks and standards, but because of the changes in the decision processes, the whole interactive routine of getting leases approved was thrown out of synchronization.

According to Schon (1983), organizational learning is for future action rather than immediate organizational coordination. Certainly, on an individual learning level, physical coordination in new tasks is learned, and that skill learning is understood to be more time consuming than is learning facts about the skill. If, therefore, the process of recoordinating in the face of these changes is thought of as a special kind of organizational learning, then it is clear that organizations (like individuals) learn skills more slowly than they learn facts. It also appears that organizations learn more slowly than their constituent learning individuals. This is because learning a new interactive routine requires all individual participants to master a new individual task with new standards and then learn to coordinate the interaction necessary to complete the routine.

By combining the concepts of domains of learning and organizational learning, Figure 11–7 emerges. It can be seen that each domain has a corollary on an organizational level. Cognitive learning on an organizational level can be seen to be facts and ideas that are shared throughout the system (Argyris and Schon 1974; Schon 1983). These are learned through talk (including reading) and observation (Johnson 1982), especially reflective discussion as in organizational development and action research (Argyris and Schon 1974; Schon 1983). The corollary of individual skill learning is the learning of interactive skills and routines requiring social coordination. These must be learned through practice. Decision process and task processes are examples of these routines.

Returning to the learning of interactive routines in case 2, it can be seen that an aspect of the learning difficulty was that effective performance within the operating context of the "old" 2JCo generated errors in the context of the "new" 2JCo (a double loop situation), but this was not fully recognized by the top management. For example, in the old 2JCo, getting lease documents from long-time customers was a minimal proposition in which the formality of the documents was not an issue, but getting the lease executed quickly was. In the new operation, getting the form right was more important than speed. Because this incongruity was entirely new to both 2JCo and 2SCo, and because the structure of new norms and standards was not well known, members were not able to rely on any kind of organizational memory to cope with the changes (and requisite learning). Furthermore, because the employees of 2JCo conceived of their situation as performance rather than learning, they did not reflect on the experience in productive terms and did not see conflict as conflict in organizational values. Thus, the conditions for quality of organizational learning were not met by the time of later interviews in this case.

In case 3, massive changes in people, task, and decision processes were directed almost immediately. Every operating unit, as it was merged, had to adjusted to new managers and new colleagues while a major change in task was also being learned without clear standards, with decision processes that were not all new but were unclear and indistinct. None of the managers, not even the top team, was clear about how decisions should be made in the new combined organization, even though 3BCo's managers, systems, and processes seemed to dominate. The effort at trying to generate the necessary system and task changes for the new combined organization either burned the managers out or caused them to fail outright. For instance, in the personnel department, designing the new personnel systems took about eighteen months and was successfully accomplished, but most of the high- and midlevel managers and professionals decided to find employment elsewhere during the process. In other departments where there were older employees not far from retirement, the combined employees made a more-or-less conscious (but tacit and private) decision to avoid having to change until those older employees could re-

	Cognitive	Skill	Affective
Individual level	Facts and information known to an individual	Motor and other skills learned by individuals	Individual beliefs, fears, tastes, and preferences
Organizational level	Facts and information shared by constituent members of the organization (information known across the organization)	Interactive routines	Norms, standards, attitudes towards change and learning, knowledge of who the enemy is, etc.

FIGURE 11–7. *Individual and Organizational Levels of Learning.*

tire. Because of their inability to relate collegially to their new organization, the top management was insulated from the reality of burn-out and resistance lower in the organization.

In these consciously resistant units, managers were forced to fail by their employees. For example, an early departure of some key managers with technical experience who were known to at least some of the employees required replacements to be hired from the outside. The employees did not care whether these newcomers failed but cared deeply that they would not have to bother with the change. In fact, the formally recommended intraunit integration/change planning process was used, in some measure, to thwart change in these resistant units. For functions where resistance to change per se was not the central issue, often this change planning process was used to force win-lose contests between factions within the combined functional unit. In other words, the recommended intraunit process was well intended but basically inconsistent with the other decision processes. As resistant staff learned to use the process, many learned it out of context, using it for their own anti-integration ends.

Task changes, in broad form, were decided at top levels rather quickly after the official decision to merge, and managers of combined functional units were also determined rather quickly, but the operating parts of the combined organization's administration were just as quickly overloaded with change. Top management had no concrete idea of the implementation problems associated with a merger of this sort but, again, because of its focus on their own concerns, failed to recognize the operational problems below them.

The conditions for quality of double-loop organizational learning in this case were clearly unmet. First, error or anomaly detection was a problem; previous errors in both organizations no longer appeared to be errors for mid- and lower-level employees. Even stated errors, such as missing deadlines for integrating a functional unit, were not viewed as anomalous by the operating units. Because the top management had become both disconnected from and not credible to the employees, and because immediate performance (not learning) was the frame of understanding of both management and the operating employees, no valid

388

efforts were made to integrate the various ways of the organization's operating values. Although it seems clear that a double-loop learning process did take place, the individual members of the organization were able to generate norms of using the process to further their own private ends rather than the new organization's.

In cases 2 and 3, the sink-or-swim linkage between learning and survival is clearly seen in the individual reactions to the uncertainty and to learning their parts of the new task. The changes themselves were rationally understood by the respondents in both cases in action and survival terms rather than in learning terms. Many of the respondents seemed surprised that everything was not done properly, easily, and quickly in the new context. They were frustrated by their own learning difficulties and the difficulties of coordination coming from other's difficulties or resistance (which seem to look the same from the angle of

the respondents). In case 2, this took the form of interpersonal conflict within 2JCo, while in case 3, it took the form of intergroup and interlevel conflict and a kind of aversive learning. Only in case 1 were there few changes that had to be learned at any one time, and only case 1 was marked by the absence of conflict. This all points to the important conclusion that in mergers and acquisitions (and in other innovation situations) the effective management of organizational change must be viewed as the management of a learning process.

Returning to Figure 11–8, organizational learning in the affective domain (especially of attitudes toward the M/A, the other organization, and higher organizational levels) is conditioned by the processes involved with learning in the other two domains. Where learning requirements were held to a minimum, as in case 1, members of 1JCo learned to trust their new parent and to trust in their own ability to improve their organization. In case 2, where some

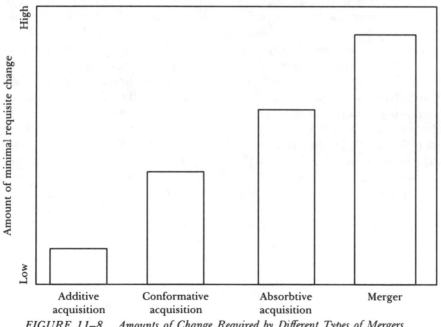

FIGURE 11–8.   *Amounts of Change Required by Different Types of Mergers and Acquisitions.*

changed processes were introduced, but where these changes were at least cognitively understood, there was conflict and confusion, but it was at a manageable level. In case 3, where there were massive changes in all aspects of organizational operations with little guidance, self-saving values abounded as did self-serving use of the change and learning processes. This double-loop learning environment also allowed people to learn contempt for the new organization and the people who had brought them into it. This brings us to the proposition that an organization moving unguided through an essentially threatening double-loop learning situation will end up with learning of poor quality (Argyris and Schon 1974).

These cases present three different kinds of inputs into (organization level) affective learning: (1) the quantity of things needing to be learned at any one time; (2) the quality of organizational learning (Argyris and Schon 1974; Schon 1983); and (3) the management of the actual learning process so that the learners felt successful in acquiring the the new skills and knowledge. From these cases, the following general findings can be stated: (1) The greater the quantity of learning required, the more resistant and hostile the learners will be, and the more difficulty they will have in successfully learning what is required; (2) the greater the quality of organizational learning (Argyris and Schon 1974; Schon 1983), the greater the acceptance of the changes and learning; and (3) the greater the attention to (and recognition of) organizational change as a learning process, the greater the acceptance of the changes. This last finding specifically relates to the delinking of performance of new routines from organizational survival.

This returns us to Bastien and Van de Ven's (1986) typology, as shown in Figure 11–9. Case 1, an additive acquisition, showed little externally (acquirer) imposed change. Case 2, a conformative acquisition, showed greater change (in two domains) induced by the

acquirer and greater learning problems and conflict. Finally, case 3, a merger, showed massive change and massive learning problems and conflict. On one hand, this tends to both confirm the rational implications of the typology. It also tends to support the finding that the greater the amount of new learning required, the greater the resistance and conflict. A further implication, however, is that when acquiring company management treats the post-M/A process as a learning process, the syndrome of problems is less intense than when the process is seen as sink-or-swim change. The typology, then, affords the acquirer to identify the domains and specific details of requisite learning for any of the four types.

SUMMARY

In this study, we have seen that formal and interpersonal (and collegial) communication serve essentially different functions: Formal communication serves to diminish uncertainty through clarifying the relationship between the two organizational systems, whereas interpersonal and collegial communication clarifies the relationship between people in the M/A as well as clarifying standards and expectations. Both were seen to be necessary for effectively managing the uncertainty (and the resultant syndrome) inherent in M/As. It should be noted that uncertainty, in this study, has dealt with issues (such as information and skills) that are not known but are learnable. The resolution to uncertainty, then, is to have learned what was not previously confidently known. The process of learning what was previously unknown must come through communication.

Second, the use of multiple sequence tracking (Poole 1983) has shown that the changes in organizational task and decision routines are causally and sequentially linked.

Institutional Affinity

		Synergistic	Exploitive
Number Of Post Acquisition Hierarchies	**Single Hierarchy**	Merger Tangible resources: all shared. Skills and knowledge: all shared, mutually taught.	Absorbtive acquisition Tangible resources: all shared. Skills and knowledge: taught by acquirer to acquired organization.
	Two Hierarchies	Additive acquisition Tangible resources: given by acquirer to acquired. Skills and knowledge: no exchange.	Conformative acquisition Tangible resources: given by acquirer to acquired. Skills and knowledge: taught by acquirer to acquired.

FIGURE 11–9.   *Resources and Knowledge by M/A Type.*

When these chains of linked changes are unresolved, there is uncertainty among the members of the organization. When there are more than one unresolved chain operating at any given time, the resultant uncertainty can overwhelm the individual members of the organization, and that this results in conflict and decreased organizational performance (the syndrome of Figure 11–1). From the tracking, two general propositions can be drawn: (1) The greater the number of unresolved chains of change at any given time, the greater the amount of uncertainty within the system and therefore the greater the likelihood of evocation of the syndrome; and (2) the longer multiple chains remain unresolved, the greater the likelihood of evocation of the syndrome. Case 1 demonstrated that a management strategy that limited the introduction of new chains of linked changes by waiting until current chains are resolved proved to be a highly effective way of managing the M/A change process.

Some important extensions and specifications of Argyris and Schon's (1974) organizational learning perspective are seen in these cases. First, the learning perspective is extended to cover the learning of actual coordination in the face of changed task, culture, and so on and to the learning of new attitudes and not just for future cognitive learning. Learning of coordination is extremely time consuming and detrimental to short-term performance. In circumstances of essential change, learning goes on whether or not the management of the organization realizes it, but it is of low quality and inefficient without the self-conscious control and reflection prescribed by Argyris and Schon (1974). Finally, when individual and organizational learning takes place in a nonreflective sink-or-swim context and when change is conceived of in terms of immediate performance, individuals orient toward learning short-term individual survival skills rather than gaining an organizational orientation. People promote their individual survival by learning how to resist change rather than productively cope with it. Furthermore, they learn to dislike both the new skills learned as well as the people who put them into the situation (the M/A decisionmakers). This has the effect of diminishing the individual's willingness to invest energy into making the M/A work and may even result in such negative behaviors as sabotage of the change process (as seen in case 3).

The learning perspective sheds interesting light on uncertainty and insecurity noted in the literature (Bastien 1987; Bastien and Van de Ven 1986; Marks and Mirvis 1986) in that it allows specification of the nature, roots, and outcomes of uncertainty. When the situation, as in cases 2 and 3, throws open questions of learning new tasks in an organizational context where there is uncertainty about norms and standards, employees found the learning task difficult to manage. In case 1, the change-limiting strategy kept the individual and organizational learning tasks at manageable levels. When employees could not be confident of what were errors/anomalies (case 3), the learning tasks for individuals became insurmountable, and employees found alternative employment. Individuals in these cases had to learn what counted (new standards and values), learn to do what counted, to do an essentially new task, and to coordinate their new and nonconfidently learned behaviors with other new nonconfident learners simultaneously and by trial and error. Furthermore, they had to learn and perform all this in a situation where "getting it right" the first time counted but what was "right" was not known. The learning tasks in case 3 frustrated employees into "dropping out" of the new organization, just a students in public high schools drop out when they find their learning tasks frustrating or overwhelming. Expending enough energy to be successful seems not only difficult but also seems pointless.

Through combination of Argyris and Schon's (1974) organizational learning perspective and the concept of domains of learning, both uncertainty and learning needs can be more specifically defined: Uncertainty can be about knowledge and standards, decision and task routines, or norms, expectations and attitudes. Synthesizing this with Bastien and Van de Ven's (1986) typology also provides a solid base for developing post–merger/acquisition accommodation plans (how to stage or limit change, how to help people and organizations learn new knowledge, practices, and so forth). This synthesis also affords M/A decisionmakers the ability to anticipate specific problems and costs for the type of M/A they envision. Certainly, whatever path is taken by organizations following a merger or acquisition, the processes of inevitable change are considerably effected by both individual and organizational learning. This brings up the central finding of this study; that effective management of post-M/A change is the management of the learning processes without which change cannot be effected. As has been shown in the three cases, change centrally involves learning, and it is the process of learning that determines both the costs and outcomes of the efforts at change and improvement.

Although this chapter has concentrated on the management of innovative change in M/As, its perspective is important and useful in examining all instances of organizational innovation. To the extent that learning to manage new processes, practices, relationships, and ideas is essential to managing the effective implementation of the innovation, then management must manage the learning process. For example, the site-based management innovation (Lindquist and Maurial, Chapter 17) depend on implementing new routines that have to be learned in both cognitive and skill domains, and on both organizational and individual levels. The issues raised by Marcus and Weber (Chapter 16) regarding internally generated versus externally induced innovations also point to both organizational and individual learning issues. Management of innovation must inevitably focus on the processes through which organizations and their individual members learn to accommodate to the changes involved in the innovation.

APPENDIX A   Chronological Listing of Events for Case 1
Column codes: C# = *chronological time (in months) in relationship to the time of actual change of ownership; I# = code letter prefix and sequence number; T# = track number.*
Chronology of Case #1

C#	I#	T#	Description
−4.5	S1	5	*Wall Street Journal* article published and seen by respondents, catching them by surprise
−4	S2	5, 6	Visit by acquiring company executives and assessment team culminates in dinner at which acquired company executives decide they will not work for this acquirer
−3.5	S3	5	*Wall Street Journal* article saying the deal is off
−2.5	S4	5	Internal formal announcement that negotiations were back on with the same acquirer
−2	S5	6	Acquiring company decision to buy
−2		5	Owners of the two companies jointly announce tentative decision to buy/sell
−1.7	S6	2, 6	Acquired company CEO goes to NYC offices of acquirer, returns with notification that he will be replaced as CEO by person from acquiring company
−1.7	S7	2, 5	Acquired company executives, upper managers prepare letters of resignation, start job searches
−1.5	G1	2	Acquiring company management team of two (new CEO and new sales executive) arrive at acquired company offices
−1	G2	5	New CEO and informant have dinner at informants house, result is major change in attitude of informant
−.7	S8	2	Executive informant sends subordinate to NYC in hopes of changing antiacquirer attitude and convincing her to not quit
−.7	G3	5	New CEO goes to Christmas parties, is very open, nonauthoritarian, communicative; starts making friends with lower-level employees
−.5	G4	6	New combined logo and major ad campaign started by acquirer for acquired company
A	X	5	Change of equity ownership, big party with cocktails and canapes
+1	G5	5	Lunch with the CEO program started
+3	G6	2	Promotion from within acquired staff to QA VP
+4	G7	1	Initiation of new QA program for sales, elimination of insurance operation

APPENDIX B Chronological Listing of Events for Case 2
Column codes: C# = chronological time (in months) in relationship to the time of actual change of ownership; I# = code letter prefix and sequence number; T# = track number.
Chronology of Case #2

C#	I#	T#	Description
−4	S1	6	New suitor looks at acquisition; suitor is well known to acquired company, as is its local rep.
−3.5	S2	6, 5	Suitor staff informs VPs of intent to buy as soon as board approves at next monthly meeting
−3	S3	6	Board fails to act
−3	S4	5	VPs act depressed, upset over failure to act
−3	S5	5	"Mole" calls and says not to worry, approval will come next month
−2	S6	6	Board fails to act
−2	S7	5	VPs act depressed, upset over failure to act, general downturn in morale
−2	S8	5	Mole again calls, says not to worry, approval will come next month
−1	S9	6	Board approves
−1	S10	5	Elation among staff
−.5	XX	5	T1 interviews
A	Z1	2	New and active CEO from acquiring company comes.
+.5	Z2	3	Two VPs differentiate functions
+.5	Z3	3	New CEO gets position in lease approval, also moves some large approvals to acquiring company
+.5	Z4	1	Addition of two new documents to requirements for lease agreement
+.5	W1	1	New bookkeeping system moves central records to acquiring company offices
+.5	Z5	4	Increased requirements for formality in documentation package
+.7	Z6	1	Credit approvals slowed by two or three days, several old customers complain, some go to competition
+1	Z7	5	Conflict between sales manager and credit manager over slowness of lease approvals
+1.5	Z8	2	New manager added from acquiring organization to help speed up credit approvals and bookkeeping communication with acquiring organization.
+2	W2	1	Accountant cannot provide management with data from central office that was previously available from local bookkeeping system
+2.5	W3	5	Accountant meets with counterpart in acquiring company, but fails to get hearing with him
+2.5	Z9	5	New credit manager informs peers and subordinates that he will work only 8–5 week days rather than hours of rest of staff
+3	V1	6	Expansion strategy moved up two and a half years to start now
+3	V2	3	Three top managers (CEO & VPs) further differentiate functions, refusing to deal with issues in the other's turf
+3	V3	1	Slowdown in credit approvals by several additional days
+3.5	V4	5	Increase in conflict between sales and credit
+3.5	U1	4	VP notices that terms used in leasing differ in meaning from same terms used in banking and that this difference in understanding has caused loss on some contracts
+3.5	U2	1	VP orders check of all lease agreements to make sure that losses do not occur in future
+4	XX	5	T2 research interviews

APPENDIX C   Chronological Listing of Events for Case 3
Column codes: C# = *chronological time (in months) in relationship to the time of actual change of ownership;* I# = *code letter prefix and sequence number;* T# = *track number.*
Chronology of Case #3

C#	I#	T#	Description
−7	S1	5	Negotiations for the merger between the two organizations actually starts
−7	S2	5	Rumors of the merger circulate through both organizations
−6	J	2	Job searches start among managers and professionals
−4	S3	5	Formal announcement that negotiations are taking place
A	A	5	Formal, legal exchange
A	M1	1, 2, 3	Reorganization plan with new organizational chart and unit-by-unit timelines for moving in together and deadlines for accomplishing the integration
+1	O1	1, 2, 3	Start on restructuring personnel plans
+1	R1	2	First management resignation
+2	M2	1, 2	First office moves
+2	K	1	Resistance pattern starts to develop
+4	O2	1, 2, 3	Actual integration planning started in most units
+4	O3	1	Normal task performance drops with added and required integration work
+8	R2	2	Resignation of informant
+10	R3	2	CEO resigns, followed by other management resignations

REFERENCES

Argyris, Chris, and Donald Schon. 1974. *Theory in Practice.* San Francisco: Jossey-Bass.

Bastien, David T. 1987. "Common Patterns of Behavior and Communication in Mergers and Acquisitions." *Human Resource Management Journal* 26:1 17–34.

Bastien, David T., and Todd Hostager. 1987. "Jazz as a Social Structure, Process, and Outcome." SMRC Discussion Paper Series. Minneapolis: University of Minnesota.

Bastien, David T., and Andrew H. Van de Ven. 1986. "Managerial and Organizational Dynamics of Mergers and Acquisitions." SMRC Discussion Paper Series. Minneapolis: University of Minnesota.

Brunner, Thomas, Thomas Krattenmaker, Robert Skitaj, and Ann Adams Webster. 1985. *Mergers in the New Antitrust Era.* Washington, D.C.: BNA.

Davidson, Kenneth. 1985. *Mega-Mergers: Corporate America's Billion-Dollar Takeovers.* Cambridge, Mass.: Ballinger.

Fogel, Alan. 1980. "The Role of Emotion in Early Childhood Education." In Lilian Katz, ed., *Current Topics in Early Childhood Education.* Vol. 3. Norwood, N.J.: Ablex. pp. xxx–xxx.

Johnson, Bonnie McDaniel. 1982. *Communication: The Process of Organizing.* Boston: American.

Kitching, John. 1967. "Why Do Mergers Miscarry?" *Harvard Business Review* 45:6 (November/December): 84–101.

Magnet, Myron. 1984. "Acquiring without Smothering." *Fortune* (November 12): 22–30.

Marks, Mitchell Lee. 1982. "Merging Human Resources." *Mergers and Acquisitions* 9(2): 38–44.

Marks, Mitchell Lee, and Philip Mirvis. 1985. "Merger Syndrome, Part I: Stress and Uncertainty." *Mergers and Acquisitions* 20(2): 50–55.

———. 1986. "Merger Syndrome, Part II: Management by Crisis." *Mergers and Acquisitions* 20(3): 70–76.

Merton, R.K., M. Fiske, and P. Kendall. 1952. *The Focused Interview.* New York: Columbia.

Poole, Marshall Scott. 1983. "Decision Development in Small Groups, III: A Multiple Sequence Model of Group Decision Development." *Communication Monographs* 50: 321–41.

Salter, Malcolm S., and Wolf A. Weinhold. 1979. *Diversification through Acquisition.* New York: Free Press.

Schon, Donald. 1983. "Organizational Learning." In Gareth Morgan, ed., *Beyond Method: Strategies for Social Research.* Beverly Hills, Calif.: Sage.

Steiner, Peter O. 1975. *Mergers: Motives, Effects, Policies.* Ann Arbor: University of Michigan.

Williams, James B., and Mark L. Feldman. 1986. "Life after the Merger: The Human Resource Factor." *Health Care Forum* (September/October): 33–37.

Yin, Robert K. 1984. *Case Study Research: Design and Methods.* Beverly Hills, Calif.: Sage.

SECTION V

Studies of Technological Innovations

The four studies in this section focus on major technological innovations. As opposed to a close-up view centered on the research and development process, these studies adopt a macro perspective. They focus on the innovation context and how it effects the development of technology in ensembles of firms and government agencies. In some cases the firms cooperate, in others they compete, and often they must simultaneously cooperate and compete with one another. Along with the macro perspective comes a long time frame: the timelines of the four studies run from five to thirty-five years. This lengthy time frame is in part responsible for the complexity observed in several of the studies. The studies make it abundantly clear that there is no definite point at which a "set" invention or product emerges. Like the studies in previous sections, these four chapters indicate that the process of technological innovation develops along a complex pathway, with many branches, deadends, promising directions that are postponed, and setbacks. Perhaps this can be appreciated best from the macro perspective adopted by these studies.

In Chapter 12 Scudder, Schroeder, Van de Ven, Seller, and Wiseman study the management of a defense innovation. The extreme complexity involved in developing two generations of sophisticated torpedoes necessitates a flexible, reactive problem-solving process. Scudder et al. argue that this can best be represented by stimulus-response patterns, some of which chain into extensive problem-solving cycles and some of which remain isolated, reactive responses. Because it grows in this incremental fashion, the innovation exhibits a complex, branching developmental path similar to the fireworks model outlined by Schroeder et al. in Chapter 4. Scudder et al. identified four complexes of issues involved in producing the torpedoes, and the interaction among these areas resulted in a proliferation of ideas and subprojects within the innovation. The developmental path was further complicated by several unexpected occurrences in relationships with the government and external vendors. The complex developmental path was the result of many incremental reactions, cumulating and sometimes having a multiplier effect on one another. As Scudder et al. note, we know relatively little about how complex

397

technological projects develop. This chapter contributes several insights for both theorists and practitioners.

The emergence of the compound semiconductor industry is the subject of Chapter 13 by Michael Rappa. Rappa studies the overall structure of the industry over a thirteen-year period. Drawing on Kuhn's theories of the structure of scientific revolutions, Rappa advances a number of propositions related to the nature of change in the community of researchers, their organizations, and the ideas and techniques they utilize as the new technology evolves. To test these propositions, Rappa used bibliometric analysis. His data largely support his framework at an industrywide level.

With a similar broad scope, Knudson and Ruttan study the development of a biological innovation, hybrid wheat. The pursuit of hybrid wheat has engaged a number of companies and hundreds of researchers and extends over several decades. A decade of high hopes was followed by a decade of setbacks and pessimism, which has been succeeded by a new optimism in the 1980s. Knudson and Ruttan use Usher's cumulative synthesis model to depict and analyze the development of experimental breeding techniques adapted to the special problems of hybrid wheat. They observed several parallel efforts in the public and private sectors, and the entry and exit of a number of different companies, agencies, and researchers. Knudson and Ruttan emphasize the importance of basic research in biological innovation and note the ephemeral nature of firms' support for such efforts.

In Chapter 15, Garud and Van de Ven recount the emergence of the cochlear implant industry. The cochlear implant is a revolutionary device that promises to restore hearing to individuals that hearing aids cannot help. Because the cochlear implant is such a new technology, a number of functions necessary to sustain the new industry did not exist. 3M, Nucleus, Storz, and other firms had to develop basic knowledge, a community of scientists, clinical standards for implants, financing, customer credibility, manufacturing applications, and other essential prerequisites for a viable industry. Garud and Van de Ven posit a model of the accumulation of industry functions based on Etzioni's epigenesis model and study the development of these functions over time. Based on their study, Garud and Van de Ven outline a number of implications and recommendations for the management of innovating firms and for national policy.

All four studies highlight the relationship between context and innovation. As we move from Chapter 12 to Chapter 15, the view of context becomes more encompassing. Scudder et al. concentrate on the relations of the defense contractor with the U.S. Navy and eternal vendors. In Chapter 13 Rappa observes the development of a research community, a critical element of the context of technological innovation. Knudson and Ruttan choose a wider lens and focus not only on the research community, but on the public and private firms involved in the research and how economic and other factors influence their participation. Finally, Garud and Van de Ven take the widest view of all: they attempt to define the full range of functions that must be met by a particular industry if it is to survive, and document how they are produced.

The approaches of the four chapters are complementary. The problems they address define a nested series of questions, each of which must be addressed in turn for a full picture of the role of context in innovation. First, managers must consider the linkages between their innovating organization and other organizations or key actors. Second is the question of how crucial communities of scientists and researchers evolve to develop basic knowledge for a technological innovation. Third, what is the role of public- and private-sector organizations in technological innovation? Finally, how does an industry emerge that provides the infrastructure necessary to commercialize a techno-logical innovation. The chapters in this part illustrate ways in which these questions might be answered. It will become clear that new technologies transgress organizational, industry, sector, and national boundaries. No single institution commands sufficient competence or resources to control techno-logical development. As a consequence, the management of technological innovation demands an appreciation of the simultaneous roles of cooperation and competition among firms building an infrastructure necessary to support both technological development and proprietary entrepreneurship.

12

MANAGING COMPLEX INNOVATIONS: THE CASE OF DEFENSE CONTRACTING

Gary D. Scudder

Roger G. Schroeder

Andrew H. Van de Ven

Gary R. Seiler

Robert M. Wiseman

Much research on innovation has addressed managing relatively small product development efforts. In these situations, identification and management of the innovation team and the resulting product is relatively straightforward. However, when the product is very technically complex, requiring the teamwork of hundreds of persons, both employees and customers, over several years, the problem of

This research was supported (in part) with a major grant from the Office of Naval Research (Contract No. N00014-84-K-0016), Honeywell, Inc., the Strategic Management Research Center, and the Operations Management Center at the University of Minnesota. This chapter has benefitted greatly from comments on an earlier draft by Professor Paul Lawrence of Harvard University.

managing innovation in a complex, uncertain environment is much more difficult. This chapter examines the development of a highly complex and uncertain innovation over time. The innovation consists of the advanced full-scale development, testing, and production of a major torpedo weapons system by a defense contractor for the U.S. Department of the Navy.

In comparison with other innovations examined in the MIRP, the development of this torpedo system is considerable more complex and uncertain in several respects. First, the organization required for the development of the innovation is large and complex. Many departments and functional areas are involved, each of which often tends to optimize its specialty resulting in suboptimization for the overall project. With one final product as the focus, integration of all of the unit efforts is required. Although many innovations occur in large organizations, the unique factor here is that a large percentage of this one organization is working on a *single* product development effort.

The second complexity encountered in this innovation is the technical sophistication of the product itself. For example, the software needed to drive the torpedo was completely different from any previously developed, so in-house expertise had to be created. In addition, the torpedo is an assembly of many different electronic components, which requires development of many differentiated parts and subsystems. Since some of these are not produced in house, there are the additional complexities of managing subcontracting and vendor relations. These "outside factories" (see Schroeder, Scudder, and Pesch 1986) must be managed by the defense contractor so that components are produced on time, within quality specifications, and within the estimated learning cost reductions contractually agreed on.

In addition to subcontractors and vendors, this innovation involves joint ventures for subsystem development. These joint ventures were with other divisions within the company as well as with other companies. Management of these intra- and interorganizational relationships creates further complexities in the number of different program managers and functional units assigned responsibilities for coordinating various system components. Indeed, coordination departments were used to coordinate coordinators.

While the above factors add to the complexity of the innovation, there are also factors of uncertainty to be managed. Uncertainty was introduced by the need to develop a system that will respond to a moving target—the "threat" usually grows over time as other countries develop their defense systems. Therefore, flexi-bility is required to modify program components while the entire complex system is being developed. In other words, "preplanning" system modules were needed not only to meet current system requirements, but also to tolerate unknown future design changes.

Another factor of uncertainty in the case examined here is the timespan over which the innovation occurs. This particular weapons development effort actually began in 1975 and production of the system is not planned until after 1990. Thus, the effort must be coordinated for fifteen years, and continuity problems arise with changes in management, workforce, and governmental regulations and personnel. Managers are promoted, some leave the company, retire, or are reassigned to other divisions of the company. The same is true of military personnel and government administrators who monitor the program for the customer. Thus, the problems of organizational learning and memory are paramount.

Another uncertainty is due to the fact that in the defense establishment the customer is very political and bureaucratic. Policies change with administrations and the defense contractor must respond to these changes. For example, the current administration is committed to multiple sourcing of weapon systems contracts in order to obtain a lower-cost, better-quality product for the taxpayer. This in turn creates tremendous cost pressures on the innovation effort for the new product. Therefore, not only is the product being designed to changing specifications, but the design must continually be targeted toward cost reduction.

Moreover, multiple sourcing of systems development contracts implies that the defense contractor is required, by the government, to reveal proprietary information and to work in a cooperative manner with its competitors who are potential second source manufacturers of the system when its development is completed. This not only requires additional documentation and training, but also very selective release of information to competitors as required, while maintaining the firm's proprietary competitive advantage.

Finally, other factors of uncertainty and complexity are introduced by changing priorities in corporate resource allocations, strategies, and profit performance over time. For example, a corporate decision was made to manufacture the new torpedo system in the same factory used to manufacture the prior generation torpedo. This requires that design and production efforts to be coordinated very closely with the current product effort. A new factory for the new product was not considered a feasible option. While the products require many similar manufacturing operations, the technology required for the new product is much more sophisticated.

LITERATURE ON MANAGING COMPLEXITY AND UNCERTAINTY

Past attempts to manage complexity and uncertainty fall generally into two categories, structure and process (Maciariello 1978). Process responses originated in the operations research field and began with the development of the PERT management system for the Polaris Missile project (Sapolsky 1972). *Network planning* or *network analysis* are terms used to identify this class of tools. The development of PERT came in response to an apparent void of management tools capable of handling large complex projects such as the Polaris. PERT and its many variations, including CPM, GERT, PEP, and NHK-SMART (Lombaers 1969), essentially involve three stages of activity: planning, scheduling, and controlling. Sophisticated computer programs create massive charts and graphs that identify all major activities and events (intermediate outcomes) and their prioritized relationships. This is followed by an estimation of cost

and time requirements for each activity. The project manager then monitors actual progress toward each event looking for abnormal deviations in cost or time from that allocated.

Though extensively promoted in the literature, it is not clear that network planning is as helpful as its sophistication would indicate or the literature seems to assume (Perry 1966). The literature in this field tends to be either normative or purely descriptive with little if any framework to guide the assessment. Sapolsky comes closest to critically exploring the use of network analysis in a large innovation and concludes that its value has been considerably overstated (1972).

Structural mechanisms to organize highly complex and uncertain programs originated almost simultaneously in both management science and organization theory literatures (Delbecq and Filley 1974). Based on norms of rationality, Thompson (1967) began by setting forth a set of propositions on structural design that organizations adopt to manage environmental uncertainty, heterogeneity, and technological interdependence. Lawrence and Lorsch (1967) further developed the relationship between organization and environment with their model of differentiation and integration. They demonstrated that increasing complexity led to both increased differentiation among the organization's subsystems and increased integration across subsystems. Expanding on these themes, Galbraith (1977) proposed the matrix organizational structure as a hybrid of product and functional departmentalization forms. Galbraith proposed the matrix structure as an organizational design that provides a high capacity to process information by a highly complex tasks environment.

The management scientists also recognized the matrix structure as an organizational arrangement for managing complexity in NASA programs (Smith 1966). Although less theoretical in its approach, management scientists realized the value of a structure designed

403

to enhance communication across functional and hierarchical levels within the organization. It was not long before attempts were made to combine matrix structure with network planning (Marquis 1969). This combination of structural and process responses to complexity resulted in a combined approach termed program management (Maciariello 1978). Program management takes the view that programs and projects are essentially a collection of subsystems that under conditions of complexity and uncertainty require special attention to the linkages among these subsystems (Sayles and Chandler 1971).

Despite the voluminous literature describing program management, few attempts have focused on actually describing the process of program or project development over time using a theoretical framework to guide the research. Works such as *Clipped Wings* (Horwitch 1982) tend to be more a narration of history than a critical assessment of the process. However, Lane, Beddows, and Lawrence (1981) and Tichy (1980) each offer insightful examinations of complex innovation utilizing a specified framework.

Lane, Beddows, and Lawrence (1981) build on the work of Lawrence and Lorsch (1967) by viewing research and development programs in three dimensions: technical, political, and organizational. Technical refers to the logic employed by an R&D unit in the research process. Political refers to the way in which relations are maintained vis-à-vis the political environment. Organizational refers to the internal relationships and administrative systems that attempt to give coherence and continuity to the organization. Each dimension relates directly to a particular context. Lane, Beddows, and Lawrence examination of nine research programs indicated that successful research programs continually adapt or coalign their organizations to their multiple contexts by (1) managing the boundaries of the program, (2) switching emphasis and resources between

technical and political dimensions as conditions demand, (3) dealing with conflict between these dimensions.

Tichy (1980) takes a more dynamic view by examining cycles among the technical, political, and cultural dimensions over time. The three cycles are political (which includes the allocation of power, determination of organization direction, and destination of benefits), technical (which includes the allocation of social and technical resources), and cultural (which includes the shared ideology and values of project members). Each of the three cycles is theorized to fluctuate through the life of a project, rising and descending in importance and criticality in a stimulus-response process.

Despite the sometimes prolific writing on research, innovation, and project management, it appears that, as Sayles and Chandler (1971) perceptively point out, "We probably know less than we think we do about the management process by which technology is converted into operating systems." One reason for this lack of knowledge is that much of the literature on program management has consisted of "arm-chair" theorizing without being anchored by systematic longitudinal research on the developmental process of program management over time. It is to these "prosaic and easily overlooked management policies" (Sapolsky 1972: 249) that this chapter addresses itself.

The next section outlines the methodology of our longitudinal study, including a description of the company, the data collection methods used, and our methods of analysis. The third section provides an overview of the entire innovation project, covering several years. In addition, four major areas of the innovation program are analyzed with respect to a stimulus-response model described in the next section. The fourth section presents the overall findings on innovation processes from this study, with the final section presenting the conclusions on managing large, complex innovations.

METHODOLOGY

The setting for this research is the Underseas Systems Division (USD) of Honeywell, Inc. Honeywell is a large, diversified company in control systems and aerospace and defense with in excess of $6.2 billion in sales per year. USD was formed in 1983 in order to concentrate on producing weapons systems that would operate underseas with the principal product and customer being torpedoes for the U.S. Navy. Currently, the Mk 46 torpedo is in the production phase, with the MK 50 currently under development as the next generation torpedo. Development of the MK 50 began in 1975 and is expected to begin full scale production in the early 1990s.

USD is organized into traditional functional departments as shown in Figure 12–1. The division is headed by a general manager who is also a vice president of the corporation. The functional departments are headed by directors who report to this general manager. The division is coordinated by frequent meetings of the operating committee which consists of the general manager and the directors. Program offices have also been established for both the MK 46 and the MK 50 systems. These program offices, which are the primary point of contact with the Navy, operate in a matrix fashion with the various functional managers.

The division operates out of one major facility that contains both the production space and the division offices. The buildings have been modified frequently as reorganization and new programs have been undertaken. Because the division is in a growth mode, there have been ample promotion opportunities for people in the division, as well as for those from other divisions within Honeywell. Few managers have been brought in from the outside. The division currently employs in excess of 1,000 people.

Data Collection

USD created a steering committee in January 1984 to assist our study in gaining access and collecting data. From February to June 1984 interviews were conducted with the general manager and most of the department directors in order to develop a case history of the MK 50 innovation program up to that point in time.

After the case history was written, the goal was to collect longitudinal data at six month intervals. The first data collection was performed in October 1984. Semi-structured interviews were conducted and ten executives filled out the Minnesota Innovation Survey questionnaire. The standard operating procedure for the interviews was to tape the interview and provide a writeup to other members of the

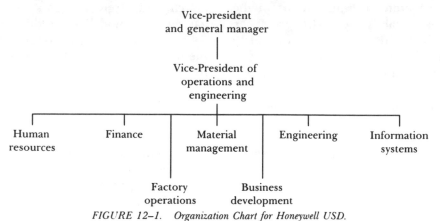

FIGURE 12–1. *Organization Chart for Honeywell USD.*

405

team within forty-eight hours. This ensured minimal information loss in the process. The data collection instruments used conformed closely to the standard set used by the MIRP studies (see Chapter 2). Similar data were also collected in May and November of 1985. (Additional data collections of this type were planned, but due to unfortunate circumstances, funding for this study was discontinued.) However, additional interviews were conducted in July 1986 with key respondents. In addition to interviews and questionnaires, data were also obtained from company documents, telephone calls, the company newsletter, and other archival data.

After each data collection, the research team entered the quantitative data into SPSS files for statistical analysis. The qualitative data from interviews, questionnaires, and archival sources were then analyzed for content and recurring themes. One method used for analysis was the categorization of the data into events. An event was described as a change in the idea of the innovation, the people involved, or the context in which it was being developed. In addition, major transactions between parties of the innovation and significant outcomes were also defined as events. For example, the promotion of the general manager out of the division was a people event. A change in context could be viewed as occuring within the Honeywell corporate environment or in the overall defense establishment—Congress, the Pentagon, and so forth. A complete listing of all of the events observed in the development of the MK 50 program is presented in the appendix to this chapter.

A graphic approach to understanding these events is presented in Figure 12–2, which shows a process map of the MK 50 development over time. This figure shows how several key events from the appendix are linked together over time and indicates the future direction of the MK 50 program and the division.

These events, when coupled with the

qualitative and quantitative data from the study, provided a rich database for identifying key themes for tracking the innovation. Following each of our data collection periods, the research team individually and collectively reviewed the data to develop feedback reports which were shared with the study participants. Each feedback report consisted of our findings, supported by qualitative and quantitative data, along with our observations on the process of managing the innovation. The feedback sessions for this project occurred in December 1984, June 1985, and February 1986.

In our initial meetings with the USD management team four major functional areas were identified as critical to the development of the overall innovation. These areas were human resource management and organization, manufacturing producibility, materials management, and strategic planning and marketing. Each of these areas is explored in greater depth in the next section of this chapter. At this point, however, a model is needed to clarify each of these areas individually, as well as collectively.

Conceptual Models for Data Analysis
Among the models proposed in the literature, two models came to the forefront in our analysis. First, the garbage can model of Cohen, March, and Olsen (1972) was considered. The garbage can model posits that problems and solutions are partially uncoupled (both exist in the organization) and they randomly intersect through fluid participation, (participants engage in activities when their interests dictate). In other words, solutions are not developed in response to particular problems but are randomly paired with problems as they arise. The basic premise here is that organizations may be organized anarchies possessing three general properties: problematic preferences, unclear technologies, and fluid participation. Although some of these characteristics are somewhat evi-

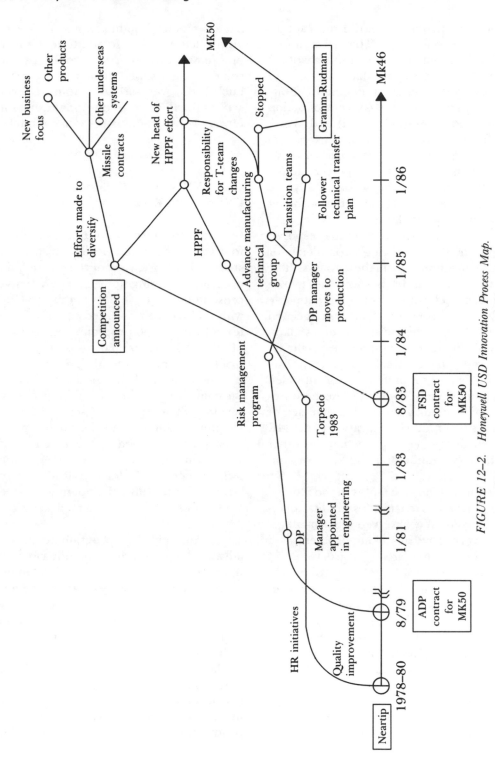

FIGURE 12–2. Honeywell USD Innovation Process Map.

407

dent in our study, this model did not capture the process we observed. (This is explored more fully in the fourth section of this chapter.)

The other model considered was that of the organization as an action generator (Starbuck 1983), particularly the concept of action loops discussed by Masuch (1985) and Weick (1979). Masuch states that an action loop occurs when an activity entails a chain of other activities, which, in turn, ultimately recreate the original situation. However, in our study, we often did not see the original situation being recreated with any regularity.

This led us to a stimulus-response model, with the stimulus being a complexity or uncertainty encountered in the process. The basic form of the model is shown in Figure 12–3. This model consists of an identifiable complexity or uncertainty that is associated with the MK 50 contract, which becomes manifest or visible in the recognition of a specific set of problems that trigger a solution or response. Over time, the solutions have both intended and unintended consequences. These consequences, in turn, may create another set of complexities that require integration of the MK 50 program or trigger other actions in other parts of the organization or in the Naval organization.

We found many examples in the Honeywell case where this chain of stimulus to response occurred. We observed USD management to be very active in seeking problems and solving others as they arose. A great deal of planning was also done and control mechanisms were widely used. In addition, the solutions had outcomes and consequences, but we did not find support for the reoccurrence or recycling of the same problem such as discussed by Masuch (1985). What we observed was more order and less anarchy than the gar-

bage can model might propose and less clearly visible action loops. In addition, in most cases, problems clearly preceded solutions, although some instances of "justifying actions" (Starbuck 1983) were noted. The next section describes the findings of this longitudinal study in terms of this stimulus-response model.

OVERVIEW OF MK 50 DEVELOPMENT

In order to more easily understand how the four areas examined here fit into the overall study, a brief overview of the entire innovation program, covering approximately ten years, is presented. First, some overall developments within USD are presented to explain the context of the MK 50 program. During the period of 1979–80, the company was engaged in a retrofit program for the Mk 46 torpedo, termed *Neartip,* which had almost disastrous results for the company. The impact of this program led to several of the changes discussed later in this chapter, including the design for production concept, preplanned improvements and a restructuring of job classifications. When Honeywell formed the Defense and Marine Systems Group in early 1983, the operation gained division status (USD) and named a new general manager.

Although the program to develop the advanced lightweight torpedo had its beginnings as early as 1975, the first identifiable event was in August 1979 when the advanced development phase was begun, with two defense contractors working in parallel on alternate systems designs. In January 1981 Honeywell was awarded a sole source contract for further systems development as their technical design had proven superior. Up to this point, the development effort was primarily technical in nature. However, partially due to earlier producibility problems, a design-to-production

Complexity/ → Problem → Solution/ → Outcomes/
uncertainty   response   consequences

FIGURE 12–3. *Stimulus-Response Model.*

408

manager, who reported to a director of engineering, was appointed in January 1983 to work on producibility issues—the earliest that production had ever been involved in a contract like this.

From the beginning of the program, USD management believed the production contract for the MK 50 would be single sourced and that they would be the leading contender for the production contract. However, in September 1984, an Industrial Briefing for Competition was held by the Navy in which basic technical information was conveyed to potential competitors. This was the first event in a long series of actions leading up to the Navy's decision to seek a second source to manufacture the torpedo. This change in the government contracting policy had many effects on the strategic planning and marketing efforts of the company as discussed below. In January 1985, the design-to-production manager was moved into the production function signifying a new focus of energies. He identified several critical tasks which needed to be addressed to ensure producibility of the torpedo and groups referred to as transition teams were formed. After a slow start, the number of teams was streamlined and they were reassigned to the production manager, with a significant improvement in results.

Another area where much effort was expended was in the development of the software to guide the torpedo. It was a new type of software, never before developed by USD and many technical and administrative difficulties were encountered in creating this software. The difficulties were eventually overcome and software became a competitive strength of the division.

The overall performance of the program was rewarded by the navy based on cost, performance, and schedule. Technically, the early results were considered outstanding, but the software effort later brought down this rating slightly. However, as the entire technical package came together, the overall performance was judged very good by the customer.

During this same development period, the organization itself was undergoing many changes. In a complex electronics product like a torpedo, many different parts and subsystems must be produced. Some of these are produced in house, but others are produced by sister divisions or outside suppliers. USD reorganized its efforts into a formal organization for material management in order to manage these materials requirements.

In addition, the strategic planning process of the company was emerging and being reshaped by the changes encountered in the MK 50 program. The primary shift seen in the five-year period was the recognition that the division could no longer be a single product company. Ultimately, the division redefined its strategy to encompass underseas weapons systems of various types.

This strategy shift was reflected by the changing charter given to task teams working on the redesign of the production facility. A decision had been made early in the MK 50 development that the new torpedo would be produced in the same factory as the previous generation torpedo. As the context changed over time, the mission of this factory was reshaped to focus on defense electronics, with maximum flexibility and low costs.

Within its management ranks, the division general manager was promoted to group vice president, and the former engineering and MK 50 program director became the new general manager. While interviewees did not consider this a major change in management context, there was a distinct shift in management style, with more emphasis placed on technical operational issues. After his promotion, the vice president promoted several USD directors to the group level to assist him in the development of group strategies. This, of course, led to several management changes for the MK 50 program in the division.

TABLE 12–1. *USD Award Fees.*

Date	Percentage
March 1985	13%
July 1985	4
October 1985	8

Note: Award fees were suspended in January 1986 after the Gramm-Rudman funding cuts. Contract negotiations were reopened, and after several months a new contract, without award fees, was signed.

During 1986 many of the earlier strategic plans came to fruition. However, a major external shock was the announcement of the Gramm-Rudman cuts in MK 50 funding in January 1986. The schedule for the program, as well as the annual funding requirements, had to be completely renegotiated, an effort which was not completed until September 1986. Much managerial time and effort was expended and resulted in significant changes in contract schedule and funding. The funding cuts were viewed differently by various members of the organization, with some persons stating the rescheduling would allow USD to meet its deadlines, which were slipping during the FY 1986 portion of the contract.

Program progress was measured in our study using two different types of outcomes criteria: (1) Navy award fees and (2) our measures in our MIRP questionnaire. Honeywell's contract with the Navy had a built-in performance

incentive system. Each quarter, the Navy conducted award fee meetings, and depending on the progress in achieving technical milestones, schedules, and cost estimates, an award percentage, between 0 and 15 percent, was determined. This translated into a dollar amount that contributed significantly to the profitability of the division. These awards varied dramatically during the course of our study, and are shown in Table 12–1.

The other outcomes monitored over time were perceived program effectiveness and degree of problems encountered, as shown in Table 12–2. The data show that perceived effectiveness declined over time while problems decreased over time as the program requirements were firmed up.

One other outcome measure used was to ask each person interviewed what grade (A–F) they would give the innovation effort for the past six months. Early in the project, the grades were unanimously As. However, in the last data collection period, a grade of C was given by one manager as recognition of the schedule slippage and funding cuts that had occurred.

With this brief overview as background, specific events related to MK 50's development are now discussed in four major areas: human resources management and organization, material management, producibility, and strategic planning and marketing. Each of these subsections contains a chronological list of the appropriate events, a stimulus-response chart showing the complexity/uncertainty factors identified, and a narrative to describe the process to the reader.

TABLE 12–2. *Survey Quantitative Data.*

	T1 N=10	T2 N=35	T3 N=30
Perceived innovation effectiveness mean	4.1	3.6	3.2
Innovation problems mean	4.6	3.2	3.1

HUMAN RESOURCES MANAGEMENT AND ORGANIZATION

Table 12–3 shows a subset of sixty events from the appendix that deal with human resources management (HRM) and organizational design of the MK 50 program from 1981 to 1986. These sixty events comprise over one-half of the total events observed in the study. For purposes of analysis, these sixty events are categorized into eight that are indirectly related and fifty-two that are directly related to the MK 50 program. The latter fifty-two events are further classified into nineteen events dealing with turnovers in key MK 50 management personnel over time, and thirty-three organizational restructurings of the MK 50 program, including seventeen that differentiated the MK 50 program into specialized units and groups, and fifteen events that clearly reflect efforts to structurally integrate various components of the MK 50 program between 1981 and 1986. We will first summarize the nature of these different groups of events and then discuss process patterns observed among these events.

Events Indirectly Related to the MK 50 Program

In the normal course of overall organizational operations in which the MK 50 program is housed, eight major events occurred between 1980 and 1986 that had significant indirect effects on the MK 50 program. These events provided an organizational context that both facilitated and inhibited the development of the MK 50 program.

Honeywell's receipt of a Navy contract to refit the Mk 46 torpedo with a *Neartip* kit in 1978, and the organization's major $70 million loss in fulfilling the contract's provisions in 1980 (event 1) was frequently referred to as a major shock and learning experience by organizational participants. The "crisis" clearly pointed out limitations in the organization's technical competence and management practices, which resulted in an extensive search for alternative ways of thinking and acting, and a sustaining strong resolve "Never to have another *Neartip!*" among all the managers interviewed in 1984.

One such change was a succession of general managers with a humanistic orientation in the Defense Systems Division—USD's parent until 1983—and in USD itself. These managers had an appreciation for the potential in utilizing the division's human resources to address business problems. For example, an "It" Steering Committee (discussed at greater length by Kanter 1983) was the brainchild of the division general manager in 1982 and was reputed to have initiated and directed a set of strategies to increase productivity, foster innovation, and create a climate of work quality and "success."

Driven by the necessity for change in the aftermath of the *Neartip* crisis and supported by the humanist perspective of top management, the personnel function began to undergo a significant metamorphosis from 1982 to 1984. Gradually, the role of HRM as a staff appendage changed to one of being a catalyst for line management. Along with this catalyst role came the relocation of responsibility for HRM from the personnel department to line management.

Working collaboratively, particularly across previously fixed organizational boundaries, was a major thrust from the "It" committee and leaders from both USD and DSD. A vivid example is the "Torpedo '83" event. Confronted with an increasingly calcified and stratified system of job titles that prevented rotation of manufacturing workers between positions, fourteen management and fourteen union representatives cooperatively negotiated over a six-month period a new job classification system that effectively consolidated thirty-five pay grades into four, accompanied by provisions for a higher wage scale overall and for multiple role assignments and training for workers.

411

TABLE 12–3.   *Events Related to MK50 HRM and Organizational Structuring.*

8 events indirectly related to MK50 program

1980	"Neartip" crisis
1979–86	Succession of humanists as division general managers
1981–82	"It" steering committee
1981–84	Human resources management from staff appendage to line management
1983	"Torpedo '83"
3/83	Underseas Systems Division (USD) spun off from Defense Systems Division as a separate profit center with single torpedo business mission
1984	"Lean, mean, and loving" division philosophy statement
3/85	Marine Systems Group created

52 events directly related to MK50 program: 1981–86

19	turnovers in key management personnel: 1 sickness, 1 death, 4 terminations, 13 internal transfers/promotions over time
33	restructurings in MK50 program organization (excluding personnel turnovers and replacements) (see detailed events below)

17 events involving organizational structural differentiation

1/81	MK50 program established in response to navy contract award
5/81	"Paperless" factory program initiated
6/81	Automatic insertion equipment installed to aid production line tasks, marking the beginning of factory automation efforts
7/82	Strategic planning and investment process begun to develop cutting-edge technologies related to growing engineering competence of USD
8/83	Division's productivity board assumes responsibility for developing a restructured materials management approach
1/84	New materials management unit announced and director appointed
7/84	Strategic plan for business diversification approved and search begun for acquisitions in order to reassign idle MK50 engineers projected in latter half of 1985 and 1986
11/84	100S software development program initiated
1/85	Advanced manufacturing technologies group formed
1/85	Software development manager position created
3/85	New general manager introduces new operating style
5/85	Factory-of-future team formed
5/85	MK50 Preproduction improvement program position formed
5/85	Reorganization of software development group into matrix
10/85	Factory automation effort in development stage
12/85	CAD/CAM group formed
Spring 1986	50 software engineers reassigned to other programs and divisions and 7 laid off

15 events involving structural integration of MK50 program

3/81	Memo from division general manager to directors of procurement and production to develop a focused approach to materials management activities; group is formed
6/81	Advanced manufacturing strategic planning unit established to integrate various factory automation innovation efforts
6/81	MK50 program manager title changed to manager of design to production to emphasize D-P transitions role
8/83	Upon Navy award of full-scale development contract, D-P unit incorporates Navy's Willoughby risk management program

TABLE 12–3. *(continued)*

2/85	22 "transitions management" teams formed
8/85	New manager of transitions teams appointed, who reduced the number from 22 to 6 teams
10/85	Transition teams evaluated and directions modified
11/85	"Tiger team" formed to help with software design problems
12/85	Transition teams terminated
2/86	Reorganization of A&D group resulting in cutbacks in division and MK50 program personnel
4/86	Strategic personnel requisition board formed as clearinghouse for internal personnel transfers.
6/86	Strategic human resources plan issued.
Spring 1986	New strategic planning system instituted that connects budgets with plans
8/86	MK50 program management unit reduced from 31 to 17 people
1986	Marine Systems Group requires USD general manager to become a more integrated player in group strategies

Another event that marked another step in the evolving organizational context of the MK 50 program was the development of a formal statement on USD's strategic management philosophy in spring 1984. This statement emphasized the central role of human resources management in the strategic direction of USD's business and articulated USD's philosophy as being "lean, mean, and loving"—meaning an efficient, competitive, and people-caring organization. This philosophy statement was printed in a brochure and widely distributed and discussed throughout USD in 1984 and 1985.

In order to manage growth into a broader and more diversified set of marine defense-related businesses, Honeywell created a new Marine Systems Group in March 1985 that included USD and a number of other marine-related Honeywell divisions and wholly owned companies. USD's general manager was promoted to group vice president, and the second person in command at USD was appointed USD general manager. In addition USD's directors of HRM and strategic planning were promoted into the new group. In September 1986 it was reported that division general managers in the group were obtaining both more and less authority. Following a "loose-tight" authority pattern, division general managers had less authority to act on their own in regard to their businesses but were given more authority in directing key groupwide committees and activities. For example, the new USD general manager became chairman of the group capital investment committee with wide discretion across divisions within the group. In return, USD is now expected to be a more "integrated player" than before with other divisions in pursuing group-level strategies.

Structural Differentiation and Integration Events in the MK 50 Program

The complex scope, uncertain technological mileposts, and long temporal duration involved in developing the MK 50 program for the Navy created significant problems and adaptations in human resources management and organizational design of the defense contractor. Although some of these problems and responses could be anticipated when contract work on the MK 50 began, most were not foreseen due to the scope, duration, and technological uncertainties of the program. As a result, an

FIGURE 12–4. Human Resources Management and Organization.

Complexity/ uncertainty ──────────⟶	Problem ──────────⟶	Solution/ response ──────────⟶	Outcomes/ consequences
Organization size	Integration between departments and levels	Task forces Transition teams Group meetings Integrative structures	Segmentation Missed Schedules Missed appointments
Technical mileposts	Technical failures, Missed deadlines	Personnel scaleup Differentiated structures	Personnel "cliff" problems Corporate ILM reassignments
Timespan of contract	Management turnover	Replacement HRM philosophy	Management discontinuity Generation of humanist managers

incremental process of structural adaptation to problems as they arose was the most frequent pattern in managing complexities. More specifically, most of the observed events related to human resources and organizational arrangements can be described by a stimulus-response model, as illustrated in Figure 12-4. Although one could systematically examine this stimulus-response model for individual sets of events in Table 12–3, due to space limitations, we will focus on the three generic strategies, shown in Figure 12–4, that appeared to be triggered by three dimensions of complexity and uncertainty: size or scope, uncertain technical requirements, and extended temporal duration of the program.

Uncertain technological tasks and mileposts required in the development of the MK 50 program often created problems of technical failures and missed deadlines. These technical problems were typically responded to (in seventeen instances) by creating new specialized units, task forces, and problem-solving teams that had the structural effect of increasingly differentiating the organization.

The extended temporal duration of the MK 50 contract created program continuity problems associated with normal personnel attrition and nineteen instances of turnovers in key management personnel between 1981 and 1986. The typical and direct response was replacement of vacated management positions. However, the ongoing and indirect socialization of USD personnel to the organizational philosophy and a long succession of humanist-oriented managers, coupled with extensive and frequent communications among levels of management, appeared to significantly help maintain program continuity as turnovers, transfers, and promotions occurred among key program managers.

The complex scope of the MK 50 program, coupled with its technological uncertainties and management turnover, led to a host of coordination and integration problems. In fifteen instances the typical response to problems of fragmentation was to institute a variety of incremental reorganizations that had the effect of integrating fragmented parts of the organization.

Although space prevents a discussion of these many individual events of structural adaptation over time, it is apparent that management of the MK 50 program involved repetitive and incremental efforts at maintaining a balance between the opposing forces of organizational differentiation and integration, as Lawrence and Lorsch (1967) proposed. However,

414

TABLE 12–4.   *Personnel Perceptions of Innovation by Level.*

	Spring 1985 Levels			Fall 1985 Levels		
	Top	Mid	Low	Top	Mid	Low
MIRP Survey Data Show That as One Compares Responses from Top to Lower Management Levels:						
Perceived effectiveness decreases	4.3	3.6	3.1	3.7	3.2	3.0
Lack of clarity about goals increases	2.3	2.8	3.3	2.0	3.1	3.3
Lack of support/resistance for key sponsors increases	2.0	2.2	3.3	2.5	2.7	3.2
Frequency of conflicts/disagreements increases	1.8	3.4	4.6	2.5	3.2	3.8
Conflict resolution by smoothing over increases	2.2	2.3	3.4	2.0	2.8	2.9
Conflict resolution by confrontation decreases	3.8	4.0	2.7	4.0	3.0	3.4
Chances of rewards for good work decrease	3.9	3.4	3.3	3.5	3.4	3.3
Overall perceived team leadership decreases	4.2	4.1	3.3	4.0	3.4	3.1

Approximate scale: 1 = none, 2 = little, 3 = some, 4 = much, 5 = very much.

the consequences of these incremental events of structural integration and differentiation are less clear. Questionnaire surveys, which measured the perceptions of cross-sections of MK 50 personnel over time, are summarized in Table 12–4. The data show that although perceptions about program coordination, control, and effectiveness by top management improved over time, they declined for lower-level personnel and an expanding gap emerged in these perceptions between organizational levels. This required managers to closely attend to communication between the different levels in the organization. Interlevel coordination problems are typical in all organizations (Tannenbaum 1968; Van de Ven and Ferry 1980), but they are attenuated in a large organization undertaking a complex innovation. Many of the HRM events described here are related to the issue of ensuring producibility of the product as discussed in the next section.

Producibility

Producibility here means that the technical design of the product can be manufactured on the factory floor within cost and performance specifications without major delays. Producibil-

ity of the MK 50 system was complicated by the fact that it is being developed over a fifteen-year time period and the resulting product must be manufactured in the same production facility as its predecessor. Table 12–5 lists the events associated with MK 50's producibility, while Figure 12–5 shows the stimulus-response diagram.

A traditional way of managing a development process would be the use of project management techniques such as PERT or CPM (Weist and Levy 1977). A PERT chart by itself can only capture a project's performance on the planned time schedule. Although the company does have a project network showing some of the major milestones, each milestone requires so much coordination and integration that additional managerial processes are needed. What is really being managed here is a transition of a complex design from engineering into the manufacturing facility. USD adapted a risk management model, along with several other transition mechanisms in order to manage the producibility of the MK 50 system.

In September 1983 the Navy requested that the company use a new, revolutionary approach to managing the design to production process. A task force of the Defense Science

TABLE 12–5.   *Events Related to Producibility.*

1975	Initial request issued for proposals for advanced lightweight torpedo
May 1979	Advanced development contract awarded to Honeywell and McDonnell-Douglas
January 1981	Navy decides Honeywell will be single source contractor for development project
January 1981	Design-to-production manager appointed, reports to engineering
October 1982	First production plan submitted
September 1983	Navy presents new process for managing risk in defense development projects, leads to formal risk management procedure
October 1983	Revised production plan submitted that incorporated new risk management procedures
January 1985	Design-to-production manager shifted to report to production
January 1985	23 tasks identified as necessary for successful transition of design into production
February 1985	Transition teams started to accomplish each of these tasks
August 1985	Transition teams reassigned and streamlined for greater effectiveness
December 1985	Transition teams put "on hold"
January 1986	Due to Gramm-Rudman funding cuts, transition activities halted for over six months

Board had developed a document for maintaining discipline in the design and production of weapons systems. This task force was comprised of both military and industrial leaders and generated a matrix of the critical events in the design, test, and production of weapons systems. These events were used to generate a set of templates or tools for managing the risk at each stage of the development process. The thirty-eight templates developed describe techniques for improving the development process to ensure a "low risk transition"—that is, minimizing the probability that design will not be producible. Each template is simply a block dia-

FIGURE 12–5.   *Producibility.*

Complexity/ uncertainty ⟶	Problem ⟶	Solution/ response ⟶	Outcomes/ consequences
Technical design	Designing a producible product	Design-to production manager Navy templates	Focus on producibility early in design phase Risk management program
	Totally new software design	Role swapping	Improved communication between navy labs and USD
		Co-location of personnel	Better communication More rapid development
Upper management decision not to build a new factory for MK 50	Merging of production processes and personnel required to integrate two products into one factory	Transition teams Task team for high performance production facility	"Outran" other areas of project after some initial problems-on hold Factory to be redesigned for future products as well as MK50 and Mk 46

gram that describes an area of risk.as well as the technical methods believed to be useful in reducing this risk. The areas of risk were drawn from the combined experiences of task force members and included the following areas: funding, design, test, production, transition plan, facilities, and management.

The transition plan template integrated the design, test, and production areas. The outline presented for decreasing risk included a revolutionary recommendation to fund a contractor transition plan *no later* than the start of the engineering development phase. The plan was continually updated until full production was started.

These templates were integrated into a formal risk management program. The risk being managed here was whether or not the design would be producible within the specifications agreed to in the contract. This risk management program consists of a monthly update of the cost, schedule, and performance of each key system or subsystem of the product in the areas specified by the templates. The observed values are compared to the planning values that were established to ensure the producibility objectives are met ten years later. The introduction of this program led to a revised production plan in October 1983. In January 1985 the emphasis of the transition effort shifted to production. The design-to-production manager reported to the head of production. At the same time, this manager identified several interdisciplinary tasks that needed to be addressed in order to make the transition, including the integration of the two torpedoes into one production facility.

To address these tasks twenty-two transition teams were created in February 1985. (See Table 12–6 for a list of the initial issues to be addressed by the teams.) Each team consisted of several people with skills relevant to the unique transition problem it was assigned. Initially, the teams reported to a management committee. Many of these teams had overlap-

TABLE 12–6. *Integration Issues Initially Assigned to Transition Teams.*

Production engineering and factory
Product assurance
Material engineering/evaluation
Production planning
Engineering support to factory
Program management
MK50 qualification
Assembly stress screening/testing
Factory model and simulation
Finance
CAD/CAM
Material management plan
Advanced manufacturing technology
Human resource/training (Support)
Capital appropriations
Mk46/MK50 hit teams
Tooling/gages
Floorspace
Configuration control
Process definition
Factory cost control (workcells)
MK50 DTC/UPC (product costs)

ping charters or addressed interdependent problems. Transition team members reported there was a lack of management commitment to their efforts. In August 1985 the number of teams was reduced from twenty-two to six and they were reassigned to the production manager. After this change, the teams became so productive in solving problems and developing plans that they were "put on hold" in December 1985 because the general manager reported they were "outrunning" the rest of the program. They were, however, viewed as having been successful in their tasks. When Gramm-Rudman funding cuts were announced in January 1986, all activities relating to transition were formally suspended for over six months.

In addition to the transition teams, a task force was created for the design of an advanced factory, referred to as the high performance production facility. Initially, sixteen goals for this factory were outlined by the torpedo factory manager. These were further refined as the company decided to design this factory for future business as well. (See the discussion in the strategic planning and marketing section of this chapter for further information on this topic.)

Other mechanisms used by the company to overcome transition problems were the use of co-location of personnel and task reassignment. Two key managers swapped roles to improve communications between two critical groups in the software development process, one within the division and the other part of a navy operation in San Diego. The co-location of analysts and programmers was used successfully in the software area to improve communications as well as the quality and production of the software.

These mechanisms and restructuring of the organization were perceived by management to be successful, and by July 1986 a prototype of the MK 50 had been developed. Software was a major bottleneck, but the problems were resolved and capabilities to develop custom software became viewed as a competitive strength of the division. The actual success of these producibility efforts can not be measured until production begins in the early 1990s, although the risk management procedure described here has now become a standard requirement in new defense contracts.

The next section describes USD's efforts to better manage the "outside factory"—its suppliers. Although this is certainly part of the producibility issue, it is important enough to warrant separate discussion.

Materials Management
As a result of the *Neartip* crisis described above, the general manager decided that the division "absolutely must produce a quality product." Since materials costs are about 70 percent of the total cost of the product and quality is similarly determined to a large degree by the sup-

pliers, it became apparent that the division had to do a better job of managing its suppliers. The management of suppliers had been very fragmented and little effort had been devoted to the task. For example, purchasing had responsibility for the cost and quality of incoming materials, but only a few purchasing agents were handling hundreds of suppliers. Various individuals were assigned to help with quality engineering or technical assistance from other parts of the organization, such as the engineering department, but these individuals were frequently not readily available or they were ineffective in solving suppliers' problems. As a result, the best purchasing could do was to hope that defective materials were detected with incoming inspections and returned to the suppliers for correction or replacement. If the situation became intolerable, another supplier was contracted to supply the parts. A preventive approach to quality was not being followed, and the division was largely reacting to quality and cost problems associated with suppliers. These problems eventually led to a new organizational structure. The events leading up to this reorganization and additional changes are listed in Table 12–7.

In 1981 the general manager decided that a more coordinated approach to suppliers was needed and that the division should devote considerable resources to managing the "outside" factory, just as they currently managed the inside factory. A committee was formed of people from production, purchasing, and engineering to look into the problem and come up with a solution. The solution that emerged was to form a new materials management department to integrate purchasing and the associated functions currently being performed by other departments. The timing was fortuitous because the purchasing manager was about to retire thereby, making possible an expanded role for materials management in the division.

The new materials management depart-

TABLE 12–7.   *Events Related to Materials Management.*

March 1981	Committee formed to develop a more coordinated approach to supplier management
August 1983	Productivity board assumes responsibility for developing new approach to material management
September 1983	Finding and recommendations presented to general manager
November 1983	New material management unit announced
January 1984	Material management department formally begins operation
January 1984	Life cycle costing begins
January 1984	Certification of suppliers begins

ment included all of those functions from other departments that were associated with managing the suppliers. These functions represented the entire chain of activities associated with obtaining and using parts from outside suppliers: purchasing, incoming inspection, quality engineering, storerooms for incoming materials, costing, and so forth. The new department was responsible for putting *good parts* into the storerooms that production could then use to make the product. It was also responsible for making these parts available on a timely basis to support the production schedule at the lowest possible total cost. This new materials management department was established January 1, 1984, and constitutes one of the many structural innovations adopted to achieve integration described earlier.

After the new department started operations, it became apparent that two additional innovations would be needed: *life-cycle costing* and *supplier certification.* Life-cycle costing was based on the idea that all the costs associated with a part should be captured. These total life-cycle costs included the purchase price of the part, as well as costs of transportation, inspection, rework, inventory storage, and the costs of eventually maintaining the part in the field in support of the finished product. The lowest-cost supplier then would be judged based on the lowest total life-cycle cost and not just the lowest purchase cost. In 1985 the division constructed and implemented a computer system that would allow it to capture the total life-cycle cost associated with the various parts. This in-

novative system was being used to track costs associated with suppliers and to provide information to management for corrective action and supplier selection decisions. Development of this system can be traced directly to navy pressures to select the "low-cost" supplier without sacrificing quality.

Another innovation was the supplier certification program. This program included visits by division materials management personnel to the supplier. During these visits the division personnel evaluated the quality control procedures used by the supplier. These evaluations included the use of statistical quality control procedures, training, measurement and testing, design procedures, organization for quality, and so forth. A supplier could be certified if the division judged that the supplier's production processes were under control and that the supplier could produce consistent parts on the basis of using statistical control principles. Under these circumstances, no incoming inspection would be used by the division and the supplier would ship parts directly to the division's production line. The "outside" factory then became, in effect, an extension of the division's production line (see Schroeder et al. 1986). Supplier quality could then be monitored on an infrequent sampling basis.

Figure 12–6 illustrates the stimulus-response model for materials management. As discussed above, there were three complexities/uncertainties in materials management. One is the inconsistent quality and delivery schedules from the suppliers. This was seen as

FIGURE 12–6. Materials Management.

Complexity/ uncertainty ⟶	Problem ⟶	Solution/ response ⟶	Outcomes/ consequences
Technical design of torpedo	Coordination of 600 unreliable Suppliers		
	Coordination among USD departments	Reorganization to material management department	Fewer suppliers Better reliability Better material control
Customer requirements	Navy requires selection of low cost bidder	Life cycle costing	Select true low-cost supplier
Supplier quality	Poor quality parts	Certified suppliers	Better quality

a problem in coordinating suppliers within USD and the unreliable nature of some suppliers. As a result, the materials management department was formed and the supplier certification program initiated. This ultimately resulted in better material control and better supplier quality and deliveries.

Another complexity was the nature of MK 50 technology, which resulted in over 600 different suppliers. A decision was made to reduce complexity by using fewer suppliers. This in turn resulted in better quality and delivery as suppliers became more integrated with the Honeywell supply system.

Another complexity was that the navy required the use of the lowest price bid on all defense contracts. This problem was resolved by using life-cycle costing, which reflected all of the costs of procurement and not just the incoming cost of the purchased parts. As a result, USD was able to select the true low-cost producer while simultaneously improving quality and reliability.

Strategic Planning and Marketing

Strategic planning in the division focused on three issues: the current torpedo product, the next generation MK 50 product, and a striving to obtain seed contracts for future defense work on different types of weapons systems. Strategic planning here is used to define the process by which future plans for the company were developed, especially in the new business area. Marketing refers to managing the customer, in this case the government complex—Capitol Hill, the Department of Defense, and various government agencies involved in the process. The events related to strategic planning and marketing are listed in Table 12–8.

The process of strategic planning for weapons systems contracts has been explored by Edwards and Larwood (1986). They point out that MacMillan's model of strategic manipulation is supported by the 250 defense establishment managers they included in their questionnaire. Our interview data, although not structured around this model, seem to support their general findings that cost, delivery, and schedule are clearly the most important factors in obtaining contracts.

As described previously, Honeywell USD is primarily involved in the manufacture of a current generation and the advanced development of the next generation of lightweight torpedoes. Although the two programs are somewhat independent, they are both affected by the government's and public's current attitudes towards defense contractors. The company was not greatly affected by the investigations of General Electric and General Dynamics, but the changes these brought to the industry were being felt and they dramatically affected strategic planning and marketing.

420

TABLE 12–8.   *Events Related to Strategic Planning and Marketing.*

Mid-1982	Initial plan to pursue nontorpedo business rejected by top management
Mid-1984	Nontorpedo strategy accepted by top management
September 1984	Industrial briefing for competition held
December 1984	Announcement of competition
January 1985	Draft request for proposal for competition issued
May 1985	Government decides to use leader/follower concept
August 1985	USD completes technical transfer plan
January 1986	Gramm-Rudman funding cuts
April 1986	Ten-year strategic plan issued
August 1986	Consultants brought in to assist in reducing product costs
August 1986	USD redefines business focus
September 1986	New development contract schedule finalized

These complexities and uncertainties led to the problems and responses shown in Figure 12–7.

Part of the fallout from these investigations is a renewed vigor on the part of the government to avoid the use of single-source contracts for weapons systems. For the current Mk 46 torpedo, the company is a sole source and has been rated highly on both delivery and performance criteria. However, in negotiating a production contract for the next three-year period, the navy imposed a price cut as a "reward" for remaining the sole supplier for the Mk 46. The government claimed that use of a second source would reduce unit cost, so the company was essentially forced to meet this projected reduction. In response, division strategic planning focused on methods to reduce its production costs so that profit margins could be maintained. The response currently being undertaken involves a complete redesign of the production facility to cut costs as well as to reposition the company to compete for future contracts in the electronics area. This response is linked to the generation of new business in that it allows the company to better pursue other contracts. As we will see, the search for new business was itself a response to the navy deciding the current development project would be multisourced.

More directly related to the MK 50 development effort, several events are of interest. When the initial interviews were conducted in early 1984, the company was convinced that it would be the sole source manufacturer for the product when development was completed. However, in September 1984 the government

FIGURE 12–7.   *Strategic Planning/Marketing.*

Complexity/ uncertainty ⟶	Problem ⟶	Solution/ response ⟶	Outcomes/ consequences
Public scrutiny of defense contracting	Mk46 price cut	Redesign of prod. facility to be more cost effective	HPPF task team Outside consultants brought in
	Competition	New business focus	Pursuing underseas electronics
Congress-Gramm-Rudman funding cuts	Budget for FY 1987 reduced in tight year	Rescheduling of rest of contract	Several months lost Primary managerial focus on new contract

announced an Industrial Briefing for Competition where the company was required to brief competitors on the product. The actual announcement of competition was made in December 1984, with a draft request for proposal issued in January 1985. At this point in time, the form of competition was still unclear—that is, whether there would be two prime contractors or whether competition would be at the subsystem level. USD undertook efforts to reverse this decision by the navy, or to keep competition at the subsystem level, with USD as the sole prime contractor.

In May 1985 the Navy decided that competition would take the form of a leader with a prime follower—that is, two companies would both produce the MK 50 torpedoes. This event triggered several other events. The company was required to develop and issue a technical transfer plan in August 1985 for use by its competitors in preparing their proposals. The company also commissioned a consulting firm in August 1986 to recommend areas where its product costs could be further reduced, so that it would be in a better position to compete on both quality *and* cost.

Competition in the next generation weapons system has been one of the driving forces behind the strategic process for obtaining nontorpedo new business. The idea for this strategy had its origin in mid-1982, but a proposal to diversify USD's products was rejected by corporate management. The USD management team kept the concept alive, however, and obtained approval for a new venture strategy in mid-1984. In April 1986 a ten-year strategic plan was issued that incorporated the new venture strategies, and by August 1986 the division redefined its business focus on three areas: core torpedo, core replacement and maintenance of torpedo products, and new product ventures.

An additional external event dramatically altered the strategic planning effort of the division. In January 1986 Gramm-Rudman funding cuts were announced. The effects of

these cuts, coupled with an expected financially weak year of the contract, led to many unplanned revisions in the development of the program. The schedule was stretched out so funding and work required could be matched. The negotiations of this schedule revision took nine months, with an agreement reached finally in September 1986. These negotiations took time, and more important they consumed scarce managerial resources that potentially could have been applied in other areas.

DISCUSSION AND ANALYSIS OF RESULTS

We will now discuss the overall findings from this case. Figure 12–8 is a combination and rearrangement of all four stimulus-response diagrams from the preceding sections. Appendix 12–B presents a "critical event" chart (see Miles and Huberman 1984) that was developed from the events in Appendix 12–A. This chart identifies, in Tichy's terms, the triggers, the cycle that is dominant or peaking, as well as the adjustments and outcomes observed. (It should be noted that in the MIRP framework, outcomes are classified as events.) These rearrangements provide the opportunity to generalize about different types of stimuli, responses, and outcomes evident in this innovation.

The major categories of stimuli fit well with Tichy's (1980) three cycles of organizational adjustment: technical, political, and cultural. For example, technical design of the torpedo, timespan of the contract, and public scrutiny of defense contracting could all be viewed as technical because they lead to complexity and interdependencies. The next four areas—Gramm-Rudman, organizational size, upper management's decision about the factory, and customer requirements—can be viewed as political cycles. Supplier quality could be considered as cultural, both between

FIGURE 12–8. *Combined Stimulus-Response Diagrams.*

Complexity/ uncertainty →	Problem →	Solution/ response →	Outcomes/ consequences
Technical design of torpedo	Coordination of 600 unreliable suppliers		
	Coordination among USD departments	Reorganization to Material management Department	Fewer suppliers Better reliability Better material control
	Technical failures Missed deadlines	Personnel scaleup Differentiated structures	Personnel "cliff" problems Corporate ILM reassignments
	Designing a producible product	Design-to Production manager Navy templates	Focus on producibility early in design phase Risk management program
	Totally new software design	Role swapping	Improved communication between navy labs and USD
		Co-location of personnel	Better communication More rapid development
Timespan of contract	Management turnover	Replacement HRM philosophy	Management discontinuity Generation of humanist managers
Public scrutiny of defense contracting	Mk46 price cut	Redesign of production facility to be more cost effective	HPPF task team Outside consultants brought in
	Competition	New business focus	Pursuing underseas electronics
Congress-Gramm-Rudman funding cuts	Budget for FY 1987 reduced in tight year	Rescheduling of rest of contract	Several months lost Primary managerial focus on new contract
Organization size	Integration between depts. and levels	Task Forces Transition teams Group meetings Integrative structures	Fragmentation Disjointed schedules Missed appointments Attention management
Upper management decision not to build a new factory for MK 50	Merging of production processes and personnel required to integrate two products into one factory	Transition teams Task team for high performance production facility	"Outran" other areas of project after some initial problems Factory to be redesigned for future products as well as MK50 and Mk 46
Customer requirements	Navy requires selection of low cost bidder	Life cycle costing	Select true low-cost supplier
Supplier quality	Poor quality parts	Certified suppliers	Better quality

organizations as well as within the USD organization, since it also required major reshuffling of responsibilities.

It should be noted here that these "stimuli," as classified into Tichy's categories, do indeed seem to eventually create peaks in one or both of the other two cycles. For example, the decision of upper management to build the MK 50 in the existing factory ultimately created a peak in the technical cycle because MK 50 and Mk 46 tasks were forced to become highly interdependent and hence more complex. On the cultural side, the transition teams that were created required the merging, in most cases, of groups with different goals and agendas. As an example of an event that caused a peak independent of the other cycles, the Gramm-Rudman funding cuts and the eventual contract renegotiation were independent of the other cycles.

The responses and outcomes may also be categorized and viewed using Tichy's model. Many of the responses were stimuli concerning organizational arrangements—changes in people, rearrangements of work groups to achieve new goals, and even changing how the work is to be performed. These are evident in such responses as role swapping, the use of the templates for risk reduction, and the reorganization of the materials management department. The desired and actual outcomes served, in most cases, to simplify a very complex process, by decreasing the supply chain, improving communication, and resolving production problems early in the process.

A closer examination of Appendix 12–B reveals some interesting "quantitative" results concerning the observed trigger events, as well

as the cycle types. In the thirty-three events listed, the triggers can be seen to be almost evenly divided between those independent of the other cycles and those dialectical events that can be traced to other previous events. In fact, four of the thirty-three triggers can be viewed as having elements of both types, an eventuality allowed by Tichy but not explicitly discussed. These "dual" triggers appear to be the culmination of a series of other events.

It is also interesting to examine the dominant cycles observed in this study—the types, their frequencies, and where overlaps occur. Table 12–9 summarizes these data for the thirty-three events in Appendix 12–B. Only five of the thirty-three cycles are viewed as cultural, while the remaining cycles are almost evenly split between technical and political. This is not an unexpected result if one examines the nature of this innovation and the organization and environment in which it was developed. The MK 50 weapon system represents a very significant technical product development challenge and hence, one would expect many peaks in the technical cycle over time. The customer complex—the U.S. Navy, Congress, the Department of Defense, and the public—creates many political cycle peaks in any defense systems development effort. The explanation for the low number of cultural peaks, at least as indicated by our data, is that this Honeywell division has a deeply ingrained culture in its management team that is highly adaptive to the changes described above.

Although it appears that our data and stimulus-response model fit well with Tichy's model, the following are additional findings from the study. Two major themes consistently

TABLE 12–9. *Cycle Analysis of Critical Events.*

Technical	Political	Cultural	Technical/ Political	Political/ Cultural	Technical/ Political/ Cultural
13	14	2	1	1	2

emerged: (1) A hybrid approach of strict control mechanisms along with adaptive, incremental micro-innovations are needed to manage complex innovations; (2) the defense contracting environment now requires managing cooperation with competitors. Each of these themes, along with the various supporting evidence, is discussed below.

Much of the innovation observed in this study consists of a series of administrative micro-innovations, small organizational innovations required to make a major technological innovation workable. Anderson (1985) describes the concept of micro-innovations for new product design. He defines micro-innovation as "a novel, useful idea that has been implemented in a product design." The implication is that the development of an innovative product depends on people working together to continually generate micro-innovations and the motivation of these people is clearly related to the product's eventual success. As Anderson suggests, team composition in the innovation process must change as project needs change. These concepts are clearly supported by our data as evidenced by the many reorganizations we observed. He posits that broad design goals may be more appropriate where many micro-innovations are desired. However, for defense contractors like USD, it appears that the design goal is very clear, but the micro-innovations are still needed so mechanisms are required, such as the risk management program, to ensure success.

At USD, many micro-innovations were evident as ways of handling complexity. Marginal and incremental innovations permit flexibility to adapt to changes as they arise in a complex organization, as pointed out by Quinn (1980). First, as noted earlier, the organization was restructuring continuously in order to meet changing requirements. Most of these micro-innovations appeared to be triggered as responses to unforeseen problems of complexity and uncertainty that arose in the product development. In this environment, long-range strategic planning is necessary, but the case shows its scope is limited due to continually changing requirements. Adjustment of plans and strategies in the short or medium time horizons, and even in the longer term, is required for continued survival in this business. Marketing efforts must be continually adapted to changing customer requirements, both in terms of price and product performance.

Another theme that emerged in this study—cooperation with competitors—is rather unusual in most traditional businesses. There were two major instances where USD had to cooperate with competitors. First, the navy forced cooperation with other potential suppliers of the MK 50 torpedo. This involved preparing a technical transfer plan, briefing the competition, and offering certain levels of assistance to competitors. Part of this cooperation was also a result of the navy's design-to-production, risk-reduction program, which affected USD as well as its suppliers. Also, the relationships with suppliers in this case are cooperative, including joint ventures with other companies and other divisions of Honeywell. In addition, divisions within the company group structure were being required to cooperate more fully with each other, where they had sometimes competed before.

Another question examined in this study is the degree to which innovation events proceed in an orderly and systematic progression, as suggested by the action loops, or in a random probabilistic mode as the garbage can model suggests; or is there some in-between level? If one examines the events in Figure 12–8, it can be seen that many of these events can not be explained by the action loops. Although some problems arose that did not trigger a response or solution, not all solutions had recognizable consequences either. In these few cases, the loose coupling behavior suggested by the garbage can model seems to hold. However, for the most part, many of the events in this innovation were not loosely coupled. In fact, some of the stimulus-response models described in the previous section are not independent of each other and can be thought of as forming a complex web. In these cases, management may intervene by introducing programs, whose effects are unpredictable, which in turn, may contribute to a vicious cycle of increasingly expanding complexity and unpredictability. Thus, we find a process that incorporates elements of both the orderly, systematic progression and random, probabilistic activities.

CONCLUSIONS

This chapter has presented one of the first attempts to track longitudinally a complex innovation with real-time measurement and formal data collection procedures. Three years of on-site data, plus several years of archival data, were used to identify issues of managing complexity and uncertainty in government contracts. Government contracts in the environment of the 1980s are one of the most difficult types of programs to manage. The customer controls almost every step of this process and has the right to change the rules by which the game is played at any point in time.

Producibility of complex products is always a complicated issue. The findings from this case history suggest several propositions on the management of complex innovations.

1. Early involvement of production personnel in the product design effort reduces the number of problems subsequently encountered in the development of complex innovations.
2. The use of interdisciplinary task teams (such as swapping of personnel) to deal with problems that cross organizational boundaries minimizes the problems encountered in the development of complex innovations.
3. Changes that arise from the development of a complex innovation require ongoing replanning and restructuring.
4. Technologically complex innovations require administrative innovations to manage the overall effort.

As these propositions suggest, advance planning cannot be relied on to determine the course of a complex innovation. Instead, a fluid view of innovation incorporates setbacks and surprises as a normal course of events. Advance planning is still needed, but it cannot be overly rigid. Adaptability to contingencies may ultimately become a competitive strength of the organization.

REFERENCES

Abernathy, W.J., and J.V. Utterback. 1982. "Patterns of Industrial Innovation." In M.L. Tushman and W. Moore, eds., *Readings in the Management of Innovation,* pp. 97–108. Boston: Pitman.

Adams, J.R., S.E., Barndt, and M.D. Martin. 1979. *Managing by Project Management.* Dayton, Ohio: Universal Technology.

Anderson, D.G. 1985. "Micro-Innovation in New Product Development." Unpublished paper, Massachusetts Institute of Technology.

Cohen, M.D., J.G., March, and J.P. Olson. 1972. "A Garbage Can Model of Organization Choice." *Administrative Science Quarterly* 17: 1–25.

Daft, R. 1978. "A Dual Core Model of Organization Innovation." *Academy of Management Journal* 21: 193–210.

Damanpour, F., and W.M. Evan. 1984. "Organizational Innovation and Performance: The Problem of 'Organizational Lag.' *Administrative Science Quarterly* 29: 392–409.

Delbecq, A., and A. Filley. 1974. "Program and Project Management in a Matrix Organization: A Case Study." Bureau of Business Research and

Service, University of Wisconsin–Madison Graduate School of Business, (January).

Edwards, F.L., and L. Larwood. 1986. "Strategic Competitive Factors in the Acquisition of Technology: The Case of the Major Weapons Systems." Academy of Management Conference, Chicago.

Frontini, G.F., and P.G. Richardson. 1984. "Design and Demonstration: The Key to Industrial Innovation." *Sloan Management Review* 25: 39–50.

Galbraith, J.R. 1977. *Organization Design.* Reading, Mass.: Addison-Wesley.

Horwitch, M. 1982. *Clipped Wings: The American SST Conflict.* Cambridge, Mass.: MIT Press.

Kanter, R.M. 1983. *The Changemasters.* New York: Simon and Schuster.

Lane, H.W., R.G. Beddows, and P.R. Lawrence. 1981. *Managing Large Research and Development Programs.* Albany: State University of New York Press.

Lawrence, P.R., and J.W. Lorsch. 1967 *Organization and Environment: Differentiation and Integration.* Cambridge, Mass.: Harvard University.

Lombaers, H.J.M. 1969. *Project Planning by Network Analysis.* Amsterdam: North-Holland.

Maciariello, J.A. 1978. *Program Management Control Systems.* New York: John Wiley.

March, J.G., and J.P. Olsen, eds. 1976. *Ambiguity and Choice in Organizations.* Bergen, Norway: Universitetsforlaget.

Marquis, D.G. 1969. "A Project Team Plus PERT Equals Success or Does It?" *Innovation* (January) 1.

Martin, M.J.C. 1984. *Managing Technological Innovation and Entrepreneurship.* Reston, Va.: Reston.

Masuch, M. 1985. "Vicious Circles in Organizations." *Administrative Science Quarterly* 30: 14–33.

Miles, M.B., and A.M. Huberman. 1984. *Qualitative Data Analysis: A Sourcebook of New Methods.* Beverly Hills, Calif.: Sage.

Perry, R.L. 1966. "The Mythography of Military R&D." Rand Corporation P-3356.

Quinn, J.B. 1980. *Strategies for Change: Logical Incrementalism.* Homewood, Ill.: Irwin.

Sapolsky, H.M. 1972. *The Polaris System Development.* Cambridge, Mass.: Harvard University.

Sayles, L.R., and M.K. Chandler. 1971. *Managing Large Systems, Organizations for the Future.* New York: Harper & Row.

Schroeder, R.G., G.D. Scudder, and M.E. Pesch. 1986. "Approaches to Managing the Cost of Materials." *International Journal of Physical Distribution and Materials Management* 16: 57–69.

Schroeder, R.G., G.D. Scudder, A.H. Van de Ven, D. Gilbert, L. Heeringa, and G. Seiler. 1984. *The Management of Innovation at Underseas System Division, Aerospace and Defense Group, Honeywell Corporation.* Unpublished proprietary working paper.

Smith, R.A. 1966. "The Matrix Organization Form: A Social Concept for Enterprise Effectiveness." NASA, George C. Marshall Space Flight Center, Management Development Office.

Starbuck, W.H. 1983. "Organizations as Action Generators." *American Sociological Review* 48: 91–115.

Tannenbaum, A.S. 1968. *Control in Organizations.* New York: McGraw Hill.

Thompson, J.D. 1967. *Organizations in Action.* New York: McGraw-Hill.

Tichy, N.M. 1980. "Problem Cycles in Organizations and the Management of Change." In J. Kimberly and R. Miles, eds., *The Organization Life Cycle: Issues in the Creation, Transformation, and Decline of Organizations,* San Francisco: Jossey-Bass.

Van de Ven, A.H., and D.L. Ferry. 1980. *Measuring and Assessing Organizations.* New York: Wiley.

Weick, K.E. 1979. *The Social Psychology of Organizing,* 2 ed. Reading, Mass.: Addison-Wesley.

Weist, J.D., and F.K. Levy. 1977. *A Management Guide to PERT/CPM.* Englewood Cliffs, N.J.: Prentice-Hall.

APPENDIX 12–A. Events in the Development of the MK 50 Program

1975	Navy issues request for technical design for next generation torpedo to replace MK46
1975	Advanced lightweight torpedo program initiated at Honeywell in response to Navy request
5/76	Torpedo development program wins torpedo R&D investment funding over other potential investment alternatives
1976–77	Restructuring at Honeywell splits the Government and Aeronautical Products Division into the Avionics and Defense Systems Divisions; USD would evolve from latter in 1983
1977	With the appointment of a general manager for the division in which USD is to be housed, a strategic vision begins to take shape; a ten-year horizon with diversification of the torpedo operations into other related fields is envisioned
1977	DSD identifies Garrett as a source of propulsion capabilities and enters into team relationship for development of AWLT
1977	Competition for the advanced development phase of ALWT is reduced to four teams of competitors
1977	Honeywell/Garrett team recognized as distinctive with their chemical propulsion design
1977	Honeywell wins the digital signal processor contract creating an advantage in the development phase.
1/78	Navy issues draft request for proposal on ALWT development.
1978	Honeywell wins retrofit contract for updating and improving MK46 torpedo *Neartip*
1978–80	*Neartip* crises, heavy financial losses in retrofit program; leads to new systems and approaches for managing
4/78	Navy cancels RFP for ALWT
4/78	Cancellation of RFP threatens development funds; Honeywell elects to maintain development team with internal funds
10/78	Navy reissues RFP and Honeywell prepares its proposal.
5/79	Navy selects two contractors for the advanced development phase of ALWT; Honeywell/Garrett one of the teams selected
1979–86	A new chief operating officer named to lead torpedo unit brings a humanist approach to management with a focus on people and technology programs; top-down management replaced with teams structures
1979–80	New general manager brings in new "team" and reorganizes torpedo unit for more innovative approaches
7/80	Efforts aimed at improving factory production result in development of innovative techniques, methods, and tools; marks beginning of factory automation/innovation efforts
Summer 1980	USD requests authorization to redesign Mk46 torpedo to improve producibility
9/80	Navy refuses design change request; instead asks USD focus on redesigning the manufacturing system
1980	General manager of DSD develops "seven principles of management" for the division—oriented toward teamwork and participation
1980–82	"IT" steering committee formed to direct task teams in implementing the "seven principles"
1981–84	Change of culture as human resource management at USD is recast in role of facilitator to help teams of line managers address human resource issues.
1/81	USD wins MK 50 development contract; navy drops second competitor
1/81	MK 50 program formally established to develop prototype

APPENDIX 12–A. (continued)

1/81	MK 50 production manager named; design-to-production approach gives early development of MK 50 a divisionwide emphasis
3/81	Materials management restructing effort begins with object of developing focused approach
5/81	Computer applications in production give rise to "paperless factory" program
6/81	Automatic insertion equipment installed to aid production line tasks giving rise to "factory of the future" efforts
6/81	Design-to-production transition program initiated to facilitate closer relationship between design and production engineers in early phases of product development
1982	Strategic business planning position created to formalize strategic thinking for torpedo unit
1982–83	Convergence of strategic thinkers as a result of the new planning unit and formalized efforts at strategic management
Mid-1982	Conceptual proposal for moving torpedo unit into new business ventures; represented early attempt to develop acquisitions planning
Mid-1982	Division management soundly rejected torpedo unit's proposal for diversification; rather, unit instructed to concentrate efforts on the torpedo business
7/82	Technology strategy for torpedo unit initiated aimed at developing cutting edge technologies related to the growing engineering competence
10/82	The first production plan for the MK 50 program is submitted to the Navy
Late 1982	Torpedo unit wins favorable funding decision from division management for an internal R&D capability
1982	"Torpedo 83" program launched for the purpose of gaining greater flexibility of human resources; joint union/management review process
1983	"Torpedo 83" reduces 35 pay grades into 4; represents a radical change in personnel policy
1/83	Design-to-production manager appointed to work on producibility issues
3/83	Underseas Systems Division (USD) is spun off from Defense Systems Division as a separate profit center with a single business mission—torpedoes
5/83	USD wins full-scale development contract for the MK 50 program
6/83	Advanced manufacturing strategic planning unit established to integrate the factory automation innovation efforts at USD
8/83	Productivity board established to develop a restructured materials management approach
9/83	Productivity board recommends a new materials management unit
9/83	Navy encourages USD to take a "risk management" approach to the MK 50 program
10/83	USD initiates program aimed at managing the risk of whether the design of USD products would be producible within navy contract limitations
10/83	A revised production plan for MK 50 program, incorporating new risk management procedures, is submitted to the Navy
11/83	New materials management unit, responsible for coordinating the material needs for both the MK46 and MK 50 torpedoes, announced
1/84	New materials management unit begins operation
5/84	USD submits the production plan for full-scale development of the MK 50 torpedo
5/84	Strategic plan for business diversification approved and a search begun for acquisitions that would permit USD to reassign its MK 50 engineers
5/84	SMRC baseline interviews completed

429

APPENDIX 12–A. (continued)

9/84	Navy conducts briefing for firms with potential interest in bidding as second contractor for the MK 50 program
10/84	SMRC case history for the MK 50 program completed and approved.
10/84	SMRC baseline data collection completed
11/84	USD's 100-S software development program to test concept is initiated
12/84	The Navy announces the MK 50 torpedo competition for production; as many as seven firms may be involved in the bidding
12/84	A key individual on the MK 50 suffers stroke; successful test of USD's management structure and succession
12/84	Change of scope as a result of an oversight by USD and navy; Timelines are drawn out to accommodate changes
12/84	First SMRC feedback to USD details results of baseline survey
1/85	Restructuring at USD reflects transition of innovation; design-to-production team changes reporting responsibilities
1/85	USD earns award fee from the Navy in excess of 13 percent
1/85	Advanced manufacturing technology group formed to guide the development of a flexible manufacturing system
1/85	Manager named to oversee the development of the MK 50 software, giving added emphasis to this area of the innovation
1/85	200S software development program launched
1/85	USD wins VHSIC contract from the Navy
1/85	Twenty-three task areas identified as critical for transition of design into production
1/85	USD completes RFP for competition
2/85	First prototype MK 50 delivered to navy for testing
2/85	Twenty-two transition teams formed for the purpose of developing interfunctional communication and a long-range proactive planning approach for MK 50
3/85	USD earns highest Navy award fee ever given a contractor
3/85	New structure for the division is announced; USD is housed within a new marine systems group
3/85	New general manager named for USD to replace manager promoted to head the new marine systems group
3/85	New manager appointed to direct the Mk46 production program
5/85	Navy confirms a leader-follower program for MK 50
5/85	USD investigates opportunities to become a multiproduct firm in other defense related product areas
5/85	HRD at USD conducts program health review identifying problem areas
5/85	Sixteen goals that will guide USD toward the "factory of the future" are identified by transition team
5/85	Draft readiness for production reports from followers are submitted to the Navy
5/85	A key MK 50 manager is elevated to the level of vice president of operations
5/85	A preplanned production improvement program (P3I) is initiated to work on the next generation of the MK 50
5/85	P3I program manager named based on familiarity with MK 50
5/85	The software development unit is reorganized into a matrix form; software delays experienced
6/85	Second SMRC feedback session details results of T-2 survey
7/85	Navy gives USD an award fee reflecting lower performance

430

APPENDIX 12–A. (continued)

8/85	New director assumes guidance for transition teams
8/85	Transition teams reduced from twenty-two to six; more focused effort sought
8/85	USD completes the technical transfer plan for follower
10/85	Task force formed to develop "high performance production facility" concept
10/85	Two MK 50 managers leave program citing high stress
10/85	Improved Navy award fee for USD; still short of the high performance level experienced in 3/85
10/85	Work of transition teams is reviewed and evaluated
11/85	Transition between facilities is accomplished through a two-way personnel transfer between the USD site and a California facility
11/85	Software development for the MK 50 is judged by its manager to be eight months behind schedule
11/85	"Tiger Teams" assigned to help with software design problems
11/85	CAD/CAM technology development effort instituted for application to design and production efforts
11/85	Manager named for CAD/CAM effort
12/85	Transition teams put on "hold"; some teams' activities are too far advanced
12/85	Key MK 50 program manager dies
12/85	Four key members of MK 50 program leave USD
12/85	USD freezes all new hiring and attempts made to "stretch" schedule
12/85	First release for 200S software program
1/86	Specter of Gramm-Rudman forces USD/Honeywell cost-saving cuts
1/86	Diversification efforts result in USD pursuit of subcontracts on missiles program
2/86	Reorganization of A&D group results in cutbacks in division and MK50 personnel; greater coordination encouraged
1986	Marine systems group forces USD into a more integrated role in group strategies
2/86	Third SMRC feedback session to USD management based on T-3 survey results
3/86	USD human resource manager promoted to group level; additional personnel soon follow as A&D is restructured and restaffed
SP/86	Fifty software engineers reassigned to other programs and divisions; seven laid off
04/86	New strategic planning system designed to connect budgets and plans is instituted
4/86	Strategic personnel requisition board is formed as a clearinghouse for internal personnel transfers
6/86	Strategic human resource plan is issued
8/86	MK 50 program management unit reduced from thirty-one to seventeen people
8/86	Team of cost reduction consultants brought in to aid USD in reducing product costs
8/86	USD redefines its business focus into three main areas—core, core replacement, and new business
9/86	Agreement reached with Navy for stretched MK 50 schedule

APPENDIX 12–B. Critical Event Cycles

Event	Trigger	Dominant Cycle	Adjustment (Tools)	Outcome(s)
1. Navy request for technical design, 1975	Independent environmental change: e.g., perception of USSR threat changes	Technical: development of next generation of torpedo	Development program needed; funds critical Gear up for new products and approaches	Beginning of HW MK50 development program (ALWT)
2. ALWT program, 1975	1975 event	Political: allocating resources	Weighing alternative investment options	HW funds ALWT; Beginning of strategic vision; Search for new methods and partners
3. Honeywell selected for ALWT development, 1977–79	5/79 event Independent: technological	Political: allocating power and resources	Planning for old and new product	Greater commitment to MK50
4. MK46 retrofit program, 1978	Independent: technological	Technical: how to produce	Reevaluation of productive system	No adjustments made: therefore, poor product quality, profitability nosedive, *Neartip* crisis
5. *Neartip* crises,	1978–80 event	Political: inadequacy of top-down management Technical: change in product leads to problems in QC	Need for new management structures and approaches Need for new technologies and manufacturing techniques	New management team. New tools and techniques introduced—e.g., computers (event 7/80)
6. New chief operating officer named, 1979	Independent: Changes in people	Technical: how organizations social and technical resources will be arranged Cultural: development of new management value system Political: Where to focus power and resources	New approaches to production system (tools-new structures; new people) Need to develop and enunciate management and organizational values Need to reapportion power and D-M mechanisms	1. New HRM role 2. Evolutionary change in culture 3. GM's seven principles 4. "IT" steering committee Greater participation; new staff-line relationships

APPENDIX 12–B. (continued)

Event	Trigger	Dominant Cycle	Adjustment (Tools)	Outcome(s)
7. Design change request	Summer, 1980 event	Technical: producibility issue	Develop better coordination systems between design and production capabilities	USD focuses on redesigning manufacturing system; efficiency and quality via operator manufacturing aids
8. USD wins MK50 development contract	1/81 event Independent: environmental complexity (new methodologies)	Technical: task complexity	Establish program capable of developing and delivering a prototype torpedo to U.S. Navy	MK50 program established; MK50 production manager named; production plan submitted (10/82)
9. Materials management restructuring effort	3/81 event Independent: technological changes (new methodologies)	Technical: production quality problem	Reevaluation of current system; developing coordinating structure	Coordinating mechanism developed (committee formed to develop coordinated approach to supplier management); Division productivity board established; new coordinated materials management unit established (11/83)
10. Future factory improvement program	5/81 event Independent: technological changes	Technical: developing new production technology	Developing new production concepts	Paperless factory program (5/81); early "factory of the future" efforts (6/81)
11. Design to production program initiated	6/81 event: shift in work means	Technical: how to design product	Integrate design/production efforts	Production engineers brought into design efforts at earlier stage
12. Strategic business planning unit created	Early 1982: shift in goals	Political	Formalization of strategic planning into organizational structure	Convergence of strategic thinkers into teams (1982–83), New business ventures explored (mid-1982), Technology strategy developed for USD (7/82), Internal R&D capability established (late 1982),

APPENDIX 12-B. (continued)

Event	Trigger	Dominant Cycle	Adjustment (Tools)	Outcome(s)
13. "Torpedo 83" program launched	1982 event: environmental; increased complexity	Political: power, resources	Development of new exchange and bargaining relationships	Review of the labor/management agreement on job descriptions and classifications; reduction of 35 pay grades into 4; new, more cooperative labor/management environment (1983).
14. USD formed	3/83 event	Political: new allocation of resources, goals, prestige	Restructuring	USD is set up as separate profit center; efforts of unit given greater visibility at Honeywell
15. Full-scale development contract awarded USD	5/83 event	Technical: how to develop prototype	Building capability	Rewards USD's groundwork efforts; intensifies development of second product focus; USD submits production plan (5/84), first prototype delivered (2/85),
16. Advanced manufacturing strategic planning unit established	6/83 event	Political: focus efforts	Consolidate and integrate factory automation innovation efforts	Automation efforts focused on MK46/50 production facility
17. Navy encourages USD risk management	9/83 event: need for change in methods	Political: how to organize efforts	Restructure approach	USD initiates a risk management program; Becomes model for defense industry (10/83), steers USD toward a structured design-to-production process; revised production plan incorporating risk management is submitted to navy (11/83).

APPENDIX 12–B. (continued)

Event	Trigger	Dominant Cycle	Adjustment (Tools)	Outcome(s)
18. Business diversification plan approved	7/84 event	Political: allocation of resources; change in goals	Reorientation of strategy	Search begun for acquisitions; attempt to identify new opportunities; contracts for diversified business are won (1/85).
19. Software development begins	11/84 event	Technical	Developing necessary technology for new generation of torpedoes	Development of 100S (11/84) and 200S (1/85) software series; software development team builds (1/85).
20. MK50 torpedo competition announced	12/84 event: Navy bidding process	Political	Establishing strong position with navy and against competitors	USD earns navy recognition and performs well on Award fees (1/85, 3/85), USD submits proposal for competition (1/85).
21. Restructuring reflecting change in emphasis	1/85 event: managing development	Political: reallocation of resources	New organizational structure	Reporting responsibility changes to reflect progress of development; organizational structure used as transition mechanism
22. Transition tasks identified	1/85 event	Technical: highly interdependent tasks	Create transition mechanisms	Twenty-three transition tasks identified Transition teams formed to guide MK50 development (2/85).
23. New divisional structure announced	3/85 event: shift in agreement over methods	Political: goals and resources Cultural: new managers/new approaches	New structural relationships	USD housed in Marine Systems Group; USD manager promoted to new Marine Systems Group (3/85); H-R manager promoted to group (3/86); personnel cut at USD as greater group integration is sought (1/86)

APPENDIX 12–B. (continued)

Event	Trigger	Dominant Cycle	Adjustment (Tools)	Outcome(s)
24. Management change	3/85 event	Political: new approach; uncertain relationships	Promotion from within	Series of promotions—new manager for MK46 program (3/85), new vice-president of operations, 1985
25. Leader/follower contract adopted by navy	5/85 event: complex/uncertain environment	Political: uncertainty of stability in bargaining and exchange relations	Adopt new approaches and tactics in relationships	USD revises planning and cost estimations; change in business strategy—seek opportunities as multiproduct defense firm (5/85), USD pursues new business outside torpedo line (1/86).
26. Factory of the future goals	5/85 event	Technical: How should firm design future production systems?	Seek flexibility	Transition team develops a set of 16 goals; high-performance production facility plan developed (10/85).
27. Preproduction improvement program announced	5/85 event	Political: How can firm stay ahead of competition?	Build in renewal process	USD begins work on next generation of MK50; manager named (5/85).
28. Software development recognized as critical	5/85 event: new technologies	Technical: complex, interdependent task	Apply dynamic organizational structures	Software development organized into matrix form; software development falls behind schedule; navy award fee declines (7/85), by 11/85 software is 8 months behind, and "Tiger Teams" applied to problems.
29. Change in transition team management	8/85 event	Cultural: change in people	Restructuring	Poorly performing teams pruned (8/85), number of teams reduced from 22 to 6; Subsequent evaluation of teams leads to cessation of activities (10/85, 12/85).

APPENDIX 12–B. (continued)

Event	Trigger	Dominant Cycle	Adjustment (Tools)	Outcome(s)
30. Key personnel changes	10/87–3/86 event: lower award fees leading to high stress; other causes	Cultural: changes in people		Two managers leave due to stress 10/85; Personnel transfer used as transition mechanism (11/85), key MK50 manager dies (12/85), four key members of MK50 leave program (12/85), hiring freeze initiated (12/85), key managers promoted to group level.
31. CAD/CAM development	11/85 event: technological changes	Technical	Application of new technology	Development effort for design and production applications; CAD/CAM program manager named.
32. Gramm-Rudman impact	1/86–8/86 event: environmental complexity	Technical: production schedule slippage Political: distribution of resources Cultural: values questioned	Retrenchments and cutbacks	50 software engineers reassigned and 7 laid off (Sp. 86) MK50 program management unit reduced from 31 to 17 (8/86), help sought from cost reduction consultants (8/86).
33. Focus on strategic planning	4/86 event	Political: goal selection and resource allocation	Development of planning structures and systems	A new strategic planning system connecting plans and budgets (4/86), strategic personnel requisition board formed as a clearinghouse for internal personnel transfers (4/86), a strategic human resources plan is issued (6/86), USD redefines its business focus into three main areas—core, core replacement, and new business (8/86).

13

ASSESSING THE EMERGENCE OF NEW TECHNOLOGIES: THE CASE OF COMPOUND SEMICONDUCTORS

Michael A. Rappa

The objective of this chapter is to develop a conceptual framework and method for monitoring systematically the emergence of radically new technologies. The intent is to improve our understanding and judgment in predicting the progress—or lack of progress—during a technology's development from its beginning as a curiousity in research laboratories to its fulfillment as a commercial reality.

Traditional methods of monitoring the emergence of new technologies can be divided into three basic approaches. The first approach is rooted in the field of technological forecasting and relies heavily on the measurement of various performance parameters to chart the progress over time for a given technology.[1]

The author greatly appreciates the comments of Professors Andrew Van de Ven and Bruce Erickson, and Raghu Garud of the Carlson School of Management, University of Minnesota. Financial support for this work has come from the University of Minnesota, the McKnight Foundation, and IBM Corporation.

The theory underlying the analysis is that a technology exhibits a sigmoid flexure in its performance parameters over its life cycle. This s-curve pattern is used to project the rise and decline of technologies by extrapolating the available data.

One limitation with this approach tends to be the lack of precise data, such that the drawing of s-curves becomes a highly stylized endeavor and the extrapolation of turning points extremely difficult. But even with satisfactory data, evidence of a well-defined s-curve pattern is likely to become apparent too late for it to help firms make an effective transition to an emerging technology. Indeed, historical experience shows that a new technology can emerge and prosper well before the established technology shows any indication of slowing down its own progress. Furthermore, the orientation of the s-curve approach toward understanding the saturation point of the established technology, and thus the need to change, neglects the need for determining which emerging technology among various alternatives will most likely transform a firm's future business.

The second approach is based on the use of large-scale surveys of the scientific and technical communities, sometimes in conjunction with expert panels and delphi methods. This approach is used frequently in Japan and with a notable degree of success.[2] The surveys are useful particularly in generating early awareness of emerging areas of science and technology as well as gaining consensus among organizations regarding which are likely to have the greatest impact in the future. But these surveys do not attempt to bring a theoretical framework to understanding the emergence of new technologies. Nor are they a practical choice for implementation by individual firms. Rather, the use of this approach is most suitable for a governmental agency or perhaps an industry association.

The third approach to monitoring the emergence of new technologies has its tradition rooted in the field of bibliometrics. The term *bibliometrics* is used to describe analyses based on the measurement of various aspects of the scientific and technical literature. The objective in most previous bibliometric studies has been the measurement of individual, organizational, or international comparisons of research performance in terms of publication and citation frequencies.[3]

Bibliometric techniques also have been used to assess the emergence of new fields of science and technology.[4] The most common approach is to scan the literature for areas of research that display a rapid growth in publications or "co-citation clusters" as a means for identifying hot topics of research. This method has proven to be of some use and as a result has gained the attention of managers and consultants in industry. But the usefulness of this approach has been largely untapped, due primarily to the fact that researchers have focused on measuring the literature (by counting papers and citations) as opposed to extracting from it the wide variety of data it contains.

This study proposes a framework for understanding and monitoring the emergence of a new technology by using the scientific and technical literature as a source of data. The literature provides a unique opportunity for understanding technological development because it is comprehensive in scope, allowing us to examine the multitude of researchers, their individual contributions, and the organizational context in which they reside. It is chronologically documented in a way that facilitates the assessment of change over time. Thus, unlike other approaches, using the science and technology literature helps to broaden our attention from examining technology in and of itself to observing technological change within the context of the community of researchers and organizations who are committed to developing a new technology.

In order to observe the kinds of changes that occur as a new technology emerges over time, the development of compound semiconductor integrated circuit (IC) technology is examined in this study. Compound semiconductor ICs are electronic and optoelectronic devices fabricated from composite materials—usually gallium arsenide, alone or in conjunction with other materials—as opposed to a single or elemental material such as silicon. Today, most semiconductor devices are made from silicon. But since the mid-1970s, an effort has been underway to make IC devices using compound semiconductors to take advantage of the material's theoretical potential to yield devices of higher speeds than silicon, among other desirable qualities. Since 1985, the technology has entered the initial stage of commercialization, with a number of firms having introduced compound semiconductor IC products to the marketplace. Thus, the development of compound semiconductors between 1976 and 1986 provides an excellent opportunity to investigate the process of technological emergence from the laboratory to the marketplace.

As a new technology, such as compound semiconductor ICs, emerges over time, what kinds of changes should we expect to observe? The analysis will focus on three interrelated components: the community of researchers dedicated to advancing the new technology; the community of organizations employing those researchers; and, lastly, the technological paradigm that the community holds in common.

First, with respect to the community of researchers, we investigate the rate at which the community grows over time, the role of graduate education as a gateway to the community, and the movement of researchers between organizations. Second, concerning the organizational community, we examine changes in the distribution of researchers among organizations, changes in the collaboration between organizations, and changes in the extent of specialization among firms. Third, with respect to the paradigm, we look at the anticipation and multiple discovery of certain ideas and techniques, the extent to which those ideas and techniques spread among firms, and the growth of an intellectual leadership in the field. In particular, we propose that as a new technology emerges over time, the following will be observed:

1. A community of researchers will form, and
   a. the community will grow at a rate sufficient to double its size within five years;
   b. the number of individuals entering the community via graduate training in the new technology will increase;
   c. the movement of researchers between firms will increase, particularly between established firms and new ventures.
2. A community of organizations will form, and
   a. the distribution of the researcher population among organizations will become less concentrated;
   b. the amount of collaboration between organizations will increase;
   c. an increasing number of industrial firms will be specialized in terms of the activities they perform.
3. A new technological paradigm will form, and
   a. anticipation and multiple discovery of new ideas and techniques will occur;
   b. certain ideas and techniques will become the focus of widespread attention;
   c. there will emerge a small group of researchers who provide intellectual leadership to the larger community.

The data used in this study were gathered from 5,905 journal articles and conference presentations on the subject of compound semiconductor materials and devices published between January 1974 and March 1987.[5] The publications were obtained by scanning fourteen scientific and technical journals and six conference series.[6] From these sources a database was constructed containing information on the work of 7,320 researchers employed by 333 different organizations in the United States, Canada, Great Britain, West Germany, France, Austria, the Netherlands, and Japan.[7]

THE COMMUNITY OF RESEARCHERS

At any particular point in time, the community of researchers includes those individuals who currently are working on compound semiconductor research and development activities and who contribute to the scientific or technical literature by having the results of their work published. The size of the research community in a

441

given year is determined by the number of individual researchers who author (or jointly author) at least one publication submitted in that year. An individual's membership in the community, however, is not related to the magnitude of their contribution to the literature in any given year, but simply to the fact that the researcher continues to publish from time to time.

The size of the community changes from year to year, increasing as new individuals begin to publish and decreasing as existing members in the community cease to publish. Since only a small fraction of researchers publish consistently from year to year, the community is examined over three, three-year time periods: 1976–78, 1980–82, and 1984–86. For example, the research community in 1976–78 includes those individuals who had at least one publication submitted in that time period. If a community member did not publish in 1980–82 as well, he or she would no longer be considered to be participating in the research community.[8]

The research community is examined according to how quickly its membership grows, the entry of new members via graduate training in the technology, and the movement of researchers between different employers.

The Rate of Growth

The success of an emerging technology depends on its ability to attract a growing number of researchers who are committed to its advancement. This community must grow at a rate significant enough to attain a "critical mass" of effort required to overcome the numerous technical obstacles it faces.

The first proposition states that the community of researchers will grow at a rate fast enough to double its size about every five years. This translates into an average annual growth rate of greater than or equal to 14 percent.

Price (1986) examined the growth of scientific fields in aggregate by measuring the growth in the number of scientific journals over three centuries. One of Price's basic conclusions is how quickly the size of science grows over time. He found that for much of its history, scientific literature as a whole has grown at an exponential rate, doubling about every fifteen years.

The community committed to a newly emerging technology should grow faster than the scientific community in general (that is, doubling in less than fifteen years), but how much faster? This question is difficult to answer, due to the fact that not all subfields grow at the same rate. Menard (1971) offers a detailed look at the growth rate of several subfields within science. Menard's data show that significant differences in the growth rate among subfields exist, and that life-cycle patterns do occur. For example, Menard found a large difference in the literature doubling rate of subfields within physics. Established research fields such as optics and acoustics grew at a slow rate during the twentieth century, with a doubling period of about forty to fifty years. In contrast, younger fields, such as nuclear physics and solid state physics, grew much faster. Menard found that the literature of nuclear physics (from 1920 to 1970) and solid state physics (from 1950 to 1970) doubled about every four to five years.

An important conclusion from Menard's work is that a newly emerging field is likely to grow at a very fast rate, with its size doubling more quickly than the whole of science. These findings suggest that a community dedicated to an emerging technology should be growing at a rate whereby it doubles every five years or sooner.

The growth in the number of researchers involved in the research and development of compound semiconductors is shown in Table 13–1. The total number of researchers participating in the community grew from 1,456 in 1976–78 to 2,853 in 1980–82 to 4,630

TABLE 13–1. *Number of Compound Semiconductor Researchers and Rate of Growth.*

Sector	Number of Researchers (Percentage of Total)			Compound Growth Rate from 1976–78 to 1984–86
	1976–78	1980–82	1984–86	
Industry	923 (63%)	1781 (62%)	2934 (63%)	15.6%
University/FFRDC	416 (29)	855 (30)	1455 (31)	16.9
Government	117 (8)	217 (8)	241 (5)	9.5
Total	1,456	2,853	4,630	15.6

in the lastest period, 1984–86. The compound growth rate in the number of researchers from 1976–78 to 1984–86 is 15.6 percent. At this rate of growth, the size of the researcher community doubled about every four and one-half years.[9]

An examination of the sectoral growth rates shows that the industrial research community grew at a compound rate of 15.6 percent. The university and FFRDC sector grew at a slightly higher rate of 16.9 percent, and the government sector grew at a much slower rate of 9.5 percent.[10]

It is interesting to note that the proportion of researchers employed in each sector remained relatively constant over the three time periods. The largest sector, industry, is comprised of slightly less than two-thirds of all researchers, followed by the univeristy/FFRDC sector with just less than one-third, and with the government sector making up the balance.

A comparison of the growth rates from the first time period (1976–78) to the second time period (1980–82) and from the second time period to the third time period (1984–86) is provided in Table 13–2.

In general, the table shows that the community of researchers grew at a faster rate in the earlier years then in the later years. The table also shows that all of the growth rates from 1976–78 to 1980–82 exceed the threshold level of 14 percent by a fairly large margin.

The growth rate in the number of compound semiconductor researchers in the United States (15.6 percent) is significantly higher than the growth rate for all U.S. researchers in the electrical and electronics industry. Data collected by the National Science Foundation show that the number of research and development scientists and engineers in the electrical and electronics industry grew from 80,300 in 1976 to 121,500 in 1985; that is a compound growth rate of 4.7 percent. Similarly, the growth rate of compound semiconductor researchers in Japan (17.4 percent) exceeds the growth rate of Japanese researchers in the field of electrical and electronic engineering in total. The Statistics Bureau of the Management and Coordination Agency in Japan reports that the number of R&D researchers in the electrical machinery industry increased at a compound growth rate of 8.9 percent, from 51,174 in 1979 to 78,427 in 1984.

In summary, the data support the proposition that the community of researchers will grow rapidly, at a rate great enough to double its size in a period of five years or less. Moreover, this result is consistent within both the industrial and university/FFRDC sectors of the

TABLE 13–2. *Growth Rates in Community of Researchers*

	T1–T2	T2–T3
Industry	17.9%	13.3%
University/FFRDC	19.7	14.2
Government	16.7	2.7
Total	18.3%	12.9%

443

research community and within all three sectors during the earlier years.

Education

The second proposition is based on the annual number of graduate students who after conducting research in the emerging technology at their university, remain in the research community after graduation. The number of these students is expected to increase over time.

The importance of a convenient pool of young recruits to a new field, as Edge and Mulkay (1976) suggest in their extensive analysis of the growth of radio astronomy, is an assertion over which few people will disagree. Both Kuhn (1970) and Constant (1980) point out that it is difficult for individuals who have been committed to working within an established research paradigm to redirect their efforts toward a new paradigm. It is more likely that the research community will assimilate young recruits who are only beginning to build a career based on a particular body of scientific and technological knowledge.

The annual number of university graduate students entering the public or private sector and who continue to conduct compound semiconductor research is shown in Table 13–3.

The data show that the number of graduate students remaining in the community after graduation increased from 53 in 1976–78 to 119 in 1980–82 to 161 in 1984–86. The number of students generally increased from year to year, averaging about ten to twenty in the early years and about forty-five to fifty-five in the later years.[11]

The increase in the number of graduates in the field of compound semiconductors is significant in light of the general downward trend in the annual number of new electrical engineering doctorates. For example, the National Science Foundation reports that the annual number of Ph.D. recipients in electrical engi-

TABLE 13–3. Annual Number of Graduating M.S. and Ph.D. Students Remaining in the Compound Semiconductor Community.

Year	Number of Researchers
1975	11
1976	16
1977	17
1978	20
1979	39
1980	38
1981	40
1982	41
1983	44
1984	55
1985	50
1986 (est.)	56

neering and solid state physics declined throughout the 1970s and recovered only slightly in the early 1980s.[12]

In conclusion, the data support the proposition that the number of individuals entering the research community via graduate training in the new technology will increase as the technology emerges. Indeed, it is likely that the data presented here underestimates considerably the flow of university graduates into the compound semiconductor field, since only a small proportion of all graduate students become authors in the kind of journals used as sources of data in this study. A more accurate estimate could be obtained from using the dissertation abstracts database. Nevertheless, the data do provide an indication that graduate training is a gateway into the compound semiconductor research community for an increasing number of individuals.

Researcher Mobility

The third and last proposition regarding the community of researchers concerns their mobility between organizations. In particular, the movement of industrial researchers from the permanent employment of one firm to another organization is examined.[13]

There are many possible reasons a researcher might have for choosing to move from one organization to another, including important individual-specific motivations. Nevertheless, there may be particular mobility patterns related to the emergence of a new technology.

One reason for the mobility of researchers is their search for a supportive management willing to make a commitment to a new technology. Early on many researchers with an interest in a new technology reside in organizations dominated by an entrenched technology, and there are real obstacles in the ability of the new technology to flourish in such environments.[14] The accumulated investment in the existing technology and the likelihood of its continued progress weigh much in its favor. Meanwhile, the lack of understanding about the new technology, difficulties in evaluating its chances for success, and its relatively high investment requirements present overwhelming odds against the wholehearted commitment on the part of many firms with a well-established technological base.

Therefore, with the growth of a community of researchers committed to developing a new technology, we should see a rise in the incidence of mobility as individuals search out firms willing to make a serious investment. In some cases, those individuals with an entre-

preneurial bent and support from venture capitalists will form new organizations where the new technology can develop in isolation from existing technologies.

Table 13–4 provides data on the number of industrial researchers who made a permanent change of employment in each of the three time periods. The total number of researchers changing employers increased from 41 in 1976–78 to 80 in 1980–82 and to 120 in 1984–86. However, although the number of industrial researchers involved in employment changes grew rapidly, the rate of mobility among industrial researchers describes a different pattern of behavior. The table shows that the number of industrial researchers as a percentage of all industrial researchers increased just barely from 4.4 percent in 1976–78 to 4.5 percent in 1980–82, and then the number declined slightly to 4.1 percent in 1984–86.

Table 13–4 also offers further insight into the nature of mobility by providing a breakdown in the data according to the type of organization researchers moved to: industrial firm, new venture, university/FFRDC, or government agency. The data indicate that the majority of employer changes tend to be intra-industry—that is, from one firm to another. In addition, the movement of industrial researchers to new ventures is the only instance in which the rate of mobility increased over each of the three periods.

It is important to note that mobility patterns differ significantly among countries. A re-

TABLE 13–4.   Number of Compound Semiconductor Researchers in Industry Involved in a Change of Employer.

Sector of New Employer	1976–78		1980–82		1984–86	
	Number of Researchers	Percentage of Total	Number of Researchers	Percentage of Total	Number of Researchers	Percentage of Total
Industry	25	2.7%	48	2.7%	50	1.8%
(New venture)	0	0.0	9	0.5	40	1.3
University/FFRDC	15	1.6	23	1.3	28	1.0
Government	1	0.1	0	0.0	2	0.1
Total	41	4.4%	80	4.5%	120	4.1%

TABLE 13–5. *Tenure with Employers of Industrial Researchers*

Number of Years with Current Employer	Percentage of U.S. Researchers	Percentage of Japanese Researchers	Percentage of European Researchers
Less than 5 years	25.8%	11.2%	10.1%
5 to 9 years	31.1	19.3	34.8
10 to 14 years	16.7	24.5	18.8
15 to 19 years	10.8	27.9	24.6
20 or more years	15.6	17.1	11.6
Total	100.0%	100.0%	100.0%

gional breakdown of the data indicates that the mobility rate for researchers in the U.S. is the highest (between approximately 6 and 7 percent from 1976–78 to 1984–86), followed by Western Europe (between 2 to 5 percent), and Japan (less than 1 percent). Furthermore, the movement of researchers to new ventures occurs predominantly in the United States.

The relatively high rate of mobility in United States is reflected in the biographical data on industrial researchers. Table 13–5 shows that close to 57 percent of the industrial researchers in the U.S. have less than ten years' tenure with their current employer. In comparison, about 45 percent of the European researchers and only about 30 percent of the Japanese researchers have less than ten years' tenure.

In summary, the data support the proposition that the number of industrial researchers changing employers will increase over time. However, the pattern of change in the rate of mobility is less certain. In only one case, the movement of researchers to new ventures, does the number of researchers increase along with the mobility rate over all three periods. The results suggest that a small, relatively fixed percentage of industrial researchers are inclined to change employers, but as a technology emerges, an increasing percentage of these "movers" join new ventures instead of other established firms.q4

THE COMMUNITY OF ORGANIZATIONS

An organization (firm, university, or government laboratory) is considered to be a member of the community in any particular year if one or more of its researchers is involved in compound semiconductor research and development activities. Each organization in the community contributes to the overall success of the community, regardless of whether or not it actively collaborates with other organizations or even perceives itself as a member. An organization's mere presence in the community enhances the processes of competition and specialization central to the emergence a new technology.

The analysis of the organizational community will be analyzed in three ways: in terms of changes in the distribution of researchers among organizations, collaboration between researchers in different organizations, and changes in the extent of specialization within the community.

Distribution of Researchers

One measure of the organizational community is the distribution of the researcher population. The distribution of researchers among organizations can be derived by means of a concentration ratio statistic (similar to that used in industrial organization economics). The concentration ratio measures the percentage of researchers employed in a given number of the largest research programs. The first proposition states that the distribution of researchers employed in industry will become less concentrated as the new technology emerges over time.

When research and development activities are more widely distributed among organizations (that is, a low concentration ratio), the risk of a new technology failing to emerge for reasons not inherent to the technology is greatly reduced. Every firm faces numerous challenges that are not necessarily related to the merit of the new technology. Considering that the period of emergence can span a decade or more, some firms will suffer setbacks in their business: cash flow problems, abrupt management changes, acquisitions or hostile takeovers, loss of key employees, and so on. Assuming that a firm succeeds in dealing with the ongoing challenges of business, it must also face many problems that are specific to the emerging technology. The novelty of a new technology means that there are very few decisions where the implications of different choices are certain. Of course, firms are not homogeneously endowed with managerial leadership, a technical group with esprit de corps, and sound business and technical judgement; consequently, it is reasonable to believe that the efforts of some firms will fail. Even firms with superb capabilities will make decisions that ultimately may not prove to be the best.[15]

The decentralization of decisionmaking resulting from a less concentrated population distribution of researchers allows for a diversity of opinions and choices. Some of these choices will prove to be very successful—not only for the individual firm but also for all firms involved with an emerging technology. In their success these firms generate market awareness and create standards that lead the way for other firms.

Finally, as the distribution of researchers becomes less concentrated, competitive rivalry begins to grow between teams in different organizations. This rivalry sparks the spirit and determination of individual researchers to succeed in making their own ideas work and, in doing so, gain the respect of the larger community.

In sum, the declining concentration of the population distribution of researchers contributes to the emergence of a new technology by reducing the impact of firm-specific failures, increasing the diversity of technical choices taken, and providing for interorganizational rivalry.

The concentration ratios for the top-five, top-ten, and top-fifteen largest industrial research programs are provided in Table 3–6. The data show that during 1976–78, the five firms with the largest programs employed 40.8 percent of all compound semiconductor researchers. By 1980–82, the level of concentration declined to the point where the five largest programs employed 34.6 percent of all researchers. This trend continued throughout

TABLE 13–6. *Distribution of Compound Semiconductor Researchers Employed in Industry and in University/ FFRDCs.*

Sector and Concentration Ratio	Percentage of Total Number of Researchers		
	1976–78	1980–82	1984–86
Industry:			
Top 5	40.8%	34.6%	28.8%
Top 10	62.5	52.0	43.1
Top 15	77.1	64.1	54.2
Univ/FFRDC:			
Top 5	35.6	28.6	22.9
Top 10	51.0	44.3	35.9
Top 15	61.1	55.0	45.8

the latest period, 1984–86, reaching a level of 28.8 percent. The concentration ratios for the top-ten and top-fifteen largest programs also declined over time.

The declining concentration in the distribution of compound semiconductor researchers is, in part, a result of the increasing number of industrial organizations that have become involved in the technology. Table 13–7 shows that the total number of firms increased from 52 in 1976–78 to 75 in 1980–82 to 123 in 1984–86. The effect of this increase on the number of firms is quite dramatic, considering that the number of researchers employed in the largest research programs also grew rapidly. For example, the size of the five largest programs grew from 377 researchers in 1976–78 to 845 researchers in 1984–86. Thus, the decline in industry concentration occurred during a period of surging activity among an increasingly larger number of firms, including the largest programs in the field.

It is interesting to note that a similar trend in the distribution of researchers occurred within the university/FFRDC sector. Table 13–7 shows that the five largest research programs among universities and FFRDCs employed 35.6 percent of all researchers in that sector during 1976–78. The concentration ratio declined to 38.6 percent in 1980–82 and to 22.9 percent in 1984–86.

In summary, the data support the proposition that the distribution of researchers among firms will become less concentrated over time. Furthermore, the data support the same proposition for the university/FFRDC sector.

Interorganizational Collaboration

The second proposition regarding the organizational community is that the emergence of a new technology is accompanied by an increasing amount of collaboration between organizations.

The measure of collaboration used here is based on the notion that co-authorship of a journal article by two or more researchers employed by different organizations is a form of interorganizational collaboration. The type of collaboration covered in these publications varies widely, ranging from independently initiated interaction among individual researchers to highly structured joint research and development programs.

Fundamentally, collaboration represents an exchange of ideas, experience, or expertise between individuals or groups who find it in their best interest to participate. Thus collaboration does not necessarily imply a lack of competition between organizations. At times competitors find that it is mutually beneficial to enter into such exchanges, even though they may otherwise have opposing interests.

It is possible that individuals in the research community identify strongly with each other, so much so that although a wider competition between organizations does exist, it does not become an obstacle to collaboration. This may be especially true when researchers were former colleagues (at the same organization) and have maintained strong personal friendships (Rogers 1982).

Interorganizational collaboration is beneficial to the emergence of a new technol-

TABLE 13–7. *Number of Organizations Contributing to the Development of Compound Semiconductors.*

Sector	Number of Organizations		
	1976–78	1980–82	1984–86
Industry	52	75	123
University/FFRDC	69	104	168
Government	6	7	7
Total	127	186	298

TABLE 13–8. *Number of Compound Semiconductor Researchers Involved in Interorganizational Collaboration.*

Sector	1976–78		1980–82		1984–86	
	Number	Percentage of Total	Number	Percentage of Total	Number	Percentage of Total
Industry	133	14.4%	579	32.5%	1,109	37.8%
Univ/FFRDC	261	62.7	476	55.7	1,074	73.8
Government	56	47.9	95	43.8	132	54.8
Total	450	30.9%	1,150	40.3%	2,315	50.0%

ogy because it can reduce duplicated effort among different groups in solving the many technical problems that arise over time. Hard as it might be to admit, sometimes after months of effort by one group to work out a technical bug, the solution comes only after a conversation with someone in another organization. Collaboration may help a group to get on the right path or reduce the number of dead ends. In doing so, the research community's efforts are continually intensified on the real problems that have yet to be solved. Collaboration also strengthens the working relationships between people in different organizations and thus may enhance the process of specialization among firms, as will be discussed in the next section.

Collaboration data were gathered for each organization to determine the number of external researchers (that is, those employed by other organizations) with whom its researchers collaborated over time. The data in Table 13–8 show that the total number of researchers involved in collaboration grew rapidly, more than doubling from one period to the next: from 450 researchers in 1976–78 to 1,150 in 1980–82 to 2,315 in 1984–86. The table also shows the number of researchers who collaborate as a percentage of the total number of researchers active in the respective community. The collaboration ratio increased by about ten percentage points each period from 30.9 percent to 40.3 percent to 50.0 percent in the latest period.[16]

The sectoral breakdown of the data in Table 13–8 shows that although industrial researchers involved in collaboration constitute the smallest group at first, their numbers grew rapidly, and they became the largest group in 1980–82 and 1984–86. The collaboration ratio for the industrial sector shows a large increase from 14.4 percent in 1976–78 to 32.5 percent in 1980–82. The ratio continued to increase, but only less so, to 37.8 percent in 1984–86. The university/FFRDC sector had the highest level of collaboration in each of the three periods. The ratio declined at first, from 62.7 percent in 1976–78 to 55.7 in 1980–82, and then increased sharply to 73.8 percent in 1984–86. The government sector shows a similar pattern, declining from 47.9 percent to 43.8 percent and then increasing to 54.8 percent in the last period.[17]

In summary, the data generally support the proposition that interorganizational collaboration will increase over time.[18] The results in the case of compound semiconductors show that both in terms of the number of researchers and as a percentage of all researchers, interorganizational collaboration has increased from 1976–78 to 1984–86.

Specialization

The third proposition concerning the organizational community is that industrial organizations will become increasingly specialized in terms of the activities they conduct during the emergence of a new technology.

Historical evidence suggests that the emergence of a new technology can be characterized in terms of changes in the degree of specialization within the industrial community. Braun and Macdonald (1982), in particular, show that the early research and development work in semiconductor technology took place mostly in vertically integrated electronics manufacturers. As the technology progressed with the passage of time, however, the industrial community became increasingly characterized by specialized firms: that is, firms solely dedicated to the development and manufacture of semiconductor materials, equipment, or devices, or systems implementing semiconductors. While specialization was not limited to new firms, many of the semiconductor startups were of a specialized character.

Stigler (1951) explains the specialization process in terms of the division of labor among firms permitted by the growth of the marketplace. The transaction cost approach espoused by Williamson (1985) is also useful in explaining this pattern of change. Williamson claims that a firm's choice between "market or hierarchy," or alternatively, "make or buy" is largely determined by the conditions of asset specificity, uncertainty, and frequency of transactions surrounding the decision. Williamson predicts that when human or physical assets are highly specific, uncertainty great, and transactions infrequent, then a firm will chose to be inte-grated. On the other hand, when assets are nonspecific, uncertainty low, and transactions frequent, then a firm can specialize and take advantage of transacting with other firms in the market.

The growth of specialization within the industrial community enhances the emergence of a new technology, because it reduces the costs and risks born by any single firm in the community. For example, those firms specializing in devices can depend on firms that specialize in materials, and as a result forgo the cost of developing their own material. By purchasing materials, the device firm can also benefit from the economies of scale achieved by materials suppliers. Lastly, rather than depend on its own success in developing materials, the specialized device firm can benefit by chosing the best materials available among various suppliers.

Table 13–9 presents the data on specialization trends among industrial firms involved in the research and development of compound semiconductor technology. The data show that the number of specialized firms increased from 7 in 1976–78 to 18 in 1980–82 to 67 in 1984–86. The percentage of researchers employed in these specialized firms also increased from 12.7 percent in 1976–78 to 17.1 percent in 1980–82 and to 23.4 percent in 1984–86. In contrast, the data also show that the number of integrated firms remained relatively stable, increasing from forty-five in 1976–78 to forty-seven in 1980–82 to fifty-six in 1984–86. The percent-

TABLE 13–9. Specialization Ratios for Compound Semiconductor Researchers Employed in Industry.

Type of Firm	1976–78	1980–82	1984–86
Integrated:			
Number of firms	45	47	56
Percentage of researchers	87.3%	82.9%	76.5%
Specialized:			
Number of firms	7	18	67
Percentage of researchers	12.7%	17.1%	23.4%

age of researchers employed in integrated firms has decreased from 87.3 percent in 1976–78 to 82.9 percent in 1980–82 to 76.5 percent in 1984–86.

In conclusion, the data support the proposition that the industry will become increasingly specialized over time. The growth of specialization is primarily the result of an increasing number of specialized firms entering the industry over time. But even though the number of specialized firms exceeds the number of integrated firms by 1984–86, it is also true that the large majority of industrial researchers are employed in vertically integrated firms.

THE TECHNOLOGICAL PARADIGM

The previous two sections described the community of researchers and their organizations committed to the development of a new technology. This section examines different dimensions of the technological paradigm—that is, the ideas and techniques that the community of researchers come to hold in common. Of particular concern are the circumstances surrounding the intitiation of new ideas and techniques, their spread among organizations, and the sources of intellectual leadership within the research community.

Many recent contributions to the academic literature on technological change, including historical accounts on the development of the turbojet (Constant 1980), microelectronics (Dosi 1982, 1984), and wireless radio (Aitken 1985) suggest the notion of technological paradigms as a conceptual analog to Kuhn's (1970) work on scientific paradigms.

The "technological paradigm perspective" views most technological change in terms of long periods of incremental advances that flow from the steady accumulation of a body of knowledge generated by a community of practitioners. This "normal" pattern of development is punctuated infrequently by the emergence of a new technological paradigm, which holds the foundation for a new technology. The new paradigm represents a body of ideas and techniques that are in some manner significantly different from an existing paradigm and are embraced by a group of individuals dedicated to its development. The incongruence between new and existing paradigms is suggestive of a technological discontinuity.[19]

Understanding the dimensions of a new technological paradigm is a particularly challenging task, so much so that there is danger in oversimplification. Indeed, the body of ideas and techniques that form the compound semiconductor paradigm is large, complex, and constantly growing. Analyzing the paradigm is made more difficult by the fact that in many instances new ideas or techniques that are fundamentally the same nevertheless go by different names. Therefore, the analysis presented in this section is limited in focus to understanding two elements of the paradigm: an idea for a new transistor device structure called the high electron mobility transistor (HEMT) and a processing technique called metalorganic chemical vapor deposition (MO-CVD).[20]

Anticipation and Multiple Discovery

The first proposition concerning the technological paradigm asserts that anticipation and multiple independent discovery of new ideas and techniques will occur among researchers in different organizations.

Although multiple independent discoveries are commonly perceived as a curiosity, Merton (1961) has shown otherwise. He suggests that once attention is turned to a particular area of research, the likelihood of similar independent insights is very high, so much so that "the pattern of independent multiple discoveries in science is in principle the dominant pattern of discovery rather than a subsidiary one" (1961: 356).

The prevalence of "multiples" as suggested by Merton, has led to a debate over why this pattern of discovery occurs. Simonton (1978), claiming that the variation in the frequency distribution of multiples can be adequately predicted by a Poisson distribution, argues that the pattern of multiple discoveries can be modeled as a chance process. The work of other researchers has countered the "chance" explanation. Brannigan and Wanner (1983) find that the frequency of multiple discoveries are consistent with a negatively contagious Poisson distribution, a model that is more suggestive of a zeitgeist explanation. The zeitgeist—or "spirit of the times"—interpretation implies that multiple independent discoveries occur when the moment is ripe within a particular community or culture.

Unlike the other propositions being examined, this one is not assessed quantitatively. Instead, the phenomenon of anticipation and multiple independent discovery is illustrated in terms of historical descriptions of the development of the HEMT device and the MO-CVD process.

The HEMT device has its theoretical roots in the late 1960s with the pioneering research of Leo Esaki and Raphael Tsu of IBM Corporation, who studied the physical properties of superlattices such as GaAs-AlGaAs.[21] Esaki and Tsu proposed that high electron mobilities could be realized in compound semiconductors if electrons were transferred from a doped layer of AlGaAs to an adjacent layer of undoped GaAs.

In parallel to the research being conducted at IBM, advances were being made by researchers at AT&T Bell Laboratories on the creation of epitaxial layers of material using a technique known as molecular beam epitaxy (MBE). By the mid-1970s, improvements in the ability to form thin layers of GaAs-AlGaAs with MBE reactors gave impetus to research within Bell Laboratories on electronic device applications. The research culminated in 1978, when a group of researchers (including Raymond Dingle, Horst Stormer, Arthur Gossard, and William Wiegmann) demonstrated the phenomenon of high electron mobility by using a process they called modulation-doping in a GaAs-AlGaAs superlattice. The group proposed the creation of a transistor based on their discovery and filed for a patent on such a device.

News of the development at Bell Laboratories sparked efforts at several other laboratories to develop transistors using the modulation-doping concept. The first device using this principle, called a high electron mobility transistor (HEMT), was announced in Japan by Fujitsu Laboratories in 1980. Shortly thereafter, the French firm of Thomson-CSF announced the development of what it called a two-dimensional electron gas field effect transistor (TEGFET). This was quickly followed by AT&T's disclosure of the development of what it called the selectively doped heterojunction transistor (SDHT). Finally, in the same year, a team of researchers from Rockwell International and the University of Illinois at Urbana/Champaign announced the development of the modulation-doped field effect transistor (MODFET). The HEMT, TEGFET, SDHT, and MODFET are all based on the same principle: a structure composed of doped AlGaAs on undoped GaAs in a field effect transistor structure in which electrons are transferred from the doped layer to the undoped layer by modulation doping.[22]

The priority in developing the HEMT is a more heated issue than the historical description suggests. Researchers at Fujitsu claim their work was independent of the research at AT&T, and view themselves as the pioneers of HEMT devices.[23] Similarly, the Thomson-CSF group also views their work as independent of the research at Bell Laboratories, and they see themselves along with the Fujitsu group as pioneers of the device since 1979.[24] Despite these claims, researchers at AT&T and the University of Illinois maintain that the HEMT device was pioneered in the United States and that Raymond Dingle's presentations in Japan in 1979 influenced both the development of the HEMT at Fujitsu Laboratories as well as the work of researchers at Thomson-CSF.[25]

The historical development of the MO-CVD provides another illustration of multiple independent discovery.[26] The MO-CVD process involves the controlled reaction of gases (such as trimethyl aluminum, arsine, and trimethyl gallium) at atmospheric pressure to form an epitaxial layer of GaAs and AlGaAs on a heated substrate.

The first description of the a MO-CVD epitaxial process using metal alkyls was provided by R.E. Ruhrwein of the Monsanto Company and subsequently patented in 1968. However, in an independent effort, a group of researchers at Rockwell International led by H.M. Manasevit conducted the pioneering research and development on the use of MO-CVD for the epitaxial growth compound semiconductors. It took a number of years for the Rockwell group to refine the process, and it is only in the past five years that serious interest in MO-CVD has spread within industry.

In the case of both HEMT and MO-CVD, the evidence suggests that the simultaneous development of similar ideas and techniques by different groups of researchers is not an uncommon occurrence in the historical development of compound semiconductor technology. The evidence affirms the proposition that anticipation and multiple independent discovery will occur during the emergence of a technological paradigm. In addition, the broader history of the development of compound semiconductor technology suggests something else of importance: namely, that the

development of fundamental ideas and techniques is seldom a discrete event, but rather an evolutionary process that unfolds over years and perhaps even decades. What is more, the process typically benefits from the intellectual contributions of numerous individuals, the names of many of whom are lost in retrospective analyses.

Diffusion of Ideas and Techniques

The second proposition concerns the extent to which the industrial research community comes to focus on a given idea or technique. It is proposed that with the emergence of a new technology, researchers in different firms will recognize certain ideas and techniques as fundamental and join in the effort to develop them at an increasing rate.

The research community's recognition of certain fundamental ideas and techniques is important to the progress of the technology toward commercialization. By reaching consensus on some ideas and techniques, the community can focus its attention on solving the limited set of problems that surround them. Moreover, without some convergence of ideas and techniques within the community, it is difficult for firms to specialize their activities. By having certain ideas and techniques in common, the community makes it profitable for individual organizations to specialize in performing a particular activity. Thus, widespread agreement throughout the research community regarding the importance of the MO-CVD process allows some organizations to specialize in the development of MO-CVD equipment.

This proposition is not necessarily intuitive. One might expect that as a new technology emerges, firms will try to limit the flow of information regarding ideas and techniques that their researchers are developing and, by doing so, limit their spread to other organizations. Or one might also conclude that the "not invented here" syndrome will rule the behavior of researchers, and thus prevent the serious consideration of ideas and techniques that did not originate in a firm's own laboratory. Indeed, consensus regarding ideas or techniques might be viewed as a hinderance to the technology's emergence, in that consensus would serve only to limit the number of different ideas and techniques pursued.

To be sure, basic agreement within the research community about a process such as MO-CVD or a device such as the HEMT does not preclude a significant degree of variation in the approaches taken by different research teams. This analysis seeks to understand the extent of acceptance by different organizations that decide to work on the development of certain ideas and techniques. How they pursue the application of these principles is a point of departure that may lead to a wide variety of approaches among firms: For example, some organizations may pursue "low-pressure" MO-CVD, others may focus on the "psuedo-morphic" HEMT. Thus, general agreement at one level does not imply a homogeneity of approaches at other levels of analysis.

Table 13–10 provides data on the number and percentage of firms that are contributing to the development of HEMT device and the MO-CVD process. The data show that the number of firms working on the development of HEMT devices grew from one in 1976–78 to five in 1980–82 to twenty-three in 1984–86. Among those firms developing compound

semiconductor devices, the percentage of industrial laboratories working on HEMT devices grew from 3.4 percent in the first period to 9.8 percent in the second period to 30.3 percent in the last period.

In the case of MO-CVD, the number of firms working with the technique worldwide grew from three in 1976–78 to twenty in 1980–82 to forty-six in 1984–86. Among firms working on the development of compound semiconductor materials, the percentage of those involved with MO-CVD increased rapidly from 8.6 percent in period one to 39.2 in period two to 61.3 percent in period three.

In summary, the data on HEMT and MO-CVD lend support to the proposition that certain ideas and techniques will become viewed increasingly by the industrial research community as fundamental, and thus an increasing number of firms will in turn contribute to their development.

Intellectual Leadership
The final proposition concerns the rise of a small group of researchers who provide intellectual leadership to the larger community by pushing ahead the frontier of the technological paradigm. The proposition states that a small group of intellectual leaders will grow simultaneously with the emergence of a new technology.

TABLE 13–10. *Diffusion of New Ideas and Techniques in Industry: High-Electron Mobility Transistors (HEMT) and Metalorganic Chemical Vapor Deposition (MO-CVD).*

| Idea/Technique | 1976–78 | | 1980–82 | | 1984–86 | |
	Number	Percentage of Total	Number	Percentage of Total	Number	Percentage of Total
HEMT	1	3.4%	5	9.8%	23	30.3%
MO-CVD	3	8.6	20	39.2	46	61.3

The main question here is whether the achievements of a small group will attain exemplary status within the research community. Gilbert (1977) and Small (1978) suggest using citation patterns in the scientific literature as a method for identifying exemplars: "Certain papers, through their repeated use as authoritative grounds for further work, begin to achieve an exceptional status, and may come to be regarded as exemplars of valuable work in the field" (Gilbert 1977). Thus, some highly cited papers may provide insight into the intellectual leadership in a research field and the emergence of a new paradigm.[27]

The growth of a group of intellectual leaders within the research community is examined by means of citation data. The analysis is conducted by isolating those researchers who have been active in the technological paradigm over all three time periods, from 1976–78 to 1984–86. This group consists of 539 researchers, which is about 7.4 percent of the total population of 7,320 researchers in the database. A random sample of 102 researchers (about 19 percent) is selected from this group. The number of citations (not including self-citations) for each researcher in this sample is calculated for three years (1975, 1980, and 1985) using the *Science Citation Index.*

Two ratios are constructed by calculating the number of citations received by the top 5 and top 10 percent of those researchers who are cited most often in the latest year (1985) as a percent of the total number of citations received by all researchers in a given year (1975, 1980, and 1985).

The ratio is structured in such a way as to determine whether the number of citations received by those researchers who currently are cited most often increases or decreases over time from 1975 to 1985. If a group of intellectual leaders emerges over time, then that group should receive a growing percentage of all citations. Consequently, the ratio should increase.

The citation data for the sample group are presented in Table 13–11. The data show that the top 5 percent of researchers most often cited in 1985 received 242 citations in 1975, 624 in 1980 and 1,284 in 1985. The proportion of citations received by the top 5 percent in 1985 grew from 22.9 percent in 1975 to 26.0 percent in 1980 to 35.5 percent in 1985.

The number of citations for the top 10 percent of the researchers cited most often in 1985 grew from 379 in 1975 to 839 in 1980 to 1,840 in 1985. The proportion of total citations received by the top 10 percent in 1985 remained relatively unchanged at about 35 percent from 1975 to 1980, but the proportion rose sharply to 50.8 percent in 1985.

TABLE 13–11. *Citations Received by the Most-Cited Researchers in 1985 Conducting Compound Semiconductor Research and Development.*

	1975		1980		1985	
	Number	Percentage of Total	Number	Percentage of Total	Number	Percentage of Total
Top 5% most-cited researchers in 1985	242	22.9%	624	26.0%	1,284	35.5%
Top 10% most-cited researchers in 1985	379	35.8	839	35.0	1,840	50.8
Citations received by all researchers in sample by year	1,056	100.0	2,396	100.0	3,622	100.0

In summary, the data support the proposition that a small group of intellectual leaders will emerge over time within the community. However, the use of citations in this analysis must be carefully considered. Numerous scholars have pointed out the limitations of using citations as a measure of the significance of an individual's or organization's research. Nevertheless, it seems reasonable to expect that with the spread of certain ideas and techniques, we should see those researchers responsible for the initial expression of these ideas and techniques receiving an increasing amount of recognition for their work.

CONCLUSION

The development of compound semiconductors illustrates that new technologies represent an expensive, long-term investment in human and physical capital. Based on the manpower estimates in this study and compensation rates for researchers, the U.S. investment in human capital alone has amounted to more than $500 million since 1976. The magnitude of such an investment necessitates that we seek to understand the emergence of radically new technologies so that individuals and organizations can deploy their intellectual and financial resources more effectively.

In following sections we summarize the specific findings of this study and discuss what we view as the major conclusions and directions for future research.

Summary of Results

An analysis of the development of compound semiconductor IC technology provides several insights into the nature of change in the community of researchers, their organizations, and the ideas and techniques they develop as a new technology progresses from its inception in research laboratories to its entrance into the marketplace. First, we found that the number of researchers contributing to the technology's development increases rapidly over time, at a rate sufficient for the research community to double in size about every four to five years. This growth is in part fueled by a small group of major research universities, where a growing number of individuals initiate their work in the new technology in conjunction with pursuing a graduate degree. Furthermore, as the community grows, so does the number of industrial researchers who chose to change employers. Although the rate of mobility within the community remains relatively unchanged over time, there is a noticeable change in the character of mobility—namely, the movement of researchers between established firms is replaced by the movement of researchers from established firms to new ventures.

The case of compound semiconductor IC technology also shows that the organizational community undergoes significant change. As the new technology emerges, the population of researchers in the community becomes more widely distributed among a growing number of organizations in both the public and private sectors. As the organizational community expands, it becomes increasingly more common for researchers in different organizations to collaborate with each other in the research and development of the technology. In addition, among those firms that are entering the field, a large number specialize in materials or devices or equipment, such that the industrial community becomes less and less dominated by vertically integrated firms.

Lastly, the emergence of compound semiconductor IC technology shows that as attention focuses on expanding a new body of knowledge, researchers begin to anticipate one another, and consequently, new ideas and techniques develop independently and sometimes simultaneously in different organizations. As certain ideas and techniques come to the forefront of the community's research agenda, an increasing percentage of firms devote resources to their development. Moreover, with the spread of particular ideas and techniques, a small group of researchers will emerge as intellectual leaders for the rest of the community.

Discussion

There are several conclusions that can be drawn from this study. First, with respect to the methodology, it is clear that the publications researchers generate can provide a wealth of empirical data, from which chronological changes in the individual, organizational, and intellectual processes that occur in the development of a technology can be measured. Because the data are readily accessible, firms and other interested parties may by able to establish a system for monitoring the emergence of new technologies without having to resort to elaborate efforts at considerable expense.

It is curious to note that researchers with a penchant for reading the literature or attending conferences, frequently come to comprehend many of the changes we associate with the emergence of a technology: The rapid growth in the number of researchers is perhaps the most easily detected change. A lesson from this study is simply that there are benefits to doing what some researchers already do, only doing it in a more systematic fashion.

Using the science and technology literature as a source of data opens up new avenues of analysis in comparison to the traditional bibliometric approaches of measuring the literature itself. Clearly, this study has just barely begun to take advantage of these opportunities and there is much to be gained from future research.

Second, the results of the study suggest that there are numerous changes that take place as a new technology emerges, such that monitoring these changes may provide firms with an early indication of its importance and future commercial impact. Indeed, in the case of compound semiconductor technology the data show that many of the expected trends occurred in the earlier years, well before the technology was introduced into the marketplace.

Nevertheless, it must be emphasized that this is a preliminary conclusion. For one, even though there are currently many companies marketing compound semiconductor devices, materials, and equipment—and soon computer systems implementing the technology—it is still premature to draw absolute conclusions about the commercial success of this technology. What is more, the value of this approach in signaling the successful emergence of new technologies can be determined only by replicating the study with other cases, including technologies that have failed to emerge as well as those that have succeeded.

We have already begun one such study, examining the development of low-temperature superconducting Josephson junction IC technology. The preliminary results from this study offer a marked contrast to that of compound semiconductors and appear to support many of our propositions. But more research is needed. One might explore other interesting cases in the field of electronics technology such as the development of silicon-based IC technology during the 1950s and 1960s or the development of magnetic bubble devices during the 1970s. Having explored the emergence of new electronics technologies, the research agenda should expand to examine other types of technologies such as those in the field of biotechnology. By doing so, we can begin to understand in a more general way the nature of emerging technologies.

Third, it is extremely important to recognize that this study focuses on the successful emergence of a new technology as an industrywide phenomenon. Therefore, it is inappropriate to draw conclusions about the success or failure of any particular firm. For example, although the findings suggest that increasing collaboration between organizations may contribute to the successful emergence of a new technology, analysis of individual firms may show that highly collaborative firms perform less well than uncollaborative firms in reaping the benefits of the new technology. Thus we may very well find a micro-macro paradox of new technology development: that is to say, what is important for the emergence of a new technology (macrolevel phenomenon) may be contrary to the success of any particular firm (microlevel phenemenon). If this is true, then the successful emergence of a new technology may depend as much on the sacrifices made by those firms that ultimately fail as well as those firms that succeed.

Fourth, with additional study we aim to uncover the causal relationships that may exist between the various phenomena associated with the emergence of new technologies. For example, one question might be whether collaboration contributes to specialization or vice versa. However, the broader question of interest might be whether collaboration and specialization, among other things, contribute to the successful progress of a technology or whether successful progress leads to more collaboration and specialization. Understanding this dimension of causality is certainly difficult, but even the findings of this study lend some clue to an answer in the temporal sequence of events displayed by the data.

Finally, it is interesting to note that many of the changes that occur with the emergence of a new technology are suggestive of the evolution of consensus among individuals and organizations about the potential importance of a new technology.[28] Is this growing consensus and the concentration of effort it creates of central importance to the emergence of a new technology?[29] If the answer is yes, then it is necessary to understand that the emergence of a new technology is beyond the capabilities of a single organization, and success in the end will depend on the degree to which others become interested in contributing to its development.

NOTES

1. See early discussions of s-curve analysis in Cetron (1970) and Martino (1972). Recent contributions have been made by Utterback and Kim (1984), Foster (1985), and Olleros (1986).
2. Irvine and Martin (1984b) and Eto (1984) discuss the use of science and technology surveys in Japan.
3. The studies of Irvine and Martin are perhaps the most rigorous examples of the use of bibliometric data for comparisons of organizational and international performance in science and technology. See Irvine and Martin (1984a), Martin and Irvine (1984a, 1984b) and Hicks, Martin, and Irvine (1986).
4. See Small (1980), Small and Greenlee (1986), Gregory (1983, 1984), and Moravcsik and Murugesan (1979).
5. Although publication dates range from January 1975 through March 1987, submission dates range from January 1974 through December 1986. The data are coded according to submission dates. The time lag between submission and publication is typically less than one year. The publication lag time has a slight affect on the data for 1986, since some articles published in the second and third quarter of 1987 will have been submitted in 1986. Therefore the data for 1986 likely underestimates the actual amount of activity in that year.

   The data coverage in this study begins with 1974 because it was about this time when the commercial feasibility of compound semiconductor integrated circuits was first taken seriously. However, there is a long history of research in compound semiconductor materials: the first III-V compound (aluminum phosphide) was described in 1827. Research was conducted throughout the first half of the twentieth century, but it was not until the 1950s, with the development of semiconductor technology in general, that research in compounds intensified. In his book *Physics of III-V Compounds,* published in 1964, Madelung states that there were already about 1,000 publications especially concerned with the subject. The publications in the database deal with a wide variety of topics in compound semiconductor research and development, including bulk materials growth, characterization, epitaxial growth, material doping, device structure, circuit logic, wafer processing, circuit and device testing, and packaging, as well as the development of the equipment associated with these processes. In addition, the publications cover research and development of integrated circuits for analog, digital, and optoelectronic applications.

6. These sources were chosen because of their high frequency of publications on compound semiconductors, their international scope, and their strong reputation.

7. This study covers only those noncommunist countries with a major research and development effort in compound semiconductors. However, there are other countries involved in the technology, including the Soviet Union, the People's Republic of China, several countries in Eastern Europe, Switzerland, Italy, Israel, South Africa, Australia, Brazil, and the Republic of China, among others.

   An analysis was conducted to determine the number of industrial firms currently working on compound semiconductors which are not covered in this study. The results show that slightly more than 70 percent of all firms working on the technology are included in the database. Most of the firms not included were relatively small in size.

8. It is important to recognize that once researchers stop publishing, they are no longer included in the estimate of the size of the research community. But this does not mean that these individuals are no longer part of the research community. On the one hand, it may simply mean that the individual moved into a management position overseeing compound semiconductor research. On the other hand, it may mean the individual has changed his research interest and perhaps was never fully committed to the study of compound semiconductors from the start. In any case, by counting only those individuals who continue to publish, the estimates provided are likely to represent a minimum figure of the annual size of the research community.

9. The growth in the number of researchers is correlated positively with the growth in the number of articles published. This leads an important issue: namely, it may be true that the growth in the number of compound semiconductor researchers as measured through the literature was influenced by changes in behavior toward publishing between 1975 and the present. Thus, a relevant question is whether or not the habit of publishing one's research has become more or less popular over this period. Data from the National Science Foundation suggest that publishing in the field of engineering and technology worldwide has declined from 1973 to 1980:

Region	Number Included in This Study		Number Not Included	
United States	79	(70.5%)	33	(29.5%)
Japan	32	(76.2%)	10	(23.8%)
Western Europe	13	(65.0%)	7	(35.0%)
Total	124	(71.3%)	50	(28.7%)

Year	Number of Engineering and Technical Publications Worldwide	U.S. Total	U.S. Universities
1973	28,617	11,955	4,715
1974	26,600	11,088	4,346
1975	25,664	10,431	4,115
1976	25,146	10,346	4,165
1977	25,003	10,081	3,870
1978	24,588	9,694	3,790
1979	22,182	9,018	3,711
1980	21,459	8,461	3,614

459

The data are based on an analysis of 2,100 journals. Thus, the correct interpretation of the data might be simply that certain journals decline in popularity over time. NSF increased the coverage on journals to 3,500 for 1981 and 1982 and reported that the number of publications in engineering and technology worldwide decreased from 35,248 in 1981 to 32,598 in 1982. Thus, the evidence suggests that the rapid growth in compound semiconductor publications came during a period when publication frequency in general was declining.

10. The industrial sector includes government-owned firms, public corporations, and those firms that are heavily regulated by government. The worldwide trend over the past ten years has been to privatize or deregulate these organizations, such as in the case of AT&T in the United States, British Telecom Laboratories (formerly the British Post Office Research Centre) in the United Kingdom, and Nippon Telegraph & Telephone (NTT) in Japan.

    The university/FFRDC sector includes independent nonprofit research institutes. Many FFRDCs are affiliated or operated by universities, such as with MIT's Lincoln Laboratory; however, some FFRDCs in the United States are operated by industrial firms, as in the cases of Sandia National Laboratories (AT&T) and Oak Ridge National Laboratory (Martin Marietta).

    The government sector includes organizations that are owned and operated by agencies of the government. The data are aggregated to the level of department or ministry. For example, in the case of the United States, the Department of Defense is taken as a single organization even though each of the three military services have research efforts (with multiple laboratories) of their own. This method of aggregation is used to allow conformity with parent firm aggregation in the industrial sector.

    Data gathered by the National Science Foundation show that the sectoral distribution of engineering and technology articles in general in the United States is significantly different than the distribution of compound semiconductor articles:

Sector	Engineering and Technical Publications		Compound S/C
	1973	1982	1984–86
Industry	41%	39%	58%
University/FFRDC	47	52	37
Government	9	7	4

11. There is a minor problem with the education data due to the manner in which the data was coded. The majority of researchers covered in the data are graduating master's and doctoral students. However, the data also includes a few postdoctoral researchers and university faculty. Although the lack of purity in the data will inflate the numbers slightly, it is likely that it will not have a substantive impact on the basic trend in the data.

12. The National Science Foundation reports the following data on the annual number of graduating Ph.Ds in electrical engineering and solid state physics:

Year	Number
1972	1,208
1973	1,187
1974	1,027
1975	933
1976	874
1977	802
1978	706
1979	776
1980	679
1981	728
1982	777

13. Mobility is defined in terms of the movement of a researcher from the permanent employment of one organization to another. Thus, the movement of visiting researchers is not in-

cluded in the data. Nor is a change in employer caused by special circumstances such as acquisitions or divestitures. For example, AT&T researchers who moved to Bell Communications Research after deregulation in 1983 are not included in the mobility data.

14. For example, see Braun and Macdonald (1982).

15. See Olleros (1986) for a discussion of the challenges facing firms that pioneer new technologies.

16. The data do not include international collaborations. The number of researchers involved in international collaboration grew from 25 in 1976–78 to 156 in 1980–82 to 229 in 1984–86.

    The collaboration data for Japan is based on analysis of joint publications as well as an examination of a government sponsored joint research effort initiated by eight major firms, known as the Optoelectronic Joint Research Laboratory. The laboratory is operated by visiting researchers from each of the member firms. The researchers conduct joint experiments and frequently publish their results. The founding members of the laboratory are Fujitsu, Hitachi, NEC Corporation, Oki Electric Industry, Sumitomo Electric Industry, Mitsubishi Electric Corporation, Toshiba Corporation, and Matsushita Electric Industrial. Other firms joining the Laboratory include Furukawa Electric, Yokogawa Electric, Fujikawa Cable, Fuji Electric Components, Nippon Sheet Glass, and Shimatsu Seisakusho.

17. The increase in collaboration among university and industry compound semiconductor researchers in the United States comes during a period when such collaboration in general appears to be more common. However, the National Science Foundation reports that between 1973 and 1980 the number of joint publications between university and industry researchers in engineering and technology remained relatively unchanged. A separate analysis shows a slight decline in collaboration from 1981 to 1982.

18. Visiting researcher appointments is a form of collaboration between organizations. The total annual number of visiting researchers working in compound semiconductors is provided below:

Year	Number of Visiting Researchers	Year	Number of Visiting Researchers
1975	1	1981	29
1976	7	1982	35
1977	12	1983	28
1978	19	1984	38
1979	18	1985	45
1980	18	1986	52

19. There are several other individuals who have contributed to the conceptual dichotomy between increment and radical change in technology, including Utterback (1979), Gomory (1983), Abernathy and Clark (1985), Rosenbloom (1985) and Olleros (1986) as well as empirical contributions by Ettlie, Bridges and O'Keefe (1984), Dewar and Dutton (1986), and Tushman and Anderson (1986).

20. The HEMT is one of the fastest-switching transistors that has been developed. When implemented in digital functions, it can switch on and off at speeds of about ten picoseconds (i.e., 10 trillionths of a second), and when implemented in analog functions, the HEMT can operate at frequencies of about 60 gigahertz. The HEMT, which represents the third generation of compound semiconductor transistors, provides several improvements over silicon technology, including higher speed, lower power consumption, and the potential for integrating both optical and electronic functions onto a single integrated circuit. Unlike previous generations, the HEMT is a complex sandwich-like structure that consists of thin layers of undoped gallium arsenide (GaAs) and doped aluminum gallium arsenide (AlGaAs) in a field-effect transistor structure. The width of the layers is critical, in some cases a thickness of less than 60 angstroms is required. Thus, the ability to control the creation of such thin layers of material is central to the task of making the transistors. Two competing methods of fabricating thin layers of AlGaAs

and GaAs are being developed intensively: molecular beam epitaxy (MBE) and metalorganic chemical vapor deposition (MO-CVD).

21. See Solomon and Morkoc (1984), Morkoc and Solomon (1984), and Dingle (1984) for a description of the HEMT device and its historical development.

22. To simplify the exposition, the HEMT acronym is used here as an umbrella term. However, there is currently little agreement among researchers regarding a single named for the device, except for the humorous proposal of calling it MAD for many-acronymed device.

23. See Mimura and Hiyamizu (1985).

24. See Jay et al. (1985).

25. See Dingle (1984).

26. See Manasevit (1981) and Coleman and Dapkus (1985) for a description of the MO-CVD process and its historical development.

27. See Gregory (1983, 1984), Small (1980), Small and Greenlee (1986), and Moravcsik and Murugesan (1979).

28. See Irvine and Martin (1984b).

29. Biographical data on 963 researchers show that ten major universities worldwide accounted for nearly 44 percent of all the researchers with doctorates.

REFERENCES

Abernathy, W.J., and K.B. Clark. 1985. "Innovation: Mapping the Winds of Creative Destruction." *Research Policy* 14(1): 3–22.

Aitken, Hugh G.J. 1985. *The Continuous Wave: Technology and American Radio, 1900–1932.* Princeton: Princeton University.

Brannigan, Augustine, and Richard A. Wanner. 1983. "Historical Distributions of Multiple Discoveries and Theories of Scientific Change." *Social Studies of Science* 13:417–35.

Braun, E., and S. Macdonald. 1982. *Revolution in Miniature: The History and Impact of Semiconductor Electronics.* Cambridge: Cambridge University.

Cetron, Marvin J. 1970. "Forecasting Technology." In M.J. Cetron and J.D. Goldhar, eds., *The Science*

of Managing Organized Technology, Vol. 2, pp. 807–24. New York: Gordon and Breach.

Coleman, J.J., and P.D. Dapkus. 1985. "Metalorganic Chemical Vapor Deposition." In D.K. Ferry, ed., *Gallium Arsenide Technology.* Indianapolis: Sams.

———. 1980. *The Origins of the Turbojet Revolution.* Baltimore: Johns Hopkins University.

Dewar, Robert D., and Jane E. Dutton. 1986. "The Adoption of Radical and Incremental Innovations: An Empirical Analysis." *Management Science* 32(11): 1422–33.

Dingle, Raymond. 1984. "New High Speed Devices for Integrated Circuits." *IEEE Transactions on Electron Devices* ED-31: 1662–67.

Dosi, Giovanni. 1982. "Technological Paradigms and Technological Trajectories." *Research Policy* 11: 147–62.

———. 1984. *Technical Change and Industrial Transformation.* New York: St. Martin's Press.

Edge, David, and Michael Mulkay. 1976. *Astronomy Transformed: The Emergence of Radio Astronomy.* New York: Wiley.

Eto, H. 1984. "Behavior of Japanese R&D Organizations." In H. Eto and K. Matsui, eds., *R&D Management Systems in Japanese Industry.* Amsterdam: North-Holland.

Ettlie, J.E., W.P. Bridges, and R.D. O'Keefe. 1984. "Organization Strategy and Structural Differences for Radical versus Incremental Innovation." *Management Science* 30(6): 682–95.

Foster, Richard N. 1985. "Timing Technological Transitions." *Technology in Society* 7: 127–41.

Gilbert, G.N. 1977. "Referencing as Persuasion." *Social Studies of Science* 7: 113–22.

Gomory, Ralph E. 1983. "Technology Development." *Science* 220: 576–80.

Gregory, J.G. 1983. "Citation Study of a Scientific Revolution, Part 1." *Scientometrics* 5: 313–27.

———. 1984. "Citation Study of a Scientific Revolution, Part 2." *Scientometrics* 6: 307–26.

Hicks, D., B. Martin, and J. Irvine. 1986. "Bibliometric Techniques for Monitoring Performance in Technologically Oriented Research." *R&D Management* 16(3): 211–23.

Irvine, John, and Ben R. Martin. 1984a. "CERN: Past Performance and Future Prospects, Part 2." *Research Policy* 13: 247–84.

———. 1984b. *Foresight in Science—Picking the Winners.* London: Francis Pinter.

Jay, P.R., et al. 1985. "High Gain Low Noise TEG-FET Devices for 18–40 GHz Use." *IEEE GaAs Symposium,* p. 172.

Koenig, Michael E.D. 1982. "A Bibliometric Analysis of Pharmaceutical Research." *Research Policy* 12: 15–36.

Kuhn, Thomas S. 1970. *The Structure of Scientific Revolutions,* 2d ed. Chicago: University of Chicago.

Manasevit, H.M. 1981. "Recollections and Reflections of MO-CVD." *Journal of Crystal Growth* 55: 1–9.

Martin, Ben R., and John Irvine. 1983. "Assessing Basic Research: Some Partial Indicators of Scientific Progress in Radio Astronomy." *Research Policy* 12: 61–90.

———. 1984a. "CERN: Past Performance and Future Prospects, Part 1." *Research Policy* 13: 183–210.

———. 1984b. "CERN: Past Performance and Future Prospects, Part 3." *Research Policy* 13: 311–42.

Martino, Joseph P. 1972. *Technological Forecasting for Decision Making.* New York: Elsevier.

Menard, Henry W. 1971. *Science: Growth and Change.* Cambridge, Mass.: Harvard University.

Merton, Robert K. 1961. "Singletons and Multiples in Scientific Discovery." *Proceedings of the American Philosophical Society* 105(5): 470–86 (reprinted in R. Merton, ed., *The Sociology of Science,* Chicago: University of Chicago, 1973).

Mimura, Takashi, and Satoshi Hiyamizu. 1985. "The HEMT—A Fujitsu First." *Fujitsu* 36: 346–54 (original in Japanese).

Moravcsik, M.J., and P. Murugesan. 1979. "Citation Patterns in Scientific Revolutions." *Scientometrics* 1:161–69.

Morkoc, H., and P.M. Solomon. 1984. "The HEMT: A Superfast Transistor." *IEEE Spectrum* (February): 28–35.

Olleros, Francisco-Javier. 1986. "Emerging Industries and the Burnout of Pioneers." *Journal of Product Innovation Management* 3(1): 5–18.

Price, Derek J. de Solla. 1986. *Little Science, Big Science . . . and Beyond.* New York: Columbia University.

Rogers, E.M. 1982. "Information Exchange and Technological Innovation." In D. Sahal, ed., *The Transfer of and Utilization of Technical Knowledge,* pp. 105–03. Lexington, Mass.: Heath.

Rosenbloom, Richard S. 1985. "Managing Technology for the Longer Term: A Managerial Perspective." In K.B. Clark, et al., eds., *The Uneasy Alliance: Managing the Productivity-Technology Dilemma,* pp. 297–327. Boston: Harvard Business School.

Simonton, Dean Keith. 1978. "Independent Discovery in Science and Technology." *Social Studies of Science* 8: 521–32.

———. 1986. "Multiple Discovery: Some Monte Carlo Simulations and Gedanken Experiments." *Scientometrics* 9: 269–80.

Small, Henry. 1978. "Cited Documents as Concept Symbols." *Social Studies of Science* 8: 327–40.

———. 1980. "Co-citation Context Analysis and the Structure of Paradigms." *Journal of Documentation* 36: 183–96.

Small, H., and E. Greenlee. 1986. "Collagen Research in the 1970s." *Scientometrics* 10(1–2): 95–117.

Solomon, P.M., and H. Morkoc. 1984. "Modulation-Doped GaAs/AlGaAs Heterojunction Field-Effect Transistors (MODFETs), Ultra-high Speed Devices for Supercomputers." *IEEE Transactions on Electron Devices* ED-31: 1015–27.

Stigler, George J. 1951. "The Division of Labor Is Limited by the Extent of the Market." *Journal of Political Economy* 59: 185–93.

Tushman, Michael L., and Philip Anderson. 1986. "Technological Discontinuities and Organizational Environments." *Administrative Sciences Quarterly* 31: 439–65.

Utterback, James M. 1979. "The Dynamics of Product and Process Innovation in Industry." In C.T. Hill and J.M. Utterback, eds., *Technological Innovation for a Dynamic Economy,* pp. 40–65. New York: Pergamon.

Utterback, James M., and Linsu Kim. 1984. "Invasion of a Stable Business by Radical Innovation," unpublished manuscript.

Williamson, O.E. 1985. *The Economic Institutions of Capitalism.* New York: Free Press.

14

THE MANAGEMENT OF RESEARCH AND DEVELOPMENT OF A BIOLOGICAL INNOVATION

Mary K. Knudson

Vernon W. Ruttan

This chapter provides a longitudinal empirical description of the development and commercialization of hybrid wheat, a biological innovation. Although in several respects it complements the perspectives taken in other chapters to study the process of innovation, study of hybrid wheat's development entails special technological considerations and also yields some unique insights on the management of innovation. As we will show, study of this biological innovation emphasizes how advances in institutional policies and technological knowledge over time change the incentives and roles of public- and private-sector organizations engaged in the development of an innovation.

Since the late 1940s substantial resources have been denoted by both the public and private sectors to the development of commercial hybrid wheat. In the early 1960s the success of hybrid wheat development appeared highly promising. By the early 1970s this hope

had turned to skepticism. A number of private and public research organizations discontinued their commercial hybrid wheat programs. By the early 1980s there was a new wave of optimism as several major seed companies either tested or introduced hybrid wheat seed on the market, but the jury is still out as to whether hybrid wheat will become profitable or have significant market impact.

Like the other MIRP innovation studies, this study focuses on the process events that occurred in the development of hybrid wheat over time. Several dramatic changes in plant breeding have occurred in the past several decades. These changes include developing new breeding schemes, incorporating exotic germplasm in parental pools, and using genetic engineering tools in ongoing breeding programs. All have served to hasten the pace of breeding programs or to broaden the genetic diversity within a crop.

Although much research has been conducted on the factor inputs and diffusion rates of these biological innovations, such as hybrid corn, rice, and sorghum (see reviews in Griliches 1960; Rogers 1982), only limited attention has been given to actually describing the process steps and events that take place over time as firms research, develop, and commercialize these innovations. An understanding of these processes may help determine what factors facilitate and inhibit the development of biological innovations. Thus, like the other MIRP studies, this chapter hopes to contribute to an empirical understanding of the innovation process by tracing the sequence of events over time in the development of a particular innovation. We will adopt the basic MIRP framework of examining how new innovative *ideas* pertaining to hybrid wheat were developed and implemented by *people* in public- and private-sector firms who engaged in a variety of *transactions* within changing institutional and technoglocal *contexts*.

As with other product, process, and administrative innovations examined in other chapters, an understanding of the development of a biological innovation such as hybrid wheat must appreciate the unique temporal laws that govern the innovation's production process. As we will see, the rate of hybrid wheat's development is governed by biological laws that required several decades in order to move from basic research through technology development to market introduction. The hybrid wheat innovation has been following this "biological time clock" for thirty years since the late 1950s. Hence, theories and methods used to track the innovation process must be adapted to be consistent with the hybrid wheat's temporal biological law of development.

One theory that was found useful for categorizing the sequence of events in hybrid wheat's development is Usher's cumulative synthesis model (Usher 1954). This theory conceptualizes the development of a technological innovation into four general sequential stages as illiustrated in Table 14–1. Stage A involves the realization that either current objectives are not being met or new established objectives need to be met. During stage B, events that may contribute to a solution are assembled. Those actors who have the proper skills or insight to work toward a solution to the problem are brought together to contribute appropriate data and expertise. Stage C involves the breakthrough to a solution. In stage D, problems that were encountered during stage C are reviewed and the solution is modified and refined.

Ruttan (1959) showed that Usher's model generalizes to a wide variety of agricultural and technological innovations and corresponds closely to Schumpeter's model of creative destruction. Usher's model is useful for innovation management because the model points to specific periods or stages in the innovation process where managerial interventions are likely to stimulate or inhibit innovation. The R&D pathway for the most part is a highly probabilistic evolutionary process.

TABLE 14–1. *Usher's Four Stages in His Cumulative Synthesis Sequence Framework.*

Stage	Title	Definition
A	Perception of the problem	Realization that current objectives are not being met or newly established objectives need to be met
B	Setting the stage	Data or elements that contribute to a solution are assembled
C	Act of insight	Breakthrough to a solution
D	Critical revision	Problems encountered are reviewed, modified, and refined so that a solution to the problem may be realized

Usher provides a useful general framework, but a more finegrained analytical perspective, such as proposed by Van den Daele (1959), was adopted in order to qualitatively examine progressions among temporal events within each of Usher's four stages. Van den Daele proposes three necessary conditions for examining temporal patterns in sequences of events. The first condition states that a temporal sequence of events may consist of any number of subsets or stages but that these subsets must occur in an ordered progression. Each of the four stages in Usher's model is a subset and because they occur in a sequential order, Usher's model satisfies this condition.

The second condition states that a temporal sequence of events may reflect more than one subset or pathway at a given time in the ordered progression. For example, in Usher's model, more than one feasible technolgical path to develop an innovation might be pursued in a given stage. These paths diverge from each other at a point in time, and subsets of events may occur to develop and complete the progression of Usher's stages in each pathway. Any developmental progression that has more than one subset of parallel paths at a time is called a multiple progression.

Two types of multiple progressions exist. Paths may diverge when more subsets of paths emerge over time in a temporally ordered collection of events. Conversely, a convergent multiple progression exists when there is a decrease in the number of parallel paths over time. Combinations of these two types of multiple progression can occur, and a description of how multiple progressions of events diverge, proceed in parallel, or converge over time provides a useful vocabulary for making process statements about the overall developmental pattern of an innovation over time. If a developmental progression has no more than one subset of events over time, it is called a unitary progression. As Ven de Ven and Poole state in Chapter 2, most models of innovation and change in the literature are represented as unitary progressions. However, as we will see, empirical examination of events in the development of hybrid wheat clearly reflect a more complex developmental pattern that begins with a unitary progression followed by period of divergent, parallel, and convergent pathways within and between Usher's overall stages of innovation.

Van den Daele's last condition states that the events that made up each pathway may be cumulative and conjunctive. If events are cumulative, then elements found in earlier events or stages are added to and built on in subsequent events or stages. For example, ideas developed in Usher's stage setting B may be used to provide the insight for a breakthrough solution in stage C. Complete cumulation means that every event from each stage is carried from its onset until the end of the developmental progression. This, of course, seldom happens, since losses of memory, mistakes and detours, and terminated pathways all imply partially cumulative progressions. Conjunctive events are causally related events, meaning that events in one pathway may trigger or influence events in other pathways of a multiple progression. Of course, what is related at one time may be viewed as unrelated at another. Therefore, strict causality among events is difficult to establish and is seldom attempted in a process analysis. Instead, process theories normally attempt to more simply identify the necessary (not sufficient) conditions of causality among temporal events.

Thus, we will rely on Usher's cumulative synthesis theory as the framework to bracket events in hybrid wheat's development into conceptually meaningful stages of innovation, while Van den Daele provides us the analytical vocabulary and tools to qualitatively identify developmental progressions and paths among events within and between these stages.

First we will describe the methodology used to conduct this research and provide a basic introduction to the technology of hybrid seed breeding versus conventional breeding of self-pollinating crops. This is followed with a description of the historical events in the development of hybrid wheat in terms of Usher's four stages of innovation. The following section discusses developmental patterns in hybrid wheat's development and interprets the processes underlying these patterns in terms of how changes in institutional policies and technological advances changed the incentives and roles of public and private organizations over time.

METHODOLOGY

Data Collection and Tabulation Methods

Since practical considerations largely eliminated the possibility of conducting a real-time process field study of hybrid wheat's development over thirty years, historical methods were used to identify the major process events through published reports, archival records, and fifteen field interviews. Fortunately, through a "snowball" sampling process, we were able to locate, visit, and interview a sample of the key scientists, breeders, and research administrators in public and private organizations who were or are directly engaged in hybrid wheat's development. Efforts were made to minimize bias in this historical case study through a triangulation method where multiple independent data sources were obtained on most of the events in hybrid wheat's development.

Events in hybrid wheat's development were identified by determining when significant changes occurred in the basic MIRP concepts of ideas, people, transactions, context, or outcomes over time. However, unlike the MIRP innovation studies reported in Chapters 8 and 10, it was recognized at the outset of this study that it was not possible nor fruitful to undertake a finegrained analysis of the events that occurred within each of the private and public organizations engaged in hybrid wheat development over thirty years. Since the primary interest of this research was to identify the major technological paths in the development of hybrid wheat over time, and how institutional policies and technological advances influenced the incentives and roles of public versus private sector organizations, we adopted coarsegrained decision rules to operationalize the basic MIRP concepts. Changes in ideas were defined as instances when major technological advances or pathways occurred. People were defined as either individuals or firms in private or public sectors that were engaged in hybrid wheat development. Transactions represented instances when firms made significant startup investments in or divestigures from hybrid wheat R&D programs. Contextual events were defined to be instances of environmental problems (such as crop failures or blights) as well as changes in government laws and investments related to hybrid wheat. Finally, outcome events were identified when new evidence became known of the technological and economic viability of hybrid wheat products.

Following these decision rules, twenty-five major events were identified in the development of hybrid wheat, and they are presented in Table 14–2 in chronological order. Column one refers to the year these events occur. Column two refers to stages A, B, C, and D of Usher's cumulative sequential sequence model. Column three denotes the number assigned to this event. Numbers are given to events in the order to which they occurred. Column four classifies whether this event occurred within the semidwarf wheat varietal or hybrid wheat pathway. Finally, column five describes these events.

468

TABLE 14–2. *Events Occurring in the Development of Commercial Hybrid Wheat and Semidwarf Wheat Varieties.*

Year	Stage[a]	Event[b]	Case[c]	Description of Event
1948	A	E1	HYW	Success of hybrid corn
Late-1940s–early 1950s	A	E2	HYW	Outbreak of stem rust wheat 15B
1949	A	E3	HYW	Wheat acreage reduction programs coupled with increased wheat prices and lower relative input prices
Late 1940s	B	E4	SDW	Semi-dwarf gene is introduced to U.S. by S.C. Salmon
Early 1950s	B	E5	HW	Kihara and Fukusawa discover a CMS-Rf system in a closely related species of cultivated wheat
Early 1950s	C	E6	SDW	Vogel, Everson, and Borlaug develop SDW program at Washington State University
Late 1950s	D	E7	SDW	Testing of SDW varieties begin
1961	B	E8	HW	Wilson and Ross from Kansas State University find a CMS in *Triticum timopheevi*
1961	D	E9	SDW	Gaines, the first SDW in the U.S., is released
1962	C	E10	HW	Johnson and Schmidt from the University of Nebraska, and Wilson and Ross find a RFS in *T. timopheevi*
1962	C	E11	HW	Private sector becomes active and takes over role of leader in the development of commercial hybrid wheat
1968	D	E12	HW	Bozzini and Scarascia-Mugnozzo at the Laboratorio per le Applicazioni in Agricolture del C.N.E.N, Rome, Italy, find a male sterile system in wheat
1969	D	E13	SDW	SDW varieties show increases of 5–50% over standard wheat varieties
1970	C	E14	HW	The Plant Variety Protection Act passes
1970	C	E15	HW	Many public-sector actors and smaller private-sector actors drop their commercial hybrid wheat programs in favor of a semidwarf wheat program
1970	D	E16	HW	All research on male sterile systems in wheat is dropped
1970	D	E17	HW	Pollen suppressors technology enters the hybrid wheat development pathway
1978	D	E18	HW	DeKalb and Pioneer each release an experimental line of their commercial hybrid wheats (via CMS-RFS)[d]
1979	D	E19	HW	These experimental lines are pulled off the market
1979	D	E20	HW	Northrup King drops its commercial hybrid wheat program
1982	D	E21	HW	DeKalb drops its commercial hybrid wheat program
1983	D	E22	HW	Cargill's Bounty, a commercial hybrid wheat (via CMS-RFS)[d] begins to perform well in the state advanced yield trials
1987	D	E23	HW	Rohm & Haas drops their commercial hybrid wheat (via PS technology)[d] program
1987	D	E24	HW	Pioneer International drops its domestic commercial hybrid wheat program
1987	D	E25	HW	Commercial hybrid wheat development still continues using CMS-RFS and PS technologies

a. A, B, C, and D refer to stages A (perception of the problem), B (setting the stage), C (act of insight), and D (critical revision) of Usher's cumulative synthesis sequence framework.

b. These event numbers correspond to numbers referred to in the text.

c. HVW, SDW, and HW are acronyms for higher-yielding wheats, semidwarf wheat, and hybrid wheat.

d. Additional acronyms found in this table are CMS, RFS and PS, which stand for cytoplasmic male-sterility, restoration factor, and pollen suppressors.

Commercial Hybrid Breeding

Commercial hybrid breeding differs fundamentally from the conventional breeding of self-pollinated crop varieties. It involves different R&D procedures, time frames, and costs than conventional breeding schemes. In the conventional breeding scheme, the development of a new variety begins with a cross made between two genetically different inbred "pure lines." A pure line is a line that has similar genes for a trait for all its genes (or that is homozygous for all its genes). Self-fertilization or backcrossing for five to seven generations follows this initial cross-fertilization. At this point the plant reaches a genetically homozygous or uniform state.[1] It is then ready for seed multiplication for commercial sale. The total length of time required for this entire procedure—that is, from the initial cross to the last self-fertilization or backcrossing—is four to five years.

However, before this system works efficiently, much time, money, and resources are put into developing good parental and germplasm pools. These two elements are the basis of a sound breeding program because it is from these resources that a breeder makes the final product. A new germplasm pool takes an average of three years to establish. However, new gene sources are continually added to it. Inbred pure-line parents take on average six years to be developed.

Commercial hybrids are the first generation (F1) from the crossing of two parents. The parents are typically inbred pure lines (but not necessarily so) that when crossed produce "hybrid" vigor (or a heterotic effect) that results in higher yields than are obtained from conventional breeding schemes.[2] An F1 hybrid usually has dissimilar genes for many of its traits (or is heterozygous for its genes). It is this F1 that is put on the market for commercial sale.

A major difference between the conventional and hybrid schemes lies in the time it takes to develop a market product. Once the germplasm pools are completed, the time of development for the hybrid takes only one to two years (versus four to five years in the conventional scheme). This advantage assumes that the technology for producing commercial hybrids is developed. The case of hybrid wheat traces the development of this technology.

STAGES A, B, AND C IN THE RESEARCH AND DEVELOPMENT OF COMMERCIAL HYBRID WHEAT

As Usher pointed out in his cumulative synthesis sequence model, the emergence of an innovation does not begin with its research and development. Instead it begins as a demand for some new state of the art—an innovation. This demand may be created by changes in the economic environment such as in relative factor prices, factor/product price ratios, and/or factor inputs in the production process of a specific crop. However, other forces such as technological advancements in similar industries, such as hybrid corn, may also create a perception of opportunity. Hybrid wheat's emergence and development will now be presented in the following sections using Usher's model as a framework to identify the factors that contributed to its R&D. The reader may want to refer to Figure 14–1 and Table 14–2 as reference guides to this part of the text. In the text, figure and table, A, B, C, and D refer to stages A, B, C, and D, respectively, of Usher's model. Also, the event E# listed in Table 14–2 is the same event referred to in the text. By using Van den Daele's conditions, what stage each event belonged to and how each event influenced other events were determined.

Perception of the Problem (Stage A)

Toward the end of the 1940s the wheat industry witnessed a change in its breeding objectives.

470

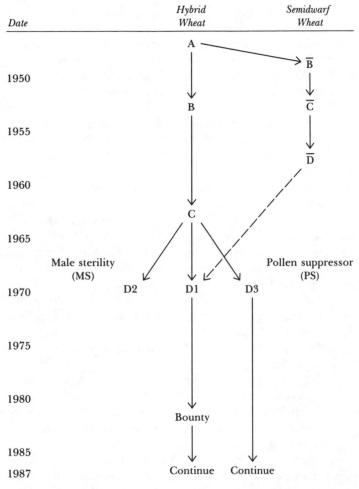

The letters A, B, C, and D refer to stages A (perception of the problem), B (setting the stage), C (act of insight), and D (critical revision), respectively, in Usher's framework.

The series A, B, C, and D1 belong to the development of commercial hybrid wheat via cytoplasmic male-sterile-restoration factor technology.

The series $\overline{B}$, $\overline{C}$, and $\overline{D}$ belong to the development of semi-dwarf wheat, which is a diverging pathway from hybrid wheat's pathway.

D2 and D3 belong to the commercial hybrid wheat development via male-sterile and pollen suppressor technology, respectively. Each of these two technologies represent the convergence of a pathway with hybrid wheat, but each remains divergent on their own pathway.

FIGURE 14–1. Usher's Cumulative Synthesis Sequence with Respect to Commercial Hybrid Wheat and Semidwarf Wheat's Development Pathways.

Primary breeding objectives changed from being primarily defensive to being mainly offensive for all five classes of wheat (see Table 14–3). Defensive breeding is breeding for drought or disease resistance or other factors designed to maintain standard yield levels and/or milling quality. Offensive breeding, on the other hand, emphasizes breeding for yield improvement. This recognition of the need to achieve higher yields set new objectives for the wheat industry and conforms to stage A, perception of the problem, in Usher's cumulative synthesis sequence framework (denoted as A in Figure 14–1).

Three events strongly influenced this change in objectives. First, the success of hybrid corn made plant breeders and farmers aware that yields could be increased more rapidly than in the past as a result of advances in biological science and in breeding technology (E1 in Table 14–2). In 1930 only 1 percent of the corn farmers grew hybrid corn, and corn yields averaged 20 bushels/acre (bu/a). By 1948, over 80 percent of the corn farmers were growing hybrid corn, and average corn yields had jumped to 40 bu/a. Other factors such as nitrogen fertilizer also contributed strongly to this yield increase (Sundquist, Menz, and Neumeyer 1982).

This increase in corn yield was one of the first major breakthroughs in increasing crop yields. Its success created a perception of a technological opportunity. That is, if technology could be used to increase corn yields, then perhaps a technology might also be developed that would increase yields in other crops. Wheat fit into this category that would benefit from a technological opportunity to increase yield as its yield had remained on a plateau of 10 bu/a from 1866 to 1940 (Dalrymple 1980).

A second development came after an outbreak of a stem rust race, 15B, that occurred in the 1950s and caused epidemic yield and quality losses to the wheat crop (E2 in Table 14–2). It was quickly realized that control measures, such as breeding for resistance to stem rust, had to be undertaken to either maintain existing yields or attain high wheat yields. If higher yields could be obtained, an outbreak of stem rust would not be as devastating. There would be more cushion to fall back on from the increased yields. This behavior is risk aversive, so the subsequent response to this even can be classified as risk induced.

Concomitant to the occurrence of these two events was a sharp reduction in land committed to wheat acreage and an increase in wheat prices (E3 in Table 14–2). The acreage allotment and crop diversion programs the federal government implemented after World War

TABLE 14–3. *Market Class Acreage, Protein Percentage, and Use for United States.*

Class	Acreage (Percentage of Total) U.S.[a]	Protein Percentage[b]	Use
Hard red winter	59%	15%	Bread
Hard red spring	13	16	Bread
Soft red winter	12	12	Pastry
White	9	11	Pastry
Durum	7	16	Macaroni

Source: Simmons (1979).
a. 1972 statistics.
b. Whole grain.

II resulted in a large reduction in the land area available for wheat production. Wheat acreage steadily dropped from 83.9 million acres in 1949 to 48.7 million acres in 1970 (Bond and Umberger 1979). Furthermore, wheat increased from $1.50 in 1940 to $3.50 and $2.50 in 1945 and 1950 (Hayami and Ruttan 1985). A profitable opportunity was created through increased wheat prices and substituting other inputs, such as new varieties, fertilizers, pesticides and irrigation, for land to increase wheat production (Binswanger and Ruttan 1978). The effect was to induce a demand for land saving innovation.

Setting the Stage (Stage B)

A new objective now existed for the wheat industry—a stronger demand for higher-yielding wheat. Because breeding for higher-yielding wheat had not been a high priority until the late 1940s, better breeding mechanisms than what existed were needed to satisfy this criterion. The pursuit of such mechanisms characterizes stage B of Usher's cumulative synthesis sequence framework.

It is not restrictive that only one mechanism to attain this goal be found. But for each mechanism found, each will continue on its own separate development pathway. This represents a divergence of the pathway for developing higher yield wheat into two separate series of fundamental studies: One opened up the possibility of hybrid wheat production (B in Figure 14–1) and the other opened up the possibility of semidwarf wheat (SDW) production (B in Figure 14–1). In this paper, we will only discuss hybrid wheat's development.[3]

The concept of hybrid wheat breeding began in the 1950s through the works of H. Kihara and H. Fukusawa of Kyoto University in Japan. Kihara found a cytoplasmic-male sterile system in an offspring from *Aegilops caudata x Triticum vulgare* in 1951. In 1953 H. Fukusawa discovered a similar system in an offspring of *Aegilops ovata x Triticum durum*. Both the *Aegilops* and *Triticum* species are closely related to cultivated conventional wheat varieties. Both Kihara and Fukusawa made their findings accidently while doing other cytogenetic analysis of these species (E5 in Table 14–2).

A cytoplasmic male-sterile system (CMS) consists of separate sterile factors in the cytoplasm and the nucleus. These two factors interact to cause male sterility. Then a plant that typically self-fertilizes is incapable of doing so if a cytoplasmic system is present. The plant can be cross-fertilized without having to hand emasculate.

Kihara and Fukusawa's interest in continuing their research on CMS was twofold. First, they were curious as to how these systems operated genetically. Second, because hybrid corn and hybrid sorghum were successfully produced by using a CMS, they thought that a similar system found in wheat might also be used successfully in producing hybrid wheat.

Unfortunately, the CMS found in the *Aegilops sp.*'s produces deleterious effects in a hybrid, such as reduced hybrid vigor, delayed maturity, pistillody, and poor germination (Sage 1976). Therefore, the CMS in these particular *Aegilops sp.*'s was of little value in a breeding program.

By the late 1950s work on finding other CMSs was under way. By 1957 some programs were set up across the United States to find and develop cytoplasmic male-sterile lines. In 1961 J.A. Wilson and W.M. Ross from the Hayes Experiment Station at Kansas State University presented findings of a new CMS that did not produce the deleterious effects in the hybrid that was yielded by the *Aegilops sp.* (E8 in Table 14–2). This new source was found in *T. timopheevi*, which is more closely related to cultivated wheat varieties than the aforementioned species are.

The pursuit of these cytoplasmic male-sterile lines was accompanied by a pursuit for a restoration factor system (RFS). A restoration

473

factor system is a genetic system carried by the male parent that restores male fertility to the resulting hybrid seed. Without this RFS, the hybrid seed would also be male-sterile. Plants from these seeds could only produce via cross-fertilization. Cross-pollination in a plant, such as wheat, does not occur at a high rate without assistance—that is, hand pollinations. These plants have evolved to rely on reproduction via self-fertilization. Grain fill is low in those plants that have a CMS but do not have a restoration factor system. It is essential then for a breeder to incorporate a RFS in the male when producing hybrids via CMS. A RFS was still needed to be found.

Within a year of Wilson and Ross's finding, V.A. Johnson and J.W. Schmidt from the University of Nebraska and Wilson and Ross separately discovered a cytoplasmic male-sterile–restoration factor system (CMS-RFS) in *Triticum timopheevi* (E10 in Table 14–2). The convergence of CMS and RFS technological paths made it feasible to produce a commercial hybrid wheat.

The Act of Insight (Stage C)

A mechanism to produce higher-yielding wheat via a hybrid technology now existed. Efforts were directed toward developing it and increasing information on it. These efforts characterize stage C of Usher's cumulative synthesis framework, the act of insight (C in Figure 14–1). Work to develop a hybrid wheat were carried out for the two wheat classes, hard red winter (HRW) and soft red winter (SRW). CMS-RFS were found only in these classes.

Both the private and public sectors were active in this stage. However, each sector had different reasons for its involvement and for its level of commitment. The private sector took over the leading role in this pursuit. Two major reasons exist for this change of hands. Recall that it was public-sector players who did most of the research in stage B, setting the stage.

First, Dekalb was aggressively investing in hybrid wheat research. Because DeKalb was the number one seed firm at the time, other firms looked to DeKalb for direction in planning their own programs. It was thought that if DeKalb was emphasizing hybrids, it presumably knew what it was doing. Other firms, including Cargill, Northrup King, and Pioneer International Hybrid, followed DeKalb's lead and developed their own hybrid wheat programs (E11 in Table 14–2).

Although DeKalb's move into hybrid wheat influenced other firms in this direction, many chose to work on hybrids rather than pure-line varieties because of the lack of legal protection for the development and marketing of pure-line varieties. Because of the genetic nature of the hybrid seed, they have their own built-in trade secret. It is highly unlikely that other firms and universities could copy a hybrid unless they know the parentage of the hybrid, and even then, they must have the inbred lines that serve as parents in their own parental pool before they can make an exact replica. Furthermore, farmers cannot use harvested grain from a hybrid for resale or replanting. It is easy to see why firms prefer producing hybrid seed varieties over pure-line varieties.

The public sector was not far behind the private sector in the pursuit of hybrid wheat development. The public sector's interest in hybrid wheat stemmed both from a need to keep pace with private firms and from a basic interest in the genetic basis of the CMS-RFSs used to produce hybrid wheat. The public institutions that initiated the most extensive hybrid wheat programs in the early 1960s were Washington State University, Kansas State University, North Dakota State University, University of Nebraska, and Texas A&M University.

Between 1962 and 1968, private- and public-sector programs conducted basic and applied research on the CMS-RFS found in wheat. More emphasis, however, was placed on basic research, which was designed to study the genetics of the CMS-RFS. Consequently, a majority of these basic experiments were quite small in scale. Applied research focused on building germplasm and parental pools for specific use in producing commercial hybrids via CMS-RFS technology.

Basic experiments brought to light many genetically based problems in using *Triticum timopheevi* as a source of CMS-RFS for hybrid production. These problems made restoring complete male fertility in the hybrid difficult. Either no restoration or incomplete restoration occurs instead of complete fertility. Without complete male fertility, grain fill in a hybrid would be less, and yields consequently would be reduced. Unless this trait could be corrected, hybrid wheat could not become a technologically successful innovation.

Many problems were encountered in developing the cytoplasmic male-sterile female parent and the restoration factor male parent. One of the major difficulties in developing these lines stems from the ploidy level of the common wheat variety. Ploidy level refers to the number of sets of chromosomes present in an organism. As the ploidy level increases, the number of genes controlling a trait increases. This complexity makes it more difficult to understand and manipulate a genetic system. A common wheat variety contains six chromosome sets as compared with corn, which is a diploid and has two chromosome sets. Consequently, parental development requires much more work in wheat than in corn. At this point of the R&D pathway, this work mainly involves developing and mastering the technology used to make the parent.

Developing the parents for commercial hybrid wheat production are further complicated by the biological construction of wheat. The parents used in developing the female cytoplasmic male-sterile parent were not closely related but were the only plants available that could produce cytoplasmic male-sterile

offspring. Deleterious effects, such as atypical floral morphology or lower yields, resulted.

Developing the male restoration factor parent was complicated by poor anther extrusion and pollen that stayed viable for only three hours. Poor anther extrusion prevents easy access to the pollen for pollen collection. Short viability time allows little time for successful cross-pollination. These traits existed to promote self-fertilization. Breeders had to undo what evolution had accomplished.

By the late 1960s these problems caused many wheat breeders to doubt that commercial hybrid wheat could ever become technologically successful. This doubt was enhanced by two concurrent events. First, the emergence and success of semidwarf wheat varieties made a readily available research alternative. The first semidwarf wheat variety, Gaines, was released in 1961. By 1969 (E13 in Table 14–2) 7 percent of the wheat acreage in the United States was devoted to semidwarf wheat varieties. Their yields were 5 to 50 percent greater than conventional varietal yields. Yield increases were due to a combination of short-stemness and heavier nitrogen application (Dalrymple 1980; Bond and Umberger 1979). Heavier nitrogen applications increase grain fill. Short-stemmed varieties could support the heavier heads of wheat plants, whereas the conventional varietal stems could not support these heads. They lodge, and subsequent yields are reduced.

A second event, the passage of the Plant Variety Protection Act (PVPA) in 1970 (E14 in Table 14–2), provided the legal protection for pure-line varieties. The PVPA gave plant breeders "(i) the exclusive right to sell or advertise and to license other persons to sell plants of the registered new variety and/or the reproductive material of those plants; (ii) the right to levy and collect royalties from persons selling or using new varieties registered under the Act" (Butler and Marion 1983). Pure-line varieties were now protected under law. Farmers could not sell their seed to anyone else as they could before the enactment of this law. But nothing prevented them from using one year's harvest for their next year's planting.

The passage of the PVPA may have added enough protection to act as an incentive for some firms to place greater emphasis on developing pure-line varieties rather than continuing their hybrid development programs. Many firms found that many of the parents that had been developed for hybrid production could compete successfully with many of the conventional commercial varieties. These parents were pure-lines, not hybrids. On passage of the PVPA, firms could release these parents as pure-line varieties without fear of losing sales because farmers were selling their seed to others.

By 1970 all public-sector actors except that of North Dakota State University elected to drop their hybrid wheat programs. Many of the smaller firms also dropped their programs (E15 in Table 14–2). The major private actors that remained to develop hybrid wheat via CMS-RFS technology were Cargill, Northrup King, DeKalb, Nickerson American Plant Breeders, and Pioneer. North Dakota State University, along with the private-sector actors, concentrated on hard red spring (HRS) and hard red winter (HRW) hybrid wheat research, respectively. All work on soft red winter (SRW) wheat for commercial hybrid production was dropped as the problem encountered in hybrid research were amplified in SRW wheat.

CRITICAL REVISION: THREE METHODS OF COMMERCIAL HYBRID SEED PRODUCTION (STAGE D)

According to Usher, in the critical revision stage, the technology worked on in the previous stage, the act of insight, is revised. This could mean that the technology is further developed and/or the new technologies that could be used to attain the goal of commercial hybrid wheat are introduced. Of course, the critical revision could also mean that any further development of the technology is terminated. This happened at the end of stage C when many of the public-sector and smaller private-sector firms decided to drop their hybrid wheat programs. What was left for the critical revision stage was the further development of CMS-RFS (D1 in Figure 14–1) and the introduction of two new technologies—male sterility (MS) and pollen suppressors (PS) (D2 and D3 in Figure 14–1)—to develop hybrid wheat. Both MS and PS were underway somewhat earlier under separate development pathways. Around 1970 it was found that each could be used in developing hybrid wheat. However, each technology's pathway still was distinct. MS and PS converged with hybrid wheat's pathway in a divergent manner. We will briefly discuss events for each competing technological pathway within the critical revision framework.

Cytoplasmic Male-Sterile–Restoration Factor Systems (CMS-RFS) Technology

The main actors in continuing to develop hybrid wheat via CMS-RFS technology were Cargill, DeKalb, Northrup King, Nickerson American Plant Breeders, Pioneer, and North Dakota State University. Their research focused on three main areas: finding new sources of CMS-RFS; building larger, more diverse parental pools; and testing for yield and quality. The first procedure, finding new sources of CMS-RFS was the focal point of these hybrid wheat programs in the 1970s. New sources were found that were easy to manipulate in a breeding program.[4] This was considered basic research.

During this same period efforts were undertaken to develop the parental pools. Existing CMS-RFS were further developed, tested, and selected for performance as a parent. This step is important as a good parental pool is a base for a sound breeding program. This was applied research.

Testing for yield and quality of hybrids for commercial release began in the mid- to late 1970s. The first hard red winter (HRW) hybrid varieties resulting from CMS-RFS technology were released to the market in 1978 by DeKalb and Pioneer International experimental lines (E18 in Table 14–2).

However, these market releases produced poorly and were found to be prematurely released for several reasons. First, these

hybrids still needed further testing to confirm their environmental stability, as they gave inconsistent yields across environments. Second, seed stock was shown to be impure, to be more susceptible to disease, and to give lower yields. In addition, not enough seed was produced to satisfy market distribution needs. These difficulties with market production prompted DeKalb and Pioneer International to pull their hybrids off the market (E19 in Table 14–2).

Unfortunately, the release of these two hybrid varieties and their subsequent recall caused significant damage to hybrid wheat's reputation. Many wheat breeders continued to be disillusioned with continual lack of success

TABLE 14–4.   *Advanced Yield Results in Kansas, 1983, 1984, and 1985.*

Name of Hybrid	Firm	Yield (bu/acre)		
		1983 16-Station Average	1984 14-Station Average	1985 16-Station Average
Hybrids from CMS-Rf:				
Bounty 100	Cargill	66.3	63	66
Bounty 201	Cargill		67.9	
Bounty 202	Cargill		65.6	66
Bounty 203	Cargill		71.3	70.3
Bounty 301	Cargill		64.4	68.9
Bounty 310	Cargill	68.3	65.6	64.4
Quantum H1260	Monsanto		58.7	
Hybrids from PS:				
Hybrex 1010	Rohm & Haas	62.1	61	
Hybrex 1019	Rohm & Haas		58.9	
Hybrex 1018	Rohm & Haas		59.8	
Top conventional varieties:				
Arkan		61.6	60.6	60.8
Hawk		59.9	56.6	59.1
Newton		56.5	57.5	56.1
Tam 105		60.8	61.8	54.8
Tam 107			67.1	57.7
Vona		61.8	56.7	55.8
Agripro			59.6	60.6

Source: Kansas Agricultural Experiment Station (1983, 1984, 1985).
Results include only nonirrigated plots.
Acronyms CMS-Rf and PS refer to hybrids produced via cytoplasmic male-sterility–restoration factor and pollen suppressor technology, respectively.

477

of commercial hybrid wheat. Consequently, many dropped their commercial hybrid wheat programs. By 1980 Northrup King had dropped all its programs (E20 in Table 14–2). In 1982 the leading seed company and first mover firm in hybrid wheat development, De-Kalb, also sold its hybrid wheat program to Monsanto after having spent an estimated $24 million on commercial hybrid wheat research (E21 in Table 14–2).[5]

Since 1981 the hope for the release of a successful hybrid variety via CMS-RFS technology has become a reality. Cargill produced several hybrids under the brand name "Bounty." 1981 marked the beginning wheat Bounty performed well in the advanced yield trials (E22 in Table 14–2). For example, in the 1984 and 1985 Kansas State Advanced Yield Trials, Bounty 203 produced an average yield of 70.8 bu/a (or a 16.6 percent advantage) versus an average yield of 60.7 bu/a by Arkan, which is considered a top conventional wheat variety (see Table 14–4). Hybrid wheats produced by CMS-RFS technology have also been released by Pioneer International, Nickerson American Plant Breeders, and Hybritech (Monsanto). Their hybrid have followed Bounty's lead by performing well in advance yield trials.

In 1987, due to the depressed farm economy, Pioneer International dropped its domestic commercial hybrid wheat program (E24 in Table 14–2). Cargill, Nickerson American Plant Breeders, Monsanto, and North Dakota State University are the remaining firms with an ongoing commercial hybrid wheat program via CMS-RFS technology in the United States (E25 in Table 14–2). Their success now rests on how well commercial hybrid wheat is accepted by the farm producer.

The Male Sterility System (MS)

While efforts in the late 1960s were being made to further advance CMS-RFS, work was also being done on a much smaller scale by these same organizations to develop a male-sterile (MS) technology that could be used to create hybrid wheat. The MS system differs from the cytoplasmic male-sterile system in that the male-sterile trait is determined by genes within the nuclear genome (versus a cytoplasmic-nuclear interaction). One needs to produce a female male-sterile for this technology to produce fertile hybrid offspring.

In the mid-1960s A. Bozzini and G.T. Scarascia-Mugnozza (1968), working at Laboratorio per le Applicazioni in Agricolture del C.N.E.N., Rome, Italy, found a simply inherited MS system in *Triticum sp.* (E12 in Table 14–2). Because this system was simply inherited (or controlled by one dominant gene), the technology could be easily managed. At this point several genetically based complications arose in using the cytoplasmic male-sterile–restoration factor systems and promising alternative technologies were sought. Ongoing research groups readily took an interest in developing the male-sterile technology, as it not only showed promise but was easily manipulated genetically. Soon after male-sterile research commenced, inherent problems with this system arose and proved to be much more severe than the problems uncovered during the CMS-RFS development. General problems in managing this genetic system stemmed from the higher ploidy levels, GxE interaction, gene instability, polygenetic system, epistasis, and modifier genes. These problems left MS-produced hybrid wheat low-yielding and more susceptible to diseases such as ergot and loose smut (Wilson 1968). Because of the severe problems associated with this technique, by the early 1970s virtually all efforts to develop this system were dropped (E16 in Table 14–2).

The Pollen Suppressor System (PS)

The third line of seed production technology development on hybrid wheat breeding was the use of gametocides, or pollen suppressors. A pollen suppressor is a growth regulator that on application is translocated to the anther locules where pollen is produced. This in turn causes the sterilization of the pollen. A wheat plant becomes male sterile without the need for (any) genetic manipulation. Hypothetically, direct lines from conventional wheat breeding programs could easily be transformed for use in a commercial hybrid wheat program.

By the time the wheat industry was witnessing all the troubles associated with CMS-RFS, this knowledge on chemical induction of male sterility via PS in plants had emerged. A chemical could possibly be developed very soon and tested for producing male-sterile wheats. At this point the private sector became active in gametocide research and in fact took the lead. In 1971 Rohm and Haas became the first actor to put effort into the R&D of hybrid wheat via PS. Shell and then Monsanto followed in 1975 and 1982 (E17 in Table 14–2).

In 1987 Rohm & Haas decided their progress was too slow and continual development on a PS was financially too risky to continue their PS hybrid wheat program. They dropped their program (E23 in Table 14–2).

Shell and Monsanto are still developing the PS technology (E25 in Table 14–2). The most recent PS are much better than any of their predecessors. Hybrid wheats produced by them are beginning to attain yield levels similar to that of the conventional varieties and hybrid wheats via genetic manipulation (Lucken 1982; Kansas Agricultural Experimental Station 1983, 1984, 1985). The only major stumbling block is that the 100 percent male sterility needed in the female parent for hybrid production has yet to be attained. Consequently, hybrid seed and self seed will be produced by the female parent after fertilization. This mixture results in nonhomogeneous crops and lower yields, neither of which is desirable by the firm or the producer.

TOWARD A COMMERCIAL SUCCESS

We have just examined the progression of hybrid wheat's R&D using the A.P. Usher framework with Van den Daele's three conditions. The Usher framework was used to classify the R&D activities into stages. Van den Deale's conditions helped to identify events that either directly or indirectly affected hybrid wheat's progress. This overall framework helped us to derive lessons about R&D in plant breeding. We can (1) identify the general steps in plant breeding research; (2) characterize incentives for public and private research; (3) differentiate between firms with respect to the type or innovational R&D they pursue; and (4) identify the external influences on the innovation. This section discusses each one of these points and ends with a brief note about lessons learned in the management of R&D leading to a biological innovation.

Stage A (perception of the problem) began in the 1940s with the increased demand for higher-yielding wheat. This demand provided a stimulus for more research to meet this demand. Up until this time, research was primarily defensive rather than directed to yield enhancement. Active research leading to higher yields was inititated in stage B (setting the stage). H. Kihara and H. Fukusawa (various interviews and surveys) serendipitously discovered CMS in wheat (event E5). Yet because of their training in cytogenetics, Kihara and Fukusawa recognized that this aberration in wheat was a CMS and was at the very least genetically interesting to investigate.

Many wheat breeders realized that this discovery meant hybrid wheat breeding was potentially feasible. Efforts were now directed toward finding CMS-RFS in wheat amenable to producing hybrid wheat. Wilson and Ross and Johnson and Schmidts' efforts were the first plant breeders to find such CMS-RFS (E8 and E10).

479

Research was very exploratory during stage B. There was no tangible product to develop. Costs were low. Not much was invested in terms of time, capital, and labor. Because of the ambiguity of this research, and because of the demand that a technology be found, the public sector was the primary actor at this stage. Public research was funded by tax dollars. The public sector was willing to devote resources to research in an area of high potential demand even if the probability of application was quite low in the short run.

The private-sector research is directed toward the same market. But the private-sector actors have a greater need for profits in order to have money available to reinvest in R&D. They can not afford to invest in very much purely exploratory R&D. Only tangible products can be sold to bring in profits. Private-sector actors play a more active role once the prospect for a more tangible product becomes apparent.

The prospect for a more tangible product became apparent on the discovery of a CMS-RFS in wheat. This marked the beginning of stage C (act of insight). Both the private sector and public sector were active at this point. However, now the private sector assumed the leading role. They had a tangible breeding technology that could provide them with a built-in trade secret. SDW technology, a competing technology (E7 and E9), did not provide them with such a guarantee. Furthermore, no legal protection for pure-bred varietal lines existed. Therefore, they did not put much effort in this area. Efforts were put in the area that could potentially bring in higher profits.

The public sector did not need to be concerned over trade secrecy. Any product they might produce was considered a public good. Their primary goal was to develop a higher-yielding wheat. They invested efforts in both SDW and hybrid wheat R&D. They were interested in hybrid wheat R&D because (1) they were not sure SDW technology would be successful and (2) if hybrid wheats proved to be very successful they did not want to be left behind. The public sector measured its success in terms of the returns that would be realized by farmers from the innovation rather than the more limited objective of profitability of the seed production enterprise.

Before any of these actors decided to invest in HW R&D, they considered the expected cost of the program. Research in stage C was still basic and relatively small in scale. Therefore, R&D costs were relatively low. An actor could invest in HW R&D without committing too many resources.

By the end of the 1960s enough information was collected concerning the likelihood of success of hybrid wheat. This information is used to decide (1) whether the technology is feasible and (b) if so, whether it is still worth the efforts to work on it (that is, whether a market still exists for this product). If the technology is not feasible, or new superior substitutes exist on the market, a firm or university would probably drop the program. In corn breeding, testing is twice as expensive as all the other R&D costs combined (Geadelmann, personnel communication, 1987). Critical evaluation of the program takes place as the stakes, via costs of production, are higher.

The success of SDW is an example of how a superior substitute appears on the market and affects the invention process of the competing technology (hybrid wheat). SDW yielded 5 to 50 percent higher than conventional varieties in the late 1960s (E13). Virtually all public-sector actors and many small private-sector actors were influenced by these yields and elected to drop their hybrid wheat programs in favor of a SDW program (E15). Actors in the private sector were also influenced by the passage of the PVPA (E14). They no longer needed to be concerned about a built-in trade secret.

Cargill, DeKalb, Nickerson American Plant Breeders, North Dakota State University,

Northrup King, and Pioneer International did not drop their programs. They questioned the technology. Because of the way in which they built up their hybrid wheat programs, however, and because they had more resources at their disposal to build up their programs, they decided not to drop their hybrid wheat programs.

From the beginning of stage C (1960), each of these actors built a strong parental pool to use in hybrid wheat development. The actors who dropped their programs had "borrowed" parents from parallel wheat programs in their respective institutions. It was more costly to build up these separate parental pools. But the genetic nature of the parents used in CMS-RFS, and the heterotic effect from the parents, called for a different set of parents than conventional programs. By building a separate parental pool they had better-suited parents available for hybrid development. The likelihood of successfully developing a competitive hybrid wheat was higher for these actors who built a separate parental pool than those actors who did not.

These same actors, those who did not drop their hybrid wheat programs in 1970, could afford to build up these parental pools. Cargill, Pioneer International, and Nickerson American Plant Breeders derive a large amount of their revenue from grain sales, corn sales, and various varietal sales, respectively. North Dakota State University has a large budget for hybrid wheat development because the university, and its experimental station believes that projects such as hybrid wheat development will help strengthen North Dakota's economy. Trio Genetics[6] could afford their program because of its profitable line of business in biogenetics.

This large amount of revenue provides each institute with dollars to reinvest or invest in the institute's R&D program without putting the future of the institute in danger. If the R&D endeavor should fail, and the actor had no resources to fall back on, current production processes and other R&D endeavors could be slowed down or cut. It is usually the larger plant breeding firms that can afford to develop risky new technologies.

Even though Cargill, DeKalb, Nickerson American Plant Breeders, North Dakota State University, Pioneer, and Northrup King had the resources to develop an extensive hybrid wheat program, someone had to decide that this program was worth investing in. In each case, a single person that was high up in the administration of the institute (the head of the department and research directors) convinced their respective funding source (the experiment station and the board of directors) of the value and feasibility of this research. These same people also personally oversaw that monies were continued to be given and increased as the R&D of their hybrid wheat program progressed.

It cannot be denied that these actors, Cargill and etc., were influenced to keep their hybrid wheat programs because they had invested many resources in it. R&D investments are irretrievable. A payoff exists only if the innovation becomes a market success. They had more to lose if they dropped their hybrid wheat programs than those who dropped their hybrid wheat programs at an earlier stage.

Roughly at the same time that many actors decided to drop their hybrid wheat programs, new actors joined the HW pathway. These new actors were from both the public and private sectors and introduced MS (E12) and PS technology (E17), respectively. Both technologies were brought in from separate development pathways (convergence). Yet the application of each technology to hybrid wheat development continued on separate pathways (divergence). MS and PS represent how invention is complemented by other inventions.

The public sector undertook the R&D of hybrid wheat technology via MS. MS in wheat was only recently discovered by 1970 and many uncertainties as to its application existed. It was soon found to be very unfeasible, so efforts in its development were dropped (E16).

Up until 1970 the public sector was also

481

the primary actor in PS development. No direct application of it yet existed. By 1970 enough work on generic gametocides had been done. Hybrid wheat development offered an opportunity to apply this technology. At this point the private sector became the only actor to develop PS for hybrid wheat development. R&D was very applied, but it was also extremely costly. Much of the cost lied with testing and meeting Environmental Protection Agency standards.

Like those actors who built a strong hybrid wheat program via CMS-RFS, the actors in developing hybrid wheat via PS had good resources available for investment. These actors were Rohm & Haas, Shell, and Monsanto. Rohm & Haas and Monsanto each derived a large revenue from successes on the chemical market. Shell received much of its revenue from its petroleum investments.

Unlike those actors using CMS-RFS, no particular individual within the firm sold the board of directors the idea of developing PS to produce hybrid wheats. PS was a relatively new technology. Administrators and scientists alike at Rohm & Haas, Shell, and Monsanto saw an opportunity of taking advantage of this new market of producing HW via PS.

Because neither Rohm & Haas or Shell had a wheat breeding program in which their PS product could be used, they had to form contractual agreements with those actors who had an ongoing wheat program and were interested in developing hybrid wheat via PS. Rohm & Haas and Shell took on different contractual approaches—that is, Rohm & Haas developed joint ventures and Shell made sale contracts. In order to avoid the problems of property rights over such an arrangement, Monsanto acquired DeKalb's hybrid wheat programs when DeKalb sold it in 1981. The path that the three firms took to sell their PS differed. Reasons for selecting their respective approaches, and the outcome each had with it, is worthy of a separate study in management and contracts.

The story of the R&D of commercial hybrid wheat is far from over. Some of the major actors suddenly dropped their programs in the mid-1980s, even though hybrid wheat now more than ever before seemed technologically feasible (E22). DeKalb sold its hybrid wheat program in 1982 to Monsanto (E21). It had made some poor decisions by expanding too fast and putting a hybrid on the market too early (E18 and E19) and consequently suffered a financial strain. It was better to sell now and save resources that may be put to better use in another R&D or production program.

Pioneer recently dropped their U.S. hybrid wheat program (E24). The poor farm financial situation made a hybrid wheat market in the United States appear much smaller. Hybrid wheat seed cost $22.50 for a 50-pound bag, versus $7 to $9 for a 60-pound bag of conventional seed. Farmers currently are emphasizing decreasing input costs in order to stabilize incomes. Rohm & Haas also dropped its PS hybrid wheat program due to financial risk of the product (E23).

The general conclusions that can be made from this study and that apply to the management of R&D are as follows:

1. Plant breeding R&D can be categorized into four categories; exploratory, basic R&D, development R&D and testing. Each category requires a different level of resources and intensity of research, usually starting low and ending high. Also, at the beginning of each category, decisions are made about whether or not one should do this research, and if so, at what intensity (that is, how many resources can be committed?).

2. Some actors followed a strategy of keeping their fingers in the pie without making a large commitment to R&D on hybrid wheat. What would have happened to them if hybrid wheat would have been easy to develop? Or if some other invention made

hybrid wheat development easier? These same reasons can be applied to keeping basic research efforts alive that have some potential but their immediate application is not apparent. What would happen to a firm or country if it did not try to keep on the frontier of R&D?

3. The public- and private-sector players have similar aspirations in developing successful products, but differences in how they view the consumer and where they get their funds for R&D cause them to make different R&D decisions.

4. Certain characteristics apply to those actors who can take on risky projects. They need to have large revenues for investment. When the endeavor represents something completely new, like in hybrid wheat development via CMS-RFS, someone with influence needs to help develop the program and see that it is properly funded.

5. New substitutes on the market, such as SDW, and new complementary technologies, such as PS, influence actors' decisions about future R&D of that innovation, such as in hybrid wheat. Invention is not a singularly isolated process.

Hopefully this study has added some insights into the R&D of a biological innovation. These insights may help provide a framework to analyze and forecast other R&D programs directed to advancing biological technology. Unfortunately, this supposition has not been tested. A test could be done by taking some other innovation, such as potatoes grown from true seed, and analyze its R&D the framework developed in this study. Did the R&D of true seed potatoes fit this framework? Did it identify variables in the management of R&D of a biological innovation? If yes, then the supposition that R&D of a biological innovation follows a specific pattern is correct. More decisionmak-

ing variables in R&D could also be identified. These variables could in turn be used to help make policy concerning R&D in the biological sciences. For now, suffice it to say, at least in the wheat industry, decisionmaking variables and R&D trends have been identified.

APPENDIX 14–A COMMERCIAL HYBRID BREEDING VIA CYTOPLASMIC MALE-STERILITY–RESTORATION FACTOR TECHNOLOGY

The following explanation demonstrates how a cytoplasmic male-sterile restoration factor system works in a diploid plant like corn. In a diploid, two sets of chromosomes (one set of genomes) are present in the plant's nucleus. Cytoplasmic male-sterility results from the interaction of recessive, nonrestorer (non-fertile) genes *(rf, rf)* in the nucleus and a sterile (S) cytoplasm.

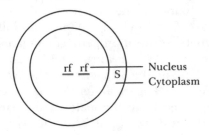

The *rf* genes are contributed by both parents, and the sterile (S) cytoplasm is from the female. Only the female transmits the cytoplasm of a cell to future generations.

Fertility is restored in subsequent generations by crossing the cytoplasmic male-sterile plant with a male plant that has the genetic restoration gene combination *(RF,____)* along with a normal cytoplasm.

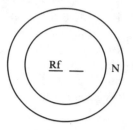

The resulting progeny will have the nucleus gene combination, *Rf rf,* which restores the male fertility system within the plant. This restoration ability is essential if the next generation is going to be able to self-fertilize and produce seed. In commercial hybrid breeding, the hybrid plant grown from the hybrid seed self-fertilizes to produce commercial seed.

In a breeding program using cytoplasmic male-sterile and restoration factor systems, A, B, and R lines are developed and maintained. A lines are female parents with *rf rf* and S. Usually the A lines are created through a series of backcrosses between an S *rf rf* female and an N *Rf rf* male in which the female denotes the desired cytoplasm and the male denotes the desired genome that will be found in the A line. All the progeny will have S cytoplasm; 50 percent will have the genetic restoration gene combination S, *rf rf,* and hence will be male-sterile. The other 50 percent of the progeny will be *Rf rf* and male-fertile. The male-fertile plants can be distinguished from the male-sterile plants because the anthers of the fertile plants are normal and those of the sterile plants are shriveled. Fertile males are discarded; the remaining progeny are crossed back (backcrossed) to the original male. Fertile males in the next generation are selected out again and discarded. This same procedure is repeated until the male-sterile progeny (S *rf rf*) have more than 90 percent of the original male parent's genome transferred into this S cytoplasm.

B lines have the genetic restoration gene combination *rf rf* and a normal cytoplasm, N. The genome of the B line is similar to the A line's genome. The main difference between the two lines is that the B line has normal cytoplasm, N, and the A line has sterile cytoplasm, S. The B line is used if any maintenance work is needed on the A line, such as increasing disease resistance. After the work is done on the B line, its genome is transferred over to an S cytoplasm through a series of backcrosses. B lines are called the maintainer lines.

The R lines have normal cytoplasm and are male-fertile (N, *Rf Rf*). They are the male parents that are used in the final hybrid crossing. They usually are selected to be used as a male parent because they offer such desirable characteristics as higher yields or genetic diversity and they can restore male fertility to the hybrid by transferring their *Rf* genes to the hybrid progeny. This system works similarly in plants with different ploidy levels than diploids.

APPENDIX 14–B. SCHEME FOR PRODUCING HYBRID WHEAT UTILIZING CYTOPLASMIC MALE STERILITY AND FERTILITY-RESTORING GENES

The male-sterile A line is maintained by pollination from the B line, which is genetically identical but is in normal cytoplasm. The hybrid seed is produced by pollinating the A line from the R line. The R line has dominant fertility-restoring genes and combines with the A line to produce a high-yielding hybrid.

FOOTNOTES

1. A homozygous condition means that the individual plant's genome contains little diversity among its genes. Subsequently, the common

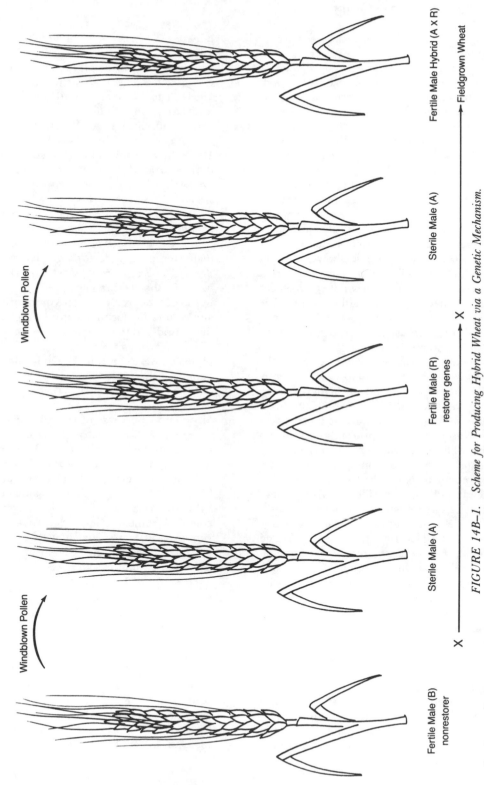

Windblown Pollen

Windblown Pollen

| Fertile Male (B) nonrestorer | Sterile Male (A) | Fertile Male (R) restorer genes | Sterile Male (A) | Fertile Male Hybrid (A x R) |

X ────────► X ────────► ────► Fieldgrown Wheat

FIGURE 14B–1. Scheme for Producing Hybrid Wheat via a Genetic Mechanism.

Adapted from: Poehlman, J.M., *Breeding Field Crops* (Westport: AVI Publishing Co. 1979), p. 181.

causal agents of genetic diversity between generations and among full-sib progeny, segregation, and recombination, do not occur at significant levels. Once this homozygous state is reached, each generation should perform similarly to each of the others. Therefore, a grower should be able to use the seed from each year's harvest for next year's planting as well as for market purposes, which means a loss in seed sales to the firms.

2. Heterosis is best defined as "increased vigor or growth of a hybrid progeny in relation" to either of the parents or to the average of the parents (Poehlman 1979). The parents' genomes must complement each other's best traits for heterosis to occur. Heterosis does not always occur in such crosses. Depending on the parents selected for the cross, they may or may not complement each other. Because of the influence of other genetic factors interacting between the parents' genomes, deleterious effects, such as lower yields, may potentially result.

3. A feasible mechanism for developing semidwarf wheat became available almost concurrently with the change in demand for higher-yielding wheat in the late 1940s. In 1946 S.C. Salmon from the USDA noticed a short-stemmed wheat variety, Norin 10, at the Morioka Branch Station in Japan. The Japanese were already growing many short-stemmed or semidwarf wheat varieties. Salmon sent Norin 10 back to the USDA research facilities at Beltsville, Maryland. Other sources of short-stemmed varieties were introduced in Beltsville within the following year— that is, Norin 16, Norin 33, Seu Suen 27, and Suweon 92. All of these plants provided a feasible mechanism by which to produce semidwarfs, which would involve crossing these sources with conventional varieties.

In the early 1950s O.A. Vogel and E.H. Everson from Washington State University began to develop a semidwarf variety, Gaines, from a cross between Norin 10 and Brevor, a U.S. wheat variety, and then crossed to two additional wheats. Breeding procedures followed a conventional scheme in which the final product is in a homozygous state. In the mid-1950s extensive varietal yield testing of Gaines commenced, and by 1961 Gaines was released as a commercial variety (Dalrymple 1980).

4. The new sources of cytoplasmic male sterility came from *Aegilops speltoides* (Lucken, in Wilson 1984); *Triticum araraticum; Triticum dilcoides* var. *nudiglumis;* and *Triticum zhukovsky* (Sasakuma and Maan, in Wilson 1984); *Aegilops kotschyi;* and *Aegilops variabilis* (Mukai and Tsunewaki 1979). New sources of restoration factor derived from *Aegilops caudata; Aegilops triaristata; Aegilops biuncialis; Aegilops columnaris; Aegilops umbellulata; Aegilops truncialis* (Mukai and Tsunewaki 1975); *Aegilops kotschyi,* and *Aegilops variabilis* (Mukai and Tsunewaki 1979).

Of all the new sources of cytoplasmic male-sterile and restoration factor systems, *Aegilops speltoides, Aegilops kotschyi,* and *Aegilops variabilis* showed the most potential for providing a new source of cytoplasmic male sterility and restoration factor and producing high-yielding hybrids. In these three species, both the sterility and fertility restoration systems were complete and were controlled by only one dominant gene. Furthermore, the cytoplasm of *Aegilops speltoides* was very similar to that of *Triticum timopheevi.*

5. Monsanto's purchase of DeKalb's hybrid wheat program indicated that Monsanto was still hopeful that hybrid wheat could become a biological and economic success. Monsanto initially was interested in purchasing the program because of its own work with gametocides and hybrids. Because Monsanto is principally a chemical company, it needed to form a joint venture with a seed company if it wanted to produce hybrids using gametocides. However, since joint ventures often end up in a debate over proprietal rights, Monsanto wanted to own its own seed company so as to avoid any such dispute. To date, Monsanto has continued developing hybrid wheat using both cytoplasmic male-sterile–restoration factor and gametocide technology.

6. On the sale of DeKalb, James Wilson, the plant breeder in charge of DeKalb's commercial hybrid wheat program, began a hybrid wheat program with Agri Genetics under the name Trio Genetics. Dr. Wilson was able to bring some of his plant material from DeKalb. This headstart,

along with Wilson's extensive knowledge about hybrid wheat, made the young program started under Trio Genetics as old as those programs begun in the early 1960s.

REFERENCES

Allard, R.W. 1960. *Principles of Plant Breeding.* New York: Wiley.

Becker, S.L. 1976. "Donald F. Jones and Hybrid Corn." Lockwood Lecture, April 9, 1976, Bulletin 763, Conneticut Agricultural Experiment Station, New Haven.

Binswanger, H.P., V.W. Ruttan, et al. 1978. *Induced Innovation. Technology, Institutions and Development.* Baltimore: Johns Hopkins University.

Bond, J.J., and D.E. Umberger. 1979. *Technical and Economic Causes of Productivity Changes in U.S. Wheat Production.* U.S. Department of Agriculture Technical Bulletin 1598.

Bozzini, A., and G.T. Scarascia-Mugnozza. 1968. "A Factor for Male Sterility Inherited as a Mendelian Recessive." *Euphytica* 17 (Supplement No. 31): 83–86.

Butler, L.J., and B.W. Marion, 1983. "The Impacts of Patent Protection on the U.S. Seed Industry and Public Plant Breeding," University of Wisconsin, mimeo.

Dalrymple, D.G. 1980. *Development and Spread of Semi-Dwarf Varieties of Wheat and Rice in the United States,* Agricultural Economic Report No. 425, USDA.

Duvick, D. 1959. "The Use of Cytoplasmic Male-Sterility in Hybrid Seed Production." *Economic Botany* 11(3): 167–95.

Geadelmann, J. 1986. Telephone interview with author.

Griliches, Z. 1960. "Hybrid Corn and the Economics of Innovation." *Science* 132 (July 29): 275–80.

Hayami, Y., and V.W. Ruttan. 1985. *Agricultural Development. An International Perspective,* 2d ed. Baltimore: Johns Hopkins University.

Kansas Agricultural Experiment Station. 1983. *Performance Tests with Winter Wheat Varieties: Report of Progress 439,* Kansas State University, Manhattan.

———. 1984. *Performance Tests with Winter Wheat Varieties: Report of Progress 459,* Kansas State University, Manhattan.

———. 1985. *Performance Tests with Winter Wheat Varieties: Report of Progress 483,* Kansas State University, Manhattan.

Kenega, C.B. 1974. *Principles of Phytopathology,* 2d ed. Lafayette: Balt.

Kirk, D. 1975. *Biology Today,* 2d ed. New York: Random House.

Lucken, K.A. 1982. "The Breeding and Production of Hybrid Wheats." Agricultural Experiment Station, North Dakota State University, mimeo.

Mukai, Y., and K. Tsunewaki. 1975. "Genetic Diversity of the Cytoplasm in *Triticum* and *Aegilops* II. Comparison of the Cytoplasms between Four $4\times$ *Aegilops* Polycides Species and the $2\times$ Relatives." *Seiken Ziho* 25–26: 67–78.

———. 1979. "Basic Studies on Hybrid Wheat Breeding VIII. A New Male Sterility. Fertility Restoration System in Common Wheat Utilizing the Cytoplasms of *Aegilops kotschyi* and *Ae. variabilis.*" *Theoretical and Applied Genetics* 54: 153–60.

Poehlman, J.M. 1979. *Breeding Field Crops.* Westport, Conn.: AVI.

Rogers, E. 1982. *Communication of Innovations.* New York: Free Press.

Ruttan, V.W. 1959. "Usher and Schumpeter on Invention, Innovation and Technological Change." *Quarterly Journal of Economics* (November): 596–606.

Sage, G.C.M. 1976. "Nucleo-Cytoplasmic Relationships in Wheat." *Advances in Agronomy* 28: 265–98.

Simmons, S. 1979. "Agronomy 3010: Adaptation, Distribution, and Production of Field Crops." University of Minnesota, St. Paul, mimeo.

Sundquist, W.B., K.M. Menz, and C.F. Neumeyer. 1982. *A Technology Assessment of Commercial Corn Production in the United States.* Statistical Bulletin 546, Agricultural Experiment Station, University of Minnesota.

Usher, A.P. 1954. *A History of Mechanical Inventions.* Cambridge, Mass.: Harvard University.

Van den Daele, L.D. 1959. "Qualitative Models in

Developmental Analysis." *Developmental Psychology* 1 (No. 4): 363–410.

Wilson, J.A. 1968. "Problems in Hybrid Wheat Breeding." *Euphytica* 17 (Supplement No. 1): 13–33.

———. 1984. "Hybrid Wheat Breeding and Commercial Seed Development." In J. Janick, ed., *Plant Breeding Reviews:* Vol. II pp. 303–19. Westport, Conn.: AVI.

15

TECHNOLOGICAL INNOVATION AND INDUSTRY EMERGENCE: THE CASE OF COCHLEAR IMPLANTS

Raghu Garud

Andrew H. Van de Ven

How do new industries emerge? What are the roles of individual firms in creating an industry? These questions not only have significant implications for national industrial policy, but they are critical to managers of technological innovations. Although most innovations represent small increments of normal change that refine and improve an established order, this chapter examines the process of commercialization of an "extraordinary innovation" that

We gratefully acknowledge useful comments on earlier drafts of this chapter from Joseph Galaskiewicz, John Mauriel, Scott Poole, Michael Rappa, Douglas Polley, Peter Ring, Vernon Ruttan, William Roering, and other colleagues involved in the Minnesota Innovation Research Program at the University of Minnesota. We also acknowledge the help of many cochlear implant industry participants for providing us with information about the industry, described here up to 1987. Support for this research program has been provided by a grant to the Strategic Management Research Center at the University of Minnesota from the Program on Organization Effectiveness, Office of Naval Research (code 442OE), under contract No. N00014-84-K-0016.

sets in motion a sequence of events that can disrupt, destroy, and make obsolete established competence or create totally new organizations and industries (Rosenbloom 1985). Such innovations have a "transilient" capacity to transform established systems of technology and markets (Abernathy and Clark 1983: 13).

Seldom can such technological innovations be developed by a single firm alone in the vacuum of a community or industrial environment. Research reviews by Mowery (1985) and Thirtle and Ruttan (1987) clearly show that the commercial success of a technological innovation is in great measure a reflection of institutional innovations that embody the social, economic, and political infrastructure that any community needs to sustain its members. Thus, the management of innovation must be concerned not only with micro developments of a particular technical device or product but also with the creation of an industry, or a macro infrastructure needed to commercialize the innovation.

In this chapter, we examine the emergence of the cochlear implant industry until 1987. The cochlear implant is a biomedical device that permits many profoundly deaf people the ability to discriminate sound by the electrical stimulation of the cochlea in the inner ear (see Figure 15–1). In granting its approval for the commercial release of the device in the United States in November 1984, the U.S. Food and Drug Administrations (FDA) announced that this was the first time that one of the five human senses had been replaced by an electronic device (Yin and Segerson 1986). Thus, the cochlear implant technology represents an extraordinary innovation that has the potential of transforming the traditional hearing-aids industry, which serves only individuals with residual hearing but not the profoundly deaf. The commercialization of cochlear implants has required the development of a totally new set of skills, knowledge, and institutional arrangements. To date, these include the development of new diagnostic and surgical procedures, otological services, trained technicians, as well as the functional competencies of R&D, manufacturing, and marketing. Commercialization of cochlear implants has also required the creation of new industry practices and regulatory procedures as well as standards of device efficacy and safety.

The chapter first introduces a new framework for viewing an industry as a social system, and an accumulation theory of change and then goes on to introduce three key questions to examine the process of industry emergence related to: (1) the sequence of development of various subsystems of an overall industry, (2) the locus of power and resources that drive industry emergence, and (3) the outcomes associated with different paths and progression of industry emergence. The next section introduces the methodology used in this longitudinal study of the cochlear implant industry. Finally, the chapter provides a qualitative description of the processes observed in the emergence of the industry and an interpretation of these processes.

CONCEPTUAL FRAMEWORK: INDUSTRY AS A SOCIAL SYSTEM

Traditionally, economists view an industry as consisting of a group of firms producing products that are close substitutes for each other. This view adopts a natural (economic) selection theory of change to explain how industry structures (ranging from perfect competition to monopoly) are transformed from within by the differential success of competing members (Kamien and Schwartz 1982:2). The analysis typically begins with a given population of organizations that are mutually susceptible to environmental vulnerability (Hannan and Freeman 1977). These firms share a "commensalistic interdependence," which implies that they engage in a competitive intraspecies struggle for economic survival because they are commonly subject to the same environmental fate (Van de Ven and Astley 1981:444). This common environmental fate includes five economic forces: entry, threat of substitution, bargaining power of buyers and suppliers, and rivalry among current competitors (Porter 1980).

But as Astley (1985) argues, by concentrating on an incremental evolutionary process of price competition among an increasingly homogeneous set of firms, this economic population ecology perspective cannot explain either the formation of entirely new industries or major extensions of old ones that are brought on by technological innovation. A punctuated equilibrium model (Tushman and Romanelli 1985) of change appears more appropriate for describing the process of creative destruction (Schumpeter 1975) and technological competition than the continuous and incremental evo-

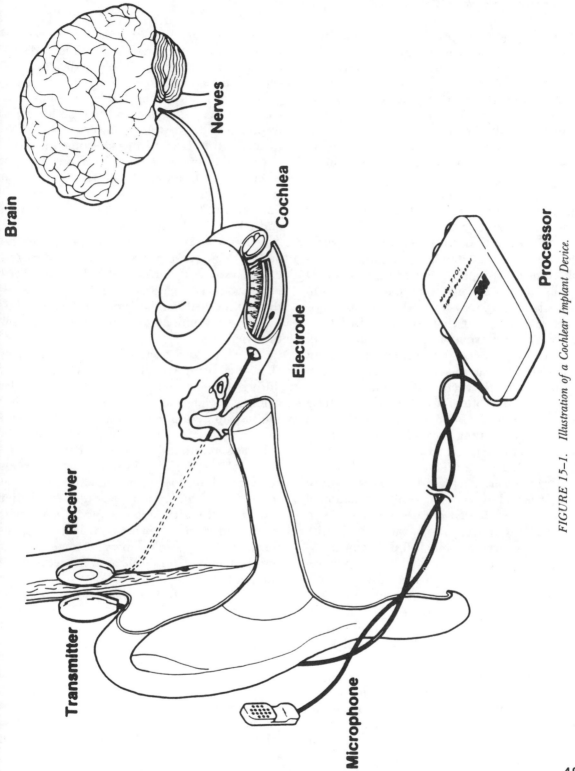

Brain

Nerves

Cochlea

Receiver

Transmitter

Electrode

Microphone

Processor

FIGURE 15–1.   *Illustration of a Cochlear Implant Device.*

491

lutionary model of price competition. In a "punctuated equilibrium" model divergent technological innovations introduce relatively short periods of discontinuous ruptures in between extended periods of convergent incremental change in the overall evolution of an industry. In other words, the random selection of a major technological variation "punctuates" extended periods of continuous equilibrium producing improvements.

For example, Piore and Sabel (1984) discuss how long periods of stability are abruptly ended by innovative breakthroughs that move industrial evolution down entirely new paths. They indicate that stability and continuity result from the adoption of a given technology because it typically entails large investments in equipment and know-how, which in turn discourage subsequent changes in industrial development. But eventually, developments with this technology become marginal, or a mutant technology appears that fosters the emergence of a new industrial sector. These extraordinary innovations are often rejected by an existing industry because of the strength and inertia built into its existing technological paradigm. Analogous to a scientific paradigm, Dosi (1982: 153) and Rappa (1987) argue that technological paradigms have a powerful exclusion effect; the efforts and the technological imagination of engineers are focused in rather precise directions while they are, so to speak, "blind" with respect to other technological possibilities. As a consequence entrepreneurs often have no recourse but to isolate themselves from existing industry branches or entire industries and "start from scratch" to create a new industry infrastructure that is necessary to commercialize their technological innovation (Dosi 1982: 154).

But an explanation of this punctuation process requires a more encompassing definition of an industry and a more finegrained theory of the punctuation process itself than the punctuated equilibrium model provides. Van de Ven and Garud (1989) propose an accumulation theory of change to explain industry emergence. As Etzioni (1963) indicates, concepts of initiation, takeoff, and startup are important for describing this accumulating change process. *Initiation* is the time when entrepreneurs decide to form a business venture (if successfully launched will become the birthday of the business unit), and *takeoff* is the time when the unit can exist without the external support of its initiators and continue growing "on its own." The period between initiation and takeoff could be called *startup,* where the new unit must draw its resources, competence, and technology from the founding leaders and external sources in order to develop the proprietary products, create a market niche, and meet the institutional standards established to legitimate the new unit as an ongoing economic enterprise.

Thus, the accumulation model views change as being stimulated by external forces during the initiation period, a transition from external to internal sources of change during the startup period, culminating at takeoff with the capability for immanent development. In comparison with the other change theories discussed above, the initiation and startup periods correspond to the "punctuation" period of technological competition, and the period after takeoff in the accumulation model converges into an incremental process of increasing refinements and evolutionary change as described by the natural selection model of price competition.

To apply this accumulation process to new industries, Van de Ven and Garud (1989) propose that an industry be viewed as a "social system" which governs, integrates, and performs all of the functions required to transform a technological innovation into a commercially viable line of products or services delivered to customers. Technological innovations stimulate discontinuous ruptures in existing industries that are centered around conventional technologies, and are replaced by clusters of new organizations that interact and isolate themselves from traditional industries by virtue of their interdependencies and growing commitments to a new technology (Astley 1985: 225). Coordination in this industry system takes place, not by a central plan or organizational hierarchy but partly through the price mechanism and mostly through interactions within relationships among industry participants (Mattsson 1986). As this cluster of organizations grows in number, a complex network of cooperative and competitive relationships evolves. This new "industry" takes on the form of a hierarchical* loosely joined system, composed of a number of subsystems, with each subsystem performing a limited range of specialized functions. Links between subsystems are only as rich or tight as is necessary to ensure the survival of the system (Aldrich and Whetten 1981: 388).

Theorists have often noted that collectivities are hierarchically stratified, with different subsystems specializing in different functions—technical instrumental functions, coordination and procurement functions, and institutional legitimation and governance (Aldrich and Whetten 1981). Although these functional subsystems are highly related, we distinguish them here to provide analytical guidance for investigating the emergence of an industry.

Instrumental Subsystem. The focus here is on the proprietary activities performed by individual firms engaged in developing and commer-

cializing products and services (Porter 1985). It includes functions such as applied R&D, manufacturing and assembly, and marketing and distribution. As Williamson's (1975) transactions costs theory suggests, some of these instrumental functions are performed within competing firms, while some are licensed or contracted by firms with outside suppliers and vendors. From a systems perspective, these make or buy decisions by individual firms produce the aggregate industry channels of raw materials, manufacturing, marketing and distribution flows (Stern and El-Ansery 1982).

Resource Procurement Subsystem. This includes the basic resources necessary to support proprietary instrumental activities. Akin to the classical economic resources of land, labor, and capital, three basic kinds of resources are critical to the emergence of most every industry: basic scientific or technological knowledge, financing, and a pool of competent human resources (Mowery and Rosenberg 1979). Separate organizations often exist to provide these necessary resources for a given industry. However, these financial, educational, and research organizations are seldom easily accessible to a new industry that is emerging to commercialize an innovation.

Institutional Subsystem. The ultimate authorities governing collective action are the legal norms of the society in which organizations function (Galaskiewicz 1985). The political context is the place for formally institutionalizing and legitimating a social system, which permits it to operate and gain access to the resources it needs (Pfeffer and Salancik 1978: 214). Thus, institutional functions include (1) establishing governance structures and procedures for the overall industry and (2) legitimizing and supporting the industry's domain in relation to other industrial, social, and political systems.

---

*Of course, hierarchy in an industry system is a matter of degree, and some industry systems may be only minimally if at all hierarchical. Hierarchy is often a consequence of institutional constraints imposed by political and governmental regulatory bodies. Hierarchy also emerges in relationships with key linking-pin organizations who either become dominant industry leaders or control access to critical resources (money, competence, technology) needed by other firms in the industry.

Few would disagree that all these system functions are essential to the emergence and maintenance of an industry. As is addressed below, from a macro viewpoint, to study industry emergence is to study how and when the three functional subsystems are organized, what firms perform these functions, and how resources flow among the firms performing different functions. Seldom can or does a single firm perform all these functions. Thus, from an individual firm viewpoint, three key decisions are made: (1) What functions will the firm perform; (2) what other organizations should the firm contract with to have the other functions performed; and consequently, (3) what organizations will the firm compete with on certain functions and cooperate with on others?

Inherent in making these decisions is the paradox of cooperation and competition. Each firm competes to establish its distinctive position in the industry; at the same time, firms must cooperate to establish the infrastructure required for the entire industry and firms within it to survive collectively. For example, it clearly benefits all firms to cooperate to set up industry standards. However, in doing so, each firm will try to ensure that standards that suit it best get institutionalized. An understanding of this paradox can offer valuable insights on how firms learn to cooperate to sustain themselves collectively, while at the same time compete to carve out their distinctive position in an emerging industry.

PROCESS QUESTIONS TO EXAMINE INDUSTRY EMERGENCE

With this framework of an industry as a social system emerging by a process of accumulation over time, we focus here on three critical issues. These three process issues have been studied in varying degrees by different disciplines; but

heretofore have been treated as "externalities" (Porter 1980), rather than central endogenous processes in industry emergence as they are treated with our social system perspective.

Sequence of Subsystem Development. First, we examine the sequence of activities that occurred in the development of the three subsystem functions during the initiation, startup, and takeoff of an industry. It is often recognized that resource endowments precede the development of instrumental and institutional activities of an industry, because basic research (which is the search for a fundamental understanding of natural phenomena) provides the foundation of knowledge that makes possible the commercial birth of a technology (Abernathy 1978; Rosenberg 1982; Mowery 1985). What is less well understood is the process by which a common pool of basic scientific or technological knowledge is transformed into proprietary innovations that can become commercial monopolies. Indeed, as Stobaugh (1985:107) points out, success at creating a monopoly by commercializing a new technology does not rest on a unique grasp of basic research, but rather on the completion of a long, complex, and uncertain journey that amounts to being an interactive search process involving large amounts of "backing and forthing" between technology and market conditions in efforts to reduce uncertainty.

Locus of Power. Next, we examine the sources of resources, influence, and constraints that directed the developmental sequence of subsystem activities in particular directions over time. Thirtle and Ruttan (1987) review the longstanding debate among economists about whether the source of technical change has been driven primarily by autonomous advances in science and technology (supply or "technology push" theories) or by economic forces ("demand pull" theories). They conclude that the extensive research on this question to date

494

has been inconclusive. Our industry framework provides a way to reconcile the debate by proposing that both models are likely at play in the emergence of an industry; "technology push" is likely to be the dominant force stimulating resource endowments activities, while "demand pull" is the major influence directing proprietary instrumental activities.

The Outcomes Associated with Different Paths and Progressions. Given that technological innovation and industry creation are inherently uncertain, one should expect many trials and errors, as witnessed by multiple divergent paths or competing technological trajectories among firms (Nelson and Winter 1977), with some leading to dead ends, stalemates, and terminations over time. Moreover, an examination of outcomes leads one to examine if and how learning occurs from previous paths and dead ends that provide guidance as to the next paths taken to create the business. By examining the outcomes of alternative paths, we hope to identify the feasible sets of paths available in the emergence of an industry. While we should expect great inter–industry differences (Mowery 1985), our study focuses on only one industry. Only by cumulative studies of these questions can we come to appreciate the endogenous dynamics of technological innovation and industry creation (Dosi 1982).

RESEARCH SETTING AND METHODOLOGY

These process questions are being examined empirically at two levels of analysis through an intensive real-time longitudinal study (begun in 1983) of the emergence of the cochlear implant industry: (1) a detailed investigation of the innovation process within an individual firm and (2) a population-level investigation of the activities of key organizations (firms, government

agencies, the medical community, etc.) in the development of the industry. Thus, our observations are about events and activities in the industry as a whole, as well as the behavior of one of the leading firms in this emerging industry.

Data on the emergence of the cochlear implant industry is being collected with multiple methods and from multiple sources in order to triangulate (Yin 1982: 50) on the major events in industry emergence over time. These sources include (1) historical cases supplemented with an extensive collection of baseline data based on interviews and archival information, and (2) field work involving attendance at trade conferences, reviews of trade literature, regular observations at management meetings of one of the firms involved in this innovation, as well as the administration of standardized questionnaires and interviews every six months with key people involved in the innovation of this firm.

The multiple data sources were content analyzed to develop a chronological list of events in the emergence of the industry. The decision rule used for designating an event was whenever a change was observed in the institutional, resource endowments, and proprietary instrumental activities related to cochlear implants. Like a motion picture, the events represent snapshots or frames, which when run together, show the progression of functional activities in the emergence of the cochlear implant industry over time.

In order to organize these events into a format conducive to processual analysis, they were next coded into three conceptual tracks consistent with the social system view of an industry. Definitions of these tracks, along with the categories developed for coding data on each track, are presented below.

1. *Institutional track.* Any activity pertaining to the establishment of governance structures and procedures for the overall industry and

which legitimizes and supports the domain of the industry in relation to to other industries. Three categories are coded in the institutional track:

A1: *Standards and rules.* Includes all efforts to establish norms of behavior, and procedures for industry participants;

A2: *Legitimation.* Includes activities undertaken to publicize, legitimize, and support the domain of the industry with respect to other industries;

A3: *Regulatory affairs.* Includes all activities involved with governmental regulatory bodies (FDA) in the setting down of uniform protocols and standards that must be met for a medical product to be approved by these regulatory bodies.

2. *Resource procurement track.* Activities that relate to the creation of fundamental knowledge and skills needed by all participants in the industry, as well as integration of instrumental activities among industry participants. Three categories are coded in this track:

B1: *Finance.* Procurement of finances both in the form of capital for large investments, as well as financing of individual implants;

B2: *Competency training.* Includes seminars, conferences, literature developed and disseminated to build and sustain required competences in large numbers of people in the industry;

B3: *Basic science.* Creation of basic knowledge that, though not applicable immediately, forms the building blocks on which the products and services are developed by individual firms.

3. *Instrumental track.* The activities performed by individual firms engaged in direct productive value creating work in the industry. The value creating work includes:

C1: *Initiation.* Includes the entry, exit of firms as well as other organizational activities such as the linkage of two or more organizations in the industry to initiate a program or to develop a specific product;

C2: *R&D.* Translates the basic technologies to knowledge that can be used to manufacture products and services for use by customers;

C3: *Clinical trials.* Includes formal experimentation to find out whether a treatment or product is therapeutically beneficial and achieves its purported claims;

C4: *Regulatory review.* Includes all activities involved in making application to and reviewing proposals with the FDA for approval and commercial release of medical products;

C5: *Manufacturing and assembly.* All activities involved in converting raw materials and "applied" knowledge into goods and services that a customer can readily use;

C6: *Marketing and distribution.* All activities related to the promotion, advertising, sales and distribution of product and services to customers;

C7: *Diagnostics.* Activities, products, and services related to determining the most appropriate treatment, if any, that a patient will benefit from most;

C8: *Rehabilitation.* Activities that enable patients to learn to use a treatment or device in the most effective way possible.

Furthermore, individual organizations involved in the occurrence of each event have also been identified.

In addition to these qualitative events, longitudinal quantitative data were obtained on other important measures of technological and industry development over time, including dates the U.S. Food and Drug Administration granted its approval for carrying out clinical investigations and for commercial sale of the device; dates, dollar amounts, and recipients of all research grants and contracts by the National Institute of Health (NIH) to conduct research on areas related to cochlear implants; citations of technical publications on cochlear implants in refereed journals; dates and recipients of all patents awarded related to cochlear implants; training programs organized by firms in the industry; and number of industry conferences held over time.

DESCRIPTION OF COCHLEAR IMPLANT INDUSTRY EMERGENCE

Following the procedures described above, a chronological list of 189 events were identified as having occurred in the development of the cochlear implant industry since the first report of a sensation of hearing due to the electrical stimulation of the acoustic nerve in 1935 until the beginning of 1987. These are summarized in Appendix 15–A. Based on this we will describe the progression of events within each of the three subsystems.

Resources Procurement Events

Basic Science. The concept of using electricity to bring hearing to the deaf goes back almost 200 years ago, when an Italian scientist, Volta, first observed the effects of electrical stimulation of the ear (ASHA 1985). More recently, experiments involving such stimulations were conducted by French researchers in 1957. The first cochlear implant in the United States was not performed until 1961 by a clinical physician, William House, founder of the House Ear Institute in Los Angeles.

Three "controversies" marked the de-velopment of basic technologies for cochlear implants. The first occurred in the late 1950s when Dr. House developed a new surgical approach, known as the mastoid facial recess, which made access to the inner ear feasible. This approach was particularly upsetting to many neurosurgeons because the surgical procedure was reported to be an invasion of their domain when it was used to approach acoustic tumors. In spite of this peer-group pressure, House continued to develop a more sophisticated cochlear implant device.

The second controversy again involved Dr. House, who in the mid-1970s was censured by colleagues in his professional association for continued development of a single-channel device in preference to a multichannel device. Many otologists believed that once a single-channel device was implanted, the ear would be unsuitable for a multiple-channel system, which was thought at the time to be technically superior to a single-channel system. This delayed commercial introduction of cochlear implants because those developing multiple-channel devices were purported to be on the verge of a major breakthrough, which did not materialize until several years later.

More recently, in 1986, the U.S. Office of Health Technology Assessment (OHTA) published a report that is surfacing a third controversy in the technical development of cochlear implants (Health Technology Assessment Report 1986). The report states that "the implantation of cochlear devices in children remains the most controversial subject in this technology as of date" as the potential for damage of the child's cochlea is very high during cochlear device implantation. In a status note on cochlear implant dated September 1986, the FDA too stated that they have limited research on cochlear implants for children because of concerns that "the procedure may damage the child's cochlea, thereby eliminating the patient from consideration for future cochlear implants with improved technology." Dr. House

497

again is the pioneer in this area and is thus a center of this third controversy.

Since the House censure in the mid-1970s, considerable research has been conducted in areas connected with the electrical stimulation of the inner ear and cochlear implants. This is illustrated in Figures 15–2 and 15–3, which plot the number of publications of technical articles on cochlear implants in leading technical journals, and patents awarded over time. This basic research has led to a proliferation of cochlear implants technologies now available in four, eight, and even twenty-two channels. Furthermore, cochlear implants can be extracochlear, which means that the electrodes do not enter the cochlea, or intracochlear; percutaneous plug, where the inner ear is accessed by a direct conduit through the cranium, or intracaneous, where the access to the inner ear is made by means of magnetic couplings. These alternative technological paths have been articulated by industry participants in a survey conducted by the American Speech and Hearing Association (ASHA 1985).

However, technological and scientific uncertainties, as well as an absence of common criteria for comparison, have made it difficult until very recently to compare the safety and efficacy of alternate technologies (Health Technology Assessment Report 1986). Instead, there have been reports of exaggerated claims made by different firms of the superiority of their device (Windmill et al. 1987). As long as technology is indefinite, it is easy to claim one has a new and better approach. Nevertheless, as the OHTA report states, the presence of such diverse strategies for processing speech is not necessarily a disadvantage, as this affords a measure of clinical flexibility that allows the selection of the most appropriate cochlear device for a particular patient.

Finance. Although comprehensive data on the funds provided by public and private organizations for the development of basic knowledge is not available, publicly available information regarding the contracts entered into and the grants awarded by the National Institute of Health has been collected and are presented as Table 15–1. As this indicates, the NIH provided a total of $29 million between 1970 and 1987 for conducting research on areas connected with cochlear implants.

For developing proprietary products and competences, the 3M cochlear implant program generated its resources solely from within the 3M corporation. In contrast, Nucleus, the other leading firm in this industry has raised resources both internally and from outside sources including private investors (See the chronological list of events in Appendix 15–A for more details.)

One other industrywide financing arrangement unique to biomedical products such as cochlear implants is the provision of third-party reimbursement payment for the cost of the device and associated surgical expenses. For cochlear implants, these expenses are of the order of $20,000 per patient. Radcliffe (1984) reported that the high cost of the device and associated implantation costs implied that the cochlear implant was not automatically available to all deaf patients inspite of FDA's approval for commercial release of the device (Radcliffe 1984). Radcliffe also reported that "third party payors were capricious in their coverage of implant costs, with payment policies varying from state to state and from patient to patient."

Symbion in 1983 was first to initiate efforts to convince third-party insurance payors to extend coverage to cochlear implants. Other firms too sought coverage for their cochlear devices. In 1983, 3M was successful in getting coverage for its first single-channel cochlear device. In December 1986, 3M, Symbion, and Nucleus were jointly successful in obtaining a wider coverage for cochlear implants from third-party payors as well as from Medicare.

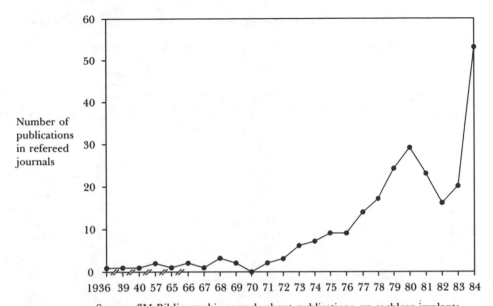

Source: 3M Bibliographic records about publications on cochlear implants.

FIGURE 15–2.   *Publications in Refereed Journals on the Clinical Development of Cochlear Implants.*

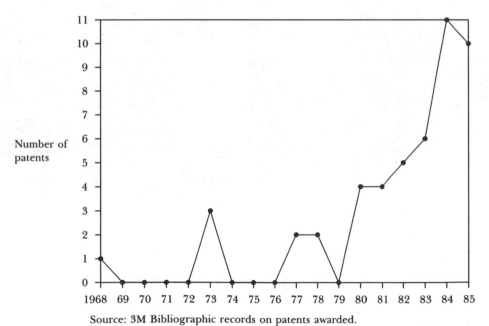

Source: 3M Bibliographic records on patents awarded.

FIGURE 15–3.   *Number of Patents Awarded for Cochlear Implant Research.*

499

TABLE 15–1.   National Institute of Health Contracts and Grants for Research on Cochlear Implants.

| Research Institute | Contract | | Duration | | Grant Amount ($m) |
	Year	Amount ($m)	Years	Months	
Stanford	1973				0.15
	1975	0.7	3		0.16
	1976				0.17
	1977	0.4	3		0.16
	1978				0.17
	1979				0.30
	1980	0.6	3		0.21
	1981				0.23
	1982				0.19
	1983	0.8	3		0.27
	1984				0.30
	1985	0.6	3		0.37
	1986	___			0.36
		3.1			3.0
University of Melbourne	1985	0.2	3		
	1987	1.1	5		
		1.3			
Research Triangle Institute	1983	0.4	3		
	1985	0.8	3		
		1.2			
University of California, San Francisco	1974				0.05
	1975				0.06
	1976				0.05
	1977				0.21
	1978				0.22
	1979				0.27
	1980				0.27
	1981				0.27
	1982				0.35
	1983	0.7	3	3	0.38
	1984	___			0.42
		0.7			2.6
EIC	1979	0.3	3		
	1981	0.5	3		
	1982	0.5	3		
	1984	0.6	3		
	1985	0.7	3		
		2.6			
Huntington	1970	1.2	9	8	

TABLE 15–1. *(continued)*

| Research Institute | Contract | | Duration | | Grant Amount ($m) |
	Year	Amount ($m)	Years	Months	
	1983	1.1	3	9	
		3.0			
Giner	1982	0.5	3		
ETC	1980	0.8	6	7	
University of Missouri	1978	0.5	3		
	1981	0.7	3		
	1984	0.6	3		
		1.8			
City University of London	1978	0.1	3		
Hughes Aircraft	1981	0.3	3		
	1985	0.6	3		
		0.9			
University of Michigan					
Miller	1979				2.34
	1984				0.79
	1985				0.75
	1986				0.59
Clopton	1983				0.08
	1984				0.09
	1985				0.09
					5.7
University of Iowa	1985				0.56
	1986				0.61
					1.17
City University of New York Graduate School	1985				0.22
	1986				0.24
					0.46
Virginia Mason Research Center	1986				0.21
Total		$16.0m			$13.0m

Source: Telephonic discussions with Dr. Hambrecht of NIH (Dec. 1986) and letters from Dr. Elkins of NIH (Jan. 20, 1987, March 24, 1987).

Competency Training. Activities directed at building personnel competencies on an industrywide basis have begun only in the last five years. There have been a number of conferences (see Figure 15–4) and technical publications (see Figure 15–2), as well as the initiation of training programs for physicians on a regular basis (see Figure 15–5). These activities have resulted in a wider appreciation of cochlear implant related skills and knowledge in the medical community.

Instrumental-Subsystem Activities

Initiation. Private firms did not become actively involved in cochlear implant development until late 1970s when corporations such

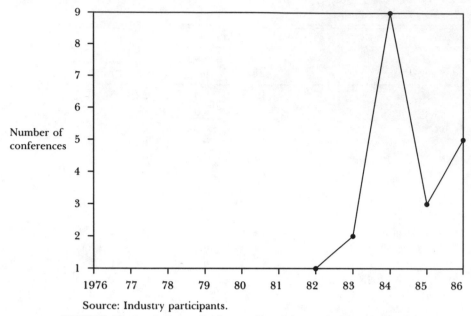

Source: Industry participants.

FIGURE 15–4.   Industry Conferences Pertaining to Cochlear Implants.

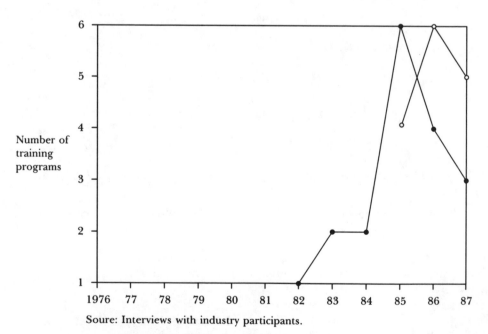

Soure: Interviews with industry participants.

● = training programs conducted by 3M.

○ = training programs conducted by Cochlear Corporation.

FIGURE 15–5.   Training Programs Conducted by 3M and Cochlear Corporation.

TABLE 15–2.   *Entry/Exit of Firms in the Cochlear Implant Industry and Their Linkages with Other Organizations.*

Year	Entry	Exit	Linkages
1977			3M and University of Melbourne explore possible linkage that does not work out
1978	Nucleus		Nucleus links up with University of Melbourne
1979	3M		3M-House begin cooperative R&D
1980			3M-UCSF work on multichannel device.
1981			Agreement between 3M and Hochmiars; formal agreement between 3M and House
1982			UCSF links up with Duke University to do patient testing; UCSF links up with RTI; UCSF-3M relationship terminated
1983	Storz, Symbion		UCSF-Storz link up; Symbion begins partnership with University of Utah
	Biostem		Biostem begins partnership with Simmons and White of Stanford; Biostem-Richards enter into a distribution agreement
1984			Cochlear corporation formed to commercialize cochlear implants
1985		Biostem	
1986		Storz	Storz first explores possible relationship with 3M and American Cynemide

Source: Chronological list of events.

as 3M, Storz, Symbion, and Nucleus initiated proprietary R&D activities to develop new businesses in cochlear implants (Table 15–2). In the case of 3M, the commitment of capital and resources to a cochlear implant program occurred in stages, starting with a preliminary exploration of the business potential and the evaluation of the technology to the formal creation of a separate cochlear implant program in 1980.

There appears to have been a period of trial and negotiations between 1978 to 1982 as firms and academicians attempted to enter into licensing agreements. A few events that occurred with 3M is evidence of this. In 1977, 3M was approached by the University of Melbourne in Australia to commercialize its cochlear implant technology with 3M. After a brief period of negotiations, the relationship between the University of Melbourne and 3M was terminated. Eventually, the University of Melbourne entered into an agreement with a new Australian firm, Nucleus, which is now 3M's major competitor. Between 1978 and 1982, 3M worked

with Dr. Robin Michelson of the University of California San Francisco (UCSF) to develop a multiple-channel implant. Through this 3M-UCSF relationship, a cochlear device was developed and implanted in two or three individuals during 1980 and 1981. On termination of this agreement in 1982, UCSF went on to license its technology with another new business startup, Storz in 1983, while 3M entered into licensing agreements with the House Ear Institute and with the Hochmiars in Vienna, Austria in 1981. Two other firms in this industry—Symbion and Biostem—entered into relationships with cochlear implant research programs underway at the University of Utah and at Stanford University, respectively, both in 1983.

R&D and Clinical Engineering.   One significant event in the emergence of this industry has been the FDA approval of the 3M-House device in November 1984. This approval was preceded by over three years of R&D and clinical trials. In October 1985 the FDA granted its approval to Nucleus to commercially market its twenty-two-channel device in the United States. Beyond these two devices, there are a number of other devices being developed by different organizations (Table 15–3; ASHA 1985). On September 1986 the FDA reported that six centers in the US were carrying out R&D and clinical trials of ten different cochlear devices after having sought "investigational device exemptions" (IDEs) from them. 3M, for instance, is in the process of developing three other devices— one an extension of the House device to be implanted in children; another an advanced single-channel device in collaboration with the Hochmiars of Austria; and third, an in-house effort to develop an advanced multichannel device. Similarly, Nucleus is in the process of developing an extension of its FDA approved twenty-two-channel device approved for children, and a four-channel device for adults.

Marketing. The FDA's approval of the 3M-House device in November 1984 and of the

Nucleus device in October 1985 marked the initial dates that cochlear implants could be sold on a commercial basis in the U.S. Although estimates indicate a current patient base of 125,000 numbers representing a potential billion dollar industry (Health Technology Assessment Report 1986), patient access has not been easy in spite of the availability of commercially salable devices. An industry informant explained that one reason for this was that patients had become accustomed to a world of deafness and may fear the risk of entering the world of sound. Furthermore, there has been little, if any, marketing infrastructure in place to access these patients. To compensate for this, in 1985, 3M launched a campaign to encourage physicians to promote cochlear implants in social and other forums.

The marketing of cochlear implants has proved to be all the more difficult because of the high technological uncertainty regarding the safety and efficacy of the device (Health

TABLE 15–3. *R&D/Product Initiatives to Develop Cochlear Devices by Academicians/Firms.*

Research Institute	1975	1978	1979	1980	1981	1982	1983	1984	1985	1986
National:										
HEI		Sigma								On the head
3M					Alpha		Vienna EC			Vienna Sprint
Nucleus				22-channel				Advanced Processor	Single channel	Mini
UCSF		Michelson		3M/ Michelson	Adv. 4 Chan.					
Symbion				Perc. plug						
Stanford/		Simmons Modular		Res.			Pulsatile			
Biostem				Pulsatile						
Foreign:										
Hochmiar	8-channel			4-channel 1-channel						
Douek						Promontory stimulator				
Banfai					8-channel	Extra-Cochlear				
Chourad		12-channel								
Portmann							1-channel			

Source: Industry participants and published documents.

Technology Assessment Report 1986). This has posed a marketing dilemma to the firms in the industry (Windmill et al. 1987). On the one hand, business firms need to promote their product in as favorable a light as possible. On the other, as researchers are still in the process of understanding this technology, it becomes important to project a realistic picture and not make exaggerated claims. At the Thirteenth International Otorhynolaryngology Conference, held in 1985, representatives of each firm that the researchers spoke to stated that others were making exaggerated claims about their cochlear devices. To the extent that these claims are indeed exaggerated, it made marketing activities all the more difficult because advertising claims are not judged to be credible.

Related to this is the fact that the implanted electrodes cannot be replaced easily without significant risk of damaging the cochlea (Health Technology Assessment Report 1986). This implies that patients once implanted with a cochlear implant may not be able to take advantage of improvements in electrode technologies. It is for this reason that the FDA reported that it was limiting research on children because of concerns that the procedure may damage the cochlea, thereby eliminating the patient from consideration for future cochlear implants with improved technologies.

By mid-1986 there were many indications to suggest that the cochlear implant market was not growing as fast as had been earlier anticipated by firms in the industry. First, in 1986 market reports indicated that the sales of Nucleus's device was not "taking off" in spite of FDA's approval. Second, Storz, informed 3M that it was considering reducing its financial commitments to its cochlear implant efforts. Storz was reported to perceive the market for cochlear implants growing at a much slower pace than they had earlier anticipated. Third, Symbion sought other partners in June 1986 for similar reasons. In October 1986, 3M too decided to shift its focus from cochlear im-

plants to the development of advanced hearing aids in the short term. However, 3M reported plans to continue long-term developmental efforts of cochlear implants.

An industry participant offered two reasons for the poor sales takeoff of cochlear devices. First, the level of technology has not yet reached a stage where true speech discrimination has been made possible with cochlear implants. Second, Medicare coverage has only recently (December 1986) been extended to cover device and implantation costs, thereby enabling many senior hearing impaired access to this new technology.

In October 1985, following FDA product recall guidelines, 3M voluntarily recalled its 3M-House device from the market due to technical defects. At 3M's initiative, the product recall prompted a number of competing firms to voluntarily join together and discuss ways to reduce the negative impact to the industry of the recall.

Diagnostics. Diagnosing specific hearing defects is a particularly important activity before surgically implanting a cochlear device because it is important to determine whether a patient will benefit from an implant (Health Technology Assessment Report 1986). In 1980 Owens and his colleagues at the University of California San Francisco developed an auditory measuring instrument for the evaluation of profound deafness called the "minimum auditory capability" battery (Owens, Kessler, and Schubert 1982). However, it was reported that different firms were using different heuristics or criteria for patient selection and evaluation. Simmons (1985) stated that there was no way to be certain about the presence of nerve fiber for stimulation, regardless of the etiology of deafness.

In 1985, 3M acquired a firm specializing in diagnostics with plans to develop this activity. However, finding that it was not commercially viable to support this activity,

3M divested from diagnostics by the end of 1986.

Rehabilitation. Another activity that was recommended by industry audiologists is postsurgical rehabilitation, which requires the skills of audiologists or speech/language pathologist, psychologists, and otologists (Health Technology Assessment Report 1986). To appreciate the importance of rehabilitation, we quote Richard Laurie, one of the cochlear implant recipients who was profoundly deaf before the implant. At a 1985 professional conference Laurie stated,

> We deaf do not hear the same way as people with normal hearing do. However, each sound evokes something specific in our minds, and consequently we can "hear" intelligibly. To accomplish this, we need all the help from our family, friends and physicians to use the cochlear implant in an effective manner.

This unique sociopsychological aspect of cochlear implants makes rehabilitation very important. In 1985, 3M initiated an experimental rehabilitation center but closed the operations in 1986 as it could not generate the required level of funding. Surgeons interviewed at the 1985 international conferences reported that they had taken the initiative to form rehabilitation centers in order to provide the needed industry downstream service activity for implanted patients. Now, in 1987, 3M has once again initiated rehabilitation activities, as have other firms. It was reported that researchers and clinicians have begun to realize the importance of rehabilitation to achieve patient performance.

Institutional Subsystem Activities

Legitimation. Hearing aids and vibrotactile devices have historically been the nearest substitute to cochlear implants (Health Technology Assessment Report 1986). Hearing aids work on the principle of amplification of sound (compared to cochlear implants, which electrically stimulate the acoustic nerve) and are not useful for alleviating the hearing loss of profoundly deaf patients. Vibrotactile devices, which transmit pressure pulses through the skin, also are poor substitutes for the real sensation of sound. On the other hand, both hearing aids and vibrotactile devices are harmless, while it is possible to damage the inner ear during surgery while implanting cochlear devices. The unsuccessful usage of hearing aids on profoundly deaf people and the limited benefits of using vibrotactile devices is one reason why potential beneficiaries have been reluctant to try cochlear implants. At the same time, the potential for damage to the cochlea has been one reason for the slow acceptance of cochlear implants by the technical community (*Hearing Journal* 1984).

The first major effort to recognize and legitimize the emergence of cochlear implants was in 1973 when an international conference was held exclusively dedicated to the electrical stimulation of the acoustic nerve. However, it was not until ten years later that three events were undertaken within a year of each other by the two most influential professional associations to indicate that momentum was building to legitimize cochlear implants. The first was an official endorsement of cochlear implants by the American Medical Association in 1983. The second was the creation of a special ad-hoc committee on cochlear implants by the American Speech Language and Hearing Association (ASHA) in 1984. In May 1985, ASHA published responses to a survey of firms in the cochlear implant industry which is reported as being read widely by the medical community. In 1985 the American Academy of Otolaryngology–Head and Neck Surgery endorsed the cochlear implant device to the OHTA. Based on this, the OHTA published a booklet in 1986

TABLE 15–4.   *Endorsement of Cochlear Implants by Institutional Bodies.*

Year	Event
1974	First international conference on the electrical stimulation of the acoustic nerve
1983	Council of Scientific Affairs of the American Medical Association endorsed the practice of cochlear implants as an acceptable procedure for postlingually profoundly deaf adults
1984	ASHA develops special committee on Cochlear implants
	FDA sponsors a press campaign on the occasion of the first approval of cochlear devices
1985	American Academy of Otolaryngology–Head and Neck Surgery provided a statement to the Office of the Health technology Assessment (OHTA) with a statement of policy on cochlear implants stating that cochlear implants is an acceptable procedure for postlingually profoundly deaf adults
1985	Thirteenth international conference on otolaryngology where cochlear implants was a key theme
	Formation of Cochlear Implant Industry Council and CHIBA

Source: Chronological list of events.

endorsing the safety and efficacy of cochlear implants, a step essential for receiving Medicare coverage for the implants.

Recently, commercial firms have also begun to play a role in institutionalizing the emerging industry. In December 1985 a Cochlear Implant Industries Council, consisting of representatives from 3M, Nucleus, Storz, and Symbion, was formed under the auspices of the Hearing Industries Manufacturers' Association. Originally, the purpose of this council was to create a united proposal to the Prospective Payment Assessment Committee of the Public Health System to obtain Medicare coverage for cochlear implants. Now, with negotiations for Medicare coverage having been completed, the association is reviewing other issues for joint action that will benefit the growth of the industry. For instance, council members are considering submitting its recommendations to the FDA on simplifying clinical testing requirements in order to reduce the costs involved and undertaking public service announcements about cochlear implants in order to increase public general awareness. In addition, the American Association of Otolaryngology has initiated a committee of representatives from industry, clinics, audiology, psychoacoustics, and other disciplines to study and recommend technical standards for this industry. Table 15–4 is a list of endorsement of cochlear implants by major institutional organizations.

Regulatory.   All medical products, including cochlear devices, are subject to review and approval by the FDA. The essential steps in the approval process have been summarized by Yin and Segerson (1986). In order to conduct clinical tests on humans, an "investigational device exemption" (IDE) must be obtained from the FDA based on clinical tests of the device on animals. Next, each of the clinical sites has to obtain an "institutional review board" clearance to certify its capability to conduct clinical tests on Humans. After test results indicate that a minimum level of safety and effectiveness has been achieved, the device has to be submitted to the FDA panel for a premarket approval (PMA). If the FDA finds that the device is safe and effective, it grants its approval for commercial sale after having approved the prevalence of "good manufacturing practices." This entire procedure of obtaining FDA approval from the initiation of clinical trials on animals can take anywhere from three to five years (Grabowski and Vernon 1982).

In 1981, when 3M applied to the FDA for an IDE status for its first cochlear implant device, it was reported that FDA personnel and panel members did not possess the necessary

TABLE 15–5.   *Regulatory Approvals Granted by the FDA for Cochlear Implants.*

Year	Organization	Investigational Device Exemptions	PMA Details
1980	HEI	Adults (now terminated)	
1981	Biostem	Adults (terminated recently)	
1982	HEI	Children (ages 2–17)	
	Nucleus	Adults	
	UCSF	Adults	
1983	Nucleus	Children (ages 10–17)	
	Symbion	Adults	
	3M	Vienna-EC-adults	
1984	Storz	Adults	3M-House–adults
1985	3M	Vienna IC–adults	Nucleus–adults
1986	Nucleus	Children (ages 2–9)	
	3M	Advanced multichannel-adults	

Source: Dr. David Segerson of the U.S. Food and Drug administration (December 1986).

knowledge to evaluate the application. As a result, 3M was requested to prepare addition documents and information in order to educate FDA personnel and scientific reviewpanels about the nature of cochlear implants and the safety of electrical stimulation of the cochlea. Since then the FDA has granted a number of approvals for clinical investigation and commercial marketing as indicated by Table 15–5. Correspondingly, the FDA has been viewed by industry analysts as having become more knowledgeable about cochlear implants, and as exercising more of its authority and knowledge by prescribing what firms must do to obtain regulatory approval for their cochlear implants.

One major incident in 1985 indicates that self-regulation by participants in the industry had also begun to emerge. This self-regulation was triggered by the product recall of the House device by 3M because of technical problems in October 1985. Realizing that the recall could irreparably tarnish the image of the infant industry, 3M initiated joint discussions with other firms in the industry in order to attempt to minimize the negative impact to the industry due to the recall.

Standards and Rules.   Industry standards pertaining to testing and selection of patients,

safety and efficacy of devices, and the reporting and comparison of results are useful because they (1) reduce technological risks by enabling comparison of different technologies, (2) restrict firms from becoming suboptimal by preventing exaggerated claims, (3) speed the regulatory review process, and (4) are essential for obtaining third-party payment for devices. At the international otorhynolaryngology conference held May 1985, participants from firms and research institutes alike were unanimous in their opinion that technical standards would benefit the industry. A number of events demonstrate the emergence of cochlear implant industry standards:

In March 1984 Yale University began work with Veterans Administration to establish medical guidelines for cochlear implants.

In mid-1985 the University of Iowa created an independent testing institute for comparing the performance of various cochlear implants. This effort was legitimized when the University of Iowa received a grant from the NIH in 1985.

At the international conference held in May 1985, coalition forming between firms was observed where different firms banded together into clusters based upon the particular technological path they had chosen.

In December of 1985, the American Association of Otolaryngology formed a committee of representatives from industry, clinics, audiology, psychoacoustics and other disciplines to study and recommend technical standards for the industry.

At the end of 1986, 3M initiated plans to promote standards for the device along with Storz.

In addition, tacit coordination in the development of industry standards have been observed. Firms are gravitating to a common set of standards through the FDA review process. Furthermore, the performance of the new generations of cochlear devices approved by the FDA have progressively improved over the previous generations.

FINDINGS ON COCHLEAR IMPLANT INDUSTRY EMERGENCE PROCESS

Based on the data described above on each industry subsystem level, we now examine developmental patterns in the emergence of the cochlear implant industry. It should be recognized, however, that this industry is still in its infancy, as witnessed by the fact that only two cochlear devices have received FDA approval for commercial distribution as recently as November 1984 and October 1985. Furthermore, it is not at all clear what directions the cochlear implant industry will take to become commercially viable, or if it will be economically successful. As a result, only preliminary inferences can be drawn from process patterns observed to date.

Sequence of Industry Development

The data presented above suggest four distinguishable periods in the developments in the cochlear implant to date. These periods are (1) basic knowledge development before 1976, (2) entrance of private firms between the period 1977 and 1983, (3) a period of proprietary product development along with the development of an industry infrastructure between 1980 and 1985, and (4) a period of commercialization and market diffusion begun in 1985. We discuss each period with respect to the issues stimulated by the accumulation model described above: the sequence of industry development, the sources of power that directed each period, and the outcomes of different paths and progressions pursued by various players in the emergence of the cochlear implant industry.

First Period: Basic Knowledge Development in the Resources Subsystem before 1976.

As Appendix 15–A indicates, basic research and technology development preceded efforts to develop other resource endowments of financing arrangements and competence pools by over ten years. Basic research was undertaken largely by an "invisible college" of researchers and physicians in different parts of the world who were often associated with universities and teaching clinics. None of these researchers or clinicians could be identified as working exclusively on cochlear implants. Most of this early work was done with no other objective than to further basic knowledge in the science of hearing. Furthermore, many breakthroughs in different disciplines were needed before they could be put together to provide this device. Individually, very few of these discoveries in each of the disciplines constituted a commercially viable innovation. The energy and dedication of "technology champions" such as Dr. William House was required to marshal all these innovations together and to galvanize resistance and censure from the neurosurgical and otological communities in 1956 and again in 1975. Since the

House censures, experimental work has been undertaken by researchers on a proliferating set of technological paths to develop cochlear implants. This research and experimental clinical work created knowledge of a wide array of alternative technologies, from among which private firms began to develop proprietary cochlear devices in the early 1980s.

Second Period: Entrance of Private Firms: 1977–83.

As Table 15–2 shows, the period between 1977–80 was marked by the entry of five private firms into the cochlear implant industry and the initiation of proprietary instrumental subsystem activities. In order to gain access to and acquire the basic scientific knowledge needed to undertake applied R&D for proprietary and commercial use, entrant firms frequently explored linkages with different academic institutions and teaching clinics. As the resource dependence theory (Pfeffer and Salancik 1978; Galaskiewicz 1985) suggests, when an organization does not possess all the capabilities necessary to develop an innovation by itself (which it seldom does), the firm enters into interorganizational relationships with others to obtain the needed resources.

As a consequence, this period was marked by much emphasis placed on *relationship management* by new entrants. In-depth observations of the first mover firm found that tremendous efforts were required to enter into and maintain enduring relationships with other firms or research units. They were observed to change constantly over time, and they entropy if left unattended. In particular, three specific insights on the development of these interorganizational relationships were observed.

First, a basic unanswered question in industry analysis is, "How do competitors arise?" Generally, the literature tends to assume that competitors are profit-seeking entrepreneurs who somehow recognize and seize commercial opportunities by entering lucrative markets. Observations in the cochlear implant industry provide a somewhat different insight: *Aborted efforts at establishing cooperative relationships turn out to become competitive relationships.* In two instances, the efforts of the first mover to initiate cooperative relationships or joint efforts with other research clinics failed, leading to the birth of the firm's competitors. Initial negotiations of possible relationships with a foreign university and a domestic university did not materialize. Otological scientists and clinicians in each of these two universities subsequently entered into licensing arrangements with two other firms (one a new company startup, the other a subsidiary of a large manufacturer) that are now the first-movers major competitors.

Second, *initial negotiations of the terms and parties to relationships were seldom sufficiently specified to accomplish business intentions.* In particular, in two interorganizational relationships the parties judged that their relationships were initially negotiated too narrowly, or without the involvement of all the key parties necessary for building a successful relationship. Moreover, organizations are pluralistic. Agreements entered into with one group of representatives from an organization were found to be contrary to and rejected by other members or associates of that organization.

Third, *the technology, business, and environment changed rapidly as the innovation developed, requiring the interorganizational relationships be renegotiated and modified over time.* For example, within a period of three years, firms involved in a partnership claimed that their relationship was no longer a "marriage" but simply a business relationship. It was reported that it was not possible to foresee at the outset these changing conditions and safeguards that were subsequently needed to maintain a relationship. Perhaps, as a result of this realization, one executive stressed the need to "always negotiate an exit clause that is fair and acceptable to both parties" when initially entering into a relationship.

Third Period: Instrumental-Resource Subsystem Linkages: 1980–85. This period, (which overlaps with the earlier period of entry by business firms), witnessed developments in the instrumental and the resource endowment subsystem.

Instrumental subsystem activities.

In the instrumental subsystem, functions of R&D, clinical trials, regulatory affairs, manufacturing, marketing, and client services emerged sequentially over time for one of the firms involved in the industry (see Figure 15–6). The FDA regulatory guidelines explain the sequential development of these instrumental functions and the direction and timing of market entry of cochlear devices. However, instances of opportunistic "leap-frogging" of functions by firms were also observed.

For example, work on the first commercially approved cochlear implant initiated with R&D, then progressed to clinical trials and regulatory review by the FDA, and then manufacturing, marketing, and service. Although this basic sequential pattern still predominates in the development of this firm's second-generation cochlear device, it is less clearly followed than it was in the development of the first generation device. Two instances of opportunistic "leap-frogging" were observed. At the 1985 otolaryngology conference the firm promoted the introduction of its second-generation device, even though it had not yet completed clinical trials, obtained FDA regulatory approval, and scaled up manufacturing on this device. In other words, the marketing function had begun before the three prior functions in the value-creation chain were completed. The second instance is when one of the firms submitted an informal application to the FDA for a premarket approval (PMA) for its cochlear implant device before the necessary preceding research and clinical trials were truly completed. It was reported that the application was submitted in order to identify FDA's standard of how many

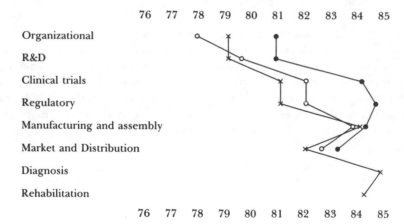

Note: The dates of initiation of product functions by only the first two firms in the industry are shown. The dates are abstracted from the chronological list of events on the emergence of the cochlear implant industry.

× 3M–First device
● 3M–Second device
o Cochlear Corporation–First device.

FIGURE 15–6. Development of Product Functions by Firms.

clinical trials were necessary to obtain regulatory approval. The FDA declined the informal application.

These instances suggest that efforts to create value for future products or an overall line of products occurred even before the chain of functions necessary to complete the initial product was completed. As observed in Chapter 8, building a new sustainable business, which is the dominant motivation by firms in developing cochlear implants, may be the major reason for this apparent leap-frogging behavior. First, market presence and viability is a systemic evolutionary process, requiring the introduction and packaging of a family of related products. It was observed that it is unlikely that a business can be created and sustained with a single product in the marketplace. An ongoing business requires the creation of synergy and economies of scale across functions, which is obtained from undertaking R&D, testing, manufacturing, marketing, and service on a family of related products and services over time.

Timely orchestration of resources among interdependent functions within and between products in the development of a business becomes a major challenge in the management of innovation. If it is true that functions are established sequentially for the first generation of products, then each preceding functional step becomes a bottleneck for succeeding steps in the product development sequence. Errors (such as a product design flaw) not detected or corrected in preceding steps are passed along to the next functional step. The consequence is that these errors may either compound developmental work for subsequent functions, or at the extreme if undetected until the product enters the market (as in the recall of the House device), may not only entail costly revisions in all steps of the product development cycle but also require counterproductive efforts by marketing and sales at "damage control."

In a multiproduct development effort, the management task is compounded by another set of "bottlenecks" in allocating and redeploying resources from the development of one product to the next generation of products. When a functional step is completed on a preceding product, those specialized resources are freed up and must be redeployed or they remain idle. In theory, these specialized resources are therefore redeployed to begin work on the next generation of products by performing tasks in their functional or disciplinary specialties.

Each generation of products usually has its own unique mix of required skills, interdependent processes, and timetable for development. However, the organization does not have that mix of skills; it has the mix of skills and technological resources that it learned and used for the last product development effort. Simple redeployment of specialized personnel to the next product when the prior one is completed, without significant retraining, invariably results in replicating or extending the previous product development effort and compromising the design specifications and timetable for the new generation product. Alternatively, the excitement and timetable for a new generation product may result in a "flee" from the predecessor product without adequately performing all the functions required to complete its development and market entry process.

Resource Endowment activities.

While firms were attempting to develop proprietary products and competencies, they had to also ensure the development of an industry infrastructure necessary for the successful emergence of the industry. For instance, commercial firms initiated training programs to create a pool of competent audiologists skilled in implanting cochlear devices. As the cochlear implant industry grows, we anticipate the emergence of additional educational programs for new specialized disciplines and services. For example, a new specialty of psychoacoustics services is emerging to provide patient therapy on sound discrimination after a surgical cochlear implant procedure—a service now still largely provided by surgeons.

Similarly, a significant level of activities at the financial category have been directed at requesting third-party payors to include cochlear implant surgery and rehabilitation procedures in their payment reimbursement systems. These efforts have been undertaken by private firms who have individually and jointly approached third-party payors to extend coverage for cochlear implants.

The creation of these resource for the industry represent "common goods" that can be freely drawn on by industry participants. It has long been recognized that the creation of these free common goods is problematic because of the "free rider" problem (Olson 1965). It is rational for an individual firm not to make investments in creating these resource endowments when it can freely draw on them. However, if all individual parties behaved in this rational way no resource endowments would emerge that are critical to sustain the overall viability of the industry and its members. Do the data provide any clues on how this common goods problem has been addressed in the cochlear implant industry thus far?

The data indicates that both patterns of self-interest and collective-regarding behavior are displayed. Self-interested behavior and free riding by private firms appear to be occurring in the development of basic research and technology for the industry, where no private firm has been found to play any significant role. Collective behavior seems to be occurring when incentives are present for individual firms to join together and cooperate to achieve an outcome they would find difficult to achieve by themselves—such as gaining coverage from third-party payors to finance sales of their

products to end-users. In addition, professional associations and industry councils (which are funded by and represent individual private firms) are developing training programs to disseminate a common stock of knowledge on cochlear implants. Finally, it appears that the burden of creating other common goods rests with the first mover in the industry.

Indeed, the first-mover firm has had to deploy considerably greater resources than any of its competitors in the industry. For example, it spent considerable efforts in educating the FDA about safety-related issues that are now taken for granted for subsequent applications. The first mover initiated cochlear implant training programs for physicians (costs that subsequent competitors did not need to incur). Finally, the first mover has been the major force in persuading third-party payors to include cochlear implants in their payment reimbursement systems. Although much of the literature has emphasized first-mover advantages in an industry, it has largely ignored Veblen's (1917) analysis of first-mover burdens of creating such common resource endowments that permit free-riding by other industry participants.

Fourth Period: Commercialization and Market Diffusion Begun in 1985. By 1985 the FDA had approved two devices for large-scale commercial sale to customers. The stage was therefore set for widespread diffusion and adoption of cochlear implants by the hearing impaired. However, the process of diffusion and adoption among potential beneficiaries has been slow and sales of cochlear implants have been far lower than anticipated. Although the driving force for firms within the emerging industry is the quest for profits, an unexpectedly weak market "demand pull" is limiting the commercial growth of the industry. Predictably, issues related to marketing, including diffusion and adoption of products approved by the FDA for commercial release, have become the dominant concerns of firms during this period of industry emergence. As a result, the commercial viability of cochlear implants as a profitable industry has become unpredictable and is being questioned by industry analysts. Indeed, as Table 15–4 indicates, two firms in the industry—Storz and Symbion—announced plans to reduce their financial commitments to their cochlear implant programs reportedly because they did not perceive the cochlear implant market growing at a fast enough pace.

Relationships between and within Tracks

Finally, we will examine the relationship of events and activities between the three industry functions, including points of convergence/divergence and possible dead-ends/terminations in the emergence of the cochlear implant industry. To address this issue, we make use of an event network chart, as suggested by Miles and Huberman (1984). This representation scheme displays the interconnections between events and activities required to be completed over time for the accomplishment of a project. Figure 15–7 presents our interpretation of the overall paths in the progressive development of the cochlear implant industry to date.

Points of Convergence. We define a convergent point as the confluence of multiple activities resulting in the occurrence of a particular event. Two such points of convergence can be identified in Figure 15–7. These are marked as events numbered [39] and [44]. These correspond to the extension of medicare coverage, and obtaining patient access respectively. For the extension of medicare coverage, three separate activities had to occur: (1) the formation of a joint set of standards by firms in the industry for recommendation to the Health Care Financing Administration [34], (2) FDA approval of a device for commercial sale [25], and (3) prior approval of other insurance payors such as the Blue Cross and Blue Shield [18].

Obtaining patient access requires the completion of the following events/activities: (1) FDA approval [25], (2) medicare coverage [39], and (3) the wide spread diffusion and adoption of cochlear implants by patients [27–44].

Points of Divergence. We define a divergent point as one that results in the initiation of many other activities. Two such points of divergence can be identified by the events numbered [3], corresponding to the development of a device with a minimum degree of safety and efficacy, and numbered [5], which represents the point at which a large business firm became involved with cochlear implants. The development of a device with a minimum safety and effectiveness level led to the endorsement of cochlear implants by the Council of Scientific Affairs [20]. It also led to business firms becoming interested in commercializing cochlear implants [4 & 5]. The involvement of business firms in the industry in turn led to the occurrence of many activities, such as education of the FDA leading to the institution of the FDA panel [9]; efforts at convincing insurance payors to cover cochlear implants leading to organizations such as the Blue Cross and Blue Shield covering cochlear implants [18]; initiation of the marketing function [8]; initiation of efforts to train physicians to implant cochlear devices leading to the formation of a core group of trained physicians [17]; initiation of R&D activities [6 & 7]; and the initiation of efforts at establishing rehabilitation [21] and diagnostics [22] functions.

Loops and Cycles. A loop, or a cycle, is a set of recurring events and activities over time. Figure 15–12 shows that each time an organization commercializes a device, it is required to seek the approval of the FDA, which ensures the safety and efficacy of the device. The two products that have received FDA approval proceeded through these iterative events. These

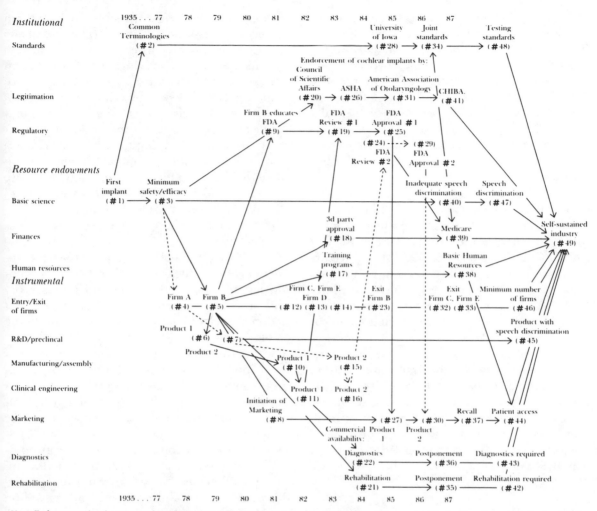

FIGURE 15–7. *Cochlear Implant Industry Emergence Chart.*

Note: Each new product has to pass through the FDA route twice before receiving FDA approval for commercial sale. First, for an investigational device exemption (IDE) which permits the researcher to conduct clinical trials, and second, for approval for marketing the device on a commercial basis (PMAA). So far, FDA has granted 6 IDEs and 2 PMAs. The paths connecting R&D and FDA for the IDEs have not been shown as this complicates the chart enormously. The PMA route is the most significant and important of the two and has been shown for the two products with PMA.

are represented in Figure 15–12 by the paths [5], [6], [10], [11], [19], [25], [27], and [4], [7], [15], [16], [24], [29], [30], for products 1 and 2, respectively.

Critical Paths. A critical path (a concept borrowed from the Project Evaluation Review Technique) is one that takes the longest period of time for the completion of a necessary set of activities for achieving the end objective; in this case, the emergence of a self-sustained industry [49]. Although there are many paths that represent logical connectivities between events and activities, there is only one "critical" path. In the case of the cochlear implant industry, a critical path is represented by [1], [3], [5], [7], [10],

[11], [19], [25], [39], [44], and [49]. Basic scientific knowledge had to first ensure safety/efficacy of the device for use in humans before business firms would become involved. The presence of firms wanting to commercialize cochlear implants was necessary as a thrust for the creation of the FDA panel for cochlear devices. FDA's approval of cochlear devices was necessary for Medicare to extend its coverage for cochlear implants, which in turn is necessary for accessing a wider patient base.

Terminations or Postponements. Three paths have shown signs of termination or postponement. The first is the exit of three of the five firms in the industry [23, 32, 33], reportedly because of their inability to raise further capital to continue their programs because of the long lead time now anticipated for the emergence of the industry. The other two terminations or postponements corresponds to diagnostics [35] and rehabilitation [36]. These two functions were initiated by the first mover firm on an experimental basis but were dropped by them as these were consuming more resources than they were willing to expend considering that firms in the cochlear implant industry have not yet been able to access the market to the extent possible.

CONCLUSION

This chapter suggested a social system framework and an accumulation process model for understanding how a new industry emerges over time with the commercialization of a revolutionary technological innovation. The social system framework incorporates and provides a more inclusive perspective of an industry than does the traditional industrial economics definition of the group of firms competing to produce similar or substitute products. In addition to these proprietary instrumental subsystem activities of competing firms, the social system framework examines the infrastructure that any industry requires in order to survive and sustain its members. The critical subsystems of an industry's infrastructure include its resource endowments of basic scientific and technological knowledge, financing, and a competent labor pool, as well as an institutional governance structure that legitimates and circumscribes the activities of industry members. Although these components of infrastructure have heretofore been treated as "externalities" (Porter 1980), they are critical functions to incorporate directly in any explanation of industry emergence.

A longitudinal case study of the development of the cochlear implant industry was conducted to empirically apply and evaluate the industry social system framework. An accumulation theory of change was used to guide the process study, for it directs attention to examining system inputs (the resources and roles contributed by public and private organizations), outputs (the instrumental, resource endowments, and institutional functions developed), and transformation processes (the temporal sequence and progressions in developing these functions) during three conceptually bracketed periods of industry initiation, startup, and takeoff. A tracking of events in the historical and real-time development of the cochlear implant technology identified four temporal progressions in the development of instrumental, resource endowments, and institutional subsystems of the industry. These are summarized below.

During the initiation period there was an extended gestation period lasting about forty years (1935–76) of "technology push" during which all events focused on developing basic research and technology by an "invisible college" of researchers and physicians associated with universities and research clinics throughout the world. This work was largely supported by university and government funding, and preceded by ten years efforts to develop other resource endowments of financing arrangements and a competent labor pool. These data suggest that proprietary activities of private firms to develop and commercialize cochlear implants would not have been possible without the prior creation of this infrastructure of resource endowments. Indeed, no private firms were engaged in or funded any of this basic science and technology development work.

Between 1977 to 1980, five private firms initiated proprietary instrumental subsystem activities by each establishing relationships with different academic institutions and teaching clinics in order to gain access to basic scientific knowledge needed to undertake applied R&D on cochlear implants. During this period, a major challenge for the first-mover firms was relationship management. Firm rivalry and competition was observed to emerge when aborted efforts at establishing cooperative relationships turned out to become competitive relationships. Further, the different technological paths that these firms adopted can be largely explained to result from the particular scientific path being followed by the universities and clinics that the private firms entered into relationships with.

The industry startup period appears to begin in 1980 and includes events focused on: developing proprietary products by competing private firms, cooperative efforts by the competing firms to establish third-party financing arrangements and otological training centers that provide common benefits to the industry participants, and institutional legitimations and endorsements by professional associations, FDA regulatory reviews, and diagnostic tests and standards for the cochlear implant industry. As these events suggest, multiple parallel and related progressions of events are being observed during this period to simultaneously and incrementally establish the instrumental, resource endowments, and institutional infrastructure of the emerging industry.

While industry startup still constitute most of the events to date, product market entry and diffusion events occurring in 1985 signify the beginning of the industry takeoff period. Industry takeoff appears to be largely governed by "demand pull" factors, and an unexpectedly "weak market" is currently limiting the commercial takeoff of the industry. In addition, tests of competing technologies provide initial evidence for the emergence of a superior technological path (multiple channel devices). Combined with the weak market demand for this product, the emerging dominant design has caused an early industry shakeout as witnessed by the withdrawal of two firms from the industry.

Of course, these temporal progressions in the emergence of the cochlear implant industry may not be found in the development of other industries. We expect that the actual process of industry emergence to be a function of technology (varying in degrees of novelty, risk, complexity, and capital intensity), market (also varying in risk, entry barriers, and demand pull), resource endowments (varying in levels of scientific advance and basic knowledge, labor competencies, and resource munificence), and institutional governance structures (varying in degrees of legitimacy, regulation, and standards). Examination of these variations represents a significant agenda for future research on industry emergence. Chakravarthy (1985) provides a useful start in this direction by proposing a typology for examining how industry emergence strategies may differ with variations in technological and market risks.

However, in order to undertake this research agenda on patterns of industry emergence, one must first have a conceptual framework for mapping the critical components of an industry. We believe that the empirical data presented and analyzed in this chapter on the emergence of the cochlear implant technology demonstrates the utility of the industry social system framework. We have found that technological innovation encompasses a wide array of issues and events that can be addressed systematically with the social system framework—such as the creation of an infrastructure of basic resources to support industry startup and take-off, questions of "technology push" and "demand pull," patterns in the development of cooperative, competitive, and multiplex relationships among firms, and the interdependencies that exist between instrumental proprietary activities of individual firms, resource endowments activities of public and private organizations, and institutional governance and regulation of industries.

Moreover, we believe that the social system framework directs attention to a number of important applied issues dealing both with firm-specific problems of product development and business creation, and public policy problems of industry structure and technological competitiveness. We will conclude by drawing out some of the practical inferences and speculations from this research about the management of industry emergence, first at the level of the firm, and then at the national policy level. Of course, evidence to substantiate these speculations will require much further theorizing and research.

Implications for the Management of the Firm

Gestation. The long gestation period of basic scientific knowledge creation observed in the emergence of the cochlear implant industry suggests that there are many opportunities for firms to monitor developments of such potentially commercializable technologies. As found in Chapter 8, Pasteur's adage that "chance favors the prepared mind" is also appropriate here. An early familiarity with the developments of a particular technology offers many benefits to the firms that undertake such monitoring activities. First, it enables them to determine when a particular technology is ready for commercialization. Second, it prepares them to choose from a variety of technological routes; a choice that appears to disappear quickly as

other firms enter into relationships to develop and commercialize the product with specific research institutes. Third, it facilitates a speedy transfer of technology once relationships have been established with research units.

Such monitoring strategies can range from simply reviewing patents over time to undertaking basic research, either within or outside the firm. Firms must decide on the most appropriate monitoring mode at different periods of gestation as each approach involves different levels of investments and commitments. Hamilton (1986) addresses precisely this issue and offers a useful framework to identify the appropriate monitoring strategy on the basis of decreasing technological and market uncertainty as a technology develops over time.

Initiation. The choice of a particular technological route to purse from the alternatives available requires more detailed knowledge about each route than a monitoring strategy may provide. The inevitability of developmental paths and the irreversibility of the investments required for each technological route makes this choice a critical one for firms, especially because of the uncertainty as to which route will eventually win out. Furthermore, the decision to enter into a licensing agreement with one research unit may restrict the possibility of entering into a licensing agreement with another research unit as each route may require nontrivial idiosyncratic investments. The information asymmetry that exists in the initiation stage between basic units and the business firm, combined with the uncertainty and the need for future technology specific investments render this situation ripe for opportunism (Williamson 1975). If the firm is not careful, it may get saddled with an inferior project (Arrow 1962).

As data from the case indicate, firms attempted to overcome this problem by exploring various relationships with a number of research institutes. In one instance, an exploratory technical collaboration was even established between a research unit and a business firm. This relationship finally fell apart when the business firm decided to pursue a different technological route.

The task of simultaneously interacting with more than one research unit during the process of exploring different technological alternatives is a delicate one for firms, for it may entail sharing of considerable proprietary information with the other organization with the expectation that they will enter into a collaborative relationship in the future. However, if such relationships do not materialize, feelings of distrust and inequity may result, which may then impact the likelihood of any future collaboration between the two organizations. Indeed, as the data shows, two of the leading firm's competitors emerged when attempts to establish a collaborative relationship did not materialize.

An important practical question then for innovation managers is, How should should such tentative relationships be negotiated and structured so that expectations are not raised prematurely and proprietary information not disclosed to potential future competitor? Ring and Van de Ven suggest in Chapter 6 that firms should engage in a process of sensemaking, understanding, and commitment where negotiations are kept deliberately informal until risk and trust become reasonably clear to permit an equitable formal transaction.

Startup. As the observations from the case suggest, the startup period was very challenging for firms as they had to develop proprietary goods and services while simultaneously ensuring the creation of an industry infrastructure. This micro-macro tension raises two managerial implications. First, firms have to decide the allocation of their resources between instrumental level activities and activities that may benefit all industry participants. Second, firms have to learn to handle a multiplexity of ties (Galaskiewicz 1985). While competing with

each other to establish a foothold in the emerging industry, each firm may need to cooperate with others to create the infrastructure that no one firm may be able to create by itself.

Takeoff. Although takeoff of this industry has not yet occurred, the data suggests a shift in activities from research and development to product market entry and diffusion. Correspondingly, this involves a greater emphasis on manufacturing activities as product price becomes increasingly important. As Abernathy (1978) suggests, this shift from a research to a marketing and manufacturing mode entails significant organizational changes that may not be easily accomplished. Indeed, an in-depth analysis of the activities of the leading firm in the industry as presented in Chapter 8, highlights the difficulties that organizations face in making such transitions. This suggests that the choice of a "right" technology, while necessary, is not a sufficient condition for commercial success. Successful companies will also need to have the ability to make a transition from a research to a marketing and manufacturing mode at the appropriate time of the industry emergence.

Implications for National Policy

Gestation. There has been considerable debate about the efficacy of basic scientific research and as to whether basic research precedes commercially successfully products or not (Myers and Marquis 1969; Mowrey 1985; Comroe and Dripps 1976).

At least for the cochlear implant technology, there has been a long period of basic research as researchers from various disciplines slowly unraveled the mysteries of providing a sensation of sound to the deaf. Such basic scientific knowledge is very costly to produce relative to its cost of diffusion and imitation (Mansfield 1985). In addition, it builds in a cumulative fashion and exhibits properties of public goods. From a national policy perspective, the important question is, Who provides the necessary resources for the creation of such basic knowledge? Arrow (1962) argues that the social returns to research investment exceed the private returns to individual firms; a condition leading to underinvestment by the firm (from a social point of view) in research. As a consequence, a variety of studies have shown that firms rely on outside sources of knowledge and technical inventions for the vast majority of their commercially significant new products (Rosenbloom 1966; Mueller 1962; Utterback 1974; Stobaugh 1985). This suggests the importance of federal support for conducting basic research for the development of technologies such as the cochlear implant.

Initiation. The data on NIH grants and contracts awarded to different research units and firms conducting basic scientific research on cochlear implants shows an interesting pattern (see Table 15–1). The leading firm (unlike other firms in this industry) did not enter into a relationship with a research unit that had obtained funds from the NIH. Nor did this firm apply for any NIH funding for conducting its research on cochlear implant. The reason offered by the laboratory manager of this firm for not seeking NIH grants was that knowledge created with federal funds was in the public domain, while this firm wanted to keep its knowledge proprietary.

This raises an interesting dilemma. Federal funding that facilitates basic knowledge creation in the gestation period may potentially cause problems during the initiation period; basic knowledge created with federal funds may still be in the public domain, while firms are motivated to convert such publicly accessible knowledge into proprietary goods and products.

However, other firms in the cochlear implant industry did choose to pursue technologies that were funded with NIH grants without any apparent detrimental effects on their pro-

prietary product development activities. How such public knowledge is used for developing proprietary goods and services is an issue worth further study.

Startup. There is a growing concern that a significant gap has arisen between the ability of the U.S. to create and implement new ideas (Ouchi 1981; Peters and Waterman 1982; Kanter 1983; Lawrence and Dyer 1983). The United States is generally regarded as the most inventive nation in the world. However other nations are surpassing the United States in developing and implementing its ideas.

One possible reason for this gap between invention and implementation could be the difficulties associated with the creation of a new industry infrastructure. No one firm may have the necessary incentive or the resource to create the infrastructure, while collective action may be difficult because each firm may "free-ride" on the others. Under these circumstances, external intervention is required to facilitate and sustain collective action when required.

As the data on the emergence of the cochlear implant industry shows, competing firms did voluntarily come together to establish Medicare payment for patients being implanted with cochlear devices, as well as to establish industry wide testing and product standards. In both instances however, firms proceeded very carefully to establish linkages with other firms because of a fear of violating antitrust laws. Thus, it appears that antitrust laws may make it even more difficult for collective action to occur, even though the creation of a part of the industry infrastructure may require the collective resources and action of all the firms involved.

Furthermore, theory on public goods suggests that certain kinds of public goods will be suboptimally provided for (from a collective perspective), while others require an exogenous force to initiate action for their creation (Oliver, Marwell and Teixeira 1985). Indeed, cochlear implant industry participants feel that certain functions such as rehabilitation services for patients have not been adequately created; a fact that may be effecting the emergence of the industry. Under these circumstances, it would be useful if public agencies could initiate and invest in the creation of such goods.

Takeoff. As mentioned earlier, there are some early indications of the emergence of a superior technological approach (the multichannel device). The role of quasi-governmental agencies (the FDA and the American Academy of Otolaryngology in the case of cochlear implants) in facilitating the identification of the superior technological path by instituting testing and product standards is an important national policy issue for new products such as the cochlear implant. For instance, one interesting issue is the timing of such intervention. Premature attempts at instituting standards may either be resisted, or if successful, may curb the development of the most appropriate technological device. At the same time, if left too late, dysfunctional side effects may arise due to the inability to compare the efficacy and safety of each product alternative.

The identification of a "winning" technological path from among alternatives may lead to a domination of the industry by those firms that happen to possess the winning technology. Indeed, in the cochlear implant industry, Nucleus is currently enjoying such a dominant position.

Some would argue for governmental intervention to rectify such a situation. However, as Schumpeter (1975) suggests, this dominance is a reward to risk-seeking entrepreneurs. Premature attempts at rectifying such situations may remove any incentives for firms to be innovative and take on risks, thereby adversely affecting the economic growth and vitality of an economy.

However, Williamson (1975) argues that such firms may continue to enjoy *indefinite* monopoly profits if market forces cannot rectify

such situations. The dominant firm is able to erect significant entry barriers to protect its monopoly position over time. Williamson therefore suggests that governmental intervention should take place to rectify such situations *after* the growth and maturity stages of an industry when market forces cannot rectify such imperfections. We would like to qualify Williamson's suggestion. We would like to suggest first a process solution and, only if this fails, a structural one. For if Schumpeter's process of creative destruction enfolds over time, no firm will be able to continue to dominate in an industry on the basis of products based on one technology. Thus, facilitating a process of creative destruction will do away with the problem of indefinite concentration. We are advocating a process of Schumpeterian competition as a first solution to issues of dominance, rather than solely relying on competition that is based on the industrial organization paradigm where competition is engendered by adhering to rigid structural rules that prevent concentration.

REFERENCES

Abernathy, W.J. 1978. *The Productivity Dilemma: Roadblock to Innovations in the Automobile Industry.* Baltimore: John Hopkins University.

Abernathy, W.J., and K.B. Clark, 1985. "Innovation: Mapping the Winds of Creative Destruction." *Research Policy.* 14: 3–22.

Aldrich, H., and D. Whetten 1981. "Organization-Sets, Action Sets, and Networks: Making the Most of Simplicity." In P.C. Nystrom and W.H. Starbuck, eds., *Handbook of Organizational Decision,* pp. 385–407. Oxford: Oxford University.

Allison, G.T. 1971. *The Essence of Decision: Explaining the Cuban Missile Crisis.* Boston: Little, Brown.

Arrow, K.J. 1962. "Economic Welfare and the Allocation of Resources for Innovative Activity." In R.R. Nelson, ed., *The Rate and Direction of Inventive Activity.* Princeton, N.J.: Princeton University.

Astley, W.G. 1985. "The Two Ecologies: Population and Community Perspectives on Organizational Evolution." *Administrative Sciences Quarterly.* 30: 224–41.

Astley, W.G., and A.H. Van de Ven. 1983. "Central Perspectives and Debates in Organization Theory." *Administrative Sciences Quarterly* 28: 245–73.

Chakravarthy, B.S. 1985. "Business-Government Partnership in Emerging Industries: Lessons from the American Synfuels Experience." In R. Lamb and P. Shrivastava, eds., *Advances in Strategic Management.* 3:257–275 Greenwich, Conn.: JAI.

"Cochlear Implants: Five Companies Respond to ASHA Survey" 1985. *American Speech Hearing Association.* pp. 27–34.

Comroe, J., and R. Dripps. 1976. "Scientific Basis for the Support of Biomedical Sciences." *Science* 192: 105–11.

"Current Status of Cochlear Implants." 1986. *U.S. Food and Drug Administration, Office of Device Evaluation, Division of OB/GYN, ENT, and Dental Devices.*

Dosi, G. 1982. "Technological Paradigms and Technological Trajectories." *Research Policy* 11:147–62.

Etzioni, A. 1963. "The Epigenesis of Political Communities at the International Level." *American Journal of Sociology* 68: 407–21.

Galaskiewicz, J. 1985. "Interorganizational Relations." *Annual Review of Sociology* 11: 281–304.

Grabowski, H.G., and J.M. Vernon. 1982. "The pharmaceutical industry." In R. Nelson, ed. *Government and Technical Progress: A Cross-Industry Analysis.* 283–362. New York: Pergamon.

Hamilton, W.F. 1986. "Corporate Strategies for Managing Emerging Technologies." In M. Horwitch, ed., *Technology in the Modern Corporation.* 103–118 New York: Pergamon.

Hannan, M., and J. Freeman. 1977. "The Population Ecology of Organizations." *American Journal of Sociology* 82: 929–64.

Health Technology Assessment Report. 1986. *Cochlear Implant Devices for the Profoundly Impaired.* Baltimore, Md.: Office of Health Technology Assessment, National Center for Health Services Research and Health Care Technology Assessment, Public Health Service.

Hearing Journal. 1984. 37 (11): 7–15.

Kamien, M.I., and N.L. Schwartz. 1982. *Market Structure and Innovation.* Cambridge: Cambridge University.

Kanter, R.M. 1983. *The Change Masters: Innovation for Productivity in American Corporation.* New York: Simon and Schuster.

Lawrence, P.R., and P. Dyer. 1983. *Renewing American Industry.* New York: Free Press.

Mansfield, E. 1985. "How Rapidly Does New Industrial Technology Leak Out?" *Journal of Industrial Economics* 34 (2): 217–23.

Mattsson, L.G. 1986. *Management of Strategic Change in a "Markets-as-Networks" Perspective.* Paper presented at "Management of Strategic Change" seminar at the University of Warwick, U.K.

McKelvey, B. 1982. *Organizational Systematics: Taxonomy, Evolution, Classification.* Berkeley: University of California.

Miles, M.B., and A.M. Huberman. 1984. *Qualitative Data Analysis: A Sourcebook of New Methods.* Beverly Hills, Calif.: Sage.

Mowery, D.C. 1985. *Market Structure and Innovation: A Critical Survey.* Paper presented at the conference on "New Technology as Organizational Innovation" at the Netherlands Institute for Advanced Studies in Humanities, Wassenaar.

Mowery, D.C., and N. Rosenberg. 1979. "The Influence of Market Demand upon Innovation: A Critical Review of Some Recent Empirical Studies." *Research Policy* (April): 103–153.

Mueller, W.F. 1962. "The Origins of the Basic Inventions Underlying DuPont's Major Product and Process Innovation, 1920–1950." In R.R. Nelson (ed.), *The Rate and Direction of Inventive Activity.* Princeton, New Jersey: Princeton University Press.

Myers, S., and D.G. Marquis. 1969 *Successful Industrial Innovations.* National Science Foundation, NSF 69–17, Washington, D.C.

Nelson, R.N., and S.G. Winter. 1977. "In Search of a Useful Theory of Innovation." *Research Policy* 6: 36–76.

Oliver, P., G. Marwell, and R. Teixeira, 1985. "A Theory of the Critical Mass: Interdependence, Group Heterogeneity, and the Production of the Collective Action." *American Journal of Sociology* 91 (3): 522–66.

Olson, M. 1965. *The Logic of Collective Action: Public*

Goods and the Theory of Groups. Cambridge: Harvard University.

Ouchi, W.G. 1981. *Theory Z.* Reading, Mass.: Addison-Wesley.

Owens, E., D.K. Kessler, E.D. Schubert, 1982. "Interim Assessment of Candidates for Cochlear Implants. *Arch Otolaryngol* 108: 478–83.

Peters, T.J., and R.H. Waterman, Jr. 1982. *In Search of Excellence.* New York: Harper & Row.

Pfeffer, J., and G. Salancik. 1978. *The External Control of Organizations.* New York: Harper & Row.

Piore, M.J., and C.F. Sabel. 1984. *The Second Industrial Divide: Possibilities for Prosperity.* New York: Basic Books.

Porter, M.E. 1980. *Competitive Strategy: Techniques for Analyzing Industries and Competitors.* New York: Free Press.

———. 1985. *Competitive Advantage: Creating and Sustaining Superior Performance.* New York: Free Press.

Radcliffe, D. 1984. "The Cochlear Implant: Its Time Has Come." *The Hearing Journal* 37 (11): 7–15.

Rappa, M.A., *The Structure of Technological Innovations: An Empirical Study of the Development of III–V Compound Semiconductor Technology.* Minneapolis: University of Minnesota Unpublished Doctoral Dissertation. 1987.

Rosenberg, N. 1982. *Inside the Black Box: Technology and Economics.* Cambridge: Cambridge University.

Rosenbloom, R.S. 1966. "Product Innovation in a Scientific Age." *New Ideas for Successful Marketing,* ch. 23. Chicago: Proceedings of the 1966 World Congress, American Marketing Association.

———. 1985. "Managing Technology for the Longer Term: A Managerial Perspective." In K.B. Clark, R. Hayes, and C. Lorenz, eds., *The Uneasy Alliance: Managing the Productivity Technology Dilemma,* pp. 297–317. Cambridge, Mass.: Harvard Business School.

Schumpeter, J.A. 1975. *Capitalism, Socialism, and Democracy.* New York: Harper & Row.

Simmons, F.B. 1985. "Some Medical, Social, and Psychological Considerations in Cochlear Implants." *Seminars in Hearing* 6:1–6.

Stern, N., and El-Ansery, A.I. 1982. *Marketing Channels.* Englewood Cliffs, N.J.: Prentice-Hall.

Stobaugh, R. 1985. "Creating a Monopoly: Product Innovation in Petrochemicals." In R. Rosenbloom, ed., *Research on Technological Innovation, Management and Policy* 2: 81–112. Greenwich, Conn.: JAI.

Thirtle C.G., and V.W. Ruttan, 1987. *The Role of Demand and Supply in the Generation and Diffusion of Technical Change.* New York: Harwood Academic Publishers.

Tushman, M.L., and E. Romanelli. 1985. "Organizational Evolution: A Metamorphosis Model of Convergence and Reorientation." In B. Staw and L. Cummings, eds., *Research in Organizational Behavior* 7: 171–222.

Utterback, J.M. 1974. "Innovation in Industry and the Diffusion of Technology." *Science* 183: 620–26.

———. 1987. *Dynamics of Innovation in Industry.* New York: Pergamon.

Utterback, J.M., and W.J. Abernathy. 1975. "A Dynamic Model of Process and Product Innovation." *OMEGA* 3(6): 639–56.

Van de Ven, A.H., and W.G. Astley. 1981. "Mapping the Field to Create a Dynamic Perspective on Organization Design and Behavior." In A. Van de Ven and W. Joyce, eds., *Perspectives on organization design and behavior,* pp. 427–68. New York: Wiley.

Van de Ven, A.H., and R. Garud. 1989. "A Framework for Understanding the Emergence of New Industries." In R.S. Rosenbloom and R. Burgelman, eds., *Research on Technological Innovation, Management and Policy,* Vol. 4. Greenwich, Conn.: JAI.

Veblen, T. 1917. *Imperial Germany and the Industrial Revolution.* London: MacMillan.

Williamson, O.E. 1975. *Markets and Hierarchies.* New York: Free Press.

Windmill, I.M., S.A. Martinez, M.B. Nolph, and B.A. Eisenmenger. 1987. "The Downside of Cochlear Implants." *The Hearing Journal* (January): 18–21.

Yin, L., and P.E. Segerson. 1986. "Cochlear Implants: Overview of Safety and Effectiveness." *Otolaryngologic Clinics of North America* 19 (2): 423–33.

Yin, R.K. 1982. "Studying the Implementation of Public Programs." In W. Williams et al. *Studying Implementation: Methodological and Administrative Issues.* 36–72. Chatham, N.J.: Chatham House.

APPENDIX 15–A.   Chronological Summary of Events in the Development of the Cochlear Implant Industry.

Date	Event	Source	Key Word
1935	First report of electrical stimulation of nerve leading to hearing	AORL, May–June 1976	Basic science
1956	House criticized for developing new surgical procedure	AORL, May–June 1976	Basic science
1957	Djourno and Eyries implant electrode on patient nerve	AORL, May–June 1976	Basic science
1961	House and Doyle implant in the United States	ASHA, May, 1985	Basic science
1962	HEI receive grants from George and Dolores Eccles	AORL, May–June 1976	Finance
1964	Simmons et al. begin implanting using multichannel Cochlear implant	AORL, May–June 1976	Basic science
1967	University of Melbourne initiates work on Cochlear implant	Letter of July 1986	Basic science
1968	Michelson's animal studies shows support for intra-Cochlear electrode	AORL, May–June 1976	Basic science
1969	House and Urban implant patient with multihard wire system	AORL, May–June 1976	Basic science
1970	NIH contract: Huntington	NIH, Dec. 1986	Finance
1971	Michelson and Mirzenich develop wearable transmitter system	AORL, May–June 1976	Basic science
	Clark at the University of Melbourne initiates R&D on multichannel Cochlear implant	Cochlear Corp. brochure	Basic science
1972	House and Urban develop first "take-home" stimulater package	AORL, May–June 1976	Basic science
	House and Urban Implant first single-wire system	AORL, May–June 1976	Basic science
1973	House began coordinated program of implantation and rehabilitation in United States	AORL, May–June 1976	Basic science
	Eddington begins work on Ineraid	AORL, Nov.–Dec. 1978	Basic science
1974	First International conference on electrical stimulation of the acoustic nerve; objective evaluation of subjects fitted with CI recommended and guidelines provided	AORL, May–June 1976	Legitimation; standards and rules
	Eddington obtains research grant from AT&T	*Surgery of the Ear and Skull,* 1982	Finance
	10 subjects implanted; rehabilitation of first large sample of subjects by House, Urban, Norton, Crary, Wxler, Berliner, and Luckey	AORL, May–June 1976	Basic science

APPENDIX 15–A. (continued)

Date	Event	Source	Key Word
1975	NIH requests various laboratories to submit proposal describing procedure for evaluating patients fitted with CI	RFP-NIH-NINCDS-75-14	Standards and rules
	NIH contract: Stanford	NIH, Dec. 1986	Finance
	HEI receives grants from Truyens	AORL, May–June 1976	Finance
	HEI receive grants from Lillian Disney	AORL, March–April 1982	Finance
	Walt Disney Hearing and Rehabilitation center opened by House	AORL, May–June 1976	Basic science
	House censured by colleagues for implanting patients with single-channel devices	Case history, p. 35	Basic science
1976	Beginning of selection of new subject sample and development of research protocol in psychoacoustics, psychology, and rehabilitation	AORL, May–June 1976	Regulatory
	University of Washington report that their research is funded by four different sources	NIH document July 1984	Finance
1977	HEI receive major grants from different foundations	AORL, March–April 1982	Finance
	University of Melbourne has a fully implantable device	Letter dated July 1986	Basic science
	Review of literature by Bilger characterized much of House's evidence as anecdotal	AORL, May–June 1977	Basic science
	University of Melbourne approached 3M for joint venture; 3M decides to pursue CI program by itself	Case history	Initiation
	Early version of Symbion's CI designed by Eddington implanted	Symbion form 10K, 1984	Basic science
	NIH contract: Stanford	NIH, Dec. 1986	Finance
1978	NIH contract: University of Missouri	NIH, Dec. 1986	Finance
	NIH contract: City University of London	NIH, Dec. 1986	Finance
	Eddington obtains research grant from Mountain Bell	*Surgery of the Ear and Skull,* 1982	Finance
	University of Melbourne implant one patient with multielectrode CI	*Journal of Laryngology and Otology,* March 1985	Basic science
	Nucleus links up with the University of Melbourne	HB, July 1986	Initiation
	3M becomes vendor for HEI	Case history	Manufacturing and assembly
1979	NIH contract: EIC	NIH, Dec. 1986	Finance
	University of Melbourne implants its 10-channel CI in three patients	Cochlear Corp. brochure	R&D

APPENDIX 15–A. (continued)

Date	Event	Source	Key Word
	3M and House begin cooperative R&D	ASHA, May 1985	R&D
	3M takes up initial research of the House device	HB, July 1986	R&D
1980	NIH contract: Stanford	NIH, Dec. 1986	Finance
	NIH contract: Huntington	NIH, Dec. 1986	Finance
	NIH contract: ETC	NIH, Dec. 1986	Finance
	HEI obtains IDE approval from FDA for their device	FDA, Dec. 1986	Regulatory review
	Separate CI program set up at 3M	Case history	Initiation
	Simmons reports 95% of his CI research started in 1960, is funded by NIH	Letter of June 1986	Finance
	NINCDS reports that it supports implant research at UCSF	Undated monogram from NINCDS	Finance
	UCSF and 3M initiate work on multichannel CI	Case history	Initiation
1981	NIH contract: EIC	NIH, Dec. 1986	Finance
	NIH contract: University of Mississippi	NIH, Dec. 1986	Finance
	NIH contract: Hughes Aircraft	NIH, Dec. 1986	Finance
	NINCDS reports that it supports implant research at Stanford University	Undated monogram from NINCDS	Finance
	Simmons/Biostem obtain IDE approval from the FDA	FDA, Dec. 1986	Regulatory review
	Agreement between 3M and Hochmiars	Case history	Initiation
	3M begins R&D work on the Vienna device	Inference	R&D
	Formal agreement between 3M and HEI	Case history	Initiation
	Clinical trials initiated by 3M for House device	RL and HB, July 1986	Clinical trials
	3M initially had to educated FDA about CI	DVDH Interview, Jan. 1986	Regulatory
	Special report on CI in *Hearing Instruments Journal*	*Hearing Instruments Journal*, June 1985	Legitimation
1982	NIH contract: EIC	NIH, Dec. 1986	Finance
	NIH contract: Giner	NIH, Dec. 1986	Finance
	UCSF have linkage with RTI to perform patient testing	HB, July 1986	Basic science
	UCSF have linkage with Duke University to carry out patient implantation	HB, July 1986	Basic science
	University of Melbourne receive grants from Deafness Foundation of Victoria	Undated monogram the University of Melbourne	Finance

APPENDIX 15–A. (continued)

Date	Event	Source	Key Word
	Eddington obtains research grant from Christian Johnson Foundation	*Surgery of the Ear and Skull,* 1982	Finance
	HEI granted IDE for their device for children by the FDA	FDA, Dec. 1986	Regulatory review
	UCSF granted IDE for their device for adults by the FDA	FDA, Dec. 1986	Regulatory review
	Nucleus granted IDE for their 22-channel device by the FDA	FDA, Dec. 1986	Regulatory review
	3M initiates training programs for physicians	RL, Feb. 1987	Competency training
	University of Melbourne/Nucleus begin clinical trials with multiple CI	*Journal of Laryngology and Otology,* March 1985	Clinical trials
	UCSF-3M relationship terminated	Case history	Initiation
	One major industry conference held	RL, Feb. 1987	Legitimation
	Marketing activities initiated by 3M for the House device	RL and HB, July 1986	Marketing and distribution
1983	NIH contract: Stanford	NIH, Dec. 1986	Finance
	NIH contract: RTI	NIH, Dec. 1986	Finance
	NIH contract: UCSF	NIH, Dec. 1986	Finance
	NIH contract: Huntington	NIH, Dec. 1986	Finance
	Biostem begins a partnership with Simmons of Stanford University	RL and HB, July 1986	Initiation
	Biostem develops a bimodal processing system	RL and HB, July 1986	R&D
	Biostem enters into a distribution agreement with Richards	RL and HB, July 1986	Initiation
	Two major industry conferences held	RL, Feb. 1987	Competency training
	UCSF and Storz link up	RL & HB, July 1986	Initiation
	Storz begins R&D work	Inference	R&D
	Wilson of RTI informs of two NIH grants (1983–86 and 1986–87)	Letter of June 1986	Finance
	Department of Science and Technology of Australia enter into a $2m contract with Nucleus to conduct research on CI	*Medical Electronics,* Nov. 1983	Finance
	Kolff Medical signs license agreement with University of Utah for CI	Kolff annual report, 1983	Initiation
	Symbion initiates R&D work	Inference	R&D
	Symbion initiates clinical trials	HB, July 1986	Clinical trials
	Council on Science Affairs of AMA recommends endorsement of CI	Symbion product brochure, 1985	Legitimation
	First efforts made by Symbion to increase third-party insurance companies awareness about CI	SCM	Finance

APPENDIX 15–A. (continued)

Date	Event	Source	Key Word
	3M convinces third-party payors to cover House device; system set up by 3M	SCM	Finance
	3M holds two training programs for physicians	RL, Feb. 1987	Competency training
	3M promotes the Vienna device at a training program	HB, July 1986	Marketing and distribution
	3M submits first PMAA for House device		Regulatory review
	3M obtains IDE approval for Vienna EC	FDA, Dec. 1986	Regulatory review
	Nucleus begins promoting its 22-channel device	RL and HB, July 1986	Marketing and distribution
	Biostem displays its products at the AAO show	RL and HB, July 1986	Marketing and distribution
1984	Symbion obtains IDE approval from the FDA	FDA, Dec. 1986	Regulatory review
	NIH contract: EIC	NIH, Dec. 1986	Finance
	NIH contract: EIC	NIH, Dec. 1986	Finance
	NIH contract: University of Mississippi	NIH, Dec. 1986	Finance
	Lawsuit pending against Biostem filed by Richards	RL and HB, July 1986	Initiation
	3M enters into a relationship a local hospital and forms CISM	SCM	Initiation
	A syndicate of eight firms invest $3m in Nucleus	Undated news clipping	Finance
	Symbion recognized first revenues associated with CI	Symbion annual report, 1984	Marketing and distribution
	3M initiate work on Rehabilitation		Rehabilitation
	Nucleus submits PMAA for their 22-channel device	Inference	Regulatory review
	Nine major industry conferences	RL, Feb. 1987	Competency training
	Two training programs for physicians by 3M	RL, Feb. 1987	Competency training
	Cochlear Corp. established to commercialize CI and enter health care field	Cochlear Corp. fact sheet	Initiation
	Storz initiates its marketing activity	HB, July 1986	Marketing and distribution
	Nucleus sets up large-scale production facilities in Australia	RL and HB, July 1986	Manufacturing and assembly
	Storz receives IDE approval from the FDA	FDA, Dec. 1986	Regulatory review
	Yale University and West Haven VA hospital work on comprehensive set of medical guidelines for cochlear implants	*N.Y. Times* 3-27-84	Standards and rules
	Kolff Medical initiates training programs	Kolff annual report, 1983	Competency training

APPENDIX 15–A. (continued)

Date	Event	Source	Key Word
	ASHA develops special committee on CI	*Hearing Instruments,* June 1985	Legitimation
	3M establishes pilot plant at St. Paul and manufacturing facilities	SCM	Manufacturing and assembly
	AIDC subscribes to $5m worth of shares in Nucleus to help in CI program	*Daily Telegraph,* April 6, 1984	Finance
	Biostim has legal problems	DJN wire, 12/27/84	Initiation
	Symbion initiates marketing of its device	RL, July 1986	Marketing and distribution
	3M receives PMA approval from the FDA the House device	FDA, Dec. 1986	Regulatory review
	FDA sponsors a press campaign on this occasion		Legitimation
	Full issue of *Hearing Journal* devoted to CI	*Hearing Journal,* Nov. 1984	Legitimation
	3M commences clinical trials commenced for the Vienna device	SCM	Clinical trials
1985	NIH contract: Stanford	NIH, Dec. 1986	Finance
	NIH contract: University of Melbourne	NIH, Dec. 1986	Finance
	NIH contract: RTI	NIH, Dec. 1986	Finance
	NIH contract: EIC	NIH, Dec. 1986	Finance
	NIH contract: Hughes Aircraft	NIH, Dec. 1986	Finance
	Biostem no longer active now	RL and HB, July 1986	Initiation
	Three major industry conferences	RL, Feb. 1987	Competency training
	Six training programs organized by 3M	RL, Feb. 1987	Competency training
	With the help of EAR foundation, CI users establish CI club	*3M Newsletter,* Sept. 1985	Competency training
	Nucleus organizes six training programs for physicians	RL, Feb. 1987	Competency training
	Symbion has the manufacturing capacity to produce 500 CIs annually	Symbion annual report, 1984	Manufacturing and assembly
	Book on CI published by Raven Press	Citations	Competency training
	Announcement of Vienna device starts cramping House sales	SCM	Marketing and distribution
	Symbion intends to develop a transcutaneous system in 1985	Symbion Form 10-K, 1984	R&D
	Establishment of titanium vendor by 3M for the House device	SCM	Manufacturing and assembly
	NASHA publishes information packet for prospective patients	NAHSA	Competency training
	U.S. government makes a grant of $200,000 to the University of Melbourne	*Australia News,* Feb. 1985	Finance
	Australian government grants $1m to the University of Melbourne	*Australia News,* Feb. 1985	Finance

APPENDIX 15–A. (continued)

Date	Event	Source	Key Word
	Landmark publication on CI for the deaf children	*Ear and Hearing*, June 1985	Basic science
	Firms respond to a survey by ASHA survey that is widely read	ASHA, May 1985	Legitimation, competency training
	Thirteenth international conference on otolaryngology	Attendance at conference	Legitimation, competency training
	At the thirteenth conference firms promote alternative technologies for the first time	Personal attendance	Marketing and distribution
	Each firm feels that others are making exaggerated claims and express the need for standards	Attendance at conference	Standards and rules
	Full issue of hearing instruments devoted to CI	Hearing Instruments, June 1985	Legitimation
	University of Iowa emerges as a testing and standards center	SCM	Standards and rules
	Surgeons come together to form rehabilitation centers	Attendance at Miami conference	Rehabilitation
	3M acquires Axonics to initiate diagnostic activities	SCM	Initiation, diagnosis
	3M starts servicing the House device	SCM	Servicing
	Nucleus initiates early clinical studies with single channel	Cochlear Corp. fact sheet	Clinical trials
	CI for children that may be extended to multichannel CI	7/85	
	PROPAC carries out a preliminary screening of CI to decide whether it should be given a separate DRG code	OTA-H-262	Finance
	Nucleus 22-channel device receives PMA approval from the FDA	FDA, Dec. 1986	Regulatory review
	Cochlear Corp. received IDE approval from the FDA for children (10–17)	Cochlear Corp. *Issues and Answers*	Regulatory review
	FDA grants 3M IDE approval for Vienna IC	FDA, Dec. 1986	Regulatory review
	3M recalls its House device from the market	3M letter to physicians	Marketing and distribution
	NIH gives University of Melbourne a grant of $300,000 to undertake advanced research for further development of CI	Australian newspaper clipping	Finance
	3M changes thrust of Vienna EC to IC	SCM	R&D
	Nucleus working on redesigning their processor for simultaneous stimulation of all the channels and on miniaturizing their device for implantation in children	HB, July 1986	R&D

APPENDIX 15–A. (continued)

Date	Event	Source	Key Word
	3M, Symbion, Storz, Cochlear Corp. discuss the best way to reduce negative impact to the industry from the recall	SCM	Standards and rules
	3M decides to divest from CISM	SCM	Initiation
	3M gets out of rehabilitation	SCM	Rehabilitation
	FDA does not accept 3M's children PMA application	SCM	Regulatory review
	Formation of Cochlear Implant industry council consisting of 3M, Cochlear Corporation, Storz, and Symbion; initially formed to attempt to get CI under medicare; subsequently, reviewing ways to simplify FDA regulations	SCM	Legitimation, finance, regulation
	American Association of OtoLaryngology initiates standards	Letter of May 1986	Standards and rules
1986	Nucleus organizes six training programs for physicians	RL, Feb. 1987	Competency Training
	3M organizes four training programs for physicians	RL, Feb. 1987	Competency Training
	Symbion expects PMA approval from the FDA in 1986	Symbion annual report, 1984	Regulatory review
	Storz approaches 3M for a possible joint-venture relationship	SCM, Jan. 1986	Initiation
	3M begins research activity for their multichannel CI	HB, July 1986	R&D
	Security analysts' report identifies the growing acceptance of the artificial ear among insurance carriers as an important factor that should facilitate future sales	Special equities rating Rothschild, Unterberg, and Townbin, Jan. 1986	Finance
	In this report, University of Iowa's test results are displayed, which shows that multichannel devices perform very much better than single-channel devices		Standards and Rules
	Symbion signed letter of intent with Bard for long-term sales and distribution rights to Ineraid	*Wall Street Journal*, 3/11/86	Initiation
	Symbion terminates its discussions with Bard and approaches 3M for a possible joint venture	KW, June 1986	Initiation
	FDA grants IDE to 3M for advanced CI	FDA, Dec. 1986	Regulatory review
	FDA grants IDE to Nucleus for children's device	FDA, Dec. 1986	Regulatory review
1986	Storz is acquired by American Cyanamide	CIP review, July 1986	Initiation

APPENDIX 15–A. (continued)

Date	Event	Source	Key Word
1986	3M decides to move into hearing aids	SCM	Initiation
1986	3M divests from diagnostics	SCM	Diagnostics
1986	OHTA publishes technical assessment report and endorses CI	OHTA report, 1986	Legitimation, finance
1986	Medicare granted	SCM	Finance
1986	Five major industry conferences	RL, Feb. 1987	Competency training
1987	NIH contract: University of Melbourne	NIH, Dec. 1986	Finance

Note: The list of events does not include publications in refereed journals or patents awarded.

SECTION
VI

Studies of Adoption of Innovation

The perspective in this section shifts to studies of innovation adoption. Previous chapters focused on processes by which innovative ideas were created, developed, and implemented within the organizations undertaking the innovations. In contrast, this section examines the processes by which organizations adopt innovations that were developed elsewhere.

In his extensive literature review, Rogers (1983) notes that no topic in the social sciences has perhaps received as much study as innovation diffusion and adoption. Whereas most of this research has focused on diffusion, which is largely concerned with the marketing, dissemination, and transfer of an innovation to individual end users, far less has dealt with adoption, or the process by which recipient user organizations select and implement an innovation in their organization. Of this smaller subset of organizational adoption studies, most have focused on statistically examining relationships between various "input" factors (characteristics of organizational context, technology, and structure) and "output" (measures of organizational innovativeness)—leaving the adoption *process* largely unexamined. Yet it is well documented that functionally similar organizations respond and perform differently in adopting similar innovations (Kimberly and Evanisko 1981; Barley 1986). In other words, the process by which organizations adopt innovations makes a difference. These adoption process differences were longitudinally examined in three MIRP studies. Each of these studies comparatively examines the processes and outcomes of adopting a given innovation in equivalent organizational settings.

Chapter 16 describes the organizational-effectiveness implications of two different reactions taken by U.S. nuclear power companies in response to a new set of nuclear safety procedures mandated by the U.S. Nuclear Power Commission. The findings indicate that the nuclear power plants with relatively poor safety records tended to repond in a rule-bound manner that perpetuated their poor safety performance, and that nuclear power plants whose safety records were relatively strong tended to retain their autonomy, a response that reinforced their strong safety performance. Marcus and Weber make a very important and generalizable inference from these findings for the management of externally induced innovations: Managers should be aware of the possible consequences of blind acceptance of external dictates, and regulators should take heed of companies that strictly obey the law. These companies may be doing so in "bad faith" and may not achieve the results the regulators intend.

Chapter 17 compares two common alternative strategies for adopting and implementing an innovation: a "depth" strategy in which an innovation is implemented and "debugged" in a demonstration site before it is generalized to other organizational units, and a "breadth" strategy in which an innovation is implemented through successive hierarchical levels across all organizational units simultaneously. Lindquist and Mauriel examine the processes and outcomes of the two adoption strategies in two public school districts that are implementing site-based management, an administrative innovation for decentralized management being adopted by many independent school districts in the country. Contrary to expectations, the school district that implemented the innovation in "breadth" (across all schools in the district) was more successful in implementing and institutionalizing more components of the site-based management program than the school district that adopted the strategy of introducing it in "depth" within a school selected as the demonstration site.

Finally, whereas Chapters 16 and 17 report of studies that simply observed the adoption process over time, in Chapter 18 Bryson and Roering take a more active participant-observer role in describing and evaluating efforts by six governmental units to adopt strategic planning systems. While strategic planning has been in vogue in private-sector organizations for over twenty years, the concept is just now gaining good currency in local units of government and municipalities as public pressure mounts to cut government spending and taxes while demand continues to grow for government services. In this political context, Bryson and Roering observe that each attempt to adopt the strategic planning innovation was prone to disintegration. External events and crises frequently occurred that distracted participants' attention and priorities and took away slack to innovate. The adoption process itself was partially cumulative; what occurred before had to be accounted for, and participants got bogged down with information overload, conflicting priorities, and divergent issues that were outside of the decision domain of the participants. Paradoxically, Bryson and Roering point out that the organizational units where adoption of the innovation was most needed is where it was the least likely to work.

REFERENCES

Barley, S. 1986. "Technology as an Occasion for Structuring: Evidence from Observations of CT Scanners and the Social Order of Radiology Departments." *Administrative Science Quarterly* 31: 78–108.

Rogers, E. 1983. *Diffusion of Innovations.* New York: Free Press.

Kimberly, J., and M. Evanisko. 1981. "Organizational Innovation: The Influence of Individual, Organizational, and Contextual Factors on Hospital Adoption of Technolgical and Administrative Innovations." *Academy of Management Journal* 24 (4): 689–713.

16

EXTERNALLY-INDUCED INNOVATION

Alfred A. Marcus

Mark J. Weber

Innovation can come into being when individuals in an organization see an opportunity that necessitates a new approach (Andrews 1971; Child 1972; Bourgeois 1984). However, it also can come into being when an external threat or challenge occurs that has not been anticipated. The stimulus to most innovation is external, according to March and Simon (1958), who attribute it primarily to necessity not to opportunity. Terreberry (1968) maintains that innovation is largely a matter of external inducement as do Downs (1967), Kelly and Kranzberg (1975), and Zaltman and Duncan (1977).

An example of an externally-induced innovation is a government requirement thrust on firms because of an external circumstance—an accident, scandal, or financial crisis—that is beyond their control. An innovation may be mandated by government agencies, but the role of market signals in inducing innovation is no less important. When the survival of an organization or some program within it is threatened because of difficulties—whether technical, legal, or economic—it is subject to increased influence from outside forces such as government agencies, industry trade associations, financial institutions, and other parties.

The difficulties can affect the organization at the outset of the innovation process or as a series of setbacks and surprises that occur during the innovation's development. In either case, outside bodies will try to introduce new ideas and practices. When recognition of the need for the new ideas and their early manifestations come from outside sources, then an innovation can be considered externally induced (Rogers and Shoemaker 1971).

Whether induced by government mandates or market signals, the new ideas will be perceived in different ways by people in the organization as complex transactions take place at each level in the organization (Wilson 1963). Conflict between different levels in the organization can occur as new managers who come on board because of the crisis impose ideas on staff who are reluctant to depart from the status quo. Externally induced innovation is relevant to the internal mandates that affect the departments and subunits and to the group mandates that affect individuals as well as to the external mandates that influence the organization as a whole. What is external and internal to the organization is largely a matter of frame of reference (see Burgelman 1983).

A classic agency problem exists (Mitnick 1980) when outside parties or interests expect members of the organization to act for, or on behalf of, the new ideas that the outsiders are trying to induce. Because the ideas may be complicated and difficult to carry out and the people in the organization have their own interests and perspectives, the tendency will be for the ideas to proliferate based on the different interpretations of their meaning and significance (Van de Ven 1986: 597; also see March and Simon 1958). If the ideas are interpreted differently, then slippage between what the inducers want and the organization does is inevitable. The monitoring costs are too great to justify the constant surveillance needed to produce perfect compliance.

In their altered form the ideas will affect the organizations, performance with a number of results being possible including formal compliance to the ends and means the ideas are intended to serve and formal disregard of both (Merton 1938; Beyer and Trice 1978; Palumbo, Maynard-Moody, and Wright 1983). The organization can achieve the goals that the outside parties are seeking while violating their form and content, or it can abandon the original goals while formally complying. Externally induced innovation affects the ideas, people

events, transactions, and the outcomes in an organization. It can thwart technological improvement or promote it, distorting the innovation process directly or indirectly (see Schweitzer 1977; Rothwell 1981; Ettlie and Rubenstein 1981).

This chapter tries to explain how organizations are likely to respond to externally induced innovation. It is based on a review of the literature and findings from a study of safety review practices introduced by the Nuclear Regulatory Commission (NRC) after the Three Mile Island (TMI) accident. In implementing externally induced innovations, two cycles are shown to exist:

1. A "vicious" one in which poorly performing organizations respond with rule-bound behavior, a response that perpetuates their poor performance; and
2. A "beneficient" one in which better performing organizations have autonomy, a response that reinforces their strong performance (see Figure 16–1).

We argue that the less rule-bound and more autonomous approaches to externally induced innovation are likely to be more effective: As opposed to doing precisely and only as much as the external body demands, better results are likely to come from customizing what an external agent requests or carrying out what it might require before it is necessary (See Marcus, 1988a and 1988b). The next section develops the rationale for why the less rule-bound and

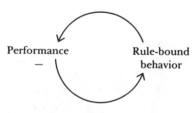 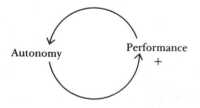

Source: Marcus 1988b, p. 396.

FIGURE 16–1. Vicious and Beneficient Cycles.

538

more autonomous approaches are appropriate. The following section presents the findings from the nuclear power study and explains what we mean by vicious and beneficent cycles. The implications and qualifications are discussed in the concluding section where comparisons are made with other MIRP studies.

WHY LESS RULE-BOUND AND MORE AUTONOMOUS APPROACHES ARE APPROPRIATE

This section relies on the literature to make four points. First, external jolts are often needed to stimulate innovation. Second, rule-bound approaches are not the appropriate response to these jolts. Third, autonomy is possible even when an organization faces the extreme demands and pressures brought on by these jolts. Finally, autonomy is not only possible under these circumstances, but it is likely to be necessary and beneficial in effectively implementing externally induced innovations.

Jolts Are Often Needed to Stimulate Innovation

Jolts, both external and internal, play a useful role in the innovation process. As Van de Ven (1986: 591) points out, people are programmed to "focus on, harvest, and protect existing practices." They therefore are likely to resist innovation. To stimulate the introduction of new programs and practices, disruptive events, which threaten a social system, are often necessary (Schon 1971).

The insight that crises, disatisfaction, tension, and significant external stresses play an important role is common. Wilson (1963: 255), for example, comments that organizations are unlikely to innovate unless there is a "crisis—an extreme change of conditions for

which there is no programmed response." Crozier (1964:196) writes:

> because of the resistance it must overcome, change in bureaucratic organizations is a deeply felt crisis. . . . Most analyses of the bureaucratic phenomenon refer only to the periods of routine. . . . But it is a partial image. Crisis is a distinctive and necessary element . . . It provides the . . . means of making . . . necessary adjustments . . . and enabling the organization to develop.

Without a severe shock or jolt, people's tendency is to unconsciously adapt to slowly changing conditions. Bateson (1979, quoted by Van de Ven 1986: 595) provides the following example:

> When frogs are placed into a boiling pail of water, they jump out—they don't want to boil to death. However, when frogs are placed into a cold pail of water, and the pail is placed on a stove with the heat turned very low, over time the frogs will boil to death.

Cyert and March (1963) argue that organizations continue with their collective routines until events require adaptation. Factors that can intervene include strong differences between expectations and aspirations, constant failure to meet objectives, and the imposition of external demands. Meyer (1983:515) refers to unwelcome surprises as "jolts" and defines them as "transient perturbations whose occurrence is difficult to foresee and whose impacts . . . are disruptive and potentially inimical." Just as "seismic tremors often disclose hidden flaws in the architecture and construction of buildings," so too environmental jolts are likely to "trigger responses that reveal how organizations adapt to their environment." Differences between expectations and aspirations and failure to meet goals can be manipulated by managers through the goal setting process. Thus, internal as well as external factors play a role in innovation. Jolts, according to Meyer (1983:-533), are "propitious opportunities for bootlegging incidental changes into organizations by camouflaging them as responses." When they are labeled as crises, they "infuse organizations with energy, legitimize unorthodox acts, and destabilize power structures."

Why Rule-Bound Approaches Are Not Appropriate

When unexpected events happen that create a perception of crisis, defense mechanisms are apt to emerge. One of the strongest is a reliance on customary rules—both the explicit ones that are codified in writing and the implicit ones that are manifested in standard procedures and routines (see Janis 1982). Starbuck and Hedberg (1977:250) have argued that successful organizations develop routines for dealing with recurring problems but as these routines become too well-entrenched, members of the organization tend to see situations as equivalent even if they are not. The result is that "programs remain in use even after the situation they fit has faded away" and that an organization's initial success breeds failure unless it rapidly revises its routines.

It is not that rules per se prevent creative adjustment but the lack of alternative rules and a triggering process to activate them. If the rules are narrow, circumscribed, and widely shared, they act as a defense against the perception of threat, blunting the impact of external challenges and preventing creative adjustment (Bromiley and Marcus 1987). The appropriate internal response to a situation of external disturbance is decreased reliance on existing rules and routines and dependence on individual coping initiatives during the crisis interim. Organizational behavior is often a habitual response to familar circumstance. Standard operating procedures seem to emerge automatically, at an "inert" level (Leibenstein 1976) according to well-learned "scripts." On the one

hand, these routines play a positive role in reducing uncertainty as they yield rapid reaction to similar situations and stimuli. On the other hand, in the face of novel circumstances, they may be radically dysfunctional. In these situations, lowered cognitive functioning and reduced responsiveness are not appropriate unless the individual coping mechanisms are activated.

Is Autonomy Possible?

There is great uncertainty about the extent to which managers must succumb to external pressures and the extent to which they can maintain their autonomy (See Van de Ven, 1979). Hannan and Freeman (1977), Aldrich (1979), and McKelvey (1982) maintain that the environment is a selection mechanism that determines their current behavior and the future performance of their organizations. In the face of external pressures, managers have few meaningful choices (see also Romanelli and Tushman 1986). According to Ashby (1958), the limiting factor is the organization's internal variety or the repetoire of responses it can generate in response to a situation. Autonomous choice is at best "problematic," according to Aldrich (1979).

Crozier (1964:156), on the other hand, maintains that managers act to preserve their autonomy, accepting dependence "only insofar as it is a safeguard against . . . [further] submission." Child (1972), Bourgeois (1984), and others have emphasized the role of managerial choice in shaping domains and the characteristics of an organizations's environment. Pfeffer and Salancik (1978:257), for example, hold that managers are involved in a "constant struggle" to maintain their "autonomy and discretion." To prevent losing autonomy, they act to reduce external dependence. They manage external demands without necessarily satisfying them. What allows them to do so is the ambiguity of most demands (on the concept of choice within constraints, see Hrebiniak and Joyce 1985). Equivocality gives managers choices. Scott (1987) shows that the demand for conformity is not categorical; it may apply to structure, procedures, *or* personnel but not every area. Similarly, Van de Ven and Drazin (1985) maintain that even if constraints operate at a macrolevel, at a microlevel managers have substantial discretion. They have the authority to employ "switching rules"—that is, to apply different designs within different subunits.

Autonomy Is Necessary and Beneficial in Implementing Externally Induced Innovations

Having autonomy does not guarantee that implementation will be a success. Once an idea has been adopted, a series of transactions take place that will affect how it is carried out. A common view is that implementation follows conception, proposal generation, and initiation and that the factors that facilitate the former inhibit the latter (Wilson 1963; 200; Duncan 1976; 172; also see Rothwell and Zegveld 1985; Strebel 1987). Rule-bound approaches, which involve central direction and highly programmed tasks, are supposed to promote implementation; that is, the number of routine tasks prescribed from above should increase as the organization moves to this stage (Wilson 1963:198). Conception, proposal generation, and initiation, on the other hand, require fewer controls and more autonomy; this is because diversity, openness, informality, and the ability to bring a variety of bases of information to bear on a problem need to be encouraged (Duncan 1976:174). Duncan (1976) suggests that making the transition from conception, proposal generation, and initiation to implementation can be difficult and that the "ambidextrous" organization that is adept at moving from stage to stage is likely to be rare.

There is a large body of research that deals with the problems of implementation.

Early case studies (see Pressman and Wildavsky 1974; Bardach 1977; Marcus 1980) support the view that excessive decisionmaking autonomy during implementation is counterproductive. When numerous decisions have to be made and many participants are involved, the probability of success decreases and the possibility of unexpected problems arising increases when autonomy is great. Critics (see Lipsky 1978; Elmore 1979; Thomas 1980; Berman 1980; Palumbo, Maynard-Moudy, and Wright 1983) of this view, however, contend that implementors have greater knowledge of multiple and contradictory demands and of conflicting legal, political, professional, and bureaucratic imperatives at the point of delivering a product or service (see Rein and Rabinowitz 1978); that the disposition of implementors is likely to be negatively affected if they are not granted a sufficient level of autonomy; and that their dispositions are often critical to policy success (see Van Meter and Van Horn 1975; Edwards 1980). According to Fidler and Johnson (1984:704), implementors may engage in "routine, mechanical operationalization" that can sabotage or impede successful implementation, if they are not given the opportunity to make modifications based on their experience.

Bourgeois and Brodwin (1984:241–42) illustrate the differences between rule-bound and autonomous styles in implementation by identifying five models of implementation including a "commander" model and a "crescive" model. The commander model, which is similar to the rule-bound approach, has a strong normative bias toward central rulemaking and enforcement, while the crescive model, which is similar to the autonomous approach, draws on managers' natural inclinations to develop opportunities as they see them in the course of day-to-day management. Bourgeois and Brodwin (1984: 260–61) maintain that although none of the approaches that they identify is correct in all situations, greater use should be made of the crescive model in environments characterized by frequent change (also see Mintzberg 1978; Nutt 1983). In a similar manner, Linder and Peters (1987: 462) summarize the implementation literature by arguing that two schools have emerged. One views implementation problems from the vantage point of a central authority that wants to see subordinates follow its rules. This is the rule-bound approach. The other approach accepts that people in the lowest echelons of the organization exhibit autonomy during the course of implementation. Rather than a command and control perspective, this approach emphasizes that bargaining takes place and that policies change during implementation (Linder and Peters 1987: 462).

Empirical studies by Beyer and Trice (1978), Guth and Macmillan (1986), and Maynard-Moody, Musheno, Palumbo, and Oliverio (1987) support the view that autonomy is often needed to facilitate successful implementation. Beyer and Trice (1978: 264) conclude that "if upper management wants better performance . . . , it will have to grant . . . more influence in decisions . . . [and] directors will have to grant more autonomy to their subordinates." Their argument is that people are unlikely to take responsibility for implementing something new unless they have discretion and are accustomed to taking responsibility. Guth and Macmillan (1986:321) find that if general management imposes its decisions, "resistance by middle management can drastically lower the efficiency with which the decisions are implemented, if it does not completely stop them from being implemented." Middle managers who believe that their self-interest is being compromised can redirect a strategy, delay its implementation, or reduce the quality of implementation. Maynard-Moody et al. (1987: 1) find that "empowered street-level workers of decentralized human service organizations play a more substantial role in the successful implementation of social policy than do less empowered workers in highly bureaucratic organizations." Only

Nutt (1986, 1987) appears to find mixed support for autonomy. In the innovations he analyzes, upper managers have the highest success rate in installing planned changes in organizations when they justify the need for change and play a critical role in formulating a plan, illustrating how performance can be improved, and showing how the plan improves performance.

THE NUCLEAR POWER STUDY

The implementation literature, however, has not been cumulative (Lester, Bowman, Goggin, and O'Toole 1987), and most of the empirical evidence comes from the study of social policies—such as alcoholism programs, half-way houses, and hospitals. No study has specifically focused on an externally induced innovation as it affects a high risk technology like nuclear power.

In *Normal Accidents* Perrow (1983a: 10) suggests that a major dilemma in the organization and management of nuclear power plants is how to balance what we call rule-bound and autonomous approaches:

> High risk systems have a double penalty; because normal accidents stem from the mysterious interaction of failures, those closest to the system, the operators, have to be able to take independent and sometimes quite creative action. But because these systems are so tightly coupled, control of operators must be centralized because there is little time to check everything out and be aware of what another part of the system is doing. An operator can't just do her own thing; tight coupling means tightly prescribed steps and invariant sequences that cannot be changed. But systems cannot be both decentralized and centralized at the same time.

In "The Organizational Context of Human Factors Engineering," Perrow (1983b) develops arguments for operator autonomy. He maintains that efforts to centralize authority and control the actions of operators by reducing their role to passive monitoring so that they no longer have significant decisions to make end up "deskilling" the operators and increasing the chances of error. These efforts encourage low system comprehension, low morale, and an inability to cope with anything but the most routine conditions. Autonomy is needed to encourage a higher level of commitment and a greater level of knowledge. Similarly, Weick (1987: 122–23) highlights the importance of autonomy while recognizing that balance between autonomy and rules is necessary to achieve reliability in high-risk technologies.

The problems of choosing between autonomous and rule-bound approaches for the organization and management of nuclear power plants are discussed here. We maintain that the more that managers exercise choice within a situation of constraints (see Hrebiniak and Joyce 1985), the better that the outcomes will be. That is, there is likely to be a relationship between the degree of autonomy nuclear power managers retain in implementing externally induced innovations and the subsequent safety outcomes that their nuclear power plants exhibit.

The concept of self-perpetuating organizational cycles (Masuch 1985) is relevant. If the prior safety record of a nuclear power plant is poor (it has had many safety events), managers will feel that they have little latitude: They have to carry out the rules precisely as they have been prescribed. Nuclear power plant managers and regulators will become more rule conscious when a plant is having more events and this may explain their rule-bound behavior. On the other hand, if the prior safety record of a nuclear power plant is good (the plant has had few safety events), its managers are likely to enjoy increased discretion. Regulators are less likely to intervene in day-to-day decisionmaking, which may partially explain the autonomy they exhibit.

In trying to avoid undesired outcomes, organizations actually can contribute to them. In implementing externally induced innovations there is likely to be (1) a vicious cycle in which organizations with poor safety records respond with rule-bound behavior, a response that perpetuates the poor outcomes, and (2) a beneficent cycle in which organizations with better safety records retain their autonomy, a response that reinforces their strong records. Evidence for the existence of these cycles comes from an examination of safety review innovations induced by the Nuclear Regulatory Commission at nuclear power plants after Three Mile Island (TMI).

Background on the Three Mile Island Study

TMI, one of the worst industrial accidents in history, has been thoroughly studied by the NRC, industry, public interest lobbyists, and academics. Some of this work is quite pessimistic about the prospects for nuclear power. Ford (1981), for example, found inertia and unwillingness to change in the nuclear industry. Perrow (1983a) suggested that accidents were inevitable and that little could be done to prevent them. Many analysts (Perrow 1983a differs) attributed what went wrong to human error (Egan 1982). Apparently, as a result of repeated assurances that the technology was safe, there was a "mindset" that the equipment was infallible and a preoccupation with the technical aspects of nuclear power as opposed to the human dimensions (Sills, Wolf, and Shelanski 1982). Institutional and organization inadequacies, however, were said to have contributed as much to the accident as mechanical breakdowns. For example, according to investigations of the accident, one of the reasons it took place was that lessons had not been learned from similar events that had occurred at other nuclear power plants (Rogovin 1979). Even before TMI, there was concern about an increase in the number of these unsafe events. Occurrence of these events had outpaced the growth in the number of new nuclear power plants, escalating from about ninety per year in 1970 to more than 3,000 per year in the late 1970s (Del Sesto 1982).

The Independent Safety Engineering Group (ISEG), the innovation that we will examine in this study, was introduced by the NRC after TMI to deal with this problem (NRC 1980). The NRC proposed that all newly licensed power plants should have an ISEG to monitor events at a plant and at other plants, to learn the appropriate lessons, and to implement prevention strategies. This innovation had not been sought by the nuclear power industry or by the utilities within it but had been thrust on them by the NRC because of the unfortunate TMI accident. The idea, as developed by the commission in revised standard technical specifications, was unique in three ways. First, there was the focus on events and their prevention—on examining events at a plant and at other similar plants that might indicate areas for improving safety. Second, the NRC proposal that newly licensed nuclear power plants have a *full-time* safety review staff was unique. Third, the NRC held that this staff be independent from nuclear power production. Thus, the five full-time engineers would be on site reporting to someone off site not in the chain of command for power production.

Four dimensions proposed by Beyer and Trice (1978) can be used for assessing the extent of this change. The new resources that were required were evidence of the *magnitude* of the task. The ISEG was an expensive addition, as five full-time engineers could cost a nuclear power plant more than a half million dollars annually. ISEG also had a *pervasive* character because as developed in the standard technical specifications, the five full-time engineers were supposed to devote exclusive attention to examining events and to suggesting ways to prevent them. The presence and functioning of an

ISEG was supposed to affect others at the plant in that all employees presumably would become more safety conscious. There was the *novelty* of safety engineers, outside the chain of command for nuclear power production, interacting with operators and production personnel and trying to influence their behavior. The only aspect about ISEG's innovativeness that may be questioned was its *duration:* How long would NRC be committed to ISEG in the form in which it was proposed? Soon after requiring that newly licensed plants implement the ISEG, the NRC initiated a study to review the ISEG and the other safety review procedures at nuclear power plants to determine if the ISEG should be extended to all power plants or if safety review systems at nuclear power plants should be revised in some other way.

Methods

In our analysis of the approaches nuclear power plants took toward implementing the ISEG, we have used both qualitative and quantitative methods. The NRC establishes what it calls *standard* technical specifications when it creates regulations for nuclear power plants; individual nuclear power plants then are allowed to customize these rules in *individual* technical specifications that the NRC must approve. A comparison of the standard and individual technical specifications along with interviews at nuclear power plants have been used to classify the implementation approaches. The implementation approaches as classified then have been related to safety outcomes and other measures of nuclear power plant performance. As NRC's desire in creating new safety review arrangements was to reduce the number of unsafe events, the safety events are the main safety measure that have been used.

Document analysis and interviews. We surveyed the situation at the end of 1981 when the United States had seventy-two licensed nuclear power plants. TMI took place in April 1979. After numerous reports about the accident were published, the NRC established the ISEG requirement in September of 1980 (NRC 1980). To add as many as five full-time engineers required a relatively long lead time because of well-documented shortages of skilled personnel in the nuclear power industry. Moreover, the adjustment of nuclear power plants to the post-TMI situation was long and complex because of the many other post-TMI changes that the NRC required (the TMI action plan had over 100 items). The interviews we conducted confirmed the impression of relatively slow adjustment in the post-TMI period and showed that in many cases that technical specifications were incomplete or inaccurate. Thus, the interviews we conducted served as a check on the document analysis.

The plants were these interviews were carried out were located in the eastern, midwestern, and southern parts of the United States. They had different reactor types (both pressurized water and boiling water), reactor suppliers (Westinghouse, Babcock and Wilcox, and GE), architect engineers, dates of initial commercial operation, and electrical power generating capabilities. The utility systems to which they belonged differed in their structure, size, and profitability. Although the sample was not entirely random, it is fairly typical of what could have been found in the nuclear power industry in 1982.

Three days were spent at most of the facilities, with visits to both the corporate office and the plant site. To ensure objectivity, interviews were conducted by a team that included Professor Marcus and at least one person with a disciplinary background different from Professor Marcus. Usually that person was an engineer with some nuclear power training. Eighty open-ended interviews with safety review staff at thirteen plants were carried out between February and September 1982.

Questions were posed about why a particular method of safety review was chosen and how this method of safety review had been implemented. The questions covered the pre-TMI requirement that plants have plant and corporate safety review groups as well as the post-TMI requirement that newly licensed plants have an ISEG. As the interviews were designed as a check on the document analysis, they followed the format of the document analysis with questions about the rationale, mission, composition, major tasks, processes, output, and workflow relations of the safety review groups. Although the questions were standardized with their precise sequence and wording determined in advance, interviewers were encouraged to probe for additional responses and to obtain other types of feedback when appropriate.

Variables

On the basis of the document analysis, the interviews, and other information that was available to us, we developed the following variables.

Implementation Approaches.

The documentary record and the interviews were used to construct a typology of implementation approaches. In further analysis we refer to this variable as IMPAPP. The primary distinction that we make is between *rule-bound* behavior that we operationalize as compliance with the standard technical specifications and *autonomy* that we operationalize as customizing these guidelines through the adoption of unique, plant specific characteristics.

To ensure coding reliability, at least three of the persons who helped with the interviews played a role in the analysis. They independently classified the safety review systems of the plants they had visited based on the documents examined (primarily the technical specifications, but during the site visits safety review staff often volunteered additional documents) and the interviews conducted. Placement in the primary groups—autonomous and rule-bound—by the analysts was virtually identical.

As a further check on this analysis, two steps were taken. First, our classification scheme was presented to the NRC officials responsible for the ISEG program. Although they contributed some refinements, they made no major modifications. Second, early drafts of the findings were shared with safety review staff who had been interviewed. As Patton (1980) remarks, analysts can learn a great deal about the "accurateness, fairness, and validity" of their findings from their subjects' comments. These checks indicated a consensus among nuclear power plant staff, the analysts, and the NRC about the primary classification in rule-bound and autonomous categories.

Safety Outcomes. The main safety outcome that we used was the number of unsafe events attributable to human error that occurred in 1982. We refer to this variable as HUM_{82}. We also looked at the total number of unsafe events that occurred in 1981 and 1982 and to the significant events that occurred in 1982. We refer to these variables as $EVENTS_{81}$, $EVENTS_{82}$, and SIG_{82}.

Unsafe events are reported to the NRC in the form of license event reports (LERs). They are one of the main methods that nuclear power plants and the NRC have for assessing safety (see Osborn and Jackson 1987 and Olson, McClaughlin, Osborn, and Jackson 1984 for a discussion of LERs and other methods of assessing nuclear power safety). Events attributable to human error constitute anywhere from a third to a quarter of the total number of LERs. Significant events are deemed by the NRC to involve serious deficiencies in major safety related systems. They may involve a call by the NRC to shut down a nuclear power reactor. While we had comparable records for the total number of events in 1981 and 1982, comparable 1981 and 1982 records for the

human factor and the significant events were not available.

Other Performance Measures. The limitations of using LERs as a measure of safety include a tendency on the part of some plants to report events more readily than others and different amounts of on-line time and other operational features that can affect a nuclear power plant's susceptibility to events. Because of these limitations, other performance measures were examined. NRC has selectively assessed the management capabilities of nuclear power plants based on various criteria. According to these assessments, if a plant is given a rating of 1, it means that management attention and involvement are "aggressive;" a rating of 2 means that management attention and involvement are "adequate"; and a rating of 3 that "weaknesses are evident." These criteria, which do not depend on self-reporting by the plants, may be less prone to manipulation by plant managers than LERs. However, they are still highly subjective inasmuch as they depend on the impressions formed by NRC staff during relatively brief site visits. NRC is aware of this limitation and largely for this reason has discontinued the management assessments. We, therefore, relied primarily on the LERs and used the management ratings only in a supplementary fashion. We refer to the management ratings as $MGMT_{82}$.

To correct for different amounts of on-line time and other operational features that can affect the number of events a plant has, we examined 1982 plant capacity ratings. Capacity ratings show the percentage of electric power that a nuclear power plant has generated in a particular period in comparison with the amount that it could generate based on its overall capacity. Downtime is expensive in a capital intensive industry like nuclear power and capacity factors are critical to a utility's financial performance. We call this variable CAP_{82}.

This indicator was very important to nuclear power plant managers and some even had instruments on their desks that provided them with up-to-the-minute reports of their progress. For our purposes, it had significance for two reasons. First, a plant can have fewer events because it has been shut down for a significant period of time; this can occur because of technical problems or it can be the result of a reduced demand for power. If plants had been shutdown for a long period of time, it would show up in low capacity ratings. The second reason for examining capacity ratings is that variations in the number of events can occur because of tradeoffs that nuclear power managers have made among different measures of performance. Conflict among competing performance goals has been noted by many scholars including Cyert and March (1963), Dill (1965), Miles and Cameron (1982), and Sonnenfeld (1982). Safety can be jeopardized to increase productive efficiency, or productive efficiency can be sacrificed for the sake of safety. If safety had been sacrificed, it would show up in a higher capacity factor combined with a lower safety rating.

Controls. Other factors besides the implementation approaches (IMPAPP) may have caused the variations in outcomes that we observed. We therefore introduced the following control variables into our analysis:

Age: Newer plants (measured by year of commercial operation in 1982) may have had more safety events because of start-up problems; or older plants may have had difficulties because of equipment obsolescence and maintenance failures.

Profitability: Profitable utilities (measured by return on equity in 1981–82) may have had the resources to be able to pay for increased safety; or less profitable utilities may have had to make sacrifices to maintain plant safety.

Size: A large commitment to nuclear power (measured by net megawattage of operational nuclear capacity in 1982) may have meant possession of the overall technical resources necessary to run safer plants; or a small commitment may have meant less bureaucracy and more flexibility and therefore an ability to manage nuclear power plants more safely.

Long-term debt: Increased debt (measured by long-term debt in 1981–82) may have meant greater spending on staff and other items related to safety, or it may have meant that a utility had less slack to pay for safety.

We called these variables AGE, PROFIT, SIZE, and LTD.

Analysis Strategy. An intercorrleation analysis of all the variables in the study was carried out first. We then compared the safety and performance outcomes (HUM_{82}, $EVENTS_{81}$, $EVENTS_{82}$, SIG_{82}, $MGMT_{82}$, and CAP_{82}) of plants having rule-bound and autonomous approaches. To test for the significance of the differences between means, t statistics were computed by running a series of regressions with IMPAPP as the independent variable and the variables listed above as dependent variables; the independent variable was a dummy variables with rule-bound = 0 and autonomy = 1.

To test for vicious and beneficent cycles, two determinations were necessary: first, if a plant's prior safety record ($EVENTS_{81}$) influenced its implementation approach (IMPAPP); and second, if the implementation approach (IMPAPP) then affected the number of safety events attributable to human error (HUM_{82}). A probit analysis was necessary to test whether $EVENTS_{81}$ affected IMPAPP because the dependent variable is dichotomous (implementation approaches are either rule-bound or autonomous). To test whether IMPAPP, controlling for age, profit, size, and

547

long-term debt, affected subsequent safety outcomes, four regressions were estimated:

$$HUM_{82} = a + bIMPAPP + bEVENTS_{81}$$

$$+ e \qquad (16.1)$$

$$HUM_{82} = a + bIMPAPP + bEVENTS_{81}$$

$$+ bAGE + bPROFIT + e \qquad (16.2)$$

$$HUM_{82} = a + bIMPAPP + bEVENTS_{81}$$

$$+ bSIZE + bLTD + e \qquad (16.3)$$

$$HUM_{82} = a + bIMPAPP + bEVENTS_{81}$$

$$+ bAGE + bPROFIT + bSIZE$$

$$+ bLTD + e \qquad (16.4)$$

Results

Two responses were classified as rule-bound behavior and four as autonomy (see Table 16–1). Those classified as rule-bound behavior were

1. Conformity (literal compliance with the external mandate; and
2. Incremental adjustment (minor alterations or adjustments).

Two plants had an ISEG exactly as NRC proposed. We therefore classified this response as *conformity*. Plants licensed prior to TMI did not have to have an ISEG or ISEG equivalent. For these plants, rule-bound behavior meant doing what the NRC expected and little more. To the extent that they modified their behavior after TMI they created subcommittees as appendages to their part-time safety groups (two plants) or added a single full-time safety review position (one plant). Thus, the response of these plants was called *"incremental adjustment"* (see Lindblom 1959 and Quinn 1980).

Responses classified as autonomy were

1. Modification (developing and applying an entirely different concept from that induced by external agents);
2. Combination (placing the five full-time engineers called for by NRC in an existing function) (quality assurance);
3. Planning (carrying out detailed studies and partially implementing NRC's standard before it was obligatory); and
4. Anticipation (acting on a plant's own initiative to implement NRC's concept before it was mandatory).

Two plants, which had been licensed after TMI, were in the process of creating a corporate nuclear safety review department with responsibility for both offsite review and onsite safety

TABLE 16–1.* Implementation Approaches after Three Mile Island (TMI).

	Licensed After TMI[a]	Licensed Prior to TMI[b]
Rule-bound behavior	Conformity: ISEG (2)	Incremental adjustment: subcommittees, full-time person (3)
Autonomy	Modification: Nuclear Safety Department (2) Combination: quality assurance (2)	Planning: technical support function (2) Anticipation: ISEG-like group (2)

Note: Number of plants pursuing these responses in parentheses.
a. ISEG or ISEG-like equivalent required.
b. ISEG or ISEG-like equivalent not required.
*This table is taken from Marcus (1988b), p. 392.

engineering. The head of this department had vice president status and reported directly to a top corporate officer. Because these modifications were intended to achieve the intent of NRC's proposal through entirely different means, this response was called *modification*.

Another response was to combine the existing quality assurance function with safety engineering. Two plants simply added the five full-time safety engineers to their existing quality assurance staff. Doing so altered the nature of what NRC intended. The distinction NRC was trying to make was between the "policeman" role that quality assurance traditionally performed and the ability *to challenge* existing procedures that the ISEG was supposed to carry out. Because these plants combined these functions, their response was called *combination*.

Plants licensed before TMI also expressed autonomy in two ways. Significant planned and actual alterations of safety review systems, when they were not required, suggested that these plants were acting on their own initiative in response to what they believed to be the lessons of TMI. Two plants simply planned for adoption, taking comprehensive steps to consider what they might do. They did detailed studies that would have created an entirely different type of safety review system. The proposed technical support group would have aided existing review groups as well as having responsibilities of its own, with full staffing taking place only if ISEG or an ISEG equivalent were mandated. Partial staffing had started even though implementation was not obligatory. Full staffing would take place if an ISEG or ISEG equivalent became mandatory. This response was called *planning*.

A different approach was to create an ISEG-like group, which was the equivalent of what NRC proposed, because management believed that such a group was necessary. To the extent that these two plants complied with NRC's proposal they did so voluntarily in a proactive manner and not because of NRC pressure or fear of NRC disapproval. The response of these plants was called *anticipation*.

A Comparison of Plants That Conformed and Plants That Anticipated: While full advantage cannot be taken of the qualitative analysis because of its length (see Marcus and Osborn 1983), a brief comparison of conforming and anticipating plants is revealing, especially in regard to the disposition of the implementors and how they shaped the responses of their plants. At the conforming plants, the ISEG offices were located in the plants' parking lot and group members had to obtain visitors' badges before entering a plant. Staff maintained that ISEG's role had not been well defined, it did not fit in with existing practices, and it was not likely to have a major impact. The ISEG was making many recommendations, but these recommendations were not accepted. The plant manager pointed to a huge stack of papers in the corner of his office and said, "Do you know how many of these [recommendations] we have acted on?" Showing a space of about a quarter of an inch between his thumb and forefinger, he answered, "That much."

In contrast, at the plants that anticipated the NRC requirements, the safety review managers we interviewed maintained that the ISEG-like group they introduced had technical potential. It was compatible with existing practices and its impact was important. The members of the ISEG-like group were reported to have had "years of operating experience." They were capable of understanding plant personnel. They had an appreciation for "what was possible," and could "put in perspective" whether something was "significant." Their recommendations, both formal and informal were accepted and were "promptly carried out."

The structure of the ISEG at these plants was similar. Both at the conforming and anticipating plants there were five engineers on-site reporting off-site. The primary emphasis of the ISEG was on events at a plant and at other plants that might indicate ways to improve

549

safety. The major difference between these plants was in the nuances of implementation. Not surprisingly, relinquishing freedom and control to an external agent (the NRC) when preferred states had been disturbed by an unwelcome surprise (TMI) created resistance, while independently tailoring a response to conditions at a plant resulted in greater understanding and acceptance. Thus, autonomy appeared to encourage a higher level of commitment, while rule-bound behavior appeared to blunt the impact of the external challenge (TMI) and prevent creative adjustment.

The statistical analysis: Table 16–2 is an intercorrelation analysis that includes all the variables in the study. As can be seen, autonomy is significantly correlated ($p < .05$) with fewer events in 1981 and 1982 and very significantly correlated ($p < .001$) with fewer human error events. This finding tends to support our view that autonomy is likely to be associated with fewer human error events.

There are also significant correlations between autonomy and higher profits, between lower profits and more events in 1982, and be-

tween lower profits and more human error events. More events are significantly correlated with more human error events, but not with significant events; the reason may be that significant events represent a situation that has dramatically deteriorated as opposed to human error events and general events that represent precursor circumstances.

Significant correlations also exist between age and events in 1981 and between long-term debt and events in 1981. There are a number of ways to understand these findings. Experience may be a factor in reducing the number of events, older technologies may be safer, or the correlation may simply represent increased reporting requirements that have been imposed on the newer plants by the NRC. Plants with utilities that have more debt also have fewer events, and in 1982 they have fewer significant events. This could reflect a safety spillover of a utility's long-term capital commitments.

Table 16–3 shows that the plants classified as autonomous outperform the plants classified as rule-bound (see Table 16–3) on nearly

TABLE 16–2.* *Descriptive Statistics and Correlations between Variables.*[a]

	Variables	Means	Standard Deviation	1	2	3	4	5	6	7	8	9	1(
1.	Implementation approach	0.62	0.51										
2.	Events 1981	66.00	29.33	−.75									
3.	Events 1982	69.85	38.84	−.83	.80								
4.	Human error events	19.92	11.85	−.98	.69	.80							
5.	Significant events	3.85	2.76	−.16	.52	.26	.03						
6.	Management rating	1.73	0.42	−.41	.05	.25	.35	−.04					
7.	Capacity rating	60.68	12.85	.10	.02	.32	−.05	.03	−.46				
8.	Profit[b]	12.59	1.57	.58	−.30	−.59	−.59	.18	−.54	−.02			
9.	Age	6.92	4.86	.43	−.61	−.53	−.46	−.38	.29	−.40	−.13		
10.	Size	1948.30	994.89	.02	−.00	.03	−.11	.16	.35	.03	.26	−.21	
11.	Long-term debt[c]	45.46	3.89	.35	−.57	−.39	−.19	−.59	−.47	.27	.30	−.10	−.:

a. n=13. Correlations coefficients above .49 are significant at p < .10; those above .57 at p < .05; those above .71 at < .01; and those above .82 at p < .001.

b. Percentage return on average common equity.

c. Percentage year-end capitalization ratios.

*This table is taken from Marcus (1988a), p. 248.

TABLE 16–3.* A Comparison of Rule-Bound and Autonomous Implementation Approaches.

Variables	Rule-Bound Mean	Rule-Bound Standard Deviation n=5	Autonomous Mean	Autonomous Standard Deviation n=8	t Statistic	Significance of t Test
Events, 1981	92.6	26.4	49.4	16.0	−3.71	.003
Events, 1982	109.2	33.1	45.3	12.7	−5.01	.000
Human error events	34.0	3.2	11.1	2.3	−15.08	.000
Significant events	4.4	3.8	3.5	2.1	−0.55	.590
Management rating	1.9	0.3	1.6	0.5	−1.50	.161
Capacity rating	59.1	17.9	61.7	9.8	0.35	.736
Profit[a]	11.5	1.6	13.3	1.2	2.35	.039
Age	4.4	2.2	8.5	5.5	1.57	.145
Size	1,922.4	568.9	1,964.5	1229.3	0.07	.944
Long-term debt[b]	43.8	4.4	46.5	3.4	1.25	.238

a. Percentage return on average common equity.
b. Percentage year-end capitalization ratios.
*This table is taken from Marcus (1988a), p. 249.

every performance indicator (the only exception is long-term debt). The smallest differences in outcomes are in the capacity ratings. Thus, productive efficiency, it appears, has not been sacrificed for the sake of safety nor has safety been jeopardized for the sake of productive efficiency.

The largest differences between the rule-bound and autonomous plants are in the number of human error events. Rule-bound plants have more than three times the number of human error events than do the autonomous plants. Significant differences also exist with respect to events in 1981 and 1982 and with re-spect to profitability. Autonomous plants have fewer events and are generally more profitable. These findings support the idea that autonomous implementation approaches outperform the rule-bound approaches with regard to safety and other indicators.

Vicious and Beneficent Cycles. To test for the existence of vicious and beneficent cycles, a determination is first made about the effect of past safety events on implementation approaches. The probit analysis shows that 1981 events correctly identify 85 percent of the implementations approaches (see Table 16–4). The R-

TABLE 16–4.* Effects of Past Safety Record on Rule-Bound and Autonomous Implementation Approaches[a]: A Probit Analysis

Independent Variables	Maximum Likelihood Estimate (MLE)	Standard Error (SE)	MLE/SE
Constant	8.14	5.34	1.52
Past safety record[b]	−0.12	0.08	−1.52

Note: Percentage predicted correctly = .85; R^2 = .91.
a. Rule-bound = 0; autonomous = 1.
b. Events in 1981.
*This table is adapted from Marcus (1988a), p. 249.

squared of .91 supports the hypothesis that a poor safety record leads to rule-bound behavior.

To determine the effect of the implementation approaches on subsequent performance, the four regression analyses have been run. The relationship between implementation approaches and human error events is strong even after the introduction of the control variables. IMPAPP (see Table 16–5) is the only variable with a significant t value in each regression. Because none of the independent variables are correlated at a level greater than .75 and only implementation approaches and events in 1981 are correlated at this level (the other correlations are at .40 or below; see Table 16–2), we do not consider multicollinearity to be a problem (Kennedy 1979: 131).

Thus, the probit analysis and regression results provide support for the existence of vicious and beneficent cycles. Plants with a poor safety record tend to respond in a rule-bound manner, a response that only perpetuates their poor safety performance; while plants with a strong safety record tend to respond in an autonomous way, a response which only reinforces their strong safety record.

Discussion of the Findings

After TMI, nuclear power plants became pervious to outside forces; the NRC introduced new organizational arrangements for safety review management. Some power plants followed the guidelines the NRC established; others customized these guidelines to fit their individual circumstances. The former approach has been termed rule-bound and the latter approach autonomous. We have related these approaches to safety outcomes and various other measures of nuclear power plant performance. Prior safety outcomes affected the implementation approaches. Poorer safety performance restricted choice. It yielded rule-bound ap-

proaches that perpetuated the poor safety outcomes. A good record, on the other hand, opened a zone of discretion. It preserved autonomy, which resulted in continued strong safety performance. Autonomy is the outcome of a good safety record and contributes to a good safety record. That is the essence of a self-perpetuating cycle—the cycle is hard to break. If poor performers are given more autonomy, our analysis suggests that their safety record is likely to improve; but our analysis also suggests that they are not likely to be given more autonomy precisely because they are poor performers. This is the essence of a vicious cycle.

Thus, we have evidence of a vicious cycle in which poorly performing nuclear power plants had their choices narrowed, which led to continued poor performance, and evidence of a beneficent cycle in which stronger performing nuclear power plants retained their autonomy, which perpetuated their strong safety performance. These findings suggest that the most potentially dangerous plants were the least likely to benefit from the innovations induced by the NRC after TMI and that the least dangerous plants were the most likely to benefit. In the short run, the performance gap between the strong and weak plants increased.

Implications

Many studies of the implementation process have been carried out, most of which focus on social policies. While some (Wilson 1963; Pressman and Wildavsky 1974; Duncan 1976) suggest that rule-bound behavior is necessary during implementation, most recent studies (Lipsky 1978; Bourgeois and Brodwin 1984; Guth and Macmillan 1986) put greater emphasis on autonomy. Our study takes the case of an externally induced innovation as it affects the organization and management of a high-risk technology and shows that autonomy is needed. The more that managers exercised

TABLE 16–5.* *Determinants of Human Error Events: OLS Regression Results.*

Independent Variables	All Six Variables	Implementation Approach/Events, 1981	Implementation Approach/Age/ Profit	Implementation Approach/Size/ Long-Term Debt
Constant:				
b	28.65	37.60	43.83	14.93
s.e.	22.60	3.83	8.78	6.84
t	1.27	9.81	5.00	2.18
Implementation approach:				
b	−22.71	−24.54	−20.48	−24.09
s.e.	2.21	2.28	2.42	1.10
t	−10.23[a]	−10.78[a]	−8.45[a]	−21.83[a]
Events, 1981:				
b	−0.18	−0.39		
s.e.	0.65	0.39		
t	−0.28	−0.99		
Age:				
b	−0.21		−0.24	
s.e.	0.28		.21	
t	−0.74		−1.17	
Profit:				
b	−0.57		−0.76	
s.e.	0.54		0.71	
t	−1.06		−1.07	
Size:				
b	−0.82			−0.74
s.e.	0.81			0.54
t	−1.01			−1.37
Long-term debt:				
b	0.36			0.47
s.e.	0.34			0.15
t	1.08			3.17*
Error term:				
b	1.98	2.66	2.70	1.80
s.e.	0.39	0.52	0.53	0.35
t	5.10	5.10	5.10	5.10
Adjusted R^2, N = 13	.97	.95	.95	.98
F (entire equation)	70.73[a]	114.07[a]	74.01[a]	170.29[a]

a. $p < .01$
*This table is adapted from Marcus (1988a), p. 250.

choice within a situation of constraints the better that the outcomes were.

Implementation is likely to be more effective if policy implementors are free to design and determine its specifics because:

1. Policy formulators may not possess sufficient information at the level where policy is carried out. Implementors are likely to have greater knowledge at the point of delivery where multiple and contradictory demands are felt.
2. Efforts to centralize authority and control the actions of implementors may deskill those who carry out policy and increase the chances of error. These efforts may encourage low system comprehension, low morale, and an inability to cope with anything but the most routine conditions. Autonomy is needed to encourage a higher level of commitment and a greater level of knowledge.
3. In particular, the disposition of implementors is likely to be negatively affected if they are not granted a sufficient level of autonomy, and it is their dispositions that are often critical to assuring program success.

On the other hand, rule-bound approaches are not appropriate for the following reasons:

1. When a situation deteriorates, a strong perception that something has gone wrong is needed.
2. Rule-bound responses are a defense against the perception of threat, blunting the impact of external challenges and preventing creative adjustment.
3. They are likely to lead to lowered cognitive functioning and reduced responsiveness.

Of course, there are important limitations to our findings, such as small sample size, use of judgment in coding implementation ap-

proaches, and the possibility that events have not been accurately reported. Additional research on the implementation of externally induced innovations after crises like TMI and on the organization and management of high-risk technologies like nuclear power is needed.

EVIDENCE FROM THE OTHER MIRP STUDIES

As a step toward further analysis we chose four MIRP studies that could be classified as externally induced innovations because they began with a substantial shock that involved external parties or they were beset by substantial setbacks and surprises during development (see Chapter 4). These studies involve site-based management of public schools, naval system, and medical devices (the cochlear implant and therapeutic apheresis). The nuclear power and school-site innovations were process innovations, the medical device innovations product innovations, and the naval system innovation involved elements of both process and product. For each of the studies, the same basic survey was administered to respondents in August and September of 1985 (see Chapter 3). In the analysis that follows we will be using an N of 117 completed surveys from five studies: twenty-three from nuclear power, nineteen from school management, thirty-five from naval systems, and forty from medical devices.

The variables were that were used were derived from measures tested by Van de Ven and Chou that passed, with minor qualifications, tests of reliability and convergent, discriminant, and concurrent validity. The dependent variable was *perceived effectiveness*. The independent variables are (1) *standardization of procedures,* (2) *decision influence,* (3) *freedom to express doubts,* (4) *autonomy encouraged,* (5) *workload pressure,* (6) *competition for obtaining resources,* and (7) *regulatory restrictiveness.* The dependent varia-

TABLE 16–6.   *Regression Analysis: Perceived Effectiveness.*

Independent Variables	Standardized Regression Coefficients
Competition for resources	$-.28$[a]
Autonomy encouraged	.23[a]
Freedom to express doubts	.22[a]
Regulatory restrictiveness	.03
Decision influence	.02
Standardization of procedures	.01
Adjusted R squared = .29	
Standard error = .21	
F = 6.00	
Significance = .00	

Note: N = 117.
a. $p < .05$.

ble and the first three independent variables are defined in Chapter 3.

Autonomy encouraged refers to the degree to which leaders encourage individual initiative and place trust in individual group members. It is the average response to the these questions:

1.   Initiative encouraged;
2.   Leader puts trust in members.

Workload pressure refers to how much work the idea imposes. It is measured by the response to these questions:

1.   Heaviness of workload;
2.   Lack of advance time.

Competition for obtaining resources refers to competitiveness in obtaining resources to develop the idea. It is the average response to these questions:

1.   Competition for finances;
2.   Competition for materials;
3.   Competition for management attention;
4.   Competition for personnel.

Regulatory restrictiveness is the restrictiveness and hostility of the regulators. It is mea-sured by the average response to these questions:

1.   Restrictiveness of regulations;
2.   Hostility of regulators.

To assess which variables had the greatest impact, we carried out a multiple regression with perceived effectiveness as the dependent variable and the other variables as the independent variables (see Table 16–6). The standardized betas reported in Table 16–6 show that competition for resources, autonomy encouraged, and freedom to express doubts had a significant impact and that no other variable significantly affected perceived effectiveness. These survey results therefore tend to support the earlier findings. Encouraging autonomy again, here defined somewhat differently, had a positive effect on organizational effectiveness, while other possible variables did not.

CONCLUSION

The results of these studies suggest that if innovation is externally induced, autonomy is likely to be the best approach. It is appropriate to customize what an external agent requests or to carry out what it might require before it is necessary, as opposed to a rule-bound approach where the organization does precisely and only as much as the external body demands. Autonomy is needed for organizations to go beyond mere formal compliance to identification and internalization (see Kelman 1961). In this respect, it resembles market-driven processes that rely on individual initiative and competence to achieve objectives that cannot be accomplished by central direction. The peculiar advantage of market-like processes is their dependence on search, trial and error, and experimentation at the point of delivery where specialized knowledge and skills are needed (Schultze 1983). If implementors have flexibility to customize external demands, implementation is likely to be with the spirit and not letter of the law, and performance outcomes are likely to be enhanced.

One question from our nuclear power study that is particularly perplexing is that the highly successful French program seems to be based on central direction and conformity. Perhaps the relationship between autonomy and performance noted here applies only in the American context. On the other hand, maybe, as Crozier (1964) suggests, autonomy is enhanced in France, albeit in a perverse way, by conformity. French officials are rule-bound because this keeps superiors from interfering in their day-to-day activities. The conformity is symbolic and designed to maintain a high level of autonomy. The specific tactics used for maintaining autonomy in different cultural contexts needs to be investigated.

This chapter points to the variety of ways autonomy can be expressed even when innovation is induced externally under extremely stressful circumstances such as a major and unprecedented industrial accident. Managers can *combine* an external demand with existing organizational procedures and practices. Another possibility is to *modify* the demand, thereby changing the external dictate and the internal environment simultaneously. Before a regulation is officially imposed, managers can *anticipate* what might have to be done. They can also *plan* for adoption but take few steps to carry out their plan. These approaches are to be contrasted with simple *conformity* and *incremental adjustment.*

Such choices are not made in a vacuum. Prior performance affects the decisions managers take. Poor performance restricts choice. It leads to rule-bound compliance, which perpetuates weak performance. Good performance, on the other hand, opens a zone of discretion. It preserves the ability to act autonomously, which should result in continued strong performance. Thus, there is evidence of vicious cycles in which poorly performing organizations cannot escape external control and evidence of beneficial cycles in which strongly performing organizations have their right to choose protected.

Managers, therefore, should be aware of the possible consequences of "blind" acceptance of external dictates and regulators should take heed of companies that strictly obey the law. These companies may not achieve the results that the external inducers intend.

REFERENCES

Aldrich, H.E. 1979. *Organizations and Environments.* Englewood Cliffs, N.J.: Prentice-Hall.

Andrews, K.R. 1971. *The Concept of Corporate Strategy.* New York: Dow Jones-Irwin.

Ashby, W.R. 1958. *Introduction to Cybernetics.* Andover, Mass.: Chapman and Hall.

Bardach, E. 1977. *The Implementation Game.* Cambridge, Mass.: MIT.

Bateson, G. 1979. *Mind and Nature.* New York: Dutton.

Berman, P. 1980. "Thinking about Programmed and

Adaptive Implementation: Matching Strategies to Situations." In Helen Ingram and Dean Mann, eds., *Why Policies Succeed or Fail*, pp. 205–31. Beverly Hills, Calif.: Sage.

Beyer, J.M., and H.M. Trice. 1978. *Implementing Change: Alcoholism Policies in Work Organizations.* London: Free Press.

Bourgeois, L.J. 1984. "Strategic Management and Determinism." *Academy of management Review* 9: 586–96.

Bourgeois, L.J., and D.R. Brodwin, 1984. "Strategic Implementation: Five Approaches to an Elusive Phenomenon." *Strategic Management Journal* 5: 241–64.

Bromiley, P., and A. Marcus. 1987. "Deadlines, Routines, and Change." *Policy Sciences.*

Burgelman, R.A. 1983. "A Model of the Interaction of Strategic Behavior, Corporate Context, and the Concept of Strategy." 20: 85–103. *Academy of Management Review* 8 (1) (January):61–70.

———. 1984. "Managing the Internal Corporate Venturing Process." *Sloan Management Review* 25 (Winter): 33–48.

Child, J. 1972. "Organization Structure, Environment and Performance: The Role of Strategic Choice." *Sociology* 6: 2–22.

Crozier, M. 1964. *The Bureaucratic Phenomenon.* Chicago: University of Chicago.

Cyert, R.M., and J.G. March, 1963. *A Behavioral Theory of the Firm.* Englewood Cliffs, N.J.: Prentice-Hall.

Del Sesto, S.L. 1982. "The Rise and Fall of Nuclear Power in the United States and the Limits of Regulation." *Technology in Society* 4: 295–314.

Dill, W.R. 1965. "Business Organizations." In March, J.C., ed., *Handbook of Organizations,* pp. 1071–1114. Chicago: Rand-McNally.

Downs, A. 1967. *Inside Bureaucracy.* Boston: Little, Brown.

Duncan R.B. 1976. "The Ambidextrous Organization: Designing Dual Structures for Innovation." In R.H. Kilmann, L.R. Pandy, and D.P. Slevin, eds., *The Management of Organization Design: Strategies and Implementation.* New York: Elsevier, North Holland.

Edwards, G.C. 1980. *Implementing Public Policy.* Washington, D.C.: Congressional Quarterly Press.

Egan, J.R. 1982. "To Err Is Human Factors." *Technology Review* (February/March): 23–31.

Elmore, R.F. 1979. "Mapping Backward: Using Implementation Analysis to Structure Policy Decisions." Paper presented at the annual meeting of the American Political Science Association, Washington, D.C., September.

Ettlie, J.E., and A.H. Rubenstein. 1981. "Stimulating the Flow of Innovations to the United States Automotive Industry." *Technological Forecasting and Social Change* 19: 33–55.

Fidler, L.A., and J.D. Johnson. 1984. "Communication and Innovation Implementation." *Academy of management Review* 9(4): 704–11.

Ford, D.F. 1981. *Three Mile Island.* New York: Penguin.

Guth, W.D., and I.C. Macmillan. 1986. "Strategy Implementation versus Middle Management Self-interest." *Strategic Management Journal* 7: 313–27.

Hannan, M.T., and J.H. Freeman. 1977. "The Population Ecology of Organizations." *American Journal of Sociology* 82: 929–64.

Hrebiniak, L.G., and W.F. Joyce. 1985. "Organizational Adaptation: Strategic Choice and Environmental Determinism." *Administrative Science Quarterly* 30 (3): 336–49.

Janis, I. 1982. *Groupthink: Psychological Studies of Policy Decisions and Fiascoes.* Boston: Houghton Mifflin.

Kelly, P., and M. Kranzberg. 1975. *Technological Innovation: A Critical Review of Current Knowledge.* National Science Foundation. Atlanta: Georgia Tech.

Kelman, H.C. 1961. "Processes of Opinion Change." *Public Opinion Quarterly* 25: 608–15.

Kennedy, P. 1979. *A Guide to Econometrics.* Cambridge, Mass.: MIT. Press.

Leibenstein, H. 1976. *Beyond Economic Man.* Cambridge, Mass.: Harvard University.

Lester, J.P., A.M. Bowman, M.L. Goggin, and L. O'-Toole, Jr. 1987. "Public Policy Implementation: Evolution of the Field and Agenda for Future Research." Prepared for delivery at the 1987 Annual Meeting of the American Political Science Association, Palmer House, September 3–6.

Lindblom, C.E. 1959. "The Science of Muddling Through." *Public Administration Review* 19: 79–88.

Linder, S.H. and B.G. Peters. 1987. "A Design Perspective on Policy Implementation: The Fallacies of Misplaced Prescription." *Policy Studies Review* 6(3): 459–75.

Lipsky, M. 1978. "Standing the Study of Public Policy Implementation on Its Head." In W.D. Burnham and M.W. Weinberg, eds., *American Politics and Public Policy,* pp. 391–402. Cambridge, Mass.: MIT.

Marcus, A.A. 1980. *Promise and Performance: Choosing and Implementing an Environmental Policy.* Westport, Conn.: Greenwood.

Marcus, A.A., and R.N. Osborn. 1983. *Safety Review at Nuclear Power Plants.* Report prepared for the Nuclear Regulatory Commission Office of Human Factors Safety, Washington, D.C.

Marcus, A.A. 1988a. "Implementing Externally Induced Innovations: A Comparison of Rulebound and Autonomous Approaches." *Academy of Management Journal* 30(2) (June): 235–57.

Marcus, A.A. 1988b. "Responses to Externally Induced Innovation: Their Effects on Organizational Performance." *Strategic Management Journal* 9(3) (Summer): 387–402.

March, James G., and Herbert A. Simon. 1958. *Organizations.* New York: Wiley.

Masuch, M. 1985. "Vicious Circles in Organizations." *Administrative Science Quarterly* 30(1) (March): 14–33.

Maynard-Moody, S., *M. Musheno, D. Palumbo, and A. Oliverio. 1987. "Street-Wise Social Policy: Empowering Workers to Resolve the Dilemma of Discretion and Accountability." Delivered at the Annual Meeting of American Political Science Association, Chicago.*

McKelvey, W. 1982. Organizational Systematics: Taxonomy, Evolution, and Classification. Los Angeles: University of California.

Merton, Robert K. 1938. "Social Structure and Anomie." *American Sociological Review* 3: 672–82.

Meyer, A. 1983. "Adapting to Environmental Jolts." *Administrative Science Quarterly* 27: 515–37.

Miles, R.H., and K. Cameron. 1982. *Coffin Nails and Corporate Strategies.* Englewood Cliffs, N.J.: Prentice-Hall.

Mitnick, Barry M. 1980. *The Political Economy of Regulation: Creating, Designing and Removing Regulatory Forms.* New York: Columbia University.

Mintzberg, H. 1978. "Patterns in Strategy Formation." *Management Science* 24(9): 934–48.

Nuclear Regulatory Commission. 1980. *Guidelines for Utility Management.* Office of Nuclear Reactor Regulation Division of Human Factors Safety, NUREG 0731.

Nutt, P.C. 1983. "Implementation Approaches for Project Planning." *Academy of Management Review* 8(4): 600–11.

Nutt, P.C. 1986. "Tactics of Implementation." *Academy of Management Journal* 8(4): 600–11.

———. 1987. "Identifying and Appraising How Managers Install Strategy." *Strategic Management Journal* 8: 1–14.

Olson, J., S. McClaughlin, R.N. Osborn, and D.H. Jackson. 1984. "An Initial Empirical Analysis of Nuclear Power Plant Organization and Its Effect on Safety Performance." Washington, D.C.: NRC.

Osborn, R.N., and D.H. Jackson. 1987. "Leaders, Riverboat Gamblers, Or Purposeful Unintended Consequences in Complex Dangerous Technologies." Working paper.

Palumbo, D.J., S. Maynard-Moody, and P. Wright. 1983. "Measuring Degrees of Successful Implementation: Achieving Policy versus Statutory Goals." Paper prepared for presentation at the Western Political Science Association Meetings, Seattle, Wash., March 26.

Patton, M. 1980. *Qualitative Evaluation Methods.* Beverly Hills, Calif.: Sage.

Perrow, C. 1983a. "The Organizational Context of Human Factors Engineering." *Administrative Science Quarterly* 28 (December): 521–41.

———. 1983b. *Normal Accidents: Living with High-Risk Technologies.* New York: Basic Books.

Pfeffer, J., and G. Salancik. 1978. *The External Control of Organizations.* New York: Harper & Row.

Pressman, J., and A. Wildavsky. 1974. *Implementation.* Berkeley: University of California.

Quinn, J.B. 1980. *Strategies for Change: Logical Incrementalism.* Homewood, Ill.: Irwin.

Rein, M., and F.F. Rabinovitz. 1978. "Implementation: A Theoretical Perspective." In W.D. Burnham and M.W. Weinberg, eds., *American Politics and Public Policy,* pp. 307–35. Cambridge, Mass.: MIT.

Rogers, E.M., and F. Shoemaker. 1971. *Communication of Innovation.* New York: Free Press

Rogovin, M. 1979. *Report of the President's Commission on the Accident at Three Mile Island.* Washington, D.C.: U.S. Government Printing Office.

Romanelli, E., and M.L. Tushman. 1986. "Inertia,

Environments, and Strategic Choice: A Quasi-Experimental Design for Comparative-Longitudinal Research." *Management Science* 32(5) May: 608–21.

Rothman, J. 1974. *Planning and Organizing for Social Change.* New York: Columbia University.

Rothwell, R. 1981. "Some Indirect Impacts of Government Regulation on Industrial Innovation in the United States." *Technological Forecasting and Social Change* 19: 57–80.

Rothwell, R., and W. Zegveld. 1985. *Reindustrialization and Technology,* New York: Sharpe.

Schon, D.A. 1971. *Beyond the Stable State.* New York: Norton.

Schultze, C. 1983. "Industrial Policy: A Solution In Search of a Problem." *California Management Review* 4 (Summer): 27–52.

Schweitzer, G.E. 1977. "Regulations, Technological Progress, and Societal Interests. *Research Management* (March) 13–17.

Scott, W.R. 1987. *Organizations: Rational, Natural, and Open Systems.* Englewood Cliffs, N.J.: Prentice-Hall.

Sills, D., C.P. Wolf, and V. Shelanski, eds. 1982. *Accident at the Three Mile Island Nuclear Power Plant: The Human Dimensions.* Boulder, Colo.: Westview.

Sonnenfeld, J. 1982. *Corporate Views of the Public Interest.* Cambridge, Mass.: Auburn House.

Starbuck, W., and B. Hedberg. 1977. "Saving an Organization from a Stationary Environment." In H. Thorelli, ed., *Strategy + Structure = Performance,* Bloomington: Indiana University. pp. 250–287.

Strebel, P. 1987. "Organizing for Innovation over an Industry Cycle." *Strategic Management Journal* 8: 117–24.

Terreberry, S. 1971. "The Evolution of Organizational Environments." In John G. Maurer, ed., *Readings in Organizational Theory: Open System Approaches,* New York: Random House. pp. 58–75.

Thomas, R.D. 1980. "Implementing Federal Programs at the Local Level." *Political Science Quarterly* 94 (3) (Fall): 419–35.

Van de Ven, A.H. 1979. "Review of *Organizations and Environments* by Howard Aldrich." *Administrative Science Quarterly* 24: 320–26.

———. 1986. "Central Problems in the Management of Innovation." *Management Science* 32 (5) May: 590–607.

Van de Ven, A.H., and R. Drazin. 1985. "The Concept of Fit in Contingency Theory." In Barry M. Staw and L.L. Cummings, eds., *Research in Organizational Behavior* 7, pp. 333–65. Greenwich, Conn.: JAI.

Van Meter, D.S., and C.E. Van Horn. 1975. "The Policy Implementation Process: A Conceptual Framework." *Administrative and Society* 6 (February): 445–88.

———. 1987. "Organizational Culture as a Source of High Reliability." *California Management Review* (2): 112–128.

Wilson, J.Q. 1963. "Innovation in Organization: Notes toward a Theory." In J. Thompson, eds., *Approaches to Organizational Design,* pp. 193–218. Pittsburgh: University of Pittsburgh.

Zaltman, G., and R. Duncan. 1977. *Strategies for Planned Change.* New York: Wiley-Interscience.

17

DEPTH AND BREADTH IN INNOVATION IMPLEMENTATION: THE CASE OF SCHOOL-BASED MANAGEMENT

Karin M. Lindquist

John J. Mauriel

Once an innovation is created and developed (either within or outside of an organization), how should it be introduced and implemented by an adopting organization? In particular, if management has made the strategic choice to adopt an administrative innovation throughout its organization, should management begin by concentrating its implementation efforts in depth in one or two specific organizational subunits, or in breadth across all organizational subunits?

The first approach, which we will call a depth adoption strategy, assumes that after an innovation has been successfully adopted and debugged by a demonstration unit, it can be transferred and diffused to other organizational units. The second approach, called a breadth adoption strategy, assumes that a more effective way to change an organization is to introduce and implement an innovation across the board, often through successive hierarchical levels across all organizational units.

Implicitly or explicitly, this question is addressed by strategic managers every time they initiate action to adopt an innovation or introduce a change throughout their organization. Yet there has been little systematic theorizing or research conducted on this question of examining the relative effectiveness of depth versus breadth innovation adoption strategies. Even less scholarly attention has been directed at examining what events and problems are typically encountered over time as the two adoption strategies are implemented in organizations.

The purpose of this chapter is to examine the processes and relative implementation effectiveness of the depth versus breadth innovation adoption strategies by reporting the findings of a longitudinal comparative study of the adoption of a school-based management (SBM) innovation in two public school districts over the past four years. A secondary question in this study of innovation adoption is whether an emphasis on breadth or depth in the implementation of an organizational innovation (such as SBM) improves the chances of that innovation becoming institutionalized or incorporated. SBM is an administrative innovation currently being adopted by many independent school districts throughout the United States. Basically, SBM involves the transfer of authority for many school governance and curriculum decisions from the central district superintendent's office to the level of the local school building, along with the development of participatory management processes for newly created local site councils consisting of the school administrators, staff, and parents.

In this chapter, we first develop a conceptual framework for examining and comparing the depth versus breadth innovation adoption strategies. Although the depth and breadth strategies are conceptually viewed as two ends of a concentration-dispersion adoption scale, seldom do we find such clear-cut case examples. Instead, the two cases selected for this empirical comparison of depth versus breadth strategies were observed to reflect relative degrees of emphasis on these two strategies. In the second section we describe the research approach used to investigate the adoption of SBM innovation. This section discusses the innovation under investigation, briefly summarizes the two research sites, and then presents the methodology of the investigation. The third and fourth sections report the findings from longitudinal field studies of two school districts; Metro School District, which emphasized a depth strategy to adopt SBM, and River School District, which emphasized a breadth innovation adoption strategy. The fifth section presents and discusses our findings and the sixth section offers some conclusions concerning the innovation effort and how our propositions have held up in the organizations under observation.

DEPTH AND BREADTH: A CONCEPTUAL FRAMEWORK

Definitions of Depth and Breadth

For the purposes of this study, breadth is defined as the number of organizational lines crossed horizontally in the innovation adoption process. Measures of breadth include the number of organizational units affected, amount of communication across organizational lines, and the project's place in the agenda of top management (the district superintendent). An innovation that emphasizes a breadth approach to implementation would involve more similar organizational units and fewer diverse stakeholder groups.

Depth is defined as an innovation effort that focuses on a specific work group or organizational unit. Some measures of depth include the number of different stakeholder groups involved in implementation, the number of different organizational levels involved in the innovation, amount of external versus internal communication regarding the innovation, and the innovation's place in the agenda of top management (the district superintendent). This approach to innovation implementation would concentrate on one division (a school) of an organization but involve more diverse stakeholder groups in a more "top to bottom" effort.

Our definitions of breadth and depth are implicitly oriented toward organizational communications patterns and also concern the nature of the relationship between the central office and the field location where the innovation is being implemented and made operational. Our idea of breadth implies a great deal of communication across divisional lines—that is, between and among various field locations, typically involving a smaller number of external stakeholders and a larger number of internal units. Breadth implementation usually involves more people at the home office or central executive staff level, since each reporting unit is involved in some form or some aspect of the implementation process. As a result, in breadth implementation, an organization is almost forced to develop more organizational wide communications, although this may be tempered by the degree of the interdependence required in implementing the innovation.

Our conception of depth, however, involves more intradivisional communication as opposed to across divisional or operating unit lines. It typically involves more external stakeholders and less intense concern from top management, although there may be a product champion at the home office level. This was the case in our depth organization—Metro School District.

A few other studies in the literature have examined depth versus breadth innovation adoption strategies. Munson and Pelz (1982) examined the incorporation or adoption of innovation in a variety of organizations. However, their definition of depth and breadth strategies differed from ours. They refer to depth as the extent of behavioral change required by those directly affected by the innovation. Breadth, in their terms, is a measure of the percentage of all people in the organization who are in any way affected by the innovation. Their definitions of depth and breadth are thus more behaviorally oriented than those used in our research effort although not necessarily inconsistent with ours.

Pelz and Andrews (1976) used a depth versus breadth measure in studying laboratory groups of scientists and engineers. Their use of the term referred to the degree of similarity or difference in technical approach and to the preference for broad interest in new problem areas versus a narrow specialization. Depth meant a high degree of similarity among group members and a preference for narrow specialization. Breadth meant that more diversity in technical approaches were represented in the

group and there was an openness to new problems and issues among members.

Pelz and Andrews developed a depth versus breadth index and correlated this to colleague assessments of both the scientific contribution and the overall usefulness to the organization (laboratory) of the groups they studied. They found that scores on the level of breadth were more positively correlated with scientific contribution for both younger research groups and older groups (age referred to how long the groups had worked together as a team). When comparing the age of groups that all scored high on the breadth index, they also found that older groups were significantly more useful than younger groups, though in both cases breadth was positively related to usefulness (statistically significant at the .05 level only for the older groups). Thus, as might be expected, experienced generalist groups were found to make greater contributions than inexperienced specialist groups.

Although other authors have not used the terms "depth" and "breadth" in describing innovation and implementation processes, their work can be related to ours. For example, Van de Ven's (1980a, 1980b) comparison of the broad program planning model with other more narrow planning approaches on the implementation of early childhood programs in Texas from 1973 to 1976, is consistent with our breadth and depth approaches as we define them here. His findings suggest that the breadth approach implied in the PPM model (crossing more organizational and agency lines initially in obtaining input on design and planning) lead to more successful adoptions of innovations.

Research and inductive models on structuring organizational innovations suggest these findings should incorporate a temporal understanding of the stage of innovation adoption: A more open, decentralized, and organic organizational approach is needed in the initiation phase, and a more centralized, bureaucratic, and routinized process is typically used and perhaps needed in the implementation phase (Hage and Aiken 1970; Harvey and Mills 1970; Wilson 1966; Zaltman et al. 1973, 1977). These findings suggest that relative shifts in emphasis from breadth to depth strategies over time facilitate innovation adoption.

To examine these possible temporal developments in breadth and depth strategies, we searched for a conceptual framework that could guide our study of the processes and hurdles encountered in the adoption of innovations. Such a framework was proposed by Yin (1979). Yin's framework, developed from observations of several innovations in public organizations, identifies ten passages and cycles and four features that most innovations experience on the path to institutionalization. This framework can be viewed as a life-cycle process of the adoption of an organizational innovation. Yin's findings suggest that the innovations that survive a higher proportion of the ten stages and cycles have a better chance at becoming institutionalized in the organization. Yin identifies and labels three basic stages of innovation implementation on the path to institutionalization: improvisation, expansion, and disappearance. He suggests that although the improvisation stage is important, the expansion and disappearance stages are critical to the long-term incorporation of an innovation into an organization. The improvisation stage involves the organization of activities during which the innovation begins to operate following its adoption. The expansion stage involves the diffusion of an innovation to all parts of the organization and the disappearance stage involves the gradual incorporation of the innovation into the daily procedures and routines of the organization.

It is possible but not necessary for any of the ten cycles or passages to occur during the improvisation stage. However, Yin suggests that there are several features of the improvisation stage that are important in the movement of the innovation from this part of its life cycle to the expansion stage. These features include:

1. Practitioner exposure to the innovation;
2. Resource management—financial, managerial, and technical;
3. Nature of the innovating team; and
4. Strategic choices in adopting the innovation.

The ten passages and cycles that Yin found important to the institutionalization of an innovation are divided between the expansion and disappearance stages.

Expansion Stage
1. Equipment turnover (cycle);
2. Transition to support by local funds (passage);
3. Establishment of appropriate organizational status (passage);
4. Establishment of stable arrangements for supply and maintenance (passage); and
5. Establishment of personnel classification or certification in which the functions related to the innovation become part of the job description or a prerequisite (passage).

Disappearance Stage
6. Changes in organizational governance that integrates the innovation into the organization (passage);
7. Internalization of the training program (passage);
8. Survives the promotion of personnel acquainted with the innovation (cycle);
9. Survives the turnover of key personnel (cycle); and
10. Attainment of widespread use (cycle).

Propositions

From this conceptual framework and a review of related literature, the following propositions were developed regarding our expectations about the outcomes and processes associated with using depth versus breadth innovation adoption strategies:

1. A breadth approach typically has the active support of top management (school district superintendent, CEO, or division general manager). As a result, a breadth approach to innovation implementation is more easily sustained and able to overcome early barriers to implementation. In contrast a depth strategy, after an initial pronouncement of top management support, tends to be quickly overlooked by other pressing priorities of top management.
2. The degree of interdependence among organizational units will affect the importance of using a breadth versus depth approach to innovation implementation. The more independent the units, the less important is the need for top management support and the less critical it is to use a breadth approach.
3. Depth approaches tend to get off to a slower start with more rules and regulations and a more varied set of stakeholder groups to satisfy. Conversely, breadth approaches to innovation adoption will get off to a faster start with fewer rules and fewer different stakeholder groups to satisfy.
4. School-based management implies greater emphasis on depth than on breadth approaches by requiring the initial involvement of multiple internal and external stakeholders and focussing on the needs of an individual school unit rather than on a total school system. However, in relative terms, depth approaches to adopting SBM will be more difficult in the early years of the adoption process than a breadth strategy. This may be partially due to the problems of gaining the support of the many different stakeholder groups that are required to sustain it.

RESEARCH APPROACH

School-Based Management

School-based management is an education administration innovation that has grown in recent years as a response to the ongoing concern over the quality of public primary and secondary education and a desire for broader participation by both internal and external (to the school) stakeholders in decisions made in their local schools. The concern over education is not new and has led to a wide variety of innovations in curriculum, instruction, teacher education, building design, and technology. However, recent desires for more voice at the local school level have arisen because school administrators have retained a traditional hierarchical structure in which most long- and short-run decisions are made by a superintendent in a central district office. The function of the individual school, in a multischool educational system, has been to implement the curriculum decisions of the central school district office using personnel and budget limits provided on a district wide basis.

The growth of community interest in the decision making process and the growth of professional employees' power to influence decisions both question the validity of the traditional approach of central office decision making. It has been suggested that placing greater decision making authority at the school building level permits greater flexibility in the design of school programs that lead to a better match with the needs of a school's constituents. The SBM concept takes this point a step further; it suggests that the administrative authority of a school should be held by a school site council comprised of representatives of the school's major internal and external constituents or stakeholder groups. According to SBM these groups would include teachers, parents, students, noncertified school staff, school administration, and occasionally nonparent community

565

members. The development of SBM is basically a two-stage process: First, the district office delegates decision making authority to the building principal, and second, the principal delegates authority to the building site council.

There are no guidelines, however, regarding the best approach to implementing SBM. Some districts have tried the depth approach by introducing it as a pilot or demonstration project in just one school. Other districts have attempted a breadth strategy by getting all or most of their schools involved in adopting SBM at the outset. The number of different stakeholder groups included in the site council has also varied.

This chapter describes the efforts of two school districts to implement SBM. One is a suburban metropolitan district (Metro District) and the other is a district outside of the state's major metropolitan area (River District). Each district has chosen distinctly different approaches to the process of implementing and diffusing the administrative innovation of SBM. The school district is the level of analysis in this investigation. This level is important because it puts the two cases of SBM adoption at comparable organizational levels: the total school system or district.

Metro and River School Districts

This study of the adoption of SBM focuses on the efforts of two school districts that significantly differ in their relative emphasis on depth versus breadth adoption strategies. Metro School District emphasized the depth and River School District the breadth approach to innovation implementation. The case descriptions in the next two sections discuss the context and events in the adoption processes of SBM over time. Using the MIRP framework, events represent periods of significant activity, or major turning points, in the ideas, people, context, and transactions related to SBM. An event could be relatively brief in duration or could

unfold over a considerable period of time. Not all events are logically or sequentially linked. A number of events could also transpire concurrently. Important events in the process of adoption were identified through extensive interviews with participants in the adoption process. Chronological listings of the events for Metro and River School districts are included in Appendix 17–A and Appendix 17–B, respectively.

Methodology

The data on the Metro and River School districts were gathered over a three-year period in a longitudinal study that examined the adoption of SBM. Data collection included in-depth interviews, archival research, and site observations. Almost fifty interviews were conducted between the two schools and included both site council and nonsite council members and more than twenty-five site council meetings were observed over the three years.

In both districts, interviews were conducted by asking a set of eight, open-ended questions:

1. What is your background?
2. What were the key events in the development of SBM in your district?
3. Who were the people that were involved in the identified key events?
4. When did the key events occurr?
5. What specific agreements were made to further the development of SBM?
6. Did the identified events have a positive or negative impact on the adoption of SBM?
7. How have the management practices and organizational structures in your school or district been affected by SBM?
8. What do you foresee to be the issues with SBM one year from now?

In River District, interviews were conducted with eighteen individuals. The district superintendent, the junior high school principal, and two elementary principals were all interviewed three times each; these interviews were conducted on a yearly interval from fall 1984 through fall 1986. In Metro School District, interviews were conducted with sixteen site council members. Multiple interviews at approximately six-month intervals were conducted with the high school principal and with the site council coordinator. A list of interviewees is located in Table 17–1. The different mix of interviewees reflects the differences between the two school districts in the composition of the site councils and structure of SBM.

Documents relating to policies and procedures affecting staff leadership were reviewed for both research sites. Minutes of site council meetings were reviewed, and selected site council meetings were observed. At River District, two meetings at elementary schools and two meetings at the junior high school were attended. Approximately twenty site council meetings at Metro High School were observed. From these observations, comments were classified as to who initiated a new topic, who seemed to influence the discussion, and what decisions were made by a council consensus or vote pertaining to curriculum, budget, and personnel.

The criterion used to classify the Metro and River School districts into the depth and breadth adoption strategies was based on the number of internal and external stakeholders involved directly in the adoption of SBM. In order for an innovation adoption process to be

TABLE 17–1. *Interviews for School-Based Management.*

No. of Persons	Position	Frequency
River School District:		
1	District superintendent	Three times
2	High school principals	Once each
1	Administrative assistant	Once
2	Asst. high school principal	Once each
1	Junior high school principal	Three times
1	School board chair	Once
1	School board member	Once
1	Special assistant to superintendent	Once
2	Teachers	Once each
1	District curriculum director	Once
2	Parents advisory council members (school level)	Once each
2	Elementary principals	Three times
1	Elementary principal	Once
Metro School District:		
1	High school principal	Four times
1	Site council coordinator	Five times
2	Students	Once each
4	Teachers	Once each
1	Staff	Once
5	Parents/community	Once each
1	District superintendent	Once
1	School board member	Once

considered a breadth approach, it needed to be adopted in more than one organizational division, in this case, more than one school building. It also needed to directly involve fewer different stakeholder groups in the innovation adoption process with the most basic approach only including a school's staff and administration.

The criterion used for determining a depth strategy was if SBM was initially implemented in only one organizational division (such as a high school) with all the other divisions continuing with their previous approach to administration. This classification also directly involved the largest number of different internal and external stakeholder groups. The most extensive level of stakeholder involvement for the adoption of SBM would include school administration, teachers, noncertified staff, students, parents, other community members, and representatives of specific organizations such as the school board and parent-teacher associations.

METRO SCHOOL DISTRICT: DEPTH ADOPTION CASE

Context

Metro District is a large suburban district that includes all or most of seven suburban communities. There is some industry, a few remaining farms, and a large number of commercial enterprises. Most residents, however, commute to jobs outside of the school district. The district itself is a developing middle- and upper-middle-class area, but only about 20 percent of the households have children in school. As of 1985 the district had 55,000 residents and there were 6,568 students in the public school system. There were six elementary schools with 3,300 students, two middle schools with 1,400 students, and one high school with 1,800 students.

Administratively, the district has been run on a more decentralized format than is usually the case in a traditional, hierarchical system. Even before the SBM proposal, the various schools had some authority over staff hiring and building budget decisions. The superintendent himself had organized his own districtwide parent and community advisory group.

The implementation of SBM has been tried in only one building, Metro High School. This effort to try SBM has been primarily a result of the interest and initiative of the Metro High principal. It was through the efforts of the principal that external funds were secured for a three-year project to implement SBM and the initial site council was formed. The district superintendent gave his consent to the project but has not played an active role in its development.

The effort to implement SBM is now entering its fourth year. Metro High is still the only school in the district that has attempted to adopt SBM and it does not appear that any of the other schools are even considering the effort. Recently, the source of the external funds for the SBM implementation effort have ended due to a shift in the funding priorities of the granting foundation. It does not appear likely that the Metro District School Board will authorize any district monies for the project. Other external sources of funds have been investigated but have not been fruitful.

The Metro High site council is comprised of twenty-two to twenty-six members. These individuals represent teachers, students, high school administrators, noncertified school staff, parents, and community members. Originally, a profile of each position was developed with the intent of having a broad representation of the different stakeholder groups. For example, among the community members there was to be one person representing small business, one from big business, and one from a union. The site council was never quite able to fill the original specifications of member distribution but did meet their general representation requirements of teachers, students, administrators, parents, staff, and community members. There has been some turnover of membership over the four-year life span of the project, although five of the current members have been with the site council since its inception.

Arrival of the Idea

The involvement of the Metro High School principal in the Public School Incentives (PSI) programs sparked the idea of creating a site council for the high school. PSI is an ad hoc group of educators and citizens with help from a local foundation that attempts to identify creative solutions to school problems and to act as catalysts in facilitating implementation programs. The principal's involvement in SBM was further encouraged by the interest of the same Foundation, which was looking for SBM projects to fund (at that time). The informal partnership of the principal and a local consultant who was a parent and education futurist resulted in the submission of a grant proposal to the Foundation for concept development monies. A second proposal was approved by the Foundation to provide start-up funds for implementing SBM at Metro High School.

Initial Organization

A group called the Planning and Monitoring Committee was first organized by the principal and the consultant in August 1983. This group was charged with actually developing and organizing the high school site council. The Metro High School management council (their name for the site council) was finally formed and began to meet on a regular basis beginning in February 1984.

Several trips by the new management council members to other school site councils in California and Washington in February 1984 convinced them of the importance of having a council coordinator. A coordinator was hired in August 1984 for a half-time position. A Metro High School management council retreat was held at a conference center in August 1984. At this retreat the controversial issue of hall monitoring responsibilities by teachers was discussed and an alternative monitoring schedule developed. In November 1984, the Planning and Monitoring Committee was disbanded at the insistence of the Foundation, the project's funding agency. After the end of the school year in June 1985, a summer workshop was conducted to establish the management council's priorities for the 1985–86 academic year.

Implementation of School-Based Management

That following fall in November 1985, another retreat was held to discuss the implementation of the council's priorities, instruction, and staff development that were identified at the June workshop. Innovation adoption progress was slow, however. In February 1986 a routine meeting from which major advocates of the management council were all coincidentally absent, rapidly evolved into an airing of pent-up frustrations, with the official secretary being asked to stop recording the discussion. Some progress was made during the school year on the issues of school environment and quality circles, but most management council's time was taken up with efforts to secure alternative external support funds. Funding by the Foundation ended July 1986 leaving no apparent future source of funds. However, the management council was permitted to carry over unspent funds from the grant, which were sufficient to last until June 1987.

Evolution

The achievements of the Metro High site council and SBM are unclear. Many of the members of the site council believe that SBM has had an impact, but are not sure how significant that impact has been. In general, the site council believes that SBM has facilitated several program innovations in Metro High, such as the recognition awards for teachers, changes in the grading system, a staff development project, and has encouraged staff and school administration to use more consultative approaches to decision making. It is not clear, however, how many of these activities would have occurred anyway without SBM.

RIVER SCHOOL DISTRICT: BREADTH ADOPTION CASE

Context

River District encompasses most of River County that lies outside of the major metropolitan area of the state and is located in the county seat, Riverton. Most of the county's residents live within the district's boundaries. The population projections through the year 2000 show no growth in the county's population level. Situated on a major river, Riverton still derives significant economic benefit from commercial river traffic. The city is also home to a growing, diversified mix of small and midsized businesses as well as three institutions of higher education.

River District is comprised of ten schools: eight elementary schools, a junior high school, and a senior high school. Each elementary school principal is assigned administrative responsibility for two school buildings. Total school enrollment as of September 1984 was 4,400 students. Enrollment declines continue in the secondary schools although several of the elementary schools have experienced increases.

For eleven years prior to 1983, River District was administered by the same superintendent. His approach to educational adminis-

tration ran along traditional, hierarchical lines that emphasized central office decision making. On his retirement, the current superintendent was hired. The new superintendent's administrative philosophy emphasized a concept of shared involvement in decisions at both the school and the district levels. This SBM philosophy was transformed into a new program entitled "staff leadership" and was to be implemented in all schools throughout the entire district.

The purpose of staff leadership was to provide a formal leadership arrangement that would enhance the decision making processes in the district and ensure that all persons affected by a decision would have had representative input. Under staff leadership, each school has an elected School Improvement Council (SIC) comprised of teachers, certified and noncertified staff members. The school principals are formally in charge (staff leaders) of their respective councils. The members of the SIC have three primary duties: to represent the interests of their constituencies; to communicate SIC proceedings through caucus sessions in their constituencies; and to serve periodically on both standing and ad hoc district wide committees. Members serve three-year terms and are compensated with $950 per school year and necessary "time off" from other duties. The number of SIC members vary from two to twelve depending on the school, with an overall total membership of fifty-four district wide. It is possible for several staff to share one SIC position so that more people are involved in staff leadership than there are formal positions. Parents and community members are not included on any of the SICs.

At the district level, a District Improvement Council (DIC) was organized. The purpose of the DIC, according to the staff leadership plan was to "provide the communication and policy advisement for districtwide issues." As a result, the actual membership of the DIC varies with the scope and nature of the specific issues at hand. However, any member of the DIC must also be a member of his or her School Improvement Council.

River District is now in its third year of the staff leadership program. Funding for the program has been internal from the district. The avoidance of external funding for the implementation effort has been a conscious choice on the part of the superintendent. All of the schools are still involved in the program, although enthusiasm and support varies by school.

There are differences of opinion among the SIC members regarding the achievements of staff leadership. Some members have indicated in their interviews that they believe that much of their effort has been spent on less important issues such as lunch room policies and parking spaces. Others have suggested that the staff leadership system may help the schools prepare themselves better for dealing with ongoing changes in curriculum and instruction. The superintendent believes that staff leadership is a process that will take time to be fully implemented and that patience is a key ingredient to its eventual institutionalization.

Arrival of the idea

The involvement of the River District in SBM also came through one individual's inspiration, only in this case the individual was the superintendent of the district instead of the principal of a single school. The superintendent, who had held previous jobs as a teacher, guidance counsellor, and a superintendent of a smaller district, had completed his doctoral dissertation on the subject of SBM just before coming to River District.

Besides his interest in the concept of SBM, the River District superintendent has a deep belief in the value of obtaining a wide range of individual participation in decision-making. He believes that if people who are involved in carrying out decisions participate in making these decisions, then better results will be achieved.

Initial Organization

His vehicle for introducing SBM and its related participative decision processes was the development of staff leadership councils. In his first year of office, a series of transactions involving teacher negotiations occurred in fall 1983 in which the superintendent offered the staff, in lieu of the reinstatement of the department head assignment which they had requested, the opportunity to form staff leadership teams in each school building. The plan was that these teams would ultimately assume authority for the making of decisions in their respective buildings. The superintendent subsequently formed a district level curriculum committee and also involved community members in a series of newly formed committees that gave them direct input into key decision processes in their respective schools.

Implementation of School-Based Management

There has been very little turnover of participants, with the exception of the high school principal, since the innovation was introduced. The idea of staff leadership has not changed over time substantially in concept in River School District. There has been, however, a great deal of necessary clarification of terms and of the amount and kind of decision making authority that was to be invested by the SICs. In addition, the SIC members and principals have been provided training designed to help them operate effectively in a participatory manner.

The outcomes in River District are still chiefly of a process nature. The biggest changes have been in the locus of decision making. Changes in people and their skill levels also seem to have taken place as a result of the training given to SIC members.

It is still too early to speculate on school learning outcomes, but there are some process results that are important to note. Staff leaders seem to be gradually assuming more decision-making authority. The SICs also seem to be dealing with more substantive issues, such as budget and curriculum, in their second and third year of operation. This is a change from the first year of implementation when the concerns were more about how to organize, who should be on the councils, what goals the council should have, and whether or not they should have to attend so many meetings.

At the beginning of the fourth full year of operations of the SICs, the issue of which level or group (district office, school principal, school improvement council) has what kind of decision authority (provide input, recommend, decide) has surfaced. This is being dealt with by the development of a decision making matrix by the superintendent with the input of the staff. After the development of the matrix, a series of meetings will be held during the 1987–88 academic year to discuss authority levels and limits and to clarify the kind of influence a SIC might have on given decisions.

Evolution

Amidst a context that changed radically in 1983 but has since stabilized, the idea of SBM is evolving slightly over time as it is implemented. The key people are still there and many transactions have shaped the character and uses to which the innovation is being put. Process outcomes are clearly observable and becoming more central to the innovation's initial intent as implementation moves through the third and fourth years.

DISCUSSION OF FINDINGS

It appears that neither Metro nor River School District has been able to institutionalize the ad-ministrative innovation of SBM. Although the projects have been pursued for at least three years, the research suggests that both sites are still primarily in the improvisation stage with River District having expanded SBM the most. Although several of the passages and cycles from the expansion and disappearance stages of Yin's framework described earlier have been achieved, neither site can really claim that the original project has been fully expanded. Table 17–2 summarizes these findings, and the next several sections discuss findings applicable to each of Yin's passages and cycles.

Initiation of the Idea-Improvisation Stage

In both cases the idea was introduced by one person who championed it through to the first stage of implementation. In Metro District the high school principal was the person and the idea has not spread beyond the building he manages. In River District it was the superintendent who introduced the idea to all ten buildings in his system concurrently. Both have maintained the effort over the years, the Metro principal by leadership in involving a variety of stakeholders and trying to keep them interested through his active involvement; the River superintendent by his line authority over the administrators responsible for implementing the program. The latter were given no option other than continuing to support the process.

Exposure to the idea, Yin's first passage, in Metro has been limited primarily to the Metro High School site council. Other staff, students, and parents of the school have been exposed to the project through occasional mailings, but chiefly in an information giving rather than a participative manner. There has been a great deal of intraschool communication to a wide range of stakeholders, however, especially in view of the fact that all key stakeholder groups (parents, staff, students, and administration) are represented on the Metro site council. There has, however, been almost no

TABLE 17–2.   *Summary of Innovations' Degree of Institutionalization.*

	School District	
Feature or Passage/Cycle	Metro	River
Improvisation stage		
Practitioner exposure to innovation	Limited to SC members	Board
Resource management		
Fund source	External	Internal
Managerial support	Superintendent passive; principal active	Superintendent active; principals vary
Technical support	Yes (training and administrative)	Yes (meeting skills training)
Innovating team		
Needs of innovation understood and initiative taken to allocate resources	Varies	Partially (capitol budget allocations)
Coordination of innovation with existing practices	No	Yes
Technically mastering and operating innovation	Partially	Partially
Strategic choices		
Narrow scope of innovation	Yes (only in 1 school)	No (all schools)
Elimination of superseded practices	No	Yes (in most cases)
Reduction of job threat to practitioners in using innovation	Unclear	Not applicable
Expansion stage		
Equipment turnover	Not applicable	Not applicable
Transition to local funds	No	Has always used local funds
Organizational status achieved	Partially	Yes
Stable supply and maintenance arrangement	Yes—while funds were still available	Yes
Personnel classification and certification	No	Yes, for principals
Disappearance stage		
Change in organization governance	No	Unclear (not in major way)
Internalization of training program	No	Unclear
Survived promotion of personnel acquainted with innovation	Not yet an issue	Not yet an issue
Survived turnover in key personnel	Not yet an issue	High school in a transition
Attainment of widespread use	No	Always used widely

communication about the idea across organizational lines and into other schools in the system. These factors lead us to conclude that this is an example of depth emphasis in implementing the innovation process.

In River District, exposure to the idea has been limited to the very restricted membership of the site councils, consisting solely of school staff members. Communication has been extensive across school lines as district level meetings were held discussing implementation, school administrators from different buildings informally compared notes and discussed various approaches to implementing the idea, and many communications were sent out from the district superintendent to all schools. However, in River District there was very little discussion of the idea of SBM in the early years of its implementation with the various other stakeholders of the school (parents and students).

Resource management (Yin's second feature) was handled by Metro High School through the receipt of a three-year grant. However, this issue will be tested in the future due to the termination of funding by the foundation that provided their initial grant. Managerial support for the project has come almost exclusively from the building principal. The district superintendent and school board have provided passive support and exhibit a "wait and see" attitude toward the innovation. A school board member has been an active participant on the Council but primarily because of his own interest and does not represent an endorsement by the board for SBM. Some technical support has been provided to the innovation team and has included the services of both a secretary and an educational futures consultant.

In River District the source of funds has been internal from the very start, although there have been efforts by the River District school board to curtail the funding of SBM. There has been substantial managerial support for the innovation. The primary advocate of the effort has been the district superintendent. The level of support on the part of the school principals has varied without a specific pattern between secondary and primary buildings. Technical support has been provided for the site council leaders and has included training in meeting skills for all SIC members.

The Work and Choices of the Innovation Team

The innovating team, in the case of Metro High School is the Metro High site council. There is an inconsistent pattern among the council members of understanding the needs of the innovation. Interviews with and site observations

of site council members indicate a wide range of understanding of the meaning of SBM and what successful implementation may require of the members. This problem has been aggravated by the lack of coordination of the innovation with the existing school administration practices. For the first three years, none of the three assistant principals was directly involved in the site council, nor were the department chairs, the faculty senate, or the student senate. Technically mastering and operating the innovation of SBM has only been partially achieved. This again reflects the wide variation among the council members in their understanding and interpretation of the innovation.

The innovating teams, in the case of River District, are the staff leadership groups. There is an inconsistent level of understanding among the groups concerning the needs of the innovation and the amount of initiative required to allocate resources. However, each building has the authority to decide how its capital budget can be spent. The coordination of the SBM innovation with existing administrative practices has been accomplished with the help of a great deal of communication. This achievement may be attributed to the active involvement of the district superintendent in the implementation of the innovation by attending meetings and visiting SICs on numerous occasions. The needed techniques for mastering and operating SBM have been partially learned in initial training for SIC members. However the superintendent believes that additional training still is needed. There is variation among the schools in their understanding of and enthusiasm for the administrative innovation.

The strategic choices of the Metro District effort have all been made at the school building level. The scope of the project has been narrow in the sense that it has only been attempted in the high school. This approach however, is more a reflection of the strong interest of the Metro High principal and the relatively passive interest of the district superintendent. Furthermore, there has not been a strategy of consciously eliminating old decision-making practices to make way for the new SBM decision making practices. For example, there has been no clear statement of when it is appropriate for the site council to make decisions and when it may only act in an advisory capacity. Finally, it is not clear that any real efforts have been made to reduce the fears of the practitioners (teachers, staff, board members, and school administrators) that SBM could have a negative impact on their jobs. This may be related to the limited amount of exposure that most of the school staff, students, and parents have had with the Metro High site council and SBM, and they do not perceive it as a threat. In addition, the loss of district office administrative jobs was not likely since only one school was attempting to adopt SBM.

The strategic choices of the River District effort have been made primarily at the district level with input from the principals at the school building level. The scope of the innovation is very broad and involves all of the school buildings in the district. It does not involve all of the primary stakeholder groups, however. To date, the effort to implement SBM has focused on the participation of professional staff, noncertified school staff, and administrators within the building or organizational unit. In most cases, previous approaches to school administration and decision making have been modified to reduce any potential conflict with the new SBM practices. The issue of job threat was not addressed in River District. It is possible that some district office personnel may have feared diminishing importance due to the potential transfer of some of their responsibilities to the school building level. This problem, however, has not arisen so far.

The expansion stage at Metro High School has not been entered on a districtwide basis, although SBM appears to be expanding at the schoolwide level. Recently, a representative of the assistant principals has been included in the membership roster. In the near future, representatives of the faculty senate, department chairs, and the student senate also may be included. Only two of the five passages and cycles of this stage have been accomplished; organizational status and stable supply and maintenance arrangements. The Metro High site council has achieved organizational status through a system of decision making rules and mission statements. These rules and statements, however, are principally concerned with the internal operation of the site council. It has made some decisions on school direction and has disbursed funds for special projects that are approved of by the council. The site council also has had a stable supply and maintenance arrangement for its administrative operations. It has a paid half-time coordinator whose responsibility has been to manage the administrative affairs of the council. The stability of this arrangement, however, will be tested as a result of the council's loss of operating funds due to the expiration of their grant. It is possible that the paid coordinator's position will be eliminated as a cost-saving measure. The cycle of equipment turnover is not applicable in this case due to the administrative process nature of the innovation.

The expansion stage at River District has been partly achieved as a result of the original broad design of the SBM project. Most of the passages and cycles important to this stage have been achieved. The transition to local funds has not been necessary because funding for the SBM project has always been internal. However, the superintendent has had to resist efforts to have the project cut from the budget. Organizational status has been accomplished and is evident from the various written documents and the integration of staff leadership into the decisionmaking processes of the schools. There has been a stable supply and maintenance arrangement because of the internal source of funding. Any potential instability

575

in such arrangements only occurred during the efforts of the school board to reduce the project funding. The passage of the establishment of a personnel classification and certification system has been partially achieved at the level of the building principal by adding the role of staff leader to his or her job responsibilities. Again, the cycle of equipment turnover is not applicable in this case due to the administrative process nature of the innovation.

Disappearance Stage

None of the passages and cycles of the disappearance stage have been achieved in the Metro School District. However, recent changes in the focus and structure of the Metro High site council indicate that efforts are being made to further integrate the council into the governance system of the high school, to increase the training in SBM, and to achieve a wider acceptance of SBM within the high school community. The two cycles of promotion of personnel acquainted with the innovation and the turnover in key innovation personnel have not been an issue in the implementation of SBM.

The original advocates of the innovation are still participants in the project, and in the case of the Metro High principal, his departure due to promotion or turnover does not appear to be likely in the near future. As a result, it may be some time before the innovation must deal with the departure of its original advocate.

River School District does not appear to have moved into the disappearance (or institutionalization) stage either. It has not achieved most of the cycles and passages important to this stage. There has been some change in the system of organizational governance, but not in a major way. Many of the final decisions about school operations are still made in the central district office. Such decisions include the overall curriculum objectives, final staff hiring decisions (although input from the school committee is considered), and the overall budget allocation for each school. It is not clear if the

staff leadership training program for the SBM project has been internalized into the schools administration. There appears to be variation among the schools regarding the amount and regularity of training, probably related to the differences in interest and enthusiasm for the innovation. The cycle of promotion of personnel acquainted with the innovation has not yet been an issue. This will only be tested when the district superintendent leaves River District, a change that appears to be some time into the future. The cycle of turnover of key personnel has been an issue only at River High School. Since staff leadership has been implemented the high school has had three principals. The original principal, however, was the least enthusiastic of all the principals about the innovation, so it is not clear what effect the turnover in staff has had on SBM at the high school level. There has been attainment of widespread use although this is probably related more to the original broad-based implementation approach than to the internalization of the innovation.

Our findings suggest that River School District has been able to complete more of Yin's features, passages, and cycles than Metro School District. This does not imply that one district has a better chance at institutionalization than the other in implementing SBM, only that River appears to be further along in the innovation institutionalization process. The time requirements for the institutionalization of an innovation vary and a longer implementation period should not be equated with less success. From the perspective of institutionalization, neither district has achieved success yet but it is still possible for both.

CONCLUSIONS AND IMPLICATIONS

From our data it appears that the first proposition about the greater sustainability of a breadth approach seems to hold up since River

School District (breadth emphasis) appears to have had an easier time maintaining momentum and overcoming barriers. The site council in Metro District has had a great deal of difficulty moving into substantive discussions and toward decisions on issues. They also seem to exhibit frustration over this lack of movement.

Schools, especially a single high school in a district with no other high schools, are reasonably independent of the other organizational units. This means that if the second proposition concerning the interdependence of organizational units is correct, River School District should not necessarily maintain a faster pace of adoption because of its breadth emphasis. It is too early to draw a conclusion on this issue however because we have data that support or disconfirm this proposition.

Proposition three dealing with the initial pace of adoption is supported because Metro High School, with its depth approach moved more slowly and seemed to flounder partly as a result of the absence of a clearly committed top management. The diverse set of involved stakeholder groups at Metro has required extensive effort to develop a discussion and decision making process that is satisfactory to all the involved parties.

The fourth proposition about the difficulties of the depth approach with its inclusion of multiple stakeholder groups also appears to be valid. One of the challenges now facing River School District is how to begin to integrate a broader group of stakeholders into its system of SBM. So far, the SICs have been limited to school administrators and staff with parents and other community members only participating in an advisory manner. At Metro School District, top management in the form of the superintendent has not been actively involved in the adoption of SBM. The source of management energy has been the high school principal, who continues to be extensively involved in the adoption effort. In the current educational system, with its district offices, superintendents, and school boards, active man-agement support primarily based on the school principal may not be adequate, however.

The answer to our central research question of which relative emphasis—depth or breadth—is more productive in facilitating smooth and effective adoption of the innovation, is inconclusive. In our search for this answer, however, we have learned some useful things about the process of innovation adoption over time.

In our two studies, the breadth emphasis produced the completion of more of Yin's stages and cycles and thus appears to be further along the road to institutionalization. This finding is contrary to Yin's proposal that innovations that are initially implemented more narrowly (depth emphasis) and then later expanded (breadth) will have a better chance of achieving institutionalization. It is possible that the source of the innovation effort, district superintendent versus building principal, may contribute to this apparent reverse finding.

Another finding is that the time required to implement an innovation varies with the depth and breadth strategies. Although the depth approach may take more time in the improvisation stage, it takes less time in the expansion or disappearance stage. A breadth approach may require less time in the improvisation and expansion stages but more in the disappearance stage.

It can be argued that River District's current level of completed features, passages, and cycles is not an accurate indicator of long-term innovation institutionalization. Yin suggests that the more broadly implemented innovations may be overly ambitious at too early a point in their life cycle, thus reducing the likelihood of eventual institutionalization.

The material on resistance to change in the organizational behavior literature should also be considered. That research discusses the movement of authority for making key decisions from one location to another in an organization and the problems that are created when the pecking order is changed or when a given role or responsibility changes in its importance to the organization (Harvey and Mills 1970; Zaltman, Duncan, and Holbeck 1973). River District introduced a more drastic change in the central office roles but a less drastic change in the role of the principal, who manages the site council consisting only of his or her subordinates. The Metro District effort involved less drastic changes at the central office because there was little new authority for making decisions granted to the school principal that had not been previously delegated to the school. In Metro District, however, there were more changes in the role of the principal through his efforts at shared decision making with a variety of stakeholders and because he was not designated as the head of the site council. One might expect that at the central office level, the River District staff might try to influence or block the SBM process if in fact they actually lose authority. On the other hand, the central office in Metro District is more secure as long as the innovation does not spread to all the schools or as long as it does not move decision making authority that now rests in the central office to the school level.

Other variables may cloud the picture. The superintendent's (top management) strong support in River District, a major change in leadership, a static situation needing change, and the initiation of new districtwide committees on important issues such as curriculum may have produced sufficient "shocks" (Schroeder et al., Chapter 4) to stimulate acceptance of SBM in the River School District. In Metro District, key policy areas (such as personnel) were delegated to the school building before the implementation of the SBM project. Furthermore, attempts to study the impact of various innovation implementation approaches are severely hampered by the inevitable existence of widely varied and dynamically changing environment in the units under study.

Continuing research will address many of the above unanswered questions. Neither school has achieved innovation institutionalization and it is important to continue tracking their efforts. Additional questions have also been raised concerning the ability and effort needed to move from either a depth or breadth emphasis to the other. It is also important to continue to explore the roles of the superintendent and the building principal in the implementation of the SBM innovation. Process changes require substantial amounts of time to observe. Our three-year study of Metro and River School districts may simply be not enough time in which to evaluate the success of the innovation efforts. Such change requires patience, faith, and support from the top that encourages all participants to stick with the change.

APPENDIX 17–A: CHRONOLOGY OF EVENTS FOR METRO SCHOOL DISTRICT

District Level

Summer 1982: The principal obtains approval of the district superintendent and the school board to try implementing school-based management at Metro High School.

Fall 1983: District superintendent participates on the Planning and Monitoring Committee.

November 1985: District office representatives attend morning sessions of site council retreat. Superintendent is not able to attend.

High School Level

Spring-Summer 1982: High school principal attends a Public Schools Incentives (PSI) meeting at which school-based management is discussed.

Fall 1982: The principal and consultant

begin a collaboration to develop a model of school-based management for Metro High School.

February 1983: The principal submits a proposal developed by the consultant and himself to the Foundation for a planning and implementation grant for the formation of a school site council.

Spring 1983: A grant is received from the Foundation for an initial three-year period with yearly renewal clause. It is hoped and "understood," however, that the funds will be supplied for a five-year period.

August 1983: The Planning and Monitoring Committee is developed and begins to hold its first meetings.

February 1984: The management council is organized and begins to hold its first meetings. The council members have been selected by the Planning and Monitoring Committee.

March 1984: A proposal for a second year of funding for the school-based management project is submitted to the Foundation.

Spring 1984: A mission statement and council by-laws are developed. Council members make several trips to examine the implementation of school-based management at other sites.

August 1984: A management council coordinator is hired on a half-time appointment. Management council retreat at which the issue of hall monitoring responsibilities are discussed.

November 1984: The Planning and Monitoring Committee is disbanded at the insistence of the granting Foundation.

June 1985: Summer workshop held to establish the management council's priorities for the following year, 1985–86.

November 1985: November retreat to discuss ways to implement the council's identified priorities of instruction and staff development. Three main subcommittees developed: school environment, quality circles, and student feedback.

February 1986: Explosive meeting at which substantial dissatisfaction with the management council was voiced by several members. Several significant council supporters were absent coincidentally that evening. Presentation on quality circles by a local business.

March 1986: Plans and drawings developed for school improvements by the school environment committee.

Spring 1986: Ongoing budget problems. The Foundation has notified the school of its decision to end all school based management funding after the third year. Significant efforts were made to find alternative sources of funding.

April 1986: A new chair and vice chair to the council were finally elected. This was originally scheduled to have been done in February but was delayed because of internal process problems.

May 1986: The management council reviewed and approved a new grading and weighting procedure. The class ranking recommendation was delayed pending further advice from departmental staff.

September 1986: The management council formally adds an assistant principal position to the membership structure of the council.

May 1987: The council is projected to run out of external funds.

Summer 1987: The management council formally includes representatives from the department chairs and faculty senate organizations to facilitate communication and coordination among the groups. Potential restructuring of management council to compensate for loss of grant money.

579

APPENDIX 17–B: CHRONOLOGY OF EVENTS FOR RIVER SCHOOL DISTRICT

District Level

Fall 1981: Budget crisis leads to district-wide elimination of department head jobs, with incumbents reverting to full-time teaching load.

Summer 1983: New superintendent appointed. Difficult teacher negotiations completed with the new superintendent in charge. Teacher demand for reinstatement of department head funding answered by the superintendent with a proposal for a staff leadership program with new school-based councils funded by the district budget.

Fall 1983: Establishment of a district curriculum council and a district staff leadership committee by the superintendent. The leadership committee was charged with the responsibility for developing a school-based management model for River District.

January 1984: District leadership committee presents proposal to faculty and staff for the River District version of school-based management called staff leadership.

April 1984: District wide vote of certified staff members overwhelmingly endorses the staff leadership plan.

May 1984: School board approves the staff leadership plan.

August–September 1984: Selection and initial training of staff leaders.

Academic year 1984–85: Each school forms its own staff leadership council (later renamed School Improvement Council) and begins to organize, set goals, and proceed with training in leadership, group methods, and shared decision making. The building principal heads each School Improvement Council and the council membership is limited to certified staff.

Spring 1985: School Improvement Councils are allocated a stipulated amount for each building to be spent on capital items. Training in budgetary process for capital expenditures is provided.

February 1985: High School principal dies affecting the development of the School Improvement Council and the staff leadership program in the largest school in River District.

December 1985: First use of the new staff leadership process for curriculum discussions. The Junior High School Improvement Council presents proposals for eleven new courses to the District Curriculum Council. Eight of the eleven proposed courses are ultimately approved by the school board.

May 1986: Superintendent defends budget item for School Improvement Councils in the face of the school board's search for items to cut.

June 1986: Resignation of new high school principal after only one year and appointment of another principal to begin in fall 1986.

Academic year 1986–87: All schools continue to make budget decisions with respect to instructional materials and capital expenditures within the limit of the total amounts allocated to these accounts by the central office. Some appoint community members to School Improvement Councils, some add noncertified staff but teachers still predominate as council members.

August 1987: Workshops with staff leaders to develop a decision authority matrix to clarify who makes what kinds of decisions and to specify the authority of the School Improvement Councils.

High School Level

February 1985: High school principal dies.

February–June 1985: Conflict among SIC members.

June 1985: New principal appointed. Participative Decision Making orientation.

June 1986: Principal leaves after one year.

September 1986: New principal arrives.

Other Schools

Academic year 1985: Each school sets up own goals and areas for emphasis.

Academic year 1986: All Staff Improvement Councils given training in capital budgeting procedures. Two schools obtain outside funding for School Improvement Council inspired projects. Multischool Staff Improvement Councils involved in the hiring of an elementary school principal.

February 1986: Union negotiates "staff representative" term to replace "staff leader."

Spring 1986: First Staff Improvement Council curriculum proposals come forth from the junior high school and the high school. Both proposals passed by the District Improvement Council.

January 1986: Junior high school Staff Improvement Council involved in planning for a new junior high building.

REFERENCES

Baldridge, J.V., and T.E. Deal. 1975. *Managing Change in Educational Organizations.* Berkeley: McCutchan.

Chesler, M., R.A. Schmuck, and R. Lippitt. 1975. "The Principal's Role in Facilitating Innovation." In J.V. Baldridge and T.E. Deal, eds., *Managing Change in Educational Organizations,* pp. 321–27. Berkeley: McCutchan.

Conway, J. 1984. "Myth, Mystery, and Mastery of Participative Decision-making in Education." *Educational Administration Quarterly* 50(3): 11–40.

Cooke, R.A., and R.J. Coughlan. 1979. "Developing Collective Decision Making and Problem Solving Structures in Schools." *Group and Organizational Studies* 4(1): 71–92.

Etzioni, A. 1963. "The Epigenesis of Political Unification." In A. Etzioni, ed., *Social Change: Sources, Patterns, and Consequences,* ch. 55. New York: Basic Books.

Freeman, R.E. 1984. *Strategic Management: A Stakeholder Approach.* Boston: Pitman.

Fullan, M., M. Miles, and G. Taylor. 1980. "Organization Development in Schools: The State of the Art." *Review of Educational Research* 50(1): 121–83.

Gross, N., J.B. Giacquinta, and M. Bernstein. 1975. "Failure to Implement a Major Organizational Innovation." In J.V. Baldridge and T.E. Deal, eds., *Managing Change in Educational Organizations,* pp. 409–26. Berkeley: McCutchan.

Gold, B.A., and M. Miles. 1981. *Whose School Is It Anyway?* New York: Praeger.

Hage, J., and M. Aiken. 1970. *Social Change in Complex Organizations.* New York: Random House.

Harvey, Edward, and R. Mills. 1970. "Patterns of Organizational Adaptation: A Political Perspective." In Mayer N. Zald, ed., *Power in Organizations,* Nashville, Tenn.: Vanderbilt University.

Lindelow, J. 1981. "School-based Management." In S.C. Smith, J.A. Mazzarella, and P.K. Piele, eds., *School Leadership: Handbook for Survival,* ch. 4. Eugene, Ore.: University of Oregon, ERIC Clearinghouse on Educational Management.

Marburger, C.L. 1985. *One School at a Time.* Columbia, Md.: National Committee for Citizens in Education.

Merritt, R.L., and A.J. Merritt, eds. 1985. *Innovation in the Public Sector.* Beverly Hills, Calif.: Sage.

Miles, M. 1975. "Planned Change and Organizational Health: Figure and Ground." In J.V. Baldridge and T.E. Deal, eds., *Managing Change in Educational Organizations,* pp. 224–49. Berkeley: McCutchan Press.

Munson, F., and D.C. Pelz. 1982. "Innovating in Organizations: A Conceptual Framework." Unpublished working paper. Ann Arbor: University of Michigan.

Pelz, D.C., and F.M. Andrews. 1976. *Scientists in Organizations.* Institute for Social Research, University of Michigan

Roberts, N., and P. King. 1987. "The Dynamic Process of Policy Innovation: The Catalytic Function of Policy Entrepreneurs." Paper presented at the "Management of Innovation" conference, University of Minnesota, 1987.

Rogers, E.M., and J. Kim. 1985. "Diffusion of Innovations in Public Organizations." In R.L. Merritt and A.J. Merritt, eds., *Innovation in the Public Sector,* pp. 85–108. Beverly Hills, Calif.: Sage.

Rothschild-Whitt, J. 1976. "Conditions Facilitating Participatory Democratic Organizations." *Sociological Inquiry* 42(2): 75–86.

Schroeder, R.G. et al. Ch 4.

Tushman, M.L., and W.L. Moore, eds. 1982. *Readings in the Management of Innovation.* Marshfield, Mass.: Pitman.

Van de Ven, A.H. 1980a. "Problem Solving, Planning, and Innovation, Part I: Test of the Program Planning Model." *Human Relations* 33(10): 771–40.

———. 1980b. "Problem Solving, Planning, and Innovation, Part II: Speculation for Theory and Practice." *Human Relations* 33(11): 757–79.

Van de Ven, A.H., and M.S. Poole. 1986. "Paradoxical Requirements for a Theory of Organizational Change." *Strategic Management Research Center Discussion Paper 58.* Minneapolis: University of Minnesota.

Weiler, H.N. 1985. "Politics of Educational Reform." In R.L. Merritt and A.J. Merritt, eds. *Innovation in the Public Sector,* pp. 167–99. Beverly Hills, Calif.: Sage.

Wilson, James Q. 1966. James D. Thompson, ed. "Innovation in Organization: Notes toward a Theory." In *Approaches to Organizational Design,* pp. 193–218. Pittsburg, Penn.: University of Pittsbugh.

Yin, R.K. 1979. *Changing Urban Bureaucracies: How New Practices Become Routinized.* Lexington, Mass.: Lexington Books.

Yin, R.K., and D. Yates. 1974. *Street Level Government: Assessing Decentralization and Urban Services.* Santa Monica, Calif.: Rand.

Zaltman, G., R. Duncan, and J. Holbeck. 1973. *Innovations and Organizations.* New York: Wiley.

Zaltman, G., D. Florio, and L. Sikorski. 1977. *Dynamics of Educational Change: Models, Strategies, Tactics, and Management.* New York: Free Press.

18

MOBILIZING INNOVATION EFFORTS: THE CASE OF GOVERNMENT STRATEGIC PLANNING

John M. Bryson

William D. Roering

Don't you love it when a plan comes together? (Col. John "Hannibal" Smith of the A-Team)

Our normal expectation should be that new programs will fail. The cards in the world are stacked against things happening, as so much effort is required to make them move. The remarkable thing is that new programs work at all. (Pressman and Wildavsky 1973: 109)

Government leaders have become increasingly interested in strategic planning as a result of the wrenching changes that have beset the public sector in recent years. The changes have

An earlier and shorter version of this chapter appeared as John M. Bryson and William D. Roering, "The Initiation of Strategic Planning by Governments," *Public Administration Review,* vol. 48, no. 6, 1988, pp. 995–1004.

The authors wish to express their deepest appreciation to all of the strategic planning teams that participated in this study. Quite literally, the study would not have been possible without them. The authors also wish to thank Andy Van de Ven for his encouragement and steadfast support of this study every step of the way. We are grateful to him for his thoughtful critique of an earlier draft of this chapter. Scott Poole also provided detailed and helpful criticism of that early draft, as did John Kimberly. Several students in Andy's doctoral seminar on innovation offered helpful critiques: Melissa Anderson, Patrick Pak, Jim Doyle, and Hyoung Koo Moon. Our colleagues in the Minnesota Innovation Research Program, of course, have been helpful throughout the study. And special thanks are due Jeremy stemmed from oil crises, demographic shifts, changing values, tax levy limits, tax indexing, tax cuts, reductions in federal grants and mandates, the devolution of responsibilities, and a volatile economy. The changes have brought into sharp relief the need for important policy choices, and thus highlighted the potential usefulness of strategic planning. Indeed, strategic planning may be defined as a disciplined effort to produce fundamental decisions and actions that define what an organization (or other entity) is, what it does, and why it does it (Bryson 1988: 5; Olsen and Eadie 1982: 4).

In effect, strategic planning is a process that can help governments recognize where they need to make major changes—that is, innovations—in the mandates they work under, the missions they pursue, or their product or service level and mix, cost, financing, organization, or management. *Strategic planning, in other words, is an administrative process innovation designed to routinize the recognition, development, and implementation of needed innovations.* The initiation of strategic planning actually represents three interrelated innovations for most governments: (1) the gathering of key actors (preferably *key decisionmakers*), (2) to work through a reasonably structured "strategic thinking and acting" process, (3) in order to focus attention on what's important, set priorities for action, and generate those actions.

The study tracked the initiation of strategic planning by eight governmental units. All are located in the Twin Cities metropolitan area of Minnesota. Each used the same basic strategic planning process. The study followed the units until they either discontinued their strategic planning efforts, or completed a strategic plan. (Two discontinued the process, while six developed strategies to deal with their most important strategic issues; all of the completed strategies, or "plans," were at least partially adopted by the appropriate decision bodies.) Although at least some plan implementation efforts were underway during the study period, plan implementation was not the primary focus of this study. The overall effectiveness of the strategic planning process model, therefore, was not assessed. Instead, the study tried to (1) document what happens when units of government work through a strategic planning process (when that process represents an innovation for the units) and (2) uncover the conditions necessary for successful initiation of a strategic planning process by governmental units.

The rest of this chapter is organized into several major sections. First, the strategic planning process followed by the eight units is presented. Second, the relationship between the process and the Minnesota Innovation Research Program framework is discussed. Third, the sample and research methods are described. Fourth, the efforts of the eight units to undertake strategic planning are described along with the outcomes of their efforts. Fifth, the efforts of the eight units are analyzed and patterns across units discussed. Finally, the chapter presents conclusions about the initiation of strategic planning by governments.

O'Grady of the London Business School, whose persistent, effective, and constructive criticism of public-sector strategic planning has been especially helpful—although he may not have realized it. Whatever readability the chapter has is due primarily to the efforts of Barbara C. Crosby. The study itself was financed in part by contract No. N00014-84-K-0016 from the Office of Naval Research, Program on Organizational Effectiveness (Code 4420E), and by grants from the Dayton Hudson Foundation to the Government Training Service, St. Paul, MN, and from the McKnight Foundation to one of the governmental units in our study.

A PUBLIC-SECTOR STRATEGIC PLANNING PROCESS

Each of the governments studied agreed to follow an outline of a public-sector strategic planning process developed by author John Bryson (see Figure 18–1) (Bryson, Freeman, and Roer

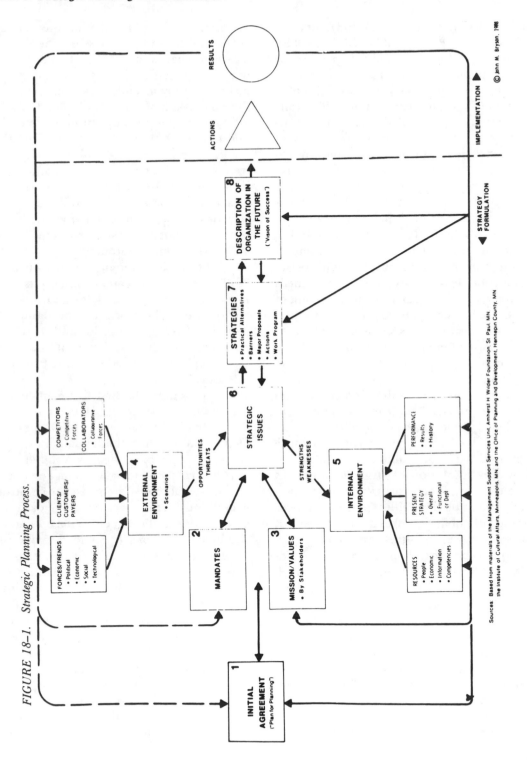

FIGURE 18–1. Strategic Planning Process.

Sources Based from materials of the Management Support Services Unit, Amherst H. Wilder Foundation, St. Paul, MN, the Institute of Cultural Affairs, Minneapolis, MN and the Office of Planning and Development, Hennepin County, MN

© John M. Bryson, 1988

ing 1986; Bryson and Roering 1987; Bryson, Van de Ven, and Roering 1987; Bryson 1988). The process begins with an *initial agreement* (or "plan for planning") among decisionmakers whose support is necessary for successful plan formulation and implementation. Typically these decisionmakers would agree on the purpose of the effort, who should be involved, what should be taken as "given," what topics should be addressed, and the form and timing of reports. Most authors agree that the support and commitment of management and the chief executive are vital if strategic planning in an organization is to succeed (Olsen and Eadie 1982). In addition, major changes in governments often must be authorized by the appropriate governing bodies—for example, city councils and county boards. Further, the involvement of key decisionmakers outside the organization usually is critical to the success of public programs if implementation will involve multiple parties and organizations (McGowan and Stevens 1983).

The second step is the *identification of the mandates,* or "musts," confronting the government. Third comes *identification of the organization's mission and values,* or "wants," because they have such a strong influence on the identification and resolution of strategic issues, as discussed below. The process draws attention in particular to similarities and differences among those who have stakes in the outcome of the process and to what the government's mission ought to be in relation to those stakeholders. "Stakeholder" is defined as any individual, group, or other organization that can place a claim on the organization's attention, resources, or output, or that is affected by that output. Examples of a government's stakeholders are citizens, taxpayers, service recipients, the governing body, employees, unions, interest groups, political parties, the financial community, and other governments.

Next come two parallel steps: identification of the *external* opportunities and threats the organization faces and identification of its *internal* strengths and weaknesses. The distinction between what is external and what is internal hinges on whether the organization controls a factor, thereby making it internal, or does not, thereby making it external (Pfeffer and Salancik 1978). To identify opportunities and threats one might monitor a variety of political, economic, social, and technological forces and trends, as well as various stakeholder groups, including clients, customers, payers, competitors, or collaborators. To identify strengths and weaknesses the organization might consider resources (inputs), present strategy (process), and performance (outputs).

Strategic planning focuses on the best "fit" between an organization and its environment. Attention to mandates and the external environment, therefore, can be thought of as planning from the "outside in." Attention to mission and values and the internal environment can be considered planning from the "inside out."

Together, the first five elements of the process lead to the sixth, *identification of strategic issues* (that is, fundamental policy questions affecting the organization's mandates, mission, values, product or service level and mix, clients or users, cost, financing, or management). Usually, it is vital that strategic issues be dealt with expeditiously and effectively if the organization is to survive and prosper. Failure to address a strategic issue typically will lead to undesirable results from a threat, failure to capitalize on an important opportunity, or both. Unfortunately, organizations often find it difficult to deal with strategic issues because of the conflicts or dilemmas involved. Every important strategic issue typically involves conflicts over *what* (ends), *why* (philosophy), *how* (means), *when* (timing), *where* (location), and *who* (winners and losers) (Hostager and Bryson 1986).

Strategy development, the seventh step in the process, begins with identification of practical alternatives for resolving the strategic issues. Then it moves to the enumeration of barriers to the achievement of those alternatives, rather than directly to development of major proposals to realize the alternatives. A focus on barriers to the achievement of those alternatives is not typical but is one way of ensuring that any strategies developed deal with implementation difficulties directly rather than haphazardly. Once alternatives and barriers are listed, major proposals are developed either to achieve the alternatives directly or else indirectly through overcoming the barriers. Then major actions and a work program are prepared to implement the proposals.

After strategy development comes an eighth step not typical of most strategic planning processes: *describing the organization in the future.* This description is the organization's "vision of success" (Taylor 1984), and outlines how the organization would look if it successfully implemented its strategies and achieved its full potential. The importance of such descriptions as a guide for performance has long been recognized by well-managed companies (Ouchi 1981; Peters and Waterman 1982; O'Toole 1985) and by organizational psychologists (Locke et al. 1981). Typically included in such descriptions are the organization's mission, its basic strategies, its performance criteria, some important decision rules, and the ethical standards expected of all employees.

Those eight steps complete the strategy formulation process. Next come *actions and decisions to implement the strategies* and, finally, the *evaluation of results.* Although the outline depicts the process as linear and sequential, it must be emphasized that the process in practice is iterative. Participants typically rethink what they have done several times before they reach final decisions. Moreover, the process does not always begin at the beginning. Instead, organizations often find themselves confronting a strategic issue that leads them to engage in strategic planning. Once engaged, an organization is then likely to go back and begin at the beginning. The steps in the process, therefore, represent more of a set of checkpoints or prompts to ask and answer specific kinds of questions than a rigid, sequential order of tasks.

RELATIONSHIP BETWEEN THE PUBLIC-SECTOR STRATEGIC PLANNING PROCESS AND THE MIRP FRAMEWORK

How does public-sector strategic planning relate to the MIRP framework? The MIRP framework defines the innovation process as the development and implementation of new *ideas* by *people* engaged in *transactions* with others in an institutional *context. Outcomes* are perceptions about performance of the process as the innovation develops over time. An *event* is defined as a change in the ideas, people involved, transactions or relations engaged in, context, or outcomes of the innovation under study. Further, the framework specifies three contingency variables that should affect the management of any innovation: its *novelty, scope or size,* and *age or stage of development.*

The Idea of Strategic Planning

Strategic planning consists of a set of concepts, procedures, and tools designed to help key decisionmakers think and act strategically on behalf of their organizations (Bryson and Roering 1987). The *idea* behind strategic planning is that if a government's key decisionmakers come together to discuss their government's mandates, mission, and situation, they will be able to produce decisions and actions that define or affect what the government is, what it does, and why it does it. Naturally there are constraints on those decisions, but the process at least implicitly challenges those constraints, alerts key decisionmakers to the need for important inno-

587

vations, and attempts to shape the creation of those innovations. In other words, the basic innovation that strategic planning represents is the idea of focusing the attention of key decisionmakers on what is truly important for the organization and then persuading them to do something about it.

Strategic planning is therefore not an ordinary idea, or set of ideas. Strategic planning is an idea about how to *organize, think about,* and *act on or toward* other ideas, people, transactions, or contexts. Said differently, *strategic planning is a process deliberately designed to produce events,* as defined by the MIRP framework.

The deliberate attempt to produce events is probably the greatest strength *and* weakness of strategic planning as a process. Changes in organizations normally occur through disjointed incrementalism, or "muddling through" (Lindblom 1959; Quinn 1980). Any process designed to force important changes therefore can be seen either as a highly desirable improvement on ordinary decision making or doomed to failure. Indeed, whatever the merits of strategic planning in the abstract, *our normal expectation has to be that most efforts to produce fundamental decisions and actions in government through strategic planning will not succeed.* At the very least, strategic decision making in public organizations should be prone to involvements by numerous actors (especially through boards, committees, task forces, and teams), variability in information, extensive negotiations, and frequent delays. Further, because of pressures for public accountability, decisions ultimately are likely to be made at the highest levels (Hickson et al. 1986: 117, 203), while political rationality dictates that top decisionmakers *not* make important decisions until forced to do so (Benveniste 1972, 1977; Quinn 1980).

Although the idea behind strategic planning may be very simple, its application therefore is not. Strategic planning processes must be tailored to the organization wishing to use it

(Barry 1986; Bryson and Roering 1987; Chakravarthy 1987).

Finally, the ideas produced by strategic planning are not all the same. We have already noted that strategic issues are likely to involve important conflicts or dilemmas. These conflicts are likely to vary in their *complexity* and in the *political interests* at stake. These differences are likely to lead to differences in the form and nature of the efforts to resolve them (Hickson et al. 1986).

People Involved

Strategic planning provides a set of concepts, procedures, and tools to help key decisionmakers focus on what is important. But strategic planning cannot work unless it is used by key decisionmakers. Strategic planning thus represents two people-related innovations for most governments. The first is the use of concepts, procedures, and tools designed to facilitate people's strategic thinking and acting, while the second is the gathering of key decisionmakers to discuss what is truly important for the organization. This second innovation is vital, as key decisionmakers in governments rarely meet to discuss matters of significance to their organization as a whole (Heclo 1977; Seidman 1986).

The initiation of strategic planning thus implies that a powerful person, or persons, must *sponsor* and *legitimize* the process. But even that is probably not enough. Some person or group also must actively push the process along, maintaining commitment and rousing enthusiasm when others would just as soon let the innovation drop. In other words, some person or group must *champion* the process.

Several authors have stressed the importance of champions (Kotler 1976; Maidique 1980; Kanter 1983). In the case of public-sector strategic planning, a special kind of champion probably is needed—a *process champion.* A process champion is a person who is not committed to any particular framing of strategic issues or answers (although they may have some good hunches about what those issues and answers are likely to be). Instead, they are deeply committed to the *process* of bringing key decisionmakers together to talk about, and act on, what is important for the organization.

Key decisionmakers are the primary audience for the information that strategic planning creates. The key decisionmakers may produce this information themselves, or they may rely on a strategic planning team to create it. (Ideally the key decisionmakers will themselves serve on the strategic planning team, since participation in the process is likely to increase commitment to the process and ownership of results.) Outside consultants may be used to facilitate the strategic planning process or to provide particular sorts of information.

A final set of people who play an important part in any strategic planning process is the organization's stakeholders (including key decisionmakers, of course). A stakeholder analysis is an important part of the process followed by the organizations in this study, as it affects development of a mission statement, strategic issue identification, and strategy development.

Transactions and Context

Because of the nature of strategic planning, it is difficult to disentangle transactions from context. This is because strategic planning is basically an exercise in informed decision making in which the rules that constitute the organization and its actions are the subject matter for decision. In other words, for purposes of decision making, the organization *is* the "rules of the game" that govern both the context and process of decisionmaking (Hickson et al. 1986: 191). Strategic planning is a decisionmaking process embedded in that context that has as its purpose the potential change of the context. As March and Olsen (1976: 31) note, "Organizations regulate the connections among problems, choice opportunities, solutions, and en-

ergy by administrative practice." Strategic planning is an administrative practice that typically seeks to change those connections.

Context is important to the strategic planning transactions that occur within it because context determines who the key decisionmakers are, what their interests are, and what might be done to satisfy those interests. Context therefore constrains the process that seeks to change it (Bryson and Delbecq 1979; Christensen 1985; Wechsler and Backoff 1987). The principal kinds of transactions, relationships, or interactions used to evaluate and change the context are discussions, conflict resolution methods, and authoritative decisions.

Contingency Factors

The MIRP framework includes three contingency variables: innovation novelty, size and scope, and age or stage. Public-sector strategic planning is not a totally novel idea, as it draws on a fund of concepts, procedures, and tools developed primarily in the private sector and applies them to public purposes (Bryson and Roering 1987). The idea of public sector strategic planning, therefore, is essentially *borrowed*.

Public-sector strategic planning basically is a *small* innovation, involving only key decisionmakers and members of a strategic planning team. However, the potential impacts on the organization using it, and its stakeholders, are large. The innovation stage of the strategic planning for the eight units in the study is basically the *design* stage, moving toward the *institutionalization* stage.

STUDY SAMPLE AND RESEARCH METHODS

Eight government units participated in the study. Five were suburban city governments, one was a county government, and two were units of the same county government (the county executive director's office and the county's public health nursing service—the Nursing Service). All units were located in the Twin Cities metropolitan area of Minnesota (see Table 18–1). The study spanned a two-and-one-half-year period between August 1984 and February 1987 (with a follow-up questionnaire in October 1987).

The authors served as consultants to the eight units as they worked through the process. Nursing Service went through the process independently from the other units. The other seven units were participants in a series of strategic planning workshops sponsored by the government training service that serves the metropolitan area. These workshops introduced the units' strategic planning teams to the strategic planning process and helped them work through the steps in the process. Workshop 1 consisted of an introduction to strategic planning and a stakeholder analysis exercise. Workshop 2 focused on mission statement development and external and internal environmental assessments. Workshop 3 involved the identification of strategic issues. The remaining four workshops were consultations with individual strategic planning teams on an as-needed basis (workshop 4, cities C, D, and E; workshop 5, cities C and D; workshop 6, counties A and B; and workshop 7, city E). Consultants' fees for services to all eight units were covered by small grants from two metropolitan area foundations.

Several kinds of data were collected. (1) Some basic background data was collected for each unit (budgetary and demographic data). (2) The applications from the seven units to participate in the government training service-sponsored exercise were reviewed. (3) Detailed process histories were prepared for city E and Nursing Service by students of the first author. (4) Periodic interviews were conducted with each strategic planning team as long as the team continued to participate in the process. (5) Separate interviews often were conducted with process champions. Notes from interviews of teams and process champions are uneven in quality; some are quite detailed, while others are sketchy. (6) Baseline MIRP questionnaire data was collected for all units. (7) Time 2, time 3, and time 4 MIRP questionnaire data, how-

TABLE 18–1.   *Characteristics of Governmental Units and Strategic Planning Processes, 1986.*

Unit	Population	Number of Full-Time Equivalent Employees	Annual Budget	Size of Team	Sponsor Present	Champion Present	Policy Board Support
City A	42,583	168	8,977,013[a]	4	Yes	Yes	?
City B	9,842	35	2,070,000[a]	10	Yes	Yes	?
City C	22,100	103	7,302,926[a]	5	Yes	Yes	Some
City D	41,207	130	7,212,612[a]	7	Yes	Yes	Some
City E	42,600	228	12,201,433[b]	9	Yes	Yes	Some
County A	127,000	550	66,783,000[b]	19	Yes	Yes	Yes
County B–Executive Director's office	471,369	2,847[c]	275,421,249[b]	5	Yes	Yes	Some
County B–Nursing Service	471,369	104	4,507,102[d]	22	Yes	Yes	Some

a. General fund.
b. Total budget.
c. Entire county government employment.
d. ½ from general fund.

ever, are available only for city E and Nursing Service. At time 2 the other strategic planning teams either had dropped out of the program (city A), had completed the process and did not want to be bothered with questionnaires (city D), refused to fill out questionnaires deemed "too academic" (city B), or else filled out too few questionnaires, too incompletely to yield useful data (city C, county A, and county B—executive director's office). (8) Finally, a short follow-up questionnaire was filled out by the heads of the strategic planning teams of units that completed the process. This questionnaire measured the success of the strategic planning efforts to supplement the authors' own assessments.

Because of the relative absence of quantifiable data, most of the description and analysis of what happened will rely on the authors' observations and interviews with strategic planning team members. The authors met frequently over the course of the study to discuss what was happening in the units. We searched our memories and data files for patterns and themes across units. We consciously tried to engage in the kind of mental activity experts usually pursue unconsciously—namely, the development and application of tentative diagnoses to individual cases and sets of cases (Johnson 1984). We were the "experts" on strategic planning but had to be much more conscious about our expertise than we would have been had we not acted as researchers as well as consultants.

We did attempt, however, to verify independently our descriptions of the eight units' efforts presented in the next section. Verification was attempted through one of two methods. First, our accounts of what happened in the two units that dropped out of the process early were checked against newspaper accounts and in telephone interviews with the government training service contact person. The accounts of what happened in the six units that completed the process were checked for accuracy with the appropriate process champions.

Although we received independent confirmation of the accuracy of our descriptive accounts, we must emphasize the speculative nature of our analyses and conclusions. We were active participant observers who wanted the eight units to succeed with their strategic planning efforts. We were active teachers, consultants, and advisors at various points through the units' efforts. Our role therefore gave us privileged access to the phenomena we were trying to study, on the one hand; but it makes our analyses and conclusions somewhat suspect, on the other hand (Sussman and Evered 1978). Further, Johnson indicates that while experts are quick to recognize patterns in their areas of expertise, they also are prone to recognize the "wrong" pattern. They frequently "jump to conclusions" based on limited data. In other words, being an expert simply may allow a person to make mistakes faster than a layperson (Johnson 1984). Once committed to a wrong diagnosis, an expert may remain committed to that diagnosis in the face of contradictory evidence (Salancik 1977; Staw 1976; Staw and Ross 1978).

The reader also will find us "intruding" much more into the eight units' efforts than other researchers did in the other innovation studies. In a very real sense, *we were an important part of the innovation process pursued by these eight units.* None of these units would have engaged in strategic planning—at least in the precise way they did—without our presence, and several of the units probably would not have completed the process had we not continuously encouraged them to finish. Unfortunately, however, we failed to collect any data on our own performance and impact throughout the study. Data on our own performance and impact would have helped us and the reader understand better what happened in the eight cases.

A final serious limitation on the study is that the criteria used to measure success (embodied in the follow-up questionnaire to process champions) are *the authors' criteria,* not necessarily the units' criteria. At the beginning of the study we did not know how success ought to be measured, other than to note whether or not units completed the strategic planning process. Toward the end of the study, we concluded that successful initiation of strategic planning ought to be assessed according to the extent the process: (1) helped focus the attention of the strategic planning team on what was important, (2) helped the team set priorities for action, (3) helped generate those actions, and (4) helped focus the attention of the organization's key decisionmakers on what was important. The follow-up questionnaire solicited the team leaders' assessments of success according to these criteria.

DESCRIPTION OF THE EIGHT UNITS' STRATEGIC PLANNING EFFORTS

Results of the strategic planning efforts of the eight units are summarized in Tables 18–2 and 18–3. Table 18–2 shows that two units discontinued the process before they identified strategic issues. The other six units completed the formulation of strategies to deal with at least some of their strategic issues. For purposes of this research, a "strategic plan" consists of one or more recommended strategies to deal with one or more strategic issues. These six units, therefore, completed preparation of strategic plans, at least according to our minimalist criterion. In addition, three units developed "visions of success."

The relevant decision bodies in all units have adopted at least part of the proposed strategic plans. Different parts of different strategies, of course, came within the purview of different decision bodies. Depending on the circumstances, one or more of the following individuals or groups was the relevant decisionmaking authority: the strategic planning team,

a single top administrator, a "cabinet" of top administrators, or an elected policymaking body (a city council or county board of commissioners).

Table 18–3 summarizes additional perceptions of process champions of the success of the process (according to our criteria for success) for the six units that completed strategic plans. All felt that the process had helped "a good bit" or "very much so" to focus their strategic planning teams focus on what was important. There was more variability in the extent to which champions felt the process helped teams set priorities for action. One-third each felt the process helped do so "a moderate amount," "a good bit," and "very much so." The process was marginally less successful in generating priority actions. Half the units thought the process helped generate those actions "a moderate amount"; one unit felt it did so "a good bit"; and two felt it helped greatly. Finally, the process was marginally even less successful in focusing the attention of key decisionmakers on what was important. One unit felt it did so only "some"; three units felt it did so "a moderate amount"; one felt it helped "a good bit"; and one felt it helped greatly.

The process was a clear success in only two of the six cases: county A and county B—Nursing Service. In the other cases success was much more mixed. The process obviously helped strategic planning teams focus on what was important, but in several cases was less successful in setting priorities for team action and generating those actions and in helping focus the attention of key decisionmakers on what was important.

A brief description of the strategic planning processes of the eight units follows. These descriptions represent a first attempt at answering our first research question: What happens when units of government work through a strategic planning process, when that process represents an innovation for the units?

TABLE 18–2. *Strategic Planning Process Steps Completed and Time Devoted to Process.*

Government Unit	Strategic Planning Process Steps[a]									Time Devoted to Process
	1	2	3	4	5	6	7	8	Adoption	
City A	Y	Y	Y	Y	Y	—	—	—	—	2 months
City B	Y	Y	Y	Y	Y	—	—	—	—	2 months
City C	Y	Y	Y	Y	Y	Y	Y	—	Partial	5 months
City D	Y	Y	Y	Y	Y	Y	Y	Y	Partial	5 months
City E	Y	Y	Y	Y	Y	Y	Y	—	Partial	24 months
County A	Y	Y	Y	Y	Y	Y	Y	Y	Y	12 months
County B–Executive Director's office	Y	Y	Y	Y	Y	Y	Y	—	Partial	7 months
County B–Nursing Service	Y	Y	Y	Y	Y	Y	Y	Y	Y	30 months

Source: Follow-up questionnaire.
a. 1 = Initial agreement   4 = External assessment   7 = Strategy development
   2 = Mandate clarification   5 = Internal assessment   8 = Vision of success
   3 = Mission development   6 = SI identification
Y = Step completed.
— = Step not completed.

TABLE 18–3. *Responses to Follow-Up Questionnaire to Strategic Planning Team Leaders.*

	City		City C	City D	City E	County A	County B Executive Director's Office	County B Nursing Service
	A[a]	B[a]						
Q1	—	—	4	4	4	5	5	5
Q2	—	—	4	3	3	5	4	5
Q3	—	—	3	3	3	5	4	5
Q4	—	—	3	3	2	5·	3	4

a. No questionnaire sent because city dropped out early in the process.
    Q1 = Did the strategic planning process help focus the attention of the *strategic planning team* on what was important?
    Q2 = Did the strategic planning process help set priorities for action by the *strategic planning team?*
    Q3 = Did the strategic planning process help generate those actions?
    Q4 = Did the strategic planning process help focus the attention of the organization's *key decisionmakers* on what was important? (Note: the organization's key decisionmakers may include a number of people not on the strategic planning team.)
    5 = Very much so
    4 = A good bit
    3 = A moderate amount
    2 = Some
    1 = Not at all

City A

City A is a middle-class, "second ring" sub-urb—an area characterized by some old housing and industry but also marked by considerable new housing and industrial development. The strategic planning team for city A was led by the city manager, a highly regarded professional who had just received the manager-of-the-year award from the state association of city managers. When the strategic planning workshops began, the team was already engaged in formulating strategies to deal with development and redevelopment of the riverfront that forms the entire northern border of the city.

The strategic planning process therefore forced the team to backtrack and to broaden its focus—that is, the team had to go back to step 1 and also had to think about the city as a whole and issues other than the riverfront. Nevertheless, the team found it difficult to step back from the riverfront issue. Their conversation during the workshops (especially coffee breaks) was dominated by the issue, and usually one or more members of the team were unable to attend the workshops because they had to deal with riverfront matters.

Team members said the workshops they attended were helpful. They felt it was very useful to discuss the city's mission and situation, particularly because doing so helped them think more clearly about the riverfront issue. City A's participation in the process came to an abrupt end, however, two months after it began, when the city manager accepted a job in a distant state, and the new city manager decided not to engage in a strategic planning process until she had acclimated herself.

City B

City B is an older, middle-class, "second ring" suburb, with a lot of older housing and not too much room for new development. City B's strategic planning team also was headed by a well-regarded, if somewhat irascible, city manager. Actually, there were two strategic planning teams—the group that attended the workshops (which included two city council members) and a much larger group back in the city that repeated the workshop process under the guidance of the city manager. (Each of the government units was to have a "core group" that attended the workshops and a larger "homework group" that repeated the workshops exercises and helped the core group think about what should be prepared for the next workshop session.) The teams developed a mission statement and performed a situation analysis, but then three events disrupted their strategic planning process.

The first event was the announcement just prior to one of the workshops that a factory, the largest employer in the city, would close. The attention of the team attending the workshop was devoted almost entirely to what to do about that issue; they participated perfunctorily in the workshop exercises.

The second event was the unexpected emergence of a major development proposal for the lakefront that was both the border of the city and its main scenic attraction. The development, which had powerful backers, called for high-rise construction that would obstruct scenic views of the lake. Again, this proposal emerged just prior to one of the workshops, and again the workshop team's attention shifted almost entirely to that issue. Of course, the workshop team was planning strategically when it devoted workshop time to discussion of these issues, even though the discussions were out of sequence with the workshop topics and exercises. Since the workshop exercises were to be repeated with the larger strategic planning team back in the city, the workshop team felt its time was better spent on discussing the two issues then and there. Unfortunately, the larger team never met.

The third event—the resignation of the city manager to take another job in the metropolitan area—terminated the process. The manager and the consultants did not part on very friendly terms. In the weeks prior to the time the manager announced that he was taking a new job, the consultants asked him several times to have his team members fill out the MIRP T2 questionnaires. (Part of the agreement under which the workshops were offered was that each unit's team would fill out four sets of questionnaires, at six-month intervals.) City B's manager asserted that he would not ask his people to fill out the questionnaires, which he deemed "too academic." We, the consultants,

594

felt our agreement with the city B's planning team had been violated.

City C

City C is an older, middle-class, "inner ring" suburb, in which most growth occurred primarily in the twenty years following World War II. Although it has experienced new development, the city's leaders mainly focus on maintaining the existing housing, industrial, and commercial base and on redeveloping certain areas of the city. The strategic planning team included the mayor, who was very supportive of the process, but not the city manager, who was more neutral. The city's planning director was the team leader and process champion. As with city B, the workshop team acted as a core group and tried to involve a larger group of decisionmakers, including the city manager, in follow-up sessions.

The workshop team pushed slowly through the process. According to the team leader, the core group was good but often hard to assemble. It was apparently impossible to get the larger team together, so the core group forgot them for most of the process. Ultimately, the core group did develop strategies to deal with what they felt were the two most important strategic issues faced by the city, and took what actions they could to deal with the issues, including raising them for discussion by the city council. The mayor assumed personal responsibility for persuading the council to address the issues, which it did at a "town meeting." The planning director also convinced the city's housing and redevelopment authority to develop a strategic plan and worked through the process with them.

City D

City D is a rapidly growing "second ring" suburb. Population and employment are rising rapidly, and the city staff were virtually overwhelmed with the tasks of handling and guiding this growth during the study period. Further, the city staff's norms of accessibility and responsiveness to citizens and other stakeholders made it difficult to assemble the strategic planning team for the workshops. In sharp contrast to the stereotype of governmental bureaucratic inertia, city D's decisionmakers and staff pride themselves on how many decisions they make, and pieces of paper (forms, message slips, letters, and so forth) they handle in a day. As a result, key decisionmakers are constantly on the phone or out of the office dealing with every item, trivial or not.

Nonetheless, city D was reasonably successful at strategic planning. This is all the more surprising because members of the strategic planning team felt that they had failed at strategic planning three months after the process began. They felt they were "failures" for several reasons. First, they misunderstood the initial agreement with the government training service and the consultants and thought that the workshops would be simply training exercises, not the "real thing." They were unprepared, they felt, for the "real thing." Second, the team leader was the assistant city manager. The city manager was a member of the team, but only a lukewarm supporter. The team members wondered if they were wasting their time on strategic planning, since their ideas might not be implemented without the city manager's involvement and support. Third, the larger team of people they had hoped to involve back in their city did not become engaged in the formal strategic planning process. Fourth, the strategic planning team identified the management of the city's drinking and recreational water resources as the city's most important strategic issue, but then realized that the team probably could not deal with that issue (since the head of the water authority and the city planning director were not team members).

The team identified other strategic issues for which they were at least partially responsible but were concerned about being the wrong team for the most important issue. Finally, they were concerned because the city council had not been directly involved in the process.

One of the authors then held two individual sessions with the strategic planning team during the last two months the team worked on the process. As "homework" for the first session, the team agreed to develop strategies for the four most important strategic issues the team had identified. At the session, the strategies were reviewed with the consultant, who concluded that all were useful and viable, as each constituted something the team members could do to help raise and resolve a strategic issue. The most important development at that first session, however, was recognition of the need for the city's key administrators to meet regularly to discuss what was important for the city government and city as a whole. Everyone was apparently "too busy" to talk with each other on a regular basis about cross-departmental and citywide matters. The team members therefore made a commitment to each other: (1) to approach the city manager as a group about the need for a "cabinet," (2) to propose that all key decisionmakers meet once a month for lunch at a local hotel to discuss what was most important for the city, and (3) to rotate responsibility for the first several months' agendas. Each month one person was to make a presentation on a different strategic issue and to lead a discussion of what to do about it.

The second individual consultation with this team was a pleasant surprise. The city manager had agreed to the team's proposal, which had been implemented within two weeks. By the time the consultant met with the team again, six weeks after the first "cabinet meeting," major progress had been made on implementing the strategies identified by the team to deal with all four strategic issues. The consul-

tant then told the team, "You've got it; that's what strategic planning is all about." Whereupon the team leader said, "Really? We thought strategic planning had to be some really formal process, that you had to have some big plan, and we always felt bad that we didn't have the time for it." City D, in other words, had many of the elements of a "strategic thinking and acting" process already in place. They simply had not put all the pieces together and called it *strategic planning*. At the close of the meeting, one of the team members, the director of public safety, likened the city government to a fast car that just needed to be pointed in the right direction.

City E

City E is another older, middle-class, "first ring" suburb. It shares a common border with city C. City E is regarded among city management professionals as one of the best-managed cities in the state. As with city D, the assistant city manager was the project leader. The city manager was a strong supporter and a member of the strategic planning team. The team that came to the workshops acted as a core group and repeated the exercises with a larger team back home. The larger team included many, but not all, of the city's key administrators. The core group and the larger team worked through the entire process, except for development of a vision of success, in five months. No severe crises disrupted the process during this period, and there were no major hitches in assembling either the core group or the larger team (unlike the experiences of some of the other units).

The only unusual feature of this period was that the team had difficulty developing a mission statement that all—and particularly the city manager—could support. Surprisingly, the disagreements did not seem to be substantive so much as stylistic. The city manager wanted a mission statement that gave him "goose bumps"—his informal criterion for a good mission statement. After several tries, the manager finally agreed to a statement. He said maybe his standards were too high—and that he thought he might have gotten at least "one bump" out of the final version.

Some actions were taken on some of the strategies developed. One major strategy was development of a city marketing and service strategy, including preparation of a budget proposal to the city council to allow pursuit of the strategy. The city council approved the budget proposal, but the process essentially came to halt for almost a year. There were three important reasons why this happened. First, the assistant city manager took a maternity leave; second, when she came back, the city administration was engaged in a new round of budget preparations and there did not seem to be time for strategic planning; and third, the council did not seem to be very interested in strategic planning.

The idea of strategic planning did not die, however, because of two events. First, the first author asked a team in his strategic planning class at the University of Minnesota to prepare a case on city E's strategic planning efforts. The city manager, assistant city manager, and other staff in city E cooperated with the team and a case was prepared. Second, the first author asked the assistant city manager to make a presentation on city E's efforts to a two-day university seminar on strategic planning. In order to prepare for the session, and for a hoped-for presentation to city E's city council, the assistant city manager prepared a strategic plan based on the efforts of the city's strategic planning team.

Eighteen months after the team began its efforts, the city manager and assistant city manager asked the city council to hear a presentation on the team's efforts and to consider the plan. The two managers were surprised and disturbed when the council continued to postpone the presentation and instead used council time to deal with matters the managers felt were either trivial or management prerogatives. They felt the council was spending too little time on policymaking and too much time on administration. Indeed, they felt the council was shirking its responsibilities as policymakers and, instead, was acting almost like a second—and unnecessary—set of managers.

Twenty months after the team began the strategic planning process, the city manager was asked to make a presentation on city E's strategic planning efforts to the state league of cities conference. He agreed and made sure one of his allies on the council also attended. One of the conference speakers was a nationally known consultant on effective governance. The city manager and council member were impressed by the speaker's presentation and both agreed that the city council could improve its performance if it adopted the speaker's advice. The city council member went back to city E and convinced a majority of his fellow council members that they had been acting too much like managers and not enough like policymakers. The council then hired the governance consultant who had spoken at the conference to run two workshops, nine months apart, on effective governance and to assist the council in becoming a more effective policymaking body. The city manager and assistant city manager are now convinced that at the end of the governance consultant's efforts, the council will be interested in strategic planning and ultimately will adopt much of the strategic plan the assistant city manager prepared.

County A

County A contains a number of urban and suburban concentrations along with substantial tracts of farmland. As with some of the other units discussed so far, county A had a core strategic planning team and a larger, back-home work group. The full strategic planning team consisted of nineteen key people, including the county administrator, planning director, and the entire county board. The county administrator was a strong supporter, while the planning director was the process champion. The process reported here actually was the county's third attempt at countywide strategic planning. The first two efforts aborted, although two departments did prepare strategic plans. The first two attempts failed because various crises disrupted the processes.

The third attempt was the most successful of any unit's efforts in this study (see Table 18–3). The third attempt, however, got off to a very shaky start, since it was difficult to get the team together until an election occurred two months after the process supposedly began. After the election, all five elected county commissioners became members of the team and showed up for the planning sessions. The process flowed reasonably smoothly. The only real hitch came when the team members identified so many strategic issues that they found it hard to set priorities and move on to strategy formulation. A session with the first author resulted in a set of priority strategic issues and the assignment of responsibility for strategy development to various individuals, committees, and task forces. A number of strategies have been developed to deal with most of these issues, many strategies have been adopted, and implementation efforts are underway.

County B: The Executive Director's Office

County B is a very populous (although not geographically large) urban county. The county executive director's office (consisting of the county executive and several staff assistants) decided to engage in strategic planning. The executive director sponsored the process but was not very supportive and did not participate in strategic planning team meetings. The executive director, in fact, was in a difficult political position. Only a small majority on the county

board supported him, and he worked purely at the pleasure of the majority and without a contract. The team leader was a person hired for two years through a foundation grant to help modernize and professionalize the county administration. The team leader was also the process champion. An influential county board member, however, did serve on the team; her participation led to a dramatic turn of events later in the process, when the county board undertook its own strategic planning process and preempted the efforts of the executive director's office. (Indeed, the more this board member understood about the intent and process of strategic planning, the more she became both the most important sponsor and most important champion of strategic planning by the county.)

The strategic planning team had difficulty agreeing on strategic issues. They also had difficulty developing strategies to deal with the issues they did identify. They just seemed to stall partway through the process. Then the board's strategic planning efforts preempted them. The county board, partly at the instigation of the board member who served on the strategic planning team, decided to hold a retreat in order to engage in strategic planning. At the retreat the board developed a mission statement for the county that later was formally adopted, and they identified eight strategic issues to be addressed. Five were already being worked on by various groups within the county government, but three were not. The strategic planning team dropped what it had been working on to attend to the remaining three issues and subsequently developed strategies and recommendations for board consideration. The board acted on several of these recommendations. The process came to a halt, however, when the county executive director was forced to resign, one of his assistants who was a strong supporter of the process found another job, and the process champion's grant ran out.

County B: Nursing Service

Nursing Service is part of the same county government as the executive director's office discussed above. The service has roughly eighty staff members and a yearly budget of approximately $3.5 million. Nursing Service was the only one of the eight units not to participate in the series of strategic planning workshops sponsored by the government training service. Instead, Nursing Service had a separate contract with the authors.

The strategic planning team was led by the director of the service, who also served as sponsor and process champion. There were other sponsors as well, including the county's executive director and the director of the department of public health, of which Nursing Service is a part. These other sponsors were not strong supporters. There also was another process champion, the department's health planner, who was an active and dedicated promoter of the process.

Unlike the other units, Nursing Service was asked by a person in a hierarchically superior position to undertake strategic planning. The county's executive director became somewhat interested in strategic planning, partly because the county had received a grant from a local foundation "to modernize and professionalize the county administration," and strategic planning looked like a way to meet the requirements of the grant. The executive director asked the county's three human services departments to undertake "pilot" efforts, so that the rest of the county government might learn from their experience. The three departments were Public Health (of which Nursing Service is the largest division), Community Services, and Corrections. They were selected because they deal with the public most directly, because the greatest changes were occurring in human services, and because they had the staff necessary to carry out the process.

The director, deputy director, and staff of Nursing Service saw strategic planning as an opportunity to rethink the service's mission and strategies in light of the rapidly changing health care environment. Some members of the team were concerned, however, that they had been selected as one of the "guinea pigs" for the executive director's experiment. The concern was twofold. First, Nursing Service has lived with the fear that it would be taken over, put out of business, or otherwise circumvented by the county government's huge medical center, a famous hospital that entered the home health care field (Nursing Service's main "business") shortly before Nursing Service began its strategic planning process. Some team members were concerned that any information or arguments created as part of the strategic planning process might be used against Nursing Service by the executive director and county board to benefit the medical center. Second, team members were concerned that the county board would not fully support the activities of the service that were perceived to compete with the private sector. Staff were fearful that strategic planning would call attention to these activities. A number of reassurances from the executive director were necessary before several Nursing Service staff members would believe the service was not being "set up." Further, additional information was sought concerning exactly what the executive director expected as output from the process.

Ironically, it was Nursing Service's strategic planning efforts that in part forced strategic planning on the county board. Nursing Service prepared its strategic issues and then was asked to make a presentation to the county board on the issues and desirable strategies to address them. The issues ultimately concerned the county government's role in the health care field and the board's willingness to pay to meet the health needs of the county's residents. County board members realized they were completely unprepared to deal with the issues raised by Nursing Service. The board also realized that they might soon be faced with similar vexing issues by the other departments engaged in strategic planning. The board felt a need to think about the county government as a whole, and about how to establish priorities, before they were presented with any more policy questions for which they had no answers. That was when the board decided to go on its retreat and identified the eight strategic issues faced by the county government as a whole. It was no accident that the same county board member who participated on the strategic planning team for the executive director's office was particularly interested in health issues and had followed Nursing Service's strategic planning efforts closely.

Also, ironically, partway through Nursing Service's planning efforts, the county board forced the county's executive director to resign. Nursing Service then saw the strategic planning process as a real opportunity to think through its position so that it could have the most impact on the thinking of the new executive director.

As a result of the process, Nursing Service identified a number of strategic issues that needed to be addressed. The principal issue was what the mission of Nursing Service should be given the changing health care environment. After rethinking their mission, the Nursing Service team rethought their strategic issues. The team identified a new set of strategic issues concerning how the new mission could be pursued. The team went on to develop a set of strategies to deal with these issues, and completed a strategic plan just before the health planner went to work for one of the service's major competitors—the county medical center. By the end of 1987 the strategies should be fully implemented.

After county A, Nursing Service has had the most success with strategic planning (see Table 18–3). Nursing Service's process, however, took over twice as long (thirty months

versus twelve months). The process was halted for long periods because of (1) the need to clarify exactly what was expected as a result of the process, (2) difficulties in assembling the team because of extraordinary workloads, (3) the health planner's maternity leave, (4) uncertainty over who the next executive director would be, (5) the need to assess his health care priorities and beliefs, (6) a county board election, (7) the need to assess the stance of the new board, and (8) the need for policy direction from the board.

ANALYSIS AND DISCUSSION

A number of patterns are apparent to us in the data drawn from the eight units' strategic planning efforts. Because the data consist primarily of interviews and participant observations, however, the articulation of these patterns remains highly speculative.

In almost all cases it appeared that something other than a "shock" stimulated the initiation of the process. As far as we can tell (from our interviews and the follow-up questionnaires), seven of the units began the process because of the inexpensive opportunity the workshops presented to focus what was important for their units. For these seven, therefore, it appeared that an event (as defined by the MIRP framework—a change in ideas, people, transactions, or context) provided the basic stimulus. Nursing Service, which was asked to undertake strategic planning and which recently had witnessed the emergence of serious competition in its core "business," is the only one that may fit Schroeder et al.'s (1988) observation that "Innovation is stimulated by shocks, either internal or external, to the organization." On the other hand, the other seven units had experienced a series of shocks over the several years preceding the workshops, and difficulties in handling these shocks may have prepared the units to take advantage of the workshop opportunity.

Also, for these seven, the precise nature of the "event" that stimulated their initiation of the process remains unclear. For some, the change was their introduction to the *idea* of strategic planning (city C, city D, county B executive director's office, and Nursing Service). For some, it was the their introduction to new *people*—the authors—who argued on behalf of strategic planning and who tried to reinforce participation through attention, instruction, advice, and encouragement (city D, county B executive director's office, and Nursing Service). For some, it was a change in *transactions* or relationships—particularly the chance to form a strategic planning team, to participate in the workshops, and to influence the organization's key decisionmakers (city C, city E, county A, county B executive director's office, and Nursing Service). And for others, the change was in the organization's *context* (city D, city E, county B executive director's office, and Nursing Service). Strategic planning is a "hot topic" in the public sector these days (Bryson and Einsweiler 1987), and some units felt they had to get "on the bandwagon" to stay ahead of, or keep abreast of, their peers, or to meet their stakeholders' expectations (Dimaggio and Powell 1983). Also, several could see changes in their units' environments on the horizon and wanted to use strategic planning to help manage those changes. For most units, in fact, some combination of the above changes probably led to their initiation of a strategic planning process.

Although shocks may not have stimulated the process directly in most cases, they certainly stopped or delayed it in a number of cases. In city A and city B the process stopped completely when their city managers (who were both sponsors and champions) left for other jobs. In county B's executive director's office, the process terminated when the executive director was forced to resign, and other key staff members left. In city E

and Nursing Service, frequent and lengthy delays occurred. The planning teams for the remaining units—city C, city D, and county A—all feared a crisis would terminate their processes prior to completion. Indeed, two prior attempts at countywide strategic planning in county A had aborted because of crises.

In each case the attempt to initiate strategic planning was prone to disintegration. In city A and city B the process did disintegrate completely when a key actor departed. In each of the others the process often appeared to be on the verge of disintegration, but not because of the *permanent* departure of key actors (although such occurrences did threaten the continuity of the process). There was always the possibility that a crisis would destroy the process, since the units would not have enough slack—particularly staff attention—to deal with the crisis *and* plan strategically. The processes therefore were always prone to the *temporary* loss of key actors. Indeed, even without major crises, planning teams rarely operated with their full complement of members, and when two process champions left for maternity leaves, their units' processes came to a temporary halt.

Beyond the loss of key staff, the process appears particularly susceptible to collapse during the strategic issue identification and strategy development steps. These two steps appear to have been difficult for three reasons. First, as the number of potential strategic issues and strategies to deal with each issue multiplied, the teams had difficulty comprehending the volume of information and its implications and had difficulty deciding what their priorities ought to be and how to set them. During these steps the process was *divergent,* and the teams had difficulty figuring out how to make it *convergent* (Van den Daele 1969).

Second, the problem of divergence was not just conceptual but political. Each issue and the associated strategies to deal with it often implied that a different *decision set* (Hickson et al. 1986) was needed to deal with the issue. The teams never incorporated the full range of possible decision sets to deal with all the issues identified. The processes were only able to converge therefore on those issues and strategies (or parts of issues and strategies) that the team was able to do something about, or else the process itself expanded out laterally or shifted to a higher hierarchical level where more of the members of the necessary decision sets were represented.

In city D, for example, the team was able to convince the city manager to form a "cabinet" in which more key decisionmakers would be represented, and in county B the county board undertook a strategic planning process that preempted the efforts of the executive director's office and set the stage for dealing with some of the key issues identified by Nursing Service. It is possible therefore that strategic planning at lower levels can force strategic planning horizontally and at higher levels, in addition to the more expected top-down progression. But when different units in the same system engage in strategic planning, each is likely to have to recycle through various parts of the process to take account of the strategic planning activities of other units.

The fact that the strategic planning teams did not encompass the appropriate decision sets for many issues probably partly accounts for some of the results noted in Table 18–3. Recall that team leaders felt the process succeeded in focusing the teams on what was important. The process did less to set priorities for team action, generate those actions, and focus the attention of the organization's key decision makers on what was important.

Third, the difficulties of divergence and partial convergence were hard enough to manage, but the partially *cumulative* nature of the processes added yet another problem (Van den Daele 1969). The processes were partially cumulative in that what happened in prior steps had to be taken into account in subsequent steps—but because more ideas and information

were created in each step than could possibly be carried on to the next, there was only a partial cumulation from step to step. Also, since teams often could not do anything about the ideas and information carried on to succeeding steps, those items, too, were dropped. As a result, teams often had difficulty deciding what to keep and what to drop as step followed step or their thinking recycled through previous steps.

The units varied widely in the amount of calendar time they devoted to their first efforts at strategic planning. For the units that completed plans the calendar time devoted to the process varied from five to thirty months. In several units a variety of stoppages, delays, and difficulties dragged out the process. This is consistent with Hickson et al.'s (1986) finding that strategic decisionmaking in public sector organizations has a tendency to be disrupted and lengthy.

Successful completion of the process appears to depend on the presence of a powerful sponsor to legitimize the process, even if that sponsor is not especially supportive. Further, multiple sponsorship is possible and probably desirable. In addition to legitimizing the process, sponsors also could make important decisions concerning the process and could push the process laterally and vertically to involve more actors and facilitate decision making. In each of the units there was multiple sponsorship. Multiple sponsorship appears to be necessary to legitimize an effort aimed at framing and informing major decisions that are cross-departmental in scope. But the sponsorship was not always strongly supportive.

Again, the fact that most units' processes were not fully sponsored by most of the key decision makers probably partly accounts for the results presented in Tables 18–2 and 18–3. Only county A's process—the most successful—appeared to enjoy relatively complete sponsorship by the unit's key decision makers. Nursing Service's process also enjoyed relatively complete sponsorship—at least ultimately—and that process was the second most successful. The other units had less complete sponsorship and also were less successful.

Successful initiation of strategic planning in governments does appear to require a strong process champion. All six units that completed the process had a strong process champion. The process champions appeared to believe that the strategic planning process would produce desirable outcomes. The champion did *not* push personally favored issues and solutions, although he or she almost always entered these into team discussions for consideration along with everyone else's ideas. It is this belief in a *process* that distinguishes process champions from other champions, such as the more general *idea* champion (Kanter 1983: 296), *product champion* (Kotler 1976: 200), or *policy entrepreneur* (Roberts and King, Chapter 9). A process champion is also different from a facilitator, although it appears to help the process if the champion is also a good facilitator. (Not all of the champions in our sample were good facilitators, although all possessed at least some facilitation skills.) Facilitators believe in the importance of process, too, but they usually give participants much more choice about the kind of process they will follow. The process champions in our sample held firmly to the importance of the steps, or "checkpoints," outlined in Figure 18–1.

Process champions appear to have been particularly important during the strategic issue identification and strategy development steps—precisely those steps when the process appeared most prone to disintegration. During these steps in the process strategic planning teams appeared to fall into serious "gumption traps" (Pirsig 1974). As the processes diverged, the teams appeared to feel the ground on which they stood was collapsing. The teams' morale, enthusiasm, and commitment waned. Only the countervailing morale, enthusiasm, and commitment of the process champions for the process appeared able to lead the teams out of the traps, toward convergence on which issues to choose, how to frame them, and what strategies to pursue in dealing with them. Even process champions, however, could fall into gumption traps. At various times over the course of the process, the consultants served as cheerleaders for the champions, helping to keep their spirits up and encouraging them to push the process along. The consultants also provided advice to all six units that completed plans on what to keep, what to drop, and how to converge on particular issues and specific strategies. Indeed, without the presence of the process champions and consultants, it is likely that the scores presented in Table 18–3 would have been lower.

Each unit relied on a strategic planning team to work through the process and prepare a strategic plan. This finding is not surprising, since the consultants required formation of a team by each unit. But the use of a team would be expected anyway in uncertain, complex, and political situations in which a variety of sources of information are needed and decisions will have cross-unit implications (Galbraith 1973; Hickson et al. 1986).

What counted as a strategic plan varied widely across units. City E and Nursing Service prepared formal strategic plans, with mission statements, situation assessments, discussions of strategic issues, and strategies. County B executive director's office ultimately addressed three strategic issues identified by the county board and prepared fairly typical governmental "decision packages" for county board consideration. The packages consisted of a discussion of the issues, strategies to deal with them, and recommendations for action. The other three units prepared much more informal plans that typically consisted of discussion papers, memoranda, and decision packages that dealt with a single issue at a time.

There appear to be several reasons for the wide variation in plans. First, throughout the process the consultants emphasized that strategic thought and action were what counted, not preparation of a formal strategic plan. The consultants argued that teams should focus on a few key issues and strategies to deal with them and prepare only such "plans" as were necessary to build coalitions around issue definitions and strategies to resolve them. (We did, however, continuously urge each team to write up the results of their discussions and analyses, so that they had adequate background materials to inform preparation of plans.) Second, the efforts at the initiation of strategic planning were almost always out of sequence with the units' normal planning and budgeting processes, so that it was difficult for most to integrate strategic planning with existent formal processes. Third, it was not necessary to integrate all, or even most, of the strategic planning efforts with normal planning and budgeting processes. The commitments recorded in formal plans and budgets were not required to deal with many strategic issues or parts of issues. Finally, the issues identified by any one unit typically were at different stages in the issue "life cycle" (Schon 1971). Separate plans for different issues therefore often seemed desirable.

Participants in the process seem to have conceptualized time in three ways: as chronos (calendar time), kyros (peak experience), and juncture (actual or potential event). Van de Ven and Poole (1988) have argued that chronos appears to be the basic metric in structural approaches to organizational change, while kyros is the basic metric in purposive approaches to human action. Time as chronos appeared to dominate participants' conception of time when the teams' efforts began. Early discussion centered on the "time commitment"—or calendar time required—for the effort, and on the scheduling of workshops or meetings, so they could be marked in appointment books. Occasionally a kyros conception of time occurred, as when teams were pleased with their efforts at workshop sessions, or a policy board adopted a team's recommendations. More typical, however, was the experience of the city E's city manager, who, after

waiting for a draft mission statement that would give him "goose bumps," finally had to settle for a "one bump" statement. Indeed, while completion of some steps in the process marked important, and often historic, "firsts," for each of the teams' units, the occurrences were hardly "peak" experiences. Instead, they called to mind Robert Sherrill's (1967) statement that "Historic moments have great difficulty escaping their intrinsic dullness."

The more prevalent conception of time that came to dominate participants' perceptions was of time as actual or potential *junctures* when things had to come together (Jakobson, personal communication, 1987). Process champions, in particular, always seemed to think about what had *to join together* for successful team meetings (such as invitations, reminders, premeeting strategy sessions, background papers, room arrangements, supplies, and briefings for consultants), presentations to relevant policy boards (such as background papers, decision packages, graphic displays, and premeeting briefings and strategy sessions), or incorporation of the results of strategic planning into ongoing unit planning and budgeting efforts.

Process champions planned both for the expected—and, to the extent they could, the unexpected—events that would occur throughout the process. Process champions, sponsors, and team members were always aware that unexpected events, or crises, could torpedo their strategic planning efforts. The conception of time as juncture therefore included both the *planned* convergence of ideas, people, transactions, and context, and the *unplanned* junctures of the "garbage can" (Cohen, March, and Olsen 1972). People realized that they had to be ready for both, that they could plan on some things happening in a reasonably predictable fashion, but that other things would not, particularly because the purpose of their strategic planning was *to produce events* the likes of which had not happened in quite the same way before. There seemed to be a sense of time among the process champions and sponsors, and team members, in other words, that corresponded to some combination of George Bernard Shaw's notion that "History may not repeat itself, but it rhymes a lot," and Robert Sherrill's (1967) comment, "History is one damn thing after another."

The consultants helped push—and facilitate—acceptance of this junctural view of time. The workshops and meetings with consultants were junctures. Process champions and teams worked hard to get their "homework" done to prepare for these occasions. Champions also prepared for interviews arranged by the consultants. Probably the best example of how the consultants facilitated acceptance of the junctural view of time is the case of city E, in which the champion prepared the city's strategic plan in order to be ready for a presentation to the first author's two-day university seminar on strategic planning.

Participants were intent on ensuring that proposed missions, strategic issues, and strategies were technically rational, politically acceptable and morally, ethically, and legally defensible. Every team appeared to apply several informal criteria to the evaluation of every statement, document, or recommendation they prepared. Discussions were never orderly, in the sense of a formally sequenced discussion of proposals against formally agreed on criteria, but they did focus on the technical workability of proposals, which stakeholders would and would not support the proposals, and whether the proposals were morally, ethically, and legally defensible. The frame of reference often would shift back and forth among criteria until finally the participants developed a proposal that satisfied all criteria. A temporal sequencing of rationalities or argumentation thus occurred: First one kind (technical, political, or moral) would predominate and then another, before the group settled on a proposal that everyone could "live with" (Schon 1979; Meyer 1984; Boland and Pondy 1986). Strate-

gic planning became a kind of planning by argumentation (Goldstein 1984).

Finally, each unit clearly adapted the process to its own situation. For example, units began the process for different reasons. Sponsors and champions of the process also varied in number, position, and influence. Units responded differently to the tendency for the process to disintegrate; some units discontinued the process altogether, while others found various ways to cope. Units varied in the time they devoted to the process. What counted as a strategic plan varied widely. The concluding section offers tentative speculations on the necessity and usefulness of these adaptations.

CONCLUSIONS

A number of conclusions emerge from this study. First, the initiation of strategic planning primarily represents three very simple, yet innovative, ideas for many governmental units—namely, (1) the gathering of key actors (preferably key decisionmakers), (2) to work through a "strategic thinking and acting" process, (3) in order to focus on what is truly important for the unit, set priorities for action, and generate those actions. Although the ideas may be conceptually simple, they are, however, quite difficult to implement. They are difficult to implement because the initiation of strategic planning is both an event, or set of events, in and of itself, but also—and more important—is a process deliberately designed to produce actual or potential events. This deliberately *disruptive* nature of the process partly explains the difficulty of implementing it.

Organizations prefer to program, routinize, and systematize as much as they can (Thompson 1967; Van de Ven 1976), yet here is a process designed to question the organizing that has occurred—along with all of the treaties that have been negotiated among stakeholders to form a coalition large enough and strong enough to govern the organization (Pfeffer 1978). Units therefore may be scared of the potential disruptions occasioned by strategic planning (as Nursing Service was), unable to pursue the disruptions they think should occur (as when city D's team was the wrong team for the city's water management issue), or unable to decide which disruptions to pursue and how (as when all teams had trouble deciding which issues and strategies to pursue). In short, it is apparently very difficult to initiate and manage a divergent, partly convergent, partly cumulative process, and hence the opening observation from Pressman and Wildavsky (1973)—that it is remarkable that anything happens at all—particularly when it comes to government strategic planning.

Said differently, nothing in the results of our study leads us to change our initial expectation: Most efforts to produce fundamental decisions and policy changes in government through strategic planning will not succeed. We might conclude that six of our eight units successfully implemented strategic planning (although there clearly were variations in *how successful* they were). But it took a lot of effort on the units' part, plus the help of two consultants, plus the efforts of the government training service (for seven units), and two small foundation grants to achieve that success. Even so, two units aborted the process early on. Strategic planning obviously is no panacea.

Second, and paradoxically, government strategic planning is probably most needed where it is least likely to work. Government strategic planning would appear to work best in units that have effective policymaking boards, strong and supportive process sponsors, superb process champions, good strategic planning teams, enough slack to handle potentially disruptive crises, experience in coping with major disruptions, and a desire to address what is truly important for the organization. Any unit with those features probably already uses some

sort of "strategic thinking and acting" process. Introducing such a unit to a formal strategic planning process probably would constitute minor tinkering with a "high-performance social system" (Vaill 1978).

Unfortunately, few governments (or organizations of any sort, for that matter) are high-performance social systems (Vaill 1978). Few, in other words, are organizational analogues of "The A-Team," whose plans always seem to come together right on time, as intended, and with positive (for them) effects. Instead most organizations tend to "muddle through" in a disjointedly incremental way from one situation (often a crisis) to the next. The introduction of strategic planning to such organizations may be doomed to failure. At the very least, the efforts of strategic planners in such situations should focus in part on how to create the conditions outlined in the previous paragraph that would make strategic planning more likely to succeed.

Third, if a government unit wants to initiate strategic planning, our study indicates that the following elements, at a minimum, should be in place: (1) a powerful process sponsor, (2) an effective process champion, (3) a strategic planning team, (4) an expectation that there will be disruptions and delays, (5) a willingness to be flexible concerning what constitutes a strategic plan, (6) an ability to think of junctures as a key temporal metric, and (7) a willingness to construct and consider arguments geared to many different evaluative criteria.

Fourth, a governmental unit must think carefully about what aspects of strategic planning it might wish to institutionalize. This study seems to indicate that at least the following elements of a strategic planning *system* (Lorange 1980, Lorange, Morton, and Ghoshal 1986) can be institutionalized: (1) a formal or informal "cabinet," (2) mission statements, (3) "policy objectives" that emerge from the sense of purpose embodied in mission and mandates (Eckhert et al. 1988) or that characterize the goals of adopted strategies, (3) periodic situation analyses, (5) periodic strategic issue identification exercises, (6) strategic issue management practices (appointment of issue managers and task forces, strategy development exercises, development of decision packages, and issue monitoring processes) (Eadie 1986; Bryson and Roering 1987), and (7) more formal multicriteria proposal evaluation procedures.

There are a number of reasons why we think these elements can be institutionalized. All of the units that completed strategic plans prepared mission statements. Several prepared policy objectives (either before, during, or after the process) to detail desired performance in key policy areas related to the mission; moreover, preparation of lists of goals and objectives is common practice in most governments. All units performed situation audits and found the exercise relatively simple, not too time consuming, and useful. All of the units identified strategic issues and developed strategies to deal with the most important issues. All of the units in effect adopted an issues management approach (Delmont and Pflaum 1983), because they realized that different issues were on different time frames and at different stages of development, required different teams to handle them, and required different decision sets to resolve them. Finally, all of the units used informal multicriteria proposal evaluation procedures. These seven aspects of strategic planning therefore seem most amenable to institutionalization by governments.

Fifth, until governmental units gain more experience in strategic planning, it seems best to judge government strategic planning according to the extent to which it (1) focuses the attention of key decisionmakers on what is important for their organizations, (2) helps set priorities for action, and (3) generates those actions.

Sixth, a number of topics clearly should be the subject of future research. Our "discovery" of the importance of the role of *process*

champions is one. The role seems to be crucial to the successful initiation of strategic planning. Are these process champions really different from other kinds of champions, and, if so, how? Can we find out exactly what they do to help the process along? Are there ways in which the champions' adherence to a particular process, or set of checkpoints, is dysfunctional? Can we train people to be effective process champions?

Another subject that should be explored further is the exact nature of the strategic planning process in practice. Our discovery that it seems to be a *divergent, partially convergent, partially cumulative process* was one of the study's findings we found most interesting. We would like to know more about the nature of such processes and more about how they can be managed.

Finally, we would like to know more about the role of strategic planning process consultants. We introduced the units to strategic planning (or, rather, we probably introduced them to the details of strategic planning, since all had already heard of it). We helped the strategic planning teams move through the process, critiqued the results of their efforts, and provided advice on how to proceed. We acted as teachers, facilitators, and cheerleaders. And yet we know very little about how—and how well—we played our role. It obviously would be helpful to know more about the role and how it should be played.

Lastly, it must be emphasized that this study concerned the *initiation* of strategic planning by governments. What was *not* studied was the effectiveness of the strategic planning process itself. That is, we studied the process through the development of strategic plans; we did not study the merits of the plans themselves or the effectiveness of efforts to implement those plans. To the authors' knowledge, only one careful study of the effectiveness of government strategic planning has been done (Boschken, 1988). (The same, of course, almost can be said about planning processes generally; see Van de Ven 1980a, 1980b, 1980c; Boal and Bryson, 1987). Each of the teams that completed plans felt that their strategic planning efforts had been worthwhile. But obviously a great deal more study is necessary before it will be possible to say exactly what does and doesn't work, under what circumstances, and why, when it comes to strategic planning by governments.

REFERENCES

Primary source material for this chapter is contained in separate files for each of the governmental units. These files are in the possession of the first author and are available for inspection—with the caveat that the names of all governmental units and study participants are to remain anonymous.

Barry, Bryan. 1986. *Strategic Planning Workbook for Non-Profit Organizations.* St. Paul, Minn.: Amherst H. Wilder Foundation.

Benveniste, Guy. 1972. *The Politics of Expertise.* Berkeley: Glendessary.

———. 1977. *Bureaucracy.* Berkeley: Glendessary.

Boal, Kimberly B., and John M. Bryson. 1987. "Representation, Testing and Policy Implications of Planning Processes." *Strategic Management Journal* 8: 211–231.

Boland, Richard J., Jr., and Louis R. Pondy. 1986. "The Micro Dynamics of a Budget Cutting Process: Modes, Models and Structure." *Accounting, Organizations and Society* 11(4/5): 403–422.

Boschken, Herman L. 1988. "Turbulent Transition and Organizational Change: Relating Policy Outcomes to Strategic Administrative Capacities." *Policy Studies Review* 7(3): 477–99.

Bryson, John M. 1988. *Strategic Planning for Public and Nonprofit Organizations.* San Francisco: Jossey-Bass.

Bryson, John M., and Andre L. Delbecq. 1979. "A Contingent Approach to Strategy and Tactics in Project Planning" *Journal of the American Planning Association* 45(2) (April): 167–79.

Bryson, John M., and Robert C. Einsweiler. 1987. "Introduction to Strategic Planning Symposium." *Journal of the American Planning Association* 53(1): 6–8.

Bryson, John M., Edward Freeman, and William Roering. 1986. "Strategic Planning in the Public Sector: Approaches and Future Directions," pp. 65–85. In Boyd Checkoyuay (ed.), *Strategic Perspectives on Planning Practice.* Lexington, Mass.: D.C. Heath.

Bryson, John M., and William D. Roering. 1987. "Applying Private-Sector Strategic Planning in the Public Sector." *Journal of the American Planning Association* 53(1) (January): 9–20.

Bryson, John M., Andrew H. Van de Ven, and William D. Roering. 1987. "Strategic Planning and the Revitalization of the Public Service." In Robert Denhardt and Edward Jennings, Jr., eds., *Toward a New Public Service,* pp. 55–75. Columbia: University of Missouri.

Chakravarthy, Balaji S. 1987. "On Tailoring a Strategic Planning System to its Context: Some Empirical Evidence." *Strategic Management Journal* 8(6): 517–34.

Christensen, Karen S. 1985. "Coping with Uncertainty in Planning." *Journal of the American Planning Association* 51(1): 63–73.

Cohen, Michael D., James G. March, and Johan P. Olsen. 1972. "A Garbage Can Model of Organizational Choice." *Administrative Science Quarterly* 17: 1–25.

Crow, Michael, and Barry Bozeman. 1988. "Strategic Public Management." In John M. Bryson and Robert C. Einsweiler, eds., *Strategic Planning in the Public and Nonprofit Sectors,* pp. 51–68. Washington, D.C. and Chicago: Planners Press.

Delmont, Timothy J., and Ann M. Pflaum. 1983. "External Scanning and Issues Management: New Planning Techniques for Colleges and Universities." St. Paul, Minn.: Association for Institutional Research.

DiMaggio, Paul J., and Walter W. Powell. 1983. "The Iron Cage Revisited: Institutional Isomorphism and Collective Rationality in Organizational Fields." *American Sociological Review* 48 (April): 147–60.

Eadie, Douglas C. 1986. "Strategic Issue Management: Improving the Council-Manager Relationship." *ICMA MIS Report* 18(6): 2–12.

Eadie, Douglas C., and Roberta Steinbacher. 1985. "Strategic Agenda Management: A Marriage of Organizational Development and Strategic Planning." *Public Administration Review* 45: 424–30.

Eckhert, Philip C., Kathleen Haines, Timothy J. Delmont, and Ann M. Pflaum. 1988. "Strategic Planning in Hennepin County, Minnesota: An Issues Management Approach." In John M. Bryson and Robert C. Einsweiler, eds., *Strategic Planning in the Public and Nonprofit Sectors*, pp. 172–83. Washington, D.C. and Chicago: Planners Press.

Galbraith, Jay. 1973. *Designing Complex Organizations.* Reading, Mass.: Addison-Wesley.

Goldstein, Harvey. 1984. "Planning as Argumentation." *Environment and Planning B: Planning and Design* 11: 297–312.

Harmon, Michael M., and Richard T. Mayer. 1986. *Organization Theory for Public Administration.* Boston: Little, Brown.

Heclo, Hugh. 1977. *A Government of Strangers.* Washington, D.C.: Brookings.

Hickson, David J., Richard J. Butler, David Cray, Geoffrey R. Mallory, and David C. Wilson. 1986. *Top Decisions: Strategic Decision-Making in Organizations.* Oxford: Basil Blackwell.

Hostager, Todd J., and John M. Bryson. 1986. "Poetics and Strategic Management." Minneapolis: University of Minnesota, Strategic Management Research Center, Discussion Paper 59.

Jakobson, Leo. Personal communication, 1987.

Johnson, Paul E. 1984. "The Expert Mind: A New Challenge for the Information Scientist." In Th. M.A. Bemelmans, ed., *Beyond Productivity: Information Systems Development for Organizational Effectiveness.* North-Holland: Elsevier Science.

Kanter, Rosabeth Moss. 1983. *The Changemasters.* New York: Simon and Schuster.

Kimberly, John P. 1981. "Managerial Innovation." In P.C. Nystrom and W.H. Starbuck, eds., *Handbook of Organizational Design*, Vol. 1, pp. 84–104. New York: Oxford University Press.

Kotler, Philip. 1976. *Marketing Management.* Englewood Cliffs, N.J.: Prentice-Hall.

Lindblom, Charles. 1959. "The Science of Muddling Through." *Public Administration Review* 19(2): 79–88.

Locke, E.A., K.W. Shaw, L.M. Saari, and G.P. Latham. 1981. "Goal Setting and Task Performance: 1969–1980." *Psychological Bulletin* 90(1): 125–52.

Lorange, P. 1980. *Corporate Planning: An Executive Viewpoint.* Englewood Cliffs, N.J.: Prentice-Hall.

Lorange, P. M.F.S. Morton, and S. Ghoshal. 1986. *Strategic Control.* St. Paul, Minn.: West.

Maidique, M.A. 1980. "Entrepreneurs, Champions, and Technological Innovation." *Sloan Management Review* 21 (Winter): 58–76.

March, James G., and Johan P. Olsen. 1976. *Ambiguity and Choice in Organizations.* Bergen, Oslo, and Tromso, Norway: Universitetsforlaget.

McGowan, R.P., and J.M. Stevens. 1983. "Local Government Initiatives in a Climate of Uncertainty." *Public Administration Review* 43(2): 127–36.

Meyer, Alan D. 1984. "Mingling Decision-Making Metaphors." *Academy of Management Review* 9(1) (January): 6–17.

Nutt, Paul. 1984. "Types of Organizational Decision Processes. *Administrative Science Quarterly* 29(3): 414–50.

Olsen, J.B., and D.C. Eadie. 1982. *The Game Plan: Governance with Foresight.* Washington, D.C.: Council of State Planning Agencies.

O'Toole, James. 1985. *Vanguard Management.* New York: Doubleday.

Ouchi, William. 1981. *Theory Z: How American Business Can Meet the Japanese Challenge.* Reading, Mass.: Addison-Wesley.

Peters, Thomas J., and Robert W. Waterman, Jr. 1982. *In Search of Excellence: Lessons From America's Best-Run Companies.* New York: Harper & Row.

Pfeffer, Jeffrey. 1978. *Organizational Design.* Arlington Heights, Ill.: AHM.

Pfeffer, Jeffrey, and Gerald Salancik. 1978. *The External Control of Organizations.* New York: Harper & Row.

Pirsig, Robert. 1974. *Zen and the Art of Motorcycle Maintenance.* New York: Basic Books.

Pressman, Jeffrey, and Aaron Wildavsky. 1973. *Implementation.* Berkeley: University of California Press.

Quinn, James B. 1980. *Logical Incrementalism.* Homewood, Ill.: Irwin.

Ring, Peter S., and James L. Perry. 1985. "Strategic Management in Public and Private Organizations: Implications of Distinctive Contexts and Constraints." *Academy of Management Review* 10(2): 276–86.

Salancik, Gerald R. 1977. "Commitment and the

Control of Organizational Behavior and Belief." In Barry M. Staw and Gerald R. Salancik, eds., *New Directions in Organizational Behavior.* Chicago: St. Clair.

Schon, Donald A. 1971. *Beyond the Stable State.* London: Temple Smith.

———. 1979. "Generative Metaphors." In Andrew Ortony, ed., *Metaphor and Thought,* pp. 254–83. Cambridge: Cambridge University Press.

Schroeder, Roger, Andrew H. Van de Ven, Gary Scudder, and Douglas Polley. 1986. "Managing Innovation and Change Processes: Findings from the Minnesota Innovation Research Program." *Agribusiness Management Journal* 2(4): 501–23.

Seidman, Harold. 1986. *Politics, Position, and Power.* New York: Oxford.

Sherrill, Robert. 1967. *Gothic Politics in the Deep South.* New York: Grossman.

Starbuck, William. 1983. "Organizations as Action Generators." *American Sociological Review* 48 (February): 91–102.

Staw, Barry M. 1976. "Knee-Deep in the Big Muddy: A Study of Escalating Commitment to a Chosen Course of Action." *Organizational Performance and Human Behavior* 16: 27–44.

Staw, Barry M., and J. Ross. 1978. "Commitment to a Policy Decision: A Multi-Theoretical Perspective." *Administrative Science Quarterly* 23: 40–64.

Sussman, Gerald, and Robert Evered. 1978. "An Assessment of the Scientific Merits of Action Research." *Administrative Science Quarterly* 23(4):582–603.

Taylor, Bernard. 1984. "Strategic Planning—Which Style Do You Need?" *Long Range Planning* 17(3): 51–62.

Thompson, James D. 1967. *Organizations in Action.* New York: McGraw-Hill.

Vaill, Peter. 1978. "High Performance Social Systems." In R. McCall and X. Lombardo, eds., *Leadership—Where Else Do We Go?,"* pp. 103–25. Durham, N.C.: Duke University.

Van den Daele, Leland D. 1969. "Qualitative Models in Developmental Analysis." *Developmental Psychology* 1(4): 303–10.

Van de Ven, Andrew H. 1976. "A Framework for Organizational Assessment." *Academy of Management Review* 1:64-78.

———. 1980a. "Problem Solving, Planning and Innovation. Part I. Test of the Program Planning Model." *Human Relations* 33: 711–40.

———. 1980b. "Problem Solving, Planning and Innovation. Part II. Speculations for Theory and Practice." *Human Relations* 33: 757–79.

———. 1980c. "Early Planning, Implementation, and Performance of New Organizations." In J.R. Kimberly, R.H. Miles, and Associates, eds., *The Organizational Life Cycle,* pp. 83–134. San Francisco: Jossey-Bass.

Van de Ven, Andrew H., and M. Scott Poole. 1988. "Paradoxical Requirements for a Theory of Planned Change." In R. Quinn and K. Cameron, eds., *Paradox and Transformation,* pp. 19–63. Cambridge, Mass.: Ballinger.

Wechsler, Bart, and Robert Backoff. 1987. "Dynamics of Strategy in Public Organizations." *Journal of the American Planning Association* 53(1) (January): 34–43.

SECTION
VII

Analyzing and Interpreting
the Studies

In Section VII we switch from a focus on individual research projects covered in Sections III through VI and return to an overview of the total Minnesota Innovation Research Program and its implications for theory and practice. In this part we take a somewhat broader view in order to draw conclusions and reach some inferences from those individual studies with respect to innovation leadership, an emerging general process theory of innovation, and prescriptions for the management of innovation.

In Chapter 19, Manz, Bastien, Hostager, and Shapiro highlight the crucial role of leadership in the process of innovation. They indicate that what constitutes good leadership may vary substantially depending on the temporal stage of development of the innovation. The chapter matches three classic modes of organizational involvement (identification, compliance, and internalization) with three fundamentally different leadership paradigms (rhetorical-visionary, transactional, and participative). Switching from a static view to a dynamic, time-phased approach, and based on a synthesis of several examples from the preceding chapters, the chapter develops a double-loop model of leadership for organizational effectiveness.

As becomes abundantly clear from the variety of studies detailed in these chapters, no single theory can begin to account for the diversity and complexity of process patterns that have been observed in MIRP. Poole and Van de Ven, in Chapter 20, set forth desiderata for developing a general theory of innovation processes. This chapter integrates many of the separate theoretical contributions of other chapters and moves toward development of a metatheory of organizational innovation. As the chapter indicates, such a metatheory would provide a classification scheme for microtheories, based on level of analysis as well as type of theory; specify conditions under which different theories within such a categorization scheme would best apply; and propose a set of switching rules for determining when to switch between models, as innovations develop over time.

Finally Chapter 21, by Angle and Van de Ven, provides the practitioner a set of recommendations for improving the management of innovation in organizations. This chapter's prescriptions are partitioned both by developmental stage and the key components or junctures commonly observed across the innovations studied by MIRP researchers. Although the chapter acknowledges the difficulty of trying to make innovation a completely "manageable" process, several suggestions aimed at improving the odds of innovation success are included in the chapter.

19

LEADERSHIP AND INNOVATION: A LONGITUDINAL PROCESS VIEW

Charles C. Manz

David T. Bastien

Todd J. Hostager

George L. Shapiro

There are countless perspectives on leadership in the literature. It is one of the most discussed topics in all of the social sciences. The definitions, perspectives, and boundaries of conceptualizations and investigative approaches have changed both over time and across cultures without necessarily disproving the previous views (Bass 1981; Yukl 1981). Rather, each perspective has tended to add a new viewpoint, situation, or set of concrete experiences, while changing the definition of the term *leadership.* Lombardo and McCall (1978) stated that

> students of leadership have discovered three things: (1) the number of unintegrated models, theories, and prescriptions, and conceptual schemes for leadership is mind boggling; (2) much of the literature is fragmentary, trivial, un-

realistic, or dull; and (3) the research results are characterized by Type III Error (solving the wrong problem precisely) and by contradictions. . . . leadership theories have been traditionally short range and atomistic, focusing on leader-group relations and passing over leader-group-systems relationships.

According to Terry (1986) there are currently at least 100 accepted academic definitions of leadership. He identifies several current Western perspectives for the study of the phenomenon: trait theories, situation-based theories, power-oriented theories, visionary theories, ethical assessment theories, and organizational theories. Of these various perspectives our focus is primarily on the organizational theories that are unique in that they emphasize position and role within formal organization hierarchies and represent situation specific adaption of other theories. Thus, they borrow from the array of existing leadership theories to study leadership processes within an organizational context.

One way to examine the deepest function or purpose of leadership is to see it as needing to balance the "habits" of individual and work group creativity, differentiation, and specialization with the need for and association with commitment to group and organization mission, or integration and community. In this chapter we examine leadership as a process that can occur at multiple levels of the organization and in multiple directions. More specifically, it can be expressed in a downward, upward, or sideways direction within the organization hierarchy, both as a force for stimulating individual as well as collective effort. Leadership can transpire as a process of one person externally influencing others or as a process in which multiple persons are encouraged and allowed to take more responsibility for influencing themselves.

LEADERSHIP WITHIN THE CONTEXT OF INNOVATION

The notion of leadership within organizations is an especially important concept for the Minnesota Innovation Research Program (MIRP). This became apparent across projects as longitudinal tracking indicated strong qualitative and quantitative support for the importance of leadership in the innovation process. Leaders, and often multiple leaders, helped sensitive and fragile innovations at various stages of development and implementation (specific cases are reviewed later in this chapter). In fact, the psychometric assessment of the innovation survey suggested that leadership (of the innovation group) was the only factor that was consistently and significantly positively correlated with perceived innovation effectiveness (Chapter 3), and this relationship was found across "originations" and "adaptations," "small" and "large" innovations, and at the "idea" generation and "implementation" stages. The positive correlations ranged from .43 to .67 and were all statistically significant to at least the .05 level.

The kinds of leadership displayed across studies were many and varied widely (this chapter focuses in particular on visionary, transactional, and participative leadership processes). It is probably safe to say that there is no one best approach or combination of approaches that represent *the way* to lead in order to facilitate effective innovation. Nevertheless, we believe some leadership perspectives are probably more appropriate than others, at least for specific situations, and that some interesting leadership themes emerged from the Minnesota studies. Our intent is to provide some preliminary insight regarding this important innovation factor and to suggest some directions for further research.

More specifically, in this chapter we attempt to develop a multiview interactive process perspective based on analysis of leadership events that occurred over time in several MIRP case studies. Brown and Lennenberg (1954) have stated that a word or perspective is a window on reality, promising that if we use a particular perspective or window we will be able to see, explain, and make useful predictions. In this light the model that we develop might best be viewed as an attempt to provide a useful window through which to view and understand leadership processes. Furthermore, our clear intent is to provide some leadership for the future study and understanding of influence processes within the context of innovation.

This chapter draws on organizational leadership views that we believe are particularly relevant to the process of innovation in organizations. The core process of leadership might be viewed as "influencing" (Gardner 1986).

This chapter takes as its perspective for examining leadership three specific influence processes: (1) the rhetorical process (specifically used in defining and gaining commitment to a vision or mission), labeled visionary leadership here; (2) the process of exchanging incentives for support and energy to pursue the vision (such as bargaining), labeled transactional leadership here; and (3) the process of facilitating employee input, involvement and self-influence and consequently ownership and commitment, labeled participative leadership here. We provide a brief introduction to each of these perspectives along with relevant research propositions and then present a more detailed overview of each perspective, followed by a cross-sectional analysis of some specific cases from the innovation studies that shed light on these propositions. Ultimately, as mentioned earlier, we present an interactive process leadership framework relevant to the unfolding of innovation processes over time and out of which flows some general questions for future research.

GENERAL PROPOSITIONS

We begin our presentation with a fairly simplistic cross-sectional view of leadership as a set of relatively distinct influence processes that are examined in a time static manner. That is, for the sake of initial clarity and simplicity we begin by examining a set of propositions as they apply to the three separate views of leadership we have chosen for analysis and provide case examples that shed light on these propositions. Later in the chapter we posit our more complex interactive process model of leadership in which all three perspectives combine and interact over time and suggest questions for future research based on this model.

Our propositions address several issues of relevance to innovation processes. In our analysis, we were especially interested in examining the types of involvement of followers. The three types of involvement that we address include the following: (1) *identification* (commitment to the leader or the leader's vision—the innovation—that in turn serves to facilitate one's self-definition as a follower); (2) *compliance* (calculative involvement—willingness to follow directives based on getting something in return); (3) *internalization* (involvement by instilling a sense of ownership of the goals and objectives being pursued and thereby facilitating some sense of self-leadership within the persons pursuing them) (Etzioni 1975; Kelman 1961; Manz 1986; Morgan 1986). Probably the primary distinguishing feature between identification and internalization is ownership. With identification perceived ownership principally rests with the leader. For internalization it is possessed more by the followers (that is, those who implement the innovation).

In addition, the leadership perspectives we address were chosen because they represent differing influence directions. That is, the process of influence associated with rhetorical/visionary leadership is mostly top-down; with participative leadership, bottom-up; and with transactional leadership, reciprocal. These ideas are integrated with a preliminary conceptual framework (presented in Table 19–1) that classifies leadership/influence processes based on the dominant type of involvement activated and the primary direction of influence employed.

TABLE 19–1. *Integrative Framework for Conceptualizing Different Leadership and Influence Processes.*

Leadership Perspective	Primary Direction of the Influence Process	Primary Type of Involvement
Rhetorical/ visionary	Top-down	Identification
Transactional	Reciprocal	Compliance
Participative	Bottom-up	Internalization

Presented below are the basic propositions that we address in our cross-sectional discrete analysis of isolated leadership processes (rhetorical, transactional, and participative) across several innovation studies and that are largely derived from the logic of our framework. First, a very brief description of the leadership perspectives we have adopted is presented (a more detailed overview is provided later in this chapter), followed by the set of general propositions to be studied for each. The logic for these propositions derives primarily from our initial framework (see Table 19–1).

A *visionary/rhetorical* view of leadership suggests a process in which various persuasive methods are employed to achieve a common view of reality in followers and to develop a vision that encompasses a common mission. The vision tends to be based primarily on the leader's insight and viewpoint.

Table 19–1 suggests that visionary leadership unfolds primarily as a mode of influence flowing from leaders to followers (a top down process). It serves as a mechanism for facilitating subordinate involvement based on identification with the leader and his or her vision. When significant changes that directly affect subordinates are initiated, facilitation of positive identification with the driving vision can be very important for overcoming natural resistances to change (see, for example, Lewin 1947). The propositions generated for this perspective are presented below.

> *Proposition 1.* When identification with an innovation (particularly one that is primarily defined and elaborated by the leader) across the wide range of employees is important for effective implementation, rhetorical visionary leadership tends to be effective.
>
> *Proposition 2.* When situations exist in which employees are forced to experience major system changes (in the basic power

structure or in the way they do their work and the task itself) rhetorical visionary leadership tends to be useful due to facilitation of employee identification.

A *transactional* view of leadership attempts to explain how the reciprocal process of influence between leaders and followers occurs over time. It is based on the idea that an exchange of some kind is made between leaders and followers (e.g., carrying out the leader's directions and allowing the leader freedom to stray from group norms is provided by the followers in exchange for competent leadership and loyalty to the group) (Hollander 1978).

Table 19–1 indicates that transactional leadership is based on a reciprocal exchange of costs and benefits between leaders and subordinates and serves to induce compliance. That is, a calculative type of involvement is fostered in which the followers comply with the leader's direction as long as desired incentives are forthcoming (see, for example, Etzioni 1975). When followers are subjected to major system changes, however, naturally occurring resistance (Lewin, 1947) makes it less likely that the normal exchange of incentives will be adequate to allow for an effective change process. Normal involvement based on compliance may simply be inadequate to overcome barriers to significant innovation. Propositions for the transactional view include the following:

> *Proposition 3.* Transactional leadership is effective in those situations in which compliance as opposed to identification and internalization is adequate for the effective development and implementation of an innovation.
>
> *Proposition 4.* Transactional leadership is not effective in situations involving major changes in the culture and/or the basic power structure because compliance is inadequate in these circumstances.

The *participative* leadership view involves a process in which part of the leadership function is passed on to followers. This perspective prescribes a situation in which followers to a greater degree become their own source of direction and influence. Typically they take on greater responsibility and become more fully involved in the organizational (especially management) process (Locke and Schweiger 1979).

Table 19–1 suggests that participative leadership operates more as a bottom-up influence process. On gaining greater self-direction followers can obtain an increased sense of ownership of the objectives and goals being pursued. Thus, involvement based on internalization is facilitated. This is especially useful when leaders desire the creative and intellectual inputs from followers (as opposed to simply compliance in implementing the leader's ideals). When employees are allowed discretion in terms of the means for furthering the innovation, participative leadership is useful for helping integrate these discretionary efforts through a process of employee internalization of organizational objectives and a sense of ownership.

Also, this participative-based influence process should be relatively free from other external control forces. That is, when other sources of influence are oriented toward compliance rather than internalization, the ability to effectively participate can be overshadowed by constraining conditions that surround the participative process. Some propositions for this leadership view include the following:

Proposition 5. Participative leadership is especially important when the intellectual input, as well as the physical implementation of the innovation, are desired for the successful completion of the innovation process.

Proposition 6. Participative leadership is not as appropriate for situations where employees are placed under external con-

straints (e.g., from the leader) that directly conflict with the area of discretion provided.

METHOD

It is clear from the above discussion that our analysis will require data concerning the *direction of influence* between organization members and the *type of involvement* fostered in these members. The Minnesota Innovation Research Program (MIRP) incorporates multiple methods for data collection in its longitudinal study of innovation, including interviews, archival data, standardized questionnaires, and research diaries of meetings (Chapter 2). Although each of these methods has strengths and weaknesses, our requirements were best served by the data generated through the interview and research diary methods.

Although the archival method did provide data on the formal structure of leadership relations, it did not furnish us with data regarding the actual direction of influence between organization members. Moveover, it did not give us information about the type of involvement instilled in members. Similarly, the questionnaire data generated by the MIS survey (see Chapter 3) did not provide us with information on direction of influence and type of involvement. The MIS findings provided clear support for the importance of leadership in general and specifically several *outcomes* of the leadership process in successful innovations (namely, member initiative, clear responsibilities, clear feedback, task emphasis, human relations emphasis, trustworthy followers). These measures of leadership outcomes did not, however, provide us with insight into the interpersonal (direction of influence) and the intrapersonal (type of involvement) *processes* through which initiative was encouraged, responsibilities were made clear, and so forth.

This chapter is an attempt first to analyze distinct perspectives on leadership within the context of innovation from an isolated cross-sectional viewpoint. Then we move toward a process model of leadership in innovation that specifies integrative ways of achieving the aforementioned outcomes; ways that differ in terms of the weights and meanings assigned to the outcomes and the mechanisms used to achieve the outcomes. For several of the MIRP cases, data from interviews and from research diaries of meetings did furnish insight into the interpersonal and intrapersonal processes of leadership. Following the recommendation of Yin (1984), we employ a pattern-matching logic in which patterns in the empirically based case data are compared with patterns predicted by our conceptual framework. In so doing, we bridge several MIRP cases and surface empirical regularities that help us to develop a more complex conceptual framework and to make preliminary recommendations for leadership within organizations in contexts of innovation.

CROSS-SECTIONAL OVERVIEW AND CASE ANALYSIS OF THE PRIMARY LEADERSHIP PERSPECTIVES

In this section we describe in some detail each of the three leadership perspectives addressed in this chapter and provide case material to shed light on our propositions. The relatively static case examples here do not address the interaction of the three perspectives over time.

Rhetorical/Visionary Leadership

Rhetoric is a discipline tracing its lineage directly to Aristotle. Most broadly, it is the study of how leaders communicate visions to motivate groups to action through persuasion. Rhetoric has as its specific goal to have the ac-

tion accomplished through the unity and depth of commitment to the action. It focuses on communication acts, examining factors in the message, the leader, and the recipients of the message. Aristotle, and many of his disciples through the millennia, have preferred to study public speaking as a means of exercising (or losing) political leadership, but rhetorical theory has been increasingly broadly applied to institutional and organizational leadership (Bormann 1980; Howell 1982).

The nature of the discipline, then, is that it is centrally concerned with relating the vision and insight of a leader, the ways that the leader expresses the vision and insight, and the ways a group responds to the expression and vision. In this long theoretical tradition, the concepts of leaders and followers are appropriate. Leaders express leadership through superior visions and persuasion, and others, identifying with the vision, follow. This is the conceptual and assumptive tradition from which much of organizational leadership theory derives, and it therefore serves as an important way of analyzing at least some organization innovation leadership situations. In particular, this set of assumptions about the nature of leadership has informed management practitioners, especially in times of necessary organizational transformation and change (House 1976; Berlew 1974).

Before moving into a direct discussion of these organizational leadership theories and their relationship to the innovations studied by the MIRP, several definitions with their attendant concepts should be specified in order to better generalize about these theories.

1. *Vision:* an individual idea of a future state used as a guide to action. It may or may not be shared with or congruent with visions of the rest of the community.
2. *Rhetorical vision:* a shared vision (Howell and Brembeck 1976) constructed through communication. It contains dramatic personae and plot lines. A vision is complex: It

618

includes the myths, themes from history, and group fantasies that members of the vision sharing community relate to and is mediated symbolically (Bormann 1972).

3. *Rhetorical community:* a group of people sharing a rhetorical vision, the object of rhetoric (Bormann 1980). Organizations are one frequently discussed type of rhetorical community (Howell 1982).

4. *Rhetorical sensitivity:* the awareness in a leader of the communication style, values, and rhetorical visions of a rhetorical community, and the adaptation of those features to the leader's communication.

Theories in this general perspective need not be theories directly descended from Aristotelian thought, although many are. Regardless of genesis, however, they all share a focus on the leader as a communicant and as the carrier of a vision (Roberts 1984). The vision must be expressed in ways that are sensitive to the rhetorical visions, myths, and styles of the organization. Bormann's (1972) work has principally focused on the process of development of rhetorical vision in small groups. As rhetorical visions diffuse through a community, some become pervasive and commonly held while others fade in the process. Those that survive and become pervasive are, in some measure, the projection of the organization's myths, style, and people into a hypothetical future. As visions are expressed and elaborated, they are compared with real events in time and serve to both explain the events and guide future action (Bormann 1972). This perspective suggests that the leader's ability to generate superior visions and to communicate them so that they dominate the future-oriented behavior of the organization is the determinant of success of the leader in getting organization members to follow. It is around these visions that groups organize for action (Bormann 1972; Roberts 1984).

Howell (1982) and Bormann (1972) assert that successful leadership is dependent on the insight of the leader into the ways that members of an organization can see a future. New visions of a leader/manager must be congruent with the current self-perceptions of the organization, must appear to guide the organization to a more desirable state, and must also be expressed in ways that are rhetorically sensitive. This does not mean that vision must come from the group, but that the leader must be sufficiently empathetic with the individuals and sensitive enough to the community to know what visions can capture the group's attention, imagination, motivation, and commitment. Finally, the vision must include high but achievable individual expectations.

When the organization is facing times of crisis and uncertainty, visions that provide the members of the organization new roles, relationships, and directions are most likely to dominate—visions that are not only new, but allow the individuals in the organization to see a personal role of importance and feel that the leader and others respect and appreciate their involvement. Theories of visionary, charismatic, or transformational leadership, all within the general rhetorical perspective, discuss the processes of generating rhetorical visions that are new, involving, and rhetorically sensitive. These theories focus on the leader's ability to know what followers think and feel through face-to-face communication. Roberts (1984) and others (House 1976; Berlew 1974) discuss participation in which leaders involve followers in elaborating the vision. In any event, these theories all involve the visionary leader's direct involvement with followers and their direct involvement with elaborating the vision. Participation in elaborating the vision (Roberts 1984) and developing empathy (Howell 1982) represent ways for rhetorical leadership to know what is possible within the visionary context of the organization, to get individuals to feel ownership of the vision and action, and to generate personal emotional

619

commitment to the new vision (and to the leader that generated it).

All theories within the general rhetorical perspective assume a leader-follower paradigm, in which the manager/leader is the carrier and communicator of superior vision. Effective managers/leaders in all cases must be sufficiently involved with and knowledgeable of the rhetorical community (organization) to be rhetorically sensitive to the rhetorical vision of the organization. The rhetorical vision of the organization necessarily includes plots and action as well as dramatic personae. In other words, the sensitive manager/leader must separate the good from the bad, and use the good to eliminate the bad and achieve the plot, and this is done through the communication of vision through persuasive communication with the personae in the plot. This communication functions both to direct the organization and to draw the organization into a greater state of cohesion (Bormann 1972; Howell 1982; Roberts 1984).

Examples from MIRP Cases Viewed from a Visionary/Rhetorical Leadership View. There are several innovation cases in which a leader's vision and communication of that vision seem to be very salient in the situation. Most of these cases are exemplars of effective rhetorical/vision and leadership, but we also discuss a case where the instability of the vision created performance problems.

Mergers and acquisitions are one general arena that (consistent with our first two propositions) tends to involve a change in commitment across an entire organization and in which at a minimum employees are forced to experience major changes in the basic power structure. Bastien's discussion of common patterns of behavior and communication in mergers and acquisitions (see Chapter 11) includes one case (case 1) that clearly shows rhetorical/visionary leadership (a more detailed analysis of this case is provided in the next section of

this chapter). Here, a new chief executive officer was installed against the wishes of the acquired organization. This leader spent a great deal of time and energy initially both trying to assimilate the style of the acquired organization and coming to understand the rhetorical vision of the community. He did this through intensive and constant personal communicative involvement with the entire staff, using the information he acquired in this communication to help shape the vision as well as using the communication to establish his personal trustworthiness as a leader. In this case, the "bad guys" were the managers of the acquiring organization and the "good guys" were all of the existing staff of the acquired organization. The new leader, recognizing this, sealed the acquired organization from his peers on the acquiring side. Changes were made through existing acquired company staff, and the CEO's strategy was to integrate his vision of the future with the rhetorical vision of the acquired organization. Clearly, this was a case where broadly based identification and commitment were needed as a major organizational change took place. As proposition 1 and 2 would predict, visionary/rhetorical leadership was an important mode of influence in this case.

Bastien and Van de Ven (1986) noted another acquisition case in which the rhetorical vision for the employees of the acquired organization included the belief that the organization was in trouble because of inept management and that the line employees were capable of much better performance than they were allowed to generate. Here, the acquiring company management recognized this rhetorical vision and supported it while bringing in technical and marketing changes to further support and enable the plot of "we could do much better if we had management that knew us and the business."

Here the acquiring company management let the old managers go and, quite literally, celebrated the line employees. They consciously confirmed the people and plots of the acquired organization's rhetorical vision through first establishing a rough vision and then becoming involved with the staff in meetings and social events. Prior to the acquisition, the hourly employees were segregated from management and received little consideration except what was guaranteed by the union contract. The new leadership socially integrated the organization by including hourly staff in all management meetings and making benefits and personal considerations the same for management as for hourly workers. Again, consistent with propositions 1 and 2 visionary/rhetorical leadership was very instrumental in this case, which involved major system changes and a need for employee identification and commitment.

Roberts (1984) studied a basic change in a school district in a time of environmental crisis (a decrease in funding). Here the vision was strongly associated with the leader, but there was extensive participation by all stakeholders in the school district in elaborating the vision. Clear and meaningful roles for all participants were specified, and the behavior of the leader was congruent with both the changes themselves and the participatory process. Unlike the first two cases, where outsiders were the new leaders, this leader came to the scene already well enculturated and already knowing the limits to the visions currently held by the organization. The changes in this case involved, again, some basic shifts in the power structure and necessitated unified and total commitment of the stakeholders in the organization.

In all three of these cases, the leader and the vision became inseparable. As commitment to the future of the vision grew, the identification with the leader grew and the leader came to personify the vision. Consistent with propositions 1 and 2, significant organizational changes were positively facilitated through the input of visionary/rhetorical leadership influ-

ence on employee commitment and identification.

In the SMRC merger and acquisition study (see Chapter 11) there were instances of emphasis on rational transactional approaches to leadership in the face of the need for broad and intense commitment. Bastien (case 2) noted a merger case in which, despite deep and basic changes in both organizations, a rational transactional approach to leadership was used with negative results. In this case, many individuals in both organizations bailed out (found other employment), others still simply resisted accommodating to the changes, and the merger suffered from poor performance for several years. In this case, all of the management of both organizations eventually left. The case was marked by low levels of acceptance of the new vision, low levels of participation in elaborating the vision (although there was participation in assigning roles), and basically no new roles developed. This case suggests that the absence of visionary/rhetorical influence in the face of major system changes can lead to failure. Consistent with propositions 1 and 2, the insufficient identification and commitment that resulted from this leadership void contributed to failure in implementation.

Transactional Leadership

The final example in the above section implicates the transactional approach as a second way to lead in situations of innovation. This section provides a general overview of transactional leadership and discusses this perspective in relation to the MIRP studies.

The transactional perspective emphasizes the mutual influence between leaders and followers and explains organizational leadership in terms of the exchange relationships among individuals in the workplace (Blau 1964; Graen and Cashman 1975; Homans 1958; Hollander 1978). From this vantage point, leadership is not a static attribute of formally specified roles or rules; rather, leadership concerns the relationships for exchanging individual costs and benefits, an interpersonal pattern of transacting that emerges out of recurrent interaction in organizational contexts. From this view, a leader is not necessarily one who occupies a superior position in a formal hierarchy. Instead, a leader is simply defined as an individual who occupies a particular position in the pattern of recurrent exchange relationships in an organization. This position may or may not coincide with a superior position in the formal hierarchy. Indeed, individuals in transactional leadership positions may come and go throughout the course of ongoing exchanges in the context of the organization.

Both leaders and followers are defined in terms of the costs and benefits traded in this setting. For example, a leader can be seen as one who is given such benefits as power, authority, status, recognition, esteem, and legitimacy *by followers* at the costs of providing task expertise, uncertainty reduction (e.g., clarity in follower responsibilities and feedback), equitable reward distribution, and direction in crises *to the followers.* Conversely, followers are those who are given such benefits as task advice, a clear work context (responsibilities and feedback), fair rewards, and guidance during crises *by the leader* at the costs of giving power, authority, status, recognition, esteem, and legitimacy *to the leader* (Hollander 1978; Yukl 1981). Leadership is thus neither the "top-down" influence of the autocratic or the rhetorical-visionary views of leadership, nor the "bottom-up" influence of participative leadership. Instead, leadership involves *reciprocal* influence among individuals in an organizational context of exchange.

The transactional view does not require leaders to coerce followers into submission (autocratic leadership), to charismatically elicit the commitment of followers to a shared view of reality and mission (rhetorical-visionary leadership), or to facilitate the involvement of follow-

ers (participative leadership). Rather, it simply requires that both leaders and followers obtain some benefit in exchange for the costs they incur in developing and implementing innovations. This requirement reveals an important characteristic of transactional leadership. Unlike rhetorical-visionary leadership, transactional leadership does not aspire to obtain convergence among individual needs and values as a basis for achieving integration in the name of developing or implementing innovations. It merely seeks to accommodate the plurality of needs and values by establishing a system of exchange among individuals in organizations. Such a system fosters a *mentality of instrumentality* among organization members and rewards *compliance* as opposed to *commitment* among individuals as they attempt to develop or implement an innovation. The focus on individual costs and benefits in organizational exchanges as a sufficient basis for integration distinguishes transactional leadership from rhetorical-visionary leadership, which instead highlights the integrating effects of individual commitment to superordinate organizational values and visions (Bass 1981).

Due to the instrumental nature and individual focus of transactional leadership, one might anticipate that it will not be as appropriate as rhetorical-visionary leadership in innovation situations characterized by radical changes in the power structure and/or culture, situations that typically engender significant uncertainty on the level of superordinate organizational values and visions. Transactional leadership is simply not equipped to deal with such leadership concerns on this level by itself.

Examples from MIRP Cases Involving Transactional Leadership. Our analysis of the MIRP cases revealed no instances in which transactional leadership was solely sufficient as a basis for developing or implementing innovations. Indeed, in the one case we found in which transactional leadership was the driving strategy for accomplishing an innovation, it resulted in poor outcomes on several indices, including turnover and financial performance (see the final example in the section on rhetorical-visionary leadership). This MIRP case suggests that compliance is not an adequate leadership basis when implementing an innovation that involves significant organizational system changes. The finding is consistent with what one would expect to see in a merger and acquisition innovation, which by definition entails major changes in power structures and culture.

Our review of the MIRP cases uncovered little or no support for propositions 3 and 4. This finding appears to be largely based on the discovery that compliance did not seem to be an adequate form of employee involvement in any of the cases we reviewed. The fact that we could not find in the MIRP cases a single instance in which transactional leadership was the dominant leadership mode for successfully accomplishing an innovation suggests to us that the MIRP cases may have all involved significant changes to the organizational system, including culture and power structure. The degree to which this inference is a function of the particular sample of innovations included in the MIRP is an interesting consideration that prompts a set of related questions that need to be further addressed. Were all the MIRP cases drawn from a population of innovations that involve significant change at the organization level? Is there a population of innovations that does not involve such change, a population consisting primarily of technical changes that can be dealt with on an individual cost-benefit basis through transactional leadership, a population that exists but merely was not tapped by the MIRP? Do all innovations involve significant changes in culture and power structure? Or does leadership in situations involving innovation, at any level, simply require more than an exchange of incentives for compliance to a leaders directives? We offer these questions as a way of qualifying our study of the MIRP cases

in regard to the transactional leadership view. Clearly, we found evidence that some form of transactional influence processes did occur in most of the innovations, but never as a sufficient or primary source of influence in any successful innovation.

Participative Leadership

Another view of leadership can be described as the participative leadership perspective. Participation as a management process has been widely studied and debated in the literature. There has, however, been little consensus in the literature regarding a common definition of participation (Locke and Schweiger 1979). Employee participation has been viewed in terms of a felt obligation to work in the best interest of the group (Schultz 1951), ego involvement (Allport 1945), a managerial style or a legally mandated approach in which employees can influence decisions (Strauss and Rosenstein 1970), elected employee representatives in management (Strauss 1982), a form of delegation (Sashkin 1976; Sorcher 1971), and a mechanism for power sharing (Leavitt 1965; Tannenbaum 1974), and in many other ways.

Because of the many different views of the participative process it can be a particularly difficult phenomenon to study. In addition, we are faced with the challenge of addressing the role of leadership in facilitating participation and the impact of this process on innovation in organizations. Van de Ven (1986), for example, has suggested that the strategic challenge faced by institutional leaders is that of creating an infrastructure that fosters innovation and organizational learning. This view is especially interesting in light of parallel logic suggested by Hackman (1986) regarding leadership under highly participative conditions in which self-managed groups are employed. Inspired by the work of McGrath (1962), Hackman suggests that the role of self-managed group leaders is to ensure that all critical functions for group maintenance and task completion are achieved within the group. Thus, similar to Van de Ven, Hackman suggests that the leadership challenge might be viewed in terms of providing a context (or infrastructure) for effective employee behavior. Another way to view this process is in terms of what Morgan (1986) calls a holographic approach in which the sense of the "whole" is incorporated into the "parts." This allows units to operate autonomously but with coordination between units. Similarly, Angle (Chapter 5) comments on the importance of enabling conditions provided by the organization. These viewpoints taken together suggest that leadership influence may involve a rather indirect process. That is, similar to Kerr and Jermier's (1978) notion of substitutes for leadership, a leader may focus largely on creating a backdrop and support system and equipping others with necessary skills and information for effective worker autonomy in the absence of immediate leadership influence.

If the participative leadership role can be viewed as involving considerably less direct influence than traditional leadership perspectives would suggest, and as centering on stimulating effective worker participation, the question still remains regarding what is meant by the concept participation. In this chapter we view participation as a varying process that falls on a continuum ranging from complete external control to total reliance on employee self-control (Manz and Angle 1986; Manz, Mossholder, and Luthans, 1987). The real issue is not so much the absolute level of employee freedom and self-control but the relative amount of employee control in relation to existing work context norms. Thus participative leadership is viewed in terms of a process of increasing worker opportunities and ability to be more fully involved in, and to influence, the leadership process than they would otherwise be. Part of this process involves facilitation of employees' self-leadership skill development and self-

confidence to better equip them to deal with increased levels of autonomy. In fact, the ultimate level of participative leadership could be described as leadership of others to lead themselves (Manz 1986; Manz and Sims 1989) through the application of various behavioral and cognitive self-leadership strategies. Nevertheless, participative leadership also takes place when workers are enabled to exercise some modest level of latitude and influence within a highly autocratic work environment. Participative leadership, then, is the facilitation of *increased* worker autonomy and influence (participation).

Much of the participation literature has posited the argument that worker (or follower) participation facilitates increased worker acceptance and commitment (Sashkin 1976; Sorcher 1971; Lawler and Hackman 1969). In addition, evidence has been found that suggests that when participation in decisionmaking is allowed, problemsolving effectiveness is often improved (Tjosvold 1987; Hill 1982; Kelly and Thibaut 1968). Probably the best known contemporary leadership theory that directly addresses follower participation, the Vroom and Yetton (1973) model, focuses on two primary factors relevant to participative leadership. Specifically, it prescribes leadership processes that protect/facilitate worker acceptance and the quality of the decision for differing situations.

Our position in this chapter is that the innovative process frequently requires the existence of worker acceptance and commitment for success. Thus, in general, participative leadership should be relevant within contexts of innovation and especially so when follower acceptance and commitment as well as creative intellectual input for improved quality are needed. On the other hand, we also expect that participative leadership will not be as effective under conditions of significant external constraints that directly conflict with the discretion allowed. In other words, when external control conditions overshadow the benefits that we have outlined here, participative leadership will be less appropriate and effective. In the following discussion we present relevant qualitative case data that shed light on our propositions.

Examples from MIRP Cases Involving Participative Leadership. In the *school-based management study* (Roberts 1984; Lindquist 1986) it is apparent that the originator of the innovation, the original leader, was strongly committed to extensive participation of others. As the school district superintendent at the time of the initiation of the innovation put it in her own words, 'Sometimes you assign but other times you just say [the issues] and have people coming up to you saying they would like to work on this or that. That's how you find out what the skill, or the talent or what the interest is" (personal interview). School-based management could be described as an ambitious and comprehensive form of participative management. It is designed to more fully capture the commitment and potential of teachers, administrators, parents, and students in the school district.

The superintendent's leadership philosophy was strongly tied to a belief in the value of tapping the talents of people. She commented with obvious conviction about how people often just cannot see what significant capabilities they possess. She described her role as helping others to see and acknowledge their own values, capabilities, and contributions. Up until the time she left the school district for her appointment as state commissioner of education, her leadership behavior reflected this philosophy.She frequently asked people in the district to work on task forces or special assignments beyond their normal responsibilities, apparently to foster their personal development as well as to tap their unique capabilities. She also was persistent in her recognition of the contributions of others. In one meeting a teacher actually took a count of how many times she acknowledged or praised others during the session. The final count came to 221.

While the details of this case can be read in more depth in the site-based management chapter (see Chapter 17) suffice it to say this leader wanted and obtained the intellectual and other inputs of district members. Consistent with proposition 5 she obtained this objective through a highly participative style of leadership, that facilitated member internalization and felt ownership of innovation objectives, and that fit nicely with the school-based management concept. During her tenure as superintendent she would have to be described as a highly effective leader—her successes ultimately leading to a prestigious appointment at the state level.

The *Human Resource Management Study* also provides some insight for our fifth proposition (Angle, Manz, and Van de Ven 1985). In this case the primary focal leader, the vice president of human resources, was interested in obtaining broad participation of line as well as staff managers in the human resource function in his company. The innovation essentially could be described as an attempt to get human resource managers to think more like line managers and to get line managers to think more like human resource managers. It was hoped that by sharing much of the human resource function with line management that human resource considerations would become more of an integral part of corporate strategy rather than an afterthought.

Clearly the V.P. of human resources was interested in a full participative process involving the insights, creativity, and especially the commitment of a broad range of players. Facilitating line managers' internalization and ownership of the innovation was a central objective. Consequently he relied on a participative leadership approach. It could be said that he was attempting, at least in part, to "give human resource management away to the divisions." As he explained, "I want to get to the point where, from the divisions viewpoint, it's indistinguishable whether [human resource staff] are reporting to them or are assigned to them. . . . We want to impart the functional human resource skills more and more to human resource staff assigned to the divisions" (Angle, Manz, and Van de Ven 1985: 62). He also initiated significant training and development initiatives that included, among other things, training in human resource management skills for line managers, again with the intention of getting them more involved in the process. Indeed, his desire to have managers more fully involved illustrates an approach for relying on participative leadership to facilitate the internalization process and ultimately the success of an innovation.

Another of the innovation studies revealed some helpful insights regarding participative leadership (Ring and Rands, Chapter 10). This particular case involved an evolving unit of research scientists with unique expertise and a unique challenge. Indeed it was a case holding considerable interest appeal simply because of its uniqueness. It also posed significant challenges for leadership practice that require unusual participation of innovation group members to get the job done. Clearly this was an innovation that could not hope to succeed without extensive intellectual participation of work unit experts in shaping the direction of the project.

It could be argued that two primary areas of influence affected work group participation in this project. First, the scientists were provided with extensive participative discretion in dealing with the challenge of designing and conducting meaningful experiments. The second area of influence concerned the administrative management of the project itself—budgetary issues, establishing and managing the time schedule of the project, and so forth. In this latter regard, tension clearly emerged between the primary position leader of the project, the director of a science research labora-

tory, and the scientists. Given his position and responsibility in the company the leader obviously felt pressure to keep the project moving along and producing output that would justify its existence. In the process of his carrying out this role it was apparent that some of the scientists came to feel subject to overly directive influence, to leadership behavior that contradicted the scientific autonomy that they felt they needed and had been conditioned to expect.

The tension that subsequently ensued can be interpreted as at least partially supporting our sixth proposition. The participatory influence that existed regarding the scientific aspects of the project significantly conflicted with constraints imposed by the administrative aspects of the project. More specifically, while the intellectual creative inputs were desired in terms of the scientific components of the project, imposition of significant administrative pressures by the primary source of administrative leadership constrained the participative process and produced significant affective reactions from the scientists. The scientists were interested in optimizing the quality of the science. The focal leader felt pressure to demonstrate that the company could put an experiment up on a timely basis. The conflicts between these areas of influence/discretion caused limited participative discretion on the part of the scientists in completing the task in a manner that met acceptable standards based on their unique expertise. Participative leadership appeared to become less effective and appropriate under these conditions.

It is interesting to note another aspect of the leadership of this project. That is that multiple leadership sources were visibly important throughout the course of the project. These sources ranged across many organizational levels, from the CEO to the direct administrative manager of the team to other related support managers that were in some way hierarchically over the team. One other person possessing some influence over the project, for example, developed as a leader more sympathetic to the concerns of the scientists throughout the project. As tensions developed between the director of the science research laboratory and the scientists, this second leader became a buffer and somewhat of an advocate for the group that provided some needed insulation for the scientists to perform their work. Apparently, in situations where autonomy is desired and expected by the innovation unit, an important leadership function may become that of an insulator from organizational pressures. This may be particularly true when significant directive leadership pressure is initiated somewhere else in the system. A consequent development can be the emergence of the classic roles of task leader and socio-emotional leader.

As a final illustration we briefly consider the externally induced innovation project concerning nuclear safety regulations (Marcus 1985; Marcus and Weber, Chapter 16). Due to the increasing occurrence and threat of nuclear accidents at nuclear power plants, the Nuclear Regulatory Commission (NRC) took actions to tighten safety controls. Specifically, new safety regulations were developed and imposed on nuclear plants across the country. The study examined responses of plants to the new regulations and resulting acceptance of the new guidelines and resulting performance.

Several patterns of reactions were identified, ranging from a proactive anticipation of the new standards adapting and implementing actions before the regulations were imposed, to passive conformity to the new standards. The results of the study indicated that those plants that took a more active participative stance, in terms of anticipating or customizing the new standards to their specific situation, demonstrated a significantly higher level of acceptance (we view this outcome as categorically similar to internalization and ownership). The plants demonstrating higher acceptance of the standards in turn were found to have significantly

higher performance in terms of fewer incidents and higher ratings of plant management by the NRC. As predicted by proposition 5, this result can be interpreted as resulting from enhanced internalization and felt ownership of the standards in those plants that more fully participated in applying them. Overall, the study provided evidence for the importance of involvement/participation for achieving higher performance in situations involving externally induced innovations.

AN INTERACTIVE LONGITUDINAL-PROCESS VIEW OF LEADERSHIP IN THE CONTEXT OF INNOVATION

In this section we present a more complex model that suggests the interactive and longitudinal processes of leadership, particularly as they relate to innovation. Then we provide case material that illustrates our model and the interaction of our three focal leadership perspectives over time.

Toward an Interactive Process Model of Leadership

In our review of several MIRP cases we discovered an interesting pattern in the leadership events that took place. First, it became readily apparent that our static, cross-sectional model of leadership, presented at the beginning of this chapter, was inadequate for capturing the process of leadership in innovation contexts. Indeed, leaders in the innovation cases tended not to rely on only one leadership approach. Rather, they nearly always fulfilled central functions of each of the three types of leadership that we have addressed. Furthermore, the more successful leaders seemed to display a significant emphasis on each of these perspectives and they appeared to do so in an interesting pattern over time.

We describe the general pattern that seemed to emerge from the studies as a double-loop leadership process (our use of the term double-loop leadership should not be confused with the distinguished work of Argyris 1982 on double-loop learning). That is, two distinct on-going feedback loops seemed to occur repeatedly (see Figure 19–1). The first loop begins with a "latent vision" (here we refer to a set of individual visions and viewpoints that tend to converge on some common pattern or theme— that is, a "latent vision") within the existing culture that the leader attempts to tap through a participative process. In this loop the leader obtains the input and participation of the players within the system. This loop enables the leader to simultaneously gain information regarding an acceptable and useful vision on which to base future leadership activity and to stimulate support and a sense of ownership for organization members. This in turn equips the leader with the necessary knowledge to "amplify" and intelligently adapt the vision and to use it as an integrating and energizing force for enhanced performance. The participative leadership phase is also a time for facilitating employee positive self-concepts and self-leadership skill development.

The first loop is then followed by a second loop that moves from amplification of the vision to transactional leadership behavior that provides incentive for followers to provide their energy and support. Thus after the leader has gathered needed input to help discover

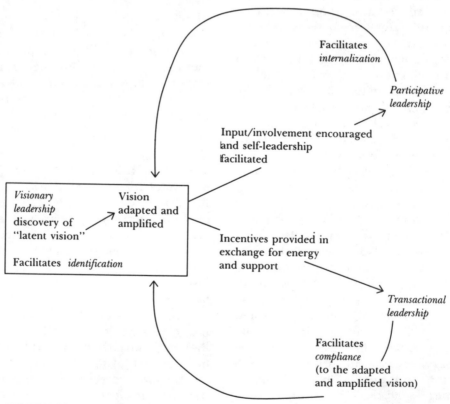

FIGURE 19–1. *An Interactive Longitudinal-Process (Double-Loop) Model of Leadership.*

vision that could be effectively amplified (one that employees are likely to accept and support and that was significantly based on their usually vast experience-based knowledge), a transactional process follows to further encourage members to "pitch in" to help pursue the vision. This "pitching in" tends to occur naturally, in part because of the previous participative process that helped establish natural motivation flowing from a sense of ownership and a better fit of the integrating vision with the various players of the innovation. Incentives are nevertheless very important because the leader typically leaves an imprint on the vision (that is, he or she typically adapts the vision) as it is amplified within the system. Also, various transactional incentives seem to serve as devices for helping to amplify the vision itself—that is, they not only support the vision but often epitomize it.

Figure 19–1 displays the essence of our double-loop model in graphic form. Note that the model connotates an ongoing process that continuously repeats each of the feedback loops. Note also that the figure encompass all three of the leadership perspectives and types of involvement (identification, compliance, and internalization) that were addressed in Table 19–1, in a single model. The participative leadership loop facilitates the internalization process for followers while helping to unleash energy (of the followers who experience a sense of ownership) to support the vision. It also provides an occasion to contribute to employee self-belief and self-leadership ability. The transactional leadership loop facilitates a degree of compliance to the vision that has become more a possession of the leader as he or she has adapted and amplified the vision. At this point the participative loop begins again, to maintain and further enhance internalization and ownership, and the cycle repeats itself. Furthermore, our observations suggested this entire double-loop process tends to take place in a very compressed fashion, with the two loops

sometimes occurring so closely together in time that they almost seem to take place simultaneously. Also, during this entire process the vision facilitates the identification process for each member in the system. Thus, the participative and transactional leader behavior feeds into and supports the vision, which as it is adapted and modified over time, reinforces identification for the followers. In the following discussion we more concretely illustrate our interactive process model with specific case material.

An Interactive Longitudinal-Process Leadership Case Example

In his discussion of mergers and acquisitions in Chapter 11 of this series, Bastien describes one case that serves as an exemplar of the double-loop leadership process that we have posited in this chapter. The leadership highlights of that case are presented here.

Prior to the acquisition, the members of the acquired company (1JCo) had widely thought of their organization as being very well and professionally managed. We view this widely shared cultural perception as a "latent vision"—that is, a view of the organization members that, while not being written down or associated with any particular leader, provided an integrating focus and sense of purpose for the membership. The corporation had been sold years previously to a holding company that had treated it and its management well. The holding company and its appointed CEO were somewhat cautious and conservative but generally well liked. During the preacquisition discussions between the management of the new acquirer (1SCo) and the owners and managers of 1JCo, the team of managers from 1SCo had thoroughly and personally offended many of 1JCo's managers and, more important, had challenged the memberships' self-image (the latent vision) of being a well and professionally managed organization.

629

Although the actual exchange of equity ownership of 1JCo was to take place in early January, the well-liked CEO was formally replaced by a person appointed by 1SCo before Thanksgiving. The newly appointed CEO was not part of the personally known management team of 1SCo and was not specifically a focus for hostility of the 1JCo management, but he did walk into a negative and potentially hostile situation. The 1JCo management did not want the acquisition to take place with 1SCo as its partner. The management of 1JCo resented the new CEO's organization because of the perceived attack on their self-image, and consequently the new CEO was a logical target for resentment, particularly since he was appointed to come in to replace a well-liked CEO.

During the six weeks between his arrival and the exchange of equity ownership, the new CEO solicited the viewpoints and participation of the organizations members (the participative loop). He spent his time getting to know the people in 1JCo and having them get to know him. During this time he learned the values and the general vision of his new organization. In general this process seemed very effective for capturing the support and energy of the people, for reinforcing their self-beliefs and abilities, and for obtaining useful information for understanding the existing latent vision as a foundation for his own leadership influence. He talked with everyone in 1JCo that he could and frequently asked for their opinions about the acquisition and the future. When asked of his plans for 1JCo, he was open and fully explained what plans he had made, but he also continued to incorporate what he had heard with his own vision for 1JCo. This talking, listening, and incorporation of existing views for the future caused the staff of 1JCo to report feeling that they were being treated collegially and that they were a central part of the future of 1JCo.

Beyond confirming the 1JCo vision of the future by seeking broad input (participa-

tion) from the membership, on the day of th
exchange of ownership of the organization
cocktail reception was held for the employee
of 1JCo. This helped to confirm their image c
themselves as mature professionals. Followin
this event, the CEO "sealed off" 1JCo throug
enforcing a strict "promote from within" pol
icy. This dramatic act can be viewed as an im
portant transactional event in which the leade
exchanged job security and opportunity for th
employees' energy and support (the transac
tional loop), and it provided formal confirma
tion of the image of being a well-managed orga
nization with enough talent so that no ne
managers needed to be added, either fror
1SCo or from the outside.

In addition to this important gesture
other transactional acts followed, includin
contribution of a substantial advertising cam
paign by 1SCo as well as some other tangibl
resources. The national ad campaign had th
effect of buoying the mood of the sales staff an
increasing their identification with the ne
CEO and organization. This infusion of re
sources ultimately increased the income of th
sales staff.

This increase in overall business als
had the longer-term effect of allowing fo
some reorganization and redefinition of th
business. This was, again, accomplishe
through the promotion of existing manager
to new positions not previously part of the or
ganization. The newly promoted manager
were given considerable latitude in how the
would manage their new functions, thereby ir
creasing the amount of ownership they felt c
those functions and the organization as
whole. Because these new promotions aug
mented rather than replaced existing manage
ment (there were no losers in the process
only winners), the staff of 1JCo viewed all c
this as confirmation of their basic vision (c
being a well-managed company); it also a
lowed them to realize the potential they saw i
themselves. Furthermore, the new manager

promoted by the CEO oversaw processes such as strategic planning and quality control that were in their basic design participatory. Thus, as time passed, through the use of participation and strategic transactional behavior, the leader succeeded in amplifying the vision as a major impetus for his own effective influence.

In summary, the new CEO, through participative and transactional behavior was able to be rhetorically sensitive and to identify and amplify the critical vision of the community of managers and employees of lJCo. Subsequently, he geared his communication to include this vision. More specifically he first relied on the active but informal participation of the managers and staff of lJCo in the process of identifying and amplifying a visionary base for his leadership. He also generated participation of lJCo managers in the subsequent reorganization. Second, through increasing both the corporate and personal incomes of lJCo staff and through promotions from within to new positions, he transacted bargains with the employees of lJCo (for their support and energy) that effectively made them winners in this acquisition.

As another illustration we briefly refer again to the school-based management innovation case discussed earlier. We discussed this case previously as an example of both visionary and participative leadership. We believe the case can be analyzed more accurately and richly with our interactive process (double-loop) model.

First, the leader had several years of experience in the school district and consequently a good sense of the "latent visions" embedded in the culture. The driving vision as amplified by the leader in this case largely rested on a belief in the capabilities of people. The school-based management concept itself encompassed a shift of power and emphasis to employees. In addition, the leader's behavior, which continually affirmed and sought the involvement of people at all levels, supported this vision by reinforcing employees confidence in themselves and ability to perform.

Continuing our analysis using our double-loop model, as the basic vision was developed and amplified the leader engaged in a highly participative leadership process (the participative loop). She widely solicited input and involvement at all levels in the district. Surrounding herself with the input of good people was a major banner of her approach. She also spent considerable time working with individuals to help them acquire the self-belief and skills to be more effectively self-leading.

As the vision was dramatically amplified, often leaving what appeared to be a charismatic tone to her activities, the leader was astute in introducing a wide range of incentives. She was lavish with her verbal recognition and reinforcement of the people involved. She also rewarded contributors with special assignments and invitations to serve on committees that further acknowledged their contributions. Perhaps most important of all, she continually rewarded people with her respect and her belief in their ability to excel and was not hesitant to communicate these positive feelings directly to persons and their peers. Thus the transactional loop of the model was vividly displayed in diverse ways.

Overall, the school-based management case appears to map well on our double-loop interactive process model, as do several other of the more successful MIRP innovation cases. In the following section we discuss some of the implications of this model and our analysis.

SUMMARY AND CONCLUSIONS

Whatever the guiding leadership and organizational assumptions of the managers and researchers are, it seems clear from the questionnaire survey data of the MIRP that leadership is a very important factor in the process of inno-

vation. Leadership was the only factor found to
be consistently positively correlated with perceived innovation effectiveness. Furthermore,
the data suggested that six fundamental issues
must be addressed: (1) Individual initiative
must be encouraged, (2) individual responsibility must be clear, (3) performance evaluation
feedback must be clear and complete, (4) there
must be a strong task emphasis, (5) human resource emphasis must be clear and strong, and
(6) the leader must demonstrably trust the
members of the organization. These six issues
correspond with the items on the innovation
survey relating to leadership of the innovation
group. Each of these issues is addressed to
some degree in each of the leadership perspectives we have discussed, but their relative
weights, meanings, and mechanisms for
achievement are different in each perspective.

A primary conclusion from our review of
the innovation studies is that multiple leadership approaches appear to be appropriate in
varying innovation contexts and at different
stages of the innovation process. The perspectives addressed in this chapter imply different
types of and levels of influence. These approaches appear to fit especially well for some
aspects of innovation and less well for others.
For example, we found no instances of a consciously applied transactional leadership process being successfully employed as the dominant influence approach. It appears that the
transactional perspective (as a principal leadership form) may be in general problematic for
innovative processes. On the other hand, after
reviewing the innovation studies, it appears
that transactional processes must be attended
to at some level in every case.

We did find numerous examples of an
emphasis on rhetorical/visionary leadership
being successfully employed, as would be expected in situations of basic organizational
change. The success of these examples, however, necessarily involved fair and equitable
transactions and high levels of participation.

Thus, the theme of multiple leadership per
spectives and styles was apparent in our analy
sis. One could argue that any of the individua
cases could be reconstructed using any of these
general leadership views. That is, we found
specific evidence that each of these perspec
tives were operative at some level for most o
the cases we reviewed, but that does not neces
sarily mean that it was the primary operative
mechanism driving the influence process
Rather, as indicated by our interactive longitu
dinal-process model (Figure 19–1) all three
types of leadership appear to be required ove
time as the innovation progresses.

Perhaps one way of looking at the prob
lem of studying leadership in the context o
innovation is to understand the tension be
tween analytic theory (the theories and per
spectives we use to drive our analysis) and op
erative theory (the theories in use that drive the
behavior of leaders/managers) (e.g., Argyri
1976, 1982). One can study an organizationa
situation from any of the three perspectives and
sets of assumptions described, but any give
manager, at a given point in time, is likely t
operate from a guiding set of beliefs abou
leadership that conform to one of the three
views. In other words, a given manager ma
believe that striking good bargains with em
ployees and other stakeholders is most impor
tant, but that manager also is likely to recogniz
employee identification facilitated by empha
sizing a vision is vital to organizational success
However, one also could study this situatio
effectively from a participation perspective an
see participation of employees as being a neces
sary condition of organizational success. W
chose to try to study the validity of all three o
these perspectives of leadership relying o
what we perceived as the guiding set of assump
tions of the managers in the cases studied.

As an integrating tool we first propose
a simple framework that draws on the perspec
tives and data reviewed in this chapter. Thi
framework centers on the direction of influenc

632

and the nature of employee involvement in the innovation. As is shown in Table 19–1 the framework specifies three directions of influence (top down, reciprocal, and bottom up) and three corresponding types of involvement (identification, compliance, and internalization). Rhetorical/visionary theories correspond with primarily a top-down influence process that fosters identification. Participative leadership is primarily a bottom-up perspective that facilitates internalization. Transactional leadership is reciprocal and compliance oriented. Our review of the cases suggested that visionary/rhetorical and participative perspectives are more prominent as dominant leadership approaches under conditions of innovation than is the transactional view, which appeared to be the least represented leadership perspective of the three viewpoints we studied.

Following our initial analysis we proposed a more complex double-loop interactive longitudinal-process model that combines all three perspectives in a temporal sequence over the course of an innovation. This model graphically forms a figure-eight pattern as the leadership process unfolds in two separate influence loops—a participative loop and a transactional loop. The pattern begins with a "latent vision" that already exists in the system and moves through a participative loop in which input and energy is sought from employees. This first loop helps the leader to determine an appropriate vision to adapt and amplify as a mechanism for leadership influence. It also promotes employee feelings of ownership of the vision and self-leadership to facilitate effective implementation. Then further support for the vision is obtained through an exchange of incentives that facilitates compliance of followers, in the second transactional leadership loop. This second loop is needed because the vision becomes more closely associated with and attributed to the leader as he or she adapts and amplifies it within the system. This double-loop process is viewed as an interactive one that repeats itself over and over again throughout the course of the innovation.

A major implication of this more complex multiple leadership perspective process model is that leaders use multiple influence procedures within innovation contexts. It suggests that our earlier more simplistic model, although useful for initial clarity in studying leadership, is inadequate for capturing the full richness of leadership processes under conditions of innovation. Rather, effective leadership combines different types of leadership influence over time as different needs arise. The perspective also suggests the need to facilitate multiple forms of employee involvement (identification, internalization, and compliance) over time as the innovation progresses. More complex and realistic analysis of several MIRP cases provided support for this process perspective.

Based on our analysis it appears that, within the context of innovation, rhetorical/visionary and participatory leadership are more frequently associated with innovation than transactional approaches. Although we did not find as much evidence for the existence of the transactional leadership view, we nevertheless observed that this process of exchange is always there at some basic but frequently subtle level. Indeed, our interactive process (double-loop) model suggests that all three leadership approaches are repeatedly combined over time to foster employee identification, internalization, and some level of compliance. The crucial contribution of leadership in the context of innovation, however, appears to be what is added beyond the exchange of incentives for compliance and effort. In particular, rhetoric and vision and the opportunity for greater participation and involvement may be the value-added contributions of a leader that best contribute to innovation effectiveness.

Overall, leadership appears to be an important variable, and specific leadership functions that enhance all three types of employee involvement (identification, internalization, and compliance) from top-down bottom-up, and reciprocal influence processes appear to be especially relevant. As a result of writing this chapter we identified some general research questions that we believe need to be addressed in future research to further our understanding of this process. They are not intended to represent an exhaustive set. Rather, the following questions are offered as preliminary suggestions for future research on the role of leadership within the unfolding of innovation processes:

1. How generalizable is our interactive longitudinal-process double-loop model of leadership under innovative conditions? Can it be replicated in studies more diverse than we have addressed here and with greater reliance on quantitative as well as qualitative data?
2. Can and should leaders combine multiple leadership styles to best facilitate the innovative process (such as rhetorical/visionary, participative, and transactional) in most all types of innovations, including small incremental changes at the work unit level? Or does this multiple perspective approach only apply to larger higher-level institutional changes?
3. To what degree are multiple leaders needed to facilitate innovations, and what are some of the more common leadership roles that need to be filled (facilitator, visionary, bargainer, insulator and protector of the innovation group, and so forth)?
4. Finally, and perhaps most important, because leadership is apparently such an important variable, what kinds of training content and procedures are needed to equip leaders to effectively contribute to innovation processes?

We believe that our chapter provides preliminary insight into each of these questions. Further, research is needed, however, to support or replace the conclusions reached in this chapter regarding the role of leadership in innovation. Clearly, the MIRP findings suggest that leadership plays a central role in determining innovation effectiveness as persons involved in the innovation process both perpetuate and experience key events over time. Consequently, further empirical examination of leadership processes within the context of innovation could significantly contribute to our understanding and ability to prescribe strategies for enhancing innovation within organizations.

REFERENCES

Allport, G.W. 1945. "The Psychology of Participation." *Psychological Review,* 53:117–31.

Angle, H., C. Manz, and A. Van de Ven. 1985. "Integrating Human Resource Management into Corporate Strategy: A Preview of the 3M Story." *Human Resource Management Journal* 24: 51–68.

Argyris, C. 1976. "Single-Loop and Double-Loop Models in Research on Decision Making." *Administrative Science Quarterly* 21: 363–375.

Argyris, C. 1982. "The Executive Mind and Double Loop Learning." *Organizational Dynamics* 11 5–22.

Bass, Bernard M. 1981. *Stogdill's Handbook of Leadership: A Survey of Theory and Research.* New York Free Press.

Bastien, David T. 1986. "Common Patterns of Communication and Behavior in Mergers and Acquisitions." *SMRC Discussion Paper Series.* Minneapolis: University of Minnesota.

Bastien, David T., and Andrew H. Van de Ven. 1986 "Managerial and Organizational Dynamics of Mergers and Acquisitions." *SMRC Discussion Paper Series.* Minneapolis: University of Minnesota.

Berlew, D.E. 1974. "Leadership and Organizational Excitement." In D.A. Kolb and J.M. McIntyre eds., *Organizational Psychology: A Book of Readings.* Englewood Cliffs, N.J.: Prentice-Hall.

Blau, P.M. 1964. *Exchange and Power in Social Life.* New York: Wiley.

Bormann, Ernest G. 1972. "Fantasy and Rhetorical Vision: The Rhetorical Criticism of Social Reality." *Quarterly Journal of Speech* 58: 596–407.

Bormann, Ernest. 1980. *Communication Theory.* New York: H.H. Rinehart, & Winston.

Brembeck, Winston L., and William S. Howell. 1976. *Persuasion: A Means of Social Influence.* Englewood Cliffs, N.J.: Prentice-Hall.

Brown, R.W., and E. H. Lennenberg. 1954. "A Study in Language and Cognition." *Journal of Abnormal and Social Psychology.* 49: 454–62.

Etzioni, A. 1975. *A Comparative Analysis of Complex Organizations: On Power Involvement and Their Correlates.* New York: Free Press.

Gardner, John. 1986. Private conversation, Greensboro, N.C., July.

Graen, G., and J.F. Cashman. 1975. "A Role Making Model of Leadership in Formal Organizations: A Developmental Approach." In J.G. Hunt and L.L. Larson, eds., *Leadership Frontiers,* Kent, Ohio: Kent State University.

Hackman, J. Richard. 1986. "The Psychology of Self-Management in Organizations." In M.S. Pollack and R.O. Perloff, eds., *Psychology and Work: Productivity Change and Employment.* Washington, D.C.: American Psychological Association.

Hill, G.W. 1982. "Group versus Individual Performance: Are N + 1 Heads Better Than One?" *Psychological Bulletin* 91: 517–39.

Hollander, E.P. 1978. *Leadership Dynamics: A Practical Guide to Effective Relationships.* New York: Free Press.

Homans, G.C. 1958. "Social Behavior as Exchange." *American Journal of Sociology* 63: 597–606.

House, R.J. 1976. "A 1976 Theory of Charismatic Leadership." In J.G. Hunt and L.L. Larson, eds., *Leadership: The Cutting Edge.* Carbondale, Ill.: Southern Illinois University.

Howell, William S. 1982. *The Empathic Communicator.* Belmont, Calif.: Wadsworth.

Howell William S., and Winston L. Brembeck. 1976. *Persuasion: A Means of Social Influence.* Englewoods Cliffs, N.J.: Prentice-Hall.

Kelly, H.H., and J.W. Thibaut. 1968. "Group Problem Solving." In G. Lindzey and E. Aronson, eds., *Handbook of Social Psychology,* vol. 3, pp. 1–105. Reading, Mass.: Addison-Wesley.

Kelman, H.C., 1961, Processes of Opinion Change, *Public Opinion Quarterly* 25: 57–78.

Kerr, S., and J. Jermier. 1978. "Substitutes for Leadership: Their Meaning and Measurement." *Organizational Behavior and Human Performance* 22: 375–403.

Lawler, E.E., and J.R. Hackman. 1969. "Impact of Employee Participation in the Development of Pay Incentive Plans: A Field Experiment." *Journal of Applied Psychology* 53: 467–71.

Lewin, K. 1947. "Frontiers in Group Dynamics." *Human Relations* 1: 5–41.

Leavitt, H.J. 1965. "Applied Organizational Change in Industry: Structural, Technological, and Humanistic Approaches." In J. March, ed., *Handbook of Organizations.* Chicago: Rand-McNally.

Lindquist, K. 1986. "School-Based Management in Orchard Valley School District." Unpublished working paper case, Strategic Management Research Center, University of Minnesota.

Locke, Edwin A., and David M. Schweiger. 1979. "Participation in Decision Making: One More Look." In L. Cummings and B. Staw, eds., *Research in Organizational Behavior,* Greenwich, Conn.: JAI. vol. 1, pp. 265–339.

Lombardo, Michael, and Morgan McCall. 1978. *Leadership: Where Else Can We Go from Here.* Durham, North Carolina: Duke University Press.

Manz, Charles C. 1986. "Self-Leadership: Toward an Expanded Theory of Self-Influence Processes in Organizations." *Academy of Management Review* 11:585–600.

Manz, Charles C., and Harold Angle. 1986. "Can Group Self-Management Mean a Loss of Personal Control: Triangulating on a Paradox." *Group and Organization Studies* 11: 309–34.

Manz, C.C., K.W. Mossholder, and F. Luthans. 1987. "An Integrated Perspective of Self-Control in Organizations." *Administration and Society* 19: 3–24.

Manz, C.C., and H.P. Sims, Jr. 1989. *SuperLeadership: Leading Others to Lead Themselves.* Englewood Cliffs, N.J.: Prentice-Hall.

Marcus, A. 1985. "Externally Induced Innovation: Response Patterns in the Nuclear Industry." Discussion Paper 41, Strategic Management research Center, University of Minnesota.

McGrath, Joseph E. 1962. "Leadership Behavior: Some Requirements for Leadership Training."

Washington, D.C.: U.S. Civil Service Commission.

Morgan, G. 1986. *Images of Organization.* Beverly Hills, Calif.: Sage.

Roberts, Nancy C. 1984. "Transforming Leadership: Sources, Process, Consequences." *SMRC Discussion Paper Series,* Minneapolis: University of Minnesota.

Sashkin, M. 1976. "Changing toward Participative Management Approaches: A Model and Methods." *Academy of Management Review* 1: 75–86.

Schultz, G.P. 1951. "Worker Participation on Production Problems: A Discussion of Experience with the 'Scanlon Plan.'" *Personnel* 28: 201–10.

Sorcher, M. 1971. "Motivation, Participation and Myth," *Personnel Administration* 34 (September/October): 20–24.

Strauss, George. 1982. "Workers Participation in Management: An International Perspective." In L. Cummings and B. Staw, eds., *Research in Organizational Behavior* 4: 173–265.

Strauss, G., and E. Rosenstein. 1970. "Workers' Participation: A Critical View." *Industrial Relations* 9 197–214.

Tannenbaum, A.S. 1974. "Systems of Formal Participation." In G. Strauss, R. Miles, C.C. Snow, and A.S. Tannenbaum, eds., *Organizational Behavior. Research and Issues.* Madison, Wis.: Industrial Relations Research Association.

Terry, R. 1986. "Leadership—Preview of a Seventh View." Working paper, Hubert H. Humphrey Institute of Public Affairs.

Tjosvold, D. 1987. "Participation: A Close Look at its Dynamics." *Journal of Management* 13: 739–750.

Van de Ven, A.H. 1986. "Central Problems in the Management of Innovation." *Management Science* 32: 590–607.

Vroom, V., and P. Yetton. 1973. *Leadership and Decision-Making.* Pittsburgh: University of Pittsburgh.

Yin, R.K. 1984. *Case Study Research: Design and Methods.* Beverly Hills, Calif.: Sage.

Yukl, Gary A. 1981. *Leadership in Organizations.* Englewood Cliffs, N.J.: Prentice-Hall.

TOWARD A GENERAL THEORY
OF INNOVATION PROCESSES

Marshall Scott Poole

Andrew H. Van de Ven

This chapter advances a theoretical framework that is emerging from the Minnesota Innovation Research Program studies. MIRP started with a set of sensitizing concepts, a few tentative models, and a research strategy that emphasized detailed, longitudinal study of a wide variety of innovations. The methodology was designed to measure and analyze innovation processes with all their real-world complexities in order to correct for the natural tendency to oversimplify and to fit observations to whatever theories a researcher is enamored with. The result is a complicated, somewhat unruly set of empirical observations that describe the multifaceted nature of innovations and that are often

beyond the explanatory capabilities of existing innovation theories. These findings include the following key observations about the innovation process:

"Shocks" trigger innovations: People initiate action when they reach a threshold of dissatisfaction or opportunity.

Ideas proliferate: After a simple unitary progression, the process proliferates into multiple divergent progressions.

Setbacks frequently arise, plans tend to be overoptimistic, commitments escalate, mixed messages about success and failure are randomly ordered over time, and mistakes snowball into vicious cycles. Vicious cycles are broken only with outside interventions.

Restructurings of the innovation unit occur often by external interventions to correct the innovation's course of action by

We gratefully appreciate the helpful comments of Harold Angle on a previous draft of this chapter. Support for this research program has been provided in part by a grant to the Strategic Management Research Center at the University of Minnesota from the Program on Organizational Effectiveness, Office of Naval Research (code 442OE), under contract no. N00014-84-K-0016, as well as other sources.

637

changes in personnel, transactions, joint ventures, or mergers.

Top management or investors are frequently involved in an innovation's development, and play four key roles: sponsor, cynic, mentor, and institutional or policy entrepreneur.

Innovation success criteria often shift over time, differ between groups, and trigger power struggles between innovation managers and resource controllers.

In-process innovation outcomes are partially a consequence of action, partially a predictor of future actions, but often unconnected to actions that occur in the development of innovations. Elements of both rational and superstitious learning occur as innovations develop over time.

An innovation that is internally developed or selected at one level of organization tends to become an externally mandated innovation to other organizational levels, creating a variety of unintended innovation adoption strategies.

In addition to these process findings common across many innovations, previous chapters have illustrated the diversity and complexity of innovation processes. They were found to vary in terms of type (product or process), novelty, degree of industry development and organization, and many other dimensions. The paths innovations take as they develop vary widely as well, ranging from rather simple sequences like those identified by Knudson and Ruttan (Chapter 14) to the more complicated "fireworks" model of Schroeder et al. (Chapter 4).

How might we make sense of these observed complexities and address the gap between data and theory? The typical strategy for sorting out this diversity is to locate the "joints" (Plato 1969; Kaplan 1963) in the innovation process and then create several new objects of study—such as invention, development, implementation, diffusion, and institutionalization—each of which has its own distinctive theory. This strategy has the advantage of stimulating development of detailed models related to innovation. However, it runs the risk of missing connections and transitions among innovation processes. Moreover, no overall unifying theory of the entire innovation process from beginning to end is likely to emerge from this strategy.

A second strategy is to develop a metatheory—that is, a theory of innovation process theories. A metatheory presumes that several theories of innovation are adequate but apply under different conditions; it attempts to specify those conditions and the relationships among the theories. For example, consider theories A, B, and C. Theory A may hold early in the innovation process, and then B after the innovating organization matures. Theory A may explain one pattern of innovation, and theory C another. A metatheory states the conditions or contingencies that govern when various theories or models hold.

This chapter adopts the latter strategy by taking some initial steps that may lead to a metatheory of innovation process. Based on the findings from the MIRP studies, we contend that a single theory cannot encompass the complexity and diversity of innovation processes. Instead, several different theories or models may explain innovation processes, and which theory holds depends on the context and conditions confronting a given innovation. This preliminary conclusion leads us to address the need for a metatheory of innovation process.

Suggestions for developing such a metatheory are presented in four major sections of this chapter. The first section discusses the desiderata for an adequate theory of innovation process by expanding and extending the discussions of Chapters 1 and 2. In the next section we discuss the basic structure of the metatheory by describing its component theories of development. This section lays the groundwork for proposing a set of contingencies in the following section that may determine the situations when different theories of development apply. These contingencies were derived from analysis of the entire set of MIRP innovation studies and represent one way of tying together some of MIRP's findings. The final section presents a set of propositions about when transitions or shifts occur between theories of innovation development. This section presents three forms that metatheoretical propositions may take.

Overall, the metatheory attempts to clarify the relationships among the various MIRP studies and to explain why some innovations followed complex paths, while others did not. This metatheory highlights several applied principles for the management of innovation. It also provides an agenda to guide future theory building and research on innovation and complex change processes in general, and the ongoing MIRP studies in particular.

METATHEORY IN THE STUDY OF INNOVATION PROCESSES

What Is a Metatheory?

A defining characteristic of a good theory is a statement of its scope. No useful theory applies to all cases, at all times, or under all circumstances. Scope conditions identify the limitations of theories, define the phenomena the theories explain, the general conditions under which their explanations apply, and factors that may interfere with theoretical predictions. So Feidler's (1964) early leadership theory is limited to (1) work groups, (2) with designated leaders, (3) in large, complex organizations. Scope conditions have been viewed most often as constraints on the application of theories. However, they can be useful in another way—as a resource for theory building.

The diverse and multifaceted nature of innovation processes has spawned a corresponding diversity of theories and models. Many have some degree of support within their scope conditions. No overarching theory has yet emerged, nor are prospects bright in the near future. However, we believe constructive initial steps toward such a theory can be undertaken by first carefully identifying the scope conditions of innovation process theories and then developing metatheory propositions that specify the situations when different theories apply and how they are related.

To proceed in this direction, it is necessary to define and distinguish several terms that are often confused. In this chapter, *theory* refers to an abstract, coherent set of principles that enables one to explain, understand, and (ideally) predict some aspect about the process of innovation. *Explanations,* then, ensue from theories. A *model* is a projection in detail of a theory that depicts a possible system of relationships, events, or actions (Suppe 1977). A model depicts the *mechanism* (the actions, processes, or techniques) by which the phenomenon under study operates and achieves its results. A mechanism usually carries the burden of explanation in a model, hence the term *explanatory mechanism.* A given theory may produce more than one model. For example, we might think in terms of a general theory that views organizations as political systems, with multiple goals, where decisions made by bargaining and power depends on the resources and allies one can muster. Incrementalism (Lindblom 1959) is one model that depicts the political process by which an innovation might emerge. An organized anarchy (Cohen, March, and Olsen 1972) is another model that builds on these assumptions.

Metatheories may incorporate both theories and models. However, for our purposes models of innovation are more useful than general theories. Models spell out in detail the processes that drive innovation, thus facilitating an evaluation of our explanations. Models also make it easy to derive recommendations for managers. So hereafter, we deal with metatheories of innovation *models;* abstract theories come into consideration only insofar as their premises are reflected in the models we discuss.

Let us consider a few examples of metatheories. Research on human development has produced several metatheories. Flavell (1972) proposed that Piaget's organismic model of development was most appropriate for childhood intelligence, while mechanistic models best described adult intelligence. This is a metatheory because it posits that different models explain the development of intellect, depending on the subject's age. In the same vein, Kohlberg (1969) suggested that ontogenetic-maturational change was the major underlying principle in early moral development, whereas experience was a more important principle in later development. Both Flavell and Kohlberg discuss processes that mediate the change from one model to its successor.

Tushman and Romanelli's (1985) punctuated equilibrium model of organizational evolution also attempts to mediate two explanatory models. Organizational development alternates between periods of convergence (which are devoted to stabilizing and elaborating existing structures and processes) and periods of reorientation (in which a transformational leader shifts the fundamental emphasis and structure of the organization). Very different explanatory mechanisms account for organizational processes in the two periods. In addition, Tushman and Romanelli describe factors that stimulate or impede transitions from one explanatory mechanism or another. This formulation mediates the tension between two well-known conceptions of change: One views change as the result of incremental shifts within a stable structure, and the other emphasizes fundamental change driven by major actors and events that alters basic organizational structures and assumptions. As such, their model

represents "paradox resolution" (Van de Ven and Poole 1988).

These are exciting moves because they attempt to reconcile seemingly opposed models and because they promise a picture commensurate with the complexity of innovation processes. However, to realize this promise, a metatheory of innovation must be more complete than the examples just reviewed. The MIRP studies suggest that an adequate metatheory must encompass multiple models cast at different levels of abstraction. If we are to cope with this complexity in a reflective manner, we must first consider the general requirements for a metatheory. This can clarify how to go about constructing adequate theories.

Desiderata of a Metatheory

What, in general terms, must a successful metatheory of innovation do? First, *a metatheory must specify the basic structure of an adequate theory of innovation.* That is, it must indicate what a good theory of innovation should do, what basic components of innovation a theory must consider, and any constraints that might rule out certain types of theory. This delimits the universe of possible models that can be brought to bear.

To describe the process of innovation is to describe how it grows, and an explanation of how things grow requires a *theory of development,* as opposed to a structural theory that explains the configurations or variations of a thing. In the next section we argue that an adequate theory of innovation must (1) incorporate models of both global (macro and long-run) and local (micro and short-run) development, (2) precisely specify the motor driving development at both levels, and (3) spell out interlevel relationships. Therefore, the domain of an innovation process metatheory should be a conceptual map of the basic alternative models of development that are relevant to explaining how innovations develop, grow, and terminate over time. To incorporate these scope requirements, the next section proposes a typology that distinguishes models of development on two dimensions: (1) level of theory, in terms of local and global models of development, and (2) *types of theory,* in terms of three motors used to explain development.

Models that focus only on local phenomena, such as individual creativity, are of limited use in explaining innovation as a whole (although they may play a key role as part of a global model of innovation processes). Consideration of local and global models leads one to examine the relations between innovation processes at different levels of analysis, such as how individuals relate to project teams, teams to organizations, organizations to a larger industry or community. An understanding of innovation processes also requires us to identify the different conceptual motors or logics that might be used to explain development at local and global levels. In particular, three motors—historical, functional, and emergent process—seem to underlie the various local and global models of development, and they will be discussed in the next section. Within this typology, we can define a number of models and motors that can serve as the basic building blocks of the proposed metatheory.

Second, *a good metatheory must specify conditions under which various models hold and when to switch between models to explain an innovation process at a given point in time.* To accomplish this, we distinguish between contingencies and switching rules in the metatheory. *Contingencies* indicate the immediate conditions in which individual models of development may apply to explain an observed innovation process. For example, we propose five such contingencies: whether natural or institutional rules prescribe the sequence of innovation development, the degrees to which innovation participants agree on the process and desired outcomes of an innovation, and innovation novelty, complexity, and resource dependence.

A more fundamental set of *switching rules* is needed to determine how and when different

developmental theories might be used and linked together to explain an overall observed innovation process from beginning to end. In the metatheory, these switching rules represent "metacontingencies"—that is, contingency-determining contingencies that govern the selection of theories to explain an innovation process within and across time. We propose three types of switching rules based on type, spatial, and temporal variations. They are fundamental parameters governing how and when to shift from one theoretical framework to another.

Finally, *the metatheory must be falsifiable.* Of course, any theory requires testing to evaluate its validity. As applied to the metatheory discussed here, there is a need to empirically test not only the fit of each explanatory model to the data, but also the contingencies and switching rules governing the appropriate applications of the models. This requires us to spell out operational criteria for each proposition so that it is possible to determine how closely each fits observed sequences and processes. Through the normal scientific process of refutation, we can narrow the range of plausible and valid models of development.

Thus, in addition to being falsifiable, we believe that a metatheory of innovation process should consist of three steps, which are illustrated in Table 20–1:

Identify and classify models of development relevant to understanding innovation process in terms of level of analysis (local or global) and type of theory (historical, functional, or emergent process motors).

Specify contingencies (institutional context, agreement on means or ends, innovation nevelty and complexity) when each type of developmental theory is likely to apply; and

Examine three switching rules (type, temporal, and spatial rules) that explain relationships between various models in the metatheory and may determine when to

TABLE 20–1. *Developmental Motors at Global and Local Levels with Examples from MIRP Studies.*

	Motor of Development		
	Historical	Functional	Emergent Process
Global level	**Logical necessity:** Cumulative synthesis (14) Transaction stages (6, 10) **Institutional requirements:** Legislative stages (9) FDA, NRC regulations (8, 16) **Biological/natural laws:** Hybrid wheat temporal stages of development (14)	**Accumulation:** Business creation (8) Industry emergence (15) **Dependency relations:** Conformative and Absorptive acquisitions (11) **Peer relations:** Strategic planning (18) Mergers and additive acquisitions (11) **Dominant relations:** Navy-defense contractor (12)	**Evolutionary:** Technology paradigm shifts (13) **Dialectical models:** Structuration (2) **Enactment:** Pushing ideas into good currency (9)
Local level	**Logical necessity:** Stimulus-response model (12) Endogenous development (14) **Institutional requirements:** Externally induced innovation adoption (16) **Biological/natural laws**	**Accumulation:** Policy entrepreneurship (9) **Dependency relations:** Rational learning (7, 11) **Peer relations:** Sensemaking, understanding, and committing (10) Strategic planning adoption process (18) **Dominant relations:** Breadth/depth adoption process (17)	**Evolutionary** **Dialectical models** **Enactment:** Success-failure action loops (7, 8) Garbage can model of choice (12)
Motor conditions	Natural or biological laws Logical necessity Institutional requirements	Clear functional goals Some institutionalization Agreement on means-ends connections	Low institutional organization Disagreement on goals and means

Note: Numbers in parentheses represent chapter numbers where model is applied.

switch between models to explain innovation processes.

The next sections represent an initial effort to undertake these steps. We believe that these steps, when performed properly, address the criteria for a good metatheory. In addition, they provide a systematic way to cope with the complexity of innovations and minimize reductionism. It may turn out that the observed complexity of innovation processes can be explained from tradeoffs and interactions among several relatively simple models of development within the metatheory.

A TYPOLOGY OF INNOVATION DEVELOPMENT MODELS

As Figure 20–1 illustrates, at its core the proposed metatheory consists of a typology of developmental models that can be drawn on to construct theories of innovation processes. The typology is based on two discriminating factors: levels and types of models. Inserted in the cells of Table 20–1 are examples of models that were used in previous chapters to explain the development of various kinds of innovations studied by MIRP researchers. Although not all the models listed fit neatly into each cell, they were classified according to their predominant level of analysis and theoretical orientation.

Local and Global Models of Development

As discussed in Chapter 2, an adequate theory of innovation development would have two levels of analysis: a model of global (macro, long-run) development and a local (micro, short-run) model of immediate action in the innovation process. The global model depicts the overall course of development of an innovation and its influences, while the local model depicts the immediate action processes that create short-run developmental patterns. Among the most well-known examples of global models in the literature are developmentalism (Nisbet 1970), epigenesis or accumulation theory (Etzioni 1963), and Darwinian evolution as well as Marxian dialectical theory. Examples of local models include stimulus-response behaviorism (Skinner 1938), path-goal theory (House 1971), and the garbage can model of decisionmaking (Cohen, March, and Olsen 1972).

Previous work has largely ignored this global/local distinction. Some theories focus on the long-run, giving the short-run little attention, while others elaborate local models in detail, giving the global short shrift. Few theories acknowledge both components equally and spell out their linkages.

Global and local theories of development differ on a number of dimensions. A global model takes as its unit of analysis the overall trajectories, paths, phases, or stages in the development of an innovation, whereas a local model focuses on the micro ideas, decisions, actions, or events of particular developmental episodes. Global models incorporate a long time frame (on the order of weeks, months, or years); local models work on the order of hours or days. In a global model, influences on development tend to come from the larger surround and include factors such as economic trends, social needs, the legal system, cultural norms, and long-term institutional arrangements. A local model focuses on influences on the immediate situation, including microlevel factors such as motivation level and group interaction processes as well as direct macro influences such as organization structure, resource control, and competition. In short, the world view in global models tends to be that of an astronaut in orbit, while it is that of the "person on street" in local models.

Both global and local perspectives are necessary for an adequate theory of innovation because innovations are extended over long periods yet managed through time by immediate action systems. When systematically conceived, a global theory would incorporate local theories of immediate action and explicate how local action unfolds in a larger temporal and social movement. In other words, global theories should address the transformation problem of showing how they can be disaggregated to describe the workings of two or more local processes, whereas local theories should be able to articulate how short-run microprocesses aggregate to produce long-term macroprocesses.

Previous work has largely ignored this global/local distinction. Some theories focus on the long-run, giving the short-run little attention, while others elaborate local models in detail, giving the global short shrift. Few theories acknowledge both components equally and spell out their linkages. In part, this is because developmental theories are still in their infancy, so to speak. As Nisbet (1970) notes, theories have depicted social and organizational development as an overly simplified progression

through a single historically and logically necessary sequence of stages. Van den Daele
(1969) termed this model a simple, unitary sequence. However desirable unitary sequences
might be for their elegent simplicity, mounting
evidence from the MIRP studies and elsewhere
suggest that innovation and change processes
are not so simple. Some changes do follow unitary sequences, but many do not. At present
most theories are at a loss to explain this complexity because they deal only with simple sequences and progressions. More precise
specification of a wider range of developmental
models is needed.

Types of Development Models

Previous chapters have often stated that innovation is an *emergent* and *goal-directed* process in
which groups of people invest their time and
resources to develop and implement new ideas
within an *institutional context.* Empirically, innovation processes were observed to differ in the
degrees to which action was constrained by institutional rules or natural programs, was rationally directed to functional outcomes, or
emerged almost randomly through a socially
constructed enactment of events. These field
observations lead us to identify developmental
models in the literature that emphasized these
differences. Three basic types of explanatory
motors or logics were found that appear to explain how and why development occurs: historical necessity, functional goal attainment, and
emergent processes. These are illustrated for
both global and local levels in Table 20–1 and
are now discussed.

Historical Motor.

Nisbet (1970) notes that by
far the most traditional and firmly entrenched
theory of development is "developmentalism,"
which is centrally based on the assumption of
immanence; that is, a social system contains
within it an underlying logic, program, or code
that regulates the process of change and moves

it from a given point of departure toward a
subsequent end that is already prefigured in the
present state. What lies latent, rudimentary, or
homogeneous in the embryo or primitive state
becomes progressively more mature, complex,
and differentiated. External events and processes can influence how the immanent form
expresses itself, but they are always mediated
by the necessity of historical logic, rules, or programs that govern development (Van de Ven
and Poole 1988: 37).

Classical theories of social development
emphasize that growth or change must necessarily take a particular course because the final
end requires a specific historical sequence of
events. Each of these events contributes a certain piece to the final product, and they must
occur in a certain prescribed order because
each piece sets the stage for the next. *Historical
necessity* dictates each stage of development as a
necessary precursor of succeeding stages. Historical necessity can result either from natural
or biological laws, logical necessity, or institutional requirements.

Initially, developmentalism relied on
natural or biological laws to explain growth from
the perspective of a gross anatomist who observes a sequence of developing fetuses and
concludes that each successive stage evolved
from the previous one. This nineteenth-cen-
tury perspective was influential in the evolution
of theories of social development (Nisbet
1970). Similar to the Aristotelian position, it
tries to view social change as development toward some final end state, and the only carrier
for such a final end must be something immanent to or genetically programmed into the developing entity. The necessary sequence of development in many animals, plants, and
humans has been shown to follow such natural
or biological laws, and genetic research has
demonstrated specific internal mechanisms
that "carry" immanence.

Nisbet's interpretation of developmentalism can be extended by considering logical

necessity and institutional requirements as two other sources of immanent prefiguration of the future in present social entities. Developments of many social entitites are not programmed or governed by natural or biological laws. Instead, *logical necessity* may dictate certain developmental sequences (Flavell 1972). One step in a sequence may simply be a logical requirement for the next, as some control over the vocal cords is a prerequisite for the development of speech. For example, in describing the process of technological development, Usher (1954) proposes a cumulative synthesis theory in which an accumulating structure of knowledge or science logically necessitates a sequence of four developmental stages, each of which logically presupposes the next: (1) setting the stage, (2) the act of insight, (3) conceptual breakthroughs, and (4) critical revision. Knudson and Ruttan use Usher's theory to explain the global development of hybrid wheat in Chapter 14.

Finally, developmental sequences of many social units are historically necessitated by *institutional requirements,* which often serve as substitutes for (and indeed sometimes contrary to) natural or biological laws and logical necessity. For example, in Chapter 9, Roberts and King provide a rich description of how legal enactment of educational reform in a state was only possible by going through an institutionally required set of lawmaking stages: idea generation and mobilization, proposal to and endorsement by a powerful elite (governor), drafting a legislative bill, transforming the bill into law, and implementing and evaluating the new law. So also, the FDA regulatory approval process of biomedical innovations in Chapters 8 and 15 shows how institutionally prescribed rules govern and explain sequences of innovation development.

Historical models were also used in preceeding chapters to explain developments at the local level. The stimulus-response model used in Scudder el al.'s study (Chapter 12) is largely based on logical necessity: The stimulus logically precedes the response. Given the global imperative by the NRC on nuclear power plants to implement new safety standards, Marcus and Weber in Chapter 16 insightfully develop an externally induced model of innovation. This model examines local adaptation strategies employed by nuclear power plants to comply with global institutional mandates. In general, the externally induced innovation model suggests that an innovation developed at a macro-organizational level often represents an institutionally prescribed innovation adoption process at the micro-organizational level.

A historical motor assumes irreversible time—that is, that development can proceed in one direction only, guided by the "arrow of time." With the final state as a criterion, we can assess progress along this arrow. Later stages cannot move back to earlier ones without a degeneration of the developing social structure.

Functional Motor. There are also models of social development that do not presume necessary sequences of events, yet which do imply standards by which change can be judged. There is no prefigured rule or logically necessary direction to these systems. However, one is still able to assess when a system is developing; it is growing more complex, or it is growing more integrated, or it is filling out a necessary set of functions. We are able to make this assessment because functional theories posit a standard of what a "developed" system is, and we are able to observe change toward a developed system as departure from some present state that is less balanced, or in tension, or not in equilibrium.

This explanation draws on the classical structural-functional tradition in biology. Functional models explain development in terms of movement toward some final state of "rest" (however temporary) that can be achieved via a number of paths, all tending toward the same endpoint. These models incorporate the systems theory assumption of equifinality. There is no assumption about historical necessity. Rather, these models rely on a *functional motor* as the explanatory principle: They posit a set of functions, goals, or forms necessary to sustain an organization, which the innovation has to acquire in order to stabilize itself. Development is movement toward functional autonomy.

Basic to a functional model is that an innovation accumulates the power and resources needed to accomplish its goals relative to its external environment. In order to do this, the innovation unit must establish its own identity, which implies creating a bounded social unit. But since no unit can ever operate totally independent of its environment, it must define its relationship to and manage transactions with external environmental units, such as resource controllers, government agencies, acquiring organizations, or markets. Based on an analysis of MIRP cases, it is useful to examine the range of power relationships between innovation units and their external environments; for these relationships largely predict the degree to which innovation units can achieve their functional goals.

In an *accumulation relationship* the unit attempts to gather the power and requirements needed to successfully develop its innovation from the external environment. Over time this involves gaining increasing control over necessary resources, personnel, skills, and organizational functions. The accumulation model is similar to Etzioni's epigenesis model. The objective of the innovation unit is to gain independent and autonomous control, sometimes in opposition to outside actors. This accumulation relationship is evident at the global level in the small business (Chapter 8) and industry emergence (Chapter 15) cases. At the local level, Roberts and King (Chapter 9) describe how policy entrepreneurs push and ride innovation ideas into sufficiently good currency to mobilize legislative action.

When the innovation unit is in a lower power position, it has a *dependent relationship* with its environment. Here the innovation unit attempts to adapt to and internalize requirements imposed from the outside. This involves determining what is required and reordering the innovation unit's developmental plans and products to satisfy outside demands. At the global level, this was observed in the conformative and absorptive acquisitions studies by Bastien (Chapter 11) and in the learning model he posits. Innovation units typically have dependent relationships with their environments in the early periods of their development or when significant setbacks or problems arise that stimulate external interventions by resource controllers and investors. Rather than controlling its own destiny, the innovation unit cedes control to external organizations.

When the power of the innovation unit is about equal to that of external units, it can engage in a *peer relationship* with others. Here the innovation unit treats external parties as equals and attempts to come to a "workable arrangement" in its transactions. At the global level, this was observed in the other two types of mergers studies by Bastien (Chapter 11) and in the planning processes implemented by Bryson and Roering (Chapter 18). The sensemaking-understanding-committing model posited by Ring et al. (Chapters 6 and 10) is a local bargaining model. Rather than achieving independence from the outside or swearing fealty to it, an innovation unit in a peer relationship assumes there must be joint or colleagial decisionmaking between itself and key external actors.

When the innovation unit is in a dominant power relationship with external parties, it tends to develop either *problem solving* or *fortification* relationships. In this case, the innovation unit maintains independence and reacts to demands or difficulties presented by external actors in its own fashion. At the global level, this relationship was observed in the relations of the defense contractor with its subcontractors reported by Scudder et al. (Chapter 12). At the local level, the implementation process of Site Based Management at River District discussed by Lindquist and Mauriel is a case of problem solving. In problem solving the innovation unit adapts, but unlike the case of a dependent relationship, the innovation unit maintains independent control of its goals. At the extreme, the unit may engage in fortification. If the innovation unit feels it has enough resources to go it alone without external support, it tends to ignore outside actors and attempts to buffer itself from possible objections or setbacks.

In each of these applications of global and local models, a functional motor of development assumes that as new social systems progress toward their functional end states, they move toward a self-sustaining equilibrium. That an organization attains the requisite functions does not mean it stays in permanent equilibrium. Varying degrees of influence of the external environment and of forces within the innovation unit itself may create instabilities that push the innovation to a new developmental path or trajectory. Theories that rely on a functional motor cannot specify what trajectory development will follow. They can at best list a set of possible paths, and outline some conditions that favor certain paths.

Emergent Process Motor. The assumption of requisite functions may be too constricting. Some models work on the premise that we can judge stability and departures from it but that there is neither a necessary prefigured direction to development nor a set of functional ends that development must satisfy. Instead, these models explain development in terms of an *emergent process motor.* They specify the process that drives development but regard this process as either socially constructed or seemingly random, in the sense that it can operate on a variety of different organizational forms and generate a variety of different developmental paths. A historical motor specifies a necessary path; a functional motor leaves the nature of the path open and specifies the end states or goals it must satisfy; and an emergent process motor leaves both the path and functions open and focuses on the process of social construction and random events by which development occurs.

Four variations of emergent process models can be distinguished in the literature. *Dialectical models* represent a well-known class of emergent process motors. They posit that an antithesis to the current state develops and that from the class between thesis and antithesis a stabilized synthesis emerges, which then becomes a new thesis as the process continues. Reigel (1975) describes a dialectical model of human development that has implications for the study of innovation. Moving away from a focus only on the developing child, Reigel argues that human development comes from interaction between the child and others. This interaction can be modeled as a dialogue in which caregiver and child develop together. The child's development is stimulated by others in a dialectic: The child's current state is inadequate for dealing with the other and the child develops further in order to achieve a more satisfactory mesh with the other. The theory of structuration (Giddens 1979, 1985) is one social theory that seems to explain change dialectically.

Evolution is another well-known model that explains change as a probabilistic process of variation, selection, and retention of innovations (Aldrich 1979; Hannan and Freeman 1977; Hull 1974). Typically, organizational evolutionary theories deal with processes of change in overall populations of organizations and technologies at the global level. For example, in Chapter 13, Rappa describes the global emergence of a community of scientists with an evolutionary model of paradigm shifts. Weick (1979) has adopted the evolutionary model at a local level to explain social-psychological processes of *enactment.*

A third variation of emergent process

theories relies on *self-reinforcing cycles* to explain development. Masuch (1985) discusses the role of interlocking action loops among interdependent people and organizational units in creating vicious cycles that promote change for worse in organizations. Recent work on dissipative structures and chaotic systems (Prigogine and Stengers (1984) has produced powerful models of self-reinforcing systems. In Chapter 8, Van de Ven et al. describe a local model of action cycles that governs new business start-ups. The model depicts cyclical relationships between the innovation organization and external resource controllers that determine the degree of autonomy the innovation achieves. Hence, it is a local process motor that drives a global functional accumulation model.

A fourth variation of the emergent process motor is the *garbage can model,* which describes decisionmaking processes on the basis of chance intersections between problems and solutions carried by distracted people who fluidly engage in opportunities for making choices as their volitions and interests dictate (Cohen, March, and Olsen 1972). Scudder et al. (Chapter 12) use the garbage can model to account for the complex patterns they observed in defense contracting.

Relations among Models in the Typology

Overall, we might think of these historical, functional, and emergent models of development as potentially composed of three parts—a required sequence of stages, a functional endpoint, and a process of change. A historical motor incorporates all three parts. By describing in some detail a particular stage sequence, a historical motor implies a final state and a process of change. Of course, these latter parts are often left implicit in the description of stages. A functional motor has two of the components. At the outset, it does not specify a sequence of events or stages, but it does describe the form of organization that is the endpoint of the development and, by implication,

the process for getting to the end. The emergent process motor describes only how development occurs, along with indicators to enable us to identify key developmental constructs (such as enactment and action loops) at any point in time.

As we move from historical to functional to emergent process perspectives, our description of the "motor" that guides development also becomes richer. A historical motor typically relies on analytic entailments to account for moves from stage to stage. For example, logically, it is necessary to define the innovation problem before we can begin searching for solutions. This logic is hard to refute, but it very certainty discourages articulation of the processes by which this logic operates in the nonideal world. A functional motor gives a richer account of developmental processes because it explains how and why certain paths lead to a desired end state better than others. Typically, an emergent process motor gives the richest and most insightful account of developmental processes because they are its exclusive focus.

Thus, one's focus of attention to different components of the overall innovation process appears to be a critical factor that connects the three explanatory motors of development. The historical motor largely directs attention to the natural, logical, or institutional rules that prescribe programs or routines of action that must be followed in developing a given innovation, and also to input conditions, which situate the innovation on its developmental track. Functional models place their central emphasis on visions of future goals or final end states of an innovation and then examine the means to achieve these desired end states. Emergent process motors center on the means of action themselves—that is, the dynamic process of social construction and transformation of an innovation from idea to "concrete" reality.

The key to using the typology described here is to select global and local models that are complementary and specify how they link together. In the ideal case we could spell out the interactions by which the local generates the "direction" of the global development and the global shapes local processes. If this is done properly, global and local models ought to be hard to separate because they will form a "seamless" whole.

Ring and Van de Ven (Chapter 6) and Ring and Rands (Chapter 10) provide a good example of a well-articulated global-local linkage. They lay out a three-stage global model of transactions: (1) negotiation, (2) agreement, (3) administration. The local model that drives negotiations through these formal phases is the interplay between understanding, sensemaking, and understanding processes oriented to attaining requirements for successful contracting. Ring and Rands have begun to work out the factors that influence the amount and proportions of these behavioral processes. These, in turn, shape the global developmental process. For example, when parties familiar with each other enter into a new deal, a specific global process unfolds, based on the local processes. Relatively little understanding, sensemaking, and committing are done in the negotiation phase because the parties already have an established relationship. Instead, parties work on understanding and committing during the agreement phase, which leads to intense emphasis on this phase. The final phase, administration, also requires little new understanding, sensemaking, or committing because the parties already know how to administer their relationship. So this phase, too, tends to be truncated. This theory is still evolving, but it is moving toward a more complete account of the relations between local and global levels of transactions.

CONDITIONS WHEN DEVELOPMENTAL MODELS APPLY

Having proposed a typology of alternative models of innovation process, we now consider the conditions when various models might apply. In so doing we develop a set of propositions that attempt to integrate and make sense of the major process patterns observed across the MIRP innovation studies. Although these propositions are not comprehensive of every pattern discovered by MIRP, they explain several notable and seemingly confusing trends across studies. In particular, we address the following observed patterns.

First, some MIRP studies found fairly straightforward unitary developmental sequences, while others found complicated, branching, recycling developmental paths. What accounts for these differences? In addition, some local developmental models found by MIRP studies generated rather lengthy chains of activity or cycles (such as Bastien's problem chains), whereas others exhibited fragmented action "molecules" often unconnected to other local actions (such as Scudder et al.'s stimulus-response episodes).

Second, a number of MIRP studies highlight the importance of the industrial infrastructure and institutional environment in innovation. This is in line with Ruttan and Hayami's (1984) theory of institutionally induced innovation. How do we incorporate these into a theory of innovation? Must all innovation theories be social or economic theories as well?

Third, the importance of key individuals emerges in most studies. As process champions, idea champions, resource controllers, scientists, regulators, and team leaders, individuals impart direction to projects, thrusting them ahead, or impeding and sometimes killing them. The MIRP studies suggest it would be a mistake to treat individual's influence simply as part of the larger movement of the innovation, determined by other, impersonal factors. Instead, the active and purposeful roles of individuals in structuring innovations must have a central place in an innovation process theory.

Fourth, in almost every innovation studied by MIRP, the innovation organization emerged as a central focus of analysis. The process of innovation is premised on building an organization that can nurture the idea, garner resources, overcome obstacles, and orchestrate development. The organization may be large and complex, as in the case of the defense contractor, or it may be a small independent business or project, as in the case of Qnetics or TAP, or it may be a temporary project team or task force, as in the case of educational reform. Whatever its form, the organization must have sufficient infrastructure and independence to launch an innovation and act as a "center" for innovating activities. In Thompson's (1967) terminology, to build an innovation organization is to engage in a process of establishing and maintaining system boundaries strong enough to preserve the integrity of the project and to protect it from disasters and shocks. Hence, internal-external relations between the innovation unit and its environment are very important in understanding the innovation process.

Finally, what is the role of shocks, crises, surprises, and "random" events? In some MIRP studies shocks played a central role; in others shocks had little effect. In some studies unplanned or surprising events were a major impetus in the innovation process. In other studies shocks terminated otherwise promising developments. An acceptable theory must come to terms with these as well as other issues.

Structuring the Innovation Process

The type of global theory that explains an innovation's development is determined in part by the existence of rules and programs in the innovation's institutional context, and by how the innovation process is structured by key actors. The structuring process operates as follows:

Key actors attempt to structure work on the innovation according to their schemata. Most schemata have a vision of the final product and the components that much be acquired

to achieve it. In addition, a schema may specify a set of stages, mileposts, or hurdles the innovation must pass through. Schemata help organize the innovation. A schema helps actors plan what must be done, who must do it, and how they can know it has been achieved. A schema also serves as a standard for evaluating progress and for making changes in courses of action.

Because there are always several key actors performing different roles relative to an innovation, several schemata may be operative, and they may differ and clash. Differences may manifest themselves in competition between schemata being applied simultaneously by different actors, or in temporal inconsistencies due to the influence of incompatible schemata at different points in the innovation. The degree to which key actors' schemata are consistent determines the degree of organization and direction in the innovation process. Of course, when actors' schemata are vague or unknown, actors with different implicit goals may join together and collaborate in pursuit of an illusory goal (Perrow 1981).

Thus, we argue that to understand how an innovation develops, it is necessary to identify the schemata that are used by actors and their degree of influence vis-à-vis one another. Hence the transactions and relationships among key actors must be taken into account. The extent of influence a given schema has depends on (1) how many key actors hold it, (2) their degree of resource control, and (3) the amount of time they spend performing different roles relative to the innovation. The innovation team's schema (or schemata) generally has great influence because team members are usually very close to the innovation and because they often control important resources.

But as shown in Chapter 7, key external resource controllers can also exert influence on the innovation's development. However, their direct influence generally is more intermittent than that of innovation team members because their attention is usually only partly focused on the innovation. Instead, resource controllers' long-term influence often stems from the alterations they make in the schemata of members of the innovation organization. Other key actors' schemata may also play a role in shaping the innovation, especially if their introduction is well timed—as illustrated by the interventions of policy entrepreneurs in the state educational reform innovation. The nature of the innovation organization, especially how it orchestrates interactions among key actors and transactions with the external environment, mediates the influence of external schemata on the innovation.

The balance of contributions by various schemata decides whether the innovation path is simple and coherent or complex and fragmented. If a single schema is accepted by all or if it is powerful enough to assert itself, its logic informs the developmental path; barring unexpected shocks, the path is coherent and clearly the product of a single plan. If, however, several schemata exert control, either simultaneously or at different times, the path is complicated and fragmented, with no clear logic. External shocks might fragment the developmental picture still further. To understand such paths, it would be necessary to decompose them into their constituent schemata and explain how the schemata interact, as done in Chapter 8.

The institutional context of the innovation also often plays a crucial role in the creation and alignment of schemata. Actors must take highly developed institutions into account in planning and managing innovations. When the institutional context is legally binding and legitimate, actors have little discretion on the developmental progression, once a choice is made to pursue it. In such cases, the institution "forces" consensus on actors' schemata; it serves as a common point of coorientation for managing the innovation. If the institutional context madates a particular sequence for approval and development, actors orchestrate the

innovation to follow that path. If the institutional context determines necessary requirements for a viable innovation, actors' schemata tend to agree on the final form of the innovation and what is needed to attain this, although they may differ about the sequence of steps that should guide the process. If a well-developed institutional context is lacking, schemata are likely to differ, and actors must find other means of attaining agreement and consistency.

Moreover, several MIRP studies point to another crucial institutional dynamic: The innovation process itself may change the institutional environment. Often, the innovation process regularizes and defines institutional environments, as Garud and Van de Ven show in Chapter 15. But innovation may also disrupt an environment, redefining its rules and organizations.

Key Contingency Propositions
Based on the structuring process just discussed, we now suggest propositions that indicate the conditions when various global models of development should apply. Evidence for these propositions is summarized in Table 20–2.

1. A historical motor holds only for innovations occurring in a highly developed institutional context that specifies a set of procedural rules to follow in developing an innovation.

Where natural laws, institutional requirements, or logical necessity dictate the sequence of development, the innovation process can be explained on the basis of programs or routines that often consist of simple unitary sequences of steps or phases of activities. As Table 20–2 indicates, all three innovations that could be explained with a historical motor occurred in the context of well-developed institutions: Roberts and King's educa-

TABLE 20–2. *Summary of Global and Local Models Used to Examine MIRP Studies, along with Scope Conditions.*

Study	Global Model	Local Model	Scope Conditions
Ch. 8, business creation: Van de Ven, Venkataraman, Garud, and Polley	Functional motor (accumulation): business unit tries to achieve resources and competence needed to take off and become self-sufficient from outside resource controllers	Emergent motor (enactment): success-failure loops determine degree of autonomy given to innovation organization	Internal and external firm markets Moderate novelty Long-run payoff Scarce resources High complexity Private sector
Ch. 9, state educational reform: Roberts and King	Historical motor (institutionally required): legislative steps	Functional motor (accumulation): policy entrepreneurs guide educational reform ideas into good currency	Legislative institution High novelty Long-run payoff High complexity Public sector
Ch. 10, NASA-3M transactions: Ring and Rands	Historical motor (logic): three sequential phases—negotiations, agreement, execution	Functional motor (bargaining): informal sensemaking, understanding, and committing	Legal institution High novelty Long-run payoff Low complexity Private sector
Ch. 11, mergers and acquisitions: Bastien	Functional motor (dependency and peer relations): adjustment of acquired firm to acquiring firm	Functional motor (dependency): learning model	Market institution Moderate novelty Intermediate payoff Low-Moderate complexity Private sector
Ch. 12, defense contracting: Scudder et al.	Functional motor (dependency and problem solving): adjustment of defense contractor to navy requirements	Historical motor (logical necessity): stimulus-response model	Restricted market High novelty Long-run payoff High complexity Private-public sector
Ch. 13, compound semiconductor development: Rappa	Emergent motor (evolutionary): gradual evolution of community of scientists and firms	None advanced	Low organization of community (institution) High novelty Moderate-long run payoff Low-moderate complexity Private-public sector

TABLE 20–2. *(continued)*

Study	Global Model	Local Model	Scope Conditions
Ch. 14, hybrid wheat development: Knudson and Ruttan	Historical motor (logic): cumulative synthesis model (usher)	Historical motor (biological and logical): steps to endogenous development	Scientific institution Moderate novelty Long-run payoff High complexity Private sector
Ch. 15, cochlear implant industry emergence: Garud and Van de Ven	Functional motor (accumulation): acquisition of requisite functions to sustain an industry	Emergent (enactment) of cooperative and competitive relations among organizations	Embryonic industry High novelty Long-run payoffs High complexity Private-public sector
Ch. 16, adoption of NRC safety regulations by power plants: Marcus and Weber	Historical motor (institutional requirements): Externally imposed Innovation	Emergent process (enactment): responses of power plants to NRC safety regulations	Structured institution Low novelty Long-run payoff (safety) High complexity Public-private sector
Ch. 17, adoption of site-based management: Lindquist and Mauriel	Functional adoption model: Yin's institutionalization model used as standard for evaluation	Functional motor: breadth versus depth adoption strategies to site-based management implementation	School system Low-medium novelty Immediate-term payoff Low complexity Public sector
Ch. 18, adoption of government strategic planning systems: Bryson and Roering	Functional motor (adoption): innovation unit tries to set up junctures for people to learn strategic planning process	Functional motor: adoption process managed by adoptors and facilitators	Structured institution Moderate novelty Medium-term payoff Small innovation Moderate complexity Public sector

tional reform had to follow a sequence necessary for legislative action; the 3M-NASA joint ventures studied by Ring and Rands occurred under the aegis of contract law; and Knudson and Ruttan's hybrid wheat strains were developed according to canons of scientific experimentation in a community of seed companies and agricultural experiment stations. In support of Ruttan and Hayami's (1984) endogenous theory of innovation, in each case institutional norms dictated a certain historical sequence of activities on the part of the innovators that satisfied the prerequisites for a successful product or outcome. For example, in the case of hybrid wheat, a series of research projects were necessary to generate the knowledge base necessary for the act of insight that could then be consolidated into a new product. Given how Western science works, no other sequence is possible. A historical motor also holds in small portions of other cases, notably the hearing health and therapeutic apheresis programs studied by Van de Ven et al. (Chapter 8). In both instances, a unitary sequence was obtained during the periods the programs pursued regulatory approval of their devices (that is, in a strong institutional context).

The nature of the historical motor varies according to the nature of the relevant institution. In the case of legislative action, the sequence of stages in the global model is the steps required for passing a law, and the local model is political negotiation. For hybrid wheat, the global model is a sequence of stages in the cumulation and emergence of scientific insight; the local model involves the development of connections in the scientific community and the exchange of knowledge and findings.

It is important to recognize that a historical motor does not regiment the innovation process, at least for the cases in MIRP studies. The authors of all three studies observed that there were many false starts and local departures from the sequence, as well as local recycling to earlier stages. However, over the long run, the historical necessity dominated the global process.

2. A functional motor holds when key innovation participants and investors are in agreement about the goals of an innovation, or the requisite qualities of the final product or end state of the innovation.

Agreement on the particular path to this product or end state is not necessarily required. For example, the various stakeholders may more or less agree on the qualities and infrastructure required to develop an effective therapeutic apheresis product but differ about how to get these qualities and infrastructure.

The level of agreement necessary for a global organizational motor to hold can obtain in at least two ways. First, and most common, institutional isomorphism (DiMaggio and Powell 1983; Granovetter 1986) of an innovation with its environment may define prerequisites, therefore creating consensus on the final form. As we noted above, the nature of the Western economy requires innovation organizations to meet certain requirements—lines of credit, accounting procedures, product development cycles. Any competent business person knows this and will plan the innovation so as to realize them. A second means is for the key actors to

come to an open agreement about requirements, perhaps renegotiating them along the way. The TAP innovation described in Chapter 8 provides an example in which this occurred.

Because a functional motor posits only an end-state or goal but no sequence, the developmental paths of innovations under this motor are often complicated and multifaceted. The five cases that reflected functional models of development (new business creation, planning processes; mergers and acquisitions; torpedo development, and industry emergence) all had the complex, branching paths of the "fireworks" model described by Schroeder et al. in Chapter 4. They exhibited branches that started and then terminated suddenly, recycling through previous stages and ramifying chains of problems and ideas. It is significant that these innovations initially began with a simple unitary sequence that subsequently diverged into complex divergent, parallel, and convergent progressions. Some of these developmental paths were conjunctive (that is, related through a division of labor among functions and interdependent alternative paths), but many appeared disjunctive (unrelated in any noticeable form of functional interdependence). Moreover, many paths that were perceived as being cumulative and conjunctive at one time were often reframed or rationalized as being disjunctive and independent at other times.

The degree of complexity of the developmental path depends on (1) the extent of agreement on the functional requirements for the innovation and (2) the extent of agreement on correct steps to take to reach these requirements. Even in cases where institutional requirements for innovation development are clear, there are likely to be different interpretations of these requirements. If the interpretations are more or less consistent, the innovation is developed by actors with a clear sense of the final end state. More often, as reported in Chapter 8, there are some inconsistencies that introduce "interference" into the developmental process. If nothing else, turnover and the introduction of new actors at various points is likely to diversify schemata. The same holds true for the steps needed to reach the final state. Whereas key actors may be in agreement that the product must provide revenue for the innovation unit, some may want the revenues from direct sales and others from licensing the product to another company. In general, the more differences there are on these means and functional ends of an innovation, the more complex the path, and the greater the probability of failure.

The innovation organization plays a particularly important role when a functional motor operates because it manages the internal-external transactions necessary to attain the final goal. In the previous section we defined five types of functional models based on internal-external relationships—accumulation, dependence, peer relations, problem solving, and fortification. At least three contingencies seem to govern which type of relationships emerge. First, is the innovation unit a well-established center of activity? If not, then accumulation or dependency relationships are necessary; in an accumulation process a center emerges and consolidates, while in a dependency relationship the organizational unit abdicates control. Second, what is the nature of the power balance between the innovation unit and its external stakeholders? If the internal unit is more powerful, then fortication or problem-solving relationships are possible. If the internal unit is weaker, then dependency or accumulation relationships occurs. If power is approximately equal, then peer relations and problem solving result. Third, is the innovation unit proactive or reactive? If it is mainly reactive, then dependency relations result; if mainly proactive, then accumulation or fortification; if a mixture, then problem solving.

These models can also be viewed as part of a temporal sequence. As stated before, typically an innovation unit does not have a well-established center of activity in its early period

of development. At first a dependency relationship exists, and the innovation unit tends to be largely reactive to external paries because it has relatively little control over the accomplishment of its final end goal. But over the course of its development, the unit, if it is successful, accumulates the resources and power for a well-established center of activity to emerge. If its relative power with external parties becomes approximately equal, then bargaining or problem solving result. If the unit gains dominance in its relationships with external parties, it can become proactive and engage problem solving and fortification of its position.

3. An emergent process motor applies when there is no strong institutional context and little agreement on an end state or on the procedural steps of an innovation.

In this case, many different schemata are likely to exist, and the interaction among key actors generates the developmental path. The interaction process determines the nature of this path. For example, if there are two schemata, a dialectical process may ensue that pits them against each other and then achieves a synthesis. In terms of the path, first one schema would hold (thesis), and then the second would attempt to "correct" for it for a period of time (antithesis), and then a hybrid schema may emerge (synthesis). Other process principles that could shape development are discussed above.

Basically, the emergent process model holds when conditions for the operation of historical and functional motors do not hold. In the MIRP studies the shift in the hearing health program from cochlear implants to hearing aids technology represents a case where an emergent process motor appeared to be at work. Because most innovations occur in at least partially organized institutional environments, functional motors are most frequent and historical motors are fairly common.

4. The greater the novelty of an innovation, the greater the probability that an emergent model of development will apply.

The lines of reasoning in the above three propositions suggest that innovation novelty and complexity may be two other key factors that may influence conditions when historical, functional, and emergent process models are likely to hold.

This is because when truly novel or new-to-the-world innovations are undertaken, an institutional infrastructure of procedural rules and resources often remains to be developed (rendering historical models inoperative), and innovation participants are sufficiently uncertain about the innovation to understand or agree on the functional means or ends of action until after development has neared completion.

Models of development are also influenced by an innovation's complexity, in terms of its size, number of stakeholders, and component parts. Where innovation processes are well understood and governed by institutional rules and procedures, complexity can be managed through historical and functional models of development. However, when institutional rules and functional ends of an innovation are vague, complexity triggers emergent processes of development even when innovation uncertainty is relatively low. This is because complexity provides multiple opportunities for partisan groups of stakeholders to exercise their competing schemata on an innovation's developmental progression.

METATHEORY SWITCHING RULES AMONG DEVELOPMENTAL MODELS

Having specified one set of contingencies when each type of developmental model is likely to apply, we now propose a more general set of switching rules for determining how and when different developmental theories might be used and linked together to explain an overall observed process of innovation from beginning to end. In the metatheory, these switching rules represent the fundamental parameters governing how and when to shift from one theoretical framework to another. They specify three general forms metatheoretical propositions might take.

The switching rules are based on an answer to the question: *Does a single model hold throughout the entire innovation process?* If the answer is yes, then models are related by *type variation.* That is, contingencies related to the nature of the innovation and its context determine that one model holds for certain types of innovations, a second for other types, a third for other types, and so forth. If the answer to the question is no, then models may be related by *temporal variation* or *spatial variation.* In the case of temporal variation, different models hold at different times during the innovation process, depending on the conditions. In the case of spatial variation different models may hold at the same time, but in different regions of the social system in which the innovation is developing. Each of these switching rules is now considered in more detail, along with examples based on the MIRP studies.

Type Variation

In the case of type variation, the relevant contingency factors that determine the switches between models pertain to the nature of the innovation or its context. Depending on the specific properties of the innovation and context, different explanatory mechanisms hold. The models subsumed by a metatheory proposition based on type variation do not interact with one another. Instead, different models are fit whole-cloth to different situations. The models are related by the scheme of contingency factors that capture the situational variations that create different innovation processes. In the case of type variation we have a stable of models, and we trot out the one that is best for a particular race.

The metatheory propositions developed in the preceding Section are based on type variation. One set of conditions favors a historical motor, another set a functional motor, and a third set an emergent process motor.

This is one type of metatheory proposition; however, the three motors may be related in two other ways, by temporal or by spatial variation. We found examples of both propositions in the MIRP studies.

Temporal Variation

For temporal variation propositions, two general types of contingency factors determine which model holds and switches between models—innovation context and temporal stage of the innovation itself. Often, forces in the innovation context mediate the transition from one model to another. For example, Tushman and Romanelli's (1985) punctuated equilibrium model proposes that changes in economic conditions might produce a shock that stimulates reorientation of a firm; this involves moving from an incremental, convergent model of development to a transformational divergent model.

Changes in the innovation and innovating organization may also stimulate a shift from one model to another. Utterback and Abernathy (1975) posit that products and the processes for producing them evolve through three stages: (1) uncoordinated process/product performance-maximizing strategies; (2) segmental process/sales-maximizing strategy; and (3) systemic process/cost-minimizing strategy. In each of the three stages the nature of the innovation process changes: the locus of innovation, the type of innovation likely to succeed, and the array of barriers to innovation shifts. As

the product develops, the immediate action model changes.

Whereas models are considered more or less independently of one another in type variation theories, in temporal variation theories, the shift or transition from one model to another must be explained. So we are interested not only in explaining innovation development but also in explaining how changes in the grounds of development come about. A temporal shift similar to that noted by Abernathy and Utterback (1978) is implied in several MIRP studies and can serve as an example here. In the biomedical innovation cases (Van de Ven et al., Chapter 8), the industry emergence case (Garud and Van de Ven, Chapter 15), and educational policy case (Roberts and King, Chapter 9) there is evidence of temporal variation.

In the initial stages, when the need for innovation is felt, the innovation concept is emerging, and only a small group supports the innovation, a "marketplace of ideas" prevails. Events on the *idea* and *people* tracks are the driving force and the innovation path is complex and disorderly. People are searching for the right idea, the right people (including an innovation champion), the right market, and resources. Insight and research are the source of novelty at this stage. Global development is dictated by an accumulation model whereby the innovators garner the necessary configuration of ideas, people, resources, and organization. For example, Dornblaser, Yin, and Van de Ven describe in Chapter 7 how local movement is accomplished via mutual adjustments between resource controllers and innovation managers subject to criteria for judging in-process innovation success. This negotiation is shaped by the self-reinforcing cycles of escalating commitments, hypervigilant learning, and mixed messages about innovation success and failure that are randomly ordered over time; cycles that often build or undermine the power and autonomy of the innovation unit.

As the innovation develops, the configurations of concepts, resources, and organization so painfully built up "take off" and acquire functional autonomy. At this point they exert an organizing force on innovation and a transition to a second model occurs. At this stage events in the *resource* and *context* tracks assume greater importance. As described in Chapters 7 and 8, those who control resources can exert great influence because once the innovation idea and team have stabilized, resource controllers manage the greatest uncertainty facing the innovation. This control may be even stronger if resources are generated by the innovation itself because it may set up a cycle whereby the innovation fuels its own growth in current directions. In this phase there is an emphasis on systematizing the innovation to stabilize it and protect it from external shocks. The logic of systems technology, regulatory approval, and product development take over. Because these are all sequential and systematic, they impose an immanent, historically necessary sequence on the global development of the innovation. The local model is still emergent negotiation, but it has also become more functional and operates through well-defined channels and within circumscribed boundaries that have emerged with history. These well-organized processes continue unless they are interrupted by a "shock" from the environment or from a major failure or disruption of resources. This shock may propel the innovation back to its initial looser stage, or it may stimulate reorganization and consolidation of the current stage, or it may destroy the project.

There is a transition from a global functional motor with a local emergent model to a global historical model driven by a local functional model. The locus of change also shifts from ideas and people to resources and context. With this shift comes increasing stabilization of the innovation. In Chapter 7, Dornblaser, Lin, and Van de Ven found this same pattern of stabilization in their analysis of MIRP longitudinal data.

Spatial Variation

Metatheory propositions based on spatial variation locate different models in different "regions" of the innovation. In this case, the key variables governing switches are the nature of different segments of innovation activity or different parts of the innovation organization. Whereas models in type variation theories may be considered more or less independent of one another, and models in temporal variation theories are consigned to separate time periods, we must consider possible interactions of models in spatial variation theories. In cases where different innovation activities are loosely coupled, models may function independently. However, there may also be cases in which two or more models interact simultaneously as the innovation develops.

The therapeutic apheresis and cochlear implant innovations (Van de Ven et al., Chapter 8) provide examples of this. In both innovations scientific research, regulatory approval, and resource development proceeded simultaneously, were managed by different (although overlapping) sets of people and appeared to be governed by different logics of development. Scientific research, following Knutson and Ruttan, is best modeled as a necessary progression of global cumulative synthesis and local endogenous development of ideas. The process of gaining regulatory approval evolves at the global level according to an institutionally prescribed historical motor, a necessary unitary sequence codified in regulators' flow charts of steps needed to win approval. Local action is driven by a process of organized negotiation subject to rules dictated by applied research procedures and the legal system. The global motor in resource development is a functional accumulation process by which the requisite materials are gathered. The local model in resource development also involves negotiation, but it is a much looser process. Although budget cycles may give resource development long-term direction, often it is a very complex process, characterized by divergent efforts to tap several resource bases. The resource development process is also subject to sudden shifts due to large fund inflows or sudden terminations or reallocations.

So, in three different arenas of these innovations, there are different global and local models. Each of these three processes has its own different internal logic. This logic is easy enough to discover if the process unfolds independently. However, consider what happens if the logics interact. As in the case of cochlear implants, we might find the regulatory approval process suddenly interrupted by either a funding cutoff that prevents completion of lab work or a new idea that renders the current idea obsolete and further development meaningless. Or as in the case of Qnetics, resource development, built on gradually evolving negotiation with investors or corporate management, might suddenly result in a major shift in funds that cause the project to cut down on research and rush "the best of what we have" to market. Alternatively, an accidental discovery during lab work preparatory to regulatory application might break the current stream of scientific work.

The result of these and other complex interactions is a shifting and tumultuous pattern of innovation processes. If we also include disruptions due to environmental shocks, the picture gets even more complex. The interactions of these models may explain the global "fireworks" model proposed by Schroeder et al. in Chapter 4. What is striking is that such a complex representation can be generated by the interaction of two or more relatively simple developmental processes. Van de Ven et al. observed how simple mechanisms could create complicated patterns in HHP, TAP, and Qnetics startup processes. Poole (1983; Poole and Doelger 1986; Poole and Roth 1988) has documented the same complexity resulting from two simple processes in group decisionmaking.

In conclusion, a repertoire of global and local models can be related in three types of propositions. Ultimately, an adequate metatheory of innovation is likely to incorporate type, temporal, and spatial propositions.

TESTING THE METATHEORY

Of course, any grounded theory that emerges out of the foregoing interaction between field observations of innovations and theories of development requires testing to evaluate its validity. As applied to the metatheory discussed above, there is a need to empirically test not only the fit of each explanatory model to the data but also the conditions and switching rules governing the appropriate applications of the models. This requires us to spell out operational criteria, conditions, and switching rules for each theory so that it is possible to determine how closely they fit observed sequences and processes. For example, each model in the metatheory can be operationalized as representing a particular template of how innovation processes unfold over time. Then an empirical comparison can be made between each model's development template and observed innovation progressions. In particular, these operational templates can incorporate a number of dimensions:

Each developmental model in the metatheory posits a particular developmental progression and its structural properties (a logical necessity model would predict a unitary sequence).

The developmental models imply causal or correlational relations among tracks (that the idea track drives the innovation process).

The conditions specify when each model should operate, and switching rules predict the nature of transitions among developmental models.

659

Some models specify the nature of temporal relationships between successive elements in the tracks (an accumulation model would imply additive or inclusive relations, with some events mediating others).

There is also often direct evidence that a given immediate action mechanism is operating (actors should be conscious of a politically motivated process if a political/negotiative (emergent) model holds).

Making predictions about these dimensions explicit in the operational template can provide a systematic way to test the fit between theory and data. Thus, through the normal scientific process of refutation, we can narrow the range of plausible and valid models of development.

CONCLUDING DISCUSSION

This chapter has taken the initial steps to develop a metatheory of innovation process. The need for such a metatheory is based on the contention that a single theory cannot encompass the complexity and diversity of process patterns observed across the MIRP innovation studies. The metatheory

Identifies and classifies models of development relevant to understanding innovation processes in terms of level of analysis (local or global) and type of theory (historical, functional, or emergent process motors); this provides a basic set of models that one can draw on to explain a particular observed innovation process;

Specifies situations or contingencies (requirements, agreement on means or ends, and innovation novelty, complexity, and power) when each type of developmental theory is likely to apply; and

Proposes three switching rules (type, temporal, and spatial rules) that may determine when to switch between models to explain innovation processes over time.

Since the metatheory was inductively derived, it requires future testing to substantiate its validity. Conceptually, however, the metatheory was found to provide a useful repertoire of developmental theories to explain innovation processes. Attention to historical, functional, and emergent developmental models at local and global levels provides a discriminating conceptual framework for comparing and contrasting different models that were used in previous chapters to explain innovation development.

It should be abundantly evident that the proposed metatheory is far from complete. The propositions we have suggested are the first steps toward a metatheory of innovation process. It is important to note that they are limited to the database provided by the MIRP studies and require further testing. They will no doubt be supplemented by additional propositions in the future. Furthermore, the propositions are presently stated at a rather high level of abstraction. For example, more detailed propositions are needed to know the conditions when ·variations of the three models of development operate. A truly general typology and theory of innovation will be a long time coming. But even though these first starts may seem disappointing five years hence, it is important to make them for they stimulate new patterns of reasoning that may promote further developments in building theories of theories.

REFERENCES

Abernathy, W.J., and J.M. Utterback. 1978. "Patterns of Industrial Innovation." *Technology Review* 80: 40–47.

Aldrich, H. 1979. *Organizations and Environments.* Englewood Cliffs, N.J.: Prentice-Hall.

Aristotle. 1941. *The Basic Works of Aristotle,* R. McKeon, ed. New York: Random House.

Cohen, M.D., J.G. March, and J.P. Olsen. 1972. "A Garbage Can Model of Organizational Choice." *Administrative Science Quarterly* 17: 1–25.

Collins, R. 1975. *Conflict Sociology.* New York: Academic Press.

DiMaggio, P., and W. Powell. 1983. "The Iron Cage Revisited: Institutional Isomorphosim and Collective Rationality in Organization Fields." *American Sociological Review* 35: 147–60.

Etzioni, A. 1963. "The Epigenesis of Political Communities at the International Level." *American Journal of Sociology* 68: 407–21.

Feidler, F.E. 1964. "A Contingency Model of Leadership Effectiveness." In L. Berkowitz, ed., *Advances in Experimental Social Psychology.* New York: Academic Press.

Flavell, J.H. 1972. "An Analysis of Cognitive-Developmental Sequences," *Genetic Psychology Monographs* 86: 279–350.

Giddens, A. 1979. *Central Problems in Social Theory.* Berkeley: University of California.

———. 1985. *The Constitution of Society.* Berkeley: University of California.

Granovetter, M. 1985. "Economic Action and Social Structure: The Problem of Embeddedness," *American Journal of Sociology* 91, 3: 481–508.

Hannan, M.T., and J. Freeman. 1977. "The Population Ecology of Organizations." *American Journal of Sociology* 82: 929–66.

House, R.J. 1971. "A Path Goal Theory of Leadership Effectiveness." *Administrative Science Quarterly* 16: 321–338.

Hull, D. 1974. *The Philosophy of Biological Science.* Englewood Cliffs, N.J.: Prentice-Hall.

Kaplan, A. 1964. *The Conduct of Inquiry.* New York: Intext.

Kohlberg, L. 1969. "Stage and Sequence: The Cognitive-Developmental Approach to Socialization." In D.A. Goslin, ed., *Handbook of Socialization Theory and Research,* pp. 347–480. San Francisco: Rand-McNally.

Lindblom, C.E. 1959. "The Science of 'Muddling Through,'" *Public Administration Review, 19:* 78–88.

March, J. 1981. "Decisions in Organizations and Theories of Choice." In A. Van de Ven and W.F. Joyce, eds., *Perspectives on Organization Design and Behavior,* pp. 205–245. Cambridge, Mass.: Ballinger.

March, J.G., and J.P. Olson. 1976. *Ambiguity and Choice in Organizations.* Bergen, Norway: Universitetsforlaget.

Masuch, M. 1985. "Vicious Circles in Organizations." *Administrative Science Quarterly* 30: 14–33.

Nisbet, R.A. 1970. "Developmentalism: A Critical Analysis." In J. McKinney and E. Tiryakin, eds., *Theoretical Sociology: Perspectives and Developments,* pp. 167–204. New York: Meredith.

Pelz, D.C. 1985. "Innovation Complexity and the Sequence of Innovating Stages." *Knowledge: Creation, Diffusion, Utilization* 6: 261–91.

Plato. 1969. "Phaedrus." Translated by R. Hackforth. In E. Hamilton and H. Cairns, eds., *The Collected Dialogues of Plato.* Princeton, N.J.: Princeton University.

Perrow, C. 1986. *Complex Organizations: A Critical Perspective.* New York: Wiley.

Poole, M.S. 1983. "Decision Development in Small Groups, III: A Multiple Sequence Model of Group Decision Development." *Communication Monographs* 50: 321–41.

Poole, M.S., and J.A. Doelger. 1986. "Developmental Processes in Group Decision-Making." In R.Y. Hirokawa and M.S. Poole, eds., *Communication and Group Decision-Making.* pp. 35–63. Beverly Hills, Calif.: Sage.

Poole, M.S., and J. Roth. 1988. "Test of a Contingency Model of Group Decision Development." University of Minnesota: Department of Speech and Communication.

Prigogine, I., and I. Stengers. 1984. *Order Out of Chaos.* New York: Bantam.

Reigel, K.F. 1975. "From Traits to Equilibrium: Toward a Developmental Dialectics." In J. Cole and W.J. Arnold, eds., *Nebraska Symposium on Motivation,* Vol. 23. Lincoln: University of Nebraska.

Ruttan, V.W. 1959. "Usher and Shumpeter on Invention, Innovation, and Technological Change." *Quarterly Journal of Economics* 73: 596–606.

Ruttan, V., and Y. Hayami. 1984. "Toward a Theory of Induced Institutional Innovation." *Journal of Development Studies* 23 (July): 203–23.

Skinner, B.F. 1938. *The Behavior of Organisms: An Experimental Analysis.* New York: Appelton-Century-Crofts.

Suppe, F. 1977. "Introduction." In F. Suppe, ed., *The Structure of Scientific Theories,* pp. 3–232. Urbana: University of Illinois.

Thompson, J.O. 1967. *Organizations in Action.* New York: McGraw-Hill.

Tushman, M., and E. Romanelli. 1985. "Organizational Evolution: A Metamorphosis Model of Convergence and Reorientation." In B. Staw and L. Cummings, eds., *Research in Organizational Behavior,* Vol. 7, pp. 171–222. Greenwich, Conn.: JAI.

Usher, A.P. 1954. *A History of Mechanical Invention.* Cambridge, Mass.: Harvard University.

Utterback, J.M., and W.J. Abernathy. 1975. "A Dynamic Model of Process and Product Innovation." *OMEGA* 3: 639–656.

Van de Ven, A.H., and R. Drazin. 1985. "The Concept of Fit in Contingency Theory." In B.M. Staw and L.L. Cummings, eds., *Research in Organizational Behavior,* Vol. 7, pp. 333–65. Greenwich, Conn.: JAI.

Van de Ven, A.H., and M.S. Poole. 1988. "Paradoxical Requirements for a Theory of Organizational Change." In R. Quinn and K. Cameron, eds., *Paradox and Transformation: Toward a Theory of Change in Organization and Management.* Cambridge, Mass.: Ballinger.

Vanden Daele, L. 1969. "Qualitative Models in Developmental Analysis." *Developmental Psychology* 1: 303–310.

Weick, K. 1979. *The Social Psychology of Organizing.* Reading, Mass.: Addison-Wesley.

Whyte, W.F. 1982. "Social Inventions for Solving Human Problems." *American Sociological Review* 47: 1–13.

SUGGESTIONS FOR MANAGING THE INNOVATION JOURNEY

Harold L. Angle

Andrew H. Van de Ven

In Chapter 1, we stated that the applied purpose for undertaking the Minnesota Innovation Research Program (MIRP) is to provide the innovation manager a road map that indicates how and why the innovation journey unfolds, and what paths are likely to lead to success or failure. Such a road map is needed because, heretofore, very few studies have directly examined the innovation process in real time. As a result, very few empirically substantiated statements are available about how and why the innovation journey emerges, develops, proceeds, and terminates over time. Yet an appreciation of the temporal sequence of events along this journey is fundamental to the management of innovation.

Based on the findings presented in the previous chapters, this final chapter provides the practitioner with a description of the major junctures encountered along an innovation journey, as well as a set of recommendations for maneuvering them, circumventing roadblocks,

and generally transiting the essentially unpaved territories along the way. In doing this, we attempt to take a balanced perspective on both the local innovation manager and the participants directly engaged in developing a particular innovation, and the investors or top managers in the organization housing the innovation. Each perspective is useful for understanding the innovation process. More importantly, as we shall see, when innovation managers and resource controllers communicate accurately with each other, we believe that they can significantly improve their joint odds of successfully managing the innovation journey.

Of course, innovation can be accomplished in a number of different ways and the journey can unfold along many different routes, as is amply demonstrated by the MIRP studies described in earlier chapters. Along with observing typical innovation pathways, therefore, we also indicate some of the alternative routes innovations might take. Many key elements of the innovation process, however, were found to be quite similar across the highly diverse set of technological, product, process, and administrative innovations studied by the MIRP. These common patterns justify discussing a generic innovation journey in this chapter.

By generic we mean that, for the most part, we focus on an innovation that consists of a purposeful, concentrated effort to develop and implement a novel idea that is of substantial technical, organizational, and market uncertainty; that entails a collective effort of considerable duration; and that requires greater resources than are held by the people undertaking the effort. This definition includes the forms of innovation that most managers and venture capitalists typically invest in and hope will produce a useful result—in other words, those that will be profitable, constructive, or will solve a problem. But this definition excludes small, quick, incremental, lone-worker innovations. It also eliminates innovations that emerge largely by chance, accident, or afterthought—although many of these elements may be contained in our description of how generic innovations develop.

This chapter is organized much like a road atlas. First we provide an overview map summarizing the key processes and junctures commonly observed in the diverse innovations studied by MIRP. Then we examine the details of each process juncture, explaining its occurrence and making some suggestions for managing it along the innovation journey. These suggestions are speculative because they have not been tested. However, they represent an attempt to draw prudent practical inferences from the knowledge gained in the MIRP and other studies of the management of innovation.

At the outset, we emphasize that we do not propose to offer a set of management principles that, if followed, will ensure innovation success. On the contrary, our experience in MIRP has convinced us that many factors that influence the success of an innovation are not within the control of the innovators. Accordingly, rather than offering a formula for successful organizational innovation, we instead suggest some approaches that we believe will increase the *likelihood* of success.

Finally, before we map the innovation journey, we must address a challenge frequently raised by innovation managers against our efforts to study the management of innovation. Various managers often stated

> Our innovation is unique. What you learn from us will not transfer to other innovations . . . You will not be able to make many generalizations that are valid or useful for managing innovation. . . . Innovation management is an art or craft requiring great intuition; it cannot be reduced to scientific analysis and principles.

These statements, of course, constitute a major challenge not only to the utility of MIRP, but also to social science research in general.

Our response to this challenge is quite simple and direct. The preceding chapters have described the detailed methods the MIRP researchers have used to make valid observations about the innovation process over time in a highly diverse set of technological, product, organizational, and administrative innovations. When comparing these observations across the MIRP studies, these chapters have reported many similarities and differences across the innovations. In this chapter we focus on the similarities in order to draw three conclusions: (1) the highly diverse innovations studied by MIRP are not unique—they share many common processes; (2) these common processes are important, because they are concerned with how people and organizations behave in related circumstances; and (3) these processes are useful for drawing some practical and concrete inferences that may improve the art or craft of innovation management.

Most artists or craftspeople do not

become masters of their professions by intuition alone. They typically begin by taking introductory courses, rigorously devoting themselves to study, and applying the basic principles and techniques of their professions. As the Nobel laureate Herbert Simon observed, "It usually takes at least ten years of intensive study and practice to become world class at anything."

We do not pretend to suggest that the MIRP findings will generate master innovation managers. Instead, we humbly submit that they may become the starting principles for an Innovation Management 101 course. If the reader concludes that the principles presented below are not new and were in fact known all along, then at least we have succeeded in puncturing the "uniqueness mystique," which represents a formidable barrier in transferring and applying management knowledge from one innovation to another.

AN OVERVIEW OF THE INNOVATION JOURNEY

Figure 21–1 illustrates the key processes or junctures commonly found in the MIRP studies of innovation. The figure expands upon the "fireworks" model that was introduced by Schroeder et al. in Chapter 4 and refered to frequently throughout the book. Imagine ongoing operations of an organization proceeding in the general direction of point A. An innovation is launched that proceeds in the new direction of point B. The overall innovation process is partitioned into three temporal periods: (1) an initiation period, in which events occur that set the stage for launching efforts to develop an innovation; (2) a developmental period, in which concentrated efforts are undertaken to transform the innovative idea into a concrete reality; and (3) an implementation or termination period, in which the innovation

either is adopted and institutionalized as an ongoing program, product, or business, or is terminated and abandoned. The process highlights of the generic innovation journey are outlined below. They are numbered to correspond to Figure 21–1. Of course, components of this innovation journey are not the same in all innovations. As discussed later, the key process elements are expected to be more pronounced for innovations of greater novelty, size, and duration.

Initiation Period

1. Innovations are not initiated on the spur of the moment, nor by a single dramatic incident, nor by a single entrepreneur. In most MIRP studies there was an extended gestation period lasting several years during which seemingly coincidental events occurred that set the stage for the initiation of innovations.

2. Concentrated efforts to initiate innovations are triggered by "shocks," from sources either internal or external to the organization, because when people reach a threshold of dissatisfaction with existing conditions, they initiate action to resolve their dissatisfaction.

3. Plans are developed and submitted to resource controllers to obtain the resources needed to launch innovation development. In most cases, the plans serve more as "sales vehicles" than as realistic scenarios of capabilities and likely hurdles ahead as the innovation develops.

The Developmental Period

4. Once developmental activities begin, the initial innovative idea soon proliferates into several ideas, which makes the innovation journey complex to manage. More

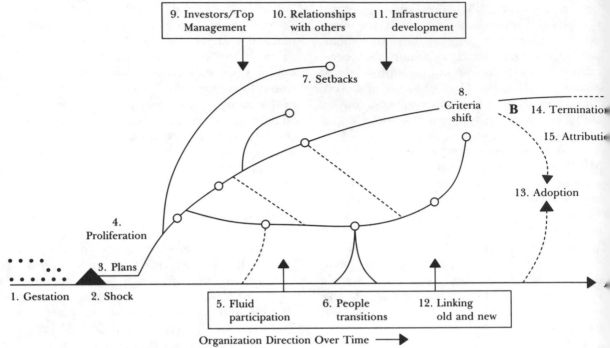

FIGURE 21–1. *Key Components of the Innovation Journey*

specifically, after the onset of a simple unitary progression of activity to develop an innovative idea, the process diverges into multiple, parallel, and interdependent paths of activities.

5.   Typically, there is part-time involvement and high turnover among innovation personnel, who (while technically competent) often lack previous experience in developing an innovation. Many of these transfers are a normal part of personnel job mobility and promotion processes. However, most of the participants involved in the initiation of an innovation are no longer involved when developmental work ends. This makes it difficult to maintain continuity and momentum and to evolve an organizational memory of innovation developmental activities.

6.   Innovation participants often experience euphoria in the beginning, frustration and

pain in the middle period, and closure at the end of the innovation journey. These changing human emotions represent some of the most gut-wrenching experiences for innovation participants and managers.

7.   Along the innovation journey, setbacks and mistakes are frequently encountered, either because plans go awry or unanticipated environmental events occur which significantly alter the ground assumptions of the innovation. As setbacks occur, resource and development timelines diverge. Initially, resource and schedule adjustments are made and provide a grace period for adapting the innovation. But with time, unattended problems often snowball into vicious cycles. Because of learning disabilities, these vicious cycles are seldom broken without outside intervention.

8. To compound the problems, criteria of success and failure shift over time, differ between resource controllers and innovation managers, and diverge in opposite directions, often triggering power struggles between insiders and outsiders.

9. Investors or top management are often involved throughout the process and play four different roles: sponsor, critic, mentor, and institutional leader.

10. Transactions or relationships initially established with other units often produce unintended consequences, and they are frequently renegotiated. These unintended consequences include: escalating risk brought on by entering into leveraged transactions with other units; cooperators who become competitors; links with scientific/technical centers that "lock in" technological alternatives; "cozy" relations that produce "groupthink"; partnerships and joint ventures that produce "hung juries"; and mergers and acquisitions that result in personnel desertions.

11. In addition to specific innovation development, managers are often involved with competitors, trade associations, and government agencies to create an industry or community infrastructure that supports technological and product innovations. Industry "team playing" minimizes first-mover costs because there are usually not one but many first movers in developing a technology or product.

Implementation/Termination Period

12. Implementation of "home-grown" innovations often occurs throughout the developmental period by linking and integrating the "new" with the "old," as opposed to substituting, transforming, or replacing the old with the new.

13. The adoption of innovations developed elsewhere is facilitated when (1) the innovation is modified to fit the local situation, (2) top management is extensively involved and committed, and (3) process facilitators help people understand and apply the innovation.

14. Innovations stop when implemented or when resources run out. This is obvious, but it explains much of the impression management games played between innovation managers and resource controllers.

15. Managers make attributions about innovation success or failure; these attributions are often misdirected but largely influence future organizational actions, as well as the careers of innovation participants.

We now examine the details of each of these key junctures in the innovation journey and make some suggestions for dealing with each one.

INITIATION PERIOD

Gestation

Innovations are not initiated on the spur of the moment, nor by a single dramatic incident, nor by a single entrepreneur. Chapters 8 to 15 show that in all the innovations studied there was an extended gestation period, often lasting three or more years, in which entrepreneurs and their organizations engaged in a variety of activities that set the stage for initiating an innovation. Most events during the gestation period were not intentionally directed toward starting an innovation. Some events triggered recognition of the need for innovation (for example, stem rust

667

blight triggering a need for improved wheat breeding), while others generated awareness of the technological feasibility of an innovation (for example, the discovery of cytoplasmic male sterility, which made hybrid wheat breeding possible). "Technology-push" and "demand-pull" events such as these often launched entrepreneurs on courses of action that, by chance, intersected with the independent actions of others. These intersections provided occasions for entrepreneurs to recognize and access new opportunities and potential resources. Where these occasions were exploited, the actors adapted their independent courses of action and undertook concerted efforts to initiate an innovation.

While the role of chance has been under-emphasized in most work on innovation, the MIRP studies clearly show how chance affects the initiation and the subsequent course of innovation development. The sheer volume of initiatives undertaken by a large number of interacting people increases the probability of stimulating innovation. The findings also reinforce the bias-for-action principle of Peters and Waterman (1982). Thus, Louis Pasteur's adage "Chance favors the prepared mind" best captures the process that sets the stage for innovation.

The important practical question then becomes, what can organizations do to increase their preparedness to capitalize on the chance of innovation? In Chapter 5, Angle provides a core suggestion for dealing with this question: *Structure the organization's context to enable and motivate innovative behavior.* As the term implies, the immediate context for most innovations is the organization within which the innovation takes place. The organization has a number of attributes, such as structure, systems, practices and culture, that increase the likelihood that innovation ideas will surface and will be developed and nurtured toward realization. Likewise, the organization is the most direct source of material, financial, and other types of resources

needed to get the innovation off and running. Finally, organizations are complex social systems that provide templates, so to speak, for playing out many distinctive roles that are important to the initiation of an innovation.

With respect to structure, there are several features that will likely have an impact on the gestation of innovative activities. Chapter 8 reported that organization structural arrangements affect the number of potential stimulants for innovation. The more complex and differentiated the organization, and the easier it is to cross boundaries, the greater are the potential number of sources of innovative ideas.

However, with increasing organizational size and complexity come segmentation (Kanter 1983) and bureaucratic procedures, which often constrain innovation unless special systems are put in place to motivate and enable innovative behavior. Key motivating factors include providing a balance of intrinsic and extrinsic rewards for innovative behaviors. Pay, in itself, is a relatively weak motivator for innovation; it more often serves as a proxy for recognition. Individualized rewards tend to increase idea generation and radical innovations, whereas group rewards tend to increase innovation implementation and incremental innovations.

Angle emphasizes, however, that the presence of motivating factors, by themselves, will not stimulate innovative behavior. The organization must also structure a context that enables innovation to happen. These enabling conditions include

resources for innovation;
frequent communications across departmental lines and among people with dissimilar viewpoints;
moderate environmental uncertainty and mechanisms for focusing attention on changing conditions;
cohesive work groups with open conflict resolution mechanisms that integrate

creative personalities into the mainstream;

structures that provide access to innovation role models and mentors;

moderate personnel turnover; and

psychological contracts that legitimate and solicit spontaneous innovative behavior.

We have often observed top managers exhorting individuals in their organizations to be more innovative in their work. This action is often counterproductive because it is the organization's context that largely enables and motivates individuals to be innovative. People generally have the potential to be creative and innovative. The actualization of this potential turns on whether management develops an organizational context that not only enables but also motivates individuals to innovate.

"Shocks" Trigger Innovation

As discussed in Chapter 4, concentrated efforts to initiate innovations were often triggered by "shocks" from sources either internal or external to the organization, because when people reach a threshold of dissatisfaction with existing conditions, they initiate action to resolve their dissatisfaction. As Johnson and Rice (1987) put it, organizations do not generally *search* for innovations but instead respond to perceived performance gaps.

The problem is that because human beings are highly adaptable, they are able to accommodate gradually changing conditions (Helson 1964) and therefore fail to notice that conditions have reached a point where they signal the appropriateness (through opportunity or threat) of a change. As a consequence, Van de Ven (1986) observes that people do not move into action to correct their situations, which over time may become deplorable. Opportunities for innovation are either not recognized or not accepted as important enough to motivate innovative action.

With regard to perception, people tend to develop cognitive routines or habits which desensitize them to novel events. Ironically, this habit-bound perception is particularly prevalent in contexts where the people are most competent. As Van de Ven put it, "what they do most is what they think about least" (1986).

People's reluctance to accept change is equally relevant. Change is often threatening because of the possibility that coping mechanisms that were at least adequate under the old situation may no longer suffice. People who were "winners" may now be lucky to break even. Adding to the relatively passive blinders that people wear because of habit or inattention are the active blinders stemming from such defense mechanisms as denial. So for two separate but related reasons, it may be difficult for people to notice the sort of change that should by all rights stimulate them to innovate. The management of attention is a central problem not only in the beginning but also throughout the innovation journey (Van de Ven 1986).

What can organizations do? While there is probably no panacea for this problem, in Chapter 5 Angle suggests that *mechanisms can be put into place for redirecting and jostling the attention of organizational members so that subtle changes and needs will be noticed.* For example, Richard Normann (1985) observed that not only are well-managed companies close to their customers, as Peters and Waterman (1982) suggest, but also they search out and focus on their *most demanding customers.* Empirically, von Hippel (1981) has shown that ideas for most new product innovations come from customers. So also has Utterback (1971) found that about 75 percent of the ideas used to develop product innovations came from outside of the organization. Being exposed face-to-face to demanding customers, to people at the cutting edge of technology, and to consultants increases the likelihood that the action thresholds of organizational participants will trigger them to pay attention to changing technological, community, and customer needs. In general, we suggest that *placing people in direct personal confrontations with sources of problems and opportunities is needed to reach the threshold of concern and appreciation required to motivate most people to act* (Van de Ven 1986).

A good example comes from Van de Ven's (1980a-c) longitudinal study of the creation of Texas child care organizations. For years, a county judge in the Texas Panhandle was aware that his county was one of the most socially and economically deprived in the state: 73 percent of the population was rural, 33 percent earned income below the poverty level, the median income was only $5600, the infant death rate was 37 per 1000 children under one year of age, and there was only one physician for every 2100 citizens in the county. The county, however, continued to allocate most of its budget to building roads and buying bulldozers. The judge recalled his personal confrontation with the problem:

One morning the county sheriff came into my office and said there was a group of children that weren't in school, but ought to be. I told him to bring them in with their families and we would talk. The next day he showed up with several children and families, and we discussed the situation with them.

The next day the sheriff and I went to the north end of town and stopped at a little house, knocked several times, and nothing happened. We opened the door, went in, and saw a red-hot wood burning stove in the middle of the room. We found four children staying by themselves. The oldest couldn't have been over four or five years. They were on an old dirty couch with covers around them and they were sick. The house was a mess. We were there for quite some time when finally the mother returned to bring some food for the children. The father was working out of town driving a tractor, making $.75 an hour. There were seven children in this family and both parents were working to support them. What I actually had done the day before was send one of the older children to school who had stayed to watch the children while their parents were at work.

When I looked at this situation, it was a little frightening to think I had sent their babysitter back to school and left these children in this house with a hot fire burning. So, I approached my commissioners and told them what we saw. We went looking for a day-care center and found there wasn't one in our small town. We spoke to state agencies, who confirmed this. Finally, we were advised to start a Community Action Program to help solve this problem, which we immediately did.

Today, this Community Action Program operates six day-care centers in the area, serving 215 children daily.

Shocks can also be purposely orchestrated to unfreeze a social system. In Chapter 9, Roberts and King describe how shocks were produced by policy entrepreneurs who helped create the perception that a crisis of education existed, thereby prompting the system to act. Ideas calling for innovation came from multiple sources. Problems were catalogued and framed to justify a need for change by outsiders; internal Department of Education bureaucrats would not voluntarily initiate educational reform.

Planning and Funding

A common juncture, signaling the end of the initiation period and the start of the developmental period, is when plans are developed and submitted to resource controllers to obtain the resources needed to launch the innovation. Most of the MIRP studies found that the plans served more as sales vehicles than realistic scenarios of capabilities and likely hurdles that would be encountered in developing the innovation. These plans served at least two "selling" functions: (1) to persuade resource controllers that the innovation idea was worth investing in and supporting, and (2) to construct a vision and a perceived need for the innovation that motivated collective action. Although such plans are functional for these purposes, Chapter 8 discussed how funding-based planning also implies overly optimistic performance plans, self-censored "lowballing" of resource requirements, and innovators who become accountable for unrealistic targets by resource controllers. These consequences precipitate the onset of a vicious cycle of impression management and "sugar-coated" administrative reviews throughout the developmental period. Indeed, we estimate that, at the outset, the typical innovation project was underfunded by about 100 percent and that the typical development schedule required at least two-thirds more time than projected.

We believe that there are two reasons for these grossly miscalculated budgets and schedules—one external, the other internal. Consider the external reason first. The innovator, in attempting to garner resource support for the innovation, is motivated to make the case for support as attractive as possible. Accordingly, cost estimates are usually optimistic, based on best-case scenarios. Whether consciously or not, the innovator is aware that, once committed to the project, funding sources are apt to add additional funding to rescue a project in trouble. It is not unlikely that the total funding (initial support plus rescue funding) could far exceed what would have been an acceptable amount up front. In a sense, it appears that even managers without much formal training in psychology are good enough street psychologists to know that the power of behavioral commitment is indeed strong (Staw and Ross 1987). Once resource allocators have invested in the project, there is a high probability that they will reinvest to save their initial outlay.

At the internal level, we suspect that there is another reason initial estimates of innovation cost are often understated. Self-deception may be operative, since innovators become

personally invested in their innovation ideas and rationalize away many of the disquieting signals that things may not work out (Aronson 1973). Part of this process includes looking at cost-benefit analyses through rose-colored glasses. Indeed, this rationalization mechanism may lie at the heart of many of the lowball budgets that are presented to resource suppliers early in the innovation process.

An obvious recommendation is that *innovation managers should develop realistic scenarios of likely courses of action.* Given the present system of funding-based plans, this recommendation is not apt to work unless (1) the process of realistic scenario-building is uncoupled from funding requests, and (2) the typically overconfident and inexperienced entrepreneurs are coached by a forceful mentor who commands respect and experience.

One might argue that, given the inherent uncertainties of innovation, it is neither possible or practical to forecast the scenarios likely to unfold along the innovation journey; real-time trial-and-error learning is a more realistic recommendation. Such an argument misses our central point, however, that the discipline of planning is prerequisite to trial-and-error learning. As former President Eisenhower once aptly stated, "Plans mean nothing; planning is everything."

Here, planning means developing and mentally rehearsing a repertoire of strategies for surmounting likely obstacles in the future. Without honing these planning skills, innovators not only have few "fall back" options available to them, but also lack the procedures for developing them when uncertainties and setbacks arise during the developmental period. Thus, the key point of our recommendation is not so much one of having an arsenal of strategies prior to launching development; instead it is developing and honing the skills of strategy development, which in turn are much the same as those required for adaptive learning. We believe this point explains why we have found in

671

our studies of organizational creation that the greater the efforts taken in planning new organizations before startup, the shorter the development time and the more successful the organizations were in their first three years after startup (Van de Ven 1980c; Van de Ven, Hudson, and Schroeder 1984).

DEVELOPMENTAL PERIOD

Proliferation

In Chapter 4, Schroeder et al. reported that in all the innovations studied by the MIRP, shortly after developmental activities began, the innovation journey became complex to manage, and the initially simple innovation process proliferated along diverse pathways. Specifically, after the onset of a simple stream of activity to develop an innovative idea, the process quickly diverged into multiple and parallel paths of developmental activities. Some of these activities were related by a division of labor among functions to develop a given alternative. Many others appeared to be unrelated and competing activities pursued by different people or organizational units. As a consequence, after a short initial period of simple unitary activities, the management of innovation soon lapses into an effort to direct controlled chaos (Quinn 1980). This mushrooming of activities over time appears to be a pervasive but little understood characteristic of the developmental process. It presents for the innovation manager the problem of "trying to grow an oak tree when there are inexorable pressures to grow a bramble bush."

In Chapter 20, Poole and Van de Ven explain this complex developmental pattern. They suggest that when we confront a complex pattern, we should look for several interacting mechanisms at work. Processes that appear complex on the surface are often the product of several simple interacting processes. Complicated developmental paths may result from pursuing alternative processes in different parts of the innovation. In particular, Poole and Van de Ven observe that historical, functional, and emergent models of development often govern different innovation activities and paths, as follows:

Activities governed by historical or institutional rules tend to follow a simple unitary sequence of stages, as prescribed.

Activities governed by a functional model tend to diverge into multiple interdependent paths and then converge into an overall cumulative sequence (much like a PERT chart).

Activities governed by an emergent process model (where institutional rules do not prevail and where significant conflict exists over innovation goals or means) tend to be divergent, quasi-independent, competitive, and not cumulative.

In most of the innovations, two or more of these models were operative simultaneously in different parts of the innovation. For example, in the development of the cochlear implant, regulatory approval, business creation, and scientific research activities proceeded simultaneously, were managed by different (although overlapping) managers, and appeared to be governed by different logics. The decision to develop the device for commercial release in the United States implied that each device must go through an institutionally prescribed sequence of hurdles to gain FDA regulatory approval. A functional model appeared to govern the developmental sequence of creating a business out of the cochlear implant technology. To become a self-sustaining business, work began with the hiring of diverse functional competencies (R&D, manufacturing, marketing, and so forth), each performing parallel and interdependent tasks to develop a first product. As these functions completed their work on the first product, they were redeployed to initiate related tasks on the next generation in a family of products that was envisioned to provide a sustainable economic business entity. Finally, an emergent model largely governed the process of scientific work (except that connected with regulatory approvals of devices). Different researchers pursued alternative technologies (single versus multiple channels) and linked up with different technology centers, which competed to become the dominant technology. This emergent process was limited only by scarce resources and curtailed by a top management decision to support only one technological route.

As Poole and Van de Ven pointed out, each of these three processes has its own internal logic, which is easy enough to discover if the process unfolds independently. However, confusion arises when these logics interact. For example, in the cochlear implant case, the regulatory approval process may suddenly be interrupted by a funding cutoff that prevents completion of lab work on a key device in the product line, or scientific evidence emerges indicating the superiority of the discontinued technological route relative to the technology chosen for further development and upon which the business creation effort is based. The result of these and other complex interactions is a shifting and tumultuous progression. This picture gets even more complex if we include disruptions from environmental events. These interactions may explain the global fireworks model illustrated in Figure 21–1 and in Chapter 4. What is impressive is that such complexity can be generated by the interaction of a few relatively simple developmental processes. As Van de Ven et al. describe in Chapter 8, a few substitutions of one simple course of action by another equally simple strategy often create exceedingly complicated and intractable action cycles.

So what can management do to cope with this proliferating complexity? Like Peters and Waterman (1982), we recommend KISS (Keep it Simple, Stupid!). We believe that much of this complexity is the result of innovation units striving to achieve too much too soon, and thereby becoming embroiled in many activities that are not necessary or essential to develop an innovation. For example, one immediate way to decrease complexity is to focus first on developing and testing the core innovation idea before efforts are undertaken to create a self-sustaining business or program. As we have seen, the impatient quest to "leapfrog" into an overall program or business spawns proliferation. It also often delays developing and market testing the core innovation idea, because innovation participants tend to focus their efforts not so much on the immediate tasks needed to develop the basic idea but more on pre-engineering systems that may be needed to develop families or generations of the innovation idea. In the process of doing so, basic problems inherent to the core idea are often masked and go unquestioned until setbacks arise (as discussed below). Administrative reviews are periodically conducted to evaluate these developing systems but tend to be poor substitutes for the acid test of the market. Restricting and simplifying developmental activities to the core innovation idea decreases cost and time to implementation. Moreover, it tends to decrease the costly mistakes of investments in innovations which do not meet the market test. Small mistakes are more tolerable and correctable than large costly mistakes.

Part-Time Involvement and Turnover

Another common process observed across the innovations studied was part-time involvement and high turnover among innovation personnel, who (while technically competent) typically lacked previous experience in developing an innovation. Most of these personnel transfers were a normal part of job mobility and promotion processes. However, most of the participants involved at the beginning of developmental work were no longer involved when it ended. Thus, contrary to the common impression that an innovation team consists of an entrepreneur who works with a fixed set of full-time people to develop an innovative idea, the staffing pattern is in fact much more temporary and fluid. It reflects the organized anarchy discussed by Cohen, March, and Olson (1972), where many people fluidly engage and disengage in the innovation process over time, as their interests and commitments dictate. This makes it difficult to maintain continuity and momentum and to develop an organizational memory of innovation developmental activities.

Where possible, we suggest it is important to *stabilize the core innovation team.* Part-time involvement usually means that participants have joint work appointments and must serve multiple masters. Given that the work demands of each job typically exceed the time available, part-time participants were often observed to be distracted and stressed, and tended to cope by "stealing from Peter to pay Paul." In many cases the innovation was "Peter," except for relatively infrequent and short intervals of peak innovation experiences or work demands. Thus, innovations were often observed to be the losers from the practice of "skunk working" (which has been the much-heralded technique to facilitate innovation by Peters and Waterman [1982]). Being an uncertain and insecure process, innovations can seldom compete with the security of employment and short-run production demands of performing clearly known routines in most permanent jobs. This situation might be even worse, except that these structural incentives of non-innovation jobs are partially counterbalanced by the greater intrinsic motivation and satisfaction most people derive from innovative tasks.

Lack of experience suggests naive planning and a dearth of fall-back positions when unanticipated events arise. Experience provides a grounding for comparing developmental processes and a repertoire of "old friends" (Simon 1947) to rely upon to diagnose and respond to situations.

Personnel turnover serves a number of useful functions (Dalton and Todor 1979; Staw 1980). Turnover contributes fresh new perspectives and competencies to an innovation team as needed to address critical problems or challenges as they arise. Also, appointment and replacements of innovation managers are key levers used by resource controllers and top management to exercise their direction and control of an innovation. Finally, one CEO observed that "job demands often outpace the people," as the level of managerial skills believed necessary to direct innovations from development to business operations often grow more rapidly than the rate at which many managers are capable of acquiring them. As a consequence, many entrepreneurs are replaced by professional managers because the former often flounder when developing the innovation into a self-sustaining business.

We have also observed, however, that high turnover creates significant continuity problems for innovations. Each person leaving takes away vital information. Not only are secrets then subject to compromise outside the group, but vital knowledge that has not been systematically recorded is then lost to the group. New persons entering the network deflect the attention of others from production to process costs, as the newcomers are "brought up to speed" and new role relationships and norms are negotiated. It is as if the group regresses somewhat along the series of tasks in group development suggested by, among others, Tuckman (1965) (that is, forming, storming, norming and performing). Much of the knowledge possessed by the innovation team is not coded in ways that permit easy transfer of information to the newcomers. As Kanter put it, "Telling about it is not only time-consuming

674

t is indeed no substitute for having been there"
(Kanter 1988).

Transitions in Innovation Personnel
Relatively little research has examined what
happens to people as they engage in an innovation over time. In Chapter 5 Angle observed
there is a need to better understand the following human dynamics that were often observed
in three periods of the development of innovation groups studied by MIRP.

During the *startup period*, the dominant
dynamics observed are individual recruitment
and engagement in an innovation team and the
problems of the "hung jury", "acquiescent
team player," and "tolerance for closure and
trust," as described by Van de Ven (1985). In
addition, the startup period is characterized by
emotional euphoria, great expectations, and
confidence among participants in the success of
the innovative undertaking.

In the *middle period* the euphoria wanes,
problems surface, and the reality of the difficulty, complexity, and high risk of success have
set in. No conclusions or solutions are in sight
to provide closure to the innovation effort. It is
here that "hen-pecking," and lack of trust or
confidence in fellow workers and in leadership
become manifest. Some people desert the effort, creating problems of continuity. New people come on board without an "organizational
memory." As a consequence, although the setting should be the most opportune time for
learning by trial and error, actual instances of
learning are remarkably infrequent.

In the *ending period*, the innovation or a
part of it terminates. An "end of the tunnel"
comes in sight and group members find closure
to their experiences. If the effort was not successful, members concoct reasons for why their
ordeal was not in vain. Attributions of failure
usually point to external "uncontrollable factors." If the effort is a "success," team celebrations emerge. Success attributions are directed

to the commitment, talent, and heroic efforts of
the team; the leader is personified as superhuman, and a variety of efforts are made to prevent or forestall termination of the group.

While this description appears to summarize the overall transitions of an *innovation
group* over time, *individual transitions* do not appear to follow this temporal sequence because,
as we have seen, individuals with diverse ambitions, frames of reference, and functional expertise often come and go as an innovation develops. As a consequence, individual
participants are often not in sync with each
other, and the overall innovation unit appears
ambivalent or incapable of taking clear courses
of action. In these instances the overall group
often gets involved in frustrating periods of internal re-evaluation and emotionally torn meetings that appear to accomplish little. Van de
Ven (1985) suggests that three contradictory
individual-group dynamics explain this ambivalence:

First, there is the *hung jury*, in which individual
group members feel strongly about specific but
opposing courses of action that should be taken
but cannot achieve a consensus. On the surface
this often appears as a technical difference of
opinion among group members. Below the surface, however, group members are pursuing a
variety of hidden agendas which are often not
clearly understood by the individuals pursuing
these agendas. They include ambiguities about
willingness to be a member [of the innovation
unit], the kind of roles the individual wishes to
perform and is capable of performing, and what
rewards the individual can receive in return for
contributions [to the effort].

Second, there is the *acquiescent team player*,
in which individuals withhold or do not forcefully argue their personal ideas and suggestions
to the group for fear that they may upset or
derail the collective effort when they deeply
want the overall group effort to succeed. This
results in a lose-lose situation. The individual
who withholds ideas loses because he/she does

675

not achieve personal ownership in the group effort and does not realize the collective effort as a way of achieving personal ambitions. As a result, the individual begins looking elsewhere to achieve personal ambitions. The group also loses, both because it loses the active involvement of a group member and because the person may have withheld suggestions that could have significantly improved the overall group product.

Third, there are varying degrees of *tolerance for ambiguity and trust* among group members. Some members feel quite comfortable with and trust others enough to be able to work productively with the uncertainties of new relationships and innovative efforts where many decisions and issues remain open. Others (often of lower hierarchical status) require and demand more closure and safeguards. In response to the latter, efforts are made to "nail down" decisions and issues, often in writing. But these decisions are sometimes premature, and "breaches" in social contracts arise when what individuals thought was "nailed down" becomes loose or is "pried open." Further, those who have a greater tolerance for ambiguity or are more secure in their relations with others begin to complain about the "needless bureaucracy" and attempt to reduce or substitute "personal trust and norms" for an "impersonal contract."

With the exception of Kanter (1983; 1988), adequate recognition and treatment of these social-psychological dynamics of group development have not been found in the innovation literature, although they have been alluded to in the literature on policy implementation (for example, Pressman and Wildavsky 1974) and organizational development (Schein 1969). Perhaps this is because they only become apparent in real-time longitudinal field research, such as those studies conducted within MIRP and reported in earlier chapters. These dynamics were observed across many MIRP studies, and they represent some of the most gut-wrenching experiences for innovation participants and managers.

These findings suggest that *it is important to be sensitive to and orchestrate the transitions human beings go through as innovations develop over time.* Kanter has provided some useful suggestions for obtaining commitment, acceptance, and compliance through persuasion, bargaining, incentives, and power relations among people both within and outside the innovation unit. She emphasizes the need for coalition building in which power is acquired by selling the project to potential allies (Kanter 1988).

We suggest that the basis for such selling may change over time as the innovation moves through various stages, because different participants and stakeholders are involved over time and their goals change. It is thus readily apparent that once sold, there is little assurance that new selling will not need to take place as the innovation moves from startup, through development to adoption or diffusion.

As Kanter (1988) indicates, there is more than one currency of exchange upon which such selling can take place. She suggests that there are at least three types of "power tools" or organizational supplies that can be invested in action: *information* (data, expertise), *resources* (funding, time), and *support* (endorsement, approval, legitimacy). Each type represents a different sort of capital that can be used to bring (and keep) various people on board. In some organizations more than others, structural or systems constraints will drastically limit how an innovation team leader can use such resources for coalition forming. In any event, the appropriateness of each type may vary considerably at different stages of an innovation.

As Manz, Bastien, Hostager, and Shapiro suggest in Chapter 19, the type of leadership that is appropriate for an innovation team will probably change over time. Use of economic and political incentives to get people to sign on to an innovation effort is needed until the innovation gets underway. Those who do come aboard at this time are apt to need some structuring of roles and reciprocal re-

676

sponsibilities (that is, through Tuckman's [1965] forming, storming, and norming stages of group development). Later on, however, when euphoria turns to disappointment and reality shock, the need for *support* becomes paramount, as people need to be bolstered and have their aspirations shored up. In general, as important as it is for an innovation manager to *have* these various resources at her or his disposal for the purpose of keeping people committed, it is equally important that the manager know *when* to use each type, in accordance with the changing priorities (and concomitantly changing receptivity) of people to different types of "currency."

Setbacks Frequently Encountered

As discussed in Chapter 4, setbacks were frequently encountered in all the innovations either because developmental plans went awry or because unanticipated environmental events occurred that significantly altered the ground assumptions and context of an innovation. As an initial response, resource and schedule adjustments were often made that provided a grace period for innovation development. But with time, many of the problems snowballed into vicious cycles because of difficulties in detecting, correcting, and learning from mistakes.

For example, while most activities in the development of initial products for business creation in Chapter 8 largely proceeded as planned, failures (which occurred in only a few critical components) resulted in slipped schedules, budget overruns, and failure of the entire first-product development efforts. These initial product failures had important spillover effects on subsequent product development efforts, made the innovations more vulnerable to subsequent unforeseen events, and triggered vicious cycles which jeopardized the survival of the overall innovation efforts.

Many of these setbacks and errors went uncorrected because of four types of learning disabilities: (1) it was difficult to discriminate substantive issues from "noise" in systems overloaded with a combination of positive, negative, and mixed signals about performance over time, (2) entrepreneurs escalated their commitments to a course of action by ignoring nay-sayers and proceeding "full scale ahead," (3) some innovation participants became hypervigilant, calling prematurely for changes in a course of action when minor, correctable problems were encountered, and (4) in-process criteria of innovation success often shifted over time (as discussed below). Thus, it was often found that while extensive errors were detected, very few were corrected and they snowballed to crisis proportions before being addressed. Many learning experiences could not be acted upon because of the time lag required to change a course of action.

An immediate recommendation that emerges from these findings is to *structure opportunities and slack resources to detect and correct mistakes as they occur before they snowball into vicious cycles.* This recommendation relates to our earlier discussion of the gross miscalculation of the time required to develop an innovation. However, it is not clear that additional slack time alone will result in adaptive learning. As pointed out in Chapter 7, trial-and-error learning is far more difficult to achieve while an innovation develops than has been assumed to be the case. With highly novel undertakings, one can seldom rely on past routines or plans to guide behavior. Moreover, the highly ambiguous information participants typically receive and the idiosyncratic experiences they encounter greatly circumscribe rational learning processes (March and Olsen 1976). Instead, these conditions spawn "superstitious learning" (Levitt and March 1988). Because information is often unreliable, ambiguous, or late, and because most innovation participants have not experienced other innovations from which to develop inferences, it becomes difficult to correctly identify cause-and-effect relationships. As we will discuss next, there is a blurring between success and failure as results are interpreted against varying personal perspectives and frames of reference.

Outcome Criteria Shift

In Chapter 7 it was reported that in all the MIRP studies, criteria of innovation success and failure often shifted over time and differed between resource controllers and innovation managers; they differed at the beginning, converged during the developmental process, and then diverged in opposite and conflicting directions as innovation implementation problems arose. These shifting criteria often triggered power struggles between insiders and outsiders.

Because of the inherent ambiguity of judging whether an innovation developmental process will turn out to be successful, and because of limitations of people's information-processing capacity, the salient criteria for evaluating innovation progress shifted over time. These shifts tended to occur when problems and setbacks arose. They also occurred with changes in the driving concerns that captured managers' attention, almost to the exclusion of other factors—a sort of sequential, single-issue coping process. Sometimes, original criteria were subordinated to new, emerging criteria in areas only indirectly related to the innovation. For example, in Chapter 18 Bryson and Roering reported an innovation in strategic planning whose positive assessments were based not on innovation outcomes per se, but on the positive spinoff effects publicity of the innovation had on other activities. Thus, not only did initially nebulous targets crystallize later into more operational criteria, but the targets themselves were often reconstructed to redirect the innovations. These changes coincided and interacted with the unanticipated developmental setbacks and problems, shifting organizational

priorities, as well as the previously discussed independent environmental events that had spillover effects on the innovations. In such an ambiguous din with frequent distractions, disruptions, and shifting targets, it is easy to understand why so little learning was observed to occur.

Under these conditions, an understanding of how judgments about in-process innovation outcomes shift over time is critical for explaining why innovations unfold and how people learn from the process. On the assumption that people act on the basis of their assessments, they will do more of what they think leads to success and less of what is perceived to lead to failure.

Chapter 7 examined some of these relationships between innovation actions and outcomes across the MIRP studies. It was found that both rational and superstitious learning occurs as innovations develop over time. In-process innovation outcomes are partially a consequence of action, partially a predictor of future actions, but often do not completely explain actions that occur in the development of innovations. These predictors and consequences of outcome assessments become clear only with time, as developmental progressions stabilize. Even when they do stabilize, unspecified and conflicting targets or changing frames of reference often produce superstitious learning and conflicts between innovation managers and resource controllers over the developmental paths of their innovations.

To explain these contradictory dynamics, a model of success and failure action loops illustrated in Figure 21-2, was proposed in Chapters 7 and 8. The model illustrates the argument that, when the course of action being pursued by an innovation unit is judged successful, external resource controllers' confidence in and willingness to delegate greater control to the entrepreneurial unit is increased, which in turn permits the innovation unit

Action Loops in Transitions of Power
between Internal and External Business Groups

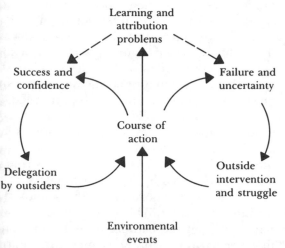

FIGURE 21–2.   Action Loops in Transitions of Power Between Internal and External Groups

greater discretion to pursue and expand its chosen course of action. However, when failure is perceived, uncertainties arise, and they trigger external resource controllers to intervene and engage in a struggle with innovation managers over the appropriateness of the innovation's present course of action. When this struggle subsides (often by the imposition of a new or modified action plan), the failure loop is completed and recycles in either positive or negative directions.

Thus, the model not only explains how innovation outcomes are both a cause and consequence of action, but also it acknowledges that outcome attributions can be produced by spurious unknown factors. In addition to an assessment of an innovation's course of action, outcome assessments are influenced by environmental events, shifting organizational priorities, as well as by mixed messages randomly ordered over time, hypervigilance, and entrepreneurial logic, as already discussed.

These research findings and inferences call into question a firmly held assumption about the management of innovation. Applied managerial research has typically searched for the actions that lead to success and treated success or failure as the bottom-line dependent variable. The MIRP findings question the wisdom of this practice since it may contribute to superstitious learning. *Innovation success might be more usefully viewed as "byproducts along the journey" than as end results.* While effectiveness judgments during innovation development can provide useful rationales for choosing subsequent actions, shifting targets and changing frames of reference produce compelling opportunities for superstitious learning, which can dominate current outcome evaluations by innovation participants and resource controllers. These practices account for many of the difficulties experienced in both the management of innovation and our ability to predict the outcome of the innovation journey.

Management Involvement and Roles

Among a broad range of innovations, MIRP research has often found investors or top management to be involved throughout the innovation process assuming a number of important roles. For example, in studying new business innovations (see Chapter 8), Van de Ven et al. suggested that investors and top management involved in an innovation enact at least four roles: *sponsor, critic, institutional leader,* and *mentor.* Also, in describing innovation in the public sector, Roberts and King identified and distinguished among the key roles of *policy entrepreneurs, champions,* and *administrators.* While these and other types of roles needed for the process of innovation may differ with respect to terminology, there are two recurrent themes: (1) innovations are not managed by one person as often implied, but by a team of managers, and (2) different managerial roles serve as checks and balances for each other in directing innovation development. Almost thirty years ago, Bavelas (1960) foretold that organizations were increasingly adopting a notion of leadership, not as a personal quality, but as an *organizational function:*

> Under this concept, it is not sensible to ask of an organization "who is the leader?" Rather we ask, "how are the leadership functions distributed in this organization?" The distribution may be wide or narrow. It may be so narrow—so many of the leadership functions may be vested in a single person—that he is the leader in the popular sense. But in modern organizations this is becoming more and more rare.

What are these "leadership functions" for managing innovation? We will answer this by describing the managerial roles suggested by Van de Ven et al. While the sponsor role is relatively well known, the roles of critic, mentor, and institutional leader have not received adequate attention.

The sponsor role is typically performed by a high-level manager who commands the power and resources to push an innovation idea into good currency. While the sponsor may or may not be the person who first thought of the innovation idea, he or she is clearly the person who "carries the ball" as an advocate for the innovation in corporate and investor circles where innovation resources are allocated. This sponsor, also called the "champion" by Pinchot (1985) and Peters and Waterman (1982), runs interference within the corporation for the innovation. Like all the leadership roles, the sponsor role may be performed by more than one person.

In addition to the sponsor, there is another partisan role—that of mentor. The mentor is typically an experienced and successful innovator, who is assigned or assumes managerial responsibility to coach (and perhaps supervise) the innovation manager or entrepreneur. This is the person or persons who serve as role models for the innovation team leader. Together with the innovation sponsor, the mentor provides encouragement, guidance and other types of support to the innovation team leader.

In the interest of checks and balances, the coalition formed by innovation sponsor and mentor needs to be offset by a devil's advocate (Janis 1983) who counterbalances their pro-innovation orientation. This role is performed by the critic, who applies dispassionate hard-nosed business criteria to the innovation idea and its development. Perhaps more than any other role we are discussing, the critic's role is likely to be shared by several persons.

At a level above all of this stands the institutional leader. This is the person who is removed from the battlefield, as it were, and therefore is presumably not subject to the partisan myopia which may afflict those closer to the innovation. It is the institutional leader's function to maintain a balance of power between the pro-innovation influences of the mentor-champion coalition and the reality-testing influences of the critic, so that the resolution of conflicts may be based on the merits of the case rather than on power alone.

In visualizing the power dynamics of the four roles we are portraying, one is drawn to an analogy of the relationships among the general manager, project manager, functional manager, and "two-boss" manager in a matrix organization as set forth by Davis and Lawrence (1977). In their framework, matrix management involves power relationships that can be diagrammed in a diamond-shaped hierarchy with the general manager at the top and the focal subordinate at the bottom. In between these levels are two equal managers who are immediately superordinate to the so-called two-boss manager. One of these supervisors is in the line chain of authority, whereas the other is in the functional chain which provided the focal subordinate to the task organization. The key matrix-management task of the general manager in this field of dynamic tension is to ensure that neither the functional nor the program interests are pursued to the detriment of the other. In other words, the key responsibility at the top is power balancing.

We can easily perceive an equivalent situation in the management of innovation. As Figure 21-3 illustrates, at the top of the diamond-shaped relationship is the institutional leader. He or she is concerned with the innovation as only one of many responsibilities. This psychological distance from the innovation, per se, allows a breadth of perspective not easily attained by the more immediate actors. At an intermediate level, we find two opposing forces, the sponsor-mentor coalition and the critic. The innovation sponsor runs interference for the project at corporate levels, while the mentor provides direct supervision, coaching, and counsel to the innovation. The counterbalancing role to this coalition is the critic, who is responsible for reality testing of the innovation against hard-nosed criteria. Without this role, the propensity of innovation managers to delude both themselves and others by seeing all ambiguity through rose-colored glasses might exhaust scarce organizational resources through a series of dubious ventures. On the other hand, were the critic able to run unchecked, no venture might be allowed a

chance to succeed because innovation is an inherently risky undertaking. Thus, the institutional leader's role is that of a power broker, ensuring that supports and restraints for the innovation are reasonably well balanced.

We believe that *the four managerial roles, in roughly equal proportions, promote the development of innovations.* In the MIRP research, innovations encountered significant hurdles in cases where one or more of the roles were absent. For example, the new company startup, reported in Chapter 8, did not enjoy the legitimacy and credibility of an institutional leader, nor the counsel of a mentor—both of which significantly hindered the ability to engage in business transactions with large customers and distributors. In addition, the board initially consisted only of inside directors, which limited the exposure of company principals to the kinds of divergent perspectives provided by critics in the corporate settings.

Relationships Frequently Altered

We have seen that as innovations develop over time, more and more players are brought into the game. A complex network of exchange relationships emerges with individuals and interest groups engaging in a series of transactions necessary to move the innovation forward. Relationships initially negotiated develop and stabilize over time, and they further shape and constrain possibilities for future interaction. These transactional relationships were often found to develop a variety of unintended consequences:

> Because of resource scarcity, innovations often enter into a leveraged set of highly interdependent and risky transactions (for example, using a customer contract as collateral to obtain a loan from a bank in order to hire employees to perform the contract). When any one of these

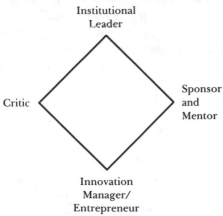

FIGURE 21–3.   *The Balance of Managerial Roles in an Innovation*

transactions falls, the entire set collapses in domino fashion.

Partnerships and joint ventures often produce "hung juries" on the strategic directions of an innovation because of the inability of parent organizations to reach agreements on concrete ways to share risks, costs, or potential payoffs of the innovation when they become apparent.

Aborted attempts to establish cooperative relationships with other organizations engaged in the development of a similar innovation result in competitive relationships later.

Close and successful relationships nurtured over the years with another organization may lead to "groupthink."

Company acquisitions undertaken to obtain technological or product competence often result in defections of the very people in the acquired organization who possess these desired competencies.

Given that innovation is a highly uncertain process, it is not surprising that these consequences should result from relationships, since innovation strategies or technological pathways change over time. What is somewhat surprising, however, is that transacting parties seldom include provisions to deal with these kinds of likely (perhaps inevitable) consequences when they initially negotiate their relationships. In no instances did MIRP researchers observe these unintended consequences to be foreseen by the parties involved when they initially entered into their relationships. And when they became apparent, the typical reaction of one party was a feeling of having been betrayed by the other party.

These findings clearly suggest the need to *continually modify and adapt the terms of relationships as innovations develop over time.* However, we have observed that organizational parties resist suggestions to reopen or renegotiate the terms of their business relationships—stating either

that this entails too much time or that it may result in an unfavorable position for their organization. Whatever the reason, the evidence is clear that the perspectives of most parties to their transactions change significantly over time and often produce unanticipated consequences if they are allowed to drift. In order to foretell and thereby minimize the negative consequences of drifting relationships, we recommend that *explicit efforts be taken to periodically evaluate the terms of relationships an innovation unit has with others.*

In particular, for business transactions with other firms, we recommend that *the terms of transactions include an exit clause and a periodic renegotiation clause.* The exit clause should set forth conditions by which parties can withdraw from the relationship in a mutually agreeable and equitable manner. The renegotiation clause should set forth periodic (perhaps yearly) deadlines in which the terms of the relationship must be renegotiated by the parties involved, or the transaction terminates by default.

We believe that many of the unanticipated and negative consequences of relationships can be minimized by modifying them as conditions change instead of allowing them to unwittingly drift over time. In Chapters 6 and 10, Ring, Van de Ven and Rands reported that among the innovations studied, the process of establishing relationships typically did not follow a simple linear set of stages of entering into negotiations with other parties, reaching agreements to the terms of relationships, and then administering or carrying them out. The more novel and complex the innovative idea, the more often trial-and-error cycles of renegotiation, recommitment, and readministration of transactions occurred. Moreover, as the risk of a deal increases and trust among parties decreases, the likelihood lessens that parties will enter into or remain in a transaction. If they do, the more costly and complex are the governance structures and contingen safeguards of the transactions, which in turn increases the likelihood of transaction failure. Ring and Rands in Chapter 10 provide a detailed and practical description of the extensive sensemaking, tacit understandings, and commitments involved in negotiating some of these interorganizational relationships over time.

Industry Team Playing

The management of innovation is not only concerned with micro or proprietary developments of a particular innovative device or service, but also often with the creation of an industry, or a macro infrastructure needed to implement or commercialize an innovation. As Rappa (Chapter 13), Knudson and Ruttan (Chapter 14), and Garud and Van de Ven (Chapter 15) report, collective action among firms in public and private sectors was found to be critical to create the social, economic, and political infrastructure a technological community needs to sustain its competing members.

For most technological innovations, this infrastructure includes institutional norms, basic scientific knowledge, financing, and a pool of competent human resources (see Chapter 15). These infrastructure resources are often initially developed as "public goods" in the public sector and appropriated by proprietary firms who transform them into "private goods" through innovation. Separate organizations often exist to provide these necessary resources for a given industry. However, these financial, educational, and research organizations are seldom easily accessible to a new industry that is emerging to commercialize an innovation. In addition, the infrastructure also requires establishing industry governance structures and procedures to regulate the behavior of competing firms and legitimating the industry's domain in relation to other industrial, social, and political systems.

Thus, the macro management of innovations demands attention to (1) the role of the public sector in stimulating or inhibiting innovations in the private sector, (2) how and when this infrastructure is organized, (3) what firms cooperate to create this infrastructure, (4) how they bundle their market transactions to establish resource distribution channels (for example, vendor-supplier-distributor relationships and joint ventures), and (5) what firms emerge as industry competitors as well as cooperators.

Inherent in these macro issues is the paradox of cooperation and competition. Each firm competes to establish its distinctive position in the industry; at the same time, firms must cooperate to establish the infrastructure required for all industry participants to survive collectively. For example, it clearly benefits all firms to cooperate to set up industry standards. In doing so, however, each firm will try to ensure that standards that best suit it become institutionalized (see Chapter 15). Another key paradox is that not only do government policies act simultaneously both to stimulate and retard industry development, but also they often change radically and unpredictably, thus creating an investment climate that can inhibit risk taking. An understanding of the processes that cause these and other paradoxes offers valuable insights on how firms and governments learn to cooperate to sustain themselves collectively, while at the same time compete, either to perform their unique roles or to carve out their distinctive positions in an emerging industry.

IMPLEMENTATION/TERMINATION PERIOD

The implementation period begins when application and adoption activities are undertaken for an innovation. If the innovation was created and developed within the organization, then implementation processes include introducing the innovation in the market, transferring it to operating sites, or distributing it to potential adopters. If the innovation was developed elsewhere, then the implementation process centers on the activities undertaken by a host organization to introduce and adopt the innovation.

It would be misleading to assume that development of an innovation is completed during the implementation period; much reinvention occurs during the implementation/termination period (Rogers 1983). More accurately, the development process simply changes in terms of (1) the goals of development and (2) who is responsible for development. This is true whether the innovation was developed within the organization (as described in Chapters 8 to 15) or was imported from the outside (as described in Chapters 16 to 18). In either event, there is often a transition of ownership and a shift from development of the innovation for its own sake to tailoring it to the organization's specific needs and constraints. (Of course, innovations may also terminate without being implemented.)

We will now discuss some of the common implementation processes of both "home-grown" and externally-developed innovations, as well as the processes observed in terminating innovations.

Implementing "Home-Grown" Innovations

In the organizations where innovations were "home grown," it was found in Chapter 4 that implementation and adoption activities often occurred throughout the developmental period, by linking and integrating the new with the old, as opposed to substituting, transforming, or replacing the old with the new. This observation implies that, because of limited organizational resources, innovations often cannot be mere additions to existing organizational programs. Neither is it practical for many innovations simply to replace existing organizational programs, because of the history of investments and the commitments to making

these older innovations work, although such a possibility is often perceived to exist. Therefore, the new innovation often represents a threat to the established order. Instead, if they are to be implemented and become institutionalized, the new innovations must overlap with and become integrated into existing organizational arrangements. Thus, in terms of the overall developmental sequence, after a period of divergent and parallel streams of activities convergent paths of activities begin to link the innovation with ongoing organizational arrangements.

The management of this convergent process appears to take several forms throughout the developmental period, including frequent restructuring of organizational arrangements, joint ventures, personnel responsibilities, use of teams, and altered control systems (see observation 5 in Chapter 4). Although the importance of integration and coordination mechanisms have long been recognized (as in Galbraith 1982), we were surprised to observe the number, fluidity, and variety of creative mechanisms used in the innovations studied by MIRP. These mechanisms provided incremental ways to make continual transitions between divergent components of the innovations and between the new innovations and existing organizational operations throughout the innovation period. Indeed, as Scudder et al. propose in Chapter 12, the earlier and the greater the involvement of interdisciplinary task teams (such as swapping of personnel) to deal with problems that cross organizational boundaries, the fewer the problems encountered in developing and implementing complex innovations.

Although these integrating mechanisms served their purpose of linking "home-grown" innovations into the operating subsystems of organizations, they did not prevent the surfacing of many problems commonly observed in organizations that adopted innovations developed elsewhere. As we will see, this is because what may be considered a "home-grown" innovation to one level or division of an organization usually represents an externally imposed innovation to another level or division of that organization. However, we have observed that *concentrated* efforts to link the new with the old throughout the developmental period provides not only more time but also more opportunities to address problems and modify the developing innovations to applied situations than is possible for organizations that adopt innovations developed elsewhere. Therefore, *we expect "home-grown" innovations normally to require less time to implement and institutionalize than externally induced innovations.*

But before we discuss the adoption of externally developed innovation, it is important to address the ending and "letting go" process that innovators must often go through when their innovation is transferred to operating units for implementation and institutionalization. In these cases there is a transfer of ownership and a shift from developing an innovation for its own sake to tailoring it to the organization's operating needs and constraints. Now the innovation must be institutionalized, and this often means it must be handed off to others who will use it. Depending on the nature of the innovation, it may be necessary to make significant changes in organizational structures and systems. There may be problems of loss of ownership, and disapproval of compromises made to the original idea in the interest of marketability, cost, or other aspects of feasibility. One is reminded of the musical complaint "Look what they've done to my song." There may be a letdown at this point for the innovation team or, more likely, a time of ambivalence. On one hand, there is a feeling of relief that the stresses and strains of the intensive innovation process are behind. On the other hand, there may be a depressing realization that there is nothing to fill the void.

An immediate problem regarding the innovation team, then, is how to deal with separation. Sutton (1987) and Albert (1984) suggest that organizations need to structure transition rituals or "funerals" in order to recognize the contributions of innovators and to facilitate the letting go process. *Just as society provides the funeral ritual to mourn the passing of loved ones, organizations need to structure ceremonies at transition times in order to help people relinquish the past and take on new assignments and ventures.* Even though it has been observed that many people involved in organizational innovations have only partial inclusion in the project (that is, they also are doing other things), it is likely that they have a heavy psychological investment in the innovation and therefore a time of mourning may be needed after their role is over.

It is likely that development of the innovation has resulted in a degree of organizational learning (although, as stated in Chapter 7, little of it could be applied to the innovation from which it was gained). This knowledge about the process of managing innovation is a valuable asset that organizations should capture, yet attempts to do so were very infrequently observed in the MIRP research. *Organizations, particularly those that are undertaking numerous innovations at any given point in time, should debrief innovators at the conclusion of their innovations by asking them what they have learned from the process and what should have been done differently.* Such information can accumulate into a storehouse of knowledge that could provide organizations with strategic competitive advantages in managing innovations.

Adopting Innovations Developed Elsewhere

Within MIRP, we have been studying the adoption of a number of innovations that were not "home grown." That is, these innovations were either suggested or imposed by outside agencies (for instance, a nuclear regulatory agency in Chapter 16, a Site-Based Management movement in Chapter 17, and a consultant on strategic planning in government agencies in Chapter 18). A key finding from these studies is that organizations within each of these settings, which appear similar in all important respects, may respond differently to innovation adoption, and these differences make a difference in implementation outcomes. Thus, the process by which innovations are adopted influences implementation success. In particular, the MIRP studies indicate that *innovation adoption is facilitated when* (1) *the adopting organization modifies and adapts the innovation to its local situation,* (2) *top management is extensively involved and commits resources to innovation adoption,* and (3) *process facilitators help people understand and apply the new innovation.*

In Chapter 16, Marcus and Weber describe the organizational effectiveness implications of two different reactions taken by 28 American nuclear power companies in response to a new set of nuclear safety procedures mandated by the U.S. Nuclear Power Commission. They find that the nuclear power plants with relatively poor safety records tended to respond in a rule-bound manner that perpetuated their poor safety performance, and that those plants whose safety records were relatively strong tended to retain their autonomy by adapting the standards to their local situations, a response that reinforced their strong safety performance. As in the case of the vicious cycles described above, those least ready or willing to adopt the innovation may be those that need them the most.

Marcus and Weber make a very important, general inference from these findings for the management of externally imposed innovations: *Be forewarned of the possible consequences of passive acceptance of external dictates by those who strictly follow the letter of the law; they may be doing so in bad faith and may not achieve the results intended.* Some autonomy is needed for an adopting unit to identify with and internalize an innovation; mere formal compliance is insufficient for innovation adoption. The disposition of innovation adopters is likely to be negatively affected if they are not granted a sufficient level of autonomy, and it is their disposition that is often critical in assuring successful adoption. This evidence points out the importance of overcoming resistance to change in imposing (or even suggesting) the adoption of innovations that did not originate in the adopting organization. The "Not Invented Here" (NIH) syndrome is well known in all sorts of organizations. Adopting agencies or organizations that have not developed any sense of commitment to those innovations may well behave bureaupathically and simply do what the letter of the law requires (Kerr 1975; Lawler and Rhode 1976).

In Chapter 17, Lindquist and Mauriel compared two common alternative strategies for adopting and implementing an innovation: a depth strategy in which the innovation is implemented and debugged in a demonstration site before it is generalized to other organizational units, and a breadth strategy in which the innovation is implemented through successive hierarchical levels across all organizational units simultaneously. It was found that the school district that implemented the Site Based Management innovation in breadth (across all schools in the district) was more successful in implementing and institutionalizing more components of the innovation than the school district that adopted the depth strategy within a school selected as the demonstration site. *This finding is contrary to conventional wisdom that successfully implemented innovations start small and spread incrementally with success* (Greiner 1970; Van de Ven 1980a). Lindquist and Mauriel provided several important generalizable explanations for this finding:

> Once the depth strategy is introduced and heralded by top management, the demonstration project loses visible attention support (and institutional legitimacy) from top-level managers, as their agen

das become preoccupied with other pressing management problems.

With a breadth strategy, top management stays in control of the innovation implementation process—thereby increasing (rather than decreasing) its power. Moreover, slack resources within the control of top management can ensure success better than can limited budgets for innovation to a demonstration site.

There is a tradeoff between implementing a few components of an innovation in breadth versus implementing all components in depth in a particular demonstration site. Fewer hurdles and resistances to change are encountered when a few (and presumably the easy) components of an innovation are implemented across the board to a few (and presumably supportive) stakeholders, than when all (both easy and hard) components of a program are implemented in depth with all partisan stakeholders involved.

With a depth strategy, it is easier for opposing forces in other parts of the organization to mobilize efforts to sabotage a "favored" demonstration site, than it is to produce positive evidence of the merits and generalizability of an innovation.

Finally, in Chapter 18 Bryson and Roering report on their participant-observer role in introducing strategic planning systems to six local governmental units. They importantly observe that each attempt to adopt the innovation was prone to disintegration because

external events and crises frequently occurred, distracting participants' attention and priorities, and taking away any slack resources available to adopt the innovation;

the adoption process itself was partially cumulative—what occurred before was re-

membered and had to be accounted for; and

participants got bogged down with information overload, conflicting priorities, and divergent issues that were outside of their decision jurisdictions or domains.

By comparing the varying adoption success of the six local government units, Bryson and Roering derive several useful recommendations that we think apply to many innovation adoption efforts. First, *have not only a powerful innovation sponsor, but also an effective process facilitator* who is committed to continuing with the adoption process, particularly when difficult hurdles and setbacks arise. Second, since disruptions and setbacks cause delays, and interest wanes with time, *structure the process into key junctures*—deadlines, conferences, and peak events. These structured junctures in the adoption process establish key deadlines to perform planned intermediate tasks, force things to come together, and facilitate unplanned intersections of key ideas, people, transactions, and outcomes. Third, *adopt a willingness to be flexible* about not only what constitutes acceptable innovation adoption, but also in constructing arguments geared to many different evaluation criteria. In short, innovation-adoption success (like development success) more often represents a socially constructed reality than an objective reality.

In conclusion, over the course of the MIRP studies we often encountered top managers who expressed frustration with the "unwillingness," "narrow-mindedness," or NIH syndrome of their subordinate unit or subsidiary managers in adopting changes they were requested to implement. For example, one corporate executive stated, "If I could only get them to take a broader view, they would see the need for change." The MIRP innovation adoption findings suggest it may be more productive to direct this frustration at corporate executives and public policymakers who fail to appreciate

687

the logical consequences of their top-down or externally imposed mandates on the behavior and performance of organizational subunits.

When Innovations Stop

Innovations terminate either when they are implemented and institutionalized (as discussed above) or when resources run out. While obvious, this is fundamentally important in explaining the tendency, throughout the developmental period, toward sugar-coated administrative reviews and impression management by innovation managers in relation to resource controllers; and the vicious cycle of conflict that is structurally inherent in the roles of resource controllers and innovation managers. That innovations stop when resources run out also provides the basis for our recommendations (1) for investors and resources controllers to develop monitoring mechanisms that don't stifle innovation management, and (2) for innovators to manage their resource controllers as well as their internal innovations.

First, *it is important not to let an innovation fall victim to the reactivity of monitoring.* Because of the necessity to perform managerial supervision, innovations are subject to progress reviews. It is obviously necessary to be able to cut one's losses early when it becomes apparent that an innovation is headed for a dead end, it is unacceptably expensive, or it cannot be completed in time to be of value. On the other hand, it is important to give the innovation a reasonable opportunity and not to expect tangible payoffs unrealistically early in the course of its development. While measurement is necessary, there are dysfunctional aspects of overreliance on concrete performance criteria (Lawler and Rhode 1976). Incentives built into managerial systems cause people to overemphasize appearances on concrete performance indicators while giving short shrift to the underlying performance that such indicators are presumed to detect. In effect, people are more

motivated to *look* good than to *be* good. Although concrete indicators, particularly financially based ones, are desirable in any business undertaking, where these indicators are robust enough to provide useful data about an enterprise's health, innovations may be so inherently subjective that, more often than not, such performance indicators only hint at the innovation effectiveness they purport to measure.

Resource controllers have two somewhat antithetical roles to play in an innovation: support and coaching on the one hand, evaluation on the other. This is quite similar to the dilemma any boss has in trying to do two drastically different things (coaching and evaluating) while undertaking a formal performance appraisal of a subordinate (Meyer, Kay, and French 1965). One solution might be for top management to get out of the review process altogether. If resource controllers are seen as making judgments as to resource support levels for an innovation, there may be too much motivation for the innovation team to sugar-coat information, thus denying upper management the factual information to make sound decisions and to help where needed.

As an alternative, we suggest that organizations look for creative ways for evaluations to be conducted by third parties who are not directly in the resource-allocation chain. For example, monitoring by a person performing a mentor role may produce more useful evaluations, because the mentor not only tends to have more direct daily contact with the innovation (which minimizes the credibility of sugar-coated reports), but also presumably commands the practical experience in evaluating realistic ways to adapt the developmental process to changing circumstances.

Another strategy that merits consideration is a peer-review process in which managers of other innovations or startup ventures are enlisted to serve as a review committee of an innovation. This strategy is modeled after the academic committees used to evaluate research proposal grants or articles submitted for competitive evaluation for publishing. (Of course, safeguards would have to be in place to prevent evaluation by persons who stand to gain from resources being cut off from the innovation being evaluated.)

A third strategy is to have periodic reviews by an unbiased external research or consulting organization. Indeed, the MIRP research itself seems to have performed this function for some of the innovation projects we studied, and periodic feedback from an outsider's perspective did seem to help the innovation teams focus their attention on issues they might otherwise have missed.

A related issue is that innovation managers need to recognize the different priorities and frames of reference held by those who allocate resources. Resource controllers simply do not see the innovation from the more personal point of view of the innovation team. Whereas the innovation may be the exclusive labor of love for the innovation team, it is but one of a set of interacting (and often competing) business considerations at the macro-organizational level. Thus, criteria for judgments of success and viability may be more dispassionate at this level. Actual criteria used to judge progress and potential may differ considerably from those used at the innovation team level. This does not necessarily mean that criteria and decision logic are superior at the organizational level—only that they are different. Given these differences, the MIRP findings clearly show that the perspective that really matters in continuing an innovation is that of the resource controllers.

The practical consequence is that the innovation team leader bears the chief responsibility for maintaining a sound relationship with those who control resources. Perhaps the key to this relationship is unambiguous mutual expectations, more often achieved in theory than in practice. Although rarely these expectations may be spelled out by one's superiors, more

often it is the responsibility of the subordinate to elicit them from the boss. All too often, it is assumed that these expectations are understood—with disastrous results.

The innovation team leader needs to develop sensitivity to the goals, criteria, and forces driving the resource controllers. These are often at odds with the immediate goals of the innovation team leader, and it may take a bit of creativity in order to reconcile the two, not unlike the process suggested to reach integrative solutions to two-party conflicts (see, for example, Filley [1975] and Ring and Rands, Chapter 10).

A key to developing a clear set of mutual expectations is the *development of trust.* In Gabarro's (1979) research on the development of "interpersonal contracts" between top managers and their immediate subordinates, it was found that the development of trust went through a series of stages (orientation, exploration, testing, and stabilization) in which the two parties sized one another up in order to learn how far, and in what areas, the other could be trusted. (In Chapter 10, Ring and Rands described these informal processes as sensemaking, understandings, and commitments.) Throughout this process, the subordinate's openness and candor seemed to be the key to developing the senior's trust. Recall, however, our discussion of the type of impression management that typically characterizes the early phases of an innovation, when the innovation team puts its best foot forward and submits sugar-coated estimates of costs and goal attainability. This type of game establishes an adversarial climate, in which "we" attempt to shape "their" perceptions—a stance antithetical to the development of trust.

One other issue deserves attention. We have noted the importance of knowing one's resource controller. It is equally important for the innovation manager to know him- or herself. Innovators are often mavericks, fitting the classic description of the entrepreneur who is not comfortable with authority relationships.

This type of personality has been termed "counterdependent" by personality psychologists. We believe that it is incumbent on innovation team leaders whose propensity is toward counterdependence to be aware of this and to take whatever steps are necessary to curb overt signs of this predisposition. If this is so deeply ingrained as to overwhelm attempt to control it, serious thought should be given to replacing the innovation manager.

Success-Failure Attributions

It would be premature to draw conclusions on the factors that lead to the ultimate success or failure of innovations, because few of the innovations being studied by MIRP have come to a natural finish (that is, been terminated or institutionalized). We have repeatedly observed, however, that judgments about in-process innovation outcomes occur throughout the innovation process, and that these attributions often appear misdirected. For example, Chapter 8 reported that the unsuccessful first-product take-off by the internal corporate venture was often attributed to problems of management implementation, although the facts indicate that many of the factors that led to failure were beyond management's control. The evidence indicates that attributing failure to mismanagement was incorrect, and resulted in making managers the scapegoats for events beyond their reasonable control. Such attributions reinforce the myth that managing innovation is fundamentally a control problem. Rather, it should be seen as one of orchestrating a highly complex, uncertain, and probabilistic process of collective action.

Why are attributions for innovation success or failure often misdirected, particularly among managers and evaluators who are closely associated with the innovations? Clearly, *there is a need to understand this process of attribution, because these attributions directly influence*

future organizational actions and the careers of innovation participants.

Mitchell, Green, and Wood (1981) provide an attributional model of the ways supervisors judge the causes of subordinates' poor performance. We believe that this model can be applied equally well to innovation. According to the Mitchell and his colleagues, deciding why something went wrong in an organization is a two-stage attribution process. First, the manager must decide whether the deficiency had internal origins (within the person), or external origins (within the situation). Another type of attribution which must be made simultaneously is whether the failure was a fluke or could be expected again under similar circumstances—in other words, whether the cause was stable or unstable. These two parallel attributions can be laid out neatly in a two-by-two table as shown in Figure 21-4.

As the table shows, an innovation failure

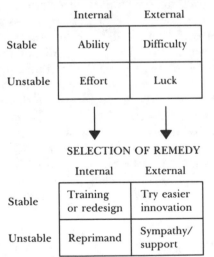

Source: Mitchell, Green, and Wood, "An Attributional Model of Leadership and the Poor Performing Subordinate" (1981).

FIGURE 21–4. *Attributions of Innovation Success or Failure*

may have occurred because of (1) internal-stable, (2) external-stable, (3) internal-unstable, or (4) external-unstable conditions.

In the first instance (internal-stable) the innovation might be seen to have failed because the innovation team and/or its leadership was not competent. The requisite skills and abilities were not present, or perhaps there was inadequate organization and leadership. In this specific attribution. It is tacitly assumed that someone else might have brought it off. The point is, however, that this particular innovation team was seen as not up to the task.

Alternatively, the task might have simply been too daunting. This type of attribution fits our second example (external-stable). In this attributional choice, it is assumed that the innovation team might be successful in other types of innovation efforts, but the one they attempted and failed at was simply too difficult. It is further assumed that other innovation teams would have met the same fate.

Both types of attributions are stable, in that the same outcomes would be expected over and over again, given either that the same innovation team was attempting the innovation, or that the same intractable problem faced the team.

The first pair of possible attributions, then, assume a reliable world; one that is essentially the same from trial to trial. In contrast, the other pair of possible attributions assume a capricious environment; one in which things may not be the same twice.

One such attribution (internal-unstable) is based on the perception that the innovation team, while perfectly competent overall, simply failed to do all that was needed to bring about success—they didn't try hard enough, or they failed to pay attention to detail. With this lesson learned, the team might do very well on another type of project, because the talent is in place.

Finally, the fourth possible attribution within our framework (unstable-external) is based on the assumption that failure resulted essentially from one or more bad breaks. Now the innovation team is not held culpable, either because of lack of effort or because of lack of ability. Furthermore, the failure is seen as one that need not have happened; the problem was solvable, but things just didn't fall into place. Better luck next time!

These four attributions do not simply occur at random. Rather, there are built-in biases in the judgment process, based on the evaluator's personality and relationship to the members of the innovation team. A more or less universal bias seems to be the tendency to make internal attributions for the failures of others; we seem to be much more even-handed in evaluating our own failures, often attributing them to external causes. With this bias operating, innovation teams associated with failed innovations may find it difficult to escape the stigma of failure. Without compelling evidence to the contrary, upper management may be predisposed to lay the blame at the feet of the innovation team. On the other hand, the same self-serving attribution bias may lead external management to take more credit than is due when things have gone well for the innovation.

This tendency to assume that all failure is the innovation team's fault can be exacerbated rather easily, if the consequences of the failure are large for the evaluator. Although a rational view of human information processing would treat evaluations of causes and of outcome severity as independent of one another, much evidence to the contrary exists (Rosen and Jerdee 1974; Shaver 1975). If a failed innovation has a particularly damaging effect on senior management, it is more likely that (1) an internal attribution of the causes of failure will be made (either low effort or lack of ability on the part of the innovation team); and (2) the remedial action will be severe, as discussed later.

Among other factors that bias the attribution process is the amount of empathy or identification between the external manager and members of the innovation team. "Essentially, any factor which makes the leader psychologically closer to the member should increase the tendency for the leader to make selflike attributions regarding the member" according to Mitchell et al. (1981). Conversely, as psychological distance increases, supervisors are more likely to make non-selflike attributions (that is, more severe). One would therefore expect that persons filling a mentor role would be biased toward making external attributions of the causes of an innovation team's failure because of the identive nature of a mentoring relationship. On the other hand, persons filling the critic role might make a clear "I–they" distinction and be inclined toward attributing failure to the team's shortcomings.

Once a cause of failure has been decided upon, it becomes upper management's job to decide what to do about it. As already mentioned, there are biases operating here, too. If the results are disastrous, the remedy will more likely be severe. One thing clearly suggested by the Mitchell et al. model is that the remedy will tend to be suggested by the specific attribution of cause made by upper management. If it was seen as a case of bad luck, perhaps supportive leadership and a chance to try again are all that is called for. If the innovation was seen as too difficult, however, this may make management more cautious in the future regarding what is to be attempted. If failure was seen as caused by lack of ability, top management may look for remedies such as better staffing of innovation teams, better training, or other enabling factors (see Chapter 5) within top management's control. If, on the other hand, failure was attributed to a lack of effort or diligence on the part of the innovation team, disciplinary action seems highly likely.

One often hears organizational maxims such as "Here, we let people fail," which indicate an espoused organizational rule that one should not stifle innovation by making people

wary that they have only one chance to make it. Yet our experience in MIRP has been that this maxim is not always operative, even in organizations in which it is commonly known that this is one of the rules. On the contrary, we have seen instances where innovation team leaders whose projects floundered are stigmatized as losers. We suspect that this label is most often applied where failure was *not* seen as an anomaly or fluke but was seen as likely to happen were a similar innovation to be attempted again. Actually, only one of the four possible attributions, bad luck, probably predisposes upper management to give the innovation team another chance. All other attributions tend to promote caution and might lead management to "place its next bet on some other horse."

CONTINGENCIES

In the Minnesota Innovation Research Program we have been struck by a paradox: innovations are alike, but innovations are different. We have seen recurrent patterns of innovation development across a wide range of product, process, and administrative innovations. Yet, we also have had to acknowledge that many variations on the innovation theme exist. This is not too surprising, because just as we learned many years ago that there is no one best way to manage, we expect that we will never find one best way to innovate. A sophisticated manager of innovation will instead try to identify those contingent factors that seem to have systematic moderating effects on what works and what does not. In particular, we propose that *many of the key process components described in the previous section tend to be more pronounced for innovations of greater novelty, size, and stage or time duration.* Systematic development of this proposition would require more space and time. We can only summarize here the major implications of this proposition. They are based on the statistical findings reported in Chapter 3, as well as on qualitative comparisons across Chapters 8 through 15 of relatively simple innovations (for example, new software company startup) with more complex innovations (for example, defense contracting).

Radical Versus Incremental Innovations.

Some innovations change the entire order of things, making obsolete the old ways and perhaps sending entire businesses the way of the slide rule or the buggy whip. Others simply build upon what is already there, requiring only modest modification of one's old world view. We expect that innovations of different levels of novelty need to be managed differently. Indeed, some organizations may be well suited to one type of innovation but not to the other. In Chapter 5, for example, it was suggested that an organization which values and rewards individualism may have the advantage in radical innovation, while a more collectivist system may do better at an incremental one. Novelty also influences an innovation's developmental pattern. Pelz (1983) found that the more novel the innovation, the more complex and overlapping the developmental sequence. In addition, in Chapter 3 it was reported that statistical relationships between perceived effectiveness and various measures of innovation ideas, people, transactions, and context were weaker for highly novel innovations than they were for less novel borrowings.

Innovation Stage and Temporal Duration

The above-stated assertion about fit between radical versus incremental innovation on one hand, and individualist versus collectivist systems and cultures on the other hand, may also be brought to bear on the changing requirements at various innovation stages of development. As mentioned by Kanter (1988), what

may be an individual undertaking, at first, soon becomes a multi-party undertaking as diffusion of the innovation becomes the primary task. In particular, in the ending stages of an innovation a shift from radical to incremental and from divergent to convergent thinking typically occurs. Empirically, Chapter 3 reported that as innovations approach the implementation stage, they become more highly structured and stabilized in their patterns, just as Zaltman, Duncan, and Holbek (1973) proposed.

The developmental pattern and eventual success of an innovation is also influenced by its temporal duration. A wide variety of business relationships and startups have been found to be time-dependent processes (Levinthal and Fichman 1988). Fichman and Leventhal (1988) argue that the initial investment to undertake an innovation represents an initial stock of assets that provides an innovation unit a honeymoon period to perform its work. These assets reduce the risk of terminating the innovation during its honeymoon period when setbacks arise and when initial outcomes are judged unfavorable. The duration of the honeymoon period varies with the degree of commitment to the innovation and the likelihood of replenishing the initial stock of assets. As we have seen, this commitment and replenishment are highly influenced by how long it takes to develop and complete the innovation. Interest and commitment wane with time. Thus, after the honeymoon period innovations terminate at disproportionately higher rates the longer the duration of development is.

Size and Scope of the Innovation

It may be that small organizations have the advantage in starting up an innovation, but that larger organizations with more slack resources have the advantage in keeping an innovation alive until it is completed. Chapter 8 reported that venture capital was more risky, of shorter term, and more difficult to obtain than was internal corporate venture funding. Larger organizations offer a more fertile ground for sustaining and nurturing spinoff innovations. Also, there may be more places to hide something in a larger organization, until such time as an innovation can stand on its own. Yet large organizations seem to need bureaucratic systems in order to manage, and this is not particularly conducive to innovation. So the message to managers is to keep finding ways to become small though large by such mechanisms as autonomous work groups, business-development units, and the like.

CONCLUDING DISCUSSIONS

This chapter has identified, explained, and made some suggestions for dealing with key junctures along a generic innovation journey. It represents our attempt to present an overall composite map of the key processes commonly observed in the innovations being tracked by the ongoing MIRP studies. These processes have been sorted into three innovation phases referred to as the initiation period, the developmental period and the implementation and termination period, because we have found the crucial tasks and challenges to differ systematically across these stages.

For the initiation period, we have highlighted the need for innovation leaders to pay attention to the creation and maintenance of both motivating and enabling conditions for innovation. Neither alone is sufficient. We have also pointed up the necessity of planning and structuring situations so that thresholds of attention are lowered and so that the organization is positioned to take advantage of favorable chance occurrences. Finally, we have noted the need for replacing dysfunctional optimism in the startup phase with a more realistic appraisal and planning process to avoid resource backlashes in the future.

In the developmental period, we have

recommended that innovation managers attempt to limit the complexity insofar as feasible by keeping their innovation effort simple. Although moderate turnover in the innovation team may actually transfuse new ideas into the innovation, it is important to limit the amount of personnel turbulence in order to limit the process costs inherent in changing team memberships. Finally, we have admonished managers to be aware of the severe changes that can be expected during this phase in such aspects as stakeholders' priorities and frames of reference, relationships, and power bases. While there are few simple prescriptions to be offered in this respect, it is clear that forewarned is forearmed. The middle period of an innovation will demand a high level of sensitivity to signals which indicate such shifts are taking place.

In the ending period, the major issues requiring resolution are dealing with endings and with transitions. Whether the innovation has been home grown or imported from outside the organization, there will be a changing of the guard. The old hands may be reluctant to let go, while the new custodians in charge of further development may not have the same sense of ownership and commitment as those who preceded them. In this phase, as in the earlier stages, the organization needs to capture the new knowledge made available through the innovation experience and to incorporate such organizational learning into the system for use in follow-on innovation efforts. It is perhaps in this final stage that such learning becomes most likely, because of the time perspective afforded.

We emphasize that the suggestions made are speculative because they have not been tested. However, they represent an attempt to derive some prudent principles for managing the innovation journey based on the knowledge gained in MIRP to date, as well as our review of other research. Moreover, we wish to restate that our purpose was not to offer a set of management principles that will ensure innovation success. The reasons for this caveat rest with a concluding lesson that we believe underlies all the key junctures along the innovation journey: *Management cannot control innovation success, only its odds.* We conclude with a discussion of this lesson, for it implies that a fundamental change is needed in conventional management philosophy and practice.

One of the authors uses a case in teaching MBA students, many of whom already have several years of managerial experience. In this case, the organization is meeting all its important goals, but management is not in control because an informal work group is managing itself by beating the system. In so doing, however, it seems to be meeting management's goals about as well as anyone could hope. The immediate supervisor in the case has essentially abdicated his role of providing motivating conditions (see Chapter 5) and is content to spend all his time providing enabling conditions. The typical reaction, on the part of students, to this scenario is outrage. The supervisor is not doing his job and top management is not in control.

We do seem to have a bias toward control in Western managerial society. Even if things are going well, we insist on being in charge. This may raise problems with the management of innovation, because the process may be inherently uncontrollable—at least at levels of intensity that show promise for making a significant new contribution.

Professor William McKelvey at UCLA tells the story of the 1976 Winter Olympics, where Franz Klammer won the men's downhill skiing competition. When interviewed after the event and asked how he managed to turn in such an incredible performance, he said that he had chosen to ski "out of control." He knew that there were many other top-level skiers entered against him, several of whom might outpace him on any given day, if he were to ski his normal speed—that is, under control. He chose, instead, to ski so fast that he abandoned any sense of control over the course. While obviously this was not alone sufficient for victory, he saw is as a necessary condition. Staying in control would virtually ensure a loss, while skiing out of control would make it at least possible to win.

Innovation managers may have an important lesson to learn from this anecdote. An innovation is, almost by definition, a leap into the unknown. It may be necessary, in order that the innovation have a chance to succeed, to relax traditional notions of managerial control. It is not that such letting go will ensure success—merely that it may be a necessary condition.

A number of practical consequences follow if innovation success is recognized to be a probabilistic process. First, innovation success or failure would more often be attributed to the external-unstable condition in Figure 21–4. This in turn will decrease the likelihood that innovation managers will be stigmatized if their innovation fails and increase the likelihood that they will be given another chance to manage future innovations. After all, one cannot become a master or professional at anything if only one trial is permitted. As we reported, relatively little trial-and-error learning occurred once the journey was begun for a given innovation. Repeated trials over many innovations are essential for learning to occur and for applying these learning experiences to subsequent innovations. Indeed, the knowledge generated by MIRP and reported in this book was possible only by carefully observing and comparing the developmental process across many diverse innovations. It is largely through repeated trials that the art or craft of innovation management can develop and thereby progressively increase the odds of innovation success as the organization's skills inventory is built up.

REFERENCES

Albert, S. 1984. "The Arithmetic of Change." Minneapolis: University of Minnesota. Strategic Management Research Center Working Paper.

Aronson, E. 1973. "The Rationalizing Animal." *Psychology Today* 6, no. 5: 46–50, 52.

Baveles, A. 1960. "Leadership: Man and Function." *Administrative Science Quarterly* 4: 491–98.

Brockhaus, R.H., Sr. 1982. "The Psychology of the Entrepreneur." In C.A. Kent, D.L. Sexton, and K.H. Vesper, eds., *Encyclopedia of Entrepreneurship.* Englewood Cliffs, N.J.: Prentice-Hall.

Cohen, M.D., J.G. March, and J.P. Olsen. 1972. "A Garbage Can Model of Organizational Choice." *Administrative Science Quarterly* 17, no. 1 (March): 1–25.

Dalton, D.R., and W.D. Todor. 1979. "Turnover Turned Over: An Expanded and Positive Perspective." *Academy of Management Review* 7:212–18.

Davis, S.M., and P.R. Lawrence. 1977. *Matrix.* Reading, Mass.: Addison-Wesley.

Fichman, M. and D.A. Levinthal. 1988. "Honeymoons and the Liability of Adolescence: A New Perspective on Duration Dependence in Social and Organizational Relationships." Unpublished paper, Graduate School of Industrial Administration, Carnegie Mellon University.

Filley, A.C. 1975. *Interpersonal Conflict Resolution.* Glenview, Ill.: Scott, Foresman.

Gabarro, J.J. 1979. "Socialization at the Top—How CEOs and Subordinates Evolve Interpersonal Conflicts." *Organizational Dynamics* 7, no. 3: 3–17.

Gabarro, J.J., and J.P. Kotter. 1983. "Managing Your Boss." In L.A. Schlesinger, R.G. Eccles, and J.J. Gabarro, eds., *Managing Behavior in Organizations.* New York: McGraw-Hill.

Galbraith, J.R. 1982. "Designing the Innovative Organization." *Organizational Dynamics* (Winter): 5–25.

Greiner, L.E. 1970. "Patterns of Organizational Change." In G. Dalton, P.R. Lawrence, and L.E. Greiner, eds., *Organizational Change and Development.* Homewood, Ill.: Irwin-Dorsey.

Helson, H. 1964. "Current Trends and Issues in Adaptation-Level Theory." *American Psychologist* 19: 23–68.

Herzberg, F., B. Mausner, and B.B. Snyderman. 1959. *The Motivation to Work.* New York: Wiley.

Janis, I. 1983. *Groupthink: Psychological Studies of Policy Decisions and Fiascoes.* Boston: Houghton-Mifflin.

Johnson, B.M. and R.E. Rice. 1987. *Managing Organizational Innovation.* New York: Columbia University.

Kanter, R.M. 1983. *The Change Masters.* New York: Simon and Schuster.

———. 1988. "When a Thousand Flowers Bloom: Structural, Collective and Social Conditions for Innovation in Organizations." In B.M. Staw and L.L. Cummings, eds., *Research in Organizational Behavior,* Vol. 10. Greenwich, Conn.: JAI.

Katz, D. 1964. "The Motivational Basis of Organizational Behavior." *Behavioral Science* 9: 131–46.

Kelman, H.C. 1961. "Processes of Opinion Change." *Public Opinion Quarterly* 25: 57–78.

Kerr, S. 1975. "On the Folly of Rewarding A, While Hoping for B." *Academy of Management Journal* 18: 769–83.

Kimberly, J., and M. Evanisko. 1981. "Organizational Innovation: The Influence of Individual, Organizational, and Contextual Factors on Hospital Adoption of Technological and Administrative Innovation." *Academy of Management Journal* 24: 689–713.

Lawler, E.E., III, and J.G. Rhode. 1976. *Information and Control in Organizations.* Pacific Palisades: Goodyear.

Levinthal, D.A., and M. Fichman. 1988. "Dynamics of Interorganizational Attachments: Auditor Client Relationships." *Administrative Science Quarterly.* Forthcoming.

Levitt, B., and J.G. March. 1988. "Organizational Learning." *Annual Review of Sociology* 14. Greenwich, Conn.: JAI.

Lewin, K. 1947. "Group Decision and Social Change." In E.E. Maccoby, T.M. Newcomb, and E.L. Hartley, eds., *Readings in Social Psychology.* New York: Holt, Rinehart and Winston.

March, J.G. and J.P. Olsen. 1976. *Ambiguity and Choice in Organizations.* Bergen: Universitetsforlaget.

Meyer, H.H., E.A. Kay, and J.R.P. French, Jr. 1965. "Split Roles in Performance Appraisal." *Harvard Business Review* 43: 123–29.

Mintzberg, H. 1973a. "Strategy-Making in Three Modes." *California Management Review* 16, no. 2: 44–53.

———. 1973b. *The Nature of Managerial Work.* New York: Harper and Row.

Mitchell, T.R., S.W. Green, and R. Wood. 1981. "An Attributional Model of Leadership and the Poor Performing Subordinate." In L.L. Cummings and B.M. Staw, eds., *Research in Organizational Behavior*, Vol. 3. Greenwich, Conn.: JAI.

Normann, R. 1985. "Towards an Action Theory of Strategic Management." In J. Pennings, ed., *Strategic Decision Making in Complex Organizations.* San Francisco: Jossey-Bass.

Pelz, D. 1985. "Innovation Complexity and the Sequence of Innovating Stages." *Knowledge: Creation, Diffusion, Utilization,* 6, no. 3 (March): 261–291.

Pelz, D., and F. Andrews. 1966. *Scientists in Organizations.* New York: Wiley.

Peters, T.J., and R.H. Waterman, Jr. 1982. *In Search of Excellence: Lessons from America's Best Known Companies.* New York: Harper and Row.

Pinchot, G., III. 1985. *Intrapreneuring: Why You Don't Have to Leave the Corporation to Become an Entrepreneur.* New York: Harper and Row.

Pressman, J. and A. Wildavsky. 1974. *Implementation.* Berkeley: University of California.

Quinn, J.B. 1980. *Strategies for Change: Logical Incrementalism.* Homewood, Ill.: Irwin.

Quinn, R.E. 1988. *Beyond Rational Management: Mastering the Paradoxes and Competing Demands of High Performance.* San Francisco: Jossey-Bass.

Rogers, E. 1983. *Diffusion of Innovations.* New York: Free Press.

Rosen, B., and T.H. Jerdee. 1974. "Factors Influencing Disciplinary Judgments." *Journal of Applied Psychology* 3: 327–31.

Schein, E. 1969. *Process Consultation: Its Role in Organization Development.* Reading, Mass.: Addison-Wesley.

Shaver, K.G. 1975. *Introduction to Attribution Processes.* Cambridge, Mass.: Winthrop.

Simon, H.A. 1947. *Administrative Behavior.* New York: Macmillan.

Staw, B.M. 1980. "The Consequences of Turnover." *Journal of Occupational Behavior* 1:253–273.

Staw, B.M., and J. Ross. 1987. "Behavior in Escalation Situations: Antecedents, Prototypes, and Solutions." In L.L. Cummings and B.M. Staw, eds., *Research in Organizational Behavior,* Vol. 9. Greenwich, Conn.: JAI.

Sutton, R.I. 1987. "The Process of Organizational Death: Disbanding and Reconnecting." *Administrative Science Quarterly* 32 (December): 542–69.

Tuckman, B. 1965. "Developmental Sequence in Small Groups." *Psychological Bulletin* 63: 384–399.

Utterback, J. 1971. "The Process of Technological Innovation Within the Firm." *Academy of Management Journal* 14: 75–88.

Van de Ven, A.H. 1980a. "Problem Solving, Planning, and Innovation. Part I. Test of the Program Planning Model." *Human Relations* 33, no. 10:711–40.

———. 1980b. "Problem Solving, Planning, and Innovation. Part II. Speculations for Theory and Practice," *Human Relations.* 33, no. 11: 757–79.

———. 1980c. "Early Planning, Implementation, and Performance of New Organizations." In J.R. Kimberly, R.H. Miles, and Associates, *The Organizational Life Cycle.* San Francisco: Jossey-Bass.

———. 1985. "Spinning on Symbolism: The Problem of Ambivalence." *Journal of Management* 11, no. 2: 101–2.

———. 1986. "Central Problems in the Management of Innovation." *Management Science* 32: 590–607.

Van de Ven, A.H., R. Hudson, and D.M. Schroeder. 1984. "Designing New Business Startups: Entrepreneurial, Organizational, and Ecological Considerations." *Journal of Management* 10, no. 1: 87–107.

Von Hippel, E. 1981. "Users as Innovators." In R.R. Rothberg, ed., *Corporate Strategy and Product Innovation.* New York: Free Press.

Wilkins, A.L. 1983. "The Culture Audit: A Tool for Understanding Organizations." *Organizational Dynamics* (Autumn): 24–38.

Zaltman, G., R. Duncan and J. Holbek. 1973. *Innovations and Organizations.* New York: Wiley.

Index

About the Contributors

HAROLD L. ANGLE is associate professor of strategic management and organization at the University of Minnesota, a position he has held since 1980. Angle received his B.A. (1956) in psychology from the University of California-Los Angeles, his M.A. (1971) in psychology from San Diego State College, and his Ph.D. (1980) in organizational behavior from the University of California-Irvine. Prior to his doctoral studies, Angle was a career officer in the United States Marine Corps where he served as director of the Office of Manpower Utilization. Angle's main research activities are in the area of member-organization relationships, and he is currently studying the effects of upheavals, such as mergers and acquisitions, reorganization, and geographic relocation, on corporate cultures and employee attitudes and performance. In addition to having published a number of research articles, Angle is co-author (with James L. Perry) of *Labor-Management Relations and Public Agency Effectiveness* (New York: Pergamon Policy Series, 1980). Angle has served on the editorial review boards of *Academy of Management Journal* and *Journal of Management*.

DAVID T. BASTIEN is assistant professor of speech-communication at the University of Wisconsin-Milwaukee. He received his B.A. (1965) in Asian studies, his M.A. (1985) and his Ph.D. (1988) in speech-communication, all from the University of Minnesota. Bastien worked for the Peace Corps early in his career where he developed and managed language training programs (1963–69). His research centers around two areas—mergers and acquisitions, and the social structures of creativity. He is currently writing on the topics of the social construction of jazz, the philosophy of social research, and the concept of management as coaching. Bastien's publications have included "Jazz as a Process of Organizational Innovation" (with Todd Hostager), in *Communication Research* (October 1988), and "Common Patterns in Mergers and Acquisitions," in *Human Resource Management Journal* (April 1987).

JOHN M. BRYSON is associate professor of planning and public affairs at the Hubert H. Humphrey Institute of Public Affairs at the University of Minnesota. He received his B.A. (1969) in economics at Cornell University, and his M.A. (1972) in public policy and administration, and his M.S. (1974) and Ph.D. (1978) in urban and regional planning, all from the University of Wisconsin-Madison. Bryson has been teaching at the University of Minnesota since 1977, and is associate director of the Strategic Management Research Center. During the 1986–87 academic year, Bryson served on the faculty of the London Business School as visiting professor. Bryson's primary area of study is strategic planning and management. He is currently beginning a series of studies on computer-assisted support systems for group-based strategic planning exercises, and also researching the applicability of the theatrical metaphor to strategic planning. Among Bryson's most recent publications are *Strategic Planning for Public and Nonprofit Organizations* (San Francisco: Jossey-Bass, 1988) and a book co-edited with Robert C. Einsweiler, *Strategic Planning— Threats and Opportunities for Planners* (Washington, D.C.: American Planning Association, 1988).

YUN-HAN CHU is assistant professor of political science at National Taiwan University in Taipei. Chu earned both his B.A. (1977) in international relations and his M.A. (1979) in political science from National Taiwan University and obtained his Ph.D. (1987) in political science from the University of Minnesota. Chu's main research interest is in the area of international and comparative political economy with emphasis on methodology and quantitative methods. Chu is currently writing on the political economy of East Asian newly industrializing countries.

BRIGHT M. DORNBLASER is professor of hospital and health care administration and coordinator for International Health in the School of Public Health at the University of Minnesota. Dornblaser received both his B.B.A. (1949) and his M.H.A. (1952) at the University of Minnesota. He has served on the faculty at the University of Minnesota since 1967 and has held a number of posts, including director of the Program in Hospital and Health Care Administration and head of the Division of Health Services Administration. Dornblaser is a past president of the Association of University Programs in Health Administration and has developed training curriculums for health administration programs in the Middle East and also a model being used in the development of training programs in Latin America. His current research focuses on international comparative studies of the management of change in health services organizations, and comparative innovativeness of investor-owned and not-for-profit hospital systems. Dornblaser has been on the editorial review board for Hospital Administration Press and currently serves as a reviewer for *Medical Care Journal* and Pluribus Press.

RAGHU GARUD is a Ph.D. candidate in the Strategic Management and Organization Department at the University of Minnesota. Garud received his bachelor of technology (1978) from the Indian Institute of Technology and his M.B.A. (1980) in marketing and finance from the Xavier Labor Relations Institute in India. Garud will join the faculty of the Leonard N. Stern School of Business at New York University in the fall of 1989. Garud's area of interest is new technology development and the interface between strategy and policy. His current research explores the patterns of cooperation and competition in the development and commercialization of a new product technology.

TODD J. HOSTAGER is assistant professor of business administration at the University of Wisconsin-Eau Claire. Hostager obtained his B.A. (1981) in psychology from St. Olaf College and is a Ph.D. candidate in the Strategic Management and Organization Department at the University of Minnesota. Hostager's research interests include communication and organizing processes in group jazz and in situations of major organizational change, metaphoric analysis as a creative approach to strategy formulation in the field of health care, and language patterns in strategy and organization. He is co-author (with David Bastien) of "Jazz as a Process of Organizational Innovation," in *Communication Research* (October 1988).

PAULA J. KING is assistant professor of management at Saint John's University in Collegeville, Minnesota. King obtained her B.A. (1973) in social work from the University of Minnesota, her M.S. (1980) in rehabilitation counseling from Mankato State University, and her Ph.D. (1988) in business administration from the University of Minnesota. King's academic specialty is public policy innovation and the policy-making process. Her specific interest is in the role of policy entrepreneurs as key participants in the innovation process. Her current research involves tracking an innovation in educational policy making at the state level. King's publications include "Policy Entrepreneurs: Catalysts in the Policy Innovation Process" in *Journal of State Government* (July/August 1987), and an article co-authored with Nancy Roberts, "Public Management Executives: Charting a Course Amidst the Political Swirl," in *Organizational Dynamics* 17 (3) (1989).

MARY K. KNUDSON is a research economist in the Technology and Resource Division of the Economic Research Service of the United States Department of Agriculture in Washington, D.C. Knudson received her B.S. (1981) in agronomy and plant genetics from the University of Minnesota, her M.S. (1983) in plant breeding and plant genetics from the University of Wisconsin-Madison, and her Ph.D. (1988) in agricultural economics from the University of Minnesota. Knudson's primary research interest is the economics of the invention and diffusion of new technology. She specializes in the fields of production economics, technological growth, and trade and development, and is researching agricultural R&D strategies in the private and public sector. Knudson is currently writing on issues of technological change in agriculture. She is co-author (with Vernon Ruttan) of "The R&D of a Biological Innovation: The Case of Hybrid Wheat," in *Food Research Institute Studies* 21 (1) (1988).

TSE-MIN LIN is a Ph.D. candidate in the Political Science Department at the University of Minnesota. He earned his B.S. (1975) in electrical engineering from National Taiwan University and his M.A. (1983) in political science from the University of Kansas. Lin will join the faculty of the Political Science Department at SUNY at Stoney Brook in the fall of 1989. Lin's research interests include the methodology and philosophy of social sciences, empirical and analytical political theory, formal modeling, and American politics. His dissertation examines the dynamics of American elections. Lin is also writing on the subjects of modeling political processes and analyzing qualitative sequences of events.

KARIN M. LINDQUIST is a Ph.D. candidate in the Strategic Management and Organization Department at the University of Minnesota. Lindquist earned her B.S. (1977) in urban planning from Michigan State University and her master of planning (1983) from the Humphrey Institute of Public Affairs at the University of Minnesota. Before beginning her doctoral studies, Lindquist worked as a planner and policy analyst for municipal, county, and state levels of government. Lindquist's dissertation focuses on the field of international strategy with an emphasis on the service sector. She is also interested in the interaction of business and government at a micropolicy level.

CHARLES C. MANZ is a Marvin Bower research fellow at Harvard Business School on leave from Arizona State University where he is associate professor of management. He received both his B.A. (1974) in general business and M.B.A. (1975) from Michigan State University, and his Ph.D. (1981) in organizational behavior from the Pennsylvania State University. His previous academic posts have included assistant professorships at the University of Minnesota (1982–88) and Auburn University (1980–82). Manz research and writing centers around the areas of leadership and self-leadership processes, power and control, vicarious learning, and self-managed work groups. He has recently completed the book, *SuperLeadership: Leading Others to Lead Themselves* (Englewood Cliffs, N.J.: Prentice-Hall, 1989), and previously authored, *The Art of Self-Leadership* (Englewood Cliffs, N.J.: Prentice-Hall, 1983). Among Manz's other publications are "Leading Workers to Lead Themselves: The External Leadership of Self-Managing Work Teams," in *Administrative Science Quarterly* 32 (1987), and "Self-Leadership: Toward an Expanded Theory of Self-Influence Processes in Organizations," in *Academy of Management Review* 11 (1986).

ALFRED A. MARCUS is associate professor of strategic management and organization at the University of Minnesota. Marcus received his B.A. (1971) in modern history at the University of Chicago, his M.A. (1973) in comparative politics and political philosophy from the University of Chicago, and his Ph.D. (1977) in public policy and administration from Harvard University. Prior to joining the faculty at the University of Minnesota, Marcus worked as a research scientist with the Battelle Human Affairs Research Centers in Seattle, Washington (1979–84). At Battelle, Marcus served as principal investigator on projects for the Nuclear Regulatory Commission, United States Department of Energy, the Environmental Protection Agency and the Industrial Forestry Association. His prior academic posts have included assistant professor of business administration at the University of Pittsburgh (1977–79) and adjunct professor of business at the University of Washington (1980). Marcus is currently researching in the areas of management response to crisis in the firm, corporate public affairs, management strategies and organizations, and the response of the stock market to auto safety recalls. He is the recipient of the 1986 Theodore Lowi Award for his article, "Airline Deregulation: Factors Affecting the Choice of Firm Political Strategy," in *Policy Studies Journal* (December 1986), and is co-editor (with Allen Kaufman and David Beam) of *Business Strategy and Public Policy* (Westport, Conn.: Greenwood Press, 1988).

JOHN J. MAURIEL is associate professor of strategic management and organization at the University of Minnesota. Mauriel obtained his B.A. (1953) in economics at the University of Michigan, and both his M.B.A. (1961) and his D.B.A. (1964) in business policy and executive development from the Harvard Business School. Mauriel has been a faculty member at the University of Minnesota since 1965 and currently directs the Bush Principal's Leadership Program and the Bush Public School Executive Fellows Program. He has also served as a visiting professor at the North European Management Institute in Oslo, Norway (1975–76). Mauriel's research specialties are in the areas of strategic management of local school districts, executive education, and cor porate strategy. His current research activities focus on strategic managemen problems of school superintendency, decentralized decisionmaking in the public schools, and site-based management. Mauriel is author of *Strategic Leadership for Local Schools* (San Francisco: Jossey–Bass, 1989) and co author (with Dan Gilbert, Edward Hartman, and R. Edward Freeman) of *The Logic of Strategy* (Cambridge, Mass.: Ballinger, 1988), and a number of man agement cases.

DOUGLAS POLLEY is a Ph.D. candidate in the Strategic Managemen and Organization Department at the University of Minnesota. Polley obtained his B.A. (1967) in mathematics at Whitman College and both his M.A. (1973) in economics and his M.A. (1975) in mathematics from the University of Min nesota. Prior to his doctoral studies, Polley spent a number of years managing the development and marketing of new telecommunications products in pri vate industry. His research interests include high technology, innovation and decisionmaking. He is currently writing on commitment to a course of action in new product development.

MARSHALL SCOTT POOLE is associate professor of speech-communi cation at the University of Minnesota. Poole received his B.A. (1973) in commu nication arts from the University of Wisconsin-Madison, his M.A. (1976) in communication from Michigan State University, and his Ph.D. (1980) in com munication arts from the University of Wisconsin-Madison. His dissertation received the Speech Communication Association's Dissertation of the Year Award in 1981. Prior to joining the faculty at the University of Minnesota Poole served as assistant professor in the Department of Speech Communica tion at the University of Illinois at Urbana-Champaign (1979–85) and was visiting assistant professor in the Communication Department at the University of Michigan (1981–82). Poole has written widely in the area of group decision making, organizational communication, conflict management, and research methodology and currently serves on the editorial boards of several academic journals, including *Human Communication Research, Communication Re search,* and *Communication Monographs.* His book *Working Through Conflic* (with Joseph Folger) (Chicago: Scott Foresman, 1983) won an award from the Center for Public Representation in Washington, D.C. He is the author (with David Seibold and Robert McPhee) of "Group Decision-Making as a Structura tional Process," in *Quarterly Journal of Speech* 71 (1985), which was recog nized by the Speech Communication Association as the best article in the field of communication for 1986. Poole is presently researching and writing in the areas of computer support for group decisionmaking, organizational climate and communication, and organizational innovation and change.

GORDON P. RANDS is a Ph.D. student in the Strategic Management and Organization Department at the University of Minnesota. Rands received his B.S. (1975) in natural resources at the University of Michigan and his M.O.B (1984) in organizational behavior at Brigham Young University. His primary interests are in the areas of organizational change and culture, corporate social responsiveness, and business ethics. Rands is currently investigating the rela tionship between individuals' values, social attitudes, and corporate social decisions.

MICHAEL A. RAPPA is assistant professor of management at the Massa chusetts Institute of Technology. Rappa earned his B.A. (1980) in economics at Union College and his Ph.D. (1987) in business administration from the University of Minnesota. Rappa is currently developing a conceptual framework and methodology for assessing the emergence of new technologies and is applying his framework to a variety of revolutionary technologies, including gallium arsenide semiconductors, superconducting devices, and neural networks.

PETER SMITH RING is associate professor of strategic management and organization at the University of Minnesota. He received his A.B. (1963) in history and government from St. Anselm College, his L.L.B. (1966) from the Law Center at Georgetown University, and his M.P.A. (1970) from the John F. Kennedy School of Government at Harvard University. Ring earned his Ph.D. (1984) in management from the University of California-Irvine following fifteen years of management experience in the public sector. Ring's present work focuses on transaction processes, strategic alliances, and comparative strategic management. His research has been published in the *Academy of Management Review, Public Administration Review* and *Administration and Society.*

NANCY C. ROBERTS is associate professor of organization behavior at the Naval Postgraduate School in Monterey, California. Professor Roberts received her diplome annuel (1966) in French at La Sorbonne in Paris, her B.A. (1967) in French and her M.A. (1968) in Latin American and South Asian history from the University of Illinois, and her Ph.D. (1983) in education from Stanford University. Prior to joining the faculty at the Naval Postgraduate School in 1986, Roberts was a visiting associate professor in the Graduate School of Business at Stanford University (1987). Her academic posts also have included an assistant professorship at the University of Minnesota (1982–85) and adjunct professorships in the Graduate School of Business at Santa Clara University (1982) and the Graduate School of Business at San Jose State University (1981). Roberts current research focuses on charismatic and transformational leadership, power—especially collective power of the whole and strategic management in the Department of Defense. Her most recent publications include, "Toward a Synergistic Model of Power," in *Shared Power* (John Bryson and Robert Einsweiler, eds.) (Lanham, Md.: University Press of America, 1988), and "Limits to Charisma" (with Raymond Trevor Bradley), in *Leadership in Management* (Jay Conger and Rabindra Kanungo, eds.) (San Francisco: Jossey-Bass, 1988). She is co-author (with Paula King) of the forthcoming book, *Public Policy Innovation* (Ballinger).

WILLIAM D. ROERING is assistant professor of strategic management at the University of Florida. Roering received his B.S. (1973) in life sciences from the University of Notre Dame, his M.S. (1976) in human systems and his M.B.A. (1980) from St. Cloud State University, and his Ph.D. (1989) in business administration from the University of Minnesota. Roering's primary areas of study include strategic management, planning, and decisionmaking. His current research investigates the decisionmaking processes of strategic managers, and his recent writing has focused on strategic management and planning in not-for-profit and public organizations. Roering's research has been published in the *Journal of Management Case Studies, Journal of the American Planning Association,* and *Public Administration Review.*

VERNON W. RUTTAN is a Regents Professor of Economics and Agricultural Economics at the University of Minnesota. Ruttan received his B.A. (1948) in economics from Yale University and both his M.A. (1952) and Ph.D. (1954) in economics from the University of Chicago. His academic positions have included professorships at Purdue University (1954–63) and at the University of Minnesota (1965–73, 1978–present), where he has served as head of the Department of Agricultural and Applied Economics (1965–70) and director of the Economic Development Center (1970–73). Ruttan has held a number of professional and political positions, among them president of the Agricultural Development Council (1973–78), president of the American Agricultural Economics Association (1971–72), economist at the International Rice Research Institute–Philippines (1963–65), and staff member of the President's Council of Economic Advisors (1961–62). Professor Ruttan's research concentrates on agricultural development, resource economics and research policy. Among his publications are *Agricultural Research Policy* (Minneapolis: University of Minnesota Press, 1982) and *Agricultural Development: An International Perspective* (Baltimore, Md.: Johns Hopkins University Press, revised 1985).

ROGER G. SCHROEDER is professor of operations management and Chair, Operations and Management Sciences Department at the University of Minnesota. Schroeder received both his B.S. (1962) in industrial engineering and his M.S.I.E. (1963) from the University of Minnesota, and his Ph.D. (1966) in operations research from Northwestern University. He joined the faculty at the University of Minnesota in 1971, where he has served as the director of the Ph.D. program in business administration. Schroeder is a past national president of the Operations Management Association (1986) and associate editor of the *Journal of Operations Management* (1984–87). His primary research interests are in the areas of operations strategy, quality improvement and the management of innovation. Schroeder is the author of *Operations Management: Decision Making in the Operations Function* (New York: McGraw-Hill, 3d ed., 1989), which has become one of the leading textbooks in its field.

GARY D. SCUDDER is associate professor of operations management at the University of Minnesota. Scudder earned both his B.S. (1974) and his M.S. (1975) in industrial engineering at Purdue University and received his Ph.D. (1981) in industrial engineering from Stanford University. He has served on the faculty of the University of Minnesota since 1980 and has also taught at the Amos Tuck School of Business Administration at Dartmouth College as visiting professor (1987–88). Scudder's research interests are in the areas of manufacturing scheduling and technology management. He is currently conducting an empirical analysis of manufacturing innovation and is examining scheduling in an automated manufacturing environment. Scudder's research has been published in numerous journals, including *Management Science, Naval Research Logistics, Journal of Operations Management,* and *International Journal of Production Research.* Scudder serves on the editorial board of the *Journal of Operations Management.*

GARY R. SEILER is associate professor of business administration at the College of St. Catherine in St. Paul, Minnesota. Seiler received his B.S. (1972) and his M.A. (1975) in business education, as well as his M.B.A. (1978) all from the University of Minnesota. He has taught at St. Catherine in the Department of Business Administration since 1973 and is currently a Ph.D. candidate in the

Strategic Management and Organization Department at the University of Minnesota. Seiler's primary areas of interest are strategic management, and business, government, and economics. His dissertation examines antitrust court decisions from a strategic and public policy perspective. Seiler has performed numerous consumer research studies for publication on a local basis.

GEORGE L. SHAPIRO is professor of speech-communication at the University of Minnesota. Shapiro received his B.A. (1953) in political science from the University of Wisconsin-Madison and both his M.A. (1957) and Ph.D. (1960) in speech-communication from the University of Minnesota. Shapiro has taught at the University of Minnesota for the last twenty-eight years. He has received the University of Minnesota Distinguished Professor Award (1977) and the Outstanding Teacher Award in the College of Liberal Arts (1966). Shapiro has held a number of extracurricular positions, including vice chair of the University Senate and chair of the Twin Cities Campus Assembly Student Affairs Committee. Over the last three years, with partial support from a Bush Sabbatical Grant, he has been researching ethical leadership both in Central America and Minnesota. Shapiro is co-author of *Interpersonal Communication in the Modern Organization* (with Earnest G. Bormann, William S. Howell, and Ralph G. Nichols) (Englewood Cliffs: Prentice-Hall, 1983).

ANDREW H. VAN DE VEN is 3M Professor of Human Systems Management at the University of Minnesota and director of the Minnesota Innovation Research Program at the University of Minnesota's Strategic Management Research Center. Van de Ven received his B.B.A. (1967) in marketing and management from St. Norbert College and both his M.B.A. (1969) in management and finance and his Ph.D. (1972) in interdisciplinary program administration from the University of Wisconsin–Madison. Prior to joining the faculty at the University of Minnesota in 1981, Van de Ven taught at the Wharton School of the University of Pennsylvania (1975–81), and Kent State University (1972–75). His publications over the years have focused on the Nominal Group Technique, organizational planning and problem-solving methods, and the Organization Assessment Framework and other instruments that evaluate the performance of organizations and interorganizational relationships. Van de Ven's current research centers on the areas of the management of innovation and change, mergers and acquisitions, and industry emergence. Van de Ven has published numerous books and articles, including *Measuring and Assessing Organizations* (with Diane Ferry) (New York: Wiley Interscience, 1980), *Group Techniques for Program Planning* (with Andrew Delbecq and David Gustafson) (Chicago: Scott Foresman, 1975), and *Perspectives on Organization Design and Behavior* (with William Joyce) (New York: Wiley Interscience, 1981).

S. VENKATARAMAN is a Ph.D. candidate in the Strategic Management and Organization Department at the University of Minnesota. Venkataraman received his M.B.A. (1982) from the Indian Institute of Management, Calcutta, India, and his M.A. (1979) in economics at the Birla Institute of Technology and Science in Pilani, India. Venkat will join the faculty at the Wharton School of the Universtiy of Pennsylvania in July, 1989. He is currently researching and writing in the area of entrepreneurship, and small business firm survival, growth, and adaptation.

MARK J. WEBER is a Ph.D. student in the Strategic Management and

719

Organization Department at the University of Minnesota. Weber received his B.A. (1974) in modern languages from the University of Notre Dame and both his J.D. (1978) in commercial law and his M.B.A. (1982) from the University of Wisconsin. Before beginning his doctoral studies at the University of Minnesota, Weber served as assistant professor at the University of Wisconsin–Superior and the University of Wisconsin–River Falls. Weber's research interest is in the relationship between government regulations and business growth. His doctoral dissertation explores the effects of product liability rules on the innovation process.

ROBERT M. WISEMAN is a Ph.D. candidate in the Strategic Management and Organization Department at the University of Minnesota. Wiseman earned his B.B.A. (1980) from the University of Wisconsin–LaCrosse and his M.B.A. (1982) at the University of Wisconsin–Milwaukee. Prior to his doctoral studies at Minnesota, Wiseman served as assistant professor of management at St. Norbert College (1982–85). His current research interests include the role of stakeholder relations, natural selection, and strategic adaptation in organizational survival. Wiseman is presently writing in the areas of strategy formulation and risk management.